GUIDELINES FOR

OPEN PIT SLOPE DESIGN

GUIDELINES FOR

OPEN PIT SLOPE DESIGN

EDITORS: JOHN READ, PETER STACEY

CSIRO
PUBLISHING

CRC

National Library of Australia Cataloguing-in-Publication entry

Guidelines for open pit slope design/editors, John Read, Peter Stacey.

9780643094697 (hbk.)
9780643095533 (ebk. : sponsors' ed.)

Includes index.
Bibliography.

Strip mining.
Slopes (Soil mechanics)
Landslides.

Read, John (John Russell Lee), 1939–
Stacey, Peter (Peter Frederick), 1942–

622.292

Published exclusively in Australia, New Zealand and South Africa by
CSIRO PUBLISHING
150 Oxford Street (PO Box 1139)
Collingwood VIC 3066
Australia

Telephone: +61 3 9662 7666
Local call: 1300 788 000 (Australia only)
Fax: +61 3 9662 7555
Email: publishing.sales@csiro.au
Website: www.publish.csiro.au

Published exclusively throughout the world (excluding Australia, New Zealand and South Africa) by CRC Press/Balkema, with ISBN 9780415874410

CRC Press/Balkema
P.O. Box 447
2300 AK Leiden
The Netherlands
Tel: +31 71 524 3080
Website: www.balkema.nl

Front cover: West Wall, Mega Pit, Sunrise Dam Gold Mine, Western Australia (Photo courtesy: AngloGold Ashanti Australia Ltd)

Set in 10/12 Adobe Minion and Optima
Edited by Adrienne de Kretser, Righting Writing
Cover and text design by James Kelly
Index by Russell Brooks
Typeset by Desktop Concepts Pty Ltd.
Printed in China by 1010 Printing International Ltd

Contents

Preface and acknowledgments *xiii*

1 Fundamentals of slope design **1**
Peter Stacey

1.1 Introduction 1

1.2 Pit slope designs 1
1.2.1 Safety/social factors 2
1.2.2 Economic factors 2
1.2.3 Environmental and regulatory factors 3

1.3 Terminology of slope design 4
1.3.1 Slope configurations 4
1.3.2 Instability 4
1.3.3 Rockfall 6

1.4 Formulation of slope designs 6
1.4.1 Introduction 6
1.4.2 Geotechnical model 6
1.4.3 Data uncertainty (Chapter 8) 8
1.4.4 Acceptance criteria (Chapter 9) 8
1.4.5 Slope design methods (Chapter 10) 9
1.4.6 Design implementation (Chapter 11) 10
1.4.7 Slope evaluation and monitoring (Chapter 12) 10
1.4.8 Risk management (Chapter 13) 11
1.4.9 Closure (Chapter 14) 11

1.5 Design requirements by project level 11
1.5.1 Project development 11
1.5.2 Study requirements 12

1.6 Review 12
1.6.1 Overview 12
1.6.2 Review levels 14
1.6.3 Geotechnically competent person 14

1.7 Conclusion 14

2 Field data collection **15**
John Read, Jarek Jakubec and Geoff Beale

2.1 Introduction 15

2.2 Outcrop mapping and logging 15
2.2.1 Introduction 15
2.2.2 General geotechnical logging 17
2.2.3 Mapping for structural analyses 19
2.2.4 Surface geophysical techniques 22

2.3 Overburden soils logging 23
2.3.1 Classification 23
2.3.2 Strength and relative density 26

2.4 Core drilling and logging 26

2.4.1 Introduction 26
2.4.2 Planning and scoping 26
2.4.3 Drill hole location and collar surveying 27
2.4.4 Core barrels 27
2.4.5 Downhole surveying 27
2.4.6 Core orientation 28
2.4.7 Core handling and documentation 29
2.4.8 Core sampling, storage and preservation 31
2.4.9 Core logging 32
2.4.10 Downhole geophysical techniques 39

2.5 Groundwater data collection **40**

2.5.1 Approach to groundwater data collection 40
2.5.2 Tests conducted during RC drilling 42
2.5.3 Piezometer installation 44
2.5.4 Guidance notes: installation of test wells for pit slope depressurisation 47
2.5.5 Hydraulic tests 49
2.5.6 Setting up pilot depressurisation trials 51

2.6 Data management **52**

Endnotes **52**

3 Geological model **53**

John Read and Luke Keeney

3.1 Introduction **53**

3.2 Physical setting **53**

3.3 Ore body environments **55**

3.3.1 Introduction 55
3.3.2 Porphyry deposits 55
3.3.3 Epithermal deposits 56
3.3.4 Kimberlites 56
3.3.5 VMS deposits 57
3.3.6 Skarn deposits 57
3.3.7 Stratabound deposits 57

3.4 Geotechnical requirements **59**

3.5 Regional seismicity **62**

3.5.1 Distribution of earthquakes 62
3.5.2 Seismic risk data 65

3.6 Regional stress **66**

4 Structural model **69**

John Read

4.1 Introduction **69**

4.2 Model components **69**

4.2.1 Major structures 69
4.2.2 Fabric 75

4.3 Geological environments **76**

4.3.1 Introduction 76
4.3.2 Intrusive 76

4.3.3 Sedimentary 76
4.3.4 Metamorphic 77

4.4 Structural modelling tools 77
4.4.1 Solid modelling 77
4.4.2 Stereographic projection 77
4.4.3 Discrete fracture network modelling 79

4.5 Structural domain definition 80
4.5.1 General guidelines 80
4.5.2 Example application 80

5 Rock mass model 83
Antonio Karzulovic and John Read

5.1 Introduction 83

5.2 Intact rock strength 83
5.2.1 Introduction 83
5.2.2 Index properties 85
5.2.3 Mechanical properties 88
5.2.4 Special conditions 92

5.3 Strength of structural defects 94
5.3.1 Terminology and classification 94
5.3.2 Defect strength 94

5.4 Rock mass classification 117
5.4.1 Introduction 117
5.4.2 RMR, Bieniawski 117
5.4.3 Laubscher IRMR and MRMR 119
5.4.4 Hoek-Brown GSI 123

5.5 Rock mass strength 127
5.5.1 Introduction 127
5.5.2 Laubscher strength criteria 127
5.5.3 Hoek-Brown strength criterion 128
5.5.4 CNI criterion 130
5.5.5 Directional rock mass strength 132
5.5.6 Synthetic rock mass model 138

6 Hydrogeological model 141
Geoff Beale

6.1 Hydrogeology and slope engineering 141
6.1.1 Introduction 141
6.1.2 Porosity and pore pressure 141
6.1.3 General mine dewatering and localised pore pressure control 146
6.1.4 Making the decision to depressurise 148
6.1.5 Developing a slope depressurisation program 151

6.2 Background to groundwater hydraulics 151
6.2.1 Groundwater flow 151
6.2.2 Porous-medium (intergranular) groundwater settings 154
6.2.3 Fracture-flow groundwater settings 156
6.2.4 Influences on fracturing and groundwater 161
6.2.5 Mechanisms controlling pore pressure reduction 163

6.3 **Developing a conceptual hydrogeological model of pit slopes** **166**
6.3.1 Integrating the pit slope model into the regional model 166
6.3.2 Conceptual mine scale hydrogeological model 166
6.3.3 Detailed hydrogeological model of pit slopes 167

6.4 **Numerical hydrogeological models** **168**
6.4.1 Introduction 168
6.4.2 Numerical hydrogeological models for mine scale dewatering applications 169
6.4.3 Pit slope scale numerical modelling 173
6.4.4 Numerical modelling for pit slope pore pressures 175
6.4.5 Coupling pore pressure and geotechnical models 179

6.5 **Implementing a slope depressurisation program** **180**
6.5.1 General mine dewatering 180
6.5.2 Specific programs for control of pit slope pressures 181
6.5.3 Selecting a slope depressurisation method 192
6.5.4 Use of blasting to open up drainage pathways 192
6.5.5 Water management and control 192

6.6 **Areas for future research** **195**
6.6.1 Introduction 195
6.6.2 Relative pore pressure behaviour between high-order and low-order fractures 195
6.6.3 Standardising the interaction between pore pressure and geotechnical models 196
6.6.4 Investigation of transient pore pressures 197
6.6.5 Coupled pore pressure and geotechnical modelling 197

7 **Geotechnical model** **201**
Alan Guest and John Read

7.1 **Introduction** **201**

7.2 **Constructing the geotechnical model** **201**
7.2.1 Required output 201
7.2.2 Model development 202
7.2.3 Building the model 202
7.2.4 Block modelling approach 205

7.3 **Applying the geotechnical model** **206**
7.3.1 Scale effects 206
7.3.2 Classification systems 210
7.3.3 Hoek-Brown rock mass strength criterion 210
7.3.4 Pore pressure considerations 211

8 **Data uncertainty** **213**
John Read

8.1 **Introduction** **213**

8.2 **Causes of data uncertainty** **213**

8.3 **Impact of data uncertainty** **213**

8.4 **Quantifying data uncertainty** **215**
8.4.1 Overview 215
8.4.2 Subjective assessment 215

8.4.3 Relative frequency concepts 216
8.5 Reporting data uncertainty 216
8.5.1 Geotechnical reporting system 216
8.5.2 Assessment criteria checklist 219
8.6 Summary and conclusions 219

9 Acceptance criteria 221
Johan Wesseloo and John Read
9.1 Introduction 221
9.2 Factor of safety 221
9.2.1 FoS as a design criterion 221
9.2.2 Tolerable factors of safety 223
9.3 Probability of failure 223
9.3.1 PoF as a design criterion 223
9.3.2 Acceptable levels of PoF 224
9.4 Risk model 225
9.4.1 Introduction 225
9.4.2 Cost–benefit analysis 226
9.4.3 Risk model process 228
9.4.4 Formulating acceptance criteria 232
9.4.5 Slope angles and levels of confidence 234
9.5 Summary 235

10 Slope design methods 237
Loren Lorig, Peter Stacey and John Read
10.1 Introduction 237
10.1.1 Design steps 237
10.1.2 Design analyses 238
10.2 Kinematic analyses 239
10.2.1 Benches 239
10.2.2 Inter-ramp slopes 244
10.3 Rock mass analyses 246
10.3.1 Overview 246
10.3.2 Empirical methods 246
10.3.3 Limit equilibrium methods 248
10.3.4 Numerical methods 253
10.3.5 Summary recommendations 263

11 Design implementation 265
Peter Williams, John Floyd, Gideon Chitombo and Trevor Maton
11.1 Introduction 265
11.2 Mine planning aspects of slope design 265
11.2.1 Introduction 265
11.2.2 Open pit design philosophy 265
11.2.3 Open pit design process 267
11.2.4 Application of slope design criteria in mine design 268
11.2.5 Summary and conclusions 276

11.3 Controlled blasting **276**
11.3.1 Introduction 276
11.3.2 Design terminology 277
11.3.3 Blast damage mechanisms 278
11.3.4 Influence of geology on blast-induced damage 279
11.3.5 Controlled blasting techniques 282
11.3.6 Delay configuration 292
11.3.7 Design implementation 294
11.3.8 Performance monitoring and analysis 296
11.3.9 Design refinement 299
11.3.10 Design platform 305
11.3.11 Planning and optimisation cycle 306

11.4 Excavation and scaling **310**
11.4.1 Excavation 310
11.4.2 Scaling and bench cleanup 312
11.4.3 Evaluation of bench design achievement 313

11.5 Artificial support **313**
11.5.1 Basic approaches 313
11.5.2 Stabilisation, repair and support methods 314
11.5.3 Design considerations 315
11.5.4 Economic considerations 316
11.5.5 Safety considerations 317
11.5.6 Specific situations 317
11.5.7 Reinforcement measures 318
11.5.8 Rockfall protection measures 325

12 Performance assessment and monitoring **327**
Mark Hawley, Scott Marisett, Geoff Beale and Peter Stacey

12.1 Assessing slope performance **327**
12.1.1 Introduction 327
12.1.2 Geotechnical model validation and refinement 327
12.1.3 Bench performance 329
12.1.4 Inter-ramp slope performance 337
12.1.5 Overall slope performance 339
12.1.6 Summary and conclusions 342

12.2 Slope monitoring **342**
12.2.1 Introduction 342
12.2.2 Movement monitoring systems 343
12.2.3 Guidelines on the execution of monitoring programs 363

12.3 Ground control management plans **370**
12.3.1 Introduction 370
12.3.2 Hazard management plan 371

13 Risk management **381**
Ted Brown and Alison Booth

13.1 Introduction **381**
13.1.1 Background 381
13.1.2 Purpose and content of this chapter 381
13.1.3 Sources of information 382

13.2 Overview of risk management 383
13.2.1 Definitions 383
13.2.2 General risk management process 383
13.2.3 Risk management in the minerals industry 384
13.3 Geotechnical risk management for open pit slopes 385
13.4 Risk assessment methodologies 389
13.4.1 Approaches to risk assessment 389
13.4.2 Risk identification 389
13.4.3 Risk analysis 391
13.4.4 Risk evaluation 395
13.5 Risk mitigation 396
13.5.1 Overview 396
13.5.2 Hierarchy of controls 398
13.5.3 Geotechnical control measures 398
13.5.4 Mitigation plans 399
13.5.5 Monitoring, review and feedback 400

14 Open pit closure 401
Dirk van Zyl
14.1 Introduction 401
14.2 Mine closure planning for open pits 403
14.2.1 Introduction 403
14.2.2 Closure planning for new mines 403
14.2.3 Closure planning for existing mines 403
14.2.4 Risk assessment and management 405
14.3 Open pit closure planning 405
14.3.1 Closure goals and criteria 405
14.3.2 Site characterisation 407
14.3.3 Ore body characteristics and mining approach 408
14.3.4 Surface water diversion 409
14.3.5 Pit water balance 409
14.3.6 Pit lake water quality 409
14.3.7 Ecological risk assessment 410
14.3.8 Pit wall stability 410
14.3.9 Pit access 412
14.3.10 Reality of open pit closure 412
14.4 Open pit closure activities and post-closure monitoring 412
14.4.1 Closure activities 412
14.4.2 Post-closure monitoring 412
14.5 Conclusions 412
Endnotes 413

Appendix 1 415
Groundwater data collection

Appendix 2 431
Essential statistical and probability theory

Appendix 3 **437**

Influence of in situ stresses on open pit design
Evert Hoek, Jean Hutchinson, Kathy Kalenchuk and Mark Diederichs

Appendix 4 **447**

Risk management: geotechnical hazard checklists

Appendix 5 **459**

Example regulations for open pit closure
Terminology and definitions *462*
References *467*
Index *487*

Preface and acknowledgments

Guidelines for Open Pit Slope Design is an outcome of the Large Open Pit (LOP) project, an international research and technology transfer project on the stability of rock slopes in open pit mines. The purpose of the book is to link innovative mining geomechanics research with best practice. It is not intended for it to be an instruction manual for geotechnical engineering in open pit mines. Rather, it aspires to be an up-to-date compendium of knowledge that creates a road map which, from the options that are available, highlights what is needed to satisfy best practice with respect to pit slope investigation, design, implementation, and performance monitoring. The fundamental objective is to provide the slope design practitioner with the tools to help meet the mine owner's requirements that the slopes should be stable, but if they do fail the predicted returns on the investment are achieved without loss of life, injury, equipment damage, or sustained losses of production.

The LOP project was initiated by and is managed on behalf of CSIRO Australia by John Read, CSIRO Exploration & Mining, Brisbane, Australia. Project planning commenced early in 2004, when a scoping document outlining a draft research plan was submitted to a number of potential sponsors and industry practitioners for appraisal. These activities were followed by a project scoping meeting in Santiago, Chile, in August 2004 and an inaugural project sponsors meeting in Santiago in April 2005. The project has been funded by 12 mining companies who are: Anglo American plc; Barrick Gold Corporation; BHP Billiton Innovation Pty Limited; Corporacion Naciónal Del Cobre De Chile ('Codelco'); Compania Minera Dona In s de Collahuasi SCM ('Collahuasi'); DeBeers Group Services (Pty) Limited; Debswana Diamond Company: Newcrest Mining Limited; Newmont Australia Limited; the Rio Tinto Group; Vale; and Xstrata Queensland Limited.

The 14 chapters in the book directly follow the life of mine sequence from project development to closure. They draw heavily on the experience of the sponsors and a number of industry and academic practitioners who have willingly shared their knowledge and experience by either preparing or contributing their knowledge to several of the chapters. In particular, the efforts of the following people are gratefully acknowledged.

- Alix Abernethy, Rio Tinto Iron Ore, Perth, Australia
- Rick Allan, Barrick Gold Corporation, Toronto, Canada
- Lee Atkinson, formerly Itasca Consulting Group, Denver, USA
- Geoff Beale, Water Management Consultants, Shrewsbury, England
- Gary Bental, BHP Billiton, Perth, Australia
- Alison Booth, formerly CSIRO Exploration & Mining, Brisbane, Australia
- Nick Brett, Nickel West, BHP Billiton, Perth, Australia
- Ted Brown, AC, Brisbane, Australia
- Gideon Chitombo, University of Queensland, Brisbane, Australia
- Paul Cicchini, Call & Nicholas Inc., Tucson, USA
- Ashley Creighton, Rio Tinto Technology & Innovation, Brisbane, Australia
- Peter Cundall, Itasca Consulting Group, Minneapolis, USA
- Mark Diederichs, Queen's University, Kingston, Canada
- Jeremy Dowling, Water Management Consultants, Tucson, USA
- John Floyd, Blast Dynamics, Steamboat Springs, USA
- Steve Fraser, CSIRO Exploration & Mining, Brisbane, Australia
- Phil de Graf, Rio Tinto Iron Ore, Perth, Australia
- Milton Harr, Longboat Key, USA
- Mark Hawley, Piteau Associates Engineering Ltd., Vancouver, Canada
- Evert Hoek, Vancouver, Canada
- Jean Hutchinson, Queen's University, Kingston, Canada
- Jarek Jakubec, SRK Consulting, Vancouver, Canada
- Mike Jefferies, Golder Associates Ltd, Calgary, Canada
- Kathy Kalenchuk, Queen's University, Kingston, Canada
- Antonio Karzulovic, Antonio Karzulovic y Asociados Ltda, Santiago, Chile
- Luke Keeney, University of Queensland, Brisbane, Australia
- Cédric Lambert, CSIRO Exploration & Mining, Brisbane, Australia
- Loren Lorig, Itasca Consulting Group, Santiago, Chile
- Mark Lorig, Itasca Consulting Group, Minneapolis, USA
- Graeme Major, Golder Associates Inc., Reno, USA
- Scott Marisett, formerly Newmont Australia, Perth, Australia
- Trevor Maton, Waihi Gold (Newmont), Waihi, NZ
- Anton Meyer, Barrick Gold Corporation, Tucson, USA
- Richard Mould, Rio Tinto Iron Ore, Peth, Australia

- Italo Onederra, University of Queensland, Brisbane, Australia
- Joergen Pilz, Rio Tinto Technology & Innovation, Salt Lake City, USA
- Frank Pothitos, OTML, Tabubil, Papua New Guinea (formerly Newcrest Mining Ltd, Orange, Australia)
- Mike Price, Water Management Consultants, Shrewsbury, England
- Martyn Robotham, Kennecott Utah Copper Company, Bingham Canyon, USA
- Eric Schwarz, Barrick Gold Corporation, La Serena, Chile
- Andrew Scott, Scottmining, Brisbane, Australia
- Joe Seery, Rio Tinto Iron Ore, Perth, Australia
- Oskar Steffen, SRK Consulting, South Africa
- Craig Stevens, Rio Tinto Technology & Innovation, Salt Lake City, USA
- Peter Terbrugge, SRK Consulting, Johannesburg, South Africa
- Julian Venter, Rio Tinto Iron Ore, Perth, Australia (formerly SRK Consulting, Johannesburg, South Africa)
- Audra Walsh, formerly Newmont Mining Corporation, Denver, USA
- Johan Wesseloo, Australian Centre for Geomechanics, Perth, Australia (formerly SRK Consulting, Johannesburg, South Africa)
- Fanie Wessels, Rio Tinto Iron Ore, Perth, Australia
- Peter Williams, Newmont Mining Corporation, Denver, USA
- Raymond Yost, Rio Tinto Minerals, Boron, USA
- Dirk van Zyl, University of British Columbia, Vancouver, Canada.

The book has been edited by John Read and Peter Stacey with the assistance of a sponsors' editorial subcommittee comprising Alan Guest (AGTC, formerly DeBeers Group Services), Warren Hitchcock (BHP Billiton), Bob Sharon (Barrick Gold Corporation) and Zip Zavodni (Rio Tinto).

John Read and Peter Stacey
May 2009

1 FUNDAMENTALS OF SLOPE DESIGN

Peter Stacey

1.1 Introduction

For an open pit mine, the design of the slopes is one of the major challenges at every stage of planning and operation. It requires specialised knowledge of the geology, which is often complex in the vicinity of orebodies where structure and/or alteration may be key factors, and of the material properties, which are frequently highly variable. It also requires an understanding of the practical aspects of design implementation.

This chapter discusses the fundamentals of creating slope designs in terms of the expectations of the various stakeholders in the mining operation, which includes the owners, management, the workforce and the regulators. It is intended to provide a framework for the detailed chapters that follow. It sets out the elements of slope design, the terminology in common usage, and the typical approaches and levels of effort to support the design requirements at different stages in the development of an open pit. Most of these elements are common to any open pit mining operation, regardless of the material to be recovered or the size of the open pit slopes.

1.2 Pit slope designs

The aim of any open pit mine design is to provide an optimal excavation configuration in the context of safety, ore recovery and financial return. Investors and operators expect the slope design to establish walls that will be stable for the life of the open pit, which may extend beyond closure. At the very least, any instability must be manageable. This applies at every scale of the walls, from the individual benches to the overall slopes.

It is essential that a degree of stability is ensured for the slopes in large open pit mines to minimise the risks related to the safety of operating personnel and equipment, and economic risks to the reserves. At the same time, to address the economic needs of the owners ore recovery must be maximised and waste stripping kept to a minimum throughout the mine life. The resulting compromise is typically a balance between formulating designs that can be safely and practicably implemented in the operating environment and establishing slope angles that are as steep as possible.

As outlined in Figure 1.1, the slope designs form an essential input in the design of an open pit at every stage of the evaluation of a mineral deposit, from the initial conceptual designs that assess the value of further work on an exploration discovery through to the short- and long-term designs for an operating pit. At each project level through this process other key components include the requirements of all stakeholders.

Unlike civil slopes, where the emphasis is on reliability and the performance of the design and cost/benefit is less of an issue, open pit slopes are normally constructed to lower levels of stability, recognising the shorter operating life spans involved and the high level of monitoring, both in terms of accuracy and frequency, that is typically available in the mine. Although this approach is fully recognised both by the mining industry and by the regulatory authorities, risk tolerance may vary between companies and between mining jurisdictions.

Uncontrolled instability, in effect failure of a slope, can have many ramifications including:

- Safety/social factors
  - → loss of life or injury;
  - → loss of worker income;
  - → loss of worker confidence;
  - → loss of corporate credibility, both externally and with shareholders.
- Economic factors
  - → disruption of operations;
  - → loss of ore;
  - → loss of equipment;
  - → increased stripping;

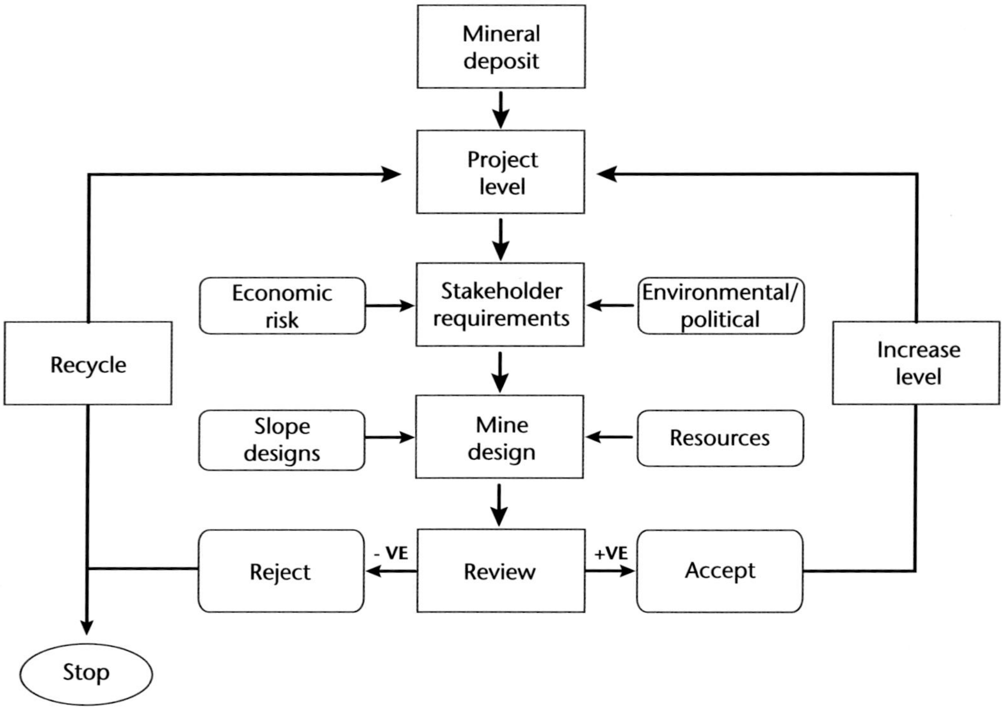

Figure 1.1: Project development flowchart

- → cost of cleanup;
- → loss of markets.
- Environmental/regulatory factors
  - → environmental impacts;
  - → increased regulation;
  - → closure considerations.

1.2.1 Safety/social factors

Safe operating conditions that protect against the danger of death or injury to personnel working in the open pit are fundamental moral and legal requirements.

While open pits have always been prone to wall instability due to the complexity of mining environments, since the adoption of formal slope design methodology in the early 1970s the number of failures has generally decreased. Even so, in recent years there have been several large failures in open pits around the world. Tragically, some of these have resulted in loss of life; most have had severe economic consequences for the operation. These failures have attracted the attention of regulators and the public. Consequently, it is becoming increasingly common for management (including executives) and technical staff to face criminal proceedings when mining codes are violated, in either the design or the operation of a mine.

While the major failures attract wide attention, it is the smaller failures, often rockfall at a bench scale, that typically result in the majority of deaths and injuries. For the mining industry to be sustainable, safety is a prime objective and must therefore be addressed at all scales of slope stability.

1.2.2 Economic factors

The main economic incentive in most open pits is to achieve the maximum slope angle commensurate with the accepted level of stability. In a large open pit, steepening a wall by only a few degrees can have a major impact on the return of the operation through increased ore recovery and/or reduced stripping (Figure 1.2).

In some instances, 'operating slopes' in initial expansion cuts may be flatter than the optimum, either to provide additional operating width or to ensure stability where data to support the designs are limited. However, this flexibility, which must be adopted with the understanding and consent of all stakeholders, almost always has negative economic consequences.

The impact of slope steepening will vary depending on the mine but, for example, it has been shown that an increase in slope angle of 1° in a 50° wall 500 m high results in a reduction of approximately 3600 m^3 (9000 t) of stripping per metre length of face.

Increasing the slope angle will generally reduce the level of stability of the slope, assuming that other factors remain constant. The degree to which steepening can be accomplished without compromising corporate and regulatory acceptance criteria, which usually reflect the safety requirements for both personnel and ore reserves,

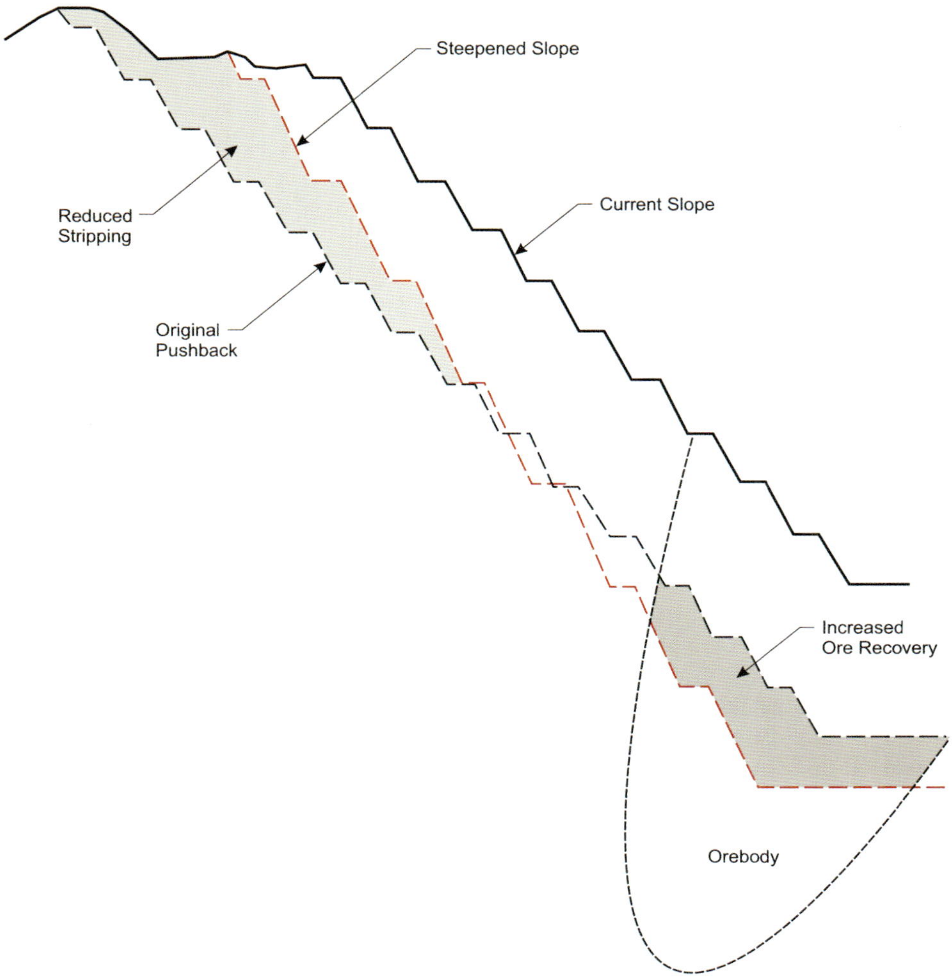

Figure 1.2: Potential impacts of slope steepening

must be the subject of stability analyses and ultimately risk assessments.

It is often no longer sufficient to present slope designs in deterministic (factor of safety) terms to a mine planner who accepts them uncritically. Increasingly, the requirement is that they be proposed within the framework of risk levels related to safety and economic outcomes for a decision-maker who may not be a technical expert in the mining field. The proposed design must be presented in a form that allows mine executives to establish acceptable levels of risk for the company and other stakeholders. In this process the slope designers must play a major role.

1.2.3 Environmental and regulatory factors

Most open pits are located in jurisdictions where there are mining regulations that specify safety and environmental requirements, including those for mine closure. The regulations may be federal, as in the case of the Mine Safety and Health Administration (MSHA) in the USA and the SNiP Codes in Russia, or local, for example the provincial mining codes in Canada and state regulations in Australia.

The regulations related to open pit slopes vary considerably between jurisdictions, as do the degrees of flexibility to modify slope configurations from those specified in the codes. However, regardless of the type of code, in most if not all jurisdictions it is the ultimate responsibility of the registered Mine Manager to maintain the 'standard of care' and regular reviews by a 'competent person' that are required.

Levels of requirements in codes can be summarised as follows.

1 Duty of Care, e.g. Western Australia, which place accountability on the registered Mine Manager to maintain appropriate design levels and safe operating procedures.
2 General Directives, e.g. MSHA, which are general in nature and do not specify minimum design criteria, although they may include definitive performance

criteria for catch benches and stable bench faces. Mines Inspectors enforce these regulations and are therefore responsible for approving the operation of a pit in terms of slope performance.

3 General Guidelines, e.g. 'Geotechnical Guidelines in Open Pit Mines – Guidelines', Western Australia, which outline the legislated background for safety in the context of the geotechnical factors that must be considered in the design and operation of open pit mines.
4 Defined General Criteria, e.g. British Columbia, Canada, which define minimum bench widths as well as maximum operating bench height, both of which are related to the capacity of the excavating equipment.
5 Detailed Criteria, e.g. the Russian SNiP Codes, which define methodologies to be used at different project levels for investigation and design of excavations.

In most jurisdictions it is possible to obtain authorisation for variations from the mining code, e.g. the use of multiple bench stacks between catch berms, provided that a clear engineering case can be presented and/or precedence for such a variation in similar conditions can be shown. For slope design practitioners, this means staying abreast of regulatory changes.

Mine closure considerations depend on regulatory requirements, company standards and/or other stakeholder interests.

1.3 Terminology of slope design

This section introduces the terminology typically used in the slope design process and presents a case for standardising this terminology, particularly with relation to slope movements and instability.

1.3.1 Slope configurations

The standard terminology used to describe the geometric arrangement of the benches and haul road ramps on the pit wall is illustrated in Figure 1.3. The terms relevant to open pit slope design as used in the manual are given in the Glossary.

It should be noted that terminology related to the slope elements varies by geographic regions. Some important examples include the following.

- Bench face (North America) = batter (Australia).
- Bench (North America) = berm (Australia). The flat area between bench faces used for rockfall catchment. The adjective 'catch' or 'safety' is often added in front of the term in either area.
- Berm (North America) = windrow (Australia). Rock piles placed along the toe of a bench face to increase rockfall catchment and/or along the crest of benches to prevent personnel and equipment falling over the face below. Note the potential confusion with the use of the term 'berm' for a flat surface.
- Bench stack. A group of benches between wider horizontal areas, e.g. ramps or wider berms left for geotechnical purposes.

Another aspect of terminology that can cause confusion is the definition of slope orientations. Slope designers usually work on the basis of the direction that the slope faces (dip direction), as this is the basis of kinematic analyses. On the other hand, mine planning programs usually require input in terms of the wall sector azimuth, which is at 180° to the direction that the slope faces, i.e. a slope facing/dipping toward 270° has an azimuth of 090° (inset, Figure 1.3). It is important that the convention adopted is clearly understood by all users and is applied consistently.

Note that the bench face angles are defined between the toe and crest of each bench, whereas the inter-ramp slope angles between the haul roads/ramps are defined by the line of the bench toes. The overall slope angle is always measured from the toe of the slope to the topmost crest (Figure 1.3).

1.3.2 Instability

Increased ability to detect small movements in slopes and manage instability gives rise to a need for greater precision in terminology. Previously, significant movement in a slope was frequently referred to in somewhat alarmist terms as 'failure', e.g. failure mode, even if the movement could be managed. It is now appropriate to be more specific about the level of movement and instability, using the definitions that recognise progression of slope movement in the following order of severity.

- Unloading response.

Initial movements in the slope are often associated with stress relaxation of the slope as it is excavated and the confinement provided by the rock has been lifted. This type of movement is linear elastic deformation. It occurs in every excavated slope and is not necessarily symptomatic of instability. It is typically small relative to the size of the slope and, although it can be detected by instruments, does not necessarily exhibit surface cracking. The deformation is generally responsive to mining, slowing or stopping when mining is suspended. In itself, unloading response does not lead to instability or large-scale movement.

- Movement or dilation.

This is considered to be the first clear evidence of instability, with associated formation of cracks and other visible signs, e.g. heaving at the toe (base) of the slope. In stronger rock, the movement generally results from

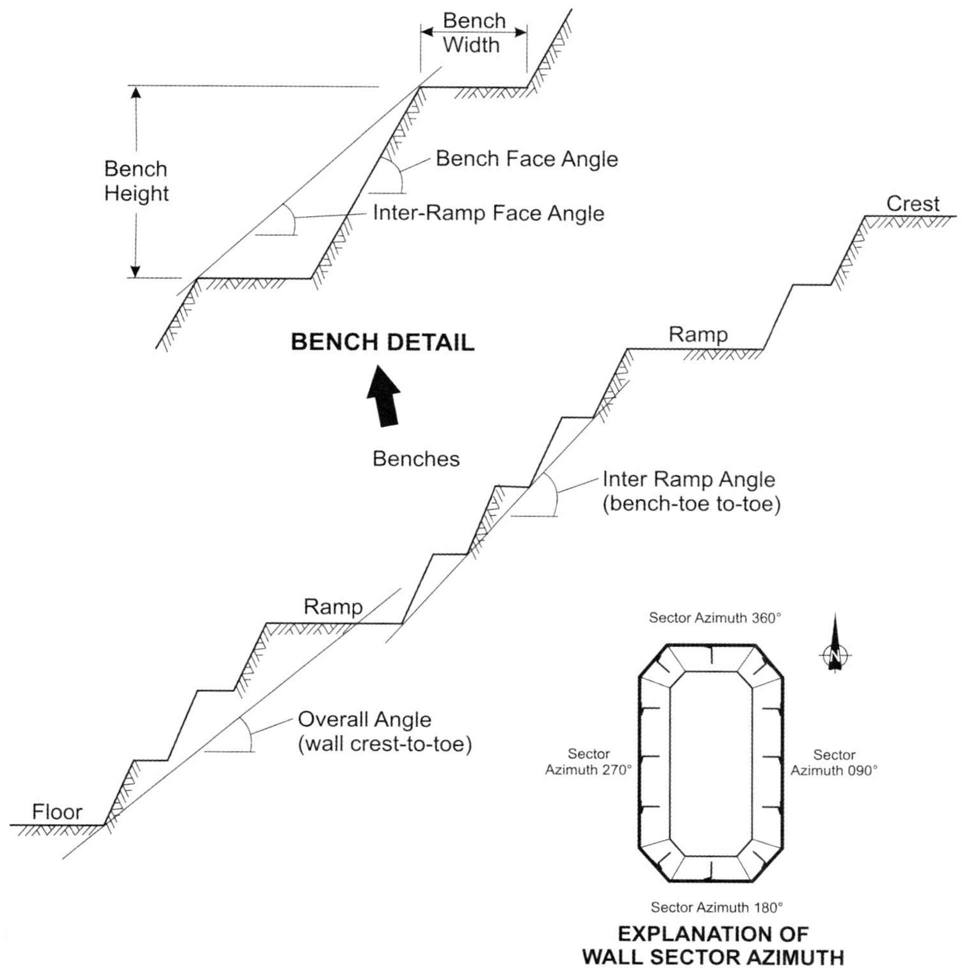

Figure 1.3: Pit wall terminology

sliding along a surface or surfaces, which may be formed by geological structures (e.g. bedding plane, fault), or a combination of these with a zone of weakness in the material forming the slope.

Slope dilation may take the form of a constant creep in which the rate of displacement is slow and constant. More frequently, there can be acceleration as the strength on the sliding surface is reduced. In certain cases the displacement may decrease with time as influencing factors (slope configuration, groundwater pressures) change. Even though it is moving, the slope retains its general original configuration, although there may be varying degrees of cracking.

Mining can often continue safely if a detailed monitoring program is established to manage the slope performance, particularly if the movement rates are low and the causes of instability can be clearly defined. However, if there is no intervention, such as depressurisation of the slope, modification of the slope configuration or cessation of mining, the movement can lead to eventual failure. This could occur as strengths along the sliding surface reduce to residual levels or if additional external factors, such as rainfall, negatively affect the stress distribution in the slope.

- Failure.

A slope can be considered to have failed when displacement has reached a level where it is no longer safe to operate or the intended function cannot be met, e.g. when ramp access across the slope is no longer possible.

The terms ‘failure’ and ‘collapse’ have been used synonymously when referring to open pit slopes, particularly when the failure occurs rapidly. In the case of a ‘progressive failure’ model, failure of a pit slope occurs when ‘the displacement will continue to accelerate to a point of collapse (or greatly accelerated movement)’ (Call et al. 2000). During and after failure or collapse of the slope, the original design configuration is normally completely destroyed. Continued mining almost always involves modification of the slope configuration, either through flattening of the wall from the crest or by stepping out at the toe. This typically results in increased stripping (removal) of waste and/or loss of ore, with significant financial repercussions.

The application of a consistent terminology such as that outlined above will also help to establish a more precise explanation of the condition of a slope for non-practitioners such as management and other stakeholders.

1.3.3 Rockfall

The term 'rockfall' is typically used for loose material that either falls or rolls from the faces. As such it is primarily a safety issue, although it could possibly be a precursor to larger-scale instability.

Rockfall can be a symptom of poor design implementation, i.e. poor blasting and/or scaling practices. However, it may also result from degradation of the slope as a result of weathering or from freeze–thaw action.

1.4 Formulation of slope designs

1.4.1 Introduction

The process of pit slope design formulation has been developed over the past 25 years and is relatively standard, although some of the methodologies vary between practioners. This section presents the general framework as an introduction to the detailed methodologies, which are discussed in the chapters that follow.

The basic process for the design of open pit slopes, regardless of size or materials, is summarised in Figure 1.4. Following this approach, the slope design process at any level of a project essentially involves the following steps:

- formulation of a geotechnical model for the pit area;
- population of the model with relevant data;
- division of the model into geotechnical domains;
- subdivision of the domains into design sectors;
- design of the slope elements in the respective sectors of the domains;
- assessment of the stability of the resulting slopes in terms of the project acceptance criteria;
- definition of implementation and monitoring requirements for the designs.

The resulting slope designs must not only be technically sound, they must also address the broader context of the mining operation as a whole, taking into account safety, the equipment available to implement the designs, mining rates and the acceptable risk levels.

The designs must be presented in a way that will allow the mine executives, who are ultimately responsible, and the operators, who implement the designs, to fully understand the basis and any shortcomings of the designs, as well as the implications of deviation from any constraints defined by the designer. In this context, a key element in the designs is the acceptance criteria against which the designs are formulated. These must be clearly defined by management working in consultation with the slope designers and mine planners.

As discussed in the following section, the available data and hence the level of confidence in the resulting designs generally improve with each successive stage in the development of an open pit mining project. However, the basic design procedures are essentially the same for all projects, with minor modification depending upon such factors as geology, groundwater conditions and proposed mine life.

The following points describe the basic elements of each step. They are discussed in following chapters, cited in parentheses.

1.4.2 Geotechnical model

The geotechnical model (Chapter 7), is the fundamental basis for all slope designs and is compiled from four component models:

- the geological model;
- the structural model;
- the rock mass model (material properties);
- the hydrogeological model.

These models also have applications for other aspects of the mining operation, for example in ore reserves and mining operations. However, particular aspects of each are critical for the slope design process.

There are other aspects of the geotechnical model that can be important in specific cases, for example in situ stress, particularly in relation to very high slopes, the presence of extensive underground openings and seismic loading.

Methods for collecting the data for each model are discussed in detail in Chapter 2.

1.4.2.1 Geological model (Chapter 3)

The geological model presents a 3D distribution of the material types that will be involved in the pit walls. The material type categories can relate not only to lithology but also to the degree and type of alteration, which can significantly change material properties, either positively (silicification) or negatively (argillisation).

In some deposits, notably those located in the tropics, geomorphology may also play a significant role in slope designs.

It is important to understand the regional geological setting and the genesis of the mineralisation. This often involves an appreciation that differs somewhat from that required by the mine geologists, who typically focus primarily on the mineralisation. Slope design studies must take a broader view of the geology of the deposit, including

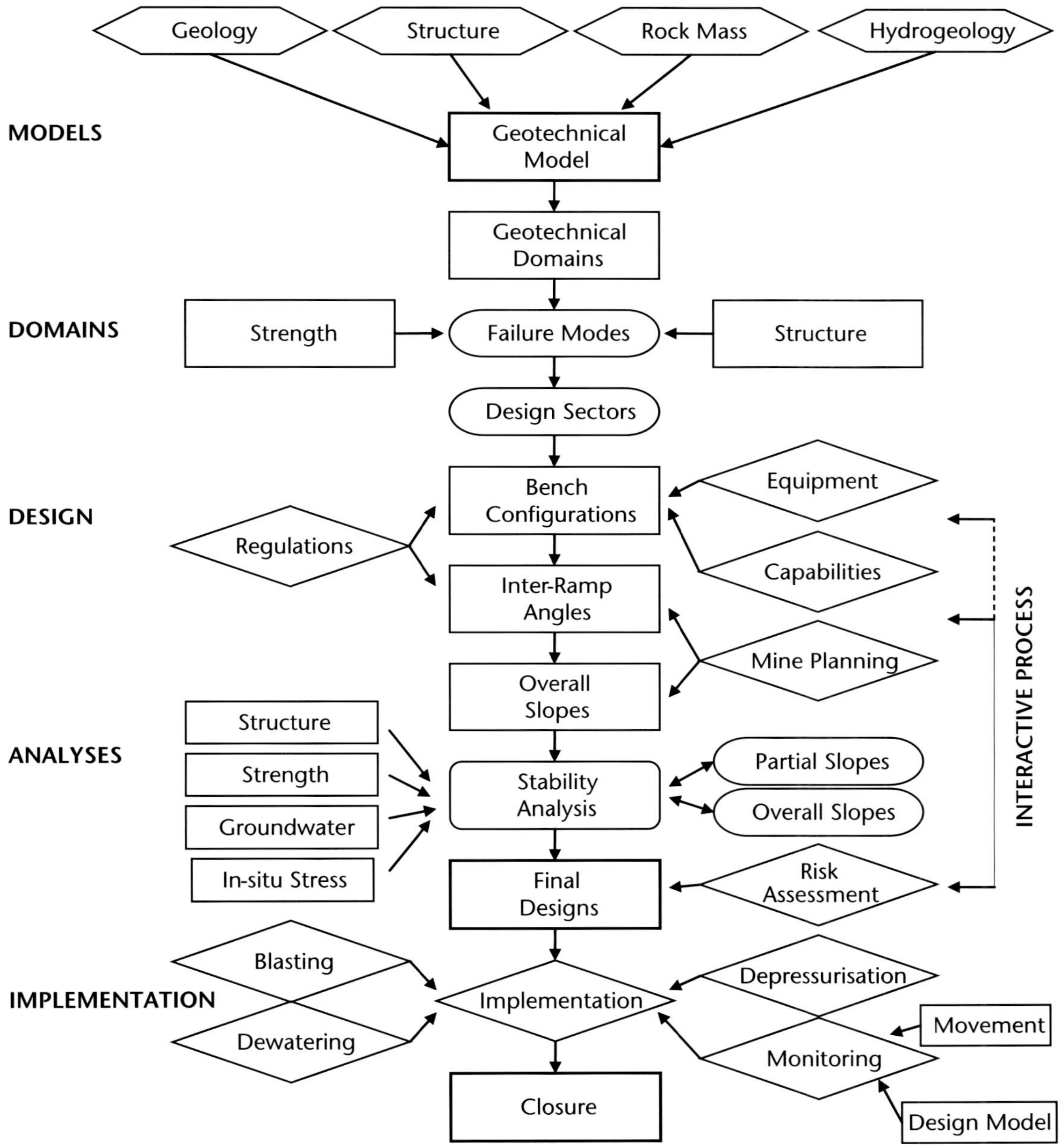

Figure 1.4: Slope design process

the surrounding waste rock, focusing on the engineering aspects.

As pit slopes become higher, the potential for impact by in situ stresses, particularly acting in combination with the high stresses created at the toe of the walls, must be considered. In situ stress assessment must be included in the geological model.

1.4.2.2 Structural model (Chapter 4)

A structural model for slope designs is typically developed at two levels:

- major structures (folds, inter-ramp and mine scale faults);
- structural fabric (joints, bench scale faults).

This differentiation relates largely to continuity of the features and the resultant impact with respect to the slope design elements. Major faults are likely to be continuous, both along strike and down dip, although they may be relatively widely spaced. Hence they could be expected to influence the design on an inter-ramp or overall slope scale. On the other hand, the structural fabric typically has limited continuity but close spacing, and therefore becomes a major consideration in design at a bench scale and possibly for inter-ramp bench stacks.

1.4.2.3 Rock mass model (Chapter 5)

The properties of the materials in which the slope will be excavated define probable performance and therefore the design approach. In strong rocks, structure is likely to be the controlling factor, even in relatively high slopes. In weaker materials and for very high slopes, the rock mass strength could be expected to play an important role, either alone or in combination with structures.

In defining the material properties, consideration must be given to the possible changes in behaviour with time. This particularly applies where there has been argillic alteration involving smectities (swelling clays) or in clay-rich shales, since the strength properties and behaviour of the material can change after exposure.

In determining the material properties, the slope designer can also provide important data for other aspects of the mining operation, for example in blast designs (Chapter 11, section 11.3). This should not be overlooked when designing the testing programs.

Back-analysis of failures and even of stable slopes can play a significant role in the determination of material properties. Detailed records of the performance of phase slopes and the initial stages of ultimate slopes can provide large-scale assessments of properties that can normally only be determined through small-scale laboratory tests during the feasibility and earlier stages of design. This is discussed in detail in Chapter 12.

1.4.2.4 Hydrogeology model (Chapter 6)

Both the groundwater pressure and the surface water flow aspects of the hydrogeological regime may have significant negative effects on the stability of a slope, and must therefore be fully understood.

These aspects are usually the only elements in a slope design that can be readily modified by artificial intervention, particularly at a large (inter-ramp and greater) scale. However, dewatering and depressurisation measures require operator commitment to be implemented effectively, and usually need significant lead time for design and implementation. Identification and characterisation of the hydrogeological regime in the early stages of any project are therefore of paramount importance.

1.4.3 Data uncertainty (Chapter 8)

With the move towards probability-based slope design methodology the need to define the reliability of the data in the geotechnical model has increased significantly. At the early stages of project development the available data are limited and hence the reliability of various model aspects will be low. This frequently leads to a situation where the

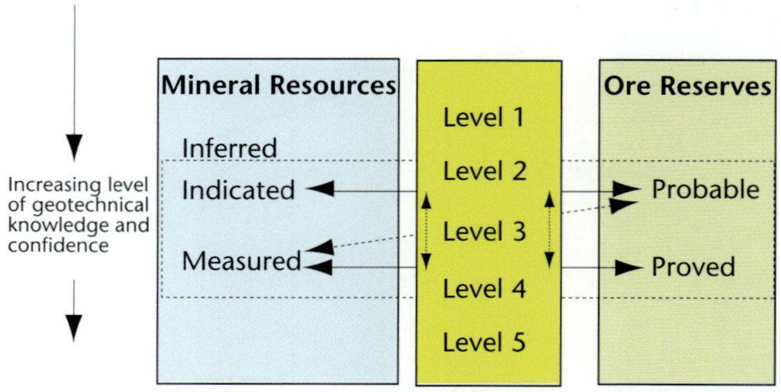

Figure 1.5: Geotechnical levels of confidence relative to the JORC code

uncertainties dominate the probabilistic results and a more deterministic approach must be used.

A high degree of uncertainty can exist even at the feasibility level, particularly where high (greater than 500 m) slopes are involved and the only available data are from drill holes and surface exposure. In this situation, either additional information obtained to reduce the uncertainties or the potential impacts must be made clear to the decision-makers.

In parallel with the introduction of codes for reporting exploration results, mineral resources and ore reserves in several countries (e.g. JORC in Australia, SAMREC in South Africa and 43-101 in Canada), the increased need to define data reliability has generated a requirement for a geotechnical reporting system related to the slope designs for the pits that define the reserves. Accordingly, a system of reporting the level of uncertainty in the geotechnical data is discussed in Chapters 8 and 9. The system is linked to the levels of effort at the various stages in the life of an open pit, outlined in section 1.5 and Table 1.2. It uses terminology to describe the different levels of uncertainty equivalent to the 'inferred', 'indicated' and 'measured' levels of confidence used by JORC (2004) to define the level of confidence in mineral resources and ore reserves (Figure 1.5).

1.4.4 Acceptance criteria (Chapter 9)

The definition of acceptance criteria allows the stakeholders, normally management or regulators, to define the level of performance required of a slope against instability and/or failure. The criteria were initially expressed in terms of a factor of safety (FoS), which compared the slope capacity (resisting forces) with the driving forces acting on the slope (gravity and water pressures). More recently, the probability of failure (PoF), i.e. the probability that the FoS will be 1 or less, has been introduced as a statistically based criterion.

The level of acceptance in either term may vary, depending upon the importance of the slope. For example, pit slopes that have no major facilities (ramps, tunnel portals, crushers) on the wall or immediately behind the

Table 1.1: Typical FoS and PoF acceptance criteria values

Slope scale	Consequences of failure	Acceptance criteria[a] FoS (min) (static)	FoS (min) (dynamic)	PoF (max) P[FoS ≤ 1]
Bench	Low–high[b]	1.1	NA	25–50%
Inter-ramp	Low	1.15–1.2	1.0	25%
	Moderate	1.2	1.0	20%
	High	1.2–1.3	1.1	10%
Overall	Low	1.2–1.3	1.0	15–20%
	Moderate	1.3	1.05	10%
	High	1.3–1.5	1.1	5%

a: Needs to meet all acceptance criteria
b: Semi-quantitatively evaluated, see Figure 13.9

crest might have an acceptable FoS of 1.2 or 1.3, or a PoF in the 10–15% range. For more critical slopes these values might be raised to 1.5 and less than 5%, respectively. Typical values are shown in Table 1.1.

Neither approach to stability assessment takes into account the consequences of instability or eventual failure or, conversely, the impacts of mitigative measures. Risk-based designs, which combine the PoF with the consequences (section 9.5), allow management to assess a slope design in terms of acceptance criteria that can easily incorporate risk in terms of safety and economic impacts, as well as societal views and legislated requirements.

1.4.5 Slope design methods (Chapter 10)

The formulation of slope design criteria fundamentally involves analysis against the predicted failure modes that could affect the slope at bench, inter-ramp and overall scales. The level of stability is assessed and compared with the acceptance criteria nominated at the various levels by the owners and/or regulators for safety levels and economic risk.

The process of slope design starts with dividing the geotechnical model for the proposed pit area into geotechnical domains with similar geological, structural and material property characteristics. For each domain, potential failure modes are assessed and designs at the respective scales (bench, inter-ramp, overall) are based on the required acceptance levels (FoS or PoF) against instability.

Once domains have been defined, their characteristics can be used to formulate the basic design approach. This involves evaluating the critical factors that will determine the potential instability mode(s) against which the slope elements will be designed. A fundamental division relates to the rock properties in that, for stronger rocks, structure is likely to be the primary control, whereas for weaker rocks strength can be the controlling factor, even down to the bench scale.

Where structure is expected to be a controlling factor, the slope orientation may exert an influence on the design criteria. In this case a subdivision of a domain into design sectors is normally required, based upon kinematic considerations related to the potential for undercutting structures (planar) or combinations (wedges), or toppling on controlling features. The 'sectorisation' can reflect controls at all levels, from bench scale, where fabric provides the main control for bench face angles, up to the overall slope, where particular major structures may be anticipated to influence a range of slope orientations with a domain.

For pits in weak rocks, where the rock mass strength is expected to be the controlling factor in slope designs, the design process commences with analyses to establish the overall and inter-ramp slope angle ranges that meet the acceptance criteria for stability. These angles are then translated down in scale into bench face configurations.

The type of stability analysis performed to support the slope design depends on several factors, including:

- the project stage (available data);
- the scale of slope under consideration;
- the properties of the materials that will form the slopes.

The main analysis types used for design include:

- kinematic analyses for bench designs in strong rock;
- limit equilibrium analysis applied to:
  - → structurally controlled failures in bench and inter-ramp design,
  - → inter-ramp and overall slopes where stability is controlled by rock mass strength, with or without structural anisotropy;
- numerical analyses for assessing failure modes and potential deformation levels in inter-ramp and overall slopes.

It should be stressed that stability analyses are tools that help formulate slope designs. The results must be evaluated in terms of other factors before they are finalised. These other factors include the mining methods and equipment that will be used to excavate the slopes, as well as the operators' capability to consistently implement such aspects as controlled blasting, surface water control and slope depressurisation.

The inter-ramp angles are normally provided to mine planners as the basic slope design criteria. Only when ramps have been added does the overall slope angle become apparent. Thus, for initial mine design and evaluation work, an overall slope angle involving the inter-ramp angle, flattened by 2–3° to account for ramps, may be used for Whittle cone analyses and other similar studies. This is discussed further in section 11.2.

1.4.6 Design implementation (Chapter 11)

Incorporating the slope design into the mine plan and implementing it requires clear understanding between all involved parties. This involves careful communication of the assumptions inherent in the design, plus the uncertainties and anticipated constraints on the construction of the slope. For the communication to be effective, the slope designer must understand the requirements and constraints influencing the other parties.

1.4.6.1 Mine planning (section 11.2)

The requirements from a slope design into the mine planning process, including the level of accuracy, depend on the project stage. At the early stages of evaluation, inter-ramp or overall angles suffice but as the project advances into the feasibility study and detailed design, more information about bench configurations and operating considerations are required. This is discussed further in section 1.5 of this chapter.

It is important at all stages that the slope designer and mine planner understand such aspects as the basis of the design, the level of accuracy, constraints and terminology. It is critical that there be regular communication between the two parties and that the slope designs be fully documented.

1.4.6.2 Operational aspects

Implementation of the slope designs typically requires the use of operating procedures that ensure minimum risk in terms of safety of personnel and recovery of reserves, including:

- the consistent application of effective controlled blasting (section 11.3);
- excavation control and face scaling (section 11.4);
- artificial support (section 11.5).

These requirements should be a fundamental part of the design definition and must be within the capability of the operators who will implement the design. It may also be necessary to consider the potential impact on production factors such as mining rate and excavation efficiency.

Where specific operating practices are required for implementing the slope design, it is critical that additional costs be incorporated into the budgets and recognised in terms of associated potential benefits to the overall revenue. For example, a mine superintendent will have little interest in implementing a controlled blasting program that allows steeper slopes unless corporate management recognises that the associated costs will be more than offset by reduced stripping costs or increased ore recovery.

The application of artificial support, either as part of the design or to stabilise a moving slope, has been in use for several decades. At a bench scale, rock bolts, mesh, shotcrete, straps and dowels are used to ensure stability or reduce degradation of the faces. Support also has a significant application where a pit slope is being mined through underground workings. These methods have largely been adapted from the underground mining environment, where the technology is well-developed. Cable bolts have been used successfully for inter-ramp slopes up to approximately 100 m in height. However, the 30 m practical length of cables is a major restriction and there have been several instances near the limit where the support has simply acted to tie together a larger mass, which subsequently failed. It is therefore important that any artificial support is carefully designed to the appropriate acceptance level, which will be partly dictated by the intended life of the supported slope and its overall importance.

1.4.7 Slope evaluation and monitoring (Chapter 12)

The performance of the slope during and after excavation must be monitored for unexpected instability and/or the potential for significant instability. Monitoring programs, which must continue throughout the life of the slope and often into closure, typically involve:

- slope performance assessment (section 12.1);
- slope displacement detection and warning (section 12.2);
- ground control management plans (section 12.3).

Assessment of slope performance focuses on validating the design model and ensuring that the operational methods for implementing the designs are appropriate and consistently applied.

It is important to validate the design model through geotechnical mapping and evaluating slope performance, particularly during the initial stages of mining. When the slope designs have been formulated on the basis of drill hole data alone, validation should include confirmation of the continuity of structures and the interpolation of geological data between holes.

Slope displacement monitoring is particularly important where instability exists and is being managed as part of the ongoing operation. A monitoring program may still be required after completion of mining, particularly if the open pit void is to be used for other purposes such as industrial (e.g. waste landfill) or recreational, where the public will have access to or below the slopes.

The ground control management plan for a pit should define responsibilities and outline the monitoring procedures and trigger points for the initiation of specified remedial measures if movement/instability is detected. It should form an integral part of the slope engineering program and the basis for the design of any required remedial measures.

1.4.8 Risk management (Chapter 13)

Certain degrees of safety, economic and financial risk have always been implicit in mining operations. In open pit mines, slope instability is one of the major sources of risk, largely due to data uncertainties, as well as the generally modest levels of stability accepted for the designs.

Factor of safety determination, which originated in the field of soil mechanics, is the traditional and widely practised slope design criterion. The uncertainty and variability of geology and rock mass properties led to increasing use of probability techniques rather than the deterministic FoS method; these provide the advantage of a linear scale for interpretation of the risks associated with slope designs. However, the concept of probability in a geotechnical sense is not easily understood by non-technical persons.

With the increasing requirement for management to be involved in the decision-making process for slope designs, a requirement for the quantification of risks has developed. To address this, risk assessment and management processes have been applied to slope designs.

Risk assessment methods range from qualitative failure modes and effects analysis (FMEA) to detailed quantitative risk/consequence analysis, depending on the level of definition favoured by management, regulators or practitioners. A fundamental requirement of all methods is that management defines acceptable levels of corporate risk against which the slope designs can be assessed. The assessment process can then be operated retroactively, with a design reviewed in relation to the acceptance criteria. Alternatively, the slope designer can proactively design a slope to meet the corporate risk profile, and the potential impacts of design variations can be assessed in terms of economic impact.

The objective of risk-based design is to provide management with quantitative information for:

- defining acceptable risks in terms of safety and economics;
- assessing relative risk levels for different slope configurations;
- benchmarking risks against industry norms and the corporate mission statement.

The risk-based design approach has been successfully applied to the design of slopes in several large open pit mines.

1.4.9 Closure (Chapter 14)

Current legislation in many jurisdictions requires mines to be designed with a view to closure and that a closure plan be in place before a mining permit is issued. Discussing the environmental aspects of closure as they relate to factors such as pit lake chemistry is outside the scope of this book, but is a critical consideration in closure.

In open pits, the closure plan should include long-term stability, particularly if the public is to have direct access to the area, for example as a recreational lake. Alternatively, if a pit lake is to be formed with outflow through a controlled surface channel, the potential for slope failures to cause waves that would overtop the channel and create a downstream flood must be considered. Other factors include aesthetics, particularly where the pit is located close to populated areas.

Stability during the closure process, for example while the pit lake is forming, could also be an issue that requires consideration and continued monitoring, particularly if slope stability has been achieved through an active slope depressurisation program. In this case, rapid repressurisation of the slopes relative to the formation of the lake could result in wall instability. This can generally be prevented by maintaining the depressurisation system until equilibrium is established.

Monitoring of slope stability can be expected to continue through the initial closure and in many cases on a continuing basis post closure, particularly if the public has access to the open pit area.

1.5 Design requirements by project level

Guidelines for the typical level of investigation and design effort expected at various stages of project development are presented in this section. It should be noted that the actual required effort can vary significantly, depending on the degree of complexity in the geotechnical model and the level of risk assurance required by the owner (sections 1.4.3. and 1.4.4).

1.5.1 Project development

There are six main levels in the development and execution of a mining project at which slope design input is required. These are:

- conceptual study (Level 1);
- pre-feasibility (Level 2);
- feasibility (Level 3);
- design and construction (Level 4);
- operations (Level 5);
- closure (Level 6).

The mine planning requirements at these levels, which are discussed in detail in section 11.2, can be summarised as follows.

At the conceptual study level, various mining methods are assessed. At this early stage the viability of open pit mining may be based on judgment or experience in similar environments. Cost estimates and slope designs are at the 'order of magnitude' level.

At the pre-feasibility level, preliminary slope designs are required to determine if the ore body is technically and economically viable to mine so that reserves and associated mining method can be defined.

The feasibility level is typically used to establish a clear picture of the anticipated costs of mine development and operation. At the completion of the study alternative interpretations may be possible, but in the view of a 'competent person' these would be unlikely to affect the potential economic viability of the project. To achieve this level of accuracy, overall slope designs in the order of ±5° are necessary.

At the design and construction level, the ore body has been shown to be potentially economic and financing has been secured for production. Confidence in the pit slope design should be increased at this stage, particularly for open pits with marginal rates of return. This stage may be skipped and initial mining may be based upon the feasibility level slope designs.

During the operations level, pit slope optimisation may be possible, based on additional data collected from the pit walls and incorporating operating experience with slope performance to refine the geotechnical model and provide revised slope design criteria for future cutbacks.

Increasingly, the slope designs must also address long-term stability associated with landforms required at closure and potential uses of the open pit void. Closure designs should be established during the operating phase, when mine staff will have experience of slope performance that may not be available post closure.

1.5.2 Study requirements

Most mining companies have specific requirements for the level of effort required to achieve the mine design at various project levels. Table 1.2 presents a summary of suggested levels of effort from the Level 1 conceptual stage through to operations (Level 5). Mine closure (Level 6) is addressed in Chapter 14. Requirements vary between companies and even between projects, therefore the table is only a guide.

The responsibility for collecting, compiling and analysing the data to establish the slope designs depends on the in-house capabilities of the mining company and on the project level. In larger companies the initial level evaluations and slope management in operating mines are typically performed by in-house staff. For larger studies (Level 3), and for most work in smaller mines, consultants play a significant role. There is an increasing requirement for independent review at the pre-feasibility and subsequent project levels (discussed further in section 1.6).

1.6 Review

1.6.1 Overview

Slope designs are increasingly subject to formal reviews, both prior to commencement of mining and during the operating phase. These reviews, which may be undertaken by in-house specialists, an external review consultant or a board of specialists, are conducted for a number of reasons. At the feasibility and mine financing stages, a review gives management and potential financiers confirmation of the viability of the proposed project. At the operating stage a review, which may involve a board addressing all geotechnical and hydrogeological aspects of the mine, gives management an independent assessment and additional confidence in the designs and the implementation procedures.

If a board is to be used, Hoek and Imrie (1995) suggested the following guidelines.

> A Review Board should be composed of a small number of internationally recognised authorities in fields relevant to the principal problems encountered on the mine. The purpose of the Board should be to provide an objective, balanced and impartial view of the overall geotechnical activities on a mine. The Board should not be used as a substitute for normal consulting services since members do not have the time to acquire all the detailed knowledge necessary to provide direct consulting opinions.
>
> The function of the Board should be to act as the technical review agency for the Mine Management. Ideally, a Board should ask the geotechnical team and associated mine planning staff 'have you considered this alternative?' rather than be asked to respond to a request such as 'please provide recommendations on a safe slope angle'.
>
> In my experience, the most effective Boards are very small (2 to 4 members) and are carefully chosen to cover each of the major disciplines involved in the

Table 1.2: Levels of geotechnical effort by project stage

	PROJECT STAGE				
Project level status	**Conceptual**	**Pre-feasibility**	**Feasibility**	**Design and Construction**	**Operations**
Geotechnical level status	**Level 1**	**Level 2**	**Level 3**	**Level 4**	**Level 5**
Geological model	Regional literature; advanced exploration mapping and core logging; database established; initial country rock model	Mine scale outcrop mapping and core logging, enhancement of geological database; initial 3D geological model	Infill drilling and mapping, further enhancement of geological database and 3D model	Targeted drilling and mapping; refinement of geological database and 3D model	Ongoing pit mapping and drilling; further refinement of geological database and 3D model
Structural model (major features)	Aerial photos and initial ground proofing	Mine scale outcrop mapping; targeted oriented drilling; initial structural model	Trench mapping; infill oriented drilling; 3D structural model	Refined interpretation of 3D structural model	Structural mapping on all pit benches; further refinement of 3D model
Structural model (fabric)	Regional outcrop mapping	Mine scale outcrop mapping; targeted oriented drilling; database established; initial stereographic assessment of fabric data; initial structural domains established	Infill trench mapping and oriented drilling; enhancement of database; advanced stereographic assessment of fabric data; confirmation of structural domains	Refined interpretation of fabric data and structural domains	Structural mapping on all pit benches; further refinement of fabric data and structural domains
Hydrogeological model	Regional groundwater survey	Mine scale airlift, pumping and packer testing to establish initial hydrogeological parameters; initial hydrogeological database and model established	Targeted pumping and airlift testing; piezometer installation; enhancement of hydrogeological database and 3D model; initial assessment of depressurisation and dewatering requirements	Installation of piezometers and dewatering wells; refinement of hydrogeological database, 3D model, depressurisation and dewatering requirements	Ongoing management of piezometer and dewatering well network; continued refinement of hydrogeological database and 3D model
Intact rock strength	Literature values supplemented by index tests on core from geological drilling	Index and laboratory testing on samples selected from targeted mine scale drilling; database established; initial assessment of lithological domains	Targeted drilling and detailed sampling and laboratory testing; enhancement of database; detailed assessment and establishment of geotechnical units for 3D geotechnical model	Infill drilling, sampling and laboratory testing; refinement of database and 3D geotechnical model	Ongoing maintenance of database and 3D geotechnical model
Strength of structural defects	Literature values supplemented by index tests on core from geological drilling	Laboratory direct shear tests of saw cut and defect samples selected from targeted mine scale drill holes and outcrops; database established; assessment of defect strength within initial structural domains	Targeted sampling and laboratory testing; enhancement of database; detailed assessment and establishment of defect strengths within structural domains	Selected sampling and laboratory testing and refinement of database	Ongoing maintenance of database
Geotechnical characterisation	Pertinent regional information; geotechnical assessment of advanced exploration data	Assessment and compilation of initial mine scale geotechnical data; preparation of initial geotechnical database and 3D model	Ongoing assessment and compilation of all new mine scale geotechnical data; enhancement of geotechnical database and 3D model	Refinement of geotechnical database and 3D model	Ongoing maintenance of geotechnical database and 3D model

project. For example, in the case of a large open pit mine, the board members could be:

- A geologist or engineering geologist with experience in the type of geological conditions that exist on the site. This is particularly important when unusual or difficult geological conditions such as very weak altered rocks or major faults are likely to be encountered.
- A rock engineering specialist with experience in rock slope stability problems in the context of open pit mining.
- A mine planning engineer with a sound understanding of rock mechanics and a strong background in scheduling, blasting and mining equipment characteristics.

Recent experience has suggested that a hydrogeologist can also play an invaluable role where large open pit slopes are concerned, since slope depressurisation is usually required.

In large projects, it is important that the reviewers be involved from the early stages and be given regular updates on progress and changes. This should avoid complications during final presentation of the design.

1.6.2 Review levels

There are three levels at which reviews are commonly performed.

1 Review at discussion level – at the discussion level the reviewer is not provided with all the relevant reports and data required for an independent assessment or independent opinion. Generally, only selective information is presented, often in meeting presentation form, and there is insufficient time to absorb and digest all the pertinent information and develop a thorough understanding of all aspects relating to the design, construction and operation. The reviewer relies on information selected by the presenter and substantially on the presenter's observations, interpretation and conclusions.

2 Review level – at this level the reviewer generally examines only key documents and carries out at least 'reasonableness of results' checks on key analyses, design values and conclusions. The reviewer generally relies on representations made by key project personnel, provided the results and representations appear reasonable and consistent with what an experienced reviewer would expect. This level of review is appropriate for all levels of project development beyond the conceptual (Level 1).

3. Audit level – an audit is a high-level review of all pertinent data and analyses in sufficient detail for an independent opinion on the general principles of design, construction and operations, and on the validity and accuracy of the key elements of the design analyses, construction control and operating methods. This level of review is often appropriate at the feasibility (Level 3) stage of investigation.

1.6.3 Geotechnically competent person

Unlike the codes in use in different countries to support ore reserve estimates (JORC in Australia, 43-101 in Canada), there is no standard definition of geotechnical competence to assess and sign off slope designs for use in reserve estimate pits. However, for slope designs it is anticipated that a definition of a 'geotechnically competent person' and/ or reviewer for slope designs will be established in the near future to complement the equivalent standards for the presentation of ore reserves. Until such a definition becomes available, the basic criteria could include:

- an appropriate graduate degree in engineering or a related earth science;
- a minimum of 10 years post-graduate experience in pit slope geotechnical design and implementation;
- an appropriate professional registration.

1.7 Conclusion

The following chapters expand on the design of large open pit slopes within the general framework outlined above. It must be a basic design premise that a slope design addresses the requirements of all stakeholders, from the owners through the operators to the regulators.

In delivering a design, technical soundness is the foundation. The slope designer must build on this, responding to the varying conditions in each phase of the mine's life. The safety of personnel and equipment is of paramount importance in all phases, and acceptable risk levels must be carefully assessed and incorporated into the designs.

By presenting the slope designs in a manner that enables mine personnel, from executives to operators, to fully understand the basis and shortcomings of the designs, practitioners provide the means of discerning the risks associated with deviation from those designs. With greater understanding, better and safer decisions can be made.

2 FIELD DATA COLLECTION

John Read, Jarek Jakubec and Geoff Beale

2.1 Introduction

The geotechnical model, together with its four components, the geological, structural, rock mass and hydrogeological models, is the cornerstone of open pit slope design. As illustrated in Figure 2.1, the model must be in place before the successive steps of setting up the geotechnical domains, allocating design sectors and preparing the final slope designs can commence.

Populating the geotechnical model with relevant field data requires not only keen observation and attention to detail, but also strict adherence to field data gathering protocols from day one in the development of the project. In this process, it is expected that the reader will be aware of the wide variety of traditional and newly developed data collection methods available to the industry. Nonetheless, it cannot be emphasised enough that those who are responsible for project site investigations must be aware of the mainstream technologies available to them, and how and when they should be applied to provide a functional engineering classification of the rock mass for slope design purposes. For geological and structural models these technologies can range from direct or digital mapping and sampling of surface outcrops, trenches and adits to direct and indirect geophysical surveys, rotary augering and core drilling. For the rock mass model they can include a plethora of field and laboratory tests. For the hydrogeological model they can include everything from historical regional hydrogeological data, to the collection of hydrogeological data 'piggy-backed' on mineral exploration and resources drilling programs and routine water level monitoring programs in specifically installed groundwater observation wells and/or piezometers.

Providing an exhaustive list of each and every technology is beyond the scope of this book. However, it is possible to outline the availability and application of the mainstream technologies used to provide a functional engineering classification of the rock mass for slope design purposes. This is the focus of this chapter and is addressed in five sections, commencing with outcrop mapping and logging in section 2.2. Section 2.3 discusses overburden soils logging, and is followed by descriptions of the applicable methods of subsurface core drilling and logging in section 2.4. Laboratory testing procedures to determine the engineering properties of the structural defects and intact rock logged and sampled during these activities are outlined in Chapter 5. Groundwater data collection is outlined in section 2.5. Finally, section 2.6 provides an overview of database management procedures.

2.2 Outcrop mapping and logging

2.2.1 Introduction

Outcrop mapping is fundamental to all the activities pursued by the teams responsible for designing and managing the pit slopes. It includes regional and mine-scale surface outcrop mapping during development prior to mining and bench mapping once mining has commenced. Preferably it should be carried out only by properly trained geologists, engineering geologists, geological engineers or specialist geotechnicians, assisted by specialists from other disciplines as needed.

Historically, the mapped data were recorded by hand on paper sheets and/or field notebooks, but advances in electronic software and hardware mean that this is increasingly replaced by electronic data recording directly into handheld tablets and/or laptop computers. Both systems have their merits, but the electronic system has the advantage that it eliminates the tedious transfer of paper data into an electronic format. It produces data that can be almost instantly transmitted for further analysis and checking in Autocad or similar systems. On the other hand, if there is not an effective file backup and saving procedure, the data are at risk of being lost in a split

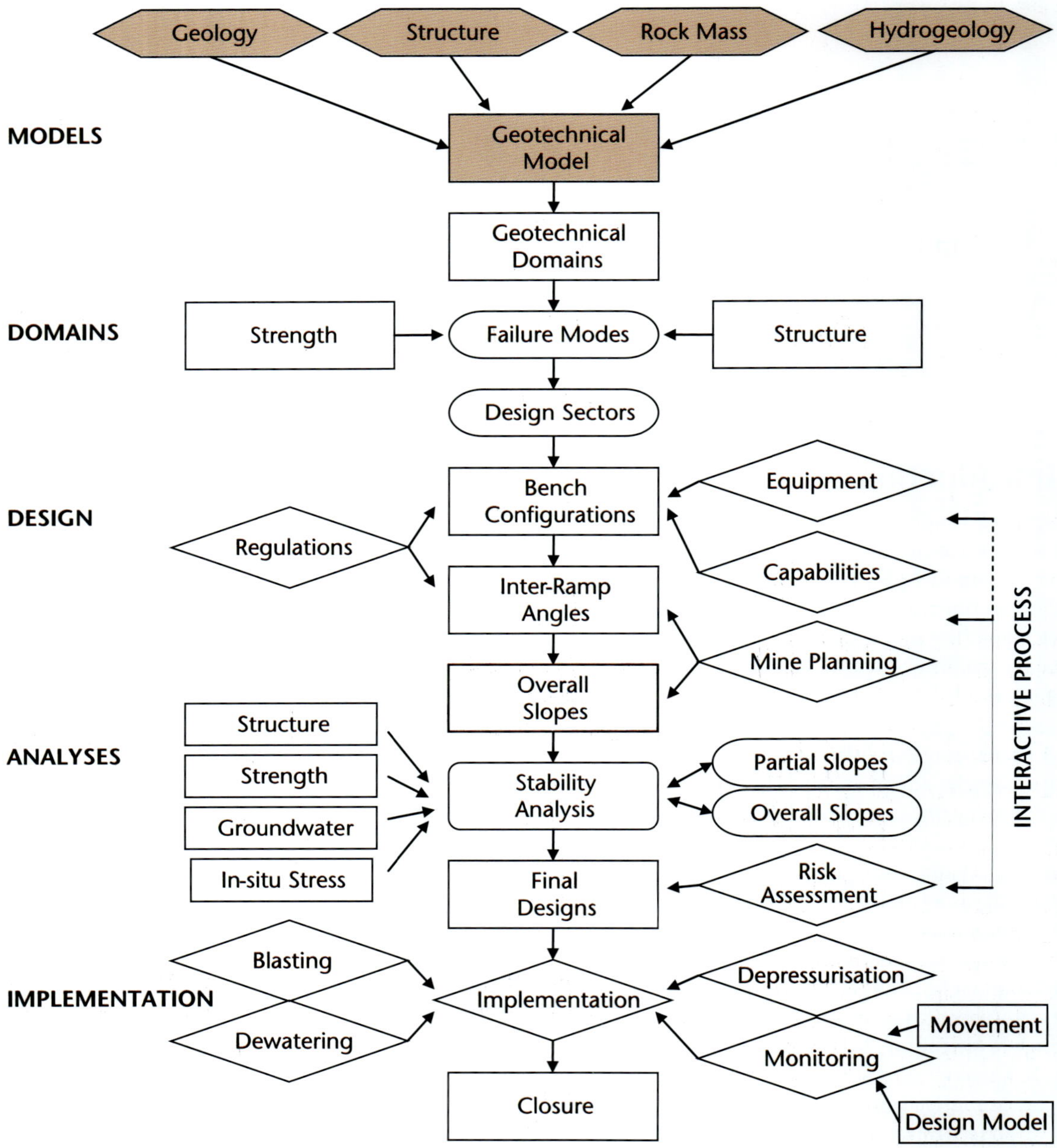

Figure 2.1: Slope design process

second. There could also be some issues with the auditing process since no field mapping sheets are available.

More recently, an area that has increased in importance is the in situ characterisation of the ore body and its surrounds by surface-based geophysical methods prior to mining. High-resolution penetrative methods can be used to assist in locating and understanding the structural setting and petrophysical properties of both the mineralised body and its surrounding materials. During this process there is an opportunity to extract valuable geotechnical information, because the petrophysical properties so determined are essentially volumetrically continuous and are from undisturbed materials. The geophysically derived determinations can be recalibrated against actual measurements taken from drill core materials or samples collected during the mining process.

Regardless of how it is recorded, it is important that all the geotechnical data captured are capable of supporting the principal rock mass classification and strength assessment methods used by the industry today. Similarly, although the level of detail captured must at least be relevant to the level of investigation, there is no reason not to collect the most comprehensive set of data even in the earliest stages of investigation. This section therefore

Table 2.1: Effect of weathering on fresh rock

Term	Symbol	Description
Fresh	Fr/W1	No visible sign of weathering
Slightly weathered	SW/W2	Partial (<5%) staining or discoloration of rock substance, usually by limonite. Colour and texture of fresh rock is recognisable. No discernible effect on the strength properties of the parent rock type.
Moderately weathered	MW/W3	Staining or discoloration extends throughout all rock substance. Original colour of the fresh rock is no longer recognisable.
Highly weathered	HW/W4	Limonite staining or bleaching affects all rock substance and other signs of chemical or physical decomposition are evident. Colour and strength of the original fresh rock no longer recognisable.
Completely weathered	CW/W5	Rock has soil properties, i.e. it can be remoulded and classified according to the USCS, although texture of the original rock can still be recognised.

outlines the data that must be collected, the procedures that are followed and the terminology and classification systems that are used.

2.2.2 General geotechnical logging

As noted above, outcrop mapping includes both regional and mine-scale surface outcrop mapping during development prior to mining and bench mapping once mining has commenced. Accordingly, the level of detail captured must not only be relevant to the level of investigation, but must also be presented at the appropriate scale. This requires thought and careful planning to set the scene before any mapping is performed. Scene setting includes understanding the geology that is to be mapped, determining what is relevant to the task in hand, setting the appropriate scale, preparing the field logging sheet, deciding on the level of data that is to be recorded and selecting the right mapping tools.

In all cases the data recorded on the field logging sheet must include at least the following items.

1 The identification of the exposure being mapped, including the northing and easting coordinates and reduced level of a reference mapping point, the mapping scale, the name of the person who carried out the logging and the date logged.
2 The rock type, the degree of weathering and/or alteration and the strength of the intact rock. Most mine sites will have a two or three letter alphanumeric code to describe the rock type. The degree of weathering and/or alteration should be estimated following the standard International Society of Rock Mechanics (ISRM 2007) classifications outlined in Tables 2.1 and 2.2. The strength of the intact rock should be estimated using the standard ISRM scale given in Table 2.3. Field estimates of the strength and relative density of soils materials are given in Tables 2.8 and 2.9 (Tomlinson 1978; AusIMM 2001).
3 The nature of the structural defects that occur in the exposure. This should include:
   → orientation (dip and dip direction);
   → frequency, spacing and persistence (observed length);
   → aperture (width of opening);
   → roughness;
   → thickness and nature of any infilling;
   → if a fault, the width of the zone of influence of the fault to either side of the fault plane.

A structural defect includes any natural defect in the rock mass that has zero or low tensile strength. This includes joints, faults, bedding planes, schistosity planes and weathered or altered zones.

An example field logging sheet is illustrated in Figure 2.2. Recommended terms for defect spacing and aperture (thickness) based on the Australian site investigation standards are given in Tables 2.4 and 2.5. A recommended classification system designed specifically to enable relevant and consistent engineering descriptions of defects, also based on the Australian standard, is given in Table 2.6 (AusIMM

Table 2.2: Effect of alteration on fresh rock

Term	Symbol	Description
Fresh	Fr/A1	No visible sign of alteration; perhaps slight discoloration on defect surfaces.
Slightly altered	SA/A2	Alteration confined to veins and/or veinlets. Little or no penetration of alteration beyond vein/veinlet boundaries. No discernible effect on the strength properties of the parent rock type.
Moderately altered	MA/A3	Alteration is controlled by veins and may penetrate wall rock as narrow vein selvages or envelopes. Alteration may also be pervasive but weakly developed. Modifications to the rock are small.
Highly altered	HA/A4	Pervasive alteration of rock-forming minerals and intact rock to assemblages that significantly change the strength properties of the parent rock type.
Completely altered	CA/A5	Intensive, pervasive, complete alteration of rock-forming minerals. The rock mass may resemble soil. For hydrothermal alteration, any alteration assemblage that results in the nearly complete or complete change of rock strength relative to the parent rock type.

ROCK MASS CELL MAPPING SHEET

General Information

Cell Location			Inspector	Blast No.
X	Y	Z		
Date	Slope Length	Slope Height	Face Orientation	Sketch / Photo

Rock Mass Conditions

Intact Rock Strength	R6		R5		R4		R3		R2		R1		<R0
Rating	15		12		7		4		2		1		0
Weathering	None		Slight			Moderate			High			Decomposed	
Av. FF/m (Spacing)	<0.1 10m	0.2 5m	0.3 3m	0.5 2m	1 1m	2	3	5	10	15	20	30	40
Rating	40	38	35	31	27	21	18	15	10	7	5	1	0

Joint Conditions

Persistence	<1m	1-3m	3-10m	10-20m	>20m
Rating	6	4	2	1	0
Fill Width	None	Stained	1mm	1-5mm	>5mm
Rating	6	5	4	1	0
Roughness	V. Rough	Rough	Sl. Rough	Smooth	Slicken
Rating	6	5	3	1	0
Fill Strength	None	Hard <5mm	Hard >5mm	Soft <5mm	Soft >5mm
Rating	6	4	2	2	0
Weathering	None	Slight	Mod	High	Decomp
Rating	6	5	3	1	0

Water

Groundwater	Dry	Damp	Wet	Dripping	Flowing
Rating	15	10	7	4	0

Defects

Set	Dip	Dip Dir	Freq.
1			
2			
3			
4			
5			
6			
7			
8			
9			
10			
11			
12			
13			
14			

Sketch

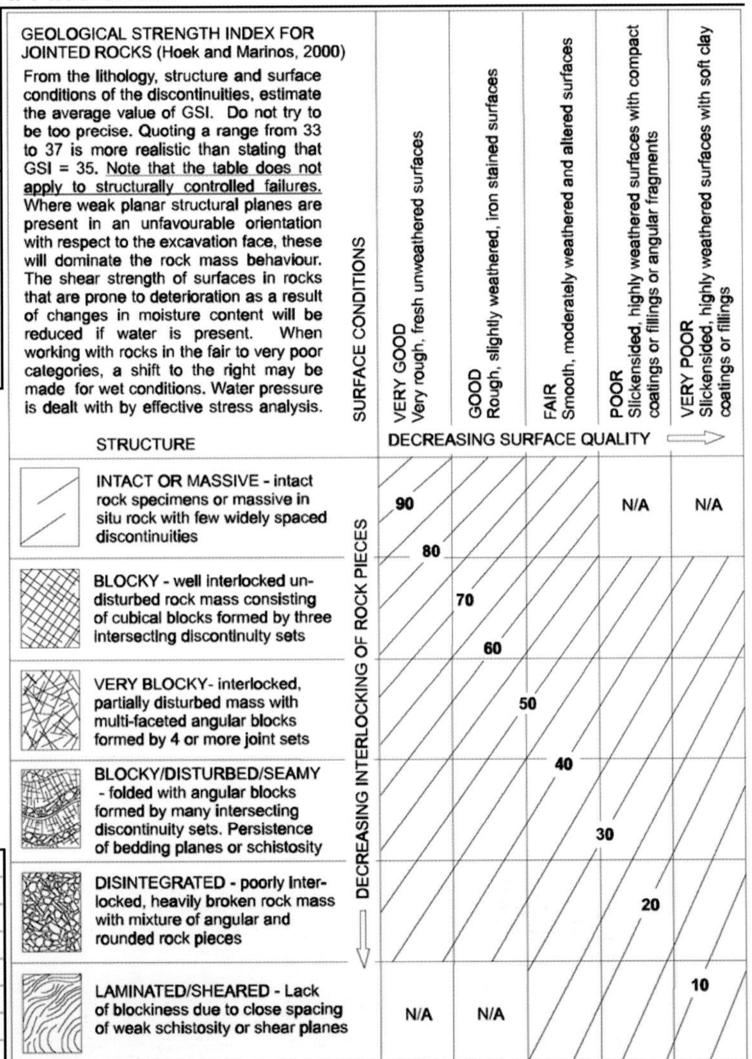

Blast Damage - Quantify

Crest Break Back		Hard Toes	
Highly Fractured		Pre-split Holes Observed	
Poor Fragmentation		Other	

Figure 2.2: Example field logging sheet for surface outcrop and bench mapping
Source: Courtesy SRK Consulting

Table 2.3: Field estimates of uniaxial compressive strength (UCS)

ISRM grade	Term	UCS (MPa)	Is_{50} (MPa)	Field estimate of strength
R6	Extremely strong	>250	>10	Rock material only chipped under repeated hammer blows, rings when struck.
R5	Very strong	100–250	4–10	Requires many blows of a geological hammer to break intact rock specimens.
R4	Strong	50–100	2–4	Handheld specimens broken by a single blow of a geological hammer.
R3	Medium strong	25–50	1–2	Firm blow with geological pick indents rock to 5 mm, knife just scrapes surface.
R2	Weak	5–25	***	Knife cuts material but too hard to shape into triaxial specimens.
R1	Very weak	1–5	***	Material crumbles under firm blows of geological pick, can be shaped with knife.
R0	Extremely weak	0.25–1	***	Indented by thumbnail.

2001). Note that the terminology used in Table 2.6 describes the actual defect, not the process that formed or might have formed it. As with the rock type descriptions, most mine sites will have a two or three letter alphanumeric code to describe mineralised infillings, but any soil-like infilling within the defects should be described using the Unified Soils Classification System (ASTM D2487, Table 2.7).

4 Moisture condition, noting any seepage zones.
5 Hoek-Brown/GSI classification.
6 A geological plan showing the distribution of the features identified in the exposure, including the rock types, the altered and/or weathered zones, the structural defects and any seepage zones.

2.2.3 Mapping for structural analyses

Structural data are a key input for kinematic, limit equilibrium and numerical slope design analyses. Gathering these data and estimating how the orientation and spatial distribution characteristics of the joint sets and faults vary across the walls of the mine is thus one of the most important structural modelling activities (Chapter 4).

Mapping techniques used for detailed structural data gathering usually fall into one of the following three types:

1 line mapping;
2 window (cell) mapping;
3 digital imaging.

Table 2.4: Terms for defect spacing

Term	Spacing (mm)
Extremely close	< 20
Very close	20–60
Close	60–200
Medium	200–600
Wide	600–2000
Very wide	>2000

2.2.3.1 Line mapping

Scanline mapping involves measuring and recording the attributes of all the structures that intersect a given sampling line. The technique has been used in mining and civil engineering for many years and has been well documented by a number of authors (Priest & Hudson 1981; Windsor & Thompson 1997; Harries 2001; Brown 2003). It is illustrated in Figure 2.3.

In Figure 2.3 the observable structures in the outcrop (usually a bench face) are shown to the left and the structures selected for mapping are shown to the right. The length of the scanline is usually matched to a prerequisite number of measurements, although there is no firm agreement on the prerequisite number. Priest (1993) suggested that 150–350 measurements should be made, with the lower number sufficient for a rock mass containing three structural sets and the larger number for a rock mass containing up to six sets. Savely (1972) suggested that a minimum of 60 measurements are required to define a set. Villaescusa (1991) suggested that at least 40 are required. For project work, it is suggested that the minimum number per set should be decided on a site-by-site basis.

2.2.3.2 Window mapping

Window mapping involves collecting all the structural data above a given cut-off size from within a specified area

Table 2.5: Terms for defect aperture (thickness)

Term	Aperture (mm)
Tight	0
Very narrow	0–6
Narrow	6–20
Moderately narrow	20–60
Moderately wide	60–200
Wide	200–600
Very wide	600–2000
Cavernous	>2000

Table 2.6: Common defects in a rock mass

TERM [1] GENERAL	LAYERING (LAYER) [2]			FRACTURES and FRACTURED ZONES		WEAK SEAMS or ZONES		
SPECIFIC	**BEDDING**	**FOLIATION**	**CLEAVAGE**	**JOINT**	**SHEARED ZONE**	**CRUSHED SEAM/ZONE**	**DECOMPOSED SEAM/ZONE**	**INFILLED SEAM/ZONE**
PHYSICAL DESCRIPTION	Arrangement in layers, of mineral grains of similar sizes or composition, and/or arrangement of elongated or tabular minerals near parallel to one another, and/or to the layers Bedding: Generally no microfractures Foliation/Cleavage: Discontinuous microfractures may be present, near parallel to the layering.			A discontinuity or crack: planar, curved or irregular, across which the rock usually has little tensile strength. The joint may be open (filled with air or water) or filled by soil substances or by rock substance which acts as a cement; joint surfaces may be rough, smooth, or slickensided	• Zone with roughly parallel planar boundaries, of rock material intersected by closely spaced (generally < 5 cm) joints and/or microscopic fracture (cleavage) planes. The joints are at small angles to the zone boundaries; they are usually slightly curved and divide the mass into unit blocks of lenticular or wedge shape; their surfaces are smooth or slickensided TYPE 'R' ranging to TYPE 'S' → Joints tightly closed cemented, but cements (usually chlorite or calcite) are weaker than the rock substance Joints not cemented but either coated with soil substances or are open, filled with air and/or water	Zone with roughly parallel planar boundaries, composed of disoriented, usually angular fragments of the host rock substance. The fragments may be of clay, silt, sand or gravel sizes, or mixtures of any of these. Some minerals may be altered or decomposed but this is not necessarily so. Boundaries commonly slickensided	Zone of any shape, but commonly with roughly parallel planar boundaries in which the rock material is discoloured and usually weakened. The boundaries with fresh rock are usually gradational. Geological structures in the fresh rock are usually preserved in the decomposed rock. 'Weathered' and 'altered' are more specific terms	Zone of any shape, but commonly with roughly parallel planar boundaries composed of soil substance. May show layering roughly parallel to the zone boundaries. Geological structures in the adjacent rock do not continue into the infill substance
ENGINEERING PROPERTIES [3 4]	• Where uniformly developed in a rock substance any of these types of structure render that rock substance anisotropic in its behaviour under stress • Compressive } Strengths { min. when β = 30° to 45° Initial shear } usually { max. when β = 0° and 90° Tensile strength usually { max. when β = 0° { min. when β = 90° Deformation modulus usually higher for β = 0° than for β = 90° Where *not* uniformly developed, these structures represent defects in the rock mass, i.e. as individual layers or layered zones			• Tensile strength low/zero • Sliding resistance depends upon properties of coatings or cement and/or condition of surfaces PARAMETERS c cohesion of coating/cement/wall-rock ϕ friction angle of coating/cement/wall-rock t angle of roughness of surface k_n normal stiffness k_e tangential stiffness	• Rock properties, very fissile rock mass. • When excavated forms GRAVEL (generally GP) SOIL properties, GRAVEL (GP, GM or GC) Both types show extreme planar anisotropy. Lowest shear strength in direction of slickensides, in plane parallel to boundaries.	• SOIL properties: either cohesive or non-cohesive • Usually shows planar anisotropy; lowest shear strength in direction of slickensides in plane parallel to boundaries	• Extremely decomposed (XD) seam has SOIL properties usually cohesive but may be non-cohesive • Mostly very compact except when soluble minerals removed • Slightly to highly decomposed substances ROCK properties but usually lower strengths than the fresh rock substance	• SOIL properties: usually cohesive (CL or CH) but may be non-cohesive
				Engineering properties commonly different from place to place especially where the defect passes through several different rock substance types				
EXTENT	Usually governed by the thickness and lateral extent of the rock substance or mass containing the defect Foliation/Cleavage: May occur in a zone continuous through several different rock substance types			From 1 cm to 50 m or more: depends on origin	Generally large (50 m to many km)		Weathered zones related to present or past land surface limited extent. Altered zones occur at/to any depth	Usually small limited to mechanically weathered zone. Can be great in rocks subject to solution
ORIGIN (USUALLY CONTROLS) EXTENT	Deposition in layers	• Viscous flow • Crystal growth at high pressures and temperatures • Shearing under high confining pressure	• Shearing during folding or faulting • Consolidation, compaction	• Shearing, extension or torsion failure; arising from faulting, folding, relief of pressure, shrinkage due to cooling or loss of fluid	**FAULTING** Shear failure by small displacements along a large number of near-parallel intersecting planes. The different strengths of Types R and S are usually due to (a) different depths of rock cover at the time of faulting or (b) Later cementation or (c) Later mechanical weathering	• Failure by large movement within narrow zone • Generally formed at shallow depth (< 3000 m)	• Decomposition of minerals, removal or rupture of cement, due to circulation of mineralized waters usually along joints sheared zones or crushed zones	• Cohesive soil carried into open joint or cavity as a suspension in water • Non-cohesive soil falls or washes in
DESCRIPTION REQUIRED	Bed thickness, grain types and sizes	Fabric description, and spacing and extent of microfractures		Shape, aperture, surface condition, coating, filling, extent	Zone width, shape and extent Pattern of joints or micro-fractures and resulting shape and size of unit blocks. Standard description of joints		Degree of decomposition	
	Ease of splitting and nature of fracture faces					Standard description of soil or rock substance		
ASSOCIATED DESCRIPTION ETC	Graded-, discord- and-, slump-bedding; other primary structures: Facing, Attitude, Lineations	Attitude of planes and of any linear structure, extent		Spacing: attitude of joint and of slickensides	Attitude of zone. Direction of slickensides and amount, direction, and sense of displacement. Type of fault. History of past movements. Any modern activity. Likelihood of future movements. The terms 'major' and 'minor' fault are defined whenever used. The definitions are made on the basis of: (a) width and nature of the fault materials, (b) significance to the profect		Attitude of zone. Classify as weathered or altered if possible and determine origin, and defect or defects influencing decomposition	Attitude of zone. Type of defect which is infilled, origin of infill substance
		Allocate to set, determine origin type						
TERMS NOT USED (FOR THESE DEFECTS) [5]	Strata, stratification, schistosity, gneissosity, micro-fissuring			Fissure, crack, slip, shear, break, fracture (except in general sense for joints, faults, cleavage planes)	Shear-, shatter-, shattered-, crush-, broken-, blocky-, zone; slip, shear, mylonite, gouge, breccia, fault-breccia, crush breccia, pug The terms 'fault' or 'fault-zone' are only used in a genetic or general sense and must be qualified by the use of the defined terms given above. 'Mylonite' is rock substance with intense planar foliation, developed due to shearing at great depth beneath the earth's crust		Rotten, disintegrated, softened, soft (unless in defined sense for clay)	Vein, fissure, pug, gouge
MAP SYMBOLS (HORIZ., VERT., DIPPING) [6]	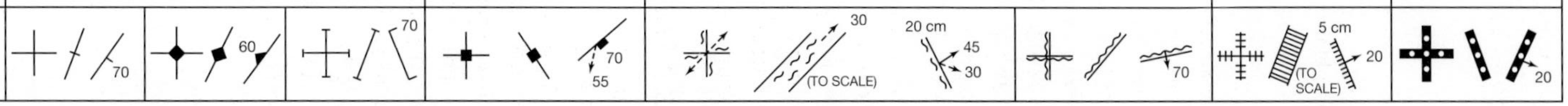							

Note that the terminology in Table 2.6 describes the actual defect, not the process that formed it. Similarly, the described properties refer to the engineering properties of the defect, not those of the rock mass containing the defect.
Source: AusIMM (2001), courtesy SAI Global.

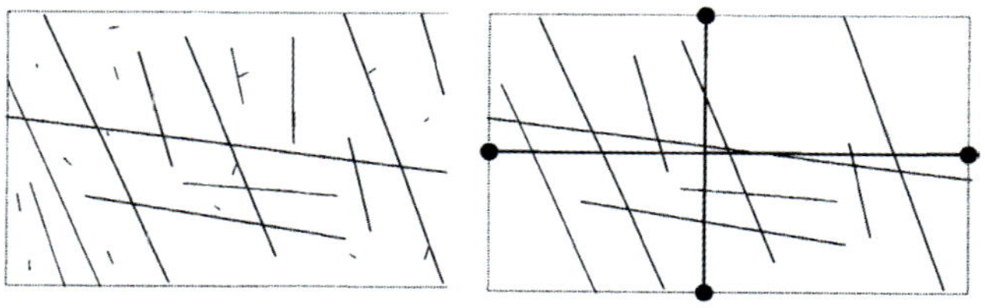

Figure 2.3: Scanline mapping technique
Source: Harries (2001)

of a rock face. Alternatively, only the attributes of each of the sets recognised within the window may be recorded (e.g. orientation, length, spacing and nature of infilling on each set), although caution is required as this procedure may introduce subjective biases into the data. In Figure 2.4 the observable structures in the outcrop (again a bench face) are shown to the left and the structures selected for mapping are shown to the right.

In an open pit mine, typically a number of windows will be located at regular intervals within each of the mapping units recognised along the benches. The spacings between windows should be decided on a site-by-site basis, but typically should provide for a 10–25% coverage of the mapping unit, depending on the geological complexity. Major structures that occur between the windows should be spot mapped individually.

2.2.3.3 *Digital imaging*

The use of 3D digital photogrammetric and laser imaging technology for structural mapping in open pit mines has increased dramatically within the last few years. The Sirojoint®[1] and 3DM Analyst®[2] digital photogrammetric systems in particular have become firmly established as routine methods of mapping exposed rock faces in both open cut and underground environments. The technology is illustrated in Figures 2.5 and 2.6.

Digital photogrammetry integrates 3D spatial data with 2D visual data to create spatially accurate representations of the surface topology of the rock. Structural properties such as orientation, length, spacing, surface roughness and distribution type can be determined remotely and accurately over long distances and in areas where access is difficult and/or unsafe. Reported accuracies range from the order of 2 cm at distances of 50 m to 10 cm at distances of up to 3 km. These features have enabled rapid, accurate, safe and low-cost geological mapping at bench and multi-bench scale using the system software or by downloading the data into mine planning software such as Vulcan™, DataMine™, MineSite™ and Surpac™. The integration of the imaging software with such mine planning systems provides the additional benefit that the data can be used in real time for mine design, mine planning and mine operating purposes.

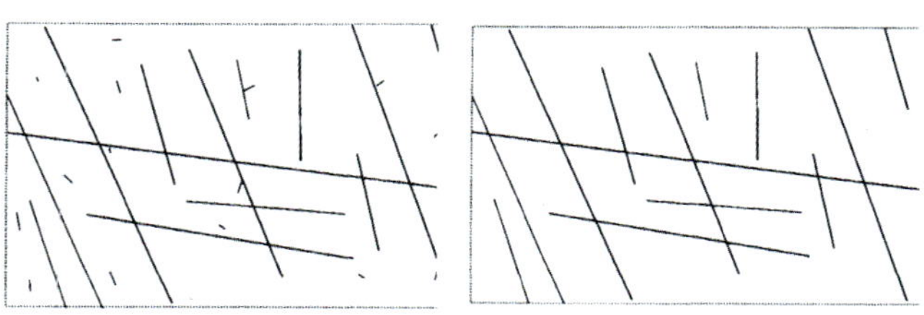

Figure 2.4: Window mapping technique
Source: Harries (2001)

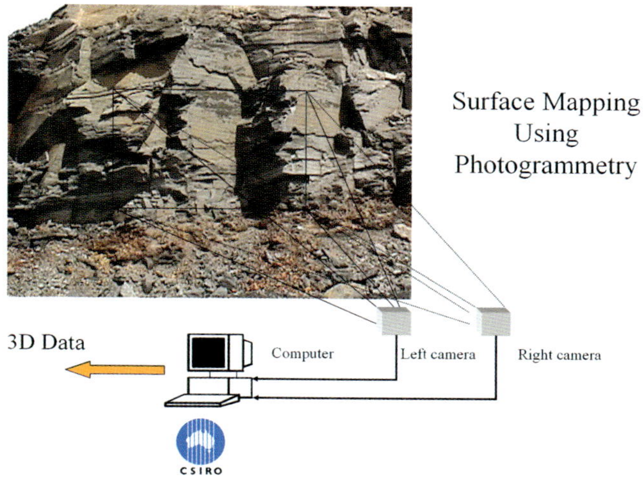

Figure 2.5: Gathering a digital photographic image of an outcrop
Source: Courtesy CSIRO

2.2.3.4 *Practical considerations*

Sampling bias and orientation measurement errors are the traditional line and window mapping issues, on the surface and underground. In open pit mining, worker safety and the time taken to map scanlines and/or windows along the benches have also become issues.

Figure 2.6: Joint orientation (equal area, lower hemisphere projection) and spacing information provided from a stereographic image of an outcrop
Source: Courtesy CSIRO

In scanlines and windows four types of sampling bias are recognised (Brown 2007):

- orientation bias;
- size bias;
- truncation or cut-off bias;
- censoring bias.

Orientation bias depends on the orientation of the scanline or window relative to the orientation of the structure. Clearly, if a structure is parallel to a scanline or window then few members of that set will be recorded. When considering size bias, the larger the structure, the more likely it is to be sampled by the scanline or window. Inversely, if a small cut-off size is used, then the size distribution of all of the structures along the scanline or inside the window may not be properly accounted for.

Understanding the nature and effect of censoring bias is important, especially when collecting data that will eventually be used in Discrete Fracture Network (DFN) modelling (section 4.4.3). The censor window is the area within which the trace lengths can be accurately measured. Joint traces that extend outside these sections are said to be censored. If both ends of the trace terminate within the window (i.e. the trace is uncensored), then something is known about the joint's persistence and size. If the trace length to termination cannot be seen or measured, then a lot less is known about its persistence. To prepare a valid DFN model there must be enough uncensored (or measured) data to arrive at a statistically viable joint size. The DFN modelling process cannot work if too many joints are censored or if censoring information is not available.

Digital photogrammetry has simplified these issues, particularly with respect to orientation accuracy, orientation bias, trace lengths (cut-off size and censoring), efficiency of mapping and worker safety.

Measurement errors in scanline and window mapping have been reported to be as much as ±10° for dip direction and ±5° for dip angle (Brown 2007). Inaccuracies of this order have been overcome by digital photogrammetry, with differences of only ±1° now being reported for dip direction and dip angle. The ability to vary the scale of mapping from bench to inter-ramp to overall pit scale from the one location is another major advantage of digital photogrammetry. This flexibility gives the user the means to examine the structural fabric at bench scale or to map large structures over multiple benches, which helps to reduce cut-off size and censoring biases. Orientation bias will always be difficult to overcome, but it is possible to address this issue by moving the camera to positions where structures visible in the ends of benches and re-entrants in the wall can be captured.

Other major benefits of digital photogrammetry are its flexibility and remote access capability, which can significantly reduce the time taken to gather the field data and remove the operators from potentially hazardous situations (Figure 2.7). In many jurisdictions, it is no longer allowable to work directly beneath open pit mine benches.

Figure 2.7: Potentially hazardous bench mapping conditions
Source: Photo courtesy 3G Software & Measurement

As noted above (section 2.2.3.3), the integration of the imaging software with mine planning software systems provides the additional benefit that the data can be used in real time for mine design, mine planning and mine operating purposes. It also provides a permanent 3D record of the mapped areas.

The disadvantages of digital imaging systems are that they still require ground proofing and cannot be used to determine the physical features of the structures, particularly surface roughness and the thickness and nature of any infillings. Their ability to accurately define flat-lying and vertically inclined structures is also questionable. However, these disadvantages can be minimised with a well-planned ground proofing and sampling program when the mapping and structural assessment process has been completed.

2.2.4 Surface geophysical techniques

2.2.4.1 Seismic methods

Seismic reflection methods have been used successfully in both sedimentary and hard rock environments for mine planning purposes (Henson & Sexton 1991; Pretorius et al. 1997). However, traditional seismic methodologies that were successful for petroleum resources have had to be extensively modified for hard rock applications. While there is a perception that seismic methods are expensive, the acoustic impedance (density and seismic velocity) information they provide can be invaluable as it essentially is a 3D image of the subsurface.

For coal mining purposes, analysis of seismic data can provide detailed structural information including the location, nature and throw of faults, definition of fracture zones and the identification of seam splitting and thickness. Also, amplitude information has been related to methane desorption (Cocker et al. 1997).

For hard rock metalliferous mining purposes, most seismic studies to date have concentrated on deposits that currently would not be considered suitable for open cut operations. However, useful pre-mining information can be gained in nearly any situation. For example, seismic studies at the Witwatersrand Basin and Bushveld Complex provided structural and lithologic information that was not viable by other means (Campbell & Crotty 1990; Campbell 1994). More recently, high-resolution imaging of near-surface deposits has been demonstrated (Urosevic et al. 2002).

Specialist near-surface seismic methodologies have been developed. For example, in sedimentary coal sequences 'Converted-Wave (PS) Seismic' can provide independent validation of mapped structures and clearer, more coherent near-surface images (Hendrick 2006). 'Surface Wave Seismic' is a seismic refraction technique that has been specifically developed to provide surface hardness and velocity information (O'Neill et al. 2003), which should be creatable with open pit mining parameters such as diggability and blastability.

2.2.4.2 *Potential field, electrical and electromagnetic methods*

At the deposit scale, a range of surface-based non-seismic geophysical methods can be used to generate subsurface parameters that can provide useful information for mine-planning purposes. Surface techniques that are amenable to inverse modelling so that voxel-volumes of petrophysical properties can be generated include time-domain electromagnetics, DC resistivity and induced polarisation, gravity and magnetics (Napier et al. 2006; Li & Oldenburg 1996, 1998, 2000).

At the Century Zinc deposit, Mutton (1997) described the use of high-resolution surface IP/resistivity survey data to map ore contacts and variations in ore quality, and for geotechnical requirements. Mutton also reported on the geotechnical use of an electromagnetic surface technique, Controlled Source Audio-Frequency Magnetotelluric (CSAMT). This technique was used to locate large blocks of detached Proterozoic shale within overlying Cambrian limestone, which were considered to be a geotechnical hazard for pit slope stability. CSAMT was also used to determine the thickness of the surrounding water-saturated limestone so as to estimate the likely water flow into the open pit during excavation. Figure 2.8 shows a plan of the resistivity model obtained from inversion of

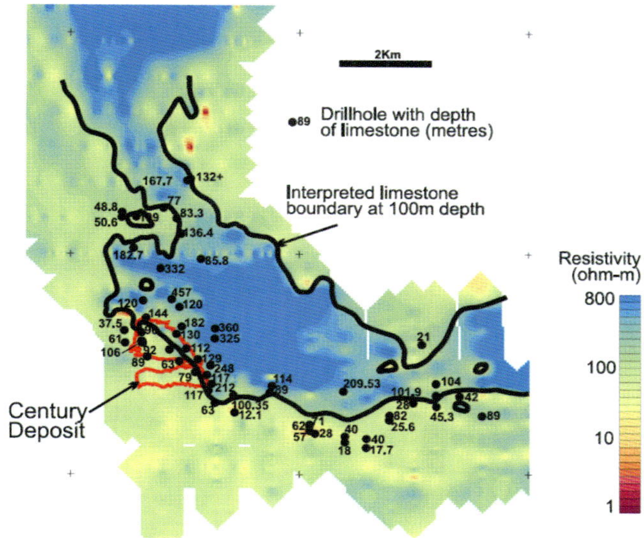

Figure 2.8: Century deposit region showing smooth-model inversion of CSAMT resistivity at 100 m depth, compared with limestone depth from drilling. The more resistive areas (blue) represent limestone greater than 100 m thick
Source: After Mutton (1997)

the CSAMT data and collated at 100 m depth. The blue colours represent the presence of resistive limestone, while the warmer colours indicate the presence of less resistive shale and siltstone.

In a landmark paper, Philips et al. (2001) detailed the compilation and interpretation of a number of 3D petrophysical property models over the San Nicholás copper-zinc deposit in Mexico. Figure 2.9 shows a simplified geological cross-section of the San Nicholás deposit as determined from drill holes for comparison with the inverted petrophysical property model sections shown on Figure 2.10. As a next step in the use of these data to derive geotechnical and mining parameters, they need to be segmented into packages with similar properties then calibrated against measured samples from strategically placed drill holes.

Ground penetrating radar (GPR) is an electromagnetic analogue of the seismic method, but with limited depth penetration. GPR in reflection mode performs best in resistive rocks as the waves are attenuated in conductive materials. GPR can be used to detect lithology and structures; it tends to be highly sensitive to clays.

2.3 Overburden soils logging

2.3.1 Classification

The global standard for the engineering logging and classification of overburden soils is the Unified Soils Classification System (USCS – ASTM D2487, Table 2.7). The basis of the system is that coarse-grained soils are logged

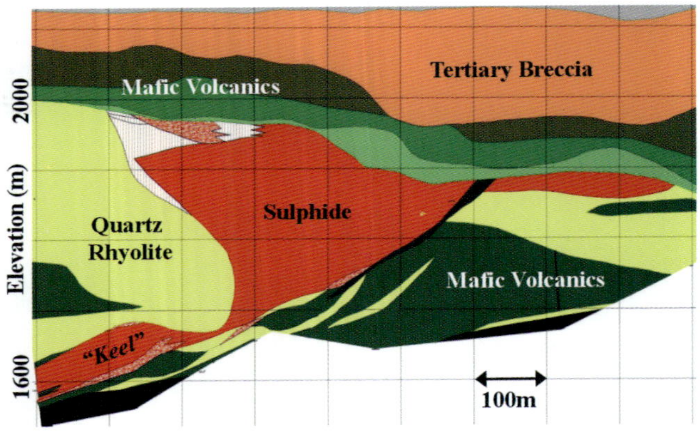

Figure 2.9: Simplified geologic cross-section of the San Nicolas deposit (line 400 south) as interpreted from drill holes (looking north)
Source: After Philips et al. (2001)

according to their grain size distributions and fine-grained soils according to their plasticity. Thus, only grain size analyses and Atterburg Limits tests are needed to completely identify and classify a soil (Holtz & Kovacs 1981).

There are four major divisions in the USCS: coarse-grained, fine-grained, organic soils and peat. The classification is performed on material passing a 75 mm sieve, with the amount of oversize being noted on the drill log. Particles greater than 300 mm equivalent diameter are termed boulders, and material between the 300 mm and 75 mm sieves are termed cobbles. Coarse-grained soils are comprised of gravels (G) and sands (S) having 50% or more material retained on the No. 200 sieve. Fine-grained soils (silt, M, and clay, C) are those having more than 50% passing the No. 200 sieve. The highly organic soils and peat can generally be divided visually.

The gravel (G) and sand (S) groups are divided into four secondary groups (GW and SW; GP and SP; GM and SM; GC and SP) depending on grain size distribution and the nature of fines in the soils. Well-graded soils have a good representation of all particles sizes; poorly graded soils do not. The distinction can be made by plotting the grain size distribution curve and computing the coefficients of uniformity (C_u) and curvature (C_c) as defined in the upper right-hand side of Table 2.7. The GW and SW groups are well-graded gravels and sands with less then 5% passing the No. 200 sieve. The GP and SP groups are poorly graded gravels and sands with little or no non-plastic fines.

The particle size limits given above are those adopted by ASTM D2487, which is published in the USA. Different limits may be adopted in different countries. For example, the Australian Standard (AS 1726-1993) adopts different limits, which are 2–60 mm for gravel, 0.062 mm for sand and less than 0.06 mm for silt and clay. As 60 mm, 2 mm and 0.06 mm sieves are not normally used, the percentage passing these sizes must be identified from a laboratory test using regular sieve sizes.

The fine-grained soils are subdivided into silt (M) and clay (C) on the basis of their liquid limit and plasticity index. Fine-grained soils are silts if the liquid limit (LL) and plasiticity index (PI) plot below the A-line on the

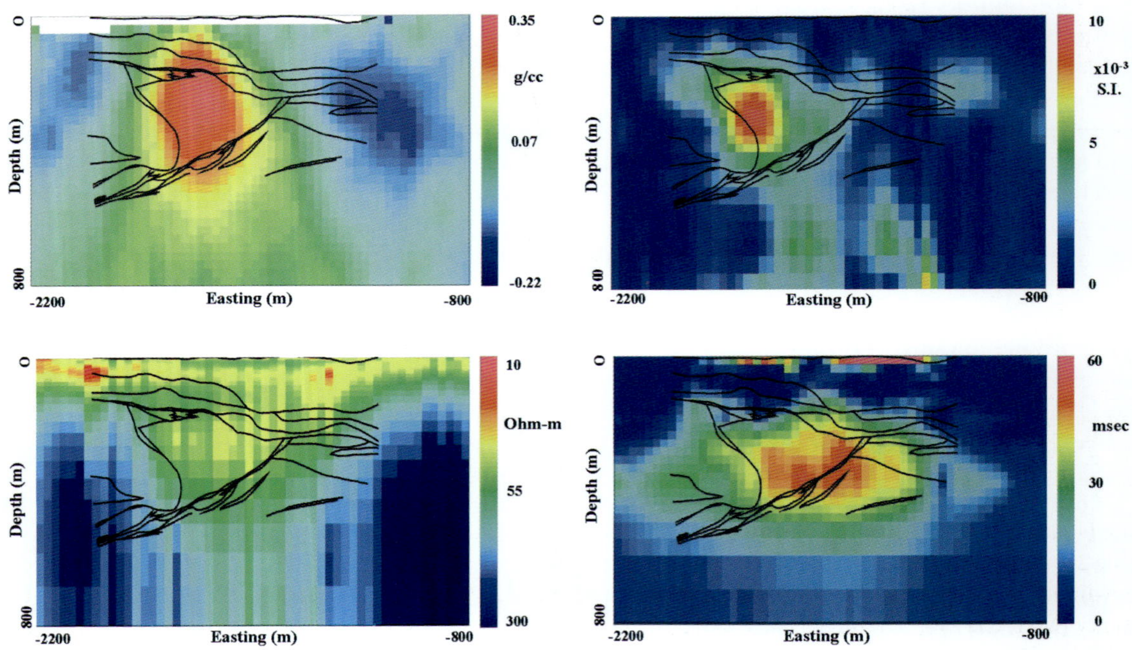

Figure 2.10: North-facing cross-section of physical property models at line 400 south with geology overlaid. (a) Density contrast model. (b) Magnetic susceptibility model. (c) Resistivity model. (d) Chargeability model
Source: After Philips et al. (2001)

Table 2.7: Unified Soils Classification System (ASTM D2487)

Major divisions			Group Symbols	Typical Names	Laboratory Classification Criteria		
Coarse-grained soils (More than half of material is larger than No. 200 sieve size)	Gravels (More than half of course fraction is larger than No. 4 sieve size)	Clean gravels (Little or no fines)	GW	Well graded gravels, gravel-sand mixtures, little or no fines	Determine percentages of sand and gravel from grain-size curve. Depending on percentage of fines (fraction smaller than No. 200 sieve size), course-grained soils are classified as follows: Less than 5 percent: GW, GP, SW, SP; More than 12 percent: GM, GC, SM, SC; 5 to 12 percent: *borderline* cases requiring dual symbols[b]	$C_u = \frac{D_{60}}{D_{10}}$ greater than 4; $C_c = \frac{(D_{30})^2}{D_{10} \times D_{60}}$ between 1 and 3	
			GP	Poorly graded gravels, gravel-sand mixtures, little or no fines		Not meeting all gradation requirements for GW	
		Gravels with fines (Appreciable amount of fines)	GM[a] d / u	Silty gravels, gravel-sand-silt mixtures		Atterberg limits below "A" line or P.I. less than 4	Above "A" line with P.I. between 4 and 7 are *borderline* cases requiring use of dual symbols
			GC	Clay gravels, gravel-sand-clay mixtures		Atterberg limits below "A" line with P.I. greater than 7	
	Sands (More than half of course fraction is smaller than No. 4 sieve size)	Clean sands (Little or no fines)	SW	Well-graded sands, gravelly sands, little or no fines		$C_u = \frac{D_{60}}{D_{10}}$ greater than 6; $C_c = \frac{(D_{30})^2}{D_{10} \times D_{60}}$ between 1 and 3	
			SP	Poorly graded sands, gravelly sands, little or no fines		Not Meeting all gradation requirements for SW	
		Sands with fines (Appreciable amount of fines)	SM[a] d / u	Silty sands, sand-silt mixtures		Atterberg limits above "A" line or P.I. less than 4	Limits plotting in hatched zone with P.I. between 4 and 7 are *borderline* cases requiring the use of dual symbols
			SC	Clayey sands, sand-silt mixtures		Atterberg limits above "A" line or P.I. greater than 7	
Fine-grained soils (More than half of material is smaller than No. 200 sieve size)	Silts and clays (liquid limit less than 50)		ML	Inorganic silts and very fines sands, rock flour, silty or clayey fine sands, or clayey silts with fine plasticity	Plasticity Chart (see below)		
			CL	Inorganic clays of low to medium plasticity, gravelly clays, sandy clays, silty clays, lean clays			
			OL	Organic silts and organic silty clays of low plasticity			
	Silts and clays (liquid limit greater than 50)		MH	Inorganic silts, micaceous or diatomaceous fine or silty soils, elastic silts			
			CH	Inorganic clays of high plasticity, fat clays			
			OH	Organic clays of medium to high plasticity, organic silts			
	Highly organic soils		Pt	Peat and other highly organic soils			

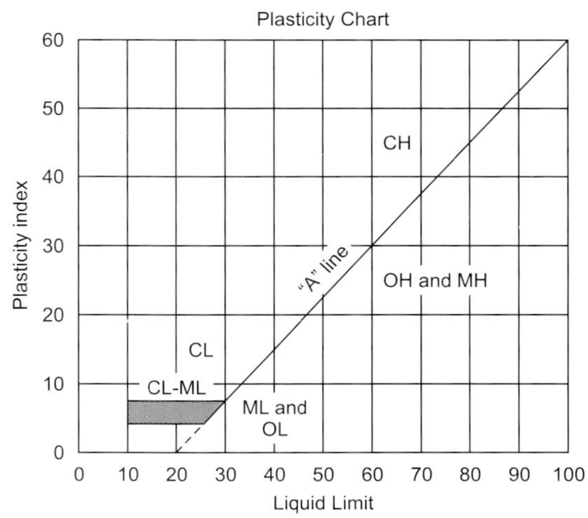

Casagrande (1948) plasticity chart in the lower right-hand side of Table 2.7. They are clays if the LL and PI values plot above the A-line. The distinction between silts and clays of high plasticity (MH, CH) and low plasticity (ML, CL) is set at a liquid limit of 50.

Coarse-grained soils with more than 12% passing the No. 200 sieve are classified as GM and SM if the fines are silty, and GC and SC if the fines are clayey. Soils with 5–12% fines are classed as borderline and have a dual symbol. The first part of the dual symbol indicates whether the soil is well-graded or poorly graded. The second part describes the nature of the fines. For example, SW-SC is a well-graded sand with some fines that plot above the A-line.

Table 2.8: Field estimates of the strength of fine-grained soils

Consistency	Term	Approximate strength (kPa)	Tactile test	SPT N-value
Very soft	S1	<25	Easily penetrated 5 cm by fist	<2
Soft	S2	25–50	Easily penetrated 5 cm by thumb	2 -4
Medium	S3	50–100	Penetrated 5 cm by thumb with moderate effort	4–8
Stiff	S4	100-200	Readily indented by thumb but penetrated only with great effort	8–15
Very stiff	S5	200–400	Readily indented by thumbnail	15–30
Hard	S6	>400	Indented with difficulty with thumbnail	>30

Fine-grained soils can also have dual symbols. The shaded zone on Table 2.2 is one example (CL-ML). It is also recommended that dual symbols (e.g CL-CH) be used if the LL and PI values fall near the A-line or near the LL = 50 line. Borderline symbols can also be used for soils with about 50% fines and coarse-grained fractions (e.g. GC-CL).

2.3.2 Strength and relative density

Field estimates of the strength and relative density of soils materials are given in Tables 2.8 and 2.9 (Tomlinson 1978; AusIMM 2001).

2.4 Core drilling and logging

2.4.1 Introduction

In open pit mining rotary core drilling is the most widely used method of subsurface investigation. For pit slope design, it helps determine within acceptable levels of confidence the geotechnical relationships and engineering properties of the rocks that will form the walls of the pit. To meet this requirement all drilling campaigns must include each of the items shown in Figure 2.11.

2.4.2 Planning and scoping

Planning and scoping the objectives of the drill hole are the most important steps of the drilling investigation. There must be clear primary and secondary objectives to extract the maximum amount of potential information. For example, geotechnical data collection may be the primary objective of the hole, but at the same time it may be possible to gain important geometallurgical and/or geohydrological information and/or use the completed hole for groundwater or other monitoring purposes. Ideally, before objectives are finalised they should be reviewed by a multidisciplinary team to ensure that all such possibilities have been taken into account.

There are other critical points.

- Before the location and orientation of the drill hole are finalised, the objectives of the hole must be checked to ensure they are consistent with the current geological, structural and hydrogeological models.
- When they have been finalised, the objectives of the drill hole must be recorded in a written memorandum that includes alternative actions in case drilling difficulties are encountered and/or it is not possible to complete the hole. The memorandum must be signed-off by all members of the team responsible for preparing the document.
- Before drilling commences, the rig site should be reviewed to ensure its location is compatible with all current and planned mining activities in the area.
- When drilling commences, it is essential that the core be photographed and logged by a properly qualified and experienced person at the rig site before it is disturbed and moved from the site to the core shed.
- Each step in the drilling process must be owned by the appropriate person. For example, the driller must

Table 2.9: Field estimates of the relative density of coarse-grained soils

Density	Relative density (%)	SPT N-value
Very loose	<15	<4
Loose	15–35	4–10
Medium	35–65	10–30
Dense	65–85	30–50
Very dense	>85	>50

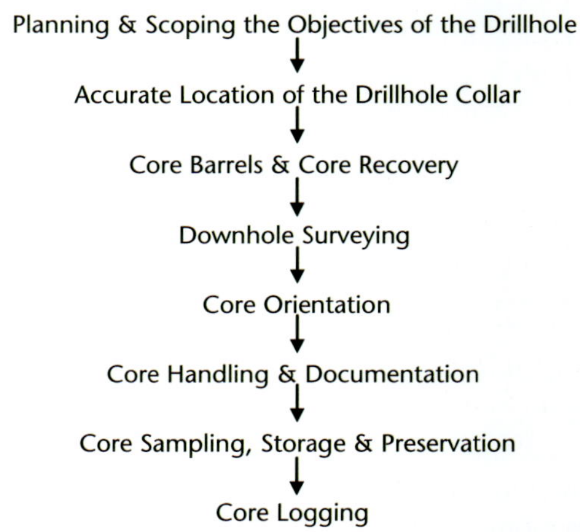

Figure 2.11: Process requirements for core drilling and logging

accept responsibility for the core recovery process, the engineering geologist for the core logging and any downhole testing, and the environmental team for decommissioning the site.

- A plan and geological section showing the drill hole trace and the expected geological/structural pierce points should be available to the drillers and loggers at the rig site.
- The drilling and logging and any downhole testing must be regularly reviewed using an appropriate QA/QC procedure.
- The potential of the drill hole for future monitoring and/or downhole testing should be continuously reviewed.

2.4.3 Drill hole location and collar surveying

Despite the introduction of sophisticated surveying techniques such as satellite guided global positioning, the seemingly simple task of providing the coordinates and elevation of the drill hole collar remains a frequent source of error at all stages of mine development. The errors are so common that it is imperative that basic checks be routinely built into every drilling campaign. These include checking for differences between the set-out pegs and the as-drilled collar locations, which frequently are quite different, and checking that the datum of the map or computer model used to plan the campaign is identical to that used at the mine site to set out the hole.

2.4.4 Core barrels

Preferably, core drilling should be performed using triple-tube core barrels where the inner tube is split. In the case of very weak and/or degradable rock the split inner tube can be replaced by a PVC sleeve that can be capped on removal and sent directly to the laboratory. In weak ground face discharge bits should be used. These steps are critical to minimise ground disturbance, core loss and core disturbance when the core is removed from the barrel (section 2.4.7). Exceptions may occur in massive competent rock, when standard double-tube systems may suffice.

2.4.5 Downhole surveying

Drill hole deviation is potentially a significant source of error in the geological and structural models. Reliable downhole surveys are therefore a must in any drilling campaign. The decision on what type of survey method(s) is appropriate for the given drilling program is critical and must be made before drilling commences.

There are two common uses for downhole surveys:

- surveys to determine the correct geometry (dip/orientation) of the drill hole trace, typically done after the drill hole is completed;
- continuous surveys performed while drilling in order to correct any drill hole deviation and reach target areas (also known as directional drilling).

Today's modern instruments employ two basic techniques – the magnetic compass and the non-magnetic gyroscope.

2.4.5.1 *Magnetic techniques*

The accuracy of the magnetic methods depends on the latitude of the drill site, the local variation of the Earth's magnetic field and the magnetic signature of the rock mass. The most widely used magnetic downhole survey techniques are:

- single-shot instruments, which are capable of one survey per trip into the drill hole. A single-shot instrument is preferred for directional drilling when successive surveys enable periodic corrections to the direction of the drill hole;
- multi-shot instruments, which can perform several readings per trip. Surveys performed with multi-shot instruments tend to be more accurate than those performed with a single-shot instrument. Multi-shot instruments are also efficient where a large number of previously drilled holes have to be surveyed and/or resurveyed.

2.4.5.2 *Non-magnetic techniques*

Where magnetic disturbances are prevalent and in high latitudes the best downhole survey results are obtained using gyroscopic tools. Three types are now commonly available:

- free-spinning gyroscopes, operating on the basis of a known direction, with changes in azimuth referenced to the starting direction, typically the azimuth of the drill hole collar;
- rate gyroscopes, which measure the point-to-point change in azimuth while the probe is in motion along the drill hole. Typically, the output of the rate gyroscope is integrated to give a change in azimuth referenced to the drill hole collar;
- north-seeking gyroscopes, which measure an absolute azimuth referenced to the Earth's geographic axis. This measurement minimises the systematic error that can be introduced from an inaccurate drill hole collar azimuth or poor calibration.

The most accurate positional survey is a combination of the rate gyroscope for a continuous measurement of azimuth and the north-seeking gyroscope for absolute accuracy, which can now be achieved using a single tool. For drilling programs where a downhole survey is critical for the accurate location of structures or geological

Table 2.10: Core orientation techniques

Technique	Complexity in use	Advantages	Disadvantages
Direct marking			
Weighted core barrel	Low	Simple to use. Clay, plasticine or spears used to form impression.	Impression may require interpretation. Unsuitable in holes inclined at <45° or >75°.
Ballmark® system	Low	Simple to use, drilling delays minimal.	Triggering mechanism may not operate in broken ground.
Scribe system	Moderate to high	Continuous scribing of core referenced to drill hole orientation.	Difficult to interpret in incompetent and/or broken ground.
EZY-Mark™ system	High	Can operate at up or down drill hole angles.	Requires an inclined hole.
Indirect marking			
ACT electronic tool	Moderate	Orientation without marking core.	Requires training in operation. Also requires an inclined hole.
Acoustic televiewers	Moderate	Geophysical log, run after drilling. Provides a continuous record of drill hole wall that can be matched to core.	Requires a stable hole. Operates in water or mud.
Optical televiewers	Moderate	Geophysical log, run after drilling. Provides a continuous record of drill hole wall that can be matched to core.	Requires a stable hole. Operates only in air or clear water.

contacts, it is recommended to use two systems and compare the results.

2.4.6 Core orientation

A number of downhole core orientation techniques are available. The choice may depend on a number of factors, including the anticipated drilling conditions and the experience of the drilling crew, but is very often guided by equipment cost and ease of operation. Some of today's most commonly used direct (physical marking) and indirect (digital) marking techniques are outlined below. Table 2.10 contains a summary highlighting the main advantages and disadvantages of each system.

2.4.6.1 Direct marking techniques

There are four main types of direct marking techniques.

- **Weighted core barrel.** As the name suggests, the weighted core barrel technique uses gravity and an impressionable substance to record the geometry of the surface or 'stub' left at the bottom of the hole after the core has been broken and returned to the surface. Typically, the core barrel is 50% weighted to help induce a consistent orientation as it free-falls down the hole. Clay, plasticine and spears have been used to form the impression. Limitations exist with drill holes inclined at shallow angles (<30°) as the weighted barrel may not reach the proper equilibrium in air or water.
- **Ballmark® system.**[3] Unlike the spear or other weighted core barrel techniques, which return to the bottom of the hole after the core has been recovered, the Ballmark® system is designed to orient the core as and when it is broken from the bottom of the hole. It does this by indent marking a soft disc with a non-magnetic free-moving ball, which gravity dictates lies at the bottom or low side of an angled hole. The indent marking process utilises the action of the inner tube back end during core-breaking, which transfers load to the outer tube via compression of a spring. Difficulties can arise in broken ground, where there is no force required to break the core. In these situations the indent triggering mechanism may fail to activate.
- **Scribe system.** Scribe orientation systems commonly provide a core which has been scribed by three tungsten carbide knives. The system's basic equipment generally consists of a multi-shot directional survey instrument which records on film the inclination, direction and orientation of the entire core, a modified double-tube core barrel, a diamond-impregnated core bit and a scribing sub situated immediately above the core bit that contains three triangular tungsten carbide knives. The recovered core is continuously scribed by the three differentially spaced knives as it enters the barrel. The reference scribe has a fixed known relation to an orienting lug which appears on the compass face of the survey so that the scribed core is continuously referenced to the drill hole azimuth and inclination data. The frequency of survey data can be varied depending upon the competency of the rock.
- **EZY-Mark™ system.**[4] The EZY-Mark system is designed to provide core orientation at most drill hole angles (up or downhole) without needing to positively break the core or run a separate tool back into the drill hole. Orientation is achieved by taking the profile shape of the bottom of the drill hole at the beginning of the core run by means of up to three

independent methods (pin profile, pencil zero-point and clay impression) while simultaneously taking three independent gravitational and non-magnetic orientations of the bottom side of the drill hole prior to starting each core run. This technique provides a measurement of true bottom dead centre to within 5°. The EZY-Mark™ system is especially applicable in conventional underground drilling operations that require constant making and breaking of long core barrels prior to the orientation data being transferred from the tool to the core. The system can also provide an audit on each orientation using a recording system which provides a record of each orientation.

2.4.6.2 *Indirect marking techniques*

- **ACT electronic core orientation tool.**[5] The ACT electronic core orientation tool is a non-marking device that provides users with oriented drill core. The ACT tool is attached directly onto the drill tube assembly. It contains three silicon accelerometers that measure individual components of the Earth's gravitational field. When coupled together, the three accelerometers behave like an electronic plumb line. Every minute the tool is downhole the accelerometers sense the low side of the core tube and make a note of its position. When the drilling run is complete the user enters the time at which the core was broken and returns the tool to the surface. The tool recalls the associated accelerometer information from its memory then guides the user to position the tool so that the same low side position is reproduced on the surface. Sophisticated processing means that accuracy is not compromised if the tool was working at an inclined angle downhole but was later laid horizontally at the surface.
- ***Televiewers.*** Acoustic (ATV) and optical (OTV) televiewers that provide continuous and oriented 360° views of the drill hole wall are principally used to determine the orientation of structures that intersect the drill hole (section 2.4.8.4), but are rapidly supplanting traditional oriented core methods in many applications. Orientation is accomplished by a three-axis fluxgate magnetometer and three accelerometers which provide the oriented image and true 3D location of the measurement. The magnetometer and accelerometers are calibrated in the laboratory. The ATV and OTV image orientation and the ATV calliper are checked in the field with a compass and oriented cylinder of known diameter (Williams & Johnson 2004). As noted in section 2.4.8.4, ATV images can be collected in water or lightly mud-filled intervals. Optimum OTV viewing conditions are provided by air or clear water.

2.4.7 Core handling and documentation

2.4.7.1 *Core recovery and labelling*

The quality of the geotechnical logging data very largely depends on the core being kept as nearly as possible in its original state. When removing the core from the split inner tube of a triple-tube core barrel the following procedure should be followed. First, the two parts of the split tube should be placed on a corrugated iron sheet or an angled iron rail. The upper split should then be removed and the core photographed and logged before it is placed in the core tray by the person responsible for logging the core. When transferring the core to the core tray, the best results are obtained by replacing the upper split with a PVC pipe that has been cut in half, rolling the combination over to transfer the core from the split tube into the cut PVC pipe, then placing the cut PVC pipe and core directly into the core tray.

To remove the core from a single-core or double-core barrel, the single-core barrel or the inner tube from the double-core barrel should be elevated to allow the core to slide out of the core barrel. The core should not be allowed to drop into the core tray; it should be captured by hand and carefully placed. Although it may be necessary to tap the core barrel with a hammer to loosen wedged core, this technique should generally be avoided to prevent artificial breaks. All the core, including the fines, must be transferred to the core tray in its correct order. The core should be pieced together as tightly as possible. Although an extruding piece of core may have to be artificially broken in order to fit it into the tray, this practice should be minimised. Where practicable, it is better to leave a small gap at the end of the core length than to break the core.

Damaged core can result in underestimates of rock mass strength and erroneous predictions of rock mass behaviour. Because of this, it cannot be emphasised strongly enough that the core should always be handled as carefully as possible, with the main objective being to minimise artificial breaks. It is very important to ensure that all artificial breaks are clearly marked. Proper core handling is the responsibility of the person logging the core (engineering geologist or geologist).

When the core has been retrieved it is the responsibility of the drillers to:

- carefully wash the core to remove any drilling mud and place a depth block at the end of each run. Washing should be done very carefully, to preserve the integrity of the core. High-pressure nozzles should not be used as these cause core misplacement and further core deterioration. Great care must be taken not to wash away the fines from any weak or broken zones;
- label the core tray with the drill hole ID, the tray serial number and an arrow pointing in the downhole direction. The proper depth should be marked on the small block after each run is recovered. Proper depth

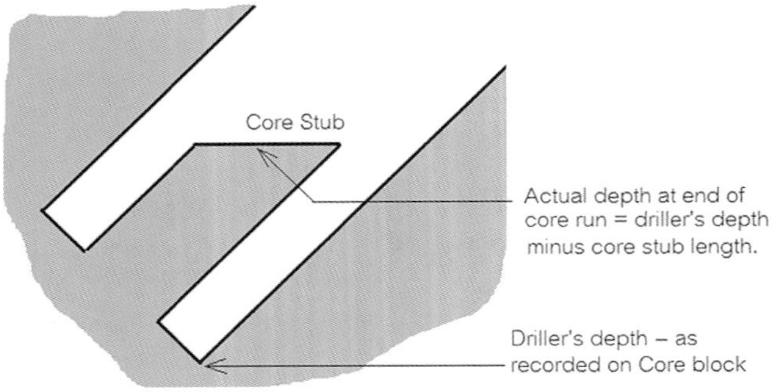

Figure 2.12: Illustration of stub length
Source: Courtesy J.L. Orpen

registering of the core is vital and the core logger should always check that it is being properly done. In this process two measurements must always be checked:

→ stick-up, which is the measurement that most often results in significant inaccuracies in core depth registration. This colloquial term arises from the fact that depth measurement of the drill hole is derived from (length of rod string plus core barrel outer tube) minus (length of rod string 'sticking up' above ground level). Since ground level is hidden under the drill rig, a convenient datum is used on the rig, usually the top of the chuck head. Measurements are made against this 'constant stick up', which has to be accurately assessed before drilling commences – a process that is simple in concept but frequently difficult in practice;

→ stub length, which is the length of the stub of core left in the bottom of the drill hole after the core has been retrieved (Figure 2.12). The stub length can vary considerably depending on the driller's skill in breaking the core at the end of each run as well as the properties of the formation being drilled. To estimate this correctly, it is necessary to inspect all breaks in the core for any grinding or similar damage. If there are no sections of grinding in a competent rock run of core, then stub length is advance minus recovery. If there is grinding, the estimate is more subjective.

2.4.7.2 *Core photography*

The entire core must be photographed as soon as possible after drilling, in colour and preferably using a digital camera with a minimum resolution of 3.0 megapixels. More sophisticated 2D and 3D core scanning techniques are being developed, but they are not yet widely used because of their cost and the need to rehandle the core.

It is important to ensure that there is a minimum of lateral distortion in the photographs. To achieve this, the plane of the core boxes must be parallel to the plane of the camera lens and the camera must be aimed at the midpoint of the core boxes. Each frame should include core from a single hole.

Core photo documentation includes three basic steps.

1 Photographing the core and verifying the required image quality. In all cases the core should be photographed in the core box prior to logging in order to minimise any bias caused by core damage. If split inner tubes have been used, the core should also be photographed in the split tube prior to handling. A single picture should cover one or two core trays at a time.
2 Labelling of the electronic file.
3 Photograph database management.

The following criteria should be observed while photographing the core.

- Lighting conditions and exposure times should be consistent throughout the project. If artificial light is being used better results can be obtained with diffused light rather than bright light.
- The core should be photographed consistently wet or dry. Experience has shown that for geotechnical purposes photographs of dry core are more informative, although in arctic conditions this may be difficult.
- The camera should be kept at a constant distance from the core. Wide-angle lenses should be avoided if possible as they will result in distortion.
- The photograph should include a label, colour bar and scale. The label should provide details of drill hole ID, date, depth, starting-point and direction of drilling.

After the core is photographed the digital file of the photograph should be renamed so that it can easily be identified in the database. The following information should be included in the file name:

- the drill hole ID;
- the depth of the interval photographed;
- an indication of whether the core is wet or dry.

The following is an example of a filename:

..\Drillhole GTS97_01\GTS97_01_204m_dry.jpg

After the core picture is properly labelled it should be stored in an appropriate database. A simple and effective solution is to store the pictures in basic folders and use a viewing program to access and review the files. This program should have the capability to view several thumbnails of photographs at the same time. An example is given in Figure 2.13.

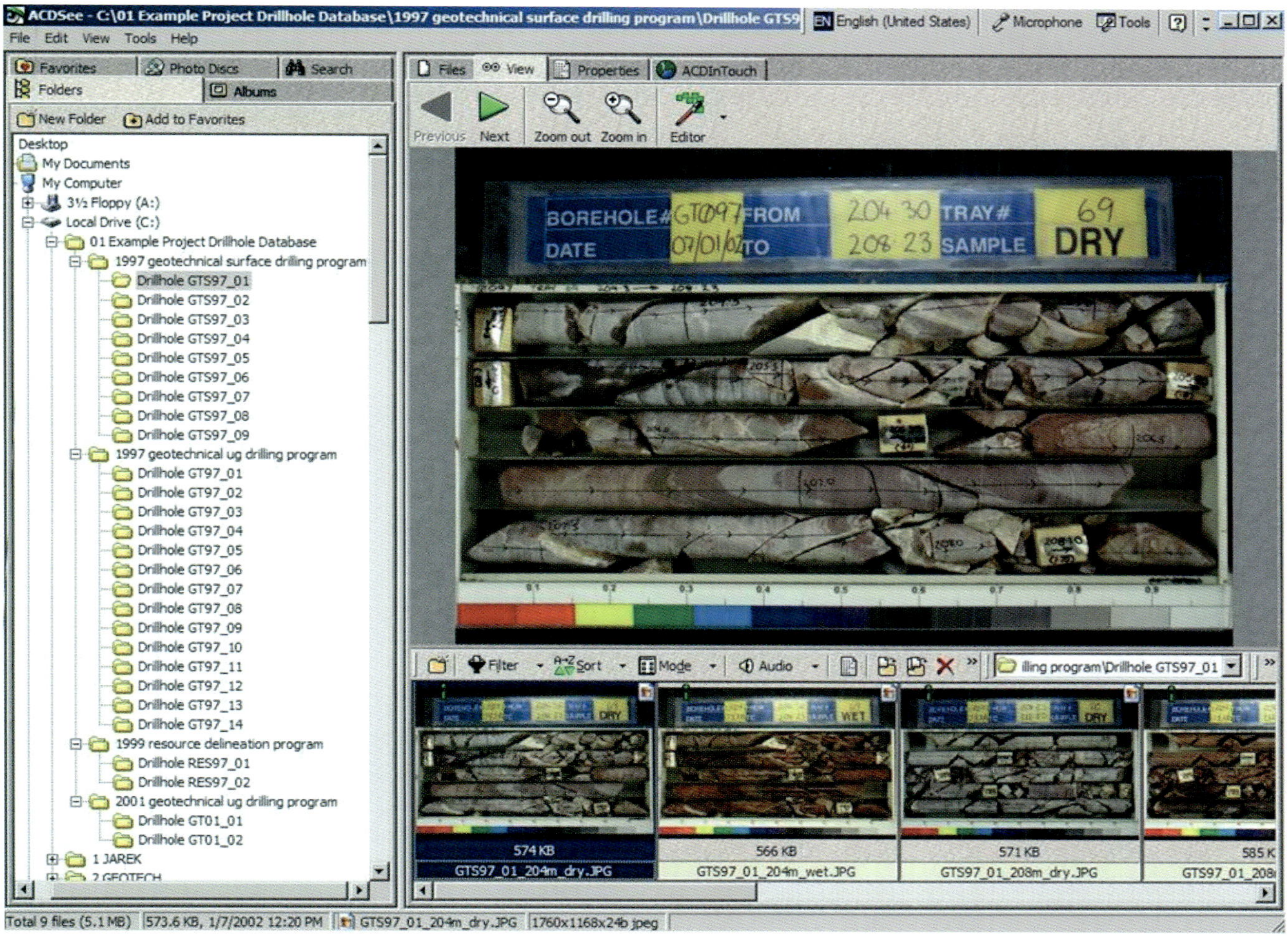

Figure 2.13: Screen snapshot of a drill core database/viewer
Source: Photo courtesy J. Jakubec.

2.4.8 Core sampling, storage and preservation

2.4.8.1 Storage and preservation

Appropriate core storage and preservation is as important as core logging. Core that is exposed to the natural elements could significantly bias the laboratory test results, especially with rocks that are susceptible to weathering. Similarly, mishandled and mislabelled core samples could have major implications on the integrity of the project.

Accordingly, proper core storage must be organised well ahead of the drilling campaign. The core should be protected from the elements, with the trays stacked in well-constructed racks that enable easy access to specific sections of the core. Storing core trays on top of each other indicates poor core management.

2.4.8.2 Laboratory test samples

The following sample handling procedure is recommended.

Proper preservation of samples for laboratory testing is critical for meaningful laboratory test results. If samples will be stored for more than 1–2 weeks before testing, or if the sampled material is susceptible to desiccation (weaker materials with moderate to high water content, e.g. clays, highly altered or weathered rock, and fault gouge) the test samples should be preserved using the following procedure.

1. Wrap the sample in plastic wrap or a plastic bag. For degradable rocks or shales that desiccate, the core should be wrapped in clingfilm (e.g. Saran/Glad Wrap) immediately after being removed from the core barrel.
2. Wrap the sample in aluminum foil.
3. Label the sample using an indelible marker on the aluminum foil and indicate the top end of core.
4. Either wrap in a plastic sample bag and seal, or coat the foil-wrapped sample completely with hot wax by dipping and rolling in a pan of molten wax.
5. Attach a permanent label to the preserved sample.

The wax coating preserves the moisture content of weak samples and provides support and protection against damage during handling and transportation. If samples to be tested are not susceptible to desiccation (stronger rock with low water content) then wrapping in two or more layers of plastic wrap or in a heavy-weight ziplock plastic bag will suffice for most samples. Fragile samples should be packed in PVC pipe or similar protective material before boxing for shipment. Samples should not be allowed to freeze.

Particularly fragile core requires more care for sample preservation and transportation. This could include shipping the samples in sealed triple-tube liners or using disposable Lexan inner tubes or tube liners for collecting and shipping. In these cases, the tubes or tube liners should be capped at both ends and sealed with wax or polypropylene tape.

2.4.8.3 Document samples

To preserve representative lithologies and enable further laboratory testing at a later stage of the project, it is advisable to regularly select representative document samples, properly label them, seal them in plastic and store them in extra core trays. Such compressed profiles provide control samples that can be used for activities such as drill hole reviews and additional laboratory tests, and may also provide the only physical evidence of the rock if the rest of the core is inadvertently destroyed, assayed or lost.

2.4.9 Core logging

2.4.9.1 Overview

Core logging is one of the fundamental techniques used to obtain geotechnical information and should only be carried out by engineering geologists, geological engineers or specialist geotechnicians. Advances in electronic technology have certainly made data capturing and manipulation more efficient but it is important to remember that the actual logging process remains a manual operation, which emphasises the need for it to be carried out by properly trained and experienced people.

Historically, the logging data was transferred by hand onto paper logging sheets, but advances in electronic software and hardware allow this procedure to be replaced by electronic data recording directly into handheld computers and/or tablets. As already noted (section 2.2.1), both systems have merit. The electronic system is much faster and typically more efficient. It eliminates the tedious transfer of data into electronic format and produces data that can be almost instantly transmitted for further analysis and checking. On the other hand, if there is no effective file backup and saving procedure, the data are at risk of being lost in a split second. Also, there could be some issues with the auditing process since no field logging sheets are available.

Regardless of how it is recorded, it is important that all the geotechnical data captured are capable of supporting the principal rock mass classification and strength assessment methods used by the industry. Similarly, although the level of detail captured must be relevant to the level of investigation, there is no reason not to collect the most comprehensive set of data even in the earliest stages of investigation. It is good geotechnical practice to record at least the intact rock strength and RQD data (step 4 below) in the exploration drill hole core logging procedures from day one. The exploration holes are often drilled well before the geotechnical holes. Hence, they can provide invaluable data for the pre-feasibility slope design and ongoing geotechnical investigation programs.

After the core has been photographed, the following geotechnical logging procedures should be carried out.

1 Recording of the basic drill hole identification data on the log, including the number, location, reduced level and orientation of the drill hole, the name of the person who carried out the logging and the date logged.
2 Recording of the basic drilling and test data, including the equipment type and drilling fluid used, the casing used, any drilling fluid loss, water levels in the drill hole before and after drilling, and the results of any downhole tests.
3 Logging of any overburden soils materials using the Unified Soils Classification System (USCS – ASTM D2487, Table 2.7).
4 Geotechnical logging of the solid core, including the rock type, the degree of weathering and/or alteration, the strength of the intact rock, total core recovery (TCR), solid core recovery (SCR) and the rock quality designation (RQD).
5 Geotechnical logging of the fractures that intersect the solid core, including details of fracture orientation relative to the core axis, and fracture spacing and infilling.
6 Large-scale structures logging.
7 Summary geotechnical logging.

At all stages of exploration or geotechnical logging, it is vital to separate artificial defects resulting from core drilling, recovery and handling processes from natural defects that exist in the rock mass. To reduce logging errors, any artificial break introduced by core handling should be clearly marked. Similarly, material such as rubble, re-drill or slough that was not in place but was recovered at the top of a core lift should not be counted as recovered core or a large-scale structure but discarded or clearly labelled to avoid subsequent misclassification (Figure 2.14).

2.4.9.2 Geotechnical logging of solid core

The main purpose of geotechnically logging the solid core is to divide the core into geotechnically similar intervals (domains) then ascribe geotechnical parameters to each

Figure 2.14: Examples of artificial defects as a result of the core recovery process. 'Rubble' at the beginning of the drilling interval (left) caused by re-drill could be mistaken for a fault zone. Severe core damage caused by the drilling process (right) should not be counted as a defect. Such defects would probably not be marked by the drillers because they were not created during the core handling
Source: Photos courtesy J. Jakubec

domain. There are, however, practical limitations to domain thickness. For example, a fault zone less than 0.5 m wide should not be selected as a separate unit, although it must be described in the large-scale structures log. Similarly, narrow highly jointed zones that are repeated in a regular pattern should not be logged as separate domains, but their character should be noted in the comments.

When logging, the following parameters should be collected.

- From, To.
- Rock type, including the effect of weathering and/or alteration. The effect of weathering and/or alteration should be judged using the ISRM-based system used when mapping outcrops (Tables 2.1 and 2.2).
- Intact rock strength, which represents the field estimate of the uniaxial compressive strength (UCS) of the intact core. The UCS should be estimated with the ISRM-based system used when mapping outcrops (Table 2.3). The estimate should represent core free of micro-defects. If the rock is anisotropic (e.g. foliation or bedding), a note should be made in the comments. A carbide scribe pen should be used for the estimate. The values should be confirmed by subsequent laboratory UCS tests.
- Total core recovery (TCR), which measures the total length of the core recovered, including broken zones, against the total length of the core drilled, expressed as a percentage. When recording the core recovery, remember that at the end of a run it is not uncommon for some core to slip through the core lifter and be dropped out of the core tube. It would then be recovered at the top of the next run, often in a crushed or ground state. If not logged properly, dropped core can result in apparent core recoveries exceeding 100%. Core recoveries should not exceed 100%. Core which was drilled in a previous run can often be identified by marks from the drilling or the core lifter.
- Solid core recovery (SCR), which measures the total length of the solid core, excluding pieces smaller than the core diameter, against the total length of the core drilled.
- Rock quality designation (RQD), which is a modified core recovery percentage in which only sound core recovered in lengths of 10 cm or greater per length of the indicated core run is counted as recovery (Deere et al. 1968).

2.4.9.3 *Geotechnical logging of fractures*

When preparing the detailed log of the fractures that intersect the solid core, the following parameters should be captured in each domain.

- Natural fracture frequency per metre, which includes all open fractures in the core except those introduced by core handling. It is sometimes difficult to distinguish between naturally open fractures and those induced by the drilling process, particularly when drilling foliated or sedimentary rocks with well-developed bedding planes. However, experience shows that drilling-induced breaks can indicate internal weakness in the rock mass, so it is important that they be captured when logging. Features indicative of drilling-induced breaks include fresh, rough and uncoated surfaces. Breaks perpendicular to the core axis may also indicate an artificial break or stress relief, for example as caused by core disking (Figure 2.15).
- Cemented joint frequency per metre, which represents the number of healed or cemented joints per metre. Defects in this category are defined as joints with more than 1 mm of infill that do not fall into the open joint category.
- Cement type and strength. The type of infill should be noted (e.g. gypsum, calcite and quartz). There are no formal procedures for estimating the joint cement strength, but a primitive 'drop test' can be used to estimate the relative strengths of the cement. Solid pieces of core including the cemented joint can be dropped from a height of 20 cm to a concrete floor. Strong joints never break on the cement floor, moderate joints sometimes break on the cement floor, weak joints always break on the cement floor. Although the test is subjective, the values indicate a relative strength compared to each other and the intact rock.
- Frequency and strength of micro-defects, which includes all non-continuous micro-defects containing less than 1 mm of infill (Figure 2.16). The frequency is expressed as follows:
  - → none;
  - → minor, spacing >100 mm;
  - → moderate, spacing ≥10 mm ≤100 mm;
  - → heavy, spacing <10 mm.

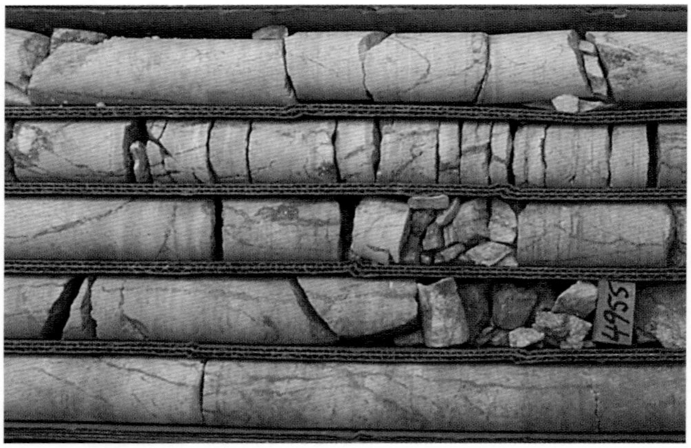

Figure 2.15: Example of core disking during drilling due to locked-in stress. The surfaces are fresh, without staining or infill and are generally perpendicular to the core axis
Source: Photo courtesy J. Jakubec

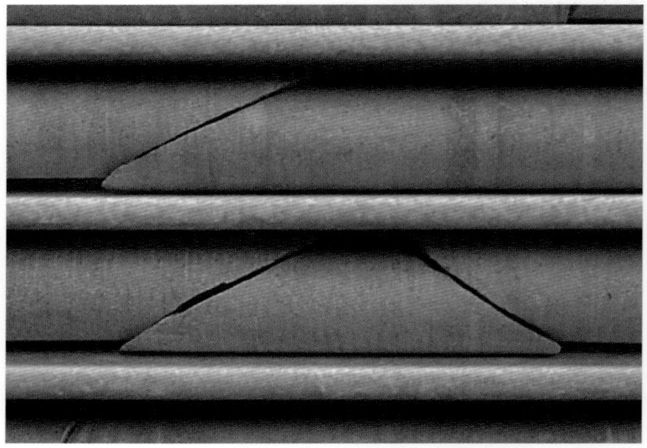

Figure 2.17: Example of two dominant joint sets intersected by the drill core. Although the Alpha angle is the same for both sets, the core pieces exhibit a different orientation
Source: Photo courtesy J. Jakubec

- Number of joint sets, Jn. This value represents the number of individual open joint sets intersected by the drill hole. Joint angles to the core axis (Alpha) and joint characteristics can help to determine the number of sets present (Figure 2.17).
- Typical angle of the individual joint set to the core axis (α). The Alpha angle captures the angle between the joint plane (the maximum dip vector) and the core axis (Figure 2.18).
- Joint conditions for the individual set (Jc). Joint conditions are expressed by small-scale irregularities (roughness) on the surface of the joint, alteration of the joint wall and the nature of the joint infill.

The nature of the joint infill should be captured using the engineering terminology of the Unified Soils Classification System (Table 2.7), which enables empirical estimates of the shear strength of the infill. Estimates of the alteration of the joint wall allow comparison of the relative strength of the joint wall against the strength of the intact rock. It should be noted whether the wall is fresh or altered. If it is altered, it should be noted if it is weaker or stronger than the intact rock.

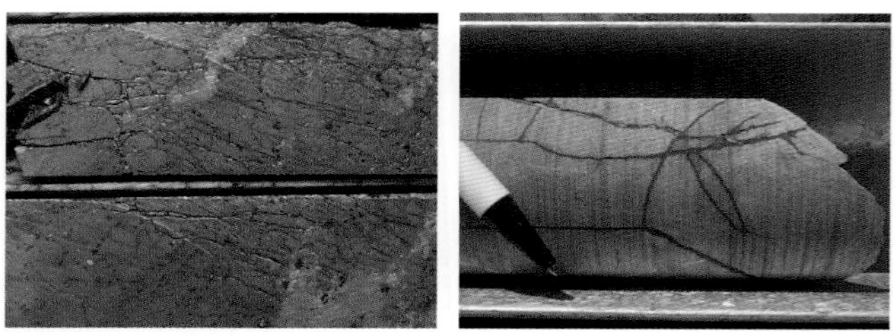

Figure 2.16: Example of micro-defect density, heavy to the left and moderate to the right
Source: Photos courtesy J. Jakubec

Joint roughness describes irregularities of the joint surface at the core scale, usually using the Barton (1987) criterion. When describing roughness, it is important to recognise that it should be assessed in the context of all possible natural variations, not just in the relative scale within one particular mineral deposit. Thus, the planar rough surface of a joint from a copper porphyry in Chile should be the approximately the same as the planar rough surface of a joint in a South African kimberlite.

The Barton criterion for small-scale roughness is shown in Table 2.11. Examples of small-scale roughness geometries are shown in Figure 2.19.

2.4.9.4 *Large-scale structures logging*

Large-scale structures represent through-going faults that extend from inter-ramp to overall pit slope and regional scale. They are weakening features that usually are widely spaced. They may form the boundaries to large-scale structural domains or define a major structural pattern within a particular structural domain.

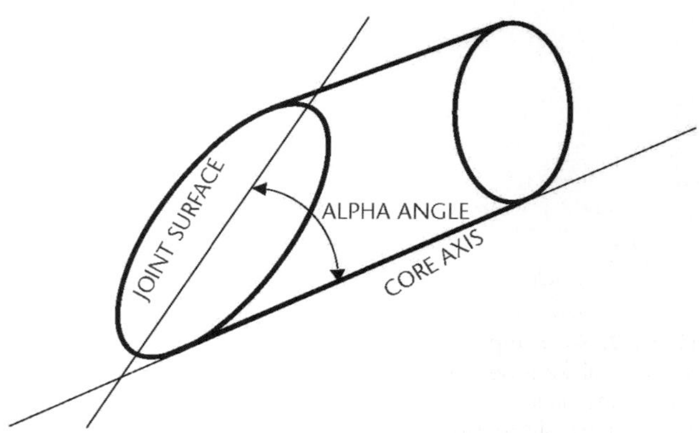

Figure 2.18: The Alpha angle (α) should be estimated for each individual joint set

Table 2.11: Small-scale roughness criterion, J_r

Description	Profile	J_r	JRC 200mm	JRC 1m
Rough		4	20	11
Smooth		3	14	9
Slickensided	Stepped	2	11	8
Rough		3	14	9
Smooth		2	11	8
Slickensided	Undulating	1.5	7	6
Rough		1.5	2.5	2.3
Smooth		1.0	1.5	0.9
Slickensided	Planar	0.5	0.5	0.4

Note that the chart is not to scale. The length of the individual surfaces illustrated should be approximately 10 cm. JRC 200 mm and JRC 1 m correspond to joint roughness coefficients when the profiles are scaled to lengths of approximately 200 mm and 1 m respectively.
Source: Barton (1987b)

Figure 2.19: Examples of small-scale joint roughness geometries. Stepped slickensided (top left), stepped rough (top right), undulating slickensided (middle left), undulating rough (middle right), planar slickensided (bottom left) and planar rough (bottom right)
Source: Photos courtesy J. Jakubec

Figure 2.20: Example of crushed material (gouge) formed by a large-scale fault. The core recovery is poor since much of the clay and silt-sized material was washed away during drilling
Source: Photo courtesy J. Jakubec

At a minimum, the following parameters should be captured on the large-scale structural logging sheet:

- rock type;
- length of intersection of fault zone along the core axis;
- quality of material within the fault zone. The quality of the material within the boundaries of the fault structure can be described using the following terms:
  - → crushed material (Table 2.6), containing angular, sand-sized fragments of rock in a matrix of silt and clay (Figure 2.20). This material is equivalent to the descriptive geological term 'gouge'. It should be described using the Unified Soils Classification System (Table 2.7) as an aid to establishing the shear strength of the material;
  - → sheared material (Table 2.6), comprised dominantly of angular sand to gravel-sized rock fragments that are smaller than the core diameter, with some silt and clay. Frequently, the angular fragments exhibit slickensided surfaces formed as the fault ruptured the rock mass (Figure 2.21). It should be possible to log this material using the Unified Soils Classification System (Table 2.7);
  - → broken material comprised almost entirely of core fragments smaller than the core diameter with only traces of silt and clay (Figure 2.22);
  - → jointed material comprising a zone of higher joint density than the rest of the core (Figure 2.23). In these zones the joint frequency per metre and the condition of the joints (Jc) should be captured on the logging sheets.

Figure 2.21: Example of a sheared zone. Note that the fragments are smaller than the core diameter. The core recovery is also poor since much of the fine material was washed away during drilling
Source: Photo courtesy J. Jakubec

Figure 2.22: Example of a broken zone. Angular fragments combine to provide almost 100% recovery
Source: Photo courtesy J. Jakubec

2.4.9.5 Determining the orientation of structures

There are at least three different means for determining the orientation of any joints or faults that may intersect the drill hole: direct measurement from the oriented core using a goniometer; direct measurement from the oriented core using digital photography and virtual 3D imagery; and downhole imaging using optical or acoustic televiewers.

Goniometry

The angle of the joint to the core axis (α, Figure 2.18) and the circumference angle (β), which represents the angle from the reference line around the core to the maximum dip vector of the joint, can be measured with a goniometer. These measurements can then be combined with the bearing and plunge of the drill hole to calculate the actual dip and dip direction of the joint (Savely & Call 1981; Brennan & Inouye 1988). Individual orientations can be quickly determined using a stereographic net, but if a large number of structures have been measured at a range of depths down the hole it is preferable to use either an Excel spreadsheet or a computer program to convert the goniometer measurements to dip and dip direction by vector mathematics and the drill hole survey data.

Caution must be used when attempting to use the Alpha angle to help orient large-scale structures. Although the Alpha angle can sometimes be used for determining the structure's orientation, it must be remembered that, because of the larger scale, the measured angle could represent only a local variation (undulation) in the geometry rather than general orientation of the structure. There are often insufficient data points to determine the true orientation of larger structures from individual drill holes.

Figure 2.23: Example of a highly jointed section of core, relative to the background joint frequency. Note the limonite staining on the joints
Source: Photo courtesy J. Jakubec

Digital photography:

Two new digital photographic systems have been developed specifically for geotechnical logging and analysis – StereoCore PhotoLog™ (Orpen 2007) and CoreProfiler™ (Sliwa et al. 2007).

Using StereoCore Photo Log™ the oriented core can be digitally photographed in the tray within a reference frame and the image processed to compensate for perspective and produce a depth-registered virtual 3D model of the core cylinders. The processed model can then be used pick the α and β angles before the system performs the structural analyses with drill hole survey data loaded from either measurement while drilling or independent drill hole survey records. Lithological logs, stereo plots and fracture frequency and RQD histograms can be reported at selected depth intervals for drill hole path depth or true vertical depth. The depth registration process also allows the driller's stick-up logs to be rigorously checked, enabling the correct identification of core loss and/or gain.

CoreProfiler™ (Figure 2.24) has been developed from CSIRO's Sirovision™ technology to reconstruct a scaled continuous image of drill core from handheld core tray or core split photographs. The development was funded by the Australian Coal Association Research Program (ACARP Project C15037) and Release 1.0[6] is freely available to the Australian coal industry.

Imported photographs of core can be digital photographs or high-quality scans of paper prints, in TIFF or JPEG format. Lens corrections can be applied to each photograph as it is imported, with perspective correction applied by identifying the corners of a rectangle of known dimensions on the image; precise depth controls can be added later. The angle of a joint or bedding to the core axis (α, Figure 2.18) can be estimated directly from the core image and the import/export logging data.

Downhole imaging

Acoustic (ATV) and optical (OTV) televiewers provide continuous and oriented 360° views of the drill hole wall from which the character, relation and orientation of lithologic and structural planar features can be defined (Figures 2.25 and 2.26).

ATVs were first developed by the petroleum industry in the late 1960s, with the optical OTVs following in the 1980s (Williams & Johnson 2004).

ATV imaging systems emit an ultrasonic pulse-echo and record the transit time and amplitude of the acoustic

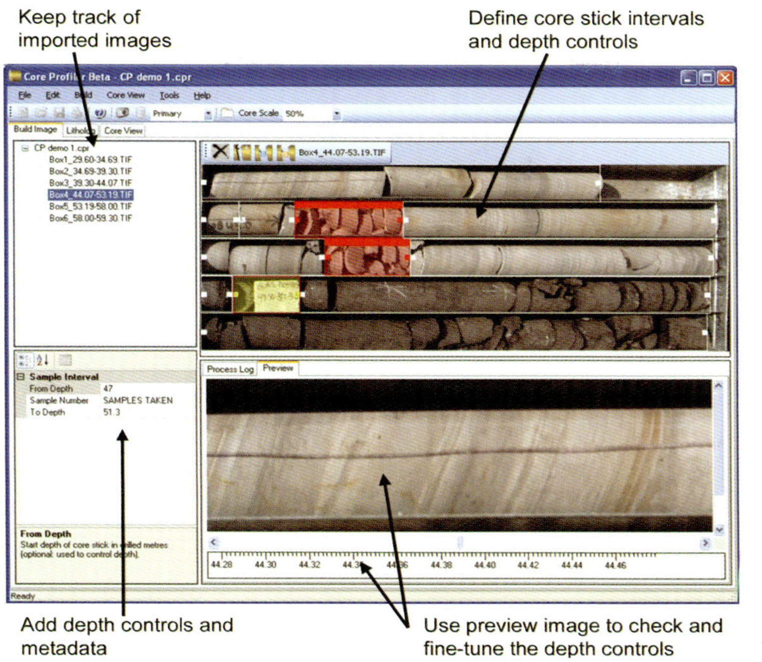

Figure 2.24: Core Profiler™ screen snapshot of the main core image builder interface used to manage the imported photographs and their division into core sticks and sample intervals
Source: After Sliwa et al. (2007)

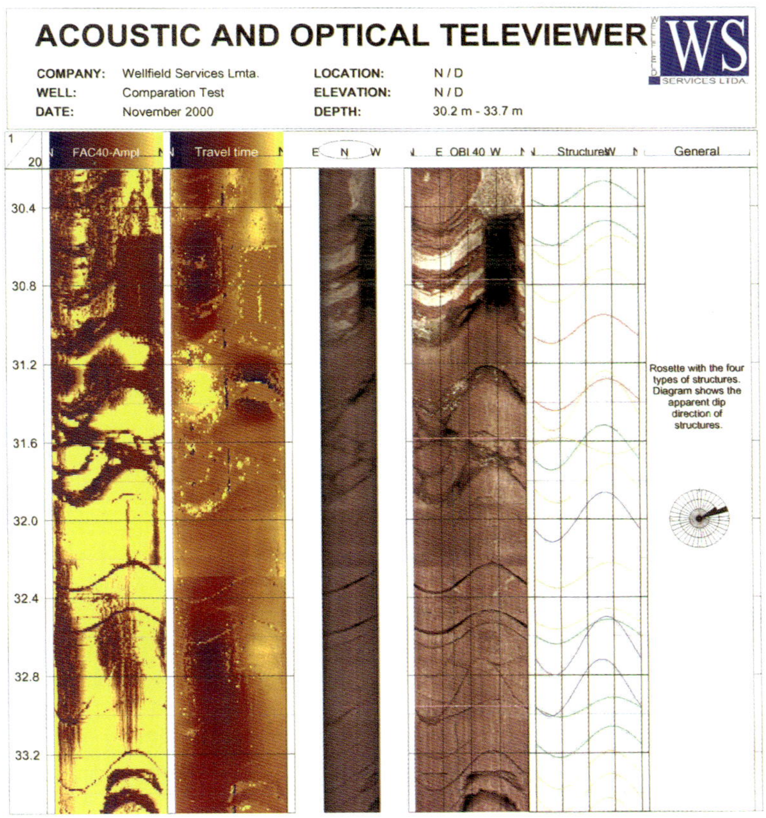

Figure 2.25: ATV and OTV images
Source: Courtesy Wellfield Services Ltda

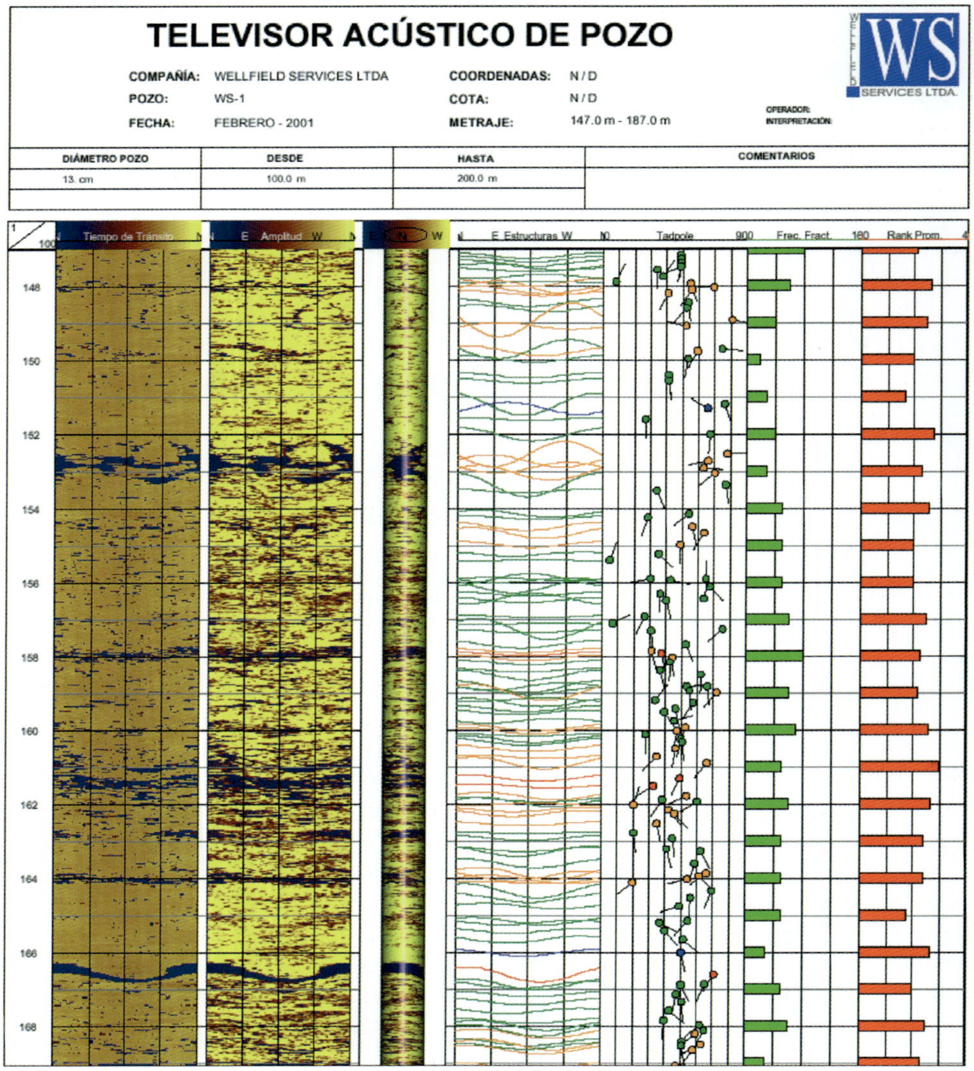

Figure 2.26: ATV images
Source: Courtesy Wellfield Services Ltda

signal reflected back from the lithological or structural features intersected by the bore hole as photographic-like images. The images can be collected in water or lightly mud-filled intervals of drill holes. Irrespective of the medium, the best images are obtained when there is sufficient drill hole relief or acoustic contrast.

OTV imaging systems use lights to illuminate the drill hole and a reflector housed in a transparent cylindrical window to focus a 360° slice of the drill hole wall in the lens of a charge-coupled device (CCD) camera that measures the intensity of the colour spectrum in red, green and blue. The lithology and structures present in the wall of the drill hole are viewed directly on the OTV images. Optimum viewing conditions are provided by air or clear water-filled drill hole intervals.

In both systems the lithological and structural planar features are classified and fitted interactively with sinusoidal traces and the true orientation calculated using the associated software. The classified features are usually displayed on 'tadpole' plots and stereo plots. On the 'tadpole' plot the dip is plotted on the x-axis of the plot and the tadpole tail points in the down dip direction. The tadpole plots are depth-dependent; the stereo plots can be reported for selected depth intervals of singular or multiple drill holes (Martel 1999).

2.4.9.6 Blind zones

When the structures that intersect the drill hole are being oriented, the occurrence of structures that have low angles of intersection (α) with the drill hole raises the issue of blind zones, which is common to all orientation methods.

The blind zone of a drill hole is the locus of the poles of the structures that are parallel to the drill hole and cannot be seen by the drill hole. Terzaghi (1965) visualised the zone by regarding the drill hole as the axis of a wheel, the blind zone as the rim of the wheel and the poles of the blind zone structures as the points of spoke attachment to the rim. Blind zone structural planes were visualised as planes tangent to the wheel rim.

Terzaghi (1965) noted that the only way to overcome the effect of blind zones is to drill a sufficient number of drill holes so oriented that no stuctural pole can lie in or near the blind zone of each hole. An appropriate layout for a single cluster of three holes was for each hole to plunge at 45º, with the orientation of the trace of each hole differing by 120º from that of the other two. A structure of any orientation would be intersected by at least one of these holes at an angle (α) equal to or greater than about 31°.

Another form of blind zone can be created by acoustic televiewers that are run down the drill hole too quickly. If the hole walls are traversed too rapidly there may be insufficient resolution of the acoustic signal from any structures that have a high angle of intersection with the drill hole. If this occurs, these structures will not be recognised.

2.4.9.7 Summary geotechnical logging

After detailed geotechnical and structural logging has been completed it is necessary to consolidate the data and confirm the geotechnical domains and the ranges of values of individual parameters. This is done in the geotechnical log summary after drilling has been completed and the whole core is laid out for review. It is good practice to review the core with the geologist and mining engineer and agree upon the larger picture and potential issues.

2.4.10 Downhole geophysical techniques

Downhole geophysical methods are commonly used at all stages of a mine's life, from exploration, through development and during production and post mine monitoring of rehabilitation activities. They are also used to investigate geophysical targets or zones of interest identified during surface geophysical surveys. Valuable geotechnical information may be gleaned from these data. Such logging data can be used to determine lithological boundaries, the in situ mechanical, physical and chemical properties of the rock mass, the locations and characteristics of the geophysical features determined from surface-based techniques, and structures that intersect the drill hole.

In hard rock mines the geophysical probe or tool is usually inserted down individual open drill holes. In this case, emphasis is placed on calibrating the geophysical data against the rock core to help determine the mechanical properties of the rock mass and the attributes of the structures that intersect the drill hole.

In coal mining, geophysical techniques are widely used for stratigraphic logging, and for estimating porosity and clay content in lieu of cored drill holes. The results from gamma–gamma and similar tools have been well correlated with coal properties such as ash content, and are used to determine stratigraphic boundaries and coal quality. The results from geophysical logs from coal measures have been used to form a rock mass rating scheme for clastic sediments (Hatherly et al. 2007).

Downhole measurements have a potential advantage over samples taken from core as they typically are continuous and, because they are measured in situ, are essentially undisturbed. If sufficient care has been taken to calibrate the measurements against known standards, downhole measurements are an efficient and effective means to rapidly assess an area. Cross-hole imaging techniques can also be used to help locate or characterise broad fracture zones and/or weathered and altered zones.

The sonic log is perhaps the most useful tool for subsurface geotechnical characterisation. Sonic velocities and attenuation are sensitive to rock composition, rock stress and strength, rock density and the degree of fracturing (Fullagher & Fallon 1997). Sonic velocity can be related or calibrated to physical properties such as hardness and rock uniaxial compressive strength (Ohkubo & Terasaki 1977; McNally 1990; Zhou et al. 2005; Hatherly 2002; Hatherly et al. 2007). Fracture frequency affects the compressional wave velocity and the full sonic waveform is attenuated, especially at higher frequencies, by rock fractures (King et al. 1978). Full waveform sonic analysis employing a dipole shear imager (DSI) is also used in the petroleum industry for evaluating the in-situ hydrological and geomechanical properties of rock fractures (Pajot 2008).

Downhole gravity and gamma–gamma methods are sensitive to rock density. Bulk density is an essential parameter for resource and reserve modelling. The gamma–gamma density is an in situ density so, depending on downhole conditions, may not correspond to a material's dry bulk density.

Downhole-induced potential (IP), resistivity and electromagnetic methods are typically used for mapping conductivity variations in resistive backgrounds. These techniques tend to be sensitive to clays, moisture, iron oxides and sulphides. In the petroleum industry, high resolution electrical resistivity imaging is used for orienting fractures and can be combined with DSI sonic logging to identify the location and orientation of the fractures that most contribute to the rock permeability (Pajot 2008).

Use of cross-hole tomographic measurements to collect continuous lithological and structural information between two drill holes is increasing. Methods with waveforms such as GPR, seismics and electromagnetic methods have been used with varying degrees of success to

Table 2.12: Principal downhole geophysical techniques

Technique	Operation	Application
Resistivity (single and multi-point)	Records the electrical resistivity of the drill hole environment and surrounding rocks and water	Geological contacts, lithology and bed thickness, variations in clays
Spontaneous potential (SP)	Records potentials or voltages developed between the drill hole fluid and the surrounding rock and fluids	Geological contacts, lithology and bed thickness, water quality
Induced polarisation (IP)	IP is observed when a steady current through two electrodes in the Earth is shut off. The voltage decays slowly, not instantaneously, indicating that charge has been stored in the rocks	Mineralisation types, variations in clays and iron oxides
Magnetic susceptibility	Records the degree of magnetization of a material in response to an applied magnetic field	Lithological and structural boundaries and zones, porosity
Electromagnetic induction (EM)	Measures variations in conductance and resistivity	Shallow soils mapping, soil-salinity mapping, ground water and subsurface contaminants, geological contacts, lithology and bed thickness, clays, ferrous mineral deposits, gravel deposits
Natural gamma	Measures the level of gamma radiation emitted by radioisotopes present in subsurface materials	Lithological boundaries; bulk density, relative porosity of soil and rock based on clay content.
Spectral gamma	Measures the energy of the gamma emissions and counts the number of gamma emissions associated with each energy level	Structural and stratigraphic features, water-rock interactions
Gamma–gamma	Density is derived from two gamma detectors that respond to variations in the specific gravity of formations. The tool emits neutrons and then measures the secondary gamma radiation that is scattered back to the detector in the instrument	Geological contacts, lithology and bed thickness, porosity and density, especially in sedimentary units.
Caliper logs	Measures the diameter of a drill hole	Useful in the analysis of other geophysical logs, including the interpretation of flow meter logs.
Drill hole gravity meter	Used to measure gravity at a sequence of depths through the zone of interest	Geological contacts, lithology and bed thickness, density. In perfectly flat uniform geology, the BHGM densities will match the normal gamma–gamma density log data
Full waveform sonic	The full wave form can be used to generate a sonic attenuation log. Abnormalities will cause attenuation of the acoustic signal and will appear on the log as variations from the normal level	Rock strength, rigidity, density, porosity, weathering, alteration, fault zone characterisation
Prompt gamma neutron activation analysis (PGNAA)	Sensitive to compositional variations such as magnetic susceptibility to iron and its many forms; natural gamma to uranium, thorium and potassium	Identifying lithologies and/or alteration zones
Fullbore formation micro-imager (FMI)	High resolution, oriented electrical resistivity imaging penetrating 250 mm beyond drill hole wall (minimum hole diameter 110 mm)	Measurement of stratigraphic and structural features in 3D space including; bedding, faults, fractures , texture/facies, fracture density, trace length and fracture porosity
Dipole shear imager (DSI)	Full waveform sonic analysis: transmits and records high frequency, multi-component acoustic/sonic (compressional, shear and Stoneley) waves	Derivation of: formation velocities, porosity and water table depth; relative fracture aperture and magnitude/extent of open fractures; Poisson's ratio, Young's and bulk modulus.

create images between drill holes or between a drill hole and the surface. Radio imaging (RIM) technology has a range of hundreds of metres in highly resistive rocks but of less than 100 m in weathered materials (Stevens et al. 2000; Thomson et al. 2003).

The principal techniques and their uses are summarised in Table 2.12.

2.5 Groundwater data collection

2.5.1 Approach to groundwater data collection

Groundwater can have a detrimental effect on slope stability. Fluid pressure acting within discontinuities and pore spaces in the rock mass reduces the effective stress, with a consequent reduction in shear strength. As noted in Chapter 1, these aspects are usually the only element of a slope design that can readily be modified by artificial intervention, usually dewatering and depressurisation. However, dewatering and depressurisation measures are usually capital-intensive, require operator commitment to be implemented effectively and require significant lead times in design and implementation. The identification and characterisation of the hydrogeological regime in the early stages of any project is therefore of paramount importance, requiring a well-organised approach to collecting the data and building the conceptual hydrogeological model.

The development of a conceptual hydrogeological model to support a slope design analysis and depressurisation program is addressed in Chapter 6. This section focuses on the field tests and procedures used to collect the raw hydrogeological data. It describes the basic approaches to data collection then outlines the nature of each appropriate test method.

2.5.1.1 *Phased approach*

A typical phased approach for collecting the required hydrogeological data to support a pit slope analysis involves the following steps.

Step 1: Initial data collection

The initial step in the hydrogeological data collection program may involve the following.

- A literature search to determine any pertinent hydrogeological information in the region surrounding the project area.
- Identification of existing observation or groundwater production wells within or surrounding the project area, with their completion and stratigraphical details; measurement of groundwater levels in any accessible observation wells; and collection of any pumping data from available production wells. In all cases, it is important to know which formations are being measured or abstracted from.
- Collection of hydrogeological data 'piggy-backed' on mineral exploration and resource drilling programs. Data collection may include:
  - → measurement of groundwater levels in the open drill holes at regular intervals during and after drilling;
  - → collection of water flow rate and other parameters at regular intervals during air drilling;
  - → noting of loss of drilling fluids during water or mud drilling;
  - → injection testing or airlift recovery testing to provide initial permeability information for individual holes.

  Note that the use of drilling additives to help stabilise the walls of the hole or increase recovery can alter the permeability around the hole and compromise the quality of the data.
- Installation of groundwater observation wells using holes drilled for the geology resource evaluation or geotechnical programs.
- Implementation of a routine water level monitoring program in all available groundwater observation wells or piezometers. For a new project, measurement of water levels on a weekly basis would be the standard procedure, possibly reducing to monthly with time as more data are collected.
- Collection of climatological information. Depending on the climatic setting, precipitation and evaporation may exert a major influence on the groundwater flow system. They are important factors for determining pit-scale hydrological conditions. The required data may already be available from existing weather stations. For remote mine sites, installation of a dedicated weather station is often required.
- Implementation of a hydrological database, which can be expanded as the project proceeds. Many operators use one database for all pit slope hydrogeological data, general dewatering and environmental management systems. In some cases the database is integrated with the geological modelling, geotechnical database and mine planning programs.

Step 2: Characterisation of the overall groundwater flow system

A good knowledge of the regional and mine-scale stratigraphy and geological structure is important for determining the hydrogeology around the mine site area. The geological information can typically be obtained from government agency reports, the regional exploration, resource drilling and condemnation drilling programs. This information is used to define the hydrostratigraphical units that may influence the pit slopes.

The focus for characterising the overall groundwater flow system is often the geological units that are more permeable, because most of the regional or mine-scale flow occurs within the permeable units.

The most appropriate test methods are as follows.

- Drilling of dedicated hydrogeological test holes, which may be completed as observation wells with broader screened intervals. These are used to characterise the components of lateral groundwater flow. Several observation wells may be required for each identified hydrostratigraphical unit, to determine lateral hydraulic gradients and groundwater flow directions. Because condemnation (or sterilisation) drill holes are often located away from the immediate pit area, they may provide an opportunity to install 'piggy-backed' observation wells.
- Injection testing or airlift recovery testing during drilling, to provide more detailed permeability information.
- Installation of test pumping wells, and carrying out pumping tests to characterise permeability and storage characteristics and to assess the groundwater flow characteristics over a wider area.
- Hydrochemical sampling and analysis of observation wells and pumping wells to characterise variations in groundwater quality which may help with the interpretation of the groundwater flow system.

Step 3: Characterisation of the local pit slopes

Once the more general data have been collected to determine the overall site setting, it will be necessary to collect data to characterise the specific conditions relevant to the immediate area of the existing or planned pit slopes.

More focused testing can then be carried out to support the detailed design of the slope depressurisation program. Information on lithology, alteration, mineralisation and geological structure will be required. It will be necessary to integrate the geology and hydrogeology information to determine structural boundaries, the influence of fault zones and groundwater flow between identified hydrostratigraphical units.

The focus for the localised area of the pit slope is often weighted towards the less permeable units in the groundwater system, because these units are typically most difficult to depressurise. However, pore pressure information will be required for each key unit in the pit slopes.

The most appropriate test methods are as follows.

- Monitoring of airlift flow rates and pressures, and static water levels during reverse circulation (RC) drilling programs.
- Injection testing and/or falling or rising head testing (slug testing) to provide local permeability information around individual piezometers.
- Monitoring of fluid loss zones and static water levels during diamond/core drilling programs.
- Installation of piezometers (vertical, horizontal, angled) with short measuring intervals, in carefully chosen locations, to characterise both the lateral and vertical variations in pore pressure and hydraulic gradients. Installation of vibrating-wire piezometers allows multi-point monitoring within a single drill hole. Depending on the nature of the slope materials and their setting, a combination of standpipe and vibrating-wire piezometers is appropriate.
- In situ permeability testing in completed standpipe piezometers.
- Measurement of water levels and pressures in piezometers at least weekly for a defined monitoring period.
- Performing an integrated geotechnical and hydrogeological drilling program, as appropriate. This may include:
  - → collection of groundwater levels at regular intervals during drilling;
  - → in situ permeability testing and/or packer testing to characterise permeability differences within discrete hydrostratigraphical units or subunits. Packer testing is appropriate for characterising the poorly permeable units in the slope;
  - → physical rock mass testing and lab testing, for porosity and density.
- Pilot testing of depressurisation methods (see section 2.5.6).

2.5.1.2 'Piggy-backed' data collection

A large amount of data can be collected from the mineral exploration and resource drilling program at minimal additional cost. In addition to basic geological information, it is possible to collect much of the information required to characterise the site-scale hydrogeology. Dual-tube reverse circulation (RC) with air is commonly used for mineral exploration drilling. This drilling method is ideal for obtaining data to characterise groundwater conditions.

The following hydrogeological data can be collected at minimal incremental cost by 'piggy-backing' on RC mineral exploration holes:

- airlift flow rate, at the end of each 6 m drill pipe. A good record of airlift flows enables the volume of water entering the hole, and the depth at which it enters, to be characterised;
- air pressure during drilling, at the end of each drill pipe, which provides supplementary data to quantify the inflow zones;
- air pressure required to unload or 'kick-off' the hole at the start of each drill pipe, which allows the recovery of the water level while adding each drill pipe to be assessed;
- static water level, measured down the inside of the drill pipe at the start of each drill shift;
- quantitative information on fracture location and intensity;
- quantitative information on the degree of clay alteration.

A standardised hydrological log form and notes for collecting data during drilling are included in Appendix 1, Attachment A.

Diamond (core) drilling is not as adaptable to groundwater exploration as RC drilling, but allows the collection of the following information from drill holes:

- static fluid level, measured down the inside of the drill pipe at the start of each drill shift;
- depth of circulation loss in the hole, which can be correlated with highly permeable zones.

2.5.2 Tests conducted during RC drilling

RC drill holes allow hydraulic testing to be carried out as the hole is advanced. When the hole is completed, it is possible to carry out controlled airlift tests at a constant rate and to measure the recovery of groundwater levels in the hole and nearby holes on completion of airlifting.

2.5.2.1 Drill-stem injection tests

Injection tests can be carried out in RC holes to characterise the fracture zones penetrated during drilling. A sketch of the set-up and procedures for injection testing are included in Appendix 1, Attachment B. Water is pumped down the annulus of the RC drill pipe using a small trash pump. The resultant rise in water level in the inner tube is measured using a sounding probe. The data allow characterisation of each main fracture zone and calculation of the specific capacity of each hole.

The drilling contractor can fabricate a suitable drill head for injection. A 50 mm inline flow metre and 3–5 kW trash pump are also required. Each test typically lasts approximately 20–30 minutes. Typically, up to three tests may be carried out on holes that show good water-bearing fractures.

Drill-stem injection tests provide a good method of characterising different intervals and different fracture zones within a drill hole. Apart from minor equipment costs, the cost of testing is limited to the cost of standing time for the drill rig. For a typical hole, 2 hours standing time may be required. The method is therefore an attractive way to rapidly obtain quantitative data from many drill holes.

2.5.2.2 *Airlift and recovery tests*

Upon completion of drilling (or after penetration of key productive fracture zones during drilling), controlled airlift testing can be carried out in RC holes or in other types of hole drilled using air circulation. In stable holes, airlift testing can be carried out for periods of 30 minutes to 4 hours to determine the potential steady-state yield of the hole.

Upon completion of airlifting, a water level sounding device can be run into the hole, and the recovery following airlift can be measured by monitoring the rising water level inside the drill pipe. The recovery data from the hole being airlifted can be analysed to estimate the permeability of the test zone.

If required, prolonged airlift testing (e.g. 8+ hours) can be carried out to observe the response in surrounding piezometers. Often, if the permeability of the slope materials is low, detailed pumping tests are not required as part of specific slope depressurisation studies. If there are other open drill holes or piezometers installed near the hole being airlifted, it is possible to monitor the groundwater level response in these observation wells. For a carefully controlled airlift test with a yield of 2 L/sec or more, the response in the observation wells can be analysed in the same quantitative manner as a conventional pumping test.

In this way, airlift testing in small-diameter RC holes can be important in helping to characterise the groundwater system in areas where the expenditure of drilling a large-diameter pumping well is not justified, or for projects where the remote or inaccessible nature of the site makes mobilisation of a large water-well drilling rig impractical. Testing multiple RC holes may also provide better spatial data coverage than testing one larger-diameter pumping well.

For example, as part of the recent site characterisation work for the remote Agua Rica copper project in Argentina, it was not feasible to mobilise a drilling rig large enough to drill water wells. Detailed testing information was obtained by controlled airlifting of RC holes for a 48 hour period, monitoring the response to airlift pumping in several purpose-installed piezometers drilled nearby. An example of the airlift data from Agua Rica is provided in Figure 2.27.

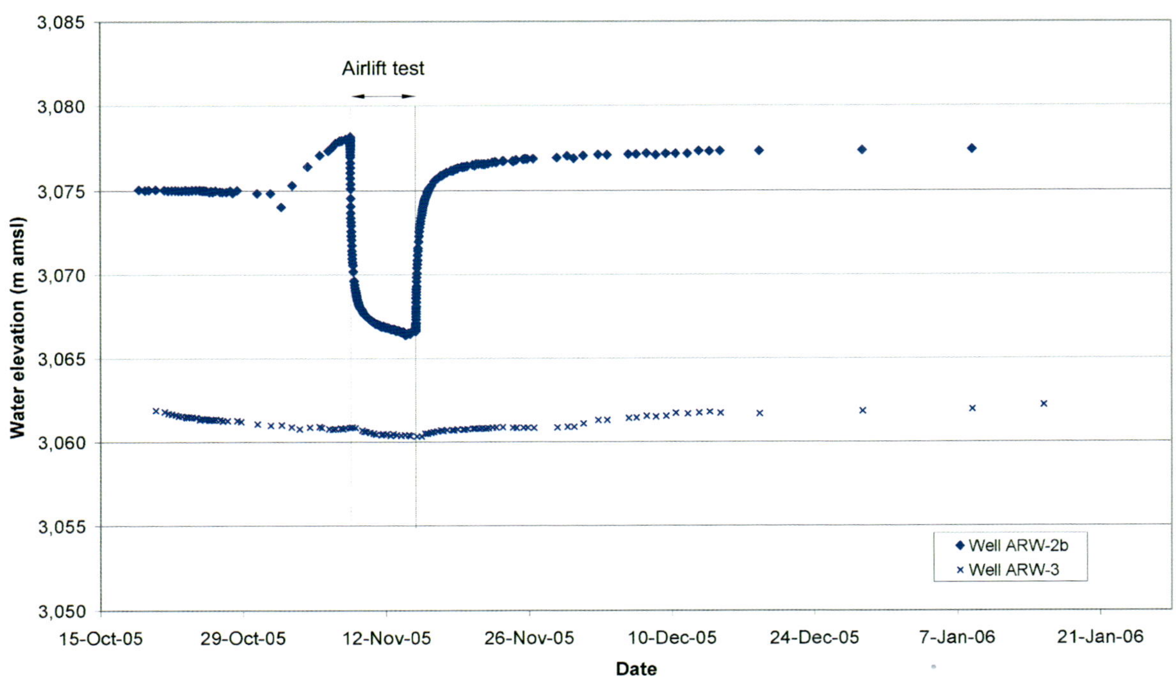

Figure 2.27: Example of drawdown response in observation wells adjacent to an airlifting test hole

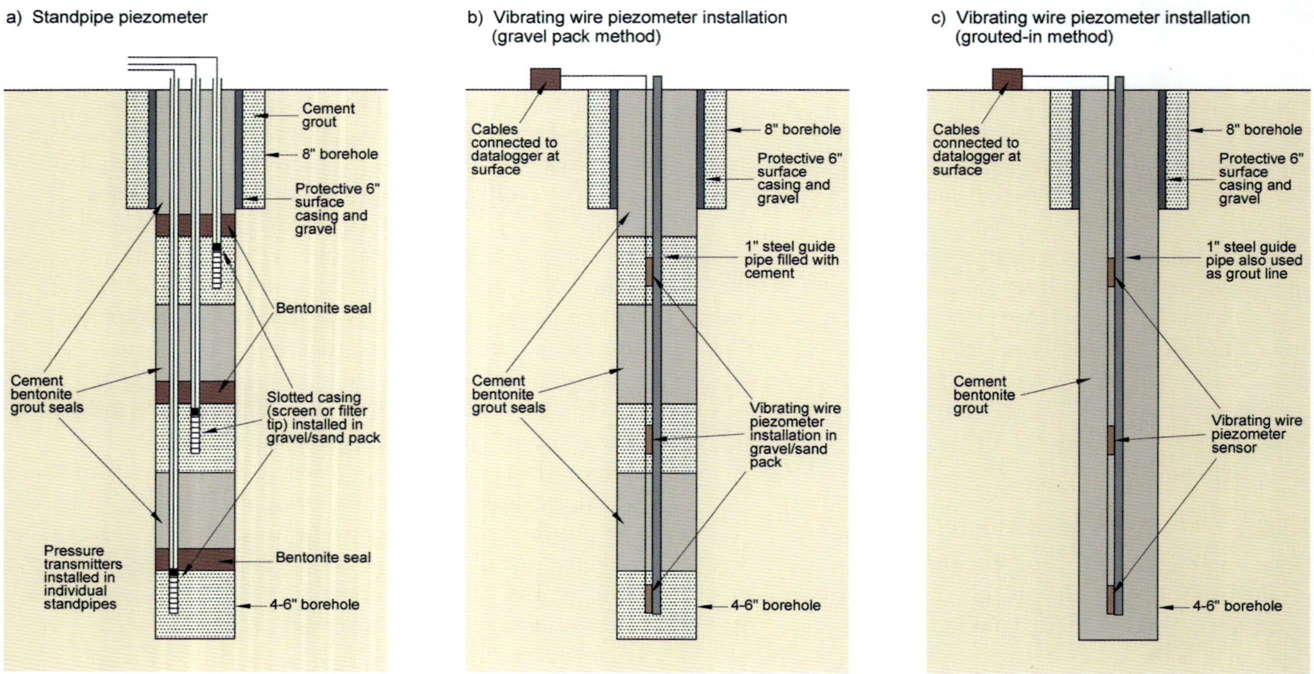

Figure 2.28: Alternative piezometer installations in core or RC drill holes

2.5.3 Piezometer installation

2.5.3.1 Terminology

The terms 'observation well' and 'piezometer' are commonly used to describe drill holes that are used for measuring water levels or pore pressures in the field. The terms are often synonymous although, in many texts and reports, 'piezometer' is often used for installations with shorter, sealed completion intervals.

Figure 2.28 shows how observation wells and piezometers may be installed as standpipes or grouted instruments. It also shows how vibrating-wire piezometers may be grouted in place to measure pore pressures at point intervals.

The term 'monitoring well' is synonymous with 'observation well' but is often used to describe wells installed primarily for water chemistry sampling.

2.5.3.2 Open observation holes

The term 'open observation hole' is used to describe an open uncompleted drill hole used for measuring groundwater levels. Open observation holes may have a small-diameter monitoring pipe installed to provide stability against hole collapse, but have no impermeable seal to isolate discrete vertical intervals in the formation.

Measurement of groundwater levels in open observation holes (with or without slotted pipe installed) can allow rapid characterisation of the piezometric surface, at low cost. The water level recorded in an open hole represents an average of the pressure over the saturated interval. This is particularly useful in the early stages of project development, where there has been little stress on the groundwater system and where significant vertical head gradients have not yet developed. Where such gradients are present, open holes may allow the vertical movement of water between intervals. Because of this, some countries have a legal requirement that open drill holes must be sealed within a specified time period.

2.5.3.3 Standpipe piezometers

The traditional piezometer is a standpipe that allows pressure to be measured as a head of water in a pipe. The piezometer consists of a filter tip or perforated interval of pipe of a desired length attached to the bottom of an impermeable casing that rises to the surface. The perforated interval is isolated by an upper seal, typically with grout placed above the seal to the surface (Figure 2.28). The level to which the water rises in the pipe indicates the average pressure over the perforated interval.

In the early stages of groundwater investigation, if piezometers are used in preference to open holes the perforated interval of the piezometer may be long (20+ m), so the average head of a formation is measured. As the investigation becomes more focused, a shorter (6 m or less) perforated interval may installed in subsequent holes to measure the pressure at more discrete depths.

General guidelines on the installation of piezometers are given in Appendix 1, Attachment C.

2.5.3.4 *Vibrating-wire piezometers*

Vibrating-wire piezometers consist of a vibrating-wire sensing element in a protective steel housing. The sensing element consists of a tensioned steel wire clamped to both ends of a hollow cylindrical body. The piezometer converts water pressure to a frequency signal via a diaphram, the tensioned steel wire and an electromagnetic coil. When excited by the electromagnetic coil the wire vibrates at its natural frequency. The piezometer is designed so that a change in pressure on the diaphragm causes a change in tension of the wire, altering its natural frequency of vibration. The vibration of the wire in the proximity of the coil generates a signal that is transmitted to the readout device. The readout device processes the signal, applies calibration factors and displays a reading.

Vibrating-wire piezometers are a good method of collecting water level data and defining vertical hydraulic gradients around active pit slopes. Multiple transducers may be set in alternating bentonite/sand packs placed using a tremie pipe (Figure 28b). Alternatively, a string of three or more vibrating-wire piezometer instruments may be grouted directly into the hole without the use of a filter pack.

The grouted-in method is a rapid way to install multiple piezometers in one drill hole, to install piezometers in holes with inclinometers or other geotechnical instruments, or to monitor pressures at a discrete point. A detailed description of grouted-in piezometer installation is given in McKenna (1995).

2.5.3.5 *Standpipe vs grouted vibrating-wire installations*

The advantage of a traditional standpipe piezometer design is that it allows the water level to be physically measured down the hole, rather than relying on the performance of an instrument. The standpipe piezometer also allows hydraulic testing and water quality sampling to be carried out. The use of grouted-in vibrating-wire piezometers relies on the instrument performing properly. If a grouted-in vibrating-wire piezometer fails, there is no way of recovering it and measuring the water level.

However, vibrating-wire piezometers typically provide a lower-cost installation, particularly where they are grouted in. It is relatively easy to install multiple vibrating-wire piezometers using one cable, so multiple piezometers can be installed in smaller-diameter holes. Another advantage is that, if access to the wellhead is lost, readings from a vibrating-wire sensor can be taken remotely (provided the cables can be identified).

Where vibrating-wire piezometers are grouted in, the pressure recorded at the piezometer is a discrete pressure at the level of the instrument. For fractured rock environments, where there is a large difference in permeability between the fracture zones and the surrounding rock mass, the lag time for piezometer response may be very large if the piezometer is not installed adjacent to a permeable fracture. For pumping tests where a more rapid response is required, it is often better to use sandpack intervals rather than grouted piezometers. Unless the vibrating-wire piezometer is accurately positioned, sandpacks are often more likely to include permeable fractures and produce a faster response.

Grouted vibrating-wire installations are more applicable in settings with low permeability. Because there is no standpipe and therefore no water in the completed hole, there are no well-bore storage effects and the initial equilibiration response time of the piezometers is much faster. Figure 2.29 shows a multiple grouted-in vibrating-wire installation where seven piezometers were installed in an HQ diamond hole.

2.5.3.6 *Joint piezometer/inclinometer completions*

At some operations it has been possible to install completions acting jointly as standpipe piezometers and inclinometers, as illustrated in Figure 2.30.

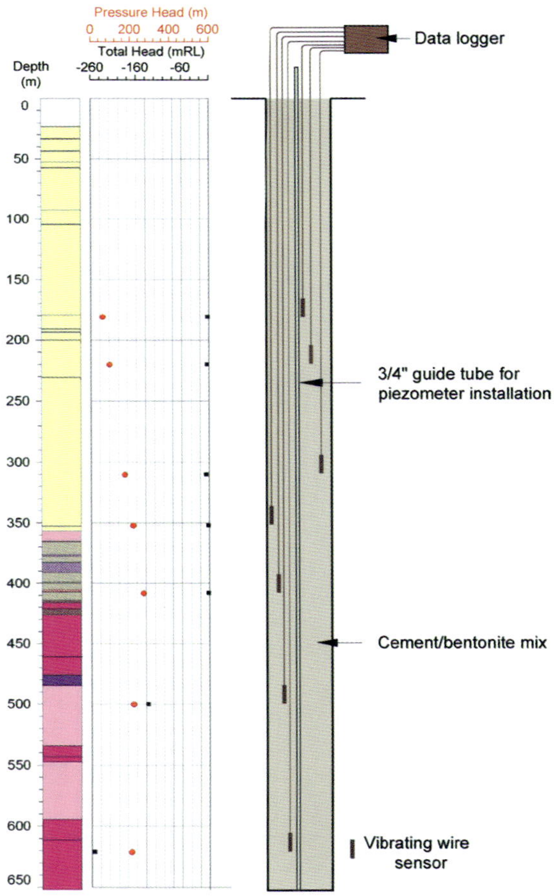

Figure 2.29: Example of a multiple vibrating-wire piezometer installation in a core hole
Source: Courtesy of Olympic Dam Expansion project

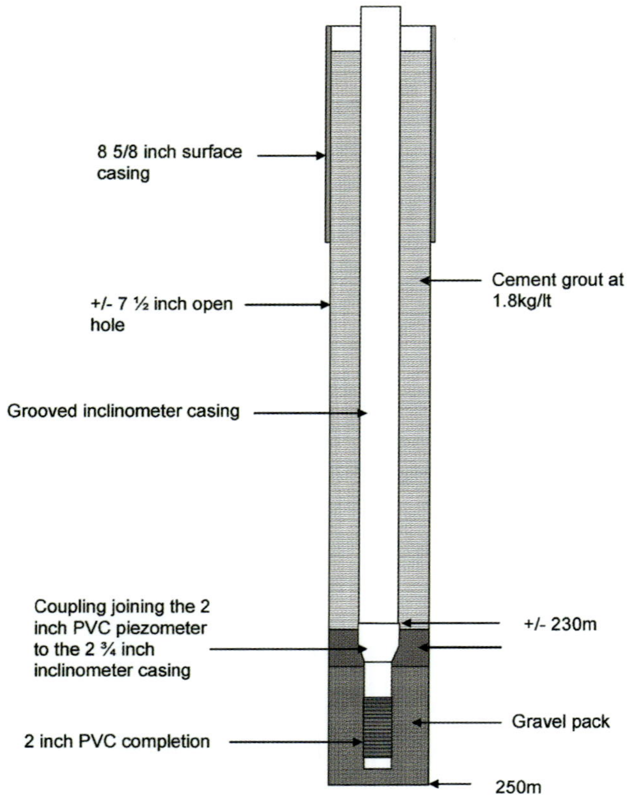

Figure 2.30: Example of dual piezometer/inclinometer installation

2.5.3.7 *Horizontal piezometers*

During active mining operations, piezometers can be installed horizontally from the pit slope, often as part a horizontal drain drilling program. However, as for vertical piezometers, the target zone for pressure monitoring needs to be isolated and sealed. There are two practical ways to achieve this. The first is to use a packer system and to grout the annulus above the packer. The second is to grout vibrating-wire transducers into place, as shown in Figure 2.31. In this process, the grout tube shown in Figure 2.31 must be extended so that the grout is introduced at the end of the hole. The second option allows multiple sensors to be installed within the hole. An issue with horizontal drilling is that the weight of the drill string often causes the holes to 'droop' (deviate downward) during drilling. This would affect subsequent vibrating-wire piezometer readings because the instruments may actually be lower than the target elevation. Therefore, if the intent is to install piezometers, it is beneficial if the hole can be directionally surveyed prior to piezometer installation.

2.5.3.8 *Piezometers installed in drainage tunnels*

The design shown in Figure 2.31 can also be applied to piezometers installed in underground drainage tunnels. Depending on the capability of the underground drilling set-up, such piezometers can be installed at any angle to target specific zones. Figure 2.32 shows a piezometer installed in a near-vertical upward hole.

2.5.3.9 *Westbay piezometers*

Another option for multi-level groundwater monitoring is the Westbay piezometer system. Westbay installations allow pore pressure monitoring from a virtually unlimited number of discrete depth zones in a single drill hole. Many existing installations include monitoring of up to 30 depth zones and several monitor up to 50 zones. The system is illustrated in Figure 2.33. It has been used worldwide for over 20 years.

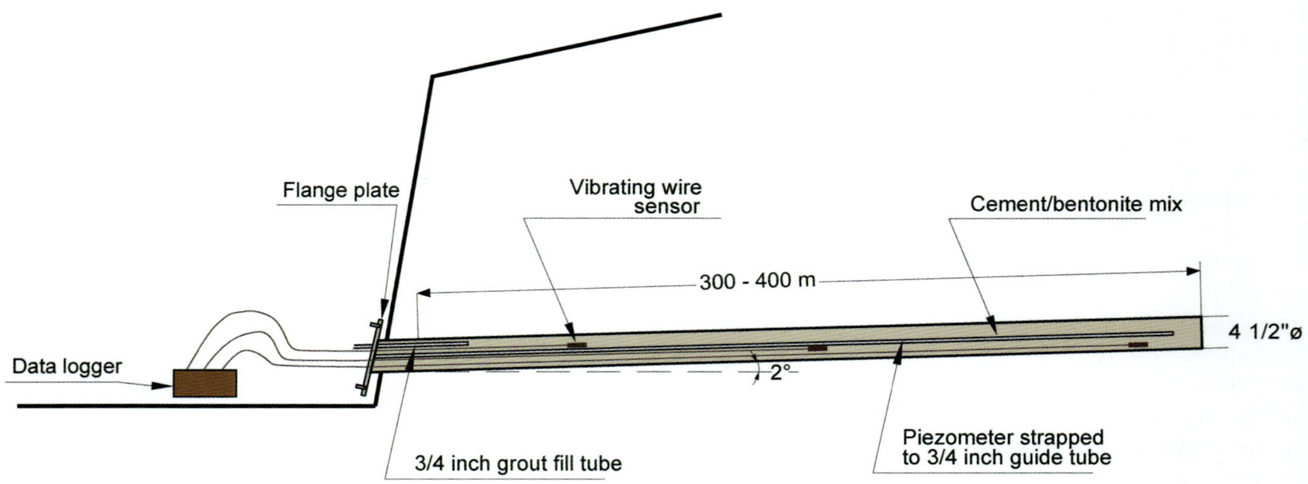

Figure 2.31: Vibrating-wire piezometers installed from the pit slope or from underground

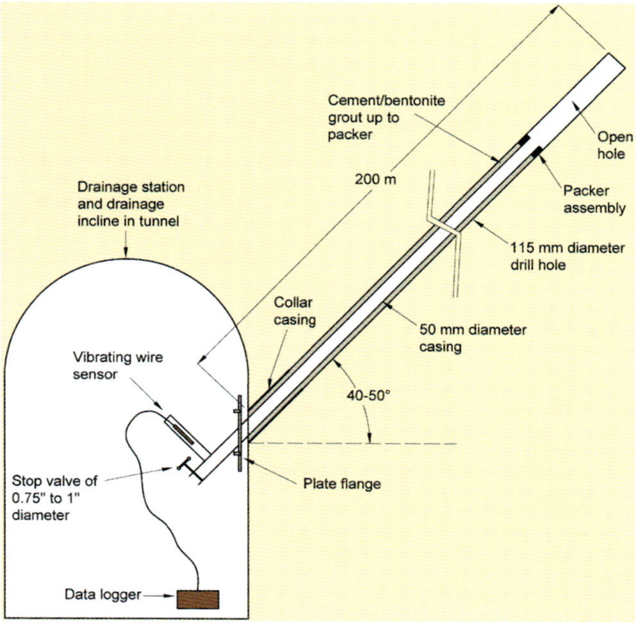

Figure 2.32: Installation of piezometers in upward holes

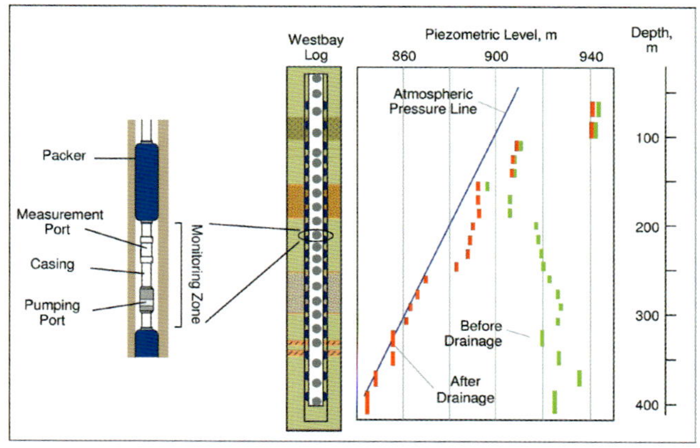

Figure 2.33: Example of multi-level pore pressure monitoring using a Westbay piezometer installation

A single casing assembly is installed into the hole, with valved 'ports' located at each required depth zone. Hydraulically inflated packers on the exterior of the casing seal the annulus of the hole between the monitoring zones. Wireline tools are lowered inside the casing to access each valved port, and the pore pressure is measured in situ. Measurements can be made manually, moving a single sensor from zone to zone, or automatically, by deploying a string of pressure sensors, each ported-in to a different monitoring zone. In low permeability environments, the Westbay system offers a similar quick response time to a vibrating wire piezometer.

An additional advantage of the Westbay system is that, if required, it also allows pulse hydraulic testing and water sampling to be carried out. Hydraulic tests can be carried out by purging within single zones, multiple zones, or cross hole testing using multiple wells. Such detailed testing can be valuable in identifying flow zones, flow barriers, and compartmentalization in complex fractured rock masses. Water samples can also be collected from each zone.

The PVC Westbay casing, ports and packers are compatible with HQ wireline drilling equipment, which allows the system to be installed inside the drill rods (with the bit removed). The system can therefore be installed to full depth without the risk of the hole collapsing. The drill rods are then withdrawn in stages and the exposed packers are inflated in sequence. This method has allowed the successful installation of the Westbay system to depths ranging from less than 100 m to over 1200 m, sometimes in poor quality rock. Pressure sensors and/or sampling devices can be retrieved for calibration and repair.

2.5.4 Guidance notes: installation of test wells for pit slope depressurisation

2.5.4.1 General

It is not the intention of this chapter to provide detailed drilling procedures for pumping wells. Suitable texts on well drilling are Driscoll (1989) and Roscoe Moss (1990). However, there are three important guidelines to follow for successful installation of test pumping wells in fractured rock settings.

1 Consider the installation of small-diameter pilot holes to establish ground conditions prior to completing high-cost production wells.
2 Minimise the use of drilling mud; if the use of mud is unavoidable, adapt the well design accordingly.
3 Use an appropriate well screen and filter-pack design for the application of the well.

2.5.4.2 Pilot holes

In fractured rock environments, most or all groundwater flow occurs within discrete fractures, joints and cavities. If holes drilled in such environments do not encounter open fractures, well yields will be low. In many fractured rock domains, well yields may vary greatly over short distances (in some cases less than 10 m) and it is not practical to use geophysics or other methods to locate open fractures prior to drilling.

In these situations, the most cost-effective approach is to install small diameter (low-cost) trial holes (pilot holes) to confirm the presence of open fractures prior to drilling larger-diameter pumping wells. There are many examples worldwide where high-cost pumping wells produced little yield because fracture zones were not intersected –

considerable savings could have been made by first proving ground conditions using pilot holes.

RC drilling provides an ideal method for drilling and testing pilot holes. The number of pilot holes required varies with location. One pilot hole may be sufficient in areas where open fracturing is pervasive or where a laterally extensive fractured horizon is present. Five or more pilot holes may be required to identify productive fracturing in tighter volcanic rock formations.

If the yield of a pilot hole is good, the hole can be directionally surveyed to determine the true location of the fractures. If the hole is determined to be vertical enough, it can be reamed into a test production well. If the hole is significantly off vertical, the collar location of the well can be adjusted to line up vertically over the fractures and a new well be drilled. If the yield of the pilot hole is low, the hole can be converted to a piezometer. Where several productive pilot holes are drilled, airlift flow and injection-test results can be quantitatively compared to assess which site will make the best production well. Locating the most productive fracture zones during pilot hole drilling can help maximise the long-term yield of subsequent production wells, ultimately leading to fewer wells being required for any given application.

Pilot holes also allow the depth of the water-producing fracture zones to be identified prior to drilling a production well. A dewatering well with higher yielding, shallow fractures may suffer a rapid loss of yield and quickly become redundant if the fractures become dewatered as the surrounding water table is lowered. A good way to determine the level of water-producing fractured intervals is to run flow (spinner), fluid conductivity and temperature logs, though these are typically run in the completed well.

In unconsolidated or sedimentary rocks where intergranular groundwater flow dominates, the use of trial holes is less important once the general water-bearing characteristics of the formation have been defined because there is typically less lateral variability in an intergranular flow setting.

2.5.4.3 Drilling mud

In porous medium-flow settings, where drilling mud is more often used, a mud cake (or filter cake) will build up on the wall. The cake develops because the pressure of mud in the hole is greater than the water pressure in the adjacent formation, so the mud liquid filters out of the hole and the mud solids form a fine-grained deposit (cake) on the hole wall. Mud cake formation can be minimised through the use of polymer drilling fluids or thin muds (or water). The more the filter cake can be cleaned subsequent to drilling, the better the yield of the hole. Procedures for removing the mud cake in porous medium-flow settings are well documented in Driscoll (1989) and other texts.

When drilling in fractured rock settings, however, it is important to realise that most of the mud losses occur into discrete fracture zones and, depending on the nature of the rock, much of the loss may occur into one or two zones. In horizons where the hole wall has low porosity and permeability, development of a filter cake may be poor to non-existent.

Depending on the fracture aperture and connection, the mud may travel a considerable distance into the fractures and may be difficult to remove by traditional techniques. If cuttings are carried into the fractures along with the drilling mud, extensive well development may be required to remove the debris from the fracture zones. If a well screen and gravel pack is installed prior to full development, debris may flow from the fracture zone into the gravel pack, permanently reducing the well yield. It is therefore important to monitor the mud level for evidence of circulation loss and to take early remedial action. However, in wells with small aperture fractures it is not always possible to identify and correct a loss of drilling fluid as the hole is being drilled.

2.5.4.4 Selection of well screen and filter pack

From the point of view of minimising head loss on water entry, the most efficient well design is to install no casing and screen – there would be nothing to impair the flow of groundwater into the well. However, there are very few formations that will stand up sufficiently (and indefinitely) to allow this; most wells in hard rock require some type of lining. Another consideration is that wells may subsequently be mined around, so will require casing and screen for future protection even if they would otherwise stand open.

For intergranular flow settings, selection processes for screen and filter pack selection are well documented in Driscoll (1989) and Roscoe Moss (1990). However, for fracture-flow settings, because much of the water enters the well over discrete intervals and therefore well entrance velocities are high, a conventional gravel pack design is often inappropriate. Under certain circumstances, installation of a well screen with no gravel pack produces the best results, particularly where debris from a fracture zone may enter the hole for a considerable time after well installation and development, and may clog the gravel pack. However, in many settings, a very coarse formation stabiliser (5–20 mm diameter grains) provides a good compromise where:

- intervals of the hole are liable to subsequently squeeze around the casing and cause damage or collapse, so some kind of hole stabilisation is required;
- it will later be necessary to mine around the well, so a formation stabiliser will dampen the effects of blasting on the well casing;

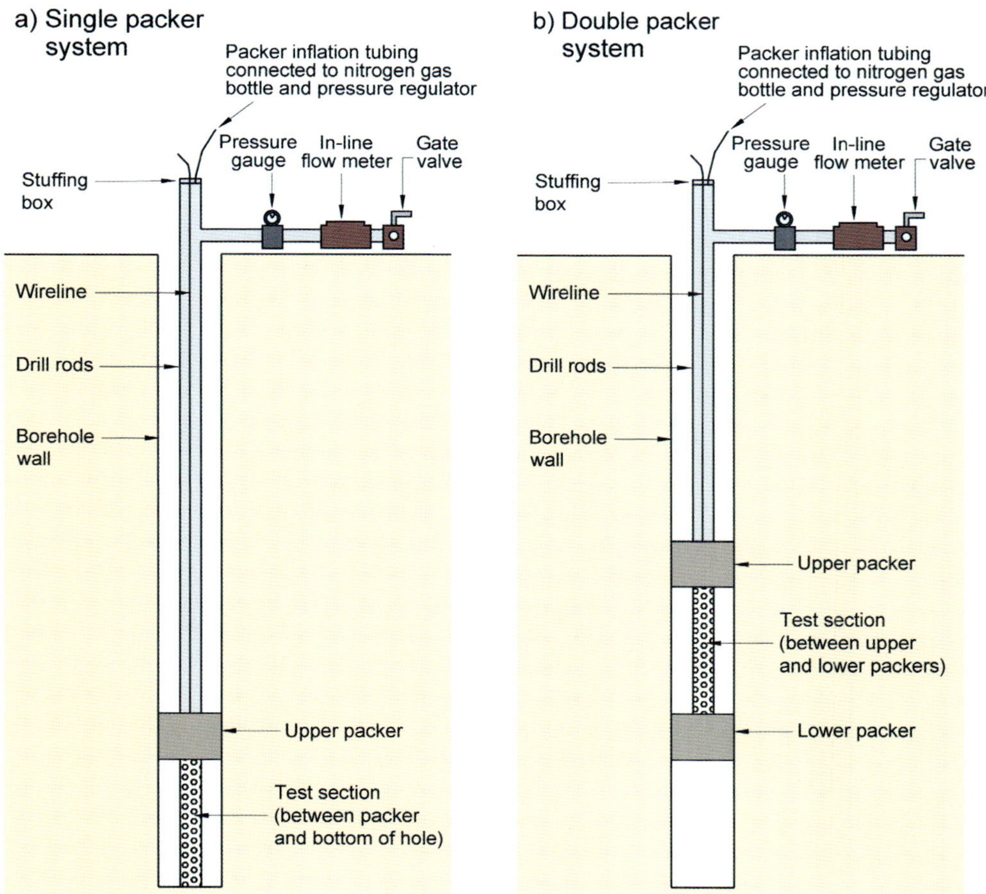

Figure 2.34: Single- and double-packer test configurations

- local regulations state that some form of filter medium is required on the outside of the well screen.

2.5.5 Hydraulic tests

2.5.5.1 Falling or rising head (slug) tests

Rising or falling head tests (slug tests) are a rapid and practical way to obtain permeability data for the formation, using standpipe piezometer installations. In a slug test, a small volume (slug) of water is quickly removed from (rising head) or introduced into (falling head) a piezometer, after which the rate of water level change in the piezometer is measured. From these measurements, the permeability of the formation in the immediate vicinity of the hole can be determined. Procedures for slug testing are provided in Appendix 1, Attachment D. Slug testing cannot be carried out in grouted-in vibrating-wire piezometer installations.

Slug testing is a highly useful and simple approach to obtaining permeability information, and is particularly useful for formations where the permeability is too low to carry out a pumping test. It is therefore a basic tool during detailed hydrogeological investigations of pit slopes. Tests can be run quickly, so there is opportunity to collect a considerable amount of data from a large number of drill sites at low cost.

2.5.5.2 Packer testing

Packer testing provides a means of testing discrete intervals of a drill hole during drilling or immediately upon completion of drilling. It is a good method of characterising vertical differences in hydraulic conditions.

Packer testing can be carried out in any type of drill hole, but smaller-diameter holes are typically better because the packers can be installed through the core barrel on the wireline and it is generally easier to obtain a seal. Packer testing in core holes (ideally HQ size) is often preferred because the packer information can be readily correlated with the core log, and because HQ-size packer equipment is readily available at many locations worldwide.

Details of procedures for packer testing are given in Appendix 1, Attachment E. Figure 2.34 shows the two main types of packer system arrangements. The first type

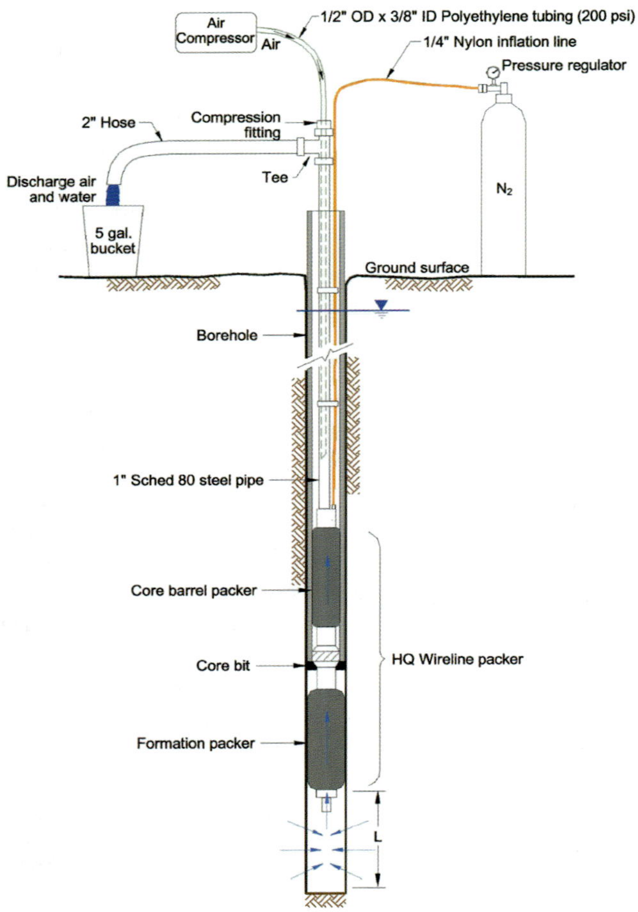

Figure 2.35: Typical set-up for airlift testing using wireline packer assembly

uses a single packer, run as the core hole is advanced, to isolate a test interval between the packer and the base of the hole. The second type is a double (straddle) packer system, typically run after the drill hole is completed to total depth.

The single-packer system has the advantage that it is only necessary to seat one packer, instead of the two required for the straddle system. For the single test procedure, the bottom seal is provided by the bottom of the hole at the time of the test. The single-packer test is easier to run because there is less risk of installation/removal problems for the packer equipment if the hole is unstable.

If drilling conditions dictate the use of mud to enhance hole stability, the use of a single-packer system may be preferable because there is less mud in the test interval. The packer system is lowered to the selected interval through the core tube, with limited exposure to the open drill hole. In addition, the longer a core hole is exposed to drilling mud the less accurate the packer test results.

However, a disadvantage of using a single-packer system is the need to stop drilling and run the equipment for each test, thus incurring additional standing charges. Also, a technician or geologist may be required at the drill site for a longer period of time to observe core, pick suitable test intervals and direct the tests.

Advantages of the double-packer method include the ability to do the tests sequentially while tripping out of the drill hole after drilling. This minimises rig standing time and allows the field supervisor to observe the entire profile of core from the hole and select all test intervals at the same time, thus reducing oversight time for the testing. If the hole is stable, it can be developed to remove mud and infill, and geophysical logging can be undertaken to help identify suitable intervals in which to seat the packers.

However, double-packer testing needs to be carried out with care by experienced staff, otherwise it is more likely to provide questionable results, especially in intervals with highly variable lithology and fracture-density conditions. This is because two packers are required to seal off the packer test interval. Difficulties may also be experienced with installation and removal of the straddle packer system in angled core holes.

Regardless of the method used, the best results are obtained by running individual tests on intervals no greater 10 m in length. The target zones to perform the packer tests are based on analysis of the core (fractures, lithology and alteration types) and any available geophysical logs. As a result, the hydraulic data obtained from the packer test can be viewed in the context of known local geological conditions. It is critical that locations for packer placement in the core hole be picked where there is a smooth drill hole wall to provide the greatest seating (formation sealing) characteristics. Again, this is easier for the single-packer system as there is only one packer to seat.

Both packer testing methods may fail due to faulty equipment or tears in the packer membrane. Poorly seated packers, poor seals, geological bypass and leaking nitrogen hoses can lead to failures and/or poor-quality test results. Close attention to detail between the driller and field supervisor is required at all times to maintain test integrity. When conducted correctly, packer tests give good results that can be correlated with other known properties of the rock obtained from geotechnical logging, particularly the locations of the fractures, the rock quality designation (RQD) and the fracture frequency (section 2.4.9)

If the tests are run properly, the hole angle is not usually important for achieving good packer test results. Single-packer testing systems may be easier to install and remove from highly inclined (more than 20° from vertical) drill holes than straddle systems are.

A more advanced alternative is to use a system that pumps water from the interval rather than injecting it (Price & Williams 1993). This adds to the expense but removes problems of clogging the interval with suspended solids in the injection water or changing the natural chemistry of the formation (if this is of concern). A typical set-up is shown in Figure 2.35.

2.5.5.3 *Pumping tests*

Pumping tests provide a method of characterising the groundwater system over a wider area than slug tests. The action of pumping a well creates a 'cone of depression' in the potentiometric surface. The shape and rate of expansion of this drawdown cone depend on the hydraulic properties of the formation. By monitoring the way an area of drawdown expands, it is possible to derive information about those properties. In a typical pumping test, a larger-diameter well is pumped and piezometers installed within the area of pumping influence are monitored for drawdown response.

Some notes and guidance for carrying out pumping tests are given in Appendix 1, Attachment F. Pumping tests are typically undertaken in permeable settings where it is possible to pump at significant rates from wells. Thus, pumping tests are more applicable for dewatering evaluations.

However, in many situations involving rock-slope depressurisation, where the permeability of the materials may be low, tests can be carried out at low rates or by airlifting as long as piezometers are closely spaced to allow discrete measurements at different vertical horizons.

2.5.6 Setting up pilot depressurisation trials

The final step for data collection is often the implementation of an instrumented pilot test area, which can be progressively expanded into a full depressurisation system. Commonly used methods for slope depressurisation are discussed in Chapter 6 (section 6.5). Figures 2.36 and 2.37 provide two examples of how pilot depressurisation trials can be used to gather design information.

Figure 2.36 shows a vertical drain trial at the Round Mountain mine in Nevada, USA. The drains were drilled to depressurise a sequence of lakebed clays and clay-altered volcanic layers (the Stebbins Hill formation). The water from the Stebbins Hill flowed down the drains and entered the underlying fractured volcanic rocks, which had already been depressurised using dewatering wells. The pilot test consisted of six test drains and two piezometers. Based on the results from the pilot area, the drain array was expanded into a field of 50 drains, each drilled to a depth of about 75 m.

Figure 2.37 shows the set-up of a horizontal drain trial. Three strings of vibrating-wire piezometers (three sensors per string) were installed in a preselected area of the pit slope. Four horizontal drains were drilled to target the level of the deepest piezometer in each string. The flow rate of the drains and the response of the piezometers were monitored. The piezometers at the level of the drains

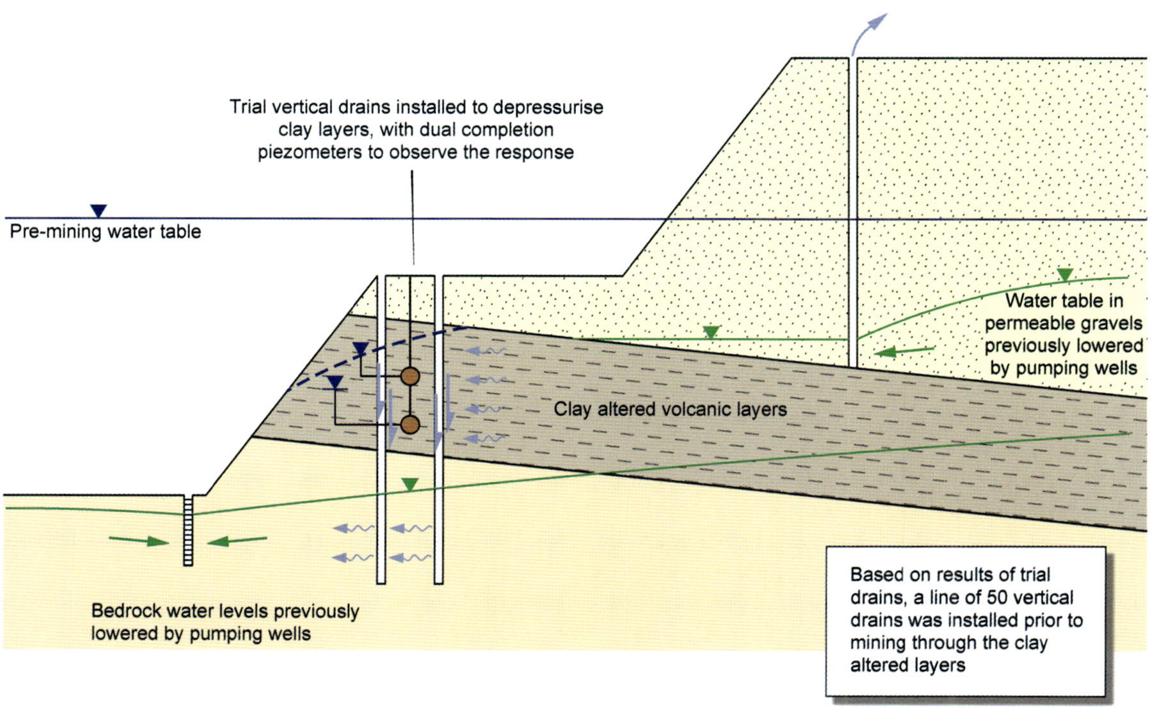

Figure 2.36: Example of depressurisation trial – vertical drains

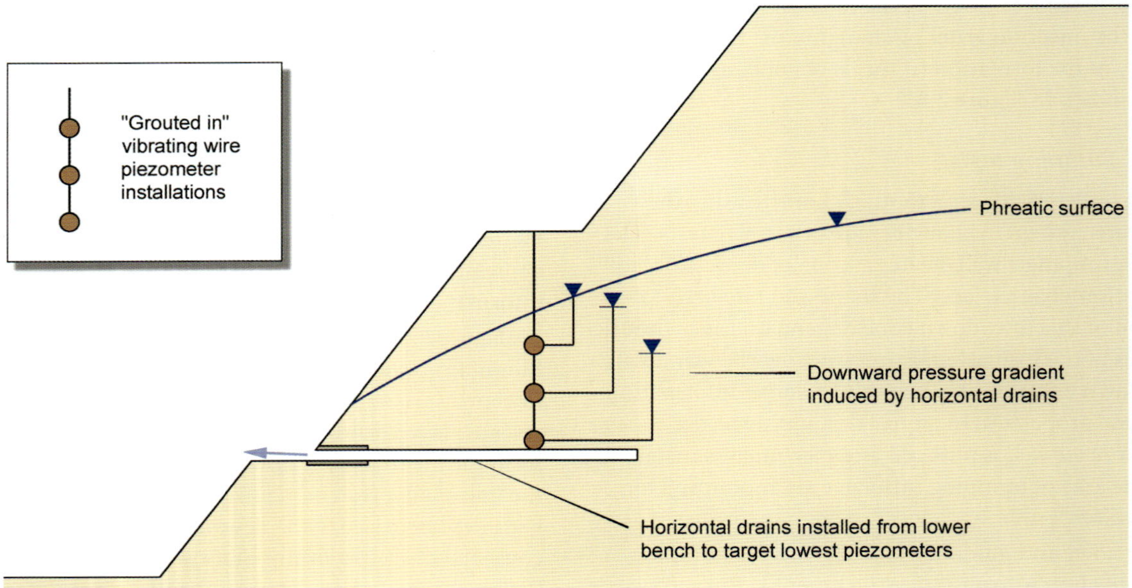

Figure 2.37: Example of depressurisation trial – horizontal drains

showed the highest response, with the shallow piezometers in each string showing a lesser response. The pilot test allowed detailed design of the depressurisation system for the slope.

2.6 Data management

A well-organised data management system to aid in the storage, interpretation, engineering and reporting of the measured field and laboratory data is just as important as the data collection process. It is imperative that a database which matches all of the anticipated needs of the project be selected and activated from day one of the project.

Historically, for specific project investigations geotechnical practitioners have mostly used back-end systems that contain the end-user application programming within the database as a single item. These include spreadsheets, systems such as Microsoft Access® and purpose-designed systems such as JointStats. The JointStats data management system was developed by the International Caving Study (Brown 2007) to accept standard structural data from face mapping or drill hole scanlines, organise it hierarchically then sort and statistically analyse it according to the standard structural attributes of orientation, length, spacing and persistence. These capabilities have now been enhanced by the LOP project to include quantitative measures of rock mass parameters (e.g. UCS, point load strength, density), enabling data uncertainty to be assessed and confidence limits determined for structural and rock mass parameters within any particular geotechnical domain.

At operating mine sites, there has been a strong corporate move towards data management systems that split the data and the programming parts into a front-end and a back-end. The back-end holds all the data, including ore reserves, grade control, geotechnical and survey data. It is accessed by users indirectly from the front-end, which holds all the application programming. Such back-end systems include Microsoft SQL®, MySQL®, Oracle® and acQuire®. The cited IT advantages of these systems include scalability, performance and concurrency. From a geotechnical viewpoint their potential advantage is that they can handle large amounts of complex data and are interoperable with a number of different geotechnical applications, including modelling systems such as Vulcan™, DataMine™, MineSite™ and Surpac™. They also have the advantage that new front-end geotechnical applications can be developed and deployed independently of the back-end database.

Endnotes

1 http://www.csiro.au/solutions/pps9a.html
2 http://www.adamtech.com.au
3 http://www.ballmark.com.au/
4 http://www.2icaustralia.com.au/core-orientation/Ezy-mark-features/
5 http://www.acedrilling.com.au
6 http://www.csiro.au/products/CoreProfiler.html

3 GEOLOGICAL MODEL

John Read and Luke Keeney

3.1 Introduction

Chapter 2 outlined the processes that should be followed to collect the data required to build the four components of the geotechnical model (Figure 3.1). Chapter 3 addresses the first of these components, the geological model.

The purpose of the geological model is to link the regional physical geology and the events that lead to the formation of the ore body to a mine-scale description of the setting, distribution and nature of the overburden soils and rock types at the site, including the effects of alteration and weathering.

Preparation of an accurate, well-understood geological model is fundamental to the slope design process. It requires a basic understanding of the essential concepts of physical geology, including how our solid earth reacts with water, air and living organisms.

Customarily, greenfields project development information is collected, consolidated and reported by the exploration team. These activities, most of which are well-described in an abundant public domain literature, are not addressed in this chapter, the main purpose of which is to outline the geotechnical requirements of the geological model. In this process it is recognised that specialist open pit mining engineering geologists or geotechnical engineers are not expected to have the type of geological expertise required of exploration or mine geologists. They are expected, however, to understand how ore bodies are formed, how they might be modified over time and which of their attributes are most likely to influence the design of the pit slopes. In the scheme of things, their main task is to work with the mine geologists to prepare a model that clearly identifies, describes geologically and outlines the 3D geometry of the different ore body and waste rock types at the mine site.

Accordingly, the chapter commences with an outline of what is required to describe the physical setting of the project site (section 3.2) and summary descriptions of the basic characteristics of the more common deposit types (section 3.3) before outlining the geotechnical requirements of building the model (section 3.4). These topics are supplemented by discussion of the causes and effects of regional seismicity (section 3.5) and the effects of regional stress (section 3.6).

For interested readers who are not trained geologists and who may not be familiar with some of the processes described and/or terminology used, a suggested supplementary text is *Physical Geology* (2007), by Plummer, Carlson and McGeary. The publication covers all aspects of physical geology, is particularly well-illustrated and contains an exhaustive list of web resources.

3.2 Physical setting

An important but often neglected part of setting up the geological model is the need to properly describe the physical setting of the project site. Many mines are situated in localities where extreme climatic and related geomorphological processes have had a profound influence not just on the layout of the mining infrastructure, but also on features such as the weathering and alteration profiles and/or the evolution of the deposit. Well-known examples of different extremes include the Ok Tedi and Lihir copper and gold mines in tropical Papua New Guinea and the copper and gold mines in the dry Atacama region in northern Chile.

The Ok Tedi mine is located in a mountainous rainforest region where the average annual rainfall is 10 m. The high rainfall and geologically rapid regional uplift and erosion have combined to develop a highly unstable terrain where 40% of the mountain ridges around the mine comprise previous landslide material (Read & Maconochie 1992). This has been accompanied by tropical weathering and alteration that has affected the

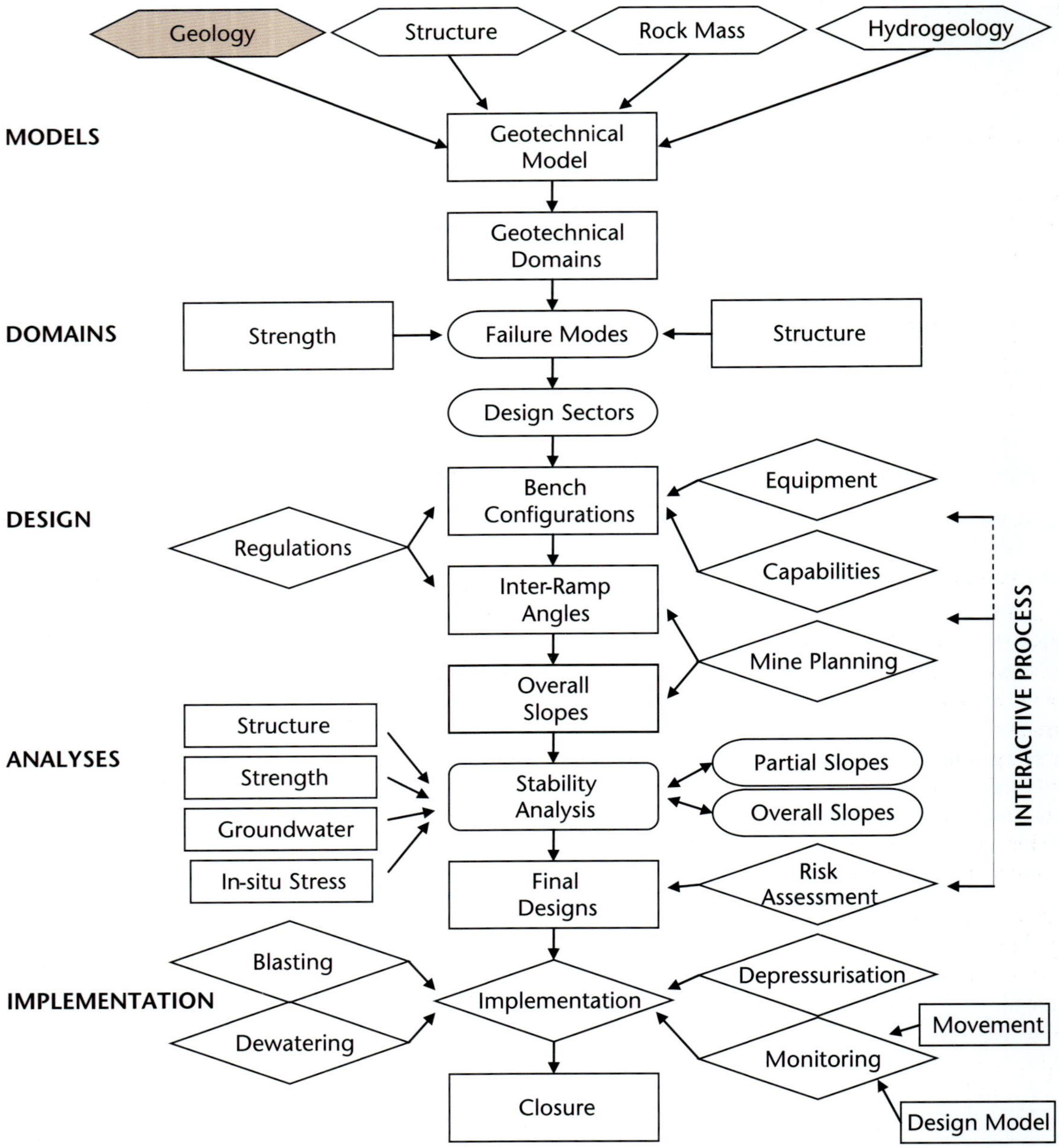

Figure 3.1: Slope design process

strength and stability of the metastable colluvial soils and the underlying bedrock. All these features have required special consideration in the pit wall design. The Lihir mine is located in an active geothermal environment – in addition to an extremely volcanically altered rock mass there is an eclectic mix of meteoric water associated with a high annual rainfall (3 m), sea water from the nearby ocean (200 m away from the seaboard wall of the pit) and geothermal water that rises from below and enters steam and gas phases as it rises towards the floor of the pit. All these must be considered in the pore pressure studies for the pit walls.

At the other extreme, the Escondida mine in northern Chile is situated in an arid and geologically mature Tertiary desert landscape where there is no vegetation and rain is a rare event. However, groundwater fluctuations during regional faulting and uplift since the ore body was emplaced produced a barren leached cap up to 180 m thick above a high-grade sequence of supergene (copper oxide) enriched ore from the deeper primary sulphide ore. Recognising the physical geologic history and distinguishing the boundaries between each zone was one of the most important aspects of project development and planned future expansions.

Many other examples could be cited, but all serve to point out that a number of physical regional features must be incorporated in the model from day one of the project. The list should include at least the following:

- geographic location;
- tectonic evolution;
- climate;
- geomorphology;
- topography;
- drainage systems.

Pit design is too often focused on the characteristics of the ore body and waste rocks in proximity. We must also take heed of the natural processes that occurred before the deposit was developed if we are to develop a reliable model.

3.3 Ore body environments

3.3.1 Introduction

There are a number of different styles of ore bodies, each with its distinctive signature. As already noted, although specialist engineering geologists or geotechnical engineers are not expected to understand the finer details of ore body formation, they should be able to recognise the characteristic geomechanical features of the different deposit types. Clearly, a detailed knowledge of the geotechnical attributes of every type of deposit is impracticable. However, a working knowledge of at least the following major ore bodies would be expected:

- porphyry deposits;
- epithermal deposits;
- kimberlites (diamonds);
- volcanic massive sulphide (VMS) deposits;
- skarn deposits;
- stratabound deposits, including Mississippi Valley type deposits, banded iron formations (BIFs), sediment-hosted copper, iron, cobalt and uranium, and coal.

In all cases, it is important to understand the regional and local geological context of individual deposits as well as the broad structure commonly associated with the deposit type.

3.3.2 Porphyry deposits

Globally, porphyry deposits such as those at the Chuquicamata and Escondida mines in Chile and the Bingham Canyon mine in the USA are probably among the best-known source of copper. Porphyry deposits occur in two main geological settings within orogenic belts (fold-belt mountain-forming areas) – island arcs and continental margins. They are large-scale low-grade ore bodies, ranging from stockworks to disseminated deposits of copper with associated primary products including molybdenum and gold.

Porphyry systems form when magmatic intrusions influence the surrounding country rock through hydrothermal interactions. The magmas associated with the intrusions can range widely in composition but are generally felsic and exhibit a porphyritic texture characterised by large crystals set in a fine-grained groundmass. The intrusions are epizonal, forming at shallow crustal depths of less than 6.5 km, and tend to be composite and disordered. They are classified in three ways:

1. plutonic porphyry copper deposits, found in batholitic settings. A batholith is a large plutonic body with an aerial extent of 100 km^2 and no known floor. Mineralisation principally occurs in one or more phases of the development of plutonic host rock;
2. volcanic porphyry copper deposits, found in the roots of volcanoes. Mineralisation occurs in both the volcanic rocks and associated deep-seated igneous intrusions, derived from the same parent magma (co-magmatic plutons);
3. classic porphyry copper deposits, which occur when high-level post-orogenic stocks intrude into unrelated host rocks. A post-orogenic stock is a small (aerial extent less than 100 km^2) plutonic body with no known floor, that has formed after a period of orogeny. In classic deposits, mineralisation may occur entirely within the stock, entirely in the country rock or in a combination of both.

Faults and strongly fractured zones are particularly important control features during intrusion emplacement and mineralisation, with intense fracturing in the intrusion and surrounding country rocks providing pathways for ore-bearing fluids to flow. Two main models describe this flow – the orthomagmatic model and the convective model. In the orthomagmatic model, metals and sulphur are derived from the magma and are concentrated in ascending residual fluids. In the convective model, metals and sulphur are derived from the surrounding host rocks by thermally driven groundwater.

Alteration is always pervasive, with strong alteration zones developing in and around the intrusion. The hydrothermal fluids involved in the alteration can be sourced from the magma or the circulating heated groundwaters. Factors that control alteration include temperature and pressure, the abundance, composition and dynamic behaviour of the fluids, and the degree of wall–rock interaction. Potassic, phyllic, argillic and propylytic alteration zones are dominant. Orthomagmatic systems are dominated by potassic and propylytic alteration with narrow zones of phyllic alteration in the area of interaction between magmatic and meteoric fluids. As a consequence, pervasive alteration and mineralisation form a series of shells around the core of the intrusion. In contrast, convective systems are dominated by phyllic alteration.

Attributes of porphyry deposits that are most likely to influence the stability of the pit slopes are the faults and strongly fractured zones that accompany the intrusion emplacement and mineralisation. Weak rock formed by the pervasive alteration that accompanies mineralisation is more likely to appear within the ore body itself, rather than in the surrounding wall rocks. However, these weak alteration zones must still be viewed with caution and carefully identified, especially during the early development stages when the toe regions of the pushbacks may well be located within the ore body being mined.

3.3.3 Epithermal deposits

Epithermal deposits form in the near-surface environment, typically within 1 km of the earth's surface in volcanic, tectonically rising areas. They are the product of low-temperature (50–300°C) hydrothermal activity generated from subvolcanic intrusions. The Manto Verde and Candelaria mines in Chile are examples.

Epithermal deposits tend to occur as small vein systems of less than 1 Mt in size, but usually exhibit good ore grades containing high unit values of precious metals with relatively low base metal content. They are referred to as either high or low sulphidation systems. High sulphidation systems are related to oxidised sulphur species. They are usually found close to volcanic vents and are associated with gold and copper, and to a lesser extent with silver, bismuth and tellurium. Low sulphidation systems are distal to volcanic vents, commonly associated with the types of fluids involved in hot springs. They contain reduced sulphur species and are associated with gold and silver, and to a lesser extent with arsenic, selenium and mercury.

Epithermal system host rocks characteristically sustain fractures over long periods of hydrothermal activity. Typically, the mineralisation occurs in siliceous vein fillings; irregular branching fissures or the close networks of veinlets that make up stockworks. Breccia pipes and vesicles formed from gas bubbles trapped in cooling lava and the pores and small fissures can also act as hosts.

Because of the high degree of fracturing and hydrothermal activity, the country rock lateral to the host rock is usually extensively altered. In felsic rocks such as rhyodacite, latite or rhyolite, the alteration is characterised by the alteration of felsic minerals to sericite, the introduction or replacement of felsic minerals by silica and the introduction of felspathoids into the system. Common alteration minerals include carbonates and clay minerals, especially kaolin and montmorillonite.

The dominant alteration process in mafic and intermediate volcanics such as basalt, andesite and dacite is propolytisation, which is a low-pressure/temperature process that produces chlorite and epidote as the most abundant alteration minerals. Other alteration products in the propylytic zone include sericite, alunite, zeolite, adularia, silica and pyrite.

Attributes of epithermal deposits that are most likely to influence the stability of the pit slopes are the high degree of fracturing and the alteration of the country rock lateral to the host rock. In these conditions, failure by rupture through the alteration weakened rock without or with only partial structural control can be as likely as structurally controlled failures.

3.3.4 Kimberlites

Diamonds form outside their host rock. They crystallise in the upper mantle under extreme pressure, available only if the lithosphere is at least 120 km thick (Evans 1993). They are then caught up by rising magma, usually kimberlite, which transports them to the crust. Evidence for their external origin comes from South African kimberlite dated at 90 million years old but containing diamonds more than 2 billion years old (Kramers 1979 in Misra 2000).

Kimberlites hosting economically important diamond deposits appear to be localised to stable shield (craton) areas older than 2.4 billion years (Misra 2000). Kimberlite itself is a potassium-rich, volatile rich, ultramafic hybrid rock which rises rapidly from the mantle depths and is emplaced explosively if it reaches the near-surface environment (Winter 2001).

Kimberlites can occur as hypabyssal dikes or sills, diatremes, crater fill or pyroclastic deposits depending on the depth of erosion and exposure (Winter 2001). Sills are less common than dikes and can be several hundred metres thick. Dike swarms 1–3 m thick occur and tend to bifurcate into stringers. Dikes pinch out near the surface and thicken with depth. They may expand locally into blows 10–20 times the dike width and 100 m long, which then feed into the root zones of diatremes.

Diatremes are 1–2 km deep, conical-shaped bodies with steeply dipping vertical sides (80–85°) tapering towards the root zone. The diatreme represents a region where decreasing pressure in rising magma results in a volatile expansion, causing an explosive eruption. At 300–400 m depth, hydrovolcanic interaction with groundwater produces gas with such violence that it breaks through to the surface. At this point, either rapid degassing and vapour exsolution or increased groundwater flow into the crater and pipe result in a downward migration of the magma, violent brecciation and mixing to form the diatreme (Mitchell 1986).

Kimberlites are preserved at the surface in a few areas where there has been little erosion. Tuff and tuff rings comprising wide low-brimmed accumulations of debris built around the volcanic vent are the main types of surface expression (Winter 2001).

Features of kimberlite deposits that are most likely to influence the stability of the pit slopes are the contact zones of the pipe itself, which usually feature highly fractured, slickensided and/or polished surfaces formed as the kimberlite pipe forced its way to the surface. Additionally, some kimberlites rapidly disaggregates when exposed to air and, particularly, water, so caution is required if benches are to be mined inside the boundaries, within the kimberlite itself.

3.3.5 VMS deposits

Volcanic massive sulphide (VMS) deposits are related to volcanically active submarine environments in major orogenic zones. The metals precipitate out of hydrothermal solutions primarily as copper, zinc, lead, silver and gold. Tin, cadmium, antimony and bismuth occur as byproducts. Individual deposits can contain anywhere from 0.1–10 million tonnes of ore, with combined copper, zinc and lead grades of more than 10% (Misra 2000).

The classic VMS deposit has a lens-shaped sulphide mound capping with a well-developed footwall alteration zone. It displays a distinct pattern of zonation, running from iron to iron and copper, then copper, lead and zinc and finally lead, zinc and barium moving up and outwards from the hydrothermal source (Robb 2005). In typical cases, the sulphide deposits may appear to be flat or tabular in shape, stacked one on top of the other, or located away from the vent.

VMS deposits can be classed as proximal or distal. A combination of temperature, salinity and degree of mixing with sea water will cause the vented submarine fluids to be more or less dense than sea water. Proximal deposits are formed when dense ore solutions precipitate close to the vent. Less dense fluids will disperse and precipitate, forming a deposit distal to the vent. The trend towards copper-dominated iron sulphide ores decreases with distance from the vent as a result of the mixing of sea water and ore-bearing hydrothermal fluids. Distal deposits are copper depleted and zinc–lead dominated, do not display systematic metal zoning and are not associated with footwall alteration (Misra 2000).

The feature of a VMS deposit that is most likely to influence the stability of a pit slope is the underlying footwall alteration zone, which can form a potentially unstable dip-slope in the footwall of the mine. Accordingly, the geometry of the ore body needs to be established early in the mine planning process. Also, many VMS deposits have been metamorphosed, in which case foliation and schistosity may influence the stability of the design.

3.3.6 Skarn deposits

The word 'skarn' comes from old Swedish. It originally referred to very hard rocks composed mainly of calc-silicate minerals associated with magnetite and chalcopyrite deposits found in Sweden (Robb 2005). The word now refers to calc-silicate mineral assemblages formed by metasomatic replacement of carbonate rocks at temperatures of 400–650°C during contact or regional metamorphic processes associated with an intruding pluton. In terms of tectonic setting, the majority of skarns appear to be located at continental margins and island arcs, forming either during or at the end of an orogenic period.

Skarns are classified as endoskarn or exoskarn, depending on the placement of the metasomatic mineral assemblages in relation to the intruding pluton. Endoskarns are created when the intrusive rocks of the pluton are replaced by fluids flowing into the pluton. An exoskarn forms when country rocks in the contact zone around the pluton are replaced by metasomatic fluid flowing up and out of the pluton. In an endo-exo skarn couplet, the skarn ore tends to be restricted to the exoskarn (Misra 2000).

The majority of skarns are not of economic quality (Evans 1993). However, skarn deposits represent some of the world's most important sources of tungsten as well as being important sources of copper, iron, molybdenum and zinc (Misra 2000). The wide variety of metal associations is a result of the differing compositions, oxidation states and metallogenic affinities of the igneous intrusion. Intrusions of mafic to intermediate compositions are generally related to iron and gold deposits. The calc-alkaline, magnetite-bearing, oxidised granitic intrusions are associated with copper, lead, zinc and tungsten deposits. The more differentiated, reduced, ilmenite-bearing granites are generally linked to molybdenum and tin (Robb 2005).

The metamorphosed endoskarn and exoskarn contacts along the margins of an intruding pluton do not usually exhibit features that will influence the stability of a pit wall. However, if the intruding pluton intersects a major regional structure such as a thrust fault, the fluids associated with the skarn formation process may react with the sheared and crushed fault material (gouge) to form a particularly weak zone along the plane of the fault. Examples are the sulphide skarns formed along a series of thrust faults at the Ok Tedi mine in Papua New Guinea, which contributed to a major landslide at the site and form controlling structural features in the design of the pit walls (Read & Maconochie 1992).

3.3.7 Stratabound deposits

Geotechnically, the major feature of stratabound deposits are the folding of bedding and the interbedding of weaker and stronger units. Attributes that can influence the stability of the pit slopes include the potential complexity of the folding, the weak zones that occur at the boundaries between stronger and weaker zones, and/or the weak zones themselves. Four different deposit types are outlined below.

Mississippi Valley

Mississippi Valley Type (MVT) deposits mostly occur in the USA, with a few also in South America, Poland and Australia. As the name suggests, they were first studied in the drainage basin of the Mississippi River in the central USA. They are strata-bound, carbonate-hosted deposits and contain significant proportions of the world's zinc and lead. Their mineralisation is epigenetic, mostly of the cavity-filling type, and shows no clear association to igneous rocks.

Tectonically, MVT deposits are located on or near the edges of intra-cratonic basins or on the shelf areas of passive continental margins. Stratigraphic evidence shows that mineralisation occurred at very shallow depths, probably between a few hundred metres and 1000 m (Misra 2000). The mineralogy is simple. Sphalerite and galena are the main sulphide minerals with fluorite, barite, dolomite and calcite making up the bulk of the gangue minerals. Most deposits contain almost no copper minerals, except in the south-east Missouri district (Robb 2005). Mineralisation can be massive or disseminated. The host rock acts mostly to provide areas for ore localisation through brecciation, faults, fractures or the permeability changes found in facies changes. Mineralisation in these areas is strata-bound and typically discordant.

Banded iron formation

Banded iron formation (BIF) deposits are a globally important source of iron. They form both proximally and distally to extrusive centres along volcanic belts, deep fault systems and rift zones. The structures they display are a result of oxygen released by cyanobacteria in association with dissolved iron in the Earth's oceans. The iron banding is a result of cyclic peaks in oxygen production causing iron oxides to precipitate out of the ocean. It can range in thickness from a few centimetres to over a metre, and form stratigraphical units that can be hundreds of metres thick and thousands of metres long (Robb 2005). The iron is found in the form of magnetite, hematite, goethite, limonite and siderite, with silica, phosphorus, aluminium and sulphur as common impurities.

The proportions of volcanic and sedimentary rocks vary between BIF deposits, of which there are three distinct types: Algoma, Lake Superior and Rapitan.

Algoma BIFs are associated with volcanic arcs and are often located in Archaean greenstone belts. They tend to be fairly small, usually only a few hundred kilometres in strike length and a few centimetres to a few hundred metres in thickness (Evans 1993).

The Lake Superior BIFs are located on stable continental platforms. They are economically important as they can extend hundreds of kilometres along strike and be anywhere from 20 m to several hundred metres thick. The type area is in the Lake Superior region of the USA. Other major deposits occur in the Hamersley Basin of Western Australia, where the Brockman BIF can be correlated over 50 000 km^2 at a scale of only 2.5 cm (Trendall & Blockley in Evans 1993), the Transvaal Basin of South Africa, the 'Quadrilatero Ferrifero' of Brazil, the Labrador trough of Canada and the Singhbhum region of India (Robb 2005).

The Rapitan ores are named for their type occurrence at the Rapitan Group in the McKenzie Mountains of north-west Canada. They are unusual because of their association with glaceogenic sediments formed during the almost global glaciation of the 'Snowball Earth' or Cryogenian period 750–580 million years ago.

Sediment-hosted ore bodies

Sediment-hosted ore bodies (copper, iron, cobalt and uranium) originated when sulphide- and oxide-rich river waters were delivered into a near-shore reducing environment of tidal flats, estuaries and bays. In time, the sedimentary rocks formed in these environments were metamorphosed to varying grades and folded into the older granitic floor rocks, with the migrating ore minerals being concentrated along stratigraphic boundaries or within folded stratigraphic traps.

The Zambian Copper Belt (ZCB) is typical of these deposits. The ZCB is one of the largest sediment-hosted stratiform copper regions and one of the great metallogenic districts in the world. Copper and cobalt ore bodies are located in shales and arenites, with most of the ore bodies situated in what is termed the shale belt. The mineralisation occurs in the form of sulphides and multi-metal iron, copper and cobalt ore minerals. Chalcopyrite is the main sulphide mineral. Distinct concentrations of minerals occur along the bedding planes with copper ore grades ranging from 3–6% and up to 15–20% between locations. At the margins of the deposit, copper and cobalt sulphides give rise to pyrrhotite and pyrite. A distinct sulphide zonation occurs with respect to the old shoreline. The shore is barren, with chalcocite occurring in the original shallow water environment, followed by bornite/chalcopyrite and carrollite farther in the basin. Pyrrhotite occurs in regions far removed from the shore line. A distinct zoning of both metal distribution and mineralogy is observed consistent with transgressive and regressive facies changes.

Coal

The first large coal deposits appeared about 400 million years ago, after the evolution of land plants in the Devonian period. Significant deposits then formed during the Carboniferous period in the northern hemisphere and during the Carboniferous and Permian periods in the southern hemisphere. Since then deposits have formed in a variety of localities around the world, principally during the Cretaceous to early Tertiary era (about 100–15 million years ago). In the slope design process for open pit coal mines the primary geotechnical interest is in:

- the rank of the coal;
- the seat-earths and other weak materials that may occur in the horizons immediately above and below the coal seam;
- the interbedded sedimentary rocks that overlie the coal seam being mined;
- the joint fabric within these sedimentary rocks and any major faults that may disrupt the stratigraphy of the deposit;
- the structural attributes of joints and/or faults that intersect the coal and the overlying sedimentary rocks;
- the strength of the coal and the overlying sedimentary rocks;
- the strength of the seat-earths and any other weak seams along the bedding.

3.4 Geotechnical requirements

When the natural regional framework discussed above has been established, each rock type at the project site should be subdivided into consistent units or domains based on a combination of the following:

- rock type (lithology);
- major structures (Chapter 4);
- mineralisation (ore and waste);
- alteration, including all pre- and post-mineralisation events;
- weathering;
- geomechanical properties.

The objective of the model is to provide a design-level understanding of the 3D boundaries and the geomechanical attributes that characterise each rock type at the site. When preparing the model it is important to recognise that at many ore bodies the overburden may have an entirely different geology from the ore and the host rock. Examples are the conglomeratic outwash gravels at Mina Sur in Chile and the regolith deposits at Olympic Dam in South Australia. There can also be thick saprolites above ore bodies in tropical environments.

To illustrate the requirement, Figure 3.2 uses mineralisation and alteration of two different lithologies to define seven basic geological units:

1 Primary mineralisation + Lithology B + Alteration A;
2 Primary mineralisation + Lithology B + Alteration B;
3 Primary mineralisation + Lithology A + Alteration B;
4 Secondary mineralisation + Lithology A + Alteration B;
5 Secondary mineralisation + Lithology A + Alteration A;

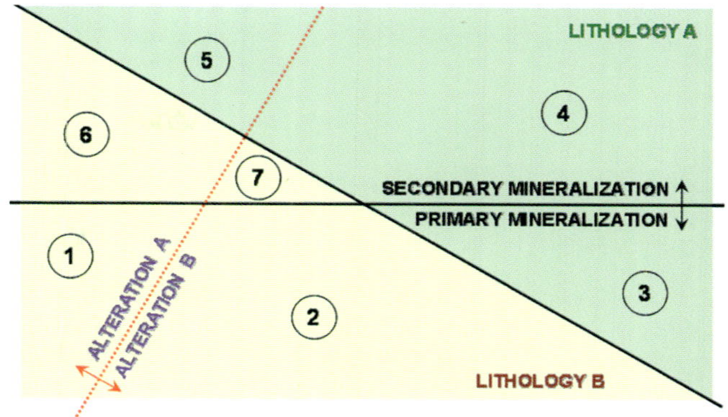

Figure 3.2: Example of the definition of basic geological units by superimposing the lithology (2 types), the mineralisation (2 types) and the alteration (2 types)
Source: Flores & Karzulovic (2003)

6 Secondary mineralisation + Lithology B + Alteration A;
7 Secondary mineralisation + Lithology B + Alteration B.

The first step in the model-building process is to compile the entire field mapping data, including core data from subsurface exploration and geotechnical drilling programs (Chapter 2), into a geological plan of the pit. This plan is then incorporated into a 3D solid geological model using a modelling system such as Vulcan™, DataMine™, MineSite™ or Surpac™. Mapped data from Autocad is usually imported as DXF files so that the geologist can connect the fault, lithological or other geological boundary traces and build on those traces in 3D to make modelled shapes or triangulations. Once the triangulations are made it is easy to cut them to pit shells or into sections. The complete process is illustrated in Figures 3.3–3.13, using examples from different mine sites.

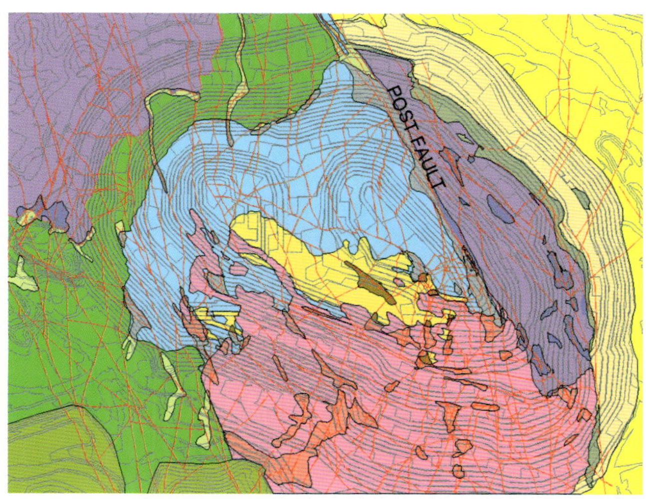

Figure 3.3: Geological model of the Goldstrike Mines Betze-Post open pit
Source: Courtesy Barrick Gold Corporation

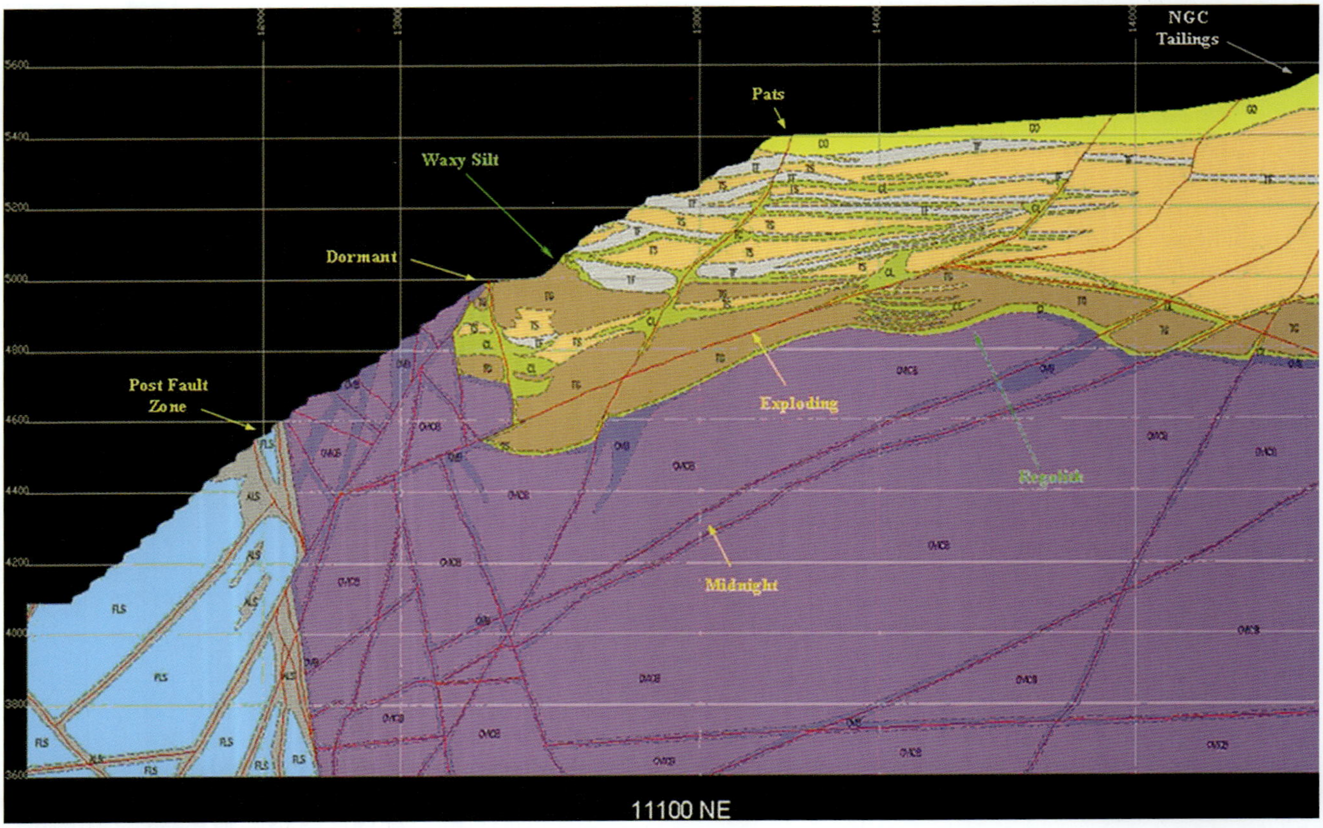

Figure 3.4: Geological cross-section through the east wall of the Goldstrike Mines Betze-Post open pit
Source: Courtesy Barrick Gold Corporation

Figure 3.3 and Figure 3.4 are from the Goldstrike Mines Betze-Post open pit. Figure 3.3 is a 2001 version of the Goldstrike geological model projected to the open pit shell. Figure 3.4 is a cross-section through the east wall of the pit.

Figure 3.4 highlights a modelling issue relating to the level of confidence in the geological information shown on the cross-section. Before computer graphics became established in the industry, geological maps and cross-sections were hand-drawn. With hand-drawn maps and cross-sections it was standard practice to designate only established or known geological boundaries and structures with solid lines. Uncertain or inferred geological boundaries and structures were shown as dashed lines or as dashed lines with question marks between the dashes. Since the introduction of computer graphics systems this practice has fallen by the wayside and all boundaries are shown as solid lines – in consequence any uncertainty in features such as lithological boundaries and major faults is not reflected in the drawing (plan or section).

In Figure 3.4 the spacing of the horizontal grid lines is 200 feet (c. 61 m). In keeping with this scale, it is reasonable to assume that the geological boundaries shown in the upper 200 feet of the cross-section are based on surface exposure and drill hole intersections and can be regarded as well-established. It is equally likely that the deeper boundaries (e.g. towards the lower right-hand side of the section) are based on projected data rather than on drill hole intersection or other real data. This introduces an element of uncertainty into the reality of their locations, which is not reflected in the solid nature of the boundaries shown on the section.

A suggested way of addressing the questions raised by Figure 3.4, based on the ideas outlined in Chapter 8, is illustrated in Figures 3.5 and 3.6. These two figures are cross-sections through a folded metasedimentary sequence that has been intruded by dolerite sills (coloured red) and a kimberlite pipe (coloured green). The figures also show the locations of the drill holes on each section and the estimated levels of certainty (80%, 66% and 33%) in the interpreted geology with depth. Although the levels of certainty shown are only estimates, they do reflect the density of drilling shown on the sections and give the reader a clear idea of which boundaries are considered to be well-established and which are not.

The impact of data uncertainty and different ways of quantifying the level uncertainty in the geological data raised by Figure 3.4 are discussed further in Chapter 8.

Figure 3.7 is an example of a solid model that has been colour-coded to show the 3D distribution of the rock types within a mine site environment. Figure 3.8 is a 3D fence

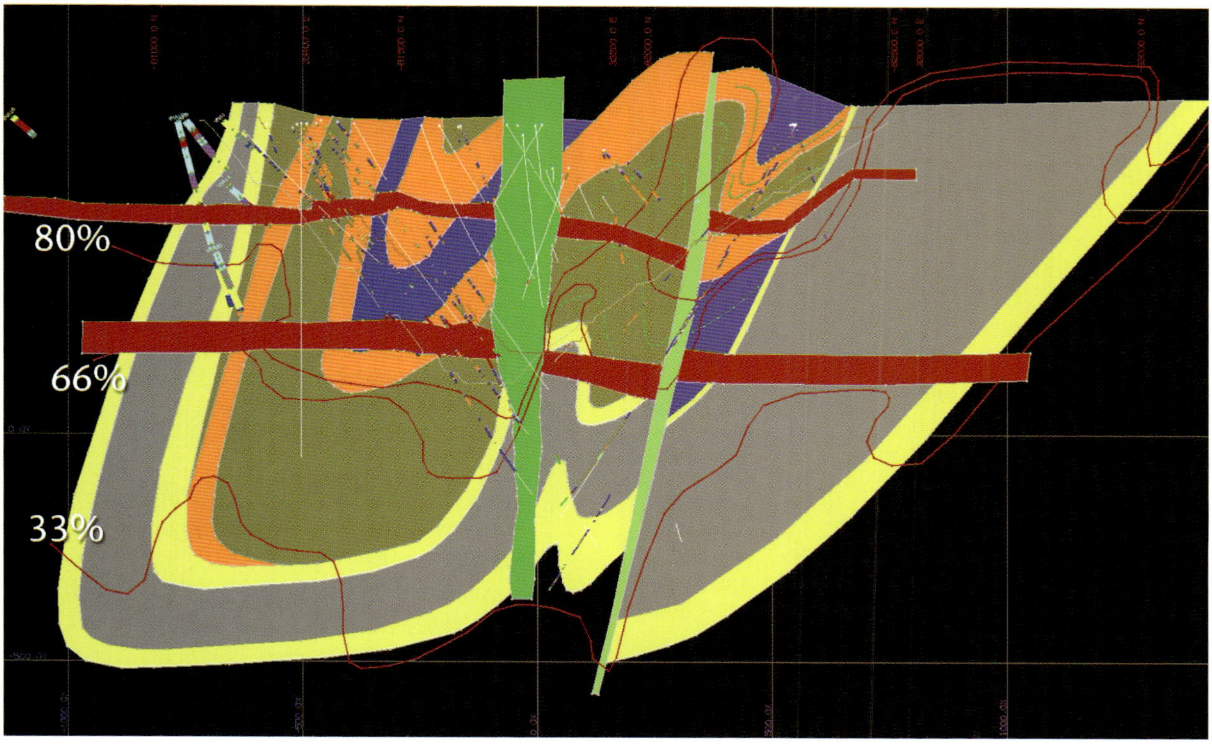

Figure 3.5: Cross-section S1 showing interpreted stratigraphic and structural boundaries, drill holes and estimated levels of data certainty with depth
Source: Courtesy DeBeers

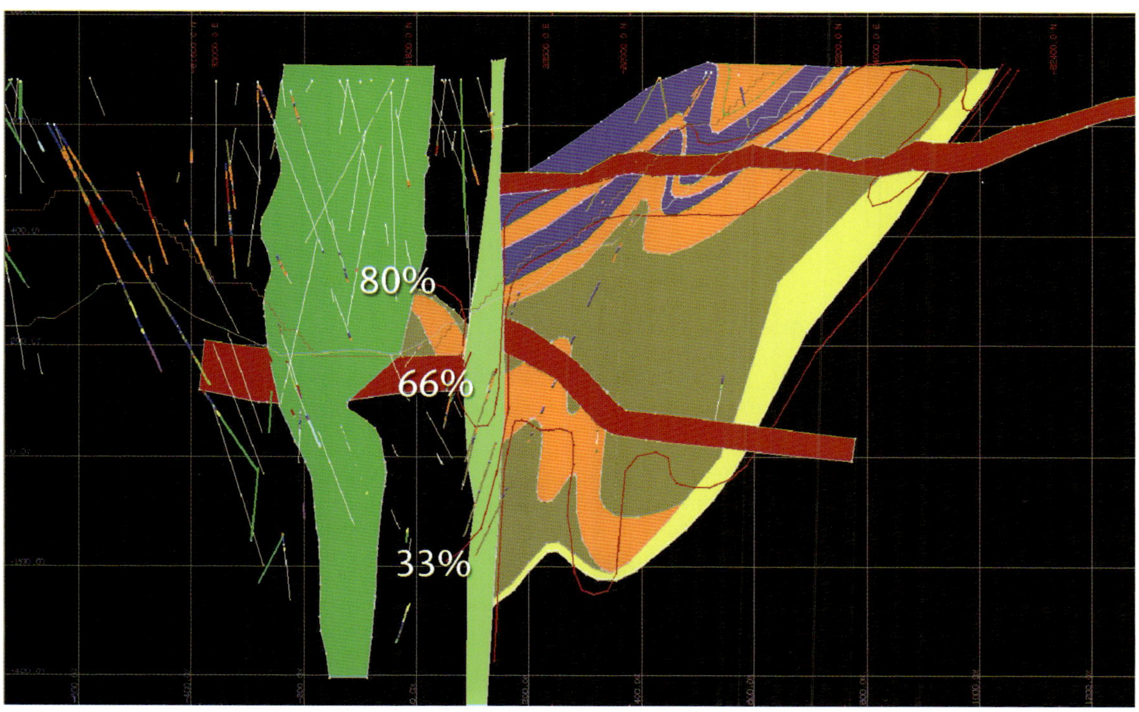

Figure 3.6: Cross-section S2 showing interpreted stratigraphic and structural boundaries, drill holes and estimated levels of data certainty with depth
Source: Courtesy DeBeers

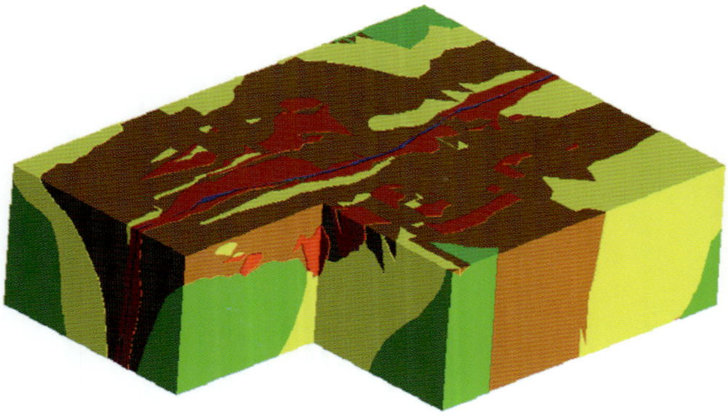

Figure 3.7: Solid model, colour-coded for rock type, showing the 3D distribution of the rock types within a mine site environment

diagram showing the stratigraphic relationships between a gabbro sill and a sedimentary pile that has been intruded by a gabbro sill and a kimberlite pipe. Figure 3.9 is a solid model of the sequence showing the boundary of the upper contact zone of the pipe. Figures 3.10, 3.11 and 3.12 are 3D solid models of twin kimberlite pipes alternately showing the contact zone and pipes removed. The final example, Figure 3.13, is another 3D solid model, again colour-coded for rock type, showing the 3D distribution of the rock types relative to the proposed pit shell at the Nickel West project in Western Australia.

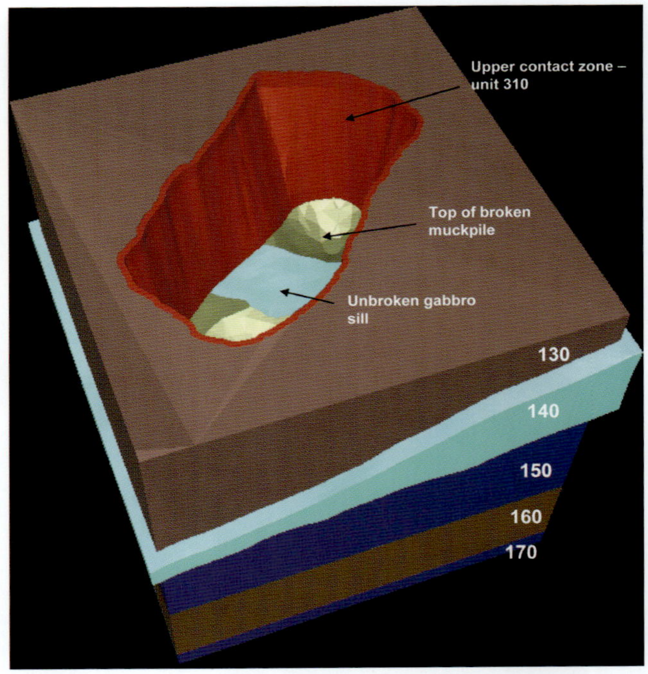

Figure 3.9: 3D solid model of sedimentary pile showing intrusive gabbro sill and the boundary of the upper contact zone of the intruding kimberlite pipe
Source: Courtesy SRK Consulting

3.5 Regional seismicity

3.5.1 Distribution of earthquakes

There are a number of instances where earthquakes have triggered landslides in natural slopes, but there are no positively identified instances of earthquakes triggering slope failures in large open pit mines. This situation has created considerable debate about the need to perform seismic analyses for open pit slopes and is the main reason why seismic loading is often ignored in open pit slope design. However, if a property was located in a seismically active region and a large earthquake were to occur near a mined slope the effects could be significant, particularly if weak soil-like materials are involved.

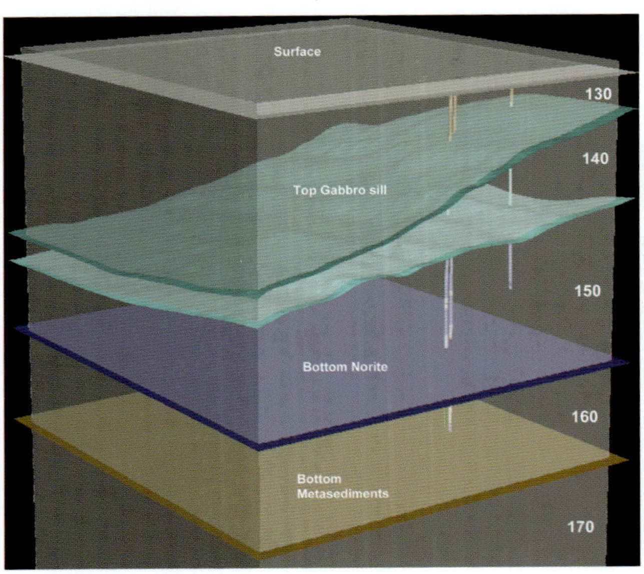

Figure 3.8: 3D fence of a stratigraphic pile intruded by a gabbro sill
Source: Courtesy SRK Consulting

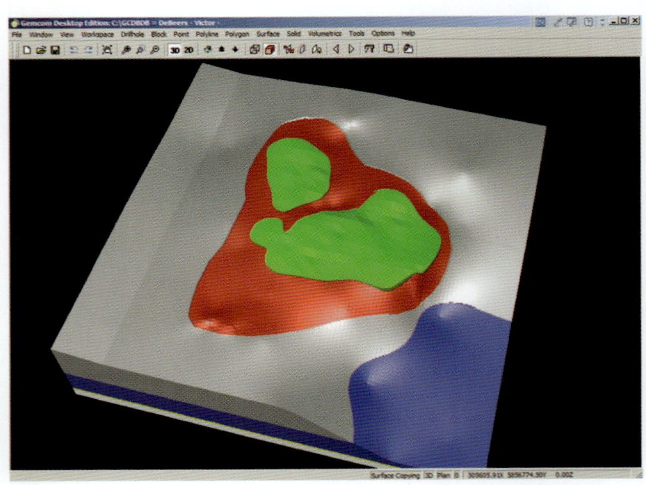

Figure 3.10: 3D solid model of twin kimberlite pipes and the surrounding contact zone
Source: Courtesy SRK Consulting and DeBeers

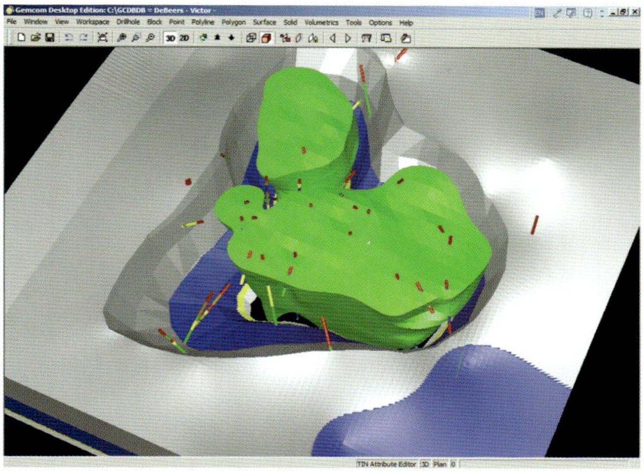

Figure 3.11: 3D solid model of twin kimberlite pipes with the surrounding contact zone removed
Source: Courtesy SRK Consulting and DeBeers

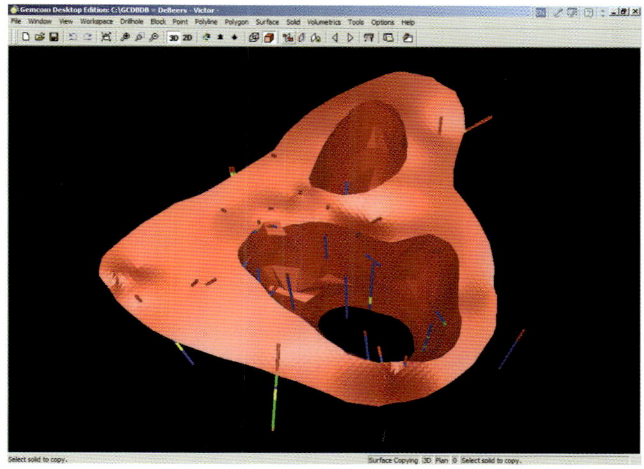

Figure 3.12: 3D solid model of twin kimberlite pipes showing the contact zone with the pipes removed
Source: Courtesy SRK Consulting and DeBeers

Additionally, other mine infrastructure, especially tailings dams, may be and have been affected by earthquakes. For this reason, documenting the regional seismicity and integrating it with the geological mode is important.

Most earthquakes are caused by interactions between two crustal plates and are concentrated in narrow geographic belts defined by the plate boundaries. Figure 3.14 outlines the world's major crustal plates. An updated digital model of plate boundaries can be obtained from Bird (2003).

Figure 3.15 shows the world distribution of earthquakes and illustrates the geographic relationship between earthquakes and plate boundaries. There are four basic types plate boundaries – divergent, transform, convergent and subductive.

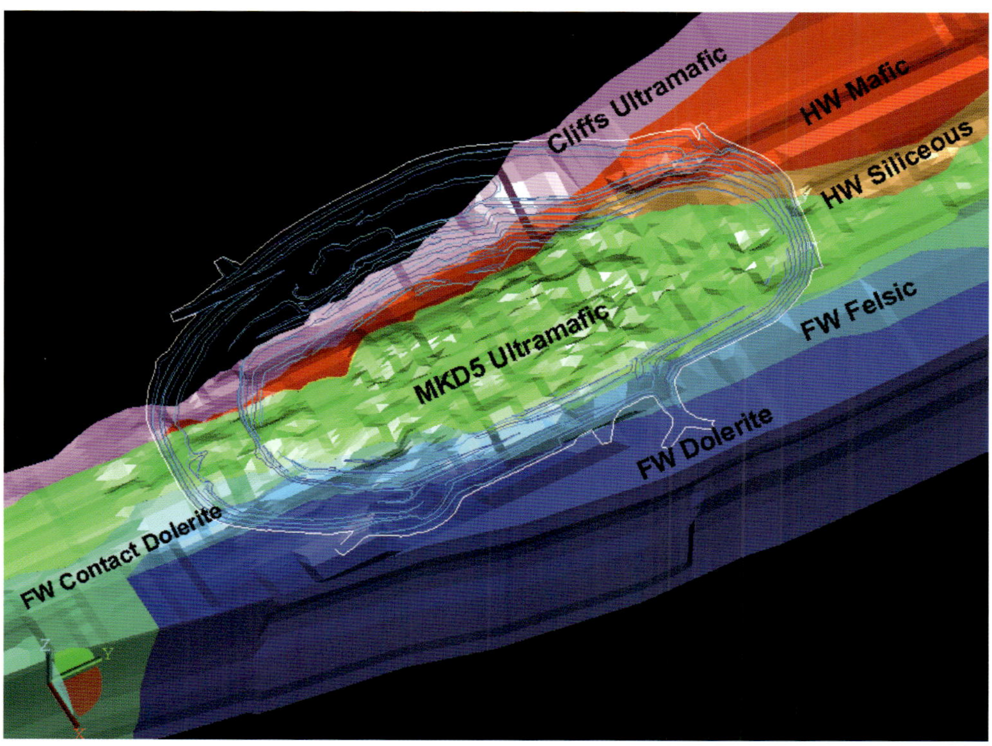

Figure 3.13: 3D solid model, colour-coded for rock type, showing the 3D distribution of the rock types relative to the proposed Nickel West pit shell
Source: Courtesy BHP Billiton, Nickel West

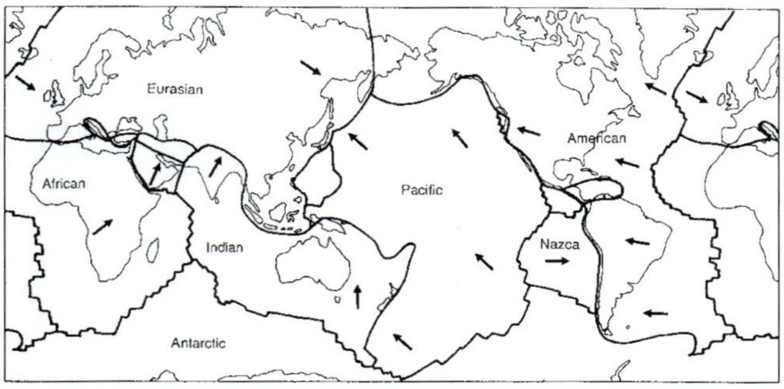

Figure 3.14: World's major crustal plates, with arrows showing the relative directions of movement
Source: Waltham (1994)

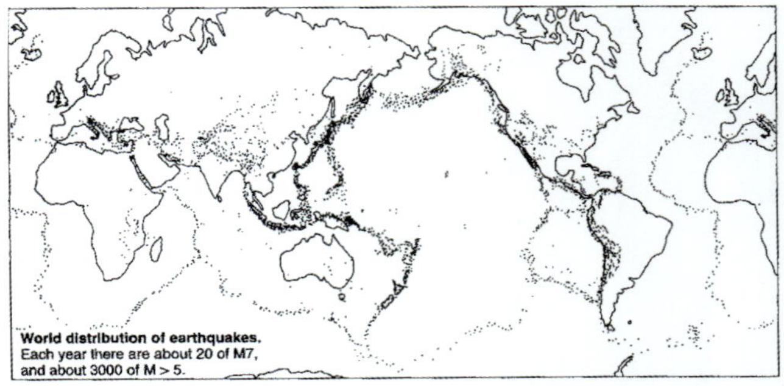

Figure 3.15: World distribution of earthquakes
Source: Waltham (1994)

Divergent plate boundaries such as the mid Atlantic ridge or the Great African Rift Valley exhibit narrow zones of shallow earthquakes along normal faults. Transform boundaries such as the San Andreas Fault in California exhibit shallow earthquakes caused by lateral strike-slip movement along the fault. Convergent boundaries where continents collide, such as in the Himalayas, exhibit broad zones of shallow earthquakes associated with complex fault systems along the line of collision. Subductive convergent boundaries, such as in the Andes where the oceanic Nazca plate is sinking beneath the continental South American plate, exhibit shallow, intermediate and deep-focus

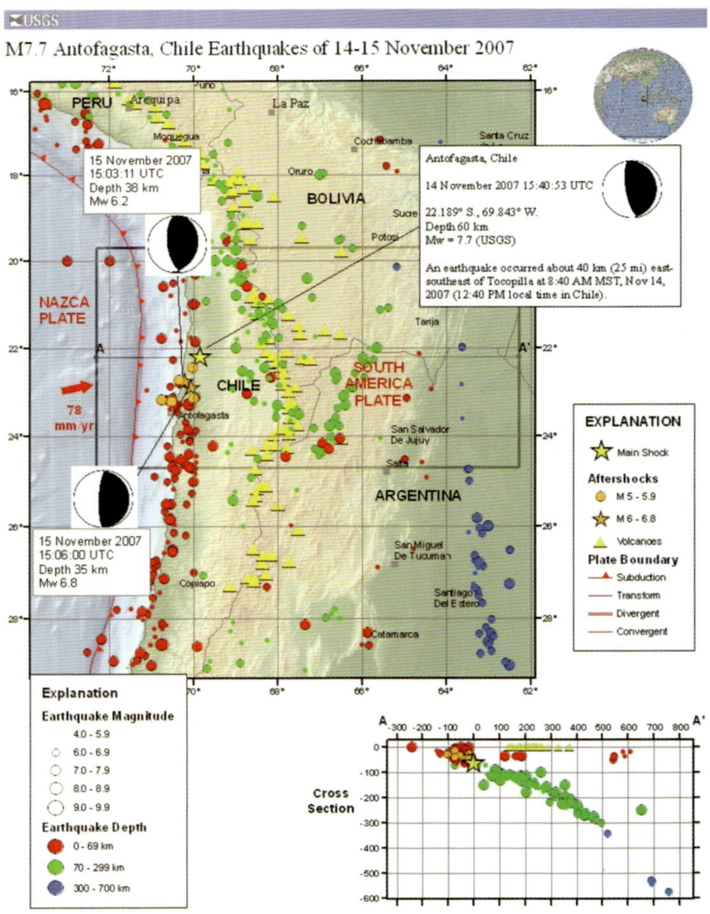

Figure 3.16: Location of earthquake epicentres associated with the Peru–Chile subduction zone in the vicinity of Antofagasta, Chile
Source: USGS (2007)

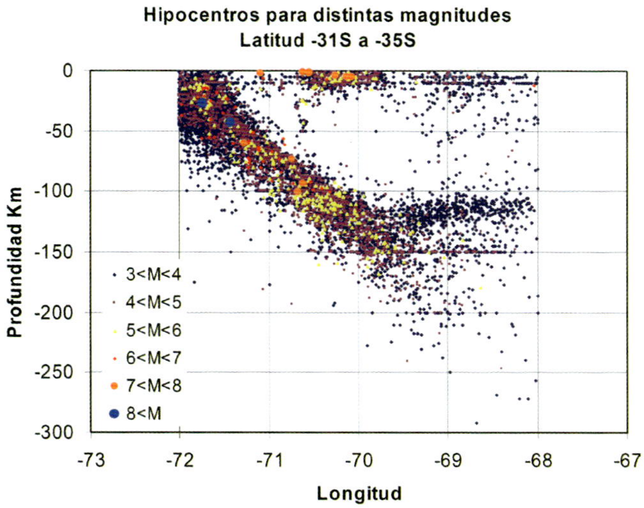

Figure 3.17: Location of earthquake epicentres associated with the Peru–Chile subduction zone between Antofagasta and Santiago, Chile
Source: Crempien (2005)

earthquakes caused by tension, underthrusting and compression.

For a well-illustrated account of the causes and distribution of earthquakes see Plummer, McGeary and Carlson (2007). Additional useful information can be gained from the following websites:

- http://quake.wr.usgs.gov/, which provides information on the latest earthquakes, historical earthquakes, how earthquakes are studied and links to other earthquake sites;
- http://seismo.unr.edu/, which provides information about recent earthquakes and links to other earthquake sites;
- http://pubs.usgs.gov/publications/text/dynamic.html, which provides general information about plate tectonics.

3.5.2 Seismic risk data

Earthquakes create four types of ground motion (Wiegel 1970):

1 ground motion that can trigger landslides or similar surface movements, which may destroy structures by simply removing their foundations;
2 sudden fault displacements that may occur at the ground surface and disrupt structures such as roads and bridges;
3 ground motion that results in soil and subsoil consolidation or settling, which may damage structures through excessive foundation deformation;
4 ground accelerations that may induce inertial forces in a structure sufficient to damage it.

The first two effects are static effects. The third effect may be static or dynamic and the fourth is dynamic.

For risk assessment purposes, the following data should be included in the model.

- The locations and magnitudes of all historic and recent earthquakes in the region of interest. The US Geological Survey (USGS) is the most common source of these data. The data should be displayed in plan and in section, as illustrated in Figures 3.16 and 3.17, which show data from the Peru–Chile subduction zone in the vicinity of Antofagasta and Santiago, Chile. In this region the Nazca plate sinks beneath the South America plate at the rate of approximately 78 mm per year. Figure 3.16 is the USGS record of a large (Mw 7.7) earthquake which occurred near Antofagasta on 14 November 2007. The map shows the locations of the earthquake and two sizeable aftershocks (Mw 6.8 and Mw 6.2), which occurred the next day. It also shows the locations of other earthquakes with magnitudes greater than Mw 4.0 that previously occurred in the region. The cross-section shows the location and depth relationship between these events and the subduction zone. Figure 3.17 widens the view of the subduction zone southward to latitude 35°, to include the recorded earthquakes between Antofagasta and Santiago. Also of interest is the comparatively large number of near-surface earthquakes associated with volcanoes in the region, not the subduction zone itself.
- The occurrence history of the earthquakes, with magnitude plotted against the number of events (Figure 3.18) and the number of events per year (Figure 3.19). Figures 3.18 and 3.19 represent the data given in Figure 3.17.
- The magnitude, return period, peak ground acceleration and distance from the project site of the maximum credible earthquake and the maximum earthquake likely to occur during the project lifetime. This information should be supplemented by the likelihood of each of these events and their corresponding peak ground accelerations being exceeded during the project life.
- Ground acceleration curves for the maximum credible and the project earthquakes, from which velocity and displacement curves can be obtained. Figure 3.20 represents the estimated ground acceleration curve for the maximum credible earthquake associated with the data given in Figures 3.17 to 3.19.

The use of seismic data in the slope design process is discussed in sections 10.3.3.5 and 10.3.4.3.

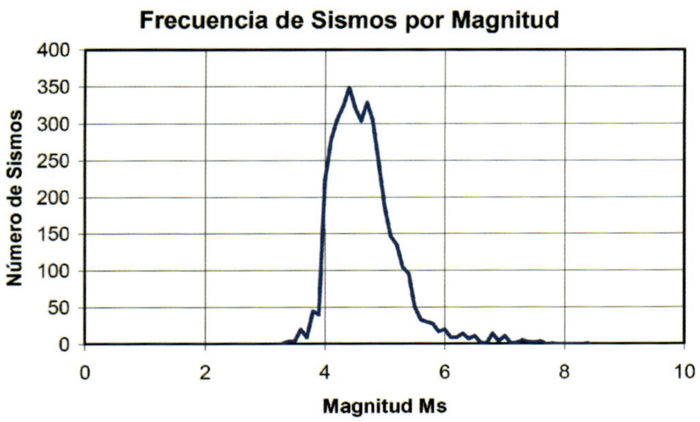

Figure 3.18: Frequency of earthquakes by magnitude (Figure 3.17 data)
Source: Crempien (2005)

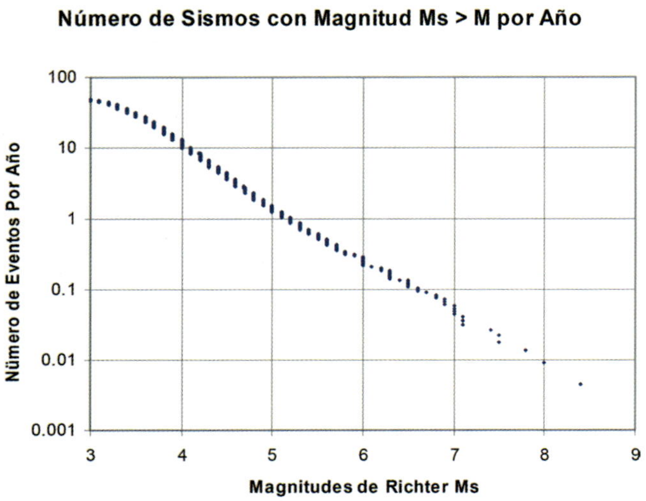

Figure 3.19: Frequency of earthquakes per year by magnitude (Figure 3.17 data)
Source: Crempien (2005)

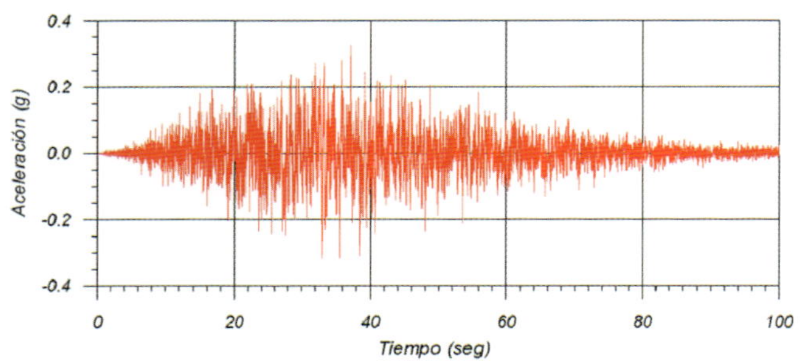

Figure 3.20: Ground acceleration curve for the maximum credible earthquake associated with the data given in Figures 3.17 to 3.19
Source: Crempien (2005)

3.6 Regional stress

The virgin stress field in an undisturbed rock mass is determined by a complex series of events controlled by gravity and active geological processes in the Earth's crust. There is a wide body of literature on the origin and measurement of virgin stress, any detailed treatment of which is well beyond the open pit mining scope of this book. Amadei and Stephansson (1997) provide a comprehensive outline of in situ stresses in the Earth's crust, methods of measuring and monitoring those stresses and their importance in rock engineering, geology and geophysics. A contemporary global database of tectonic stress of the Earth's crust is maintained by the World Stress Map Project (2005), which issues stress maps for several regions including America, Africa, Asia, Australia and Europe.

In mining, the principal application of in situ virgin stress measurements and any monitoring after the rock mass has been disturbed is underground. The knowledge is used to evaluate the stability of underground excavations and their susceptibility to stress-induced failures such as rock burst, closure, roof falls, pillar collapse and side-wall slabbing. In open pit mining however, the stress environment is dilationary not confining and most slope failures are believed to be gravity-driven. Consequently, the effects of virgin in situ stresses are thought to be minimal and are not routinely measured for slopes.

In the current environment of increasingly deeper open pits, however, regional virgin stresses should perhaps be measured, particularly if it is thought that the stresses at the bottom of the pit are likely to be higher than the strength of the rock. Also, an advantage of knowing the pre-mining stress states is that numerical analyses can be used to evaluate their relevance and whether they are likely to reinforce the ability of weak rock, adversely oriented geological structures and unfavourable pore pressures to create unstable conditions.

When selecting a test method, the available types of stress indicators are usually grouped into four categories:

1 in situ stress measurements including overcoring, hydraulic fracturing, drill hole slotting and acoustic emission (Kaiser effect) tests;
2 well bore break-outs and drilling-induced fractures;
3 earthquake focal measurements;
4 geologic data from fault-slip analyses and volcanic vent alignments.

The direct methods in the first category are those most commonly used in underground or open pit mining investigations (Appendix 3, section 4).

Detailed descriptions of the main overcoring, hydraulic fracturing and drill hole slotting techniques are given in Amadei and Stephansson (1997). The best-known overcoring methods are probably the CSIR triaxial cell (Van Heerden 1976) and the CSIRO hollow inclusion cell (Worotnicki 1993). Both have been widely used in underground mining but, because they need underground sites for the test drill holes, their usefulness in greenfields open pit investigations is restricted.

Of the other methods, drill hole slotting, hydraulic fracturing and acoustic emission testing are potentially useful in open pit investigations.

Drill hole slotting (Bock 1986) is fast and the instrument used to cut the stress relief slots is reusable, but its application is limited to 2D analyses. It has only been tested at shallow depths.

In hydraulic fracturing, pressure is applied to an isolated part of the wall of the drill hole and increased until existing fractures open or new fractures are formed. The pressures needed to sustain the opened fractures are measured and related to the in situ stress field, with the direction of the measured stresses obtained by observing and measuring the orientation of the opened fractures. The technique was developed in the petroleum industry and has been widely used in the investigation of underground nuclear waste repositories. It has not been widely used in the mining industry, probably because it requires specialist skills and equipment. However, it can be performed from the surface, which means that it can be done as an extension of the investigation drilling program. A good example of such a use is at the Ok Tedi mine, where the in situ stresses in the open pit region were determined using the hydrofrac process in routine exploration drill holes. The results were invaluable. Although plate tectonic considerations and studies of regional seismicity suggested that the most comprehensive principal stress in the region should have been north–south, the tests showed that at least for depths of up to 450 m the stress state was essentially isotropic, with the magnitude of all three principal stresses approximately equal to the overburden stress. The unexpected result was believed to have resulted from the presence of fracturing, clay-filled fault zones and numerous healed fractures in the rock mass, together with deep incisement by river erosion (Power et al. 1991).

Application of the acoustic emission test has emerged more recently. Based on the principle of the Kaiser effect (Kaiser 1953), it supposes that the immediate maximum previous stress to which a rock has been subjected by its environment may be detected by loading a rock specimen to the point of substantial increase in acoustic emission (Villaescusa et al. 2002). The attraction of the test is that it is applied to samples derived from core that can be retrieved at any depth from standard investigation drill holes. For the test, six small cylindrical samples are undercored from the retrieved core and instrumented with a pair of acoustic emission transducers. The samples are then loaded uniaxially, allowing the transducers to provide a record of the number of events that occur with increasing load. The information from the six samples is analysed to give six independent normal stresses from which the full stress tensors are obtained. There is ongoing debate, however, on whether the stress value recorded, which is taken to represent the immediate maximum previous stress to which the sample had been subjected, in fact represents the current stress condition.

4 STRUCTURAL MODEL

John Read

4.1 Introduction

The second component of the geotechnical model is the structural model (Figure 4.1). The purpose of the structural model is to describe the orientation and spatial distribution of the structural defects that are likely to influence the stability of the pit slopes. This includes features such as faults and folds that are relatively widely spaced and continuous along strike and dip across the entire mine site, and the more closely spaced joints and faults that typically do not extend for more than two or three benches.

Because of the differences in scale between the benches, inter-ramp and overall slopes, the structural model must be configured in at least two overlays that show:

1 the major structural features such as through-going faults and folds that can be used to subdivide the mine into a select number of structural domains, each of which is characterised internally by a recognisable structural fabric comprised of more closely spaced faults and joints. Lithological boundaries and the shape of the open pit may also influence the selection of the domain boundaries;

2 the attributes of the more closely spaced fault and joint fabric that occur within each structural domain.

Both these overlays should be underpinned by an outline of the regional geological setting that concisely describes the tectonic events and major faults and/or folds that have controlled or influenced the style and shape of the ore body, from evolution through to mining.

Later sections of this chapter describe the steps in this process, commencing with section 4.2, which outlines the components of the model. Section 4.3 describes the geological environments that will be encountered and section 4.4 the structural modelling tools that are most frequently used to define the domains and differentiate the structural fabric within each domain. Finally, section 4.5 provides examples from different mine sites to illustrate the process.

It is stressed that the task of erecting the structural model is one for an experienced structural geologist, rather than an exploration or a mine geologist. Exploration and mine geologists are an essential part of the modelling team, but the team leader should be a structural geologist who has the specific skills and the experience in structural geology to bring together the disparate parts of the model. This chapter is intended to provide a general picture of structural geology as applied to pit slope designs, together with the techniques used for analysis. The reader is referred to referenced texts for more detailed explanations.

4.2 Model components

4.2.1 Major structures

Major structures include the folds and faults that are continuous along strike and down dip across the mine site, and features such as the laminated structures associated with metamorphic rocks like slate, phyllite and schist. The basic terminology used to describe these features is outlined below.

4.2.1.1 Folds

Folds are one of the most commonly occurring structures in deformed rocks. They are formed when planar features such as bedding and schistosity are deflected into wavelike curviplanar or curvilinear structures. They may develop in single or multi-layers and occur mostly by bending and buckling. They may also occur by gravity slumping and may have a wide variety of geometries and sizes.

Bending is a flexuring induced by a compression acting at a high angle to the layering, as illustrated in Figure 4.2a. Buckling is a flexuring induced by compression acting at a low angle to the layering (Figure 4.2b). Bending can also occur in the form of a drape fold when, for example, sediments that cover a more rigid basement flex in response to components of vertical movement along basement faults (Figure 4.3). As the name implies, gravity

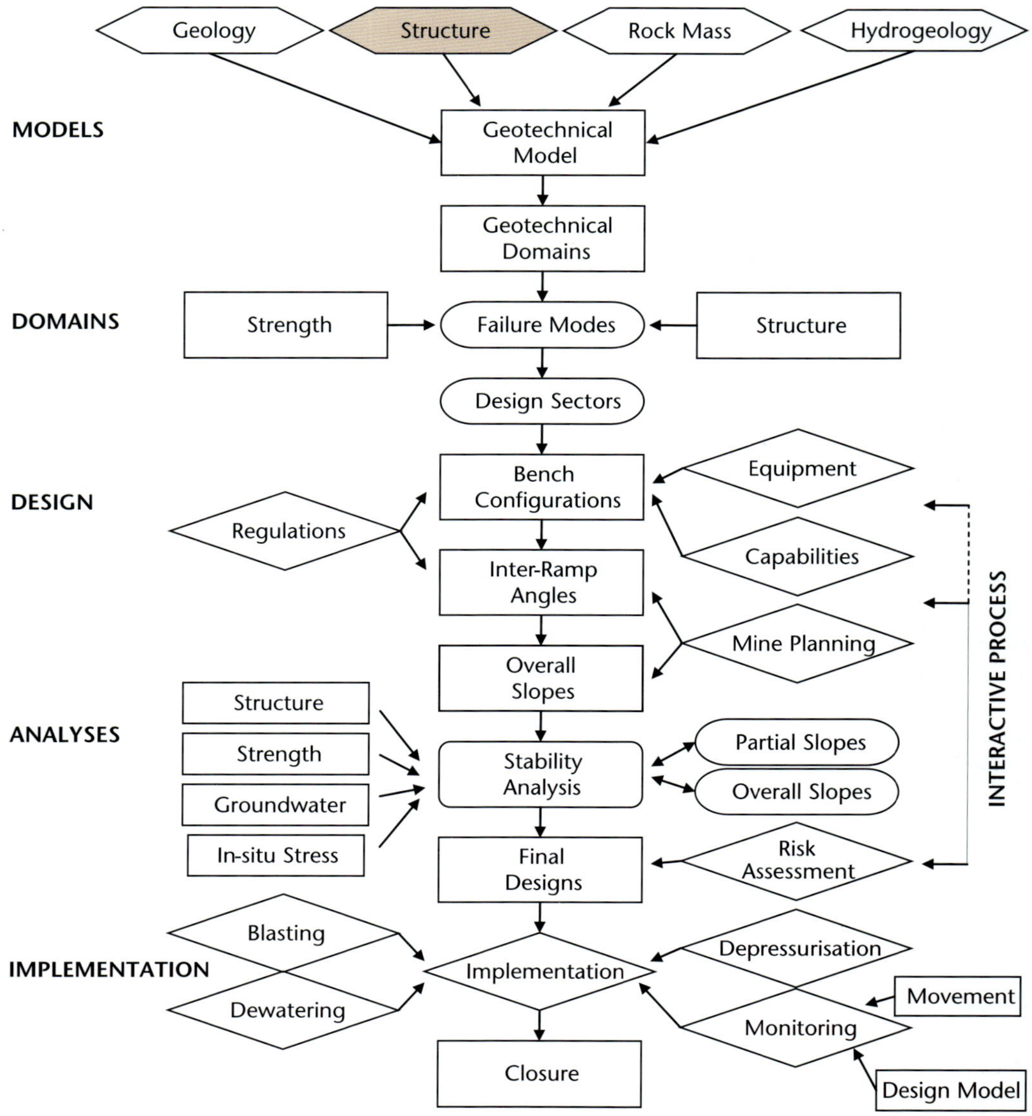

Figure 4.1: Slope design process

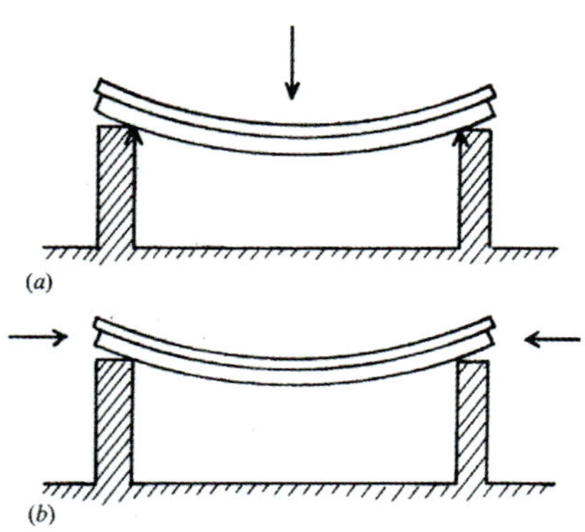

Figure 4.2: The orientation of the principal compression for (a) bending and (b) buckling of planar layers
Source: Blyth & deFreitas (1984)

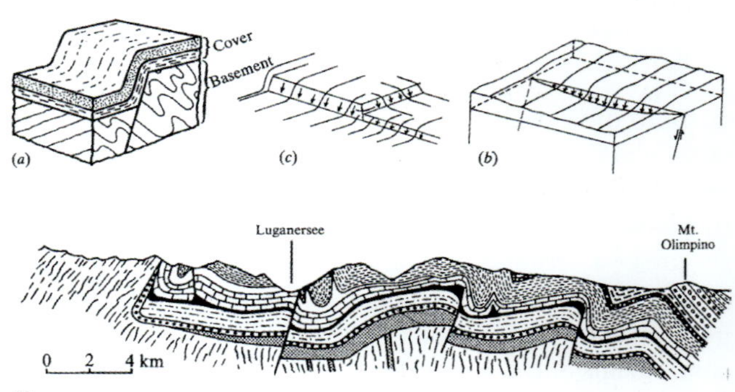

Figure 4.3: (a) and (b) Block diagrams of hypothetical drape-folds, the result of normal faulting in the basement. (c) Drape-fold geometry associated with block faulting in the basement. (d) Drape-folds over reverse faults in the basement
Source: Blyth & deFreitas (1984)

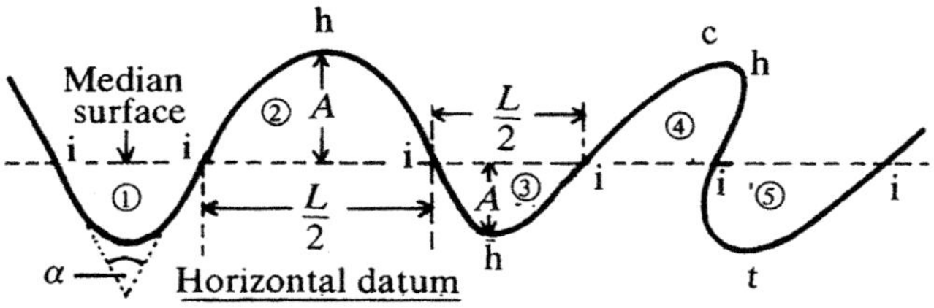

Figure 4.4: Terms used to describe the geometry of a fold profile: h = hinge; I = inflection point; c = crest; t = trough; α = interlimb angle; L = wavelength; A = amplitude
Source: Blyth & deFreitas (1984)

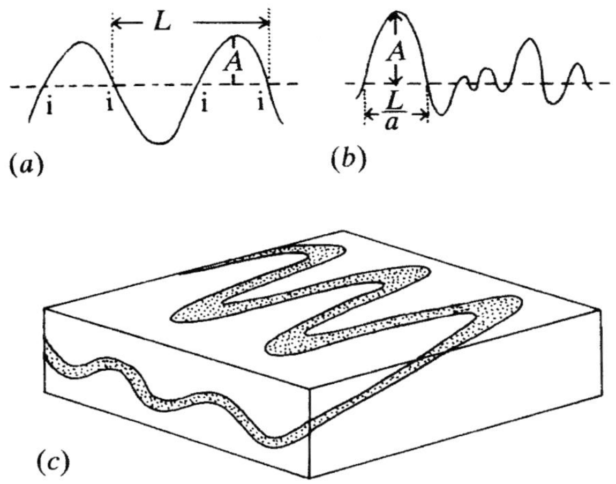

Figure 4.5: (a) and (b) Wavelength (L) and amplitude (A) of a fold. (c) Diagram showing the dependence of the fold outcrop pattern on the orientation of the plane of erosion
Source: Blyth & deFreitas (1984)

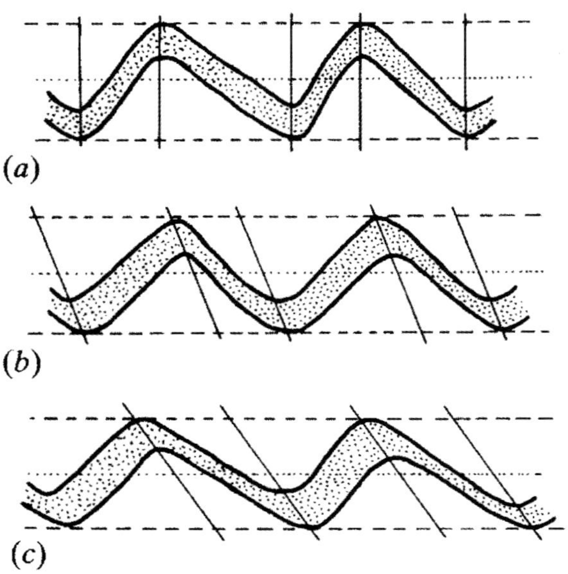

Figure 4.6: Types of asymmetric folds with differing limb lengths and positions of the hinge surface
Source: Blyth & deFreitas (1984)

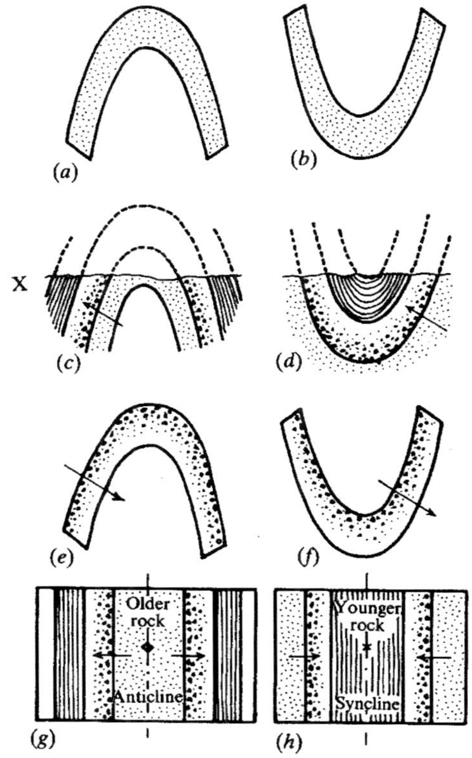

Figure 4.7: (a), (c) and (e) Upward-closing folds. (b), (d) and (f) Downward-closing folds. Arrows indicate direction of younging. Plan views of (g) eroded anticline and (h) syncline
Source: Blyth & deFreitas (1984)

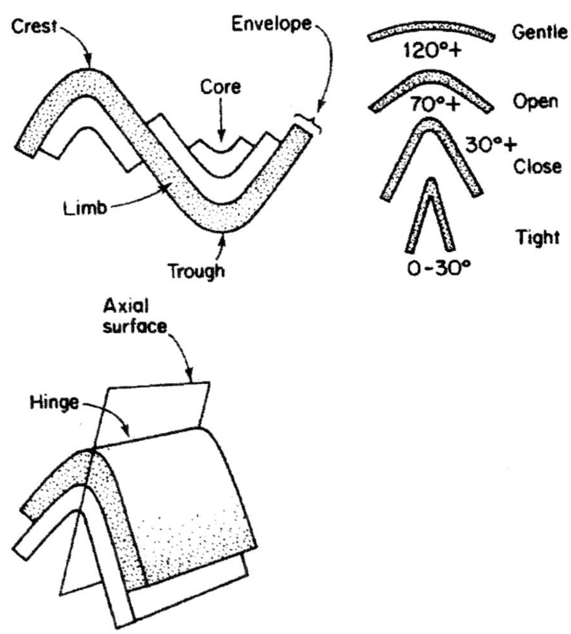

Figure 4.8: Antiform and synform in upright open folding, with corresponding degrees of acuteness in folding and the hinge of folding
Source: Blyth & deFreitas (1984)

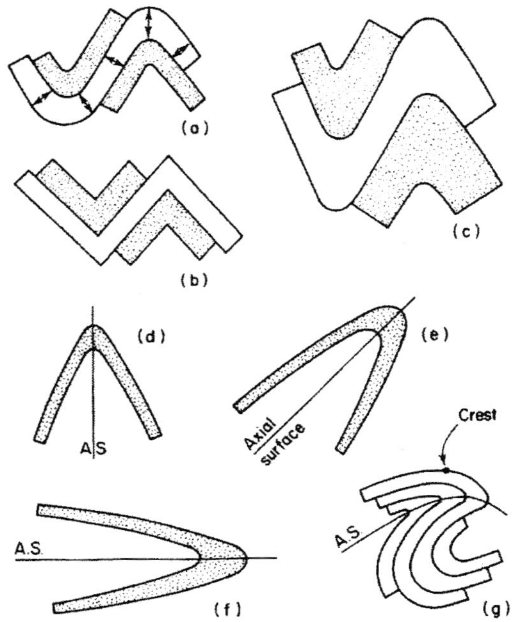

Figure 4.9: Fold forms. (a) Parallel. (b) Chevron. (c) Similar .(d) Upright. (e) Inclined. (f) Recumbent. (g) Curved axial surface
Source: Blyth & deFreitas (1984)

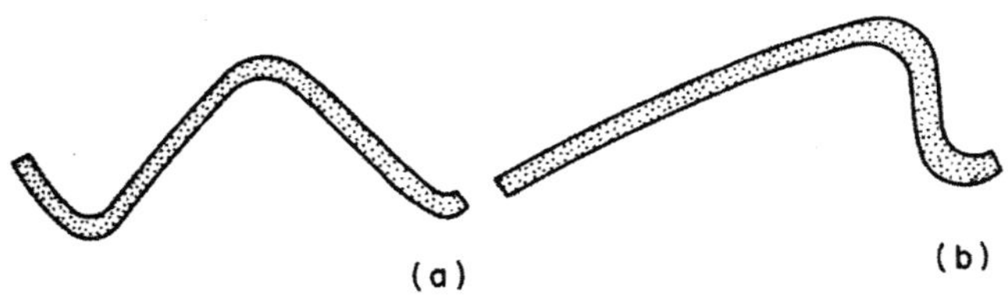

Figure 4.10: Fold symmetry. (a) Symmetric. (b) Asymmetric
Source: Blyth & deFreitas (1984)

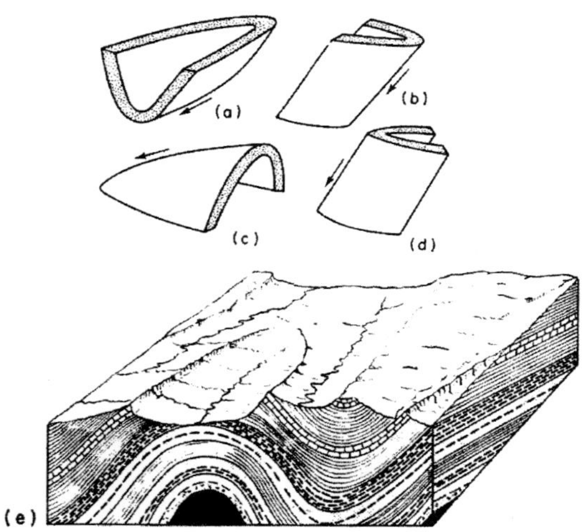

Figure 4.11: Diagrams illustrating plunge. (a) and (b) Synclinal. (c) and (d) Anticlinal. (e) Block diagram of eroded anticline and syncline, with hard beds (brick pattern) forming surface features on eroded surface
Source: Blyth & deFreitas (1984)

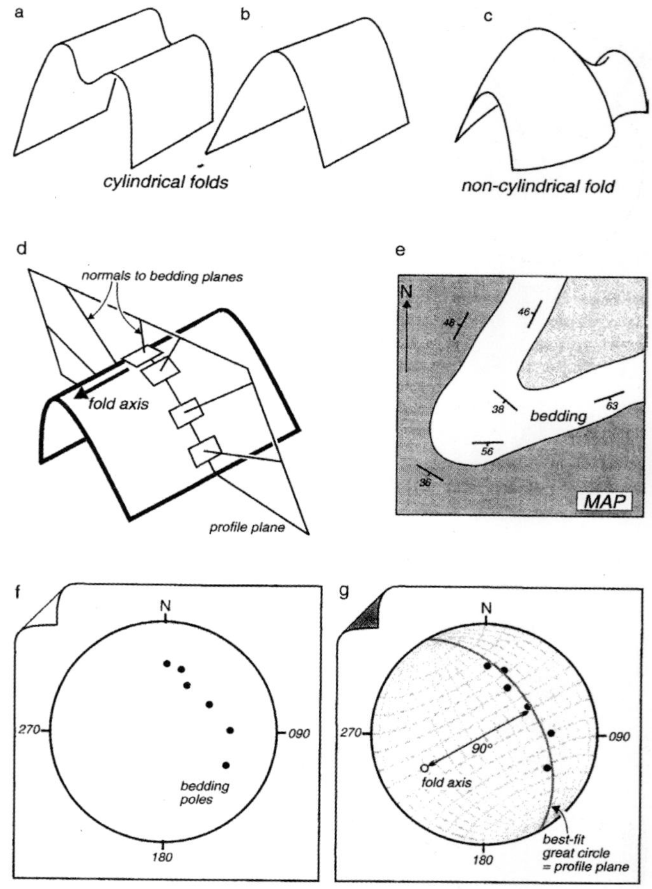

Figure 4.12: Stereonet representation of a symmetrical fold
Source: Lisle & Leyshon (2004)

slumping involves the sliding of a mass down a slope under the influence of gravity and is most common in a submarine environment.

The basic terminology used to define folds is outlined by example in Figures 4.4 and 4.5. The most common different fold forms are outlined by example in Figures 4.6 to 4.11.

Three-dimensional representations of these different styles of folding using the stereonet (section 4.4.2) are illustrated in Figures 4.12, 4.13 and 4.14.

As outlined by Lisle and Leyshon (2004), Figure 4.12 shows how the symmetry of fold can be recognised by the orientations of the normals to the folded surface taken at a succession of locations across the fold. If the fold is symmetrical, when plotted on the stereonet the poles of the normals to the fold will lie close to a single or best-fit great circle known as the profile plane. In turn, the pole of the profile plane gives the direction of the fold axis. If the poles cannot be fitted to a great circle, the fold is not symmetrical.

Figure 4.13 illustrates typical distributions of the poles on the profile plane for different degrees of openness and curvature of the fold. Typically, the degree of completeness

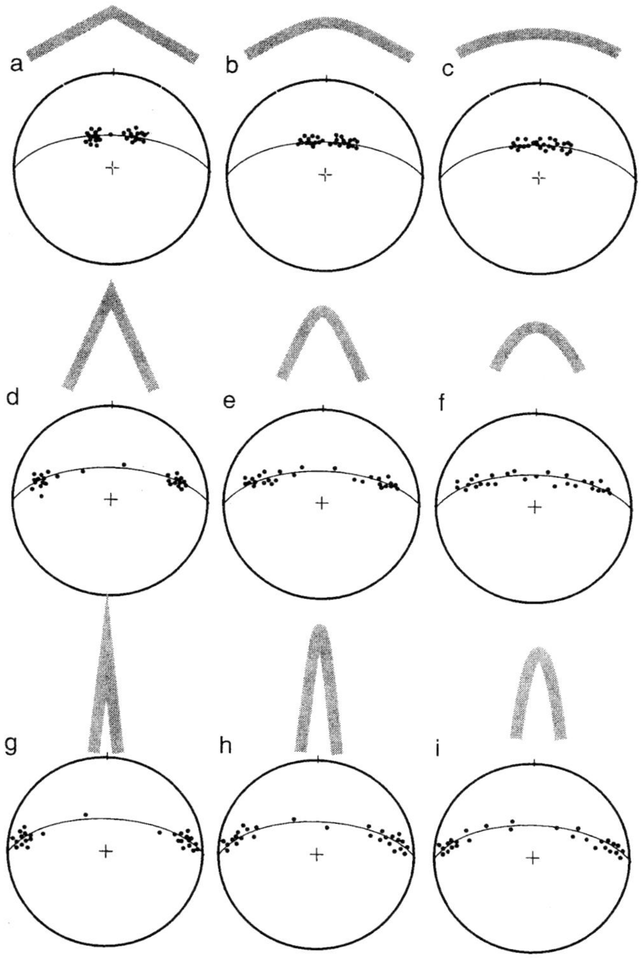

Figure 4.13: Stereonet representation of different styles of folding
Source: Lisle & Leyshon (2004)

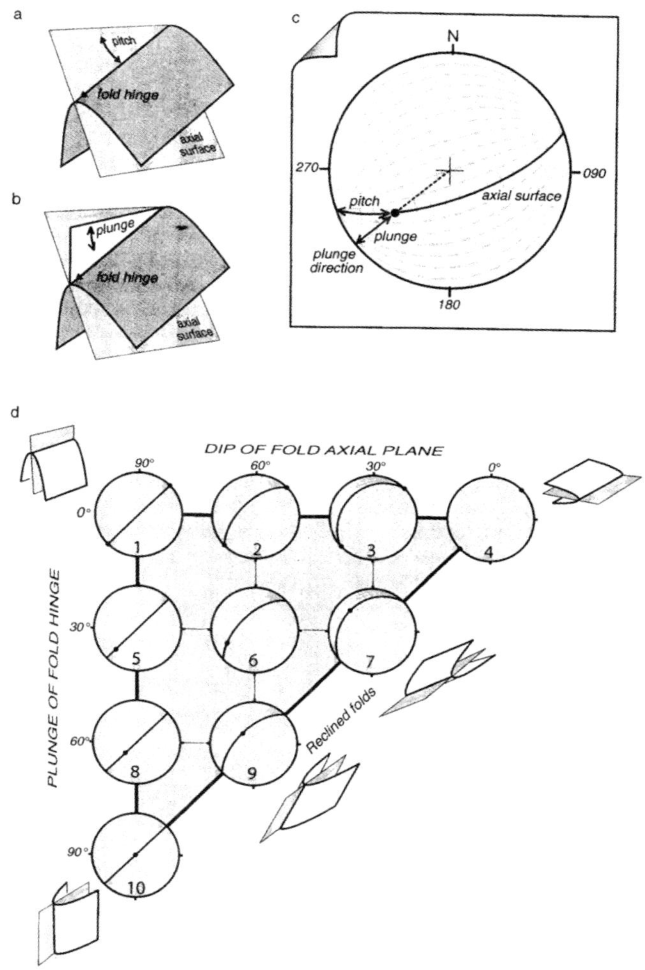

Figure 4.14: Stereonet representation of differing fold orientations
Source: Lisle & Leyshon (2004)

of the great circle reflects the tightness of the fold, with the range of orientations for a tight fold (Figure 4.13i) being greater than for an open fold (Figure 4.13c). In the same manner, planar limbs of a fold (Figure 4.13a, d and f) show two clusters of poles whereas open folds (Figure 4.13c and f) show more diffuse patterns. If the limbs of the fold have unequal lengths one cluster of poles on the profile plane is likely to be more pronounced than the other.

Figure 4.14 shows differing fold classes based on plunge (Figure 4.11) and the dip of the axial surface, both of which are independent of the openness or degree of curvature of the fold (Figure 4.13). Classifications based on plunge can range from non-plunging to vertical. Classifications based on the dip of the axial surface can range from to upright (Figure 4.9d) to recumbent (Figure 4.9f).

4.2.1.2 Faults

The dictionary definition of a fault is a fracture surface or zone along which appreciable displacement has taken place. The word 'appreciable' raises the question of how much is appreciable. For engineering purposes, however, any movement is a fault, recognising that even a minor (small-scale) fault may have considerable engineering significance in terms of strength reduction.

For slope design purposes, a suggested scale is given in Table 4.1. The components of displacement of a fault are measured in terms of throw, heave, total slip and shift (Figure 4.15).

Fault classification systems recognise a parent hydrostatic state of stress in the Earth's crust such that the magnitude of the horizontal stresses at any given depth in the crust is equal to the vertical geostatic stress induced at depth by gravitational loading. The magnitude of the horizontal stresses (σ_2 and σ_3) relative to that of the vertical stress (σ_1) can change in three ways. If the differential stress is sufficiently large these variations will give rise to three main faults – normal, thrust (reverse) and strike-slip (Figures 4.16 and 4.17).

Table 4.1: Suggested scale of fault magnitude

Length (m)	Description
<1	Minor (small-scale)
1–10	Bench
10–100	Bench to inter-ramp
100–1000	Inter-ramp to overall
>1000	Mine scale to regional

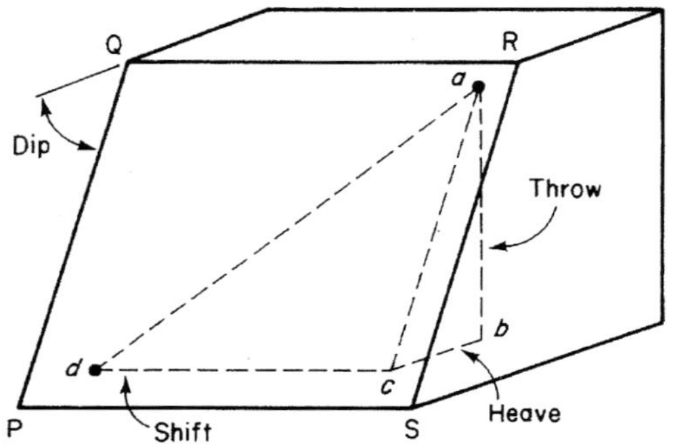

Figure 4.15: Components of fault displacement (a, c and d lie on the fault surface, PQRS)
Source: Blyth & deFreitas (1984)

1 A normal fault is a lateral extension where both the horizontal stresses decrease in magnitude, but not by the same amount (i.e. σ1 > σ2 > σ3). Normal faults can occur in any geological environment. They form grabens (Figure 4.17b), and in outcrop or drill hole exposures result in an apparent loss of strata.
2 A thrust fault results from compression. Both horizontal stresses increase in magnitude, but not by the same amount (i.e. $\sigma_2 > \sigma_3 > \sigma_1$). Thrust faults are typical of thrust and fold belt environments and result in the repetition of strata (Figure 4.18). Where the inclination of the fault surface is greater than 45° the term 'reverse fault' is used.
3 Strike-slip faults (transcurrent, tear, wrench or transform) occur where the fault plane is

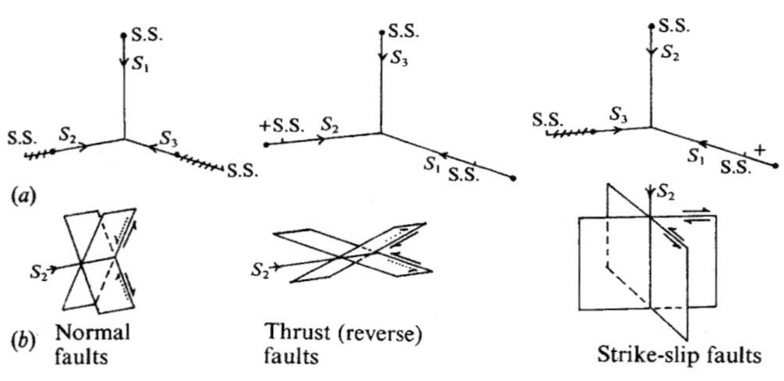

Figure 4.16: Stress directions for normal, thrust (reverse) and strike-slip faults
Source: Blyth & deFreitas (1984)

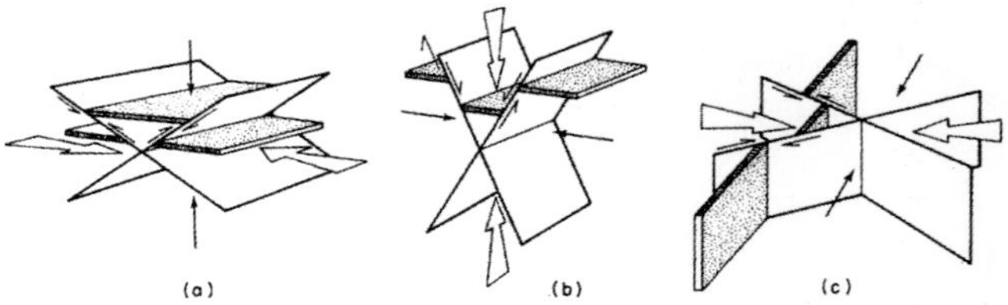

Figure 4.17: Relationship of faults to axes of principal stress. (a) Thrust. (b) Normal. (c) Strike-slip.
Source: Blyth & deFreitas (1984)

approximately vertical and movement is in the strike direction (left or right lateral). One horizontal stress increases in magnitude while the other horizontal stress decreases in magnitude (i.e. $\sigma_2 > \sigma_3 \geq \sigma_1$).

4.2.1.3 Metamorphic structures

Metamorphic rocks such as slate, phyllite and schist exhibit a planar fissility that at mine scale can have a major effect on the stability of the inter-ramp and overall pit slopes. The terminology used to describe the fissile texture of these metamorphic rocks can be confusing; it is clarified below:

- slate – a fine-grained rock with perfect schistosity;
- phyllite – a fine-grained schistose rock, sometimes with incipient segregation banding, with a lustrous sheen of mica and chlorite along the schistosity surfaces;
- schist – a strongly schistose, usually well-lineated rock, generally with well-developed segregation layering. It contains abundant micaceous minerals. The grain size is sufficient to allow easy identification of the main component minerals in hand specimens.

A feature of these descriptions is the distinction between schistosity (or foliation), segregation banding (or layering) and lineation, which can be described as follows:

- schistosity – a planar fissility in rock caused by the orientation of the mineral crystals in the rock with

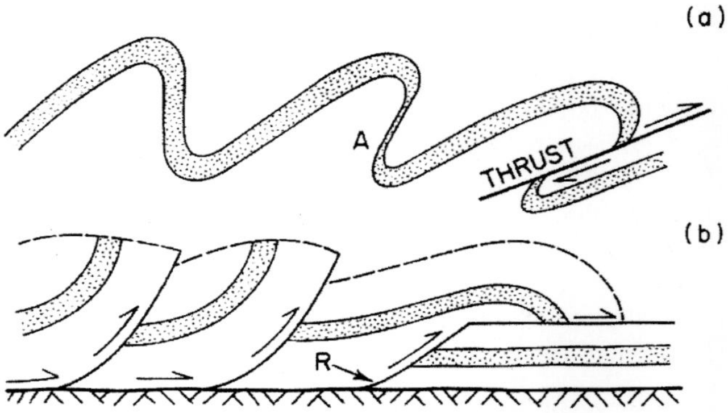

Figure 4.18: Development of (a) thrust and (b) overthrust, with repetition of strata
Source: Blyth & deFreitas (1984)

their greatest dimension subparallel to the plane of schistosity. Note that *s*-surfaces are synonymous with schistosity, but have a broader connotation in that the term is applied to any set of parallel surfaces, of metamorphic origin or not, that can be seen in the fabric of a metamorphic rock (e.g. bedding);

- segregation banding – a laminated structure resulting from the segregation of simple mineral assemblages of contrasted composition during metamorphism into alternating layers parallel to the schistosity;
- lineation - parallel alignment of linear elements in some direction within the schistosity, e.g. prismatic crystals of hornblende or epidote, rod-like aggregates of quartz, or the axes of microfolds.

4.2.2 Fabric

The bench scale structural fabric within the major domains can include micro-bedding and folding, minor faults, joints, schistosity and cleavage. The principal features of some common minor fold structures and joints are outlined below.

4.2.2.1 Minor fold structures

Common minor fold structures include fracture cleavage, tension gashes, boudinage structures and slickensides.

- Fracture cleavage consists of a series of parallel fractures (or conjugate shears) formed in an incompetent bed (e.g. shale) in response to the folding of an enclosing competent bed (e.g. sandstone), as illustrated in Figure 4.19 and the stereonet representations in Figure 4.20. Tension gashes may form by extension in the enclosing or other nearby brittle rocks in response to the folding. If the cleavage is parallel or subparallel to the axial plane of the associated fold, it is known as axial-plane cleavage. Because the amount and direction of the strains around the fold may vary,

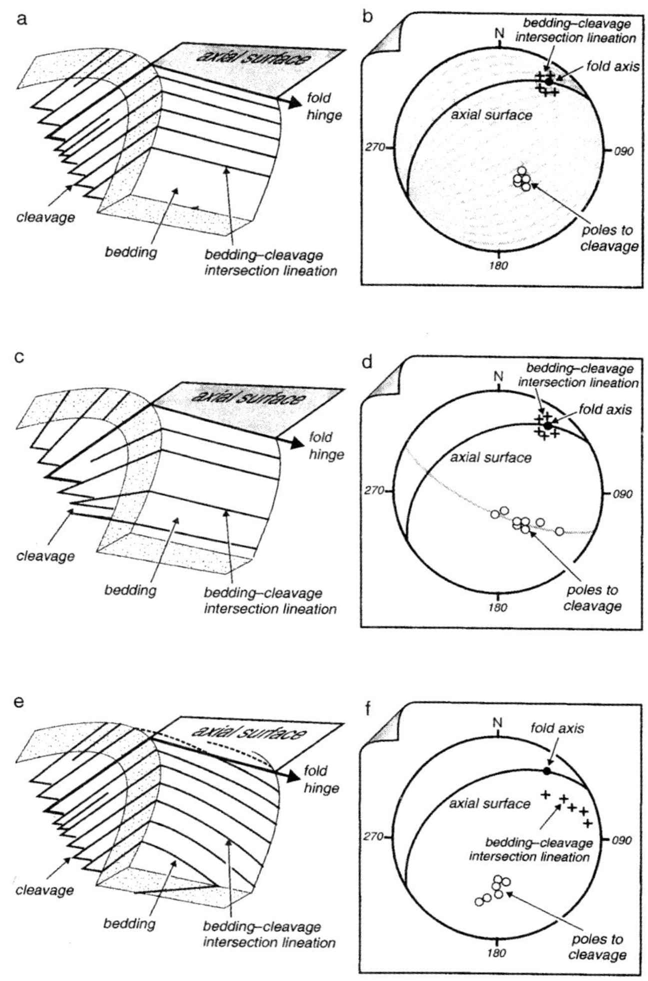

Figure 4.20: Stereonet representation of folds and cleavage
Source: Lisle & Leyshon (2004)

the axial-plane cleavage may converge (Figure 4.20c) or diverge from the inner arc of the fold. When this occurs, the poles of the cleavage planes will show a

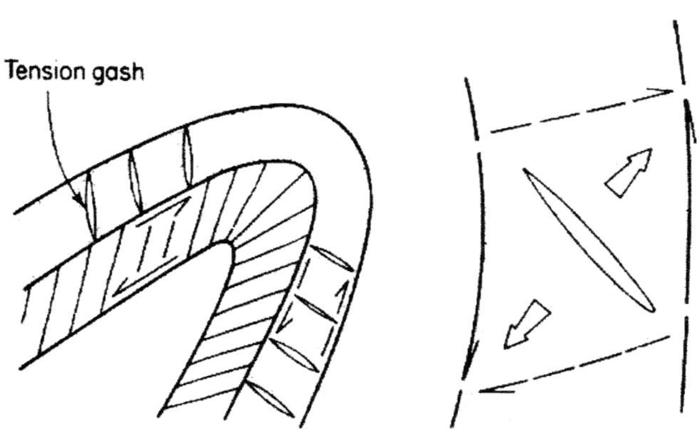

Figure 4.19: Fracture cleavage in a weaker rock folded between stronger beds, with relationship between tension gash and shear stresses
Source: Blyth & deFreitas (1984)

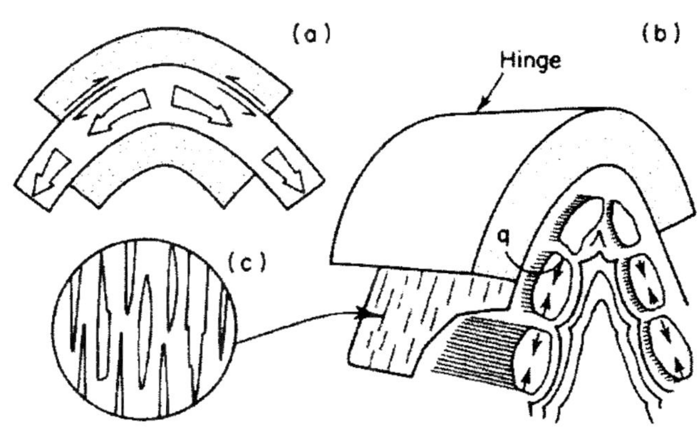

Figure 4.21: (a) Tension within competent bed. (b) Boudin structures with quartz (q) between boudins. (c) Lineations
Source: Blyth & deFreitas (1984)

greater spread, following a great circle perpendicular to the fold axis (Figure 4.20c). As noted by Lisle and Leyshon (2004), the bedding-cleavage intersections will, however, remain aligned parallel to the fold hinges (Figure 14.20c and d).
- Boudinage structures are formed by extension during the flexuring of a brittle material, totally fracturing the layer into rod-like pieces (Figure 4.21).
- Slickensides are lineations reflecting the direction of movement of adjacent beds or structures during folding or faulting (Figure 4.21).

4.2.2.2 Joints

Joints develop in response to three main geological processes:

- deformation resulting from orogenic processes;
- deformations resulting from epeirogenic (broad uplift and downlift) processes;
- shrinkage caused by cooling or desiccation.

Joints in sedimentary rocks reflect the relief of stress that remained in the rocks after (epeirogenic) deformation. The basic jointing is orthogonal with sets oriented perpendicularly to the bedding and normal to each other. However, other sets may also be present, depending on subsequent deformation events. Joints in igneous rocks can reflect contraction cooling, the contraction being taken up in extension (opening of tension joints), or deformation processes after cooling has taken place.

4.3 Geological environments

4.3.1 Introduction

As outlined in section 3.3, there are a number of different ore body styles, each with its own set of structural features that can affect the stability of the pit slopes. Many of these features are common between styles and in most cases can be related to the intrusive, sedimentary or metamorphic nature of the different geological environments.

4.3.2 Intrusive

The igneous and subvolcanic intrusive activity and mineralisation associated with porphyry and epithermal deposits and skarns are linked to faults and highly fractured zones that formed pathways for the intrusion and mineralising fluids. These structures form the basic skeleton of the structural model and may have the most impact on the slope designs. Additional questions that must be asked and items that must be added to the skeleton as the model is developed include the following.

- Does the ore body represent a single phase or multiple phases of tectonism and mineralisation? If there were multiple phases, were the existing structures remobilised or were new structures developed?
- Do the alteration zones and boundaries extend widely into the country rocks lateral to the ore body or are they confined to the faults and fractured zones? This is a particularly important question, especially in epithermal deposits, as the presence of structures and alteration-weakened rock in the walls means that failure through the weakened rock without or with only partial structural control can be as likely as structurally controlled failures.
- What is the relationship between the joints and the major structures? Were the joints and faults formed by the same stress regimes or separately at different times and under different stress conditions?

The different volcanic environments of kimberlites and VMS (volcanic massive sulphide) deposits lead to a different set of questions. Kimberlite extrusions are explosive and the geotechnical interest is highly focused on the shape and condition of the contact zones around the pipe. VMS deposits occur as lens-shaped bodies in volcanically active submarine environments. In this case the geotechnical questions concern the steepness of the footwall alteration zone and any internal layering within the ore body, which can form potentially unstable dip-slopes.

4.3.3 Sedimentary

In sedimentary environments, attributes that can influence the stability of the pit slopes include the following:

- contacts between different lithological units, including bedding planes and unconformities. Of particular interest are any weak zones at the boundaries between stronger and weaker zones (e.g. mudstone or shale overlying sandstone) and unconformities that exhibit paleo-soil horizons;
- folds, either simple or complex, which can form dip-slopes;
- joints, with sets oriented perpendicularly to the bedding and normal to each other providing release planes within unfavourably oriented beds (e.g. dip-slopes);
- cleavage, which can provide release planes within unfavourably oriented beds;
- faults, including all major regional faults. These can provide release surfaces but may also represent major failure planes, e.g. thrust faults in orogenic fold and thrust environments. Thrust faults not only repeat the beds, but geotechnically can form major planes of weakness over distances that have been measured in kilometres.

4.3.4 Metamorphic

Attributes of metamorphic rocks that can affect the stability of slopes are similar to those found in sedimentary environments, especially with respect to dip-slopes resulting from folding. Hence, the main geotechnical questions are concerned with the integrity of the planar fissility associated with slates, phyllites and schists (section 4.2.1.3). Schistosity is developed in amphibolites and gneisses, but is less obvious than in typical schists and is normally less of a concern. Other structures to be aware of include narrow zones of deformation and dislocation such as cataclasites and mylonites that have been formed by dynamic metamorphic processes during faulting and folding, and joints and cleavages.

4.4 Structural modelling tools

4.4.1 Solid modelling

Three-dimensional solid modelling of the structural geology using a commercially available modelling system such as Vulcan™, DataMine™, Gemcom™ or MineSite™ has become routine at most mine sites and design offices. Like the geological model, the first step is to compile the entire field mapping and core drilling structural data (sections 2.2 and 2.4) into a geological plan of the pit. The plan is then incorporated into a 3D solid geological model using one of the modelling systems mentioned above. Mapped data from Autocad are usually imported as DXF files so that the geologist can connect the structural or other geological boundary traces and build on those traces in 3D to make modelled shapes or triangulations. Once the triangulations are made it is easy to cut them to pit shells or into sections.

Figures 4.22 and 4.23 illustrate typical steps in this process. Figure 4.22 shows a sequence of normal faults intersecting a mapped ore body from the east (near side) to the west (far side) inside the proposed ultimate pit shell and above planned underground workings. Figure 4.23 shows major structures mapped from available drill hole and pit mapping data intersecting a proposed ultimate shell.

4.4.2 Stereographic projection

4.4.2.1 General guidelines

Structural modelling is an exercise in 3D geometry requiring the application of descriptive geometry or

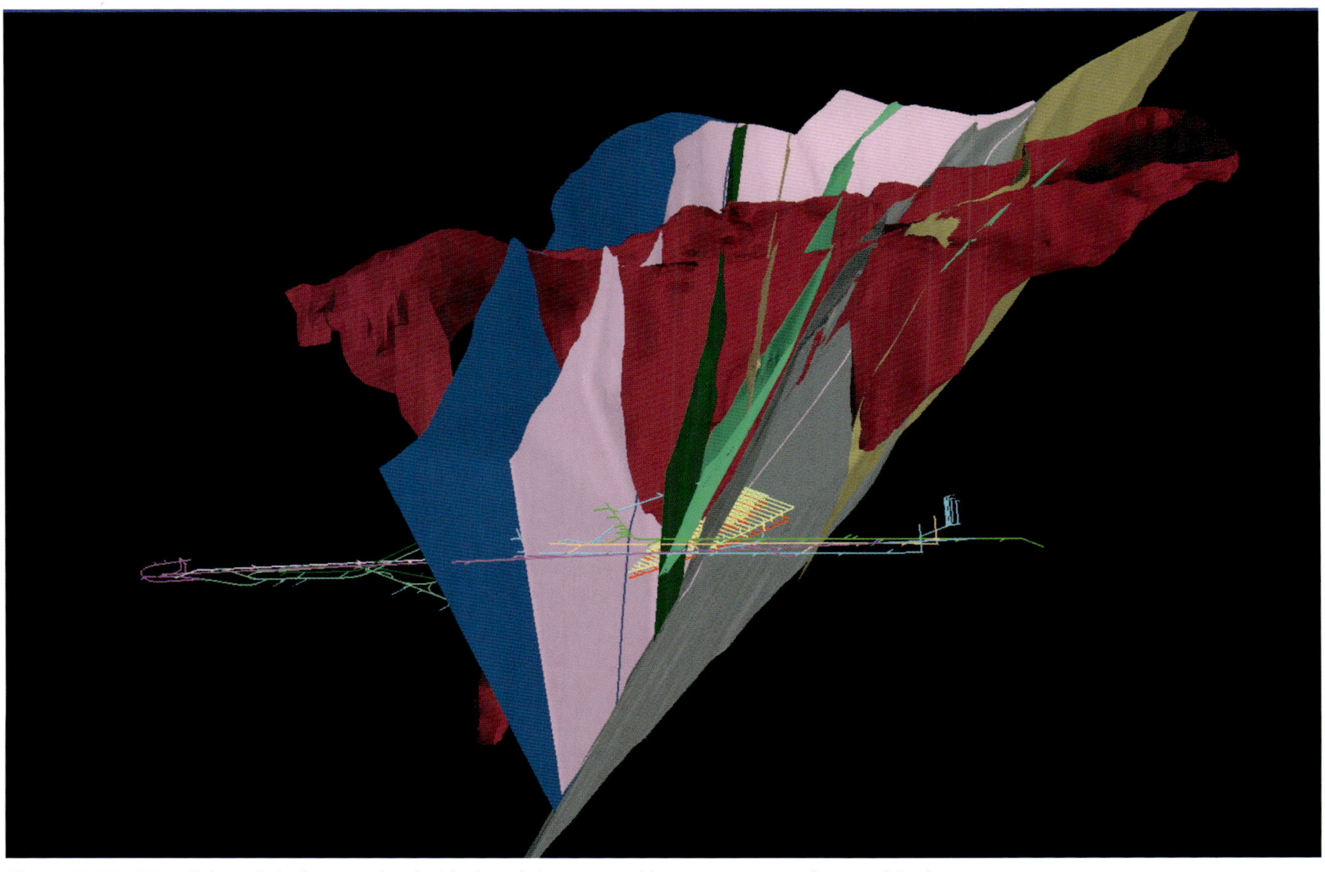

Figure 4.22: 3D solid model of an ore body (dark red) intersected by a sequence of normal faults
Source: Courtesy Argyle Diamonds

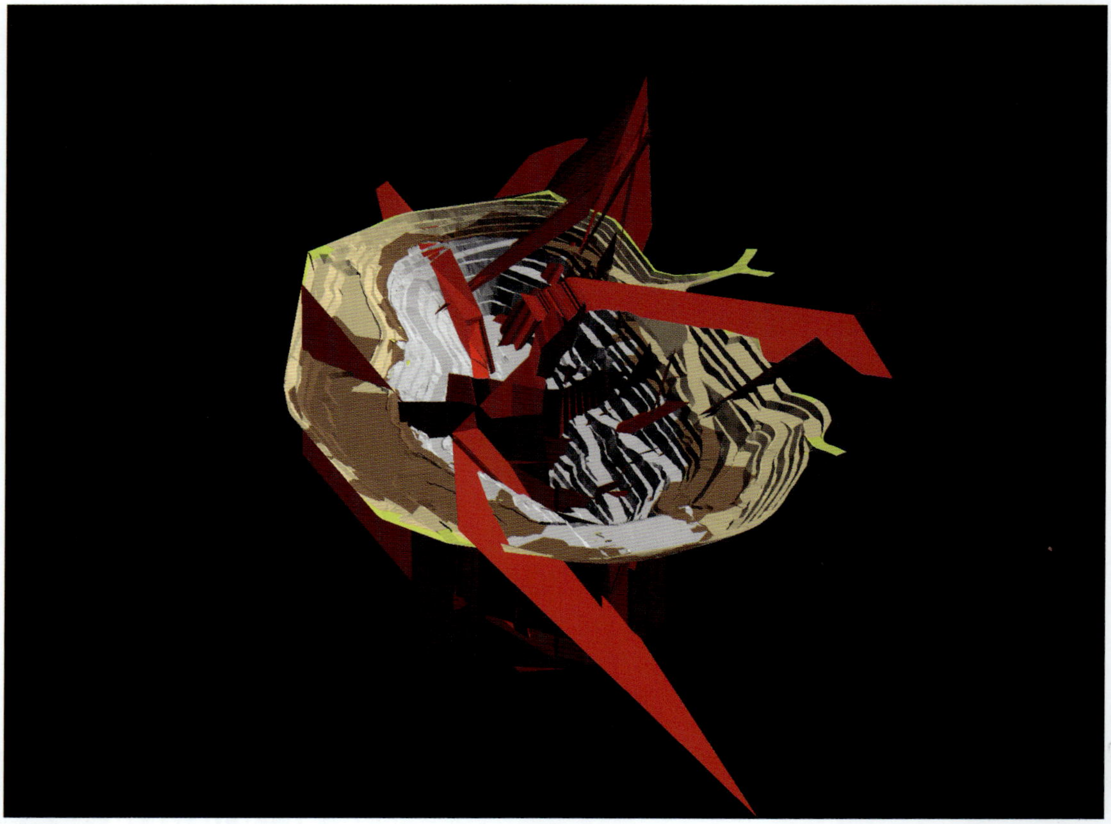

Figure 4.23: Major structures intersecting a proposed ultimate pit shell
Source: Courtesy BHP Billiton, Nickel West

trigonometry. A number of tabular and graphical aids can help construct these solutions (Badgley 1959), but they are often difficult to manipulate in three dimensions. The stereographic projection method provides the neatest solution to this difficulty. Historically the method was used mainly by crystallographers and mineralogists, but it was brought into prominence in structural geology during the 1950s by Phillips (1960). Although out of print, the Phillips publication remains the definitive stereographic projection textbook. A comprehensive outline of stereographic solutions is also given in Ragan (1985) and more recently in Lisle and Leyshon (2004). A number of basic techniques for use in slope stability engineering problems are presented in Wyllie and Mah (2004). It is vital to remember that in geotechnical engineering applications of the stereographic projection, the lower half of the hemisphere is used.

The main attraction of the stereographic projection is that it is easy to use. It can quickly provide solutions to complex geometric problems in the field or the office, and is an ideal tool for plotting and contouring sets of structural data. Because of its power and flexibility, it is recommended as the basic tool for all open pit structural modelling analyses. It is easily adapted to computer solutions and has been incorporated into a number of commercial software packages. Probably the best-known of these and certainly the most widely used in the open pit mining industry is the Rocscience Inc. program DIPS™ (Rocscience 2003), which is used illustratively in a number of figures in sections 4.2 and 4.5.

4.4.2.2 Blind zones

As outlined in section 2.4.9.6 the occurrence of structures that have low angles of intersection (α) with the drill hole raises the issue of blind zones.

All too frequently the occurrence and effect of blind zones are ignored or unrecognised when the structures in an open pit are being modelled. Most commonly they are created when the investigation drill holes along one side of the pit are angled back into the wall. Terzaghi (1965) noted that the only way to overcome their effect is to drill a sufficient number of drill holes so oriented that no structural pole can lie in or near the blind zone of each hole. An appropriate layout for a single cluster of three holes was for each hole to plunge at 45°, with the orientation of the trace of each hole differing by 120° from that of the other two. A structure of any orientation would be intersected by at least one of these holes at an angle (α) equal to or greater than about 31°.

4.4.2.3 *Terzaghi correction for joint spacing*

When the spacing of joints are measured from drill hole core (or along an outcrop scanline), the number of observations of joints of any one set is a function of the angle of intersection (inclination) between that set and the axis of the drill hole. Specifically, the number of intersection with a drill hole of given length decreases as the angle of inclination decreases such that:

$$N\alpha = \frac{L \sin \alpha}{d} \qquad \text{(eqn 4.1)}$$

where

α = inclination of the joints to the drill hole
d = the spacing between the joints
L = the length of the drill hole
$N\alpha$ = the number of joints intersected by the drill hole.

Hence, in a vertical drill hole, $N\alpha$ ranges between L/d for horizontal joints, of which α is 90°, and zero for vertical joints, of which α is zero (Terzaghi 1965).

No adequate correction can be made for joints with low angles of α. If a group of variously oriented drill holes is available, Terzaghi (1965) suggested that:

- it is generally advisable to disregard the poles of joints with an angle of inclination (α) of less than 20–30° because joints of the same set, if abundant, will be intersected at a higher angle by one or more of the other holes;
- data from the group of holes will provide a better basis for estimating the spacing of such joints.

4.4.2.4 *Terzaghi weighting*

The Terzaghi correction can also be used to establish an indication of the relative proportions of structures where a single drill hole or scan line orientation creates a bias in the structural orientation data. In this case, the relative proportions or weighting of the individual structures intersected in the scanline/hole(s) can be assessed through the equations:

R′ (true density of joint population) = 1/d = 1/d′ sinα = d′ cosecα (eqn 4.1)

W (weighting applied to individual pole for the density calculation = (1) cosecα (eqn 4.2)

where:

α = angle between plane and the drill hole or scan line
d = the true spacing of the fractures
d′ = apparent spacing along the drill hole or scan line

Since the weighting function tends to infinity as alpha (α) approaches zero, a maximum limit for this weighting must be set to prevent unreasonable results. This maximum limit corresponds to a minimum angle, which is typically set between 5° to 25°, and normally 15°.

Because the effect of applying the Terzaghi weighting to some data distributions can be quite marked, it is important to understand the weighting procedure before applying it.

4.4.3 Discrete fracture network modelling

Discrete fracture network (DFN) modelling explicitly represents how the faults and joints recognised by the structural model are spatially distributed within the rock mass. This feature has made it an important tool in helping to visualise how the rock mass deforms and slope failure mechanisms develop, particularly when the failure involves sliding along the major structures and fracture across the intact blocks of rock (rock bridges) left between these structures. Other important uses include estimating block size distributions for fragmentation analyses and determining flow conditions in hard rock masses.

The DFN modelling packages most commonly referred to in the literature include:

- FracMan (Golder Associates Inc. 2007);
- JointStats (Brown 2007);
- 3FLO (Billaux et al. 2006);
- SIMBLOC (Hamdi & du Mouza 2004).

The FracMan suite of DFN modelling tools was developed and released by Golder Associates Inc. in 1986. It was initially developed for mining and civil engineering applications and has been widely used in oil and gas and environmental projects, including radioactive waste management. More recently it has been applied to slope stability and tunnelling problems, in situ fragmentation prediction and groundwater management.

JointStats software was developed by the Julius Kruttschnitt Mineral Research Centre (JKMRC), University of Queensland, as part of the International Caving Study research and technology transfer program (Brown 2007). The original software accepts standard structural data from a face mapping or drill hole scanline but as part of the LOP project it has been enhanced to deliver a structural and a rock mass material properties database that enables data uncertainty to be assessed and confidence limits determined for specified data and/or attributes from within a single geotechnical domain. Milestones in this program included expanding the existing JointStats database to include quantitative measures of rock mass parameters and structural data collected using digital techniques.

3FLO was developed by Itasca Consultants S.A. (France) primarily for the hydrogeological analysis of fractured media. The code is capable of generating its own DFN and has many features similar to the standard Itasca codes, including the built-in programming language FISH.

FracMan, JointStats and 3FLO base their modelling on the random disc model where the size of the circular discontinuities is defined by the discontinuity radius and the locations are determined by a stochastic process, usually the Poisson process (Brown 2007). In SIMBLOC, the discontinuities are assimilated to flat discs. Each set is simulated independently of the others and the disc centres are generated in space using a uniform distribution law. The orientation of the discs is simulated following the mean and standard deviations of the distribution law that fits the actual field measurements. The radius of the disc is estimated from the trace length distribution. The joint intensity is calculated on the basis of the mean linear frequency and the radius distribution. Known applications of this code have been related mainly to block size distribution.

4.5 Structural domain definition

4.5.1 General guidelines

The information contained in the structural model is used to subdivide the rocks at the mine site into a select number of structural domains, each of which has distinct boundaries and is characterised internally by a recognisable structural fabric that clearly differentiates it from its neighbours.

All the features outlined in the sections above should be used to help define each domain. These include:

- mine-scale contacts marking changes in geology, including changes in lithology (e.g. between igneous and subvolcanic intrusive rocks and intruded sedimentary rocks), changes in weathering profiles and changes in alteration styles;
- mine-scale faults that may divide the rocks at the mine site into different structural blocks;
- mine-scale folded structures, with particular emphasis on changes in the orientation of the folds;
- mine-scale metamorphic structures, also with emphasis on changes in the orientation of the structures;
- bench and inter-ramp scale faults, folds and metamorphic structures;
- bench-scale joints, cleavage and micro-structures such as parasitic or second-order folds formed on the limbs of any inter-ramp or mine-scale folds.

All these features should have been identified from outcrop mapping and drilling, and stored in the 3D structural database.

4.5.2 Example application

4.5.2.1 Primary domain boundaries

Figure 4.24 illustrates the primary structural domains recognised at the Codelco Norté Chuquicamata mine in northern Chile. Here the domains have been given names, but more usually they are identified by numbers. The boundaries take account of lithology and

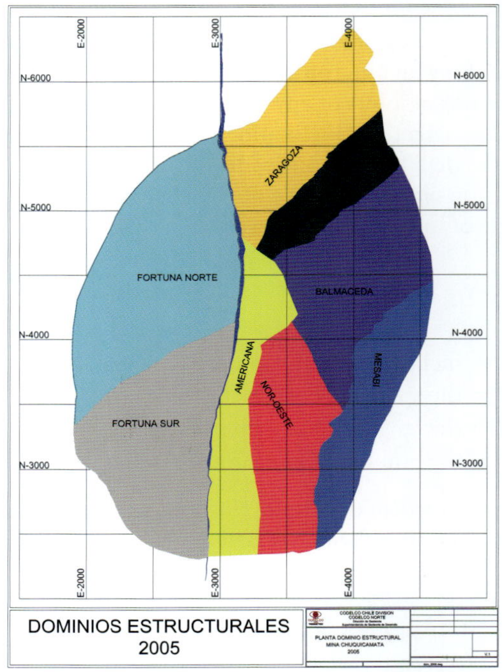

Figure 4.24: Structural domains at the Chuquicamata mine shown on a plan of the 2005 pit floor
Source: Courtesy Codelco, Division Codelco Norté

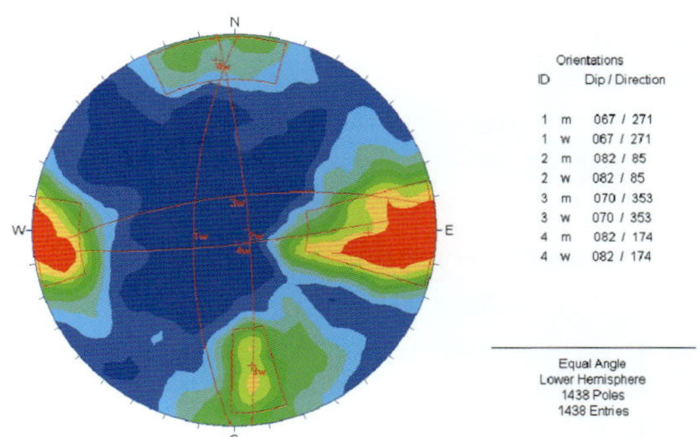

Figure 4.25: Orientation of major structures in the Fortuna North domain of the Chuquicamata mine
Source: Courtesy Codelco, Division Codelco Norté

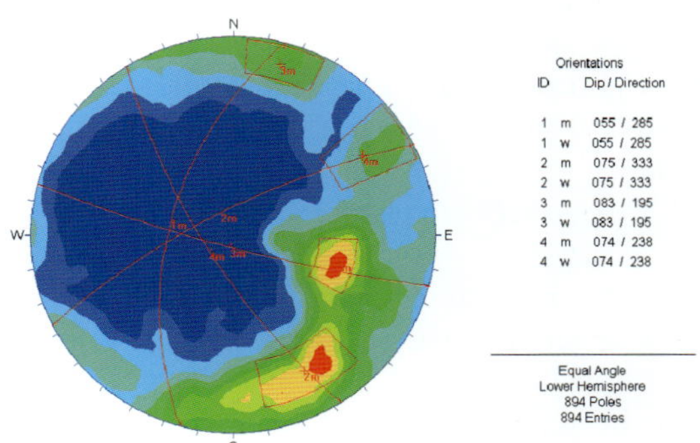

Figure 4.26: Orientation of major structures in the Fortuna South domain of the Chuquicamata mine
Source: Courtesy Codelco, Division Codelco Norté

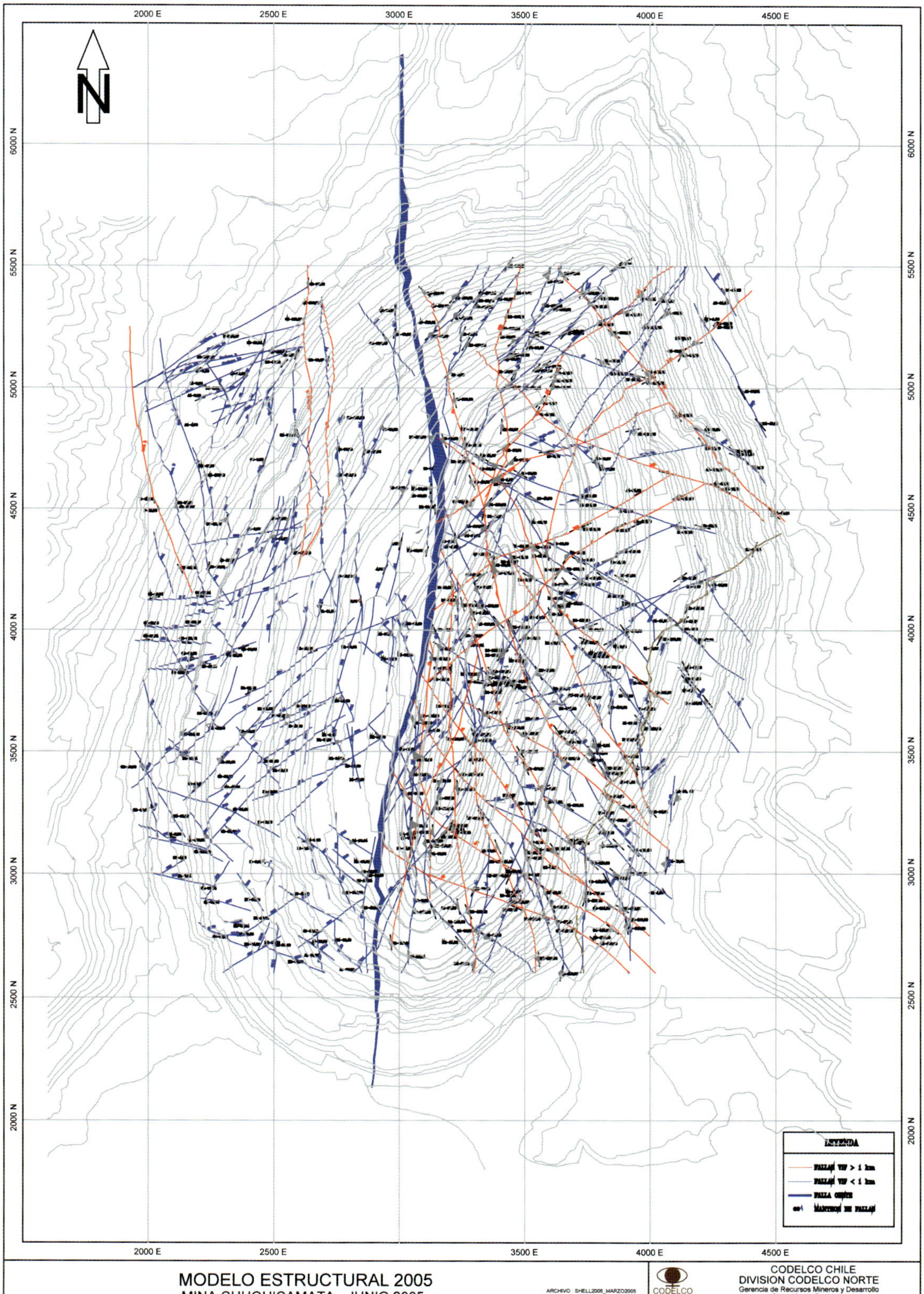

Figure 4.27: Major fault traces on the Chuquicamata mine 2005 pit shell
Source: Courtesy Codelco, Division Codelco Norté

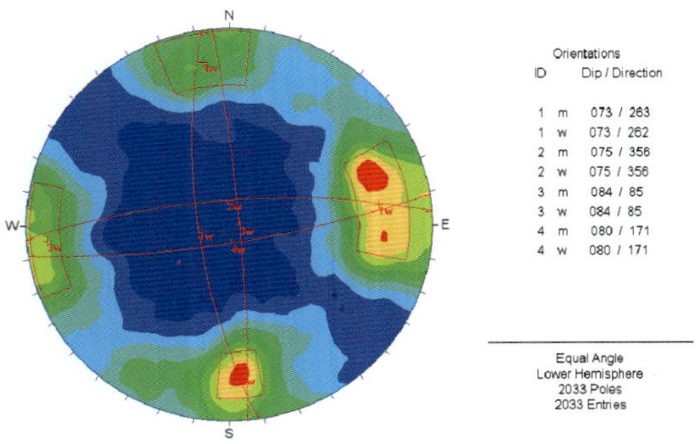

Figure 4.28: Orientation of lesser faults in the Fortuna North domain of the Chuquicamata mine
Source: Courtesy Codelco Norté

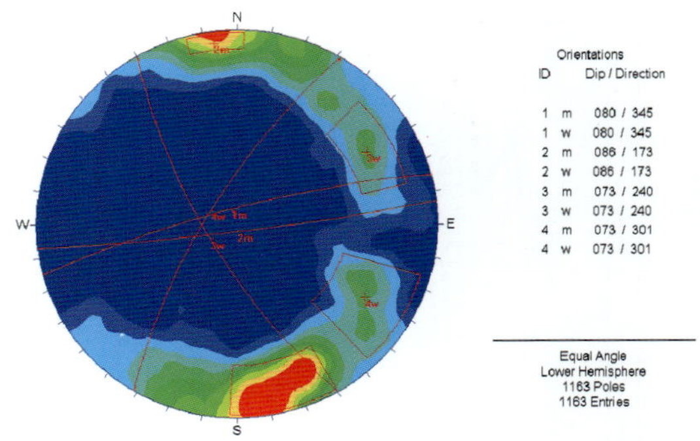

Figure 4.29: Orientation of lesser faults in the Fortuna South domain of the Chuquicamata mine
Source: Courtesy Codelco Norté

the shape of the pit but are primarily based on major faults mapped in the pit over a number of years combined with the results of surface mapping, oriented drill hole core logging and underground mapping performed between 2003 and 2005. The more recent work was done to provide additional design information for a study of the viability of steepening the pit slopes as the mine approaches a possible transition to underground mining (Calderón & Tapia 2006). The Chuquicamata mine has been used as the example because it shows the clarity that can be achieved when an established and validated 3D structural database is available for analysis. Obviously, such clarity will not be possible at the pre-feasibility and early feasibility stages of project development, but the example does illustrate the mature design objective.

Figures 4.25 and 4.26 are stereonets that illustrate the different orientation of the faults that divide the Fortuna Granodiorite in the west wall of the pit into the Fortuna North and Fortuna South domains. The differences in the orientation can be distinguished in Figure 4.27, which plots the fault traces on the 2005 pit shell. Faults shown in blue have trace lengths of up to 1 km. Faults shown in red have trace lengths greater than 1 km.

4.5.2.2 Fabric within primary domains

Once the primary domain boundaries have been selected, the bench and inter-ramp scale structures within each domain must be assessed to ensure that the internal structural fabric of the domain clearly distinguishes it from its neighbour. This process should feature an exhaustive interrogation of the structural database and may lead to changes in the primary boundaries or to subdivision of the domains. It is illustrated by the Chuquicamata mine example used in Figures 4.24–4.27.

Figures 4.28–4.31 illustrate how the lesser structures and joint fabric were used at the mine to consolidate the domains within the boundaries set by the major structures. Figure 4.28 shows the orientations of the lesser fault sets within the Fortuna North domain. These sets are quite distinct from those of the Fortuna South domain, which are shown in Figure 4.29. Similarly, Figures 4.30 and 4.31 illustrate the differences in the joint sets between the two domains.

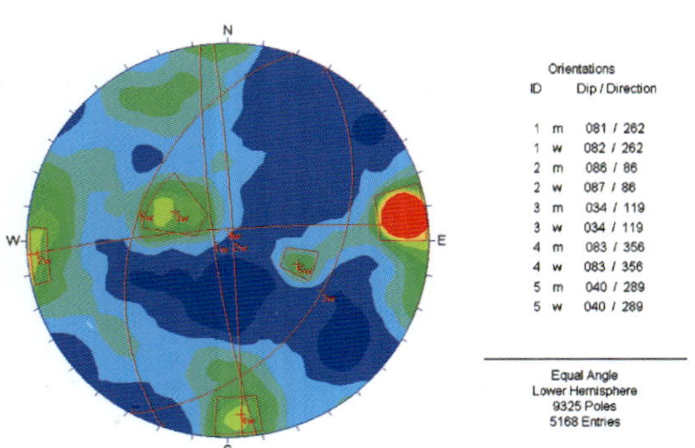

Figure 4.30: Orientation of joints in the Fortuna North Domain of the Chuquicamata mine
Source: Courtesy Codelco Norté

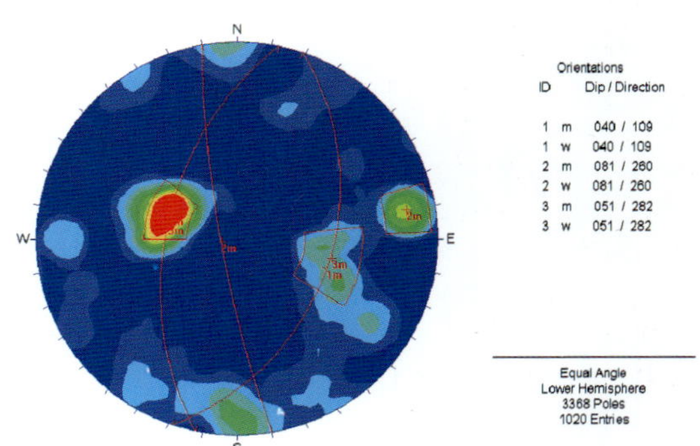

Figure 4.31: Orientation of joints in the Fortuna South domain of the Chuquicamata mine
Source: Courtesy Codelco Norté

5 ROCK MASS MODEL

Antonio Karzulovic and John Read

5.1 Introduction

Chapters 3 and 4 dealt with the geological and structural components of the geotechnical model. The third component, which must now be addressed, is the rock mass model (Figure 5.1). The purpose of this model is to database the engineering properties of the rock mass for use in the stability analyses that will be used to prepare the slope designs at each stage of project development. This includes the properties of the intact pieces of rock that constitute the anisotropic rock mass, the structures that cut through the rock mass and separate the individual pieces of intact rock from each other, and the rock mass itself.

As outlined in Chapter 10 (section 10.1.1), when assessing potential failure mechanisms of any rock mass a fundamental attribute that must always be considered is that in stronger rocks structure is likely to be the primary control, whereas in weaker rocks strength can be the controlling factor. This means that the rock mass may fail in three possible ways:

1 structurally controlled failure, where the rupture occurs only along the joints, bedding or faults. This is the case for planar and wedge slides, which are most likely to occur at bench and inter-ramp scale. In this case the strength and orientation of the structures are the most important parameters in assessing slope stability;
2 failure with partial structural control, where rupture occurs partly through the rock mass and partly through the structures, usually at inter-ramp and overall scale. In this case the strength of the rock mass and the strength and orientation of the structures are both important in assessing slope stability;
3 failure with limited structural control, where the rupture occurs predominantly through the rock mass. This can occur at inter-ramp or overall slope scale in either highly fractured or weak rock masses mostly comprising soft or altered material. In this case the strength of the rock mass is the most important parameter in assessing slope stability.

Hence, when setting out to determine the geotechnical engineering properties of the rock mass, the strength of the rock mass and the potential mechanism of failure must be considered and factored into the sampling and testing program. Data representative of the intact pieces of rock, the structures and the rock mass itself will all be required at some stage of the slope design and must be incorporated in the rock mass model. The procedures involved in gathering these data are the focus of the next four sections.

Section 5.2 deals with the properties of the intact rock. It outlines the nature of the standard index and mechanical property tests used in rock slope engineering (sections 5.2.1, 5.2.2 and 5.2.3) then outlines testing needs for special cases such as weak, saprolitic and/or highly weathered and altered rocks, degradable clay shales and permafrost conditions (section 5.2.4).

Section 5.3 deals with the strength of the mechanical defects in the rock mass, especially shear strength and the effects of surface roughness. Section 5.4 outlines the methods currently used to classify the rock mass. Section 5.5 completes the chapter, with descriptions of current and newly developed means of assessing the strength of the rock mass.

5.2 Intact rock strength

5.2.1 Introduction

The geomechanical properties of the intact rock that occurs between the structural defects in a typical rock mass are measured in the laboratory from representative samples of the intact rock. The need to obtain representative samples is important. For example, it is not uncommon that only the 'best' core samples are sent to the laboratory for uniaxial compression testing, which can

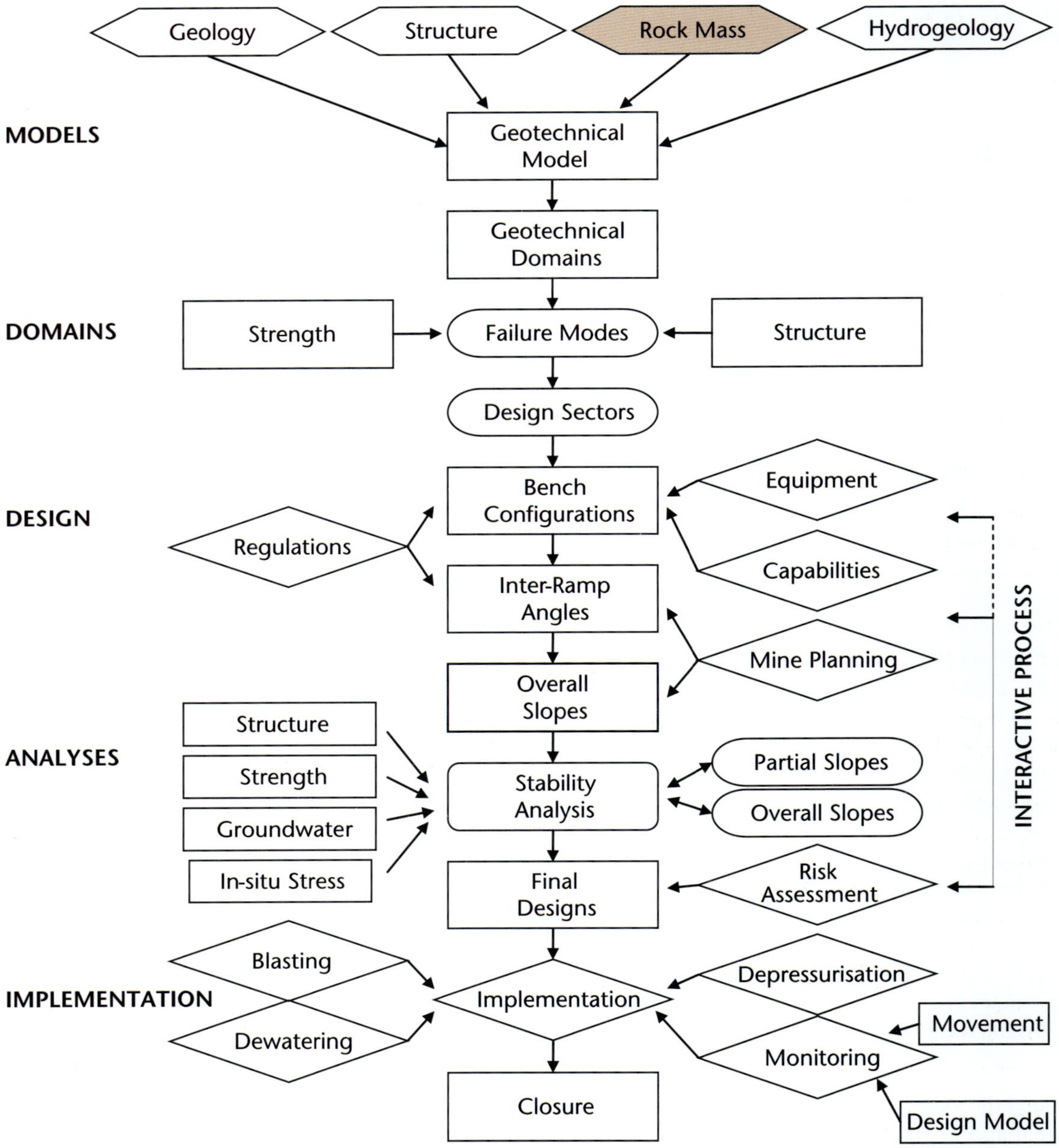

Figure 5.1: Slope design process

result in the rock strength being overestimated. If the results of the tests show a large variation or, for example, there is only partial core recovery, it may be better not to consider a unique value such as the mean or the mode, but a range defined by upper and lower values. In the case of only partial recovery, the upper bound would be represented by the uniaxial strength of the 'good' core and the lower bound, representing the zones of core loss, would represent zones of significantly reduced strength.

When sampling and testing the intact rock it is also important to differentiate between 'index', 'conductivity' and 'mechanical' properties.

- Index properties, which do not define the mechanical behaviour of the rock, but are easy to measure and provide a qualitative description of the rock and, in some cases, can be related to rock conductivity and/or mechanical properties. For example, an increase in rock porosity could explain a decrease in its strength.
- Conductivity properties are properties that describe fluid flow through the rock. An example is hydraulic conductivity.
- Mechanical properties are properties that describe quantitatively the strength and deformability of the rock. The most common example is uniaxial compres-

sive strength, which is one of the most used parameters in rock engineering.

Comprehensive discussions on rock properties and their measurement can be found in Lama et al. (1974), Lama and Vutukuri (1974), Farmer (1983), Nagaraj (1993), Bell (2000) and Zhang (2005).

In open pit slope engineering the most commonly used rock properties are the following.

- Index properties (see section 5.2.2):
  - → Point load strength index, I_s;
  - → Porosity, n;
  - → Unit weight, γ;
  - → P-wave velocity, $V_{P;}$
  - → S-wave velocity, $V_{S;}$
- Mechanical properties (see section 5.2.3):
  - → Tensile strength, *TS* or σ_t;
  - → Uniaxial compressive strength, *UCS* or σ_c;
  - → Triaxial compressive strength, *TCS*;
  - → Young's modulus, *E*, and Poisson's ratio, *v*.

5.2.2 Index properties

5.2.2.1 Point load strength index

The point load strength index, I_s, is an indirect estimate of the uniaxial compressive strength of rock. The point load test can be performed on specimens in the form of core (diametral and axial tests), cut blocks (block tests) or irregular lumps (irregular lump test). The samples are broken by a concentrated load applied through a pair of spherically truncated, conical platens. The test can be performed in the field with portable equipment, or in the laboratory. The point load strength index, I_s, is given by:

$$I_s = \frac{P}{D_e^2} \qquad \text{(eqn 5.1)}$$

where P is the load that breaks the specimen and D_e is an equivalent core diameter, given by:

$$D_e = D \qquad \text{(eqn 5.2a)}$$

$$D_e = \frac{4A}{\pi} \text{ for axial, block and lump tests} \qquad \text{(eqn 5.2b)}$$

where D is the core diameter and A is the minimum cross-sectional area of a plane through the specimen and the platen contact points. I_s varies with D_e. Hence, it is preferable to carry out diametral tests on 50–55 mm diameter specimens.

Brady and Brown (2004) indicated that the value of I_s measured for a diameter D_e can be converted into an equivalent 50 mm core I_s by the relation:

$$I_s = I_{s(De)} \times \left(\frac{D_e}{50}\right)^{0.45} \qquad \text{(eqn 5.3)}$$

where $I_{s(De)}$ is the point load strength index measured for an equivalent core diameter D_e different from 50 mm. It is not recommended to use core diameters smaller than 40 mm for point load testing (Bieniawski 1984).

Several correlations have been developed to estimate the uniaxial compressive strength of rock, σ_c, from the point load strength index (Zhang 2005), but the most commonly used is:

$$\sigma_c \approx (22 \text{ to } 24) \times I_s \qquad \text{(eqn 5.4)}$$

where I_s is the point load strength index for D_e = 50 mm.

It should be noted that the point load test is not generally applicable for rocks with a CICS value below 25 MPa (R2 and lower), since the points tend to indent the rock. Further, extreme caution must be exercised when carrying out point load tests and interpreting the results using correlations such as Equation 5.4. First, there is considerable anecdotal and documented evidence that suggests there is no unique conversion factor and that it is necessary to determine the conversion factor on a site-by-site and rock type by rock type basis (Tsiambaos & Sabatakakis 2004). Second, as noted by Brady and Brown (2004), the test is one in which the fracture is caused by induced tension and it is essential that a consistent mode of failure be produced if the results obtained from different specimens are to be comparable. Very soft rocks and highly anisotropic rocks or rocks containing marked planes of weakness such as bedding planes are likely to give spurious results. A large degree of scatter is a general feature of point load test results and large numbers of individual determinations, often in excess of 100, are required in order to obtain reliable indices. For anisotropic rocks, it is usual to determine a strength anisotropy index, I_a, defined as the ratio of the mean I_s values measured perpendicular and parallel to the planes of weakness.

ASTM Designation D5731-95 describes the standard test method for determination of the point load strength index of rock and Franklin (1985) describes the method suggested by the ISRM for determining point load strength.

5.2.2.2 Porosity

The porosity of rock, n, is defined as the proportion of the volume of voids (V_V) to the total volume (V_T) of the sample. Porosity is traditionally expressed as a percentage.

$$n = \frac{V_V}{V_T} \qquad \text{(eqn 5.5)}$$

Goodman (1989) indicates that in sedimentary rocks n varies from close to 0 to as much as 90%, depending on the degree of consolidation or cementation, with 15% being a 'typical' value for an 'average' sandstone. Chalk is among the most porous of all rocks, with porosities in

Table 5.1: Porosities of some rocks

Rock Type	Rock	Age	Depth (m)	*n* (%)
Chalk	Chalk, Great Britain	Cretaceous	Surface	28.8
Diabase	Frederick diabase	–	–	0.1
Dolomite	Beekmantown dolomite	Ordovician	3200	0.4
	Niagara dolomite	Silurian	Surface	2.9
Gabbro	San Marcos gabbro	–	–	0.2
Granite	Granite, fresh	–	Surface	0–1
	Granite, weathered	–	–	1–5
	Granite, decomposed (saprolite)	–	–	20
Limestone	Black River limestone	Ordovician	Surface	0.46
	Bedford limestone	Mississippian	Surface	12
	Bermuda limestone	Recent	Surface	43
	Dolomitic limestone	–	–	2.08
	Limestone, Great Britain	Carboniferous	Surface	5.7
	Limestone, Great Britain	Silurian	–	1.0
	Oolitic limestone	–	–	1.06
	Salem limestone	Mississippian	Surface	13.2
	Solenhoffen limestone	–	Surface	4.8
Marble	Marble	–	–	0.3
	Marble	–	–	1.1
Mudstone	Mudstone, Japan	Upper Tertiary	Near surface	22–32
Quartzite	Quartzite, Great Britain	Cambrian	–	1.7–2.2
Sandstone	Berea sandstone	Mississippian	0-610	14
	Keuper sandstone (England)	Triassic	Surface	22
	Montana sandstone	Cretaceous	Surface	34
	Mount Simon sandstone	Cambrian	3960	0.7
	Navajo sandstone	Jurassic	Surface	15.5
	Nugget sandstone (Utah)	Jurassic	–	1.9
	Potsdam sandstone	Cambrian	Surface	11
	Pottsville sandstone	Pennsylvanian	–	2.9
Shale	Shale	Pre-Cambrian	Surface	1.6
	Shale	Cretaceous	180	33.5
	Shale	Cretaceous	760	25.4
	Shale	Cretaceous	1065	21.1
	Shale	Cretaceous	1860	7.6
	Shale Oklahoma	Pennsylvanian	305	17
	Shale Oklahoma	Pennsylvanian	915	7
	Shale Oklahoma	Pennsylvanian	1525	4
	Shale, Great Britain	Silurian	–	1.3–20
Tuff	Tuff, bedded	–	–	40
	Tuff, welded	–	–	14
Tonalite	Cedar City tonalite	–	–	7

Source: Modified from Goodman (1989). Data selected from Clark (1966), Duncan (1969), Brace & Riley (1972)

Table 5.2: Dry unit weight of some rocks

Rock type	γ (kN/m3)	γ (tonne/m3)	Rock type	γ (kN/m3)	γ (tonne/m3)
Amphibolite	27.0–30.9	2.75–3.15	Dolomite	26.0–27.5	2.65–2.80
Andesite	21.6–27.5	2.20–2.80	Limestone	23.1–27.0	2.35–2.75
Basalt	21.6–27.4	2.20–2.80	Marble	24.5–28.0	2.50–2.85
Chalk	21.6–24.5	2.20–2.50	Norite	26.5–29.4	2.70–3.00
Diabase	27.5–30.4	2.80–3.10	Peridotite	30.9–32.4	3.15–3.30
Diorite	26.5–28.9	2.70–2.95	Quartzite	25.5–26.5	2.60–2.70
Gabbro	26.5–30.4	2.70–3.10	Rock salt	20.6–21.6	2.10–2.20
Gneiss	25.5–30.9	2.60–3.15	Rhyolite	23.1–26.0	2.35–2.65
Granite	24.5–27.4	2.50–2.80	Sandstone	18.6–26.5	1.90–2.70
Granodiorite	26.0–27.5	2.65–2.80	Shale	19.6–26.0	2.00–2.65
Greywacke	26.0–26.5	2.65–2.70	Schist	25.5–29.9	2.60–3.05
Gypsum	22.1–23.1	2.25–2.35	Slate	26.5–28.0	2.70–2.85
Diorite	26.5–28.9	2.70–2.95	Syenite	25.5–28.4	2.60–2.90

Source: Data selected from Krynine & Judd (1957), Lama & Vutukuri (1978), Jumikis (1983), Carmichael (1989), Goodman (1989)

some instances of more than 50%. Some volcanic materials, e.g. pumice and tuff, were well-aerated as they were formed and can also present very high porosities, but most magma-derived volcanic rocks have a low porosity. Crystalline rocks, including limestones and evaporites and most igneous and metamorphic rocks, also have low porosities, with a large proportion of the void space often being created by planar cracks or fissures. In these rocks n is usually less than 1–2% unless weathering has taken hold. As weathering progresses, n can increase well beyond 2%.

The ISRM-recommended procedures for measuring the porosity of rock are described in ISRM (2007). A detailed discussion of porosity can be found in Lama and Vutukuri (1978). The porosities of some rocks are given in Table 5.1.

5.2.2.3 Unit weight

The unit weight of rock, γ, is defined as ratio between the weight (W) and the total volume (V_T) of the sample:

$$\gamma = \frac{W}{V_T} \tag{eqn 5.6}$$

The density of rock, ρ, is defined as ratio between the mass (M) and the total volume (V_T) of rock:

$$\rho = \frac{M}{V_T} \tag{eqn 5.7}$$

The specific gravity of rock, G_s, is defined as the ratio between its unit weight (γ) and the unit weight of water (γ_w):

$$G_s = \frac{\gamma}{\gamma_w} \tag{eqn 5.8}$$

The ISRM-recommended procedures for measuring the unit weight of rock are described in ISRM (2007). A detailed discussion of unit weight can be found in Lama and Vutukuri (1978). The unit weights of some rocks are given in Table 5.2.

5.2.2.4 Wave velocity

The velocity of elastic waves in rock can be measured in the laboratory. Wave velocity is one of the most used index properties of rock and has been correlated with other index and mechanical properties of rock (Zhang 2005). Laboratory P-wave velocities vary from less than 1 km/sec in porous rocks to more than 6 km/sec in hard rocks.

Table 5.3: Average P-wave velocities in rock-forming minerals

Mineral	V_p (m/sec)	Mineral	V_p (m/sec)	Mineral	V_p (m/sec)
Amphibole	7200	Epidote	7450	Olivine	8400
Augite	7200	Gypsum	5200	Orthoclase	5800
Biotite	5260	Hornblende	6810	Plagioclase	6250
Calcite	6600	Magnetite	7400	Pyrite	8000
Dolomite	7500	Muscovite	5800	Quartz	6050

Source: Data selected from Fourmaintraux (1976), Carmichael (1989)

Table 5.4: P-wave and S-wave velocities of some rocks

Rock	V_P (m/sec)	V_S (m/sec)	Rock	V_P (m/sec)	V_S (m/sec)
Basalt	4550–6150	2550–3550	Limestone	4550–6200	2750–3600
Chalk	1550–4300	1600–2500	Norite	5950–6950	3300–3900
Diabase	3300–3750	5150–6750	Peridotite	6400–8450	3300–4400
Diorite	4750–6350	2900–3550	Quartzite	2750–5550	1600–3450
Dolomite	4850–6600	2950–3750	Rhyolite	3200–3300	1900–2000
Gabbro	5950–6950	3300–3900	Sandstones	2550–5000	1400–3100
Gneiss	2850–5450	1950–3350	Schist	2950–4950	1750–3250
Granite	4200–5900	2550–3350	Tuff	1400–1500	800–900

Source: Data selected from Carmichael (1989), Schön (1996), Mavko et al. (1998)

Wave velocities are significantly lower for micro-cracked rock than for porous rocks without cracks but with the same total void space. Hence, Fourmaintraux (1976) proposed a procedure based on comparing the theoretical and measured values of V_P to evaluate the degree of fissuring in rock specimens in terms of a quality index *IQ*:

$$IQ\% = \frac{V_P}{V_P^T} \times 100\% \qquad \text{(eqn 5.9)}$$

where V_P is the measured P-wave velocity and V_P^T is the theoretical P-wave velocity, which can be calculated from:

$$\frac{1}{V_P^T} = \sum_i \frac{C_i}{V_{P,i}} \qquad \text{(eqn 5.10)}$$

where $V_{P,i}$ is the P-wave velocity of mineral constituent *i*, which has a volume proportion C_i in the rock. Average P-wave velocities in rock-forming minerals are given in Table 5.3.

Experiments by Fourmaintraux established that *IQ* is affected by the pores in the rock sample according to:

$$IQ\% = 100 - 1.6n_p \qquad \text{(eqn 5.11)}$$

where n_p is the porosity of non-fissured rock expressed as a percentage. However, if there is even a small fraction of flat cracks or fissures, Equation 5.7 breaks down. Because of the extreme sensitivity of *IQ* to fissuring, and based upon laboratory measurements and microscopic observation of fissures, Fourmaintraux proposed a chart (Figure 5.2) as a basis for describing the degree of fissuring of a rock specimen.

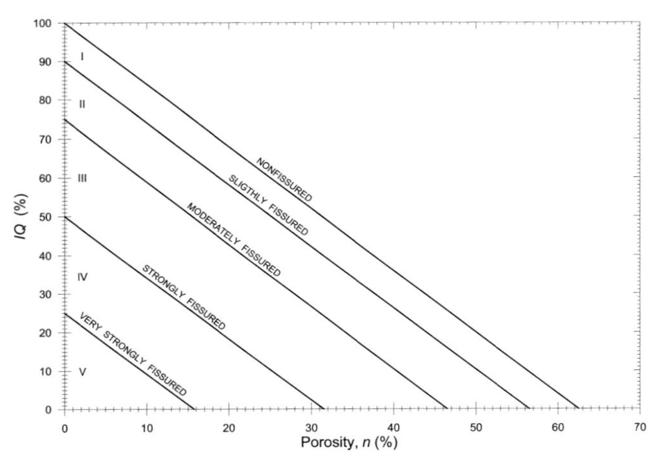

Figure 5.2: Classification of scheme for fissuring in rock specimens considering the quality index *IQ* and the porosity of the rock
Source: Fourmaintraux (1976)

Both the P-wave velocity (V_P) and the S-wave velocity (V_S) can be determined in the laboratory, with V_P the easiest to measure. ASTM D2845-95 described the laboratory determination of pulse velocities and ultrasonic elastic constants of rock, and ISRM (2007) described the methods suggested by the ISRM for determining sound velocity in rock. The P-wave and S-wave velocities of some rocks are given in Table 5.4.

5.2.3 Mechanical properties

5.2.3.1 Tensile strength

The tensile strength of rock, σ_t, is measured by indirect tensile strength tests because it is very difficult to perform a true direct tension test (Lama et al. 1974). These indirect tensile strength tests apply compression to generate combined tension and compression in the centre of the rock specimen. A crack starting in this region propagates parallel to the axis of loading and causes the failure of the specimen (Fairhurst 1964, Mellor & Hawkes 1971).

The Brazilian test is the most used method to measure the tensile strength of rock. The specimens are disks with flat and parallel faces. They are loaded diametrically along line contacts (unlike the point contacts of the otherwise similar diametral point load test). The disk diameter should be at least 50 mm and the ratio of the diameter *D* to the thickness *t* about 2:1. A constant loading rate of 0.2 kN/sec is recommended, such that the specimen ruptures within 15–30 sec, usually along a single tensile-type fracture aligned with the axis of loading.

The Brazilian tensile strength, σ_{tB}, is given by:

Table 5.5: Tensile strength of some rocks

Rock	σ_t (MPa)	Rock	σ_t (MPa)	Rock	σ_t (MPa)
Andesite	6–21	Gneiss	4–20	Sandstone	1–20
Anhydrite	6–12	Granite	4–25	Schist	2–6
Basalt	6–25	Greywacke	5–15	Shale	0.2–10
Diabase	6–24	Gypsum	1–3	Siltstone	1–5
Diorite	8–30	Limestone	1–30	Slate	7–20
Dolerite	15–35	Marble	1–10	Tonalite	5–7
Dolomite	2–6	Porphyry	8–23	Trachyte	8–12
Gabbro	5–30	Quartzite	3–30	Tuff	0.1–1

Source: Data selected from Lama et al. (1974), Jaeger & Cook (1979), Jumikis (1983), Goodman (1989), Gonzalez de Vallejo (2002)

$$\sigma_{tB} = \frac{2P}{\pi D t} \qquad \text{(eqn 5.12)}$$

where P is the compression load, and D and t are the diameter and thickness of the disk. The Brazilian test has been found to give a tensile strength higher than that of a direct tension test, probably owing to the effect of fissures as short fissures weaken a direct tension specimen more severely than they weaken a splitting tension specimen. In spite of this, Brazilian tests are widely used and it is commonly assumed that the Brazilian tensile strength is a good approximation of the tensile strength of the rock.

ASTM D3967-95a describes the standard test method for splitting tensile strength of rock specimens and ISRM (2007) describes the methods suggested by the ISRM for determining indirect tensile strength by the Brazilian tests. The tensile strengths of some rocks are given in Table 5.5.

In addition to the Brazilian test, several correlations have been developed for estimating the tensile strength of rock, σ_t. Two of the most common are (Zhang, 2005):

$$\sigma_t \approx \frac{\sigma_c}{10} \qquad \text{(eqn 5.13)}$$

$$\sigma_t \approx 1.5 I_s \qquad \text{(eqn 5.14)}$$

where σ_c is the uniaxial compressive strength and I_s is the point load strength index of the rock. These correlations must be used with caution.

5.2.3.2 *Uniaxial compressive strength*

Uniaxial compression of cylindrical rock samples prepared from drill core is probably the most widely performed test on rock. It is used to determine the uniaxial compressive strength (unconfined compressive strength), σ_c, the Young's modulus, E, and Poisson's ratio, ν:

The uniaxial compressive strength, σ_c, is given by:

$$\sigma_c = \frac{P}{A} = \frac{4P}{\pi D^2} \qquad \text{(eqn 5.15)}$$

where P is the load that causes the failure of the cylindrical rock sample, D is the specimen diameter and A its cross-sectional area. Corrections to account for the increase in cross-sectional area are commonly negligible if rupture occurs before 2–3% strain is reached.

ASTM D2938-95 and D3148-96 describe the standard test methods for uniaxial compressive strength and elastic moduli of rock specimens. ISRM (2007) describes the methods suggested by the ISRM for determining the uniaxial compressive strength and deformability of rock. Brady and Brown (2004) summarised the essential features of this recommended procedure.

- The samples should be right circular cylinders having a height:diameter ratio of 2.5:3.0 and a diameter preferably of not less than NMLC core size (51 mm). The sample diameter should be at least 10 times the largest grain in the rock.
- The ends of the sample should be flat within 0.02 mm. They should depart not more than 0.001 radians or 0.05 mm in 50 mm from being perpendicular to the axis of the sample.
- The use of capping materials or end surface treatments other than machining is not permitted.
- The samples should be stored for no more than 30 days and tested at their natural moisture content. This requires adequate protection from damage and moisture loss during transportation and storage.
- The uniaxial load should be applied to the specimen at a constant stress rate of 0.5 MPa/sec to 1.0 MPa/sec.
- Axial load and axial and radial or circumferential strains should be recorded throughout the test.
- There should be at least five replications of each test.

Additionally, all samples should be photographed and all visible defects logged before testing. After testing, the sample should be rephotographed and all failure planes logged. Only the test results where it can be demonstrated that failure occurred through the intact rock rather than along defects in the sample should be accepted.

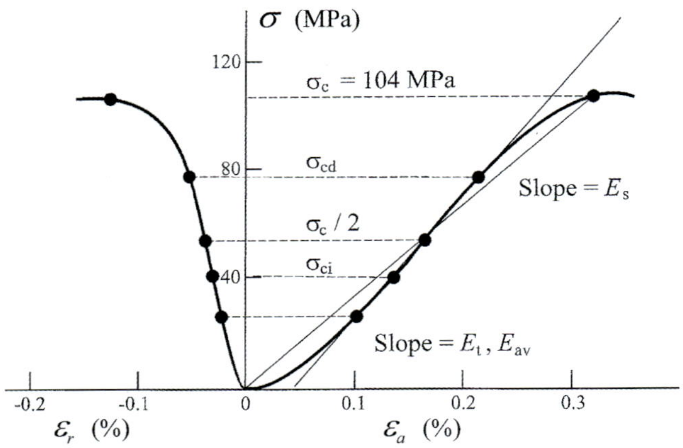

Figure 5.3: Results from a uniaxial compression test on rock
Source: Brady & Brown (2004)

An example of the results from a uniaxial compression test is shown in Figure 5.3.

An initial bedding-down and crack-closure stage is followed by a stage of elastic deformation until an axial stress of σ_{ci} is reached, at which stage stable crack propagation is initiated. This continues until the axial stress reaches σ_{cd} when unstable crack growth and irrecoverable deformations begin. This continues until the peak or uniaxial compressive strength, σ_c, is reached.

The uniaxial strength of rock decreases with increasing specimen size, as shown in Figure 5.4. It is commonly assumed that σ_c refers to a 50 mm diameter sample. An approximate relationship between uniaxial compressive

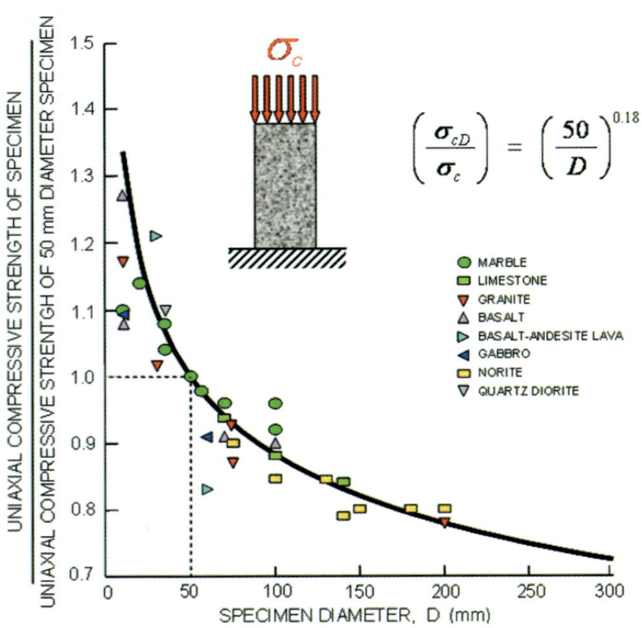

Figure 5.4: Influence of sample size on the uniaxial compressive strength of rock
Source: Hoek & Brown (1980a)

strength and sample diameter for specimens between 10 mm and 200 mm diameter is given by Hoek and Brown (1980):

$$\sigma_c = \sigma_{cD} \left(\frac{D}{50} \right)^{0.18} \qquad \text{(eqn 5.16)}$$

where σ_c is the uniaxial compressive strength of a 50 mm diameter specimen and σ_{cD} is the uniaxial compressive strength measured in a specimen with a diameter D (in mm).

In the case of anisotropic rocks (e.g. phyllite, schist, shale and slate), several uniaxial compression tests are performed on core oriented at various angles to any foliation or other plane of weakness. Strength is usually least when the foliation or weak planes make an angle of about 30° to the direction of loading and greatest when the weak planes are parallel or perpendicular to the axis. This allows the definition of lower and upper limits for σ_c and enables decisions, using engineering judgment, as to which value is the most appropriate.

For a detailed discussion on rock behaviour under uniaxial compression see Jaeger (1960), Donath (1964), McLamore (1966) and Brady and Brown (2004). For a particularly comprehensive discussion on uniaxial testing of rock see Hawkes and Mellor (1970).

5.2.3.3 *Triaxial compressive strength*

The triaxial compressive strength test defines the Mohr-Coulomb failure envelope (Figure 5.5) and hence provides the means of determining the friction (Ø) and cohesion (*c*) shear strength parameters for intact rock.

In triaxial compression, when the rock sample is not only loaded axially but also radially by a confining pressure kept constant during the test, failure occurs only when the combination of normal stress and shear stress is such that the Mohr circle is tangential to the failure envelope. Thus, in Figure 5.5, Circle A represents a stable condition; Circle B cannot exist.

The triaxial compression test is carried out on a cylindrical sample prepared as for the uniaxial compression test. The specimen is placed inside a pressure vessel (Figure 5.6) and a fluid pressure, S_3, is applied to its

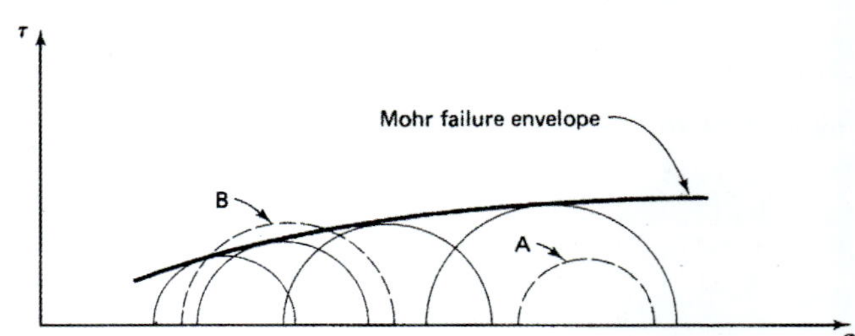

Figure 5.5: Mohr failure envelope defined by the Mohr circles at failure
Source: Holtz & Kovacs (1981)

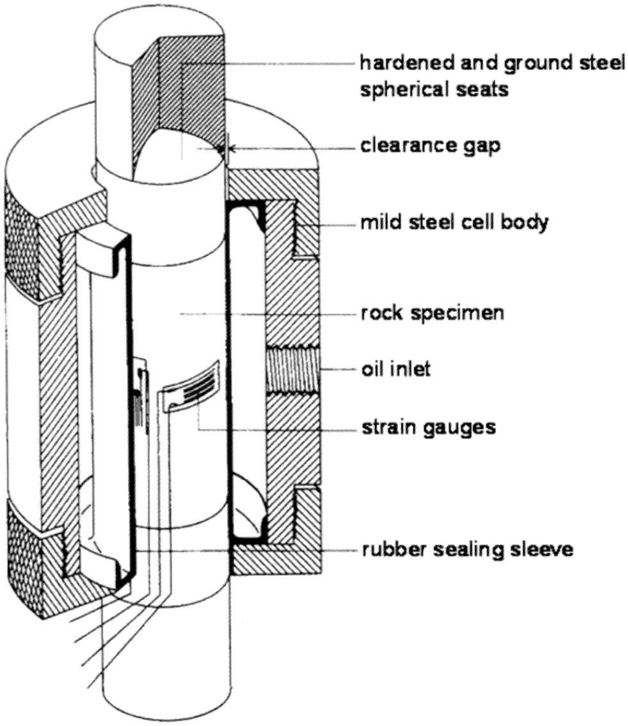

Figure 5.6: Cut-away view of the rock triaxial cell designed by Hoek & Franklin (1968)
Source: Brady & Brown (2004)

surface. A jacket, usually made of a rubber compound, is used to isolate the rock specimen from the confining fluid. The axial stress, S_1, is applied to the specimen by a ram passing through a bush in the top of the cell and hardened steel caps. Pore pressure, u, may be applied or measured through a duct which generally connects with the specimen through the base of the cell. Axial deformation of the rock specimen may be most conveniently monitored by linear variable differential transformers (LVDTs) mounted inside (preferably) or outside the cell. Local axial and circumferential strains may be measured by electric resistance strain gauges attached to the surface of the rock specimen (Brady & Brown 2004).

The confining pressure is maintained constant and the axial pressure increased until the sample fails. In addition to the friction (Ø) and cohesion (c) values defined by the Mohr failure envelope, the triaxial compression test can provide the following results: the major (S_1) and minor (S_3) principal effective stresses at failure, pore pressures (u), a stress–axial strain curve and a stress–radial strain curve.

Pore pressures are hardly ever measured when testing rock samples. These measurements are very difficult and imprecise in rocks with porosity smaller than 5%. Instead, the samples are usually tested at a moisture content as close to the field condition as possible. They are also loaded slowly enough to prevent excess pore pressures that may generate premature rupture and unrealistically low strength values.

ASTM Designation D2664-95a describes the standard test method for triaxial compressive strength of undrained rock specimens without pore pressure measurements. ISRM (2007) describes the methods suggested by the ISRM for determining the strength of rock in triaxial compression.

For all triaxial compression tests on rock, the following procedures are recommended.

- The maximum confining pressure should range from zero to half of the unconfined compressive strength (σ_c) of the sample. For example, if the value of σ_c is 120 MPa then the maximum confining pressure should not exceed 60 MPa (Hoek & Brown 1997).
- Results should be obtained for at least five different confining pressures, e.g. 5, 10, 20, 40 and 60 MPa if the maximum confining pressure is 60 MPa.
- At least two tests should be carried out for each confining pressure.

5.2.3.4 *Elastic constants, Young's modulus and Poisson's ratio*

As shown in Figure 5.3, the Young's modulus of the specimen varies throughout the loading process and is not a unique constant. This modulus can be defined in several ways, the most common being:

- tangent Young's modulus, E_t, defined as the slope of the stress–strain curve at some fixed percentage, generally 50% of the uniaxial compressive strength;
- average Young's modulus, E_{av}, defined as the average slope of the more-or-less straight line portion of the stress–strain curve;
- secant Young's modulus, E_s, defined as the slope of a straight line joining the origin of the stress–strain curve to a point on the curve at a fixed percentage of the uniaxial compressive strength.

The first definition is the most widely used and in this text it is considered that E is equal to E_t. Corresponding to any value of the Young's modulus, a value of Poisson's ratio may be calculated as:

$$\nu = \frac{-(\Delta\sigma/\Delta\varepsilon_a)}{(\Delta\sigma/\Delta\varepsilon_r)} \quad \text{(eqn 5.17)}$$

where σ is the axial stress, ε_a is the axial strain and ε_r is the radial strain. Because of the axial symmetry of the specimen, the volumetric strain, ε_v, at any stage of the test can be calculated as:

$$\varepsilon_v = \varepsilon_a + 2\varepsilon_r \quad \text{(eqn 5.18)}$$

Table 5.6: Uniaxial compressive strength, Young's modulus and Poisson's ratio for some rocks

Rock	σ_c (MPa)	E (GPa)	ν	Rock	σ_c (MPa)	E (GPa)	ν
Andesite	120–320	30–40	0.20–0.30	Granodiorite	100–200	30–70	0.15–0.30
Amphibolite	250–300	30–90	0.15–0.25	Greywacke	75–220	20–60	0.05–0.15
Anhydrite	80–130	50–85	0.20–0.35	Gypsum	10–40	15–35	0.20–0.35
Basalt	145–355	35–100	0.20–0.35	Limestone	50–245	30–65	0.25–0.35
Diabase	240–485	70–100	0.25–0.30	Marble	60–155	30–65	0.25–0.40
Diorite	180–245	25–105	0.25–0.35	Quartzite	200–460	75–90	0.10–0.15
Dolerite	200–330	30–85	0.20–0.35	Sandstone	35–215	10–60	0.10–0.45
Dolomite	85–90	44–51	0.10–0.35	Shale	35–170	5–65	0.20–0.30
Gabbro	210–280	30–65	0.10–0.20	Siltstone	35–250	25–70	0.20–0.25
Gneiss	160–200	40–60	0.20–0.30	Slate	100–180	20–80	0.15–0.35
Granite	140–230	30–75	0.10–0.25	Tuff	10–45	3–20	0.20–0.30

Source: Data selected from Jaeger & Cook (1979), Goodman (1989), Bell (2000), Gonzalez de Vallejo (2002)

The uniaxial compressive strength, Young's modulus and Poisson's ratio for some rocks are given in Table 5.6.

Using the values of E and ν the shear modulus (G) and the bulk modulus (K) of rock can be computed as:

$$G = \frac{E}{2(1+\nu)} \quad \text{(eqn 5.19)}$$

$$K = \frac{E}{3(1-2\nu)} \quad \text{(eqn 5.20)}$$

P-wave and S-wave velocities can be used to calculate the dynamic elastic properties:

$$E_d = \frac{\rho\left(3V_P^2 - 4V_S^2\right)}{\left(\frac{V_P^2}{V_S^2} - 1\right)} \quad \text{(eqn 5.21)}$$

$$G_d = \rho V_S^2 \quad \text{(eqn 5.22)}$$

$$\nu_d = \frac{\left(\frac{V_P^2}{2V_S^2} - 1\right)}{\left(\frac{V_P^2}{V_S^2} - 1\right)} \quad \text{(eqn 5.23)}$$

where ρ is the rock density, E_d is the dynamic Young's modulus, G_d is the dynamic shear modulus and ν_d is the dynamic Poisson's ratio. Typically E_d is larger than E and the ratio E_d/E varies from 1 to 3. Some correlations between E and E_d have been derived for different rock types, as shown in Table 5.7.

Moisture content can have a large effect on the compressibility of some rocks, decreasing E with increasing water content. Vasarhelyi (2003, 2005) indicated that the ratio between E in saturated and dry conditions is about 0.75 for some British sandstones and about 0.65 for some British Miocene limestones. In the case of clayey rocks or rocks with argillic alteration the effect could be larger.

Table 5.7: Correlation between static (E) and dynamic (E_d) Young's modulus of rock

Correlation	Rock type	Reference
$E = 1.137 \times E_d - 9.685$	Granite	Belikov et al. (1970)
$E = 1.263 \times E_d - 29.5$	Igneous and metamorphic rocks	King (1983)
$E = 0.64 \times E_d - 0.32$	Different rocks	Eissa & Kazi (1988)
$E = 0.69 \times E_d + 6.40$	Granite	McCann & Entwisle (1992)
$E = 0.48 \times E_d - 3.26$	Crystalline rocks	McCann & Entwisle (1992)

Both E and E_d are in GPa units
Source: Zhang (2005)

A number of classifications featuring rock uniaxial compressive strength and Young's modulus have been proposed. Probably the most used is the strength-modulus classification proposed by Deere and Miller (1966). This classification is shown in Figure 5.7 and defines rock classes in terms of the uniaxial compressive strength and the modulus ratio, E/σ_c:

- if $E/\sigma_c < 200$, the rock has a low modulus ratio (L region in Figure 5.7);
- if $200 \le E/\sigma_c \le 500$, the rock has a medium modulus ratio (M region in Figure 5.7);
- if $500 < E/\sigma_c$, the rock has a high modulus ratio (H region in chart of Figure 5.7)

5.2.4 Special conditions

5.2.4.1 Weak rocks and residual soils

Slopes containing highly weathered and altered rocks, argillic rocks and residual soils such as saprolites may fail in a 'soil-like' manner rather than a 'rock-like' manner. In

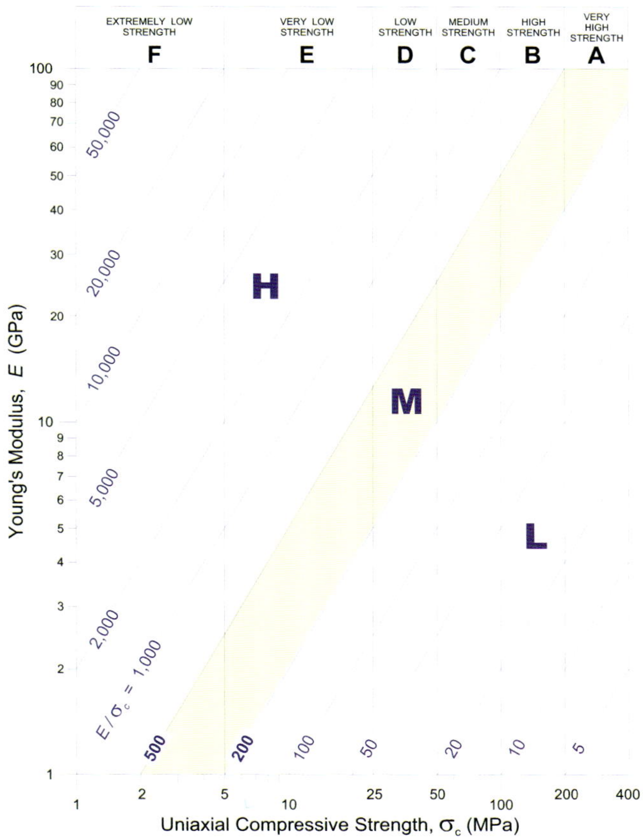

Figure 5.7: Rock classification in terms of uniaxial compressive strength and Young's modulus
Source: Modified from Deere & Miller (1966)

these cases the testing procedures outlined above may not be adequate, especially if the rock has high moisture content. If so, it may be necessary to perform soil-type tests that take account of pore pressures and effective stresses rather than rock-type tests. The sampling and testing decisions must be cognisant of the nature of the parent material and the climatic conditions at the project site. When planning the investigation, the following points must be kept in mind.

1 Usually, soil slope stability analyses are effective stress analyses. Effective stress analyses assume that the material is fully consolidated and at equilibrium with the existing stress system and that failure occurs when, for some reason, additional stresses are applied quickly and little or no drainage occurs. Typically, the additional stresses are pore pressures generated by sudden or prolonged rainfall. For these analyses the appropriate laboratory strength test is the consolidated undrained (CU) triaxial test, during which pore pressures are measured (Holtz & Kovacs 1981).
2 Classical soil mechanics theory and laboratory testing procedures have been developed almost exclusively using transported materials that have lost their original form. In contrast, residual soils frequently retain some features of the parent rock from which they were derived. Notably, these can include relict structures and anomalous void ratios brought on by cemented bonds in the parent rock matrix preventing changes associated with loading and unloading or by the leaching of particular elements from the matrix.
3 In situations where the stability analyses have been performed simply on the basis of 'representative' CU triaxial test results, persistent relict structures in residual or highly weathered and hydrothermally (argillic) altered profiles can and frequently have provided unexpected sources of instability, especially in wet tropical climates. Although relict structures can be difficult to recognise, even if only part of the slope is comprised of a residual or highly weathered and/or altered profile, they should be sought out and characterised. They may have lower shear strengths than the surrounding soils and may promote the inflow of water into the slope. Hence, common sense dictates that they must be accounted for.
4 High void ratio, collapsible materials such as saprolites, leached, soft iron ore deposits and fine-grained rubblised rock masses invariably raise the issue of rapid strain softening, which can lead to sudden collapse if there are rapid positive or negative changes in stress. Sudden transient increases in pore pressure can also lead to rapid failure, a condition known as static liquefaction.
5 Another peculiarity of materials with high void ratios (e.g. saprolites), which should not be overlooked, is the effect of soil suction on the effective stress and available shear strength. With saprolites, strong negative pore pressures (soil suction) are developed when the saturation falls below about 85%, which explains why many saprolite slopes remain stable at slope angles and heights greater than would be expected from a routine effective stress analysis. It also explains why these slopes may fail after prolonged rainfall even without the development of excess pore pressures. Without necessarily reaching 100%, the associated increase in the moisture content can reduce the soil suction, reducing the additional strength component and resulting in slope failure (Fourie & Haines 2007).
6 Sampling of weak rocks and high void ratio soil materials should be planned and executed with great care. For these types of material, high-quality block samples rather than thin-walled tube samples should be considered in order to reduce the effects of compressive strains and consequent disturbance of the sample.
7 Particular care also needs to be taken when preparing argillic, saprolitic and halloysite-bearing volcanic soils and/or weathered and altered rocks for Atterberg Limits tests (Table 2.7). Oven-drying of these materials

Figure 5.8: Degradation test of exposed core
Source: Courtesy Anglo Chile Ltda

can change the structure of the clay minerals, which will provide incorrect test results. This can be avoided if the samples are air-dried.

5.2.4.2 Degradable rocks

Certain materials degrade when exposed to air and/or water. These include clay-rich, low-strength materials such as smectitic shales and fault gouge and some kimberlites.

Standard tests of degradability such as slake durability and static durability can indicate the susceptibility of these materials to degradation. However, it is has been found that simply leaving core samples exposed to the elements is a direct and practical way of assessing degradability (see Figure 5.8). This information is required to establish catch bench design requirements (Chapter 10, section 10.2.1).

Where there is a high gypsum or anhydrite content in the rock mass, the potential for the solution of these minerals and consequent degradation must be considered when assessing its long-term strength.

5.2.4.3 Permafrost

Slope stability is typically improved where the rock mass is permanently frozen. However, in thawing conditions, the active layer will be weakened. Hence, for design purposes in permafrost environments it is necessary to determine the shear strength parameters (friction and cohesion) and moisture content for the rock and soil units in both the frozen and unfrozen states. It is also necessary to know:

- the thickness and depth of the frozen zone, including the thickness and depth of the active freeze and thaw layer;
- the ice content, whether rich or poor;
- the annual and monthly air temperatures – differences in the annual and monthly air temperatures lead to different permafrost behaviour in different regions;
- nearby water flow that can damage the permafrost;
- the snow cover and precipitation;
- the geothermal gradient;
- how the ice behaves at the free surface – whether it melts and flows, or stays in place.

Strength testing of permafrost materials requires specialised handling, storage and laboratory facilities. The samples must be maintained in a frozen state from collection to testing.

5.3 Strength of structural defects

5.3.1 Terminology and classification

A structural defect includes any mechanical defect in a rock mass that has zero or low tensile strength. This includes defects such as joints, faults, bedding planes, schistosity planes and weathered or altered zones.

Recommended terms for defect spacing and aperture (thickness) are given in Chapter 2, Tables 2.4 and 2.5. A recommended classification system designed specifically to enable relevant and consistent engineering descriptions of defects is given in Chapter 2, Table 2.6. Note that the terminology used in Table 2.6 describes the actual defect, not the process that formed or might have formed it. The materials contained within the defects are described using the Unified Soils Classification System (ASTM D2487; Chapter 2, Table 2.7).

5.3.2 Defect strength

In open pit slope engineering, the most commonly used defect properties are the Mohr-Coulomb shear parameters of the defect (friction angle, ϕ, and cohesion, c). For numerical modelling purposes the stiffness of the defects must be also be assessed. Comprehensive discussions of how these parameters are determined and applied in rock slope engineering and underground can be found in Goodman (1976), Barton and Choubey (1977), Barton (1987), Bandis (1990), Wittke (1990), Bandis (1993), Priest (1993), Hoek (2002) and Wyllie and Mah (2004).

Shear strength can be measured by laboratory and in situ tests, assessed from back-analyses of structurally controlled failures or assessed from a number of empirical methods. Both laboratory and in situ tests have the problem of scale effects as the surface area tested is usually much smaller than the one that could occur in the field. On the other hand, back-analyses of structurally controlled slope instabilities require a very careful interpretation of the conditions that trigger the failure, and judgment to assess the most probable value for the shear strength parameters. Values assessed from empirical methods also require careful evaluation and judgment.

5.3.2.1 Measuring shear strength

The shear strength of smooth discontinuities can be evaluated using the Mohr-Coulomb failure criterion, in which the peak shear strength is given by:

$$\tau_{max} = c_j + \sigma_n \tan \phi_j \quad \text{(eqn 5.24)}$$

where ϕ_j and c_j are the friction angle and the cohesion of the discontinuity for the peak strength condition (representing the peak value of the shear stress for a given confining pressure, which usually takes place at small displacements in the plane of the structure) and σ_n is the average value of the normal effective stress acting on the plane of the structure. The criterion is illustrated in Figure 5.9.

In a residual condition, or when the peak strength has been exceeded and relevant displacements have taken place in the plane of the structure, the shear strength is given by:

$$\tau_{res} = c_{jres} + \sigma_n \tan \phi_{jres} \quad \text{(eqn 5.25)}$$

where ϕ_{jres} and c_{jres} are the friction angle and the cohesion for the residual condition, and σ_n is the mean value of the effective normal stress acting on the plane of the structure. It must be pointed out that in most cases c_{jres} is small or zero, which means that:

$$\tau_{res} = \sigma_n \tan \phi_{jres} \quad \text{(eqn 5.26)}$$

ASTM Designation D4554-90 (reapproved 1995) describes the standard test method for the in situ determination of direct shear strength of rock defects and ASTM Designation D5607-95 described the standard test method for performing laboratory direct shear strength tests of rock specimens that contain defects. ISRM (2007) described the methods suggested by the ISRM for determining direct shear strength in the laboratory and in situ.

Ideally, shear strength testing should be done by large-scale in situ testing on isolated discontinuities, but these tests are expensive and not commonly carried out. In addition to the high cost, the following factors often preclude in situ direct shear testing (Simons et al. 2001):

- exposing the test discontinuity;
- providing a suitable reaction for the application of the normal and shear loads;
- ensuring that the normal stress is maintained safely as shear displacement takes place.

The alternative is to carry out laboratory direct shear tests. However, it is not possible to test representative samples of discontinuities in the laboratory and a scale effect is unavoidable. Nevertheless, the defect's basic friction angle (ϕ_b) can be measured on saw cut discontinuities using laboratory direct shear tests.

Sometimes the direct shear box equipment used for testing soil specimens is used for testing rock specimens containing discontinuities, but testing with these machines has the following disadvantages (Simons et al., 2001):

- difficulty in mounting rock discontinuity specimens in the apparatus;

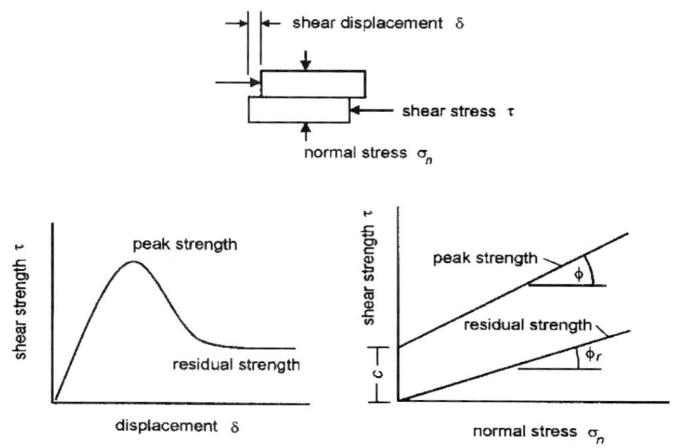

Figure 5.9: Mohr-Coulomb shear strength of defects from direct shear tests
Source: Hoek (2002)

- difficulty maintaining the necessary clearances between the upper and lower halves of the box during shearing;
- the load capacity of most machines designed for testing soils is likely to be inadequate for rock testing.

The most commonly used device for direct shear testing of discontinuities is a portable direct shear box (see Figure 5.10). Although very versatile, this device has the following problems (Simons et al. 2001):

- the normal load is applied through a hydraulic jack on the upper box and acts against a cable loop attached to the lower box. This system results in the normal load increasing in response to dilation of rough discontinui-

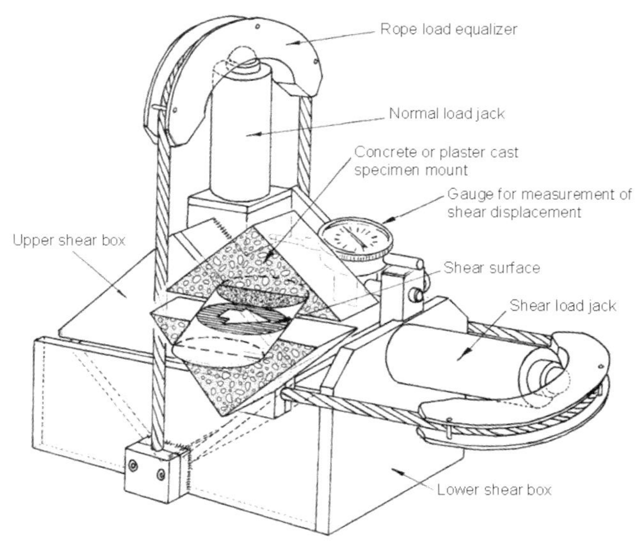

Figure 5.10: Portable direct shear equipment showing the position of the specimen and the shear surface
Source: Hoek & Bray (1981)

ties during shear. Adjustment of the normal load is required throughout the test;
- as the shear displacements increase the applied normal load moves away from the vertical and corrections for this may be required;
- the constraints on horizontal and vertical movement during shearing are such that displacements need to be measured at a relatively large number of locations if accurate shear and normal displacements are required;
- the shear box is somewhat insensitive and difficult to use with the relatively low applied stresses in most slope stability applications since it was designed to operate over a range of normal stresses from 0 to 154 MPa.

The direct shear testing equipment used by Hencher and Richards (1982) (see Figure 5.11) is more suitable for direct shear testing of discontinuities. The equipment is portable and can be used in the field. It is capable of testing specimens up to about 75 mm (i.e. NQ and HQ drill core).

The typical direct shear test procedure consists of using plaster to set the two halves of the specimen in a pair of steel boxes. Particular care is taken to ensure that the two pieces are in their original matched position and the discontinuity is parallel to the direction of the shear load.

A constant normal load is then applied using the cantilever, and the shear load gradually increased until sliding failure occurs. Measurement of the vertical and horizontal displacements of the upper block relative to the lower one can be made with dial gauges, but more precise and continuous measurements can be made with linear variable differential transformers (LVDTs) (Hencher & Richards 1989).

Where the natural fractures are coated with a clay infilling or there is significant clay alteration,

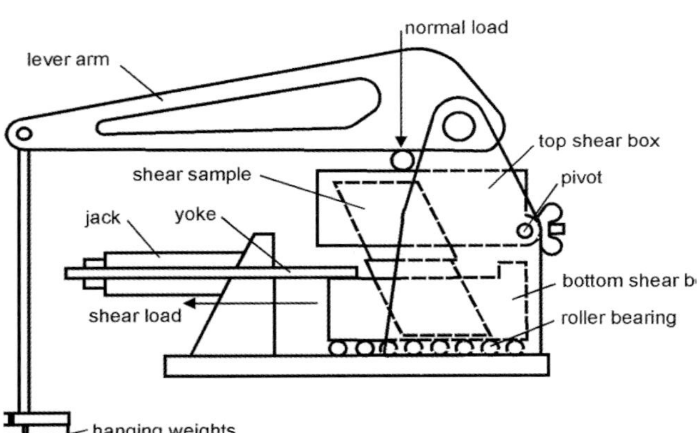

Figure 5.11: Direct shear equipment of the type used by Hencher and Richards (1982) for direct shear testing of defects
Source: Hoek (2002)

consideration should be given to performing the tests saturated. This would, however, require special apparatus.

A common practice is to test each specimen three or four times at progressively higher normal loads. When the residual shear stress has been established for a normal load the specimen is reset, the normal load increased and another direct shear tests is conducted. It must be pointed out that this multi-stage testing procedure has a cumulative damage effect on the defect surface and may not be appropriate for non-smooth defects.

The test results are usually expressed as shear displacement–shear stress curves from which the peak and residual shear stress values are determined. Each test produces a pair of shear (τ) and effective normal (σ_n) values, which are plotted to define the strength of the defect, usually as a Mohr-Coulomb failure criterion. Figure 5.12 shows a typical result of a direct shear test on a discontinuity, in this case with a 4 mm thick sandy silt infill.

It should be noted that although the Mohr-Coulomb criterion is the most commonly used in practice, it ignores the non-linearity of the shear strength failure envelope. To be valid, the shear strength parameters should be done for a range of normal stresses corresponding to the field condition. For this reason, special care must be taken when considering the 'typical' values reported in the geotechnical literature because, if

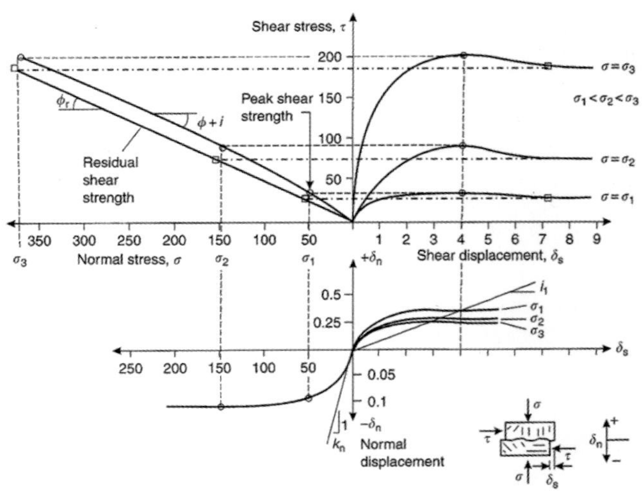

Figure 5.12: Results of a direct shear test on a defect (a 4 mm thick sandy silt infill). The shear displacement–shear stress curves on the upper right show an approximate peak shear stress as well as a slightly lower residual shear stress. The normal stress–shear stress curves on the upper left show the peak and residual shear strength envelopes. The shear displacement–normal displacement on the lower right show the dilatancy caused by the roughness of the discontinuity. The normal stress–normal displacement curves on the lower left show the closure of the discontinuity and allow the computation of its normal stiffness
Source: Modified from Erban & Gill (1988) by Wyllie & Norrish (1996)

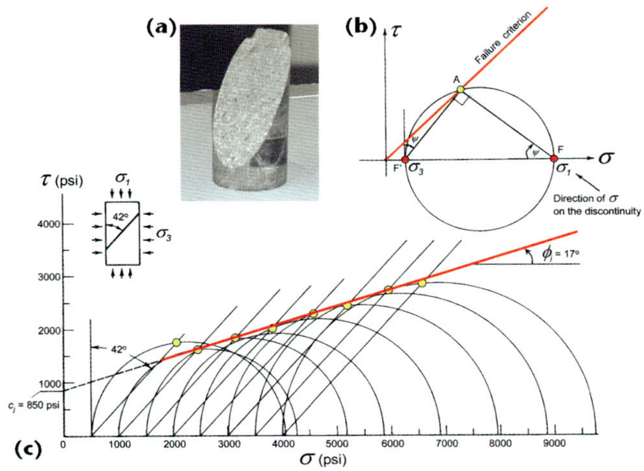

Figure 5.13: Use of triaxial compression test to define the shear strength of veins or other defects with strong infills
Source: Modified from Goodman (1989)

these values have been determined for a range of normal stresses different from the case being studied, they might be not applicable. It must be noted that many of the 'typical' values mentioned in the geotechnical literature correspond to open structures or structures with soft/weak fillings under low normal stresses. Though these 'typical' values may be useful in the case of rock slopes they may not be applicable to the case of underground mining, where the confining stresses are substantially larger than in open pit slopes.

When calculating the contact area of the defect an allowance must be made for the decrease in area as shear displacements take place. In inclined drill-core specimens the discontinuity surface has the shape of an ellipse, and the formula for calculating the contact area is as follows (Hencher & Richards 1989):

$$A_C = ab\pi - \sqrt{\left(\frac{\delta_s b(4a^2 - \delta_s^2)}{2a}\right)} - 2ab\sin^{-1}\left(\frac{\delta_s}{2a}\right) \quad \text{(eqn 5.27)}$$

where A_c is the contact area, $2a$ and $2b$ are the major and minor axes of the ellipse and δ_s is the relative shear displacement.

Triaxial compression testing of drill-core containing defects can be used to determine the shear strength of veins and other defects infills using the procedure described by Goodman (1989). If the failure plane is defined by a defect (Figure 5.13a), the normal and shear stresses on the failure plane can be computed using the pole of the Mohr circle (Figure 5.13b). If this procedure is applied, the results of several tests allow the cohesion (c_j) and friction angle ($\varnothing_j$) of the defect to be determined (Figure 5.13c).

5.3.2.2 Influence of infilling

The presence of infillings can have a very significant impact on the strength of defects. It is important that infillings be identified and appropriate strength parameters used for slope stability analysis and design. The effect of infilling on shear strength will depend on the thickness and the mechanical properties of the infilling material.

The results of direct shear tests on filled discontinuities are shown in Figure 5.14. These results show that the infillings can be divided into two groups (Wyllie & Norrish 1996).

1 Clays: montmorillonite and bentonitic clays, and clays associated with coal measures have friction angles ranging from about 8° to 20°, and cohesion values ranging from 0 kPa to about 200 kPa (some cohesion values were measured as high as 380 kPa, probably associated with very stiff clays).
2 Faults, sheared zones and breccias: the material formed in faults and sheared zones in rocks such as granite, diorite, basalt and limestone may contain clay in addition to granular fragments. These materials have friction angles ranging from about 25° to 45° and cohesion values ranging from 0 kPa to about 100 kPa. Crushed material found in faults (fault gouge) derived from coarse-grained rocks such as granites tend to have higher friction angles than those from fine-grained rocks such as limestones.

The higher friction angles found in the coarser-grained rocks reflect the frictional attributes of non-cohesive materials, which can be summarised as follows:

- in drained direct shear or triaxial tests, the higher the density (i.e. the lower the void ratio) the higher the shear strength;
- with all else held constant, the friction angle increases with increasing particle angularity;
- at the same density, the better-graded soil (e.g. SW rather than SP) has a higher friction angle.

Figure 5.15, prepared by the US Navy (1971), presents correlations between the effective friction angle in triaxial compression and the dry density and relative density of non-cohesive soils as classified by the Unified Soils Classification System (Chapter 2, Table 2.7).

Some of the tests shown in Figure 5.14 also determined residual shear strength values. The tests showed that the residual friction angle was only about 2–4° less than the peak friction angle, while the residual cohesion was essentially zero. Figure 5.16 shows an approximate relationship between the residual friction angle and the plasticity index (PI) of clayey crushed rock (gouge) from a fault. Figure 5.17 shows an empirical correlation between the effective friction angle and the plasticity index of normally consolidated undisturbed clays.

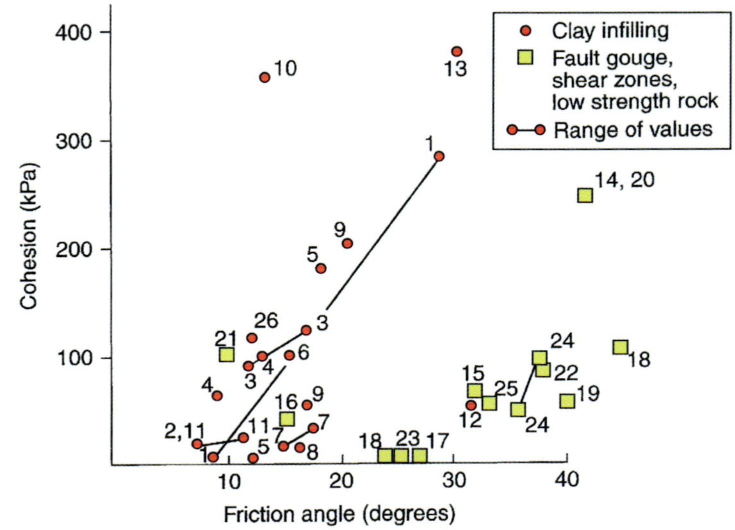

1 Bentonite shale
2 Bentonite seams in chalk
3 Bentonite; thin layers
4 Bentonite; triaxial tests
5 Clay; overconsolidated
6 Limestone; 10-20 mm clay infillings
7 Lignite and underlying clay contact
8 Coal measures; clay mylonite seams
9 Limestone; < 1 mm clay infillings
10 Montmorillonite clay
11 Montmorillonite; 80 mm clay seam in chalk
12 Schists/quartzites; stratification, thick clay
13 Schists/quartzites; stratification, thick clay
14 Basalt; clayey, basaltic breccia
15 Clay shale; triaxial tests
16 Dolomite; altered shale bed
17 Diorite/granodiorite; clay gouge
18 Granite; clay-filled faults
19 Granite; sandy-loam fault filling
20 Granite; shear zone, rock and gouge
21 Lignite/marl contact
22 Limestone/marl/lignites; lignite layers
23 Limestone; marlaceous joints
24 Quartz/kaolin/pyrolusite; remolded triaxial
25 Slates; finely laminated and altered
26 Limestone; 10-20 mm clay infillings

Figure 5.14: Peak shear strength of filled discontinuities
Source: Originally from Barton (1974), modified by Wyllie (1992)

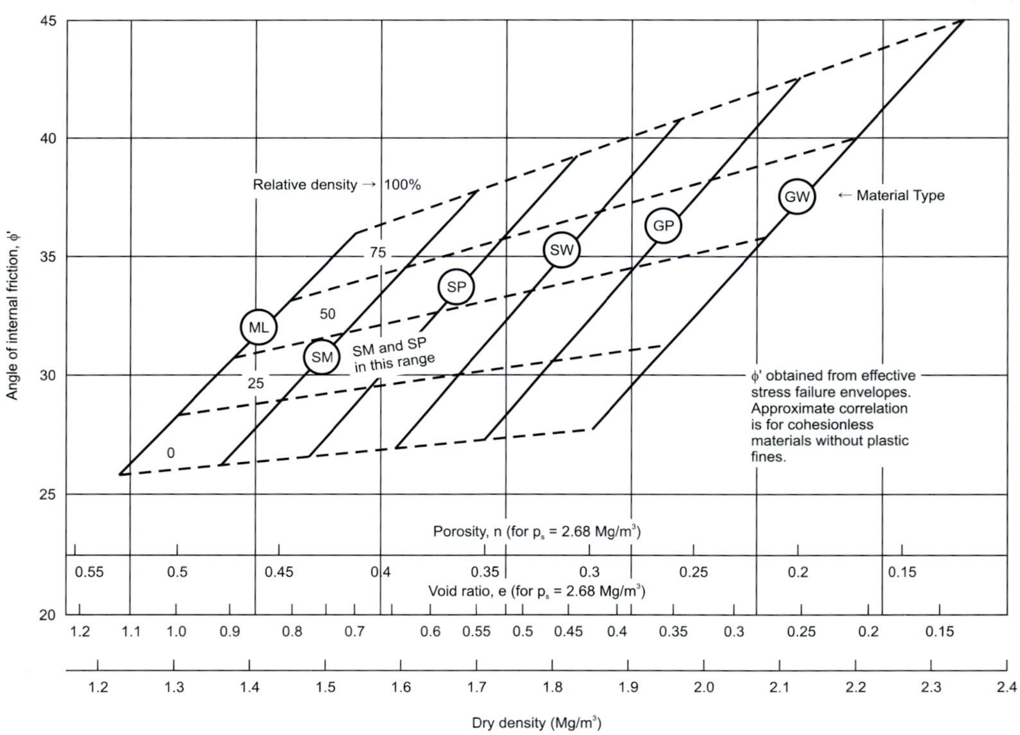

Figure 5.15: Correlations between the effective friction angle in triaxial compression and the dry density and relative density of non-cohesive soils
Source: US Navy (1971)

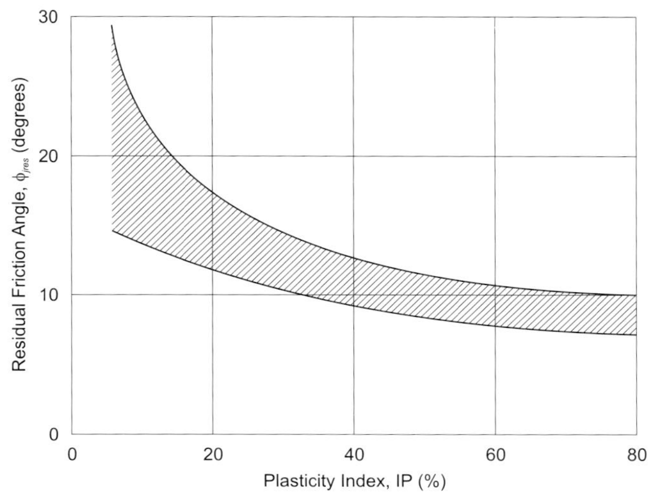

Figure 5.16: Approximate relationship between the residual friction angle (drained tests) and the plasticity index of crushed rock material (gouge) from a fault
Source: From Patton & Hendron (1974) and Kanji (1970)

A comparative list of the shear strength values of defects without infills, with thin to medium infills and with thick crushed material from faults (gouge) is provided in Tables 5.8, 5.9 and 5.10.

5.3.2.3 *Effect of defect displacement*

Wyllie and Norrish (1996) indicated that the shear strength-displacement behaviour is an additional factor to consider regarding the shear strength of filled discontinuities. In cases where there is a significant decrease in shear strength with displacement, slope failure can occur suddenly following a small amount of movement.

Barton (1974) indicated that filled discontinuities can be divided into two general categories, depending on any previous displacement of the discontinuity. These categories can be further subdivided into normally consolidated (NC) or overconsolidated (OC) materials (Figure 5.18).

Recently displaced discontinuities include faults, sheared zones, clay mylonites and bedding-surface slips. In faults and sheared zones the infilling is formed by the shearing process that may have occurred many times and produced considerable displacement. The crushed material (gouge) formed in this process may include both clay-size particles, and breccia with the particle orientations and striations of the breccia aligned parallel to the direction of shearing. In contrast, the mylonites and bedding-surface slips are defects that were originally clay-bearing and along which sliding occurred during folding or faulting. The shear strength of recently displaced discontinuities will be at, or close to, the residual strength (Graph I in Figure 5.18). Any cohesive bonds that existed in the clay due to previous overconsolidation will have been destroyed by shearing and the infilling will be equivalent to a normally consolidated (NC) material. In addition,

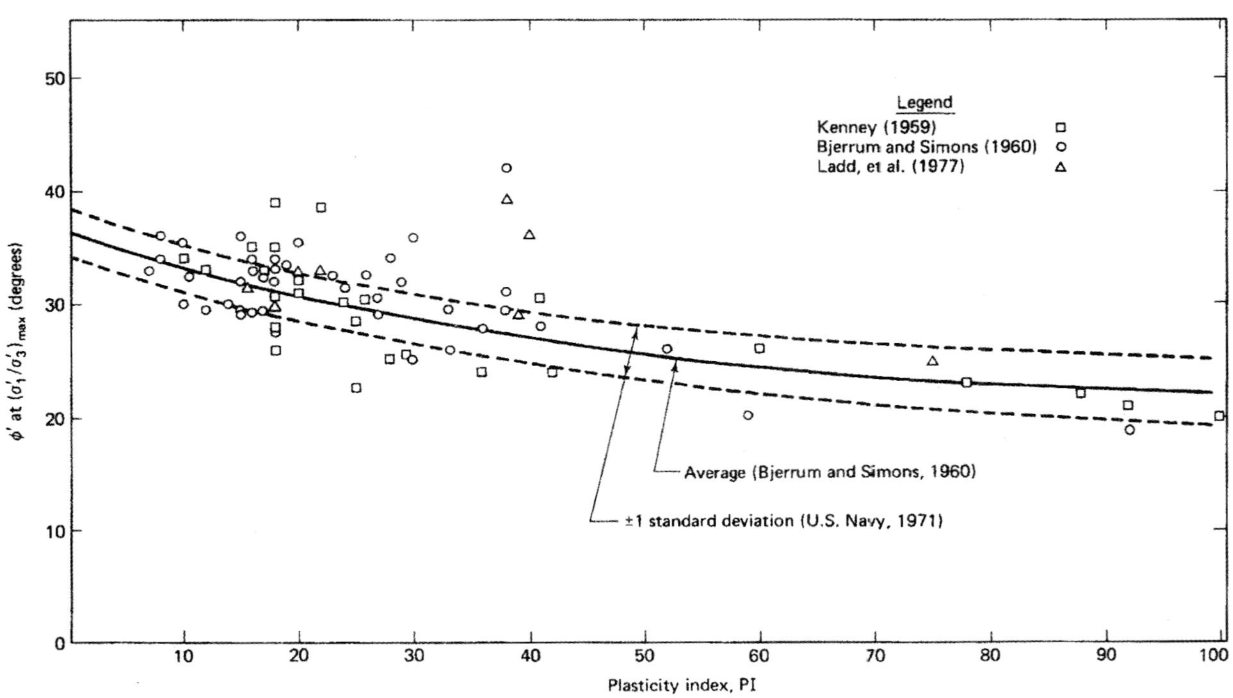

Figure 5.17: Empirical correlation between effective friction angle and plasticity index from triaxial tests on normally consolidated clays
Source: Holtz & Kovacs (1981)

Table 5.8: Shear strength of some structures without infill material

	Shear strength					
	Peak		Residual			
Rock wall/filling material	ϕ_j (°)	c_j (kPa)	ϕ_{jres} (°)	c_{jres} (kPa)	Comments	Reference
1: Structures without infills						
Crystalline limestone			42–49	0	LT (σ_n < 4 MPa?)	Franklin & Dusseault (1989)
Porous limestone			32–48	0		
Chalk			30–41	0		
Sandstones	32–37	120–660	24–35	0		
Siltstones	20–33	100–790				
Soft shales	15–39	0–460				
Shales			22–37	0		
Schists			32–40	0		
Quartzites			23–44	0		
Fine-grained igneous rocks			33–52	0		
Coarse-grained igneous rocks			31–48	0		
Basalt			40–42	0	DST-H (σ_n < 4 MPa?)	Giani (1992)
Calcite			40–42	0		
Hard sandstone			34–36	0		
Dolomite			30–38	0		
Schists			21–36	0		
Gypsum			34–35	0		
Micaceous quartzite			38–40	0		
Gneiss			39–41	0		
Copper porphyry	45–60	0			BA of bench failures at Chuquicamata	
Granite	45–50	1000–2000			IS (σ_n < 3 MPa?)	Lama & Vutukuri (1978)
Joint in biotitic schist	37–43	0			BA (DA: 120 × 100 m)	McMahon (1985)
Joint in quartzite	34–38	0			BA (DA: 20 × 10 m)	

LT Laboratory tests
DST-H Direct shear tests using a Hoek shear cell or similar
BA Back analysis of structurally controlled instabilities
DA Areal extent of the shear surface considered in the back analysis
IS In situ direct shear tests
PI Plasticity index of the clay
Source: Flores & Karzulovic (2003)

strain-softening may occur with any increase in water content, resulting in a further strength reduction (Wyllie & Mah 2004).

Undisplaced discontinuities that are infilled and have undergone no previous displacement include igneous and metamorphic rocks that have weathered along the discontinuity to form a clay layer. For example, diabase can weather to amphibolite and eventually to clay. Other undisplaced discontinuities include thin beds of clay and weak shales that are found with sandstone in interbedded sedimentary formations. Hydrothermal alteration is another process that forms infillings that can include low-strength materials such as montmorillonite, and high-strength materials such as quartz and calcite. The infillings of undisplaced discontinuities can be divided into NC and OC materials that have significant differences in peak strength (Graphs II and III in Figure 5.18). While the peak strength of OC clay infillings may be high, there can be a significant loss of strength due to softening, swelling and pore pressure changes on unloading. Strength loss also occurs on displacement in brittle materials such as calcite (Wyllie & Mah 2004).

5.3.2.4 Effect of surface roughness

In the case of clean rough defects, the roughness increases the friction angle. This was shown by Patton (1966), who

Table 5.9: Shear strength of some structures with thin to medium thick infill material

Rock wall/filling material	Shear strength: Peak ϕ_j (°)	Peak c_j (kPa)	Residual ϕ_{jres} (°)	Residual c_{jres} (kPa)	Comments	Reference
2: Structures with thin to medium thickness infills						
Bedding plane in layered sandstone and siltstone	12–14	0			BA (DA: 250 × 100 m)	McMahon (1985)
Bedding plane containing clay in a weathered shale	14–16	0			BA (DA: 30 × 30 m)	
Bedding plane containing clay in a soft shale	20–24	0			BA (DA: 200 × 600m)	
Bedding plane containing clay in a soft shale	17–21	0			BA (DA: 120 × 180 m)	
Bedding plane containing clay in a shale	19–27	0			BA (SD: 80 × 60 m)	
Foliation plane with chlorite coating in a chloritic schist	33–36	0			BA (DA: 120 × 100 m)	
Structure in basalt with fillings containing broken rock and clay	42	237			IS (σ_n: 0–2.5 MPa)	Barton (1987)
Shear zone in granite, with brecciated rock and clay gouge	45	254			IS (σ_n: 0.3-0.7 MPa)	
Bedding planes with a clay coating in a quartzite schist	41	725			IS (σ_n: 0.3-0.9 MPa)	
Bedding planes with a clay coating in a quartzite schist	41	598			IS (σ_n: 0.5-1.1 MPa)	
Bedding planes with centimetric clay fillings in a quartzite schist	31	372			IS (σ_n: 0.2-0.4 MPa)	
Limestone joint with clay coatings (<1 mm)	21–17	49–196			IS (σ_n: 0.1-2.5 MPa)	
Limestone joint with millimetric clay fillings	13–14	98				
Greywacke bedding plane with clay filling (1–2 mm)			21	0	IS (σ_n: 0-2.5 MPa)	
Clay veins (1–2.5 cm) in coal	16	12	11–12	0	IS (σ_n < 3 MPa?)	
Laminated and altered schists containing clay coatings	33	50				

LT Laboratory tests
DST-H Direct shear tests using a Hoek shear cell or similar
BA Back analysis of structurally controlled instabilities
DA Areal extent of the shear surface considered in the back analysis
IS In situ direct shear tests
PI Plasticity index of the clay
Source: Flores & Karzulovic (2003)

studied bedding plane traces in unstable limestone slopes and demonstrated that the rougher the bedding plane the steeper the slope (Figure 5.19).

Based on experimental data for shear of model joints with regular teeth, Patton proposed the following bilinear failure criterion for rough discontinuities:

$$\tau_{max} = \sigma_n \tan(\phi_b + i) \text{ if } \sigma_n \le \sigma_{ny} \quad \text{(eqn 5.28a)}$$

$$\tau_{max} = c_{jeq} + \sigma_n \tan(\phi_{jres}) \text{ if } \sigma_n \ge \sigma_{ny} \quad \text{(eqn 5.28b)}$$

where ϕ_b is the basic friction angle of a planar rock surface, i is the angle of inclination of the failure surface with respect to the direction of the shear force or roughness angle, ϕ_{jres} is the residual friction angle of the discontinuity, σ_{ny} is the effective normal stress that causes the yielding of the asperities, and c_{jeq} is the shear strength intercept derived from the asperities which defines a kind of 'equivalent' cohesion for the defect (Figure 5.20).

Patton (1966) suggested that asperities can be divided into first- and second-order asperities. First-order asperities are those corresponding to major undulations of the discontinuity. They exhibit wavelengths larger than 0.5 m and roughness angles of not more than about 10–15° (Figure 5.21).

Second-order asperities are those corresponding to small bumps and ripples of the discontinuity with wavelengths smaller than 0.1 m and roughness angles as high as 20–30° (Figure 5.21). Patton (1966) indicated that only first-order asperities have to be considered to obtain reasonable agreement with field observations, but Barton (1973) showed that at low normal stresses second-order asperities also come into play.

Table 5.10: Shear strength of crushed material (gouge) from some faults

	Shear strength					
	Peak		Residual			
Rock wall/filling material	ϕ_j (°)	c_j (kPa)	ϕ_{jres} (°)	c_{jres} (kPa)	**Comments**	**Reference**
3: Structures with thick clay gouge fillings (strength defined by gouge material)						
Smectites			5–10	0	LT (σn < 4 MPa?)	Franklin & Dusseault (1989)
Kaolinites			12–15	0		
Illites			16–22	0		
Chlorites			16–22	0		
Clays with IP < 20%			12–28	0	Correlation with the results of laboratory and in situ testing	Hunt (1986)
Clays with 20% < PI < 40%			9–16	0		
Clays with 40% < PI < 60%			8–14	0		
Clays with IP > 60%			7–12	0		
Smooth concrete and clay filling	9–16	240–425			LT (direct shear test)	Potyondy (1961)
Bentonite	9–13	60–100			LT (triaxial tests)	Barton (1974)
Consolidated clay fillings	12–19	0–180	10–16	0–3		
Limestone joint with clay filling (6 cm)			13	0	IS (σ_n: 0.8-2.5 MPa)	Barton (1987)
Shales with clay layers (10–15 cm)	32	78			IS (σ_n: 0.3-0.8 MPa)	
Structures in quartzites and siliceous schists with fillings of brecciated rock and clay gouge (10–15 cm)	32	29			IS (σ_n: 0.3-1.1 MPa)	Barton (1987)
Thick bentonite-montmorillonite vein in chalk (8 cm)	7–8	15			IS (σ_n < 1 MPa?)	Barton (1987)
Fault with clay gouge (5–10 cm)	25	75			BA (planar slide)	
4: Structures with thick non-clayey gouge fillings (strength defined by gouge material)						
Portland cement grout			16–22	0	LT (σ_n < 4 MPa?)	Franklin & Dusseault (1989)
Quartz-feldspar sand			28–40	0		
Smooth concrete with compacted silt fillings	40	0			LT (direct shear tests)	Potyondy (1961)
Rough concrete with compacted silt fillings	40	0				
Smooth concrete with dense sand fillings	44	0				
Rough concrete with dense sand fillings	44	0				

LT Laboratory tests
DST-H Direct shear tests using a Hoek shear cell or similar
BA Back analysis of structurally controlled instabilities
DA Areal extent of the shear surface considered in the back analysis
IS In situ direct shear tests
PI Plasticity index of the clay
Source: Flores & Karzulovic (2003)

Wyllie and Norrish (1996) indicated that the actual shear performance of the defects in rock slopes depends on the combined effects of the defect's roughness and wall rock strength, the applied effective normal stress and the amount of shear displacement. This is illustrated in Figure 5.22, where the asperities are sheared off and there is a consequent reduction in the friction angle with increasing normal stress. In other words, there is a transition from dilation to shearing.

The degree to which the asperities are sheared depends on the magnitude of the effective normal stress in relation to the strength of the asperities and the amount of shear displacement. A rough discontinuity that is initially undisturbed and interlocked will have a peak friction angle of $(\varnothing_b + i)$. With increasing normal stress and shear displacement, the asperities will be sheared off and the friction angle will progressively diminish to a minimum residual value. This dilation-shearing behaviour is represented by a curved strength envelope with an initial slope equal to $\tan(\varnothing_b + i)$, reducing to $\tan(\varnothing_{jres})$ at high normal stresses.

Two other important features of non-planar defects must also be considered.

1 In some cases the surface roughness may display a preferred orientation (eg, undulations, slickensides). In

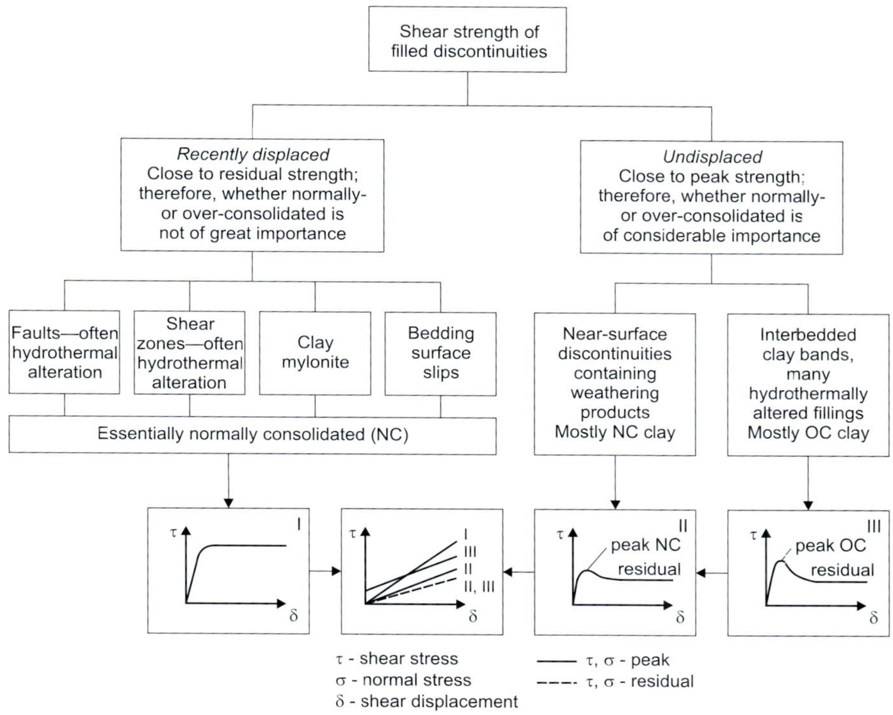

Figure 5.18: Simplified classification of filled defects into displaced and undisplaced, and normally consolidated (NC) and overconsolidated (OC) types of infill material
Source: Modified from Barton (1974) by Wyllie & Norrish (1996)

these cases, the shear strength of the defect will be affected by the direction of sliding, where the shear strength is much greater across the corrugations than along them (Figure 5.23). This effect can be very important in slope stability analyses.

2 The shear strength is affected by how the normal load is applied and how restricted the dilatancy of the defect is. This is discussed in detail by Goodman (1989), who showed that the shear strength of non-planar defects depends on the stress path, due to the interaction between the normal and tangential deformations, the dilatancy and the normal and shear stresses. This is usually ignored in practice. Usually, the shear strength criteria assume that the normal stress remains constant

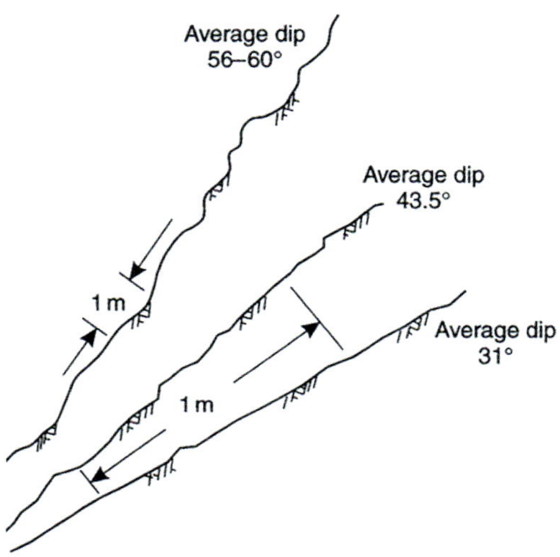

Figure 5.19: Patton's observation of bedding plane traces in unstable limestone slopes
Source: Patton (1966)

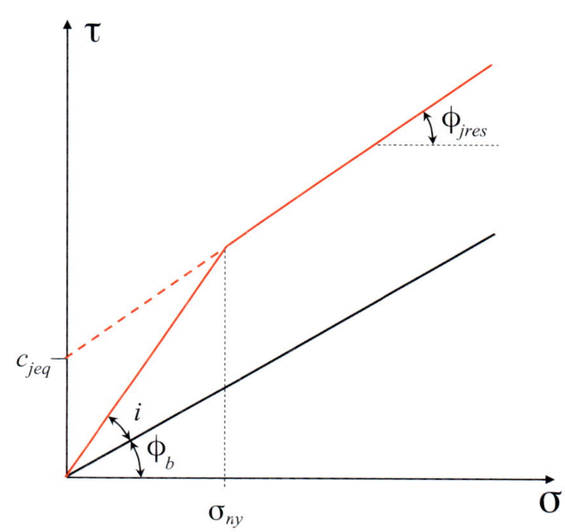

Figure 5.20: Patton's bilinear failure criterion for the shear strength of rough defects

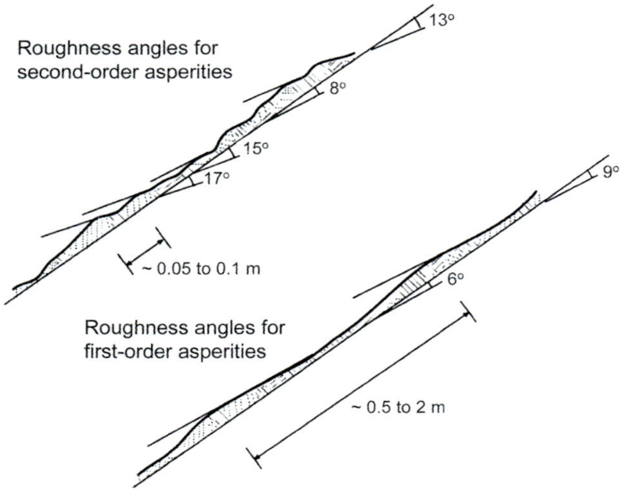

Figure 5.21: Definition of first- and second-order asperities on rough defects
Source: Wyllie & Norrish (1996)

during the shearing process even if the structure is rough. This may be permissible for open pit slopes, where a sliding block does not impose major restrictions on dilatancy. It is not necessarily permissible for an underground mine where there may be heavy restrictions on dilatancy, especially if two of the faces of a potentially instable block are parallel or quasi-parallel.

As a means of taking joint roughness and the wall rock strength into account, Barton and Bandis (1981) suggested that a first estimate of the peak friction angle can be obtained by assuming that:

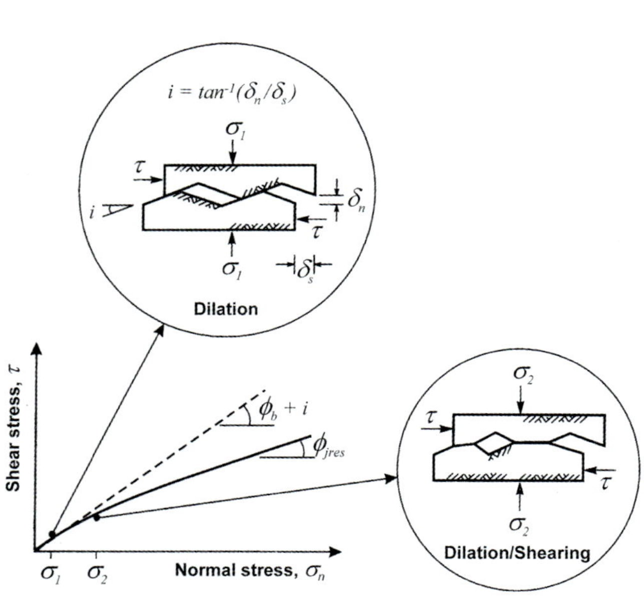

Figure 5.22: Effect of surface roughness and normal stress on the defect's friction angle
Source: Wyllie (1992)

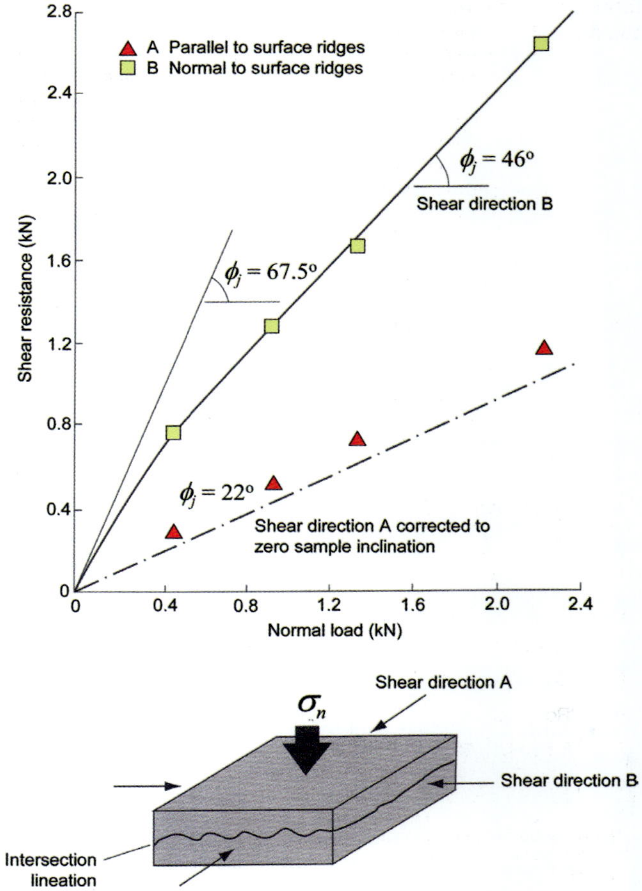

Figure 5.23: Roughness-induced shear strength anisotropy
Source: Simons et al. (2001)

$$\phi_j \approx \tan^{-1}\left(\frac{J_r}{J_a}\right) \quad \text{(eqn 5.29)}$$

where J_r is the joint roughness and J_a is the joint alteration number. Peak friction angle values obtained using this approach are given in Table 5.11 and should be compared with the values for defects either without infill material or with thin to medium thicknesses of infill material given in Tables 5.8 and 5.9.

5.3.2.5 Barton-Bandis failure criterion

Barton (1971, 1973) used the concepts of joint roughness and wall strength to introduce the non-linear empirical Barton-Bandis criterion for the shear strength of the defects in a rock mass. The criterion defines the peak shear strength of a discontinuity as:

$$\tau_{max} = \sigma_n \tan\left(JRC \log_{10}\left(\frac{JCS}{\sigma_n}\right) + \phi_b\right) \quad \text{(eqn 5.30)}$$

where ϕ_b is the basic friction angle, *JRC* is the joint roughness coefficient and *JCS* is the uniaxial compressive strength of the rock wall.

Table 5.11: First estimates of the peak friction angle of defects obtained from the joint roughness number, J_r, and the joint alteration number, J_a

		Joint alteration number, J_a				
		Tightly healed, hard, non-softening, impermeable filling, e.g. quartz or epidote	Unaltered joint walls, surface staining only	Slightly altered joint walls, non-softening mineral coatings, sandy particles, clay-free disintegrated rock etc.	Silty- or sandy-clay coatings, small clay fraction (non-softening)	Softening or low-friction clay mineral coatings, i.e. kaolinite or mica. Also chlorite, talc, gypsum, graphite etc. and small quantities of swelling clays
		A	B	C	D	E
Joint roughness number, J_r				J_a		
Description	J_r	0.75	1	2	3	4
A Discontinuous joints	4	70°		60°	55°	45°
B Rough, undulating joints	3	70°		55°	45°	35°
C Smooth, undulating joints	2	65°	60°	45°	35°	25°
D Slickensided, undulating joints	1.5	60°	55°	35°	25°	20°
E Rough or irregular, planar joints	1.5	60°	55°	35°	25°	20°
F Smooth, planar joints	1.0	50°	45°	25°	18°	15°
G Slickensided, planar joints	0.5	35°	25°	15°	<10°	

Notes
The joint roughness number assumes rock wall contact or rock wall contact before 10 cm of shear displacement.
The descriptions of different cases for J_r refer to small-scale features and intermediate-scale features, in that order.
The joint alteration number assumes rock wall contact.
These are first estimates of peak friction angle and may not be appropriate for site-specific design purposes.

As originally formulated by Barton (1973), the criterion applies only to defects of geological origin, meaning defects that were formed as a consequence of brittle failure. Defects that were subsequently modified by processes such as (a) the passage of mineralising solutions, which left behind a variety of infillings ranging from soft to weak to hard and strong such as clay, talc, gypsum, pyrite and quartz on the defect faces or (b) tectonic events, for example faulting and plastic deformation such as foliation, slaty cleavage and gniessosity, were excluded. The exclusion of all filled defects means that weathering and alteration can only be considered if the rock walls of the defect are still in direct rock/rock contact. The net effect of this exclusion means that the Barton-Bandis criterion cannot be applied to many of the geological environments found in pit slope engineering. Consequently, the criterion must be applied with great caution.

Notwithstanding these limitations, the advantage of the Barton-Bandis criterion is that it includes explicitly the effects of surface roughness, through the parameter *JRC*, and of the magnitude of the normal stress through the ratio (JCS/σ_n). Hence, equation (5.3) can be rewritten as:

$$\tau_{max} = \sigma_n \tan(\phi_j) = \sigma_n \tan(\phi_b + i) \quad \text{(eqn 5.31)}$$

where the friction angle of the defect, ϕ_j, is represented by the basic friction angle, ϕ_b, plus an increment i that depends on the roughness of the discontinuity and the magnitude of the effective normal stress relative to the uniaxial compressive strength of the wall rock. This increment is given by:

$$i = JRC \log_{10}\left(\frac{JCS}{\sigma_n}\right) \quad \text{(eqn 5.32)}$$

The values of roughness and i reach their maximum at low values of σ_n. As σ_n increases, some of the asperities will yield and the effect of roughness will decrease. As σ_n moves towards the value of *JCS*, more asperities yield and the effect of roughness diminishes. Eventually, all the asperities yield and the effect of roughness is totally overcome. When this occurs, ϕ_j equals ϕ_b.

Usually ϕ_b takes values of the order of 30°. The values given in Table 5.12 give a guide for first estimates for

Table 5.12: Typical values of the basic friction angle, ϕ_b, for some rock types

Rock type	ϕ_b dry	ϕ_b wet	Rock type	ϕ_b dry	ϕ_b wet
Amphibolite	32°		Granite, fine-grained	31–35°	29–31°
Basalt	35–38°	31–36°	Granite, coarse-grained	31–35°	31–33°
Chalk		30°	Limestone	31–37°	27–35°
Conglomerate	35°		Sandstone	26–35°	25–34°
Copper porphyry	31°		Schist		27°
Dolomite	31–37°	27–35°	Siltstone	31–33°	27–31°
Gneiss, schistose	26–29°	23–26°	Slate	25–30°	21°

Source: Data from Barton (1973), Barton & Choubrey (1977)

some rock types. In practice, ϕ_b can be determined from simple tilt-table tests or from direct shear tests on saw-cut rock samples.

The joint roughness coefficient, *JRC*, varies from 0° for smooth, planar and slickensided surfaces to as much as 20° for rough undulating surfaces. There are a number of different ways of evaluating *JRC*, but the procedure most widely used is to visually compare the surface condition with standard profiles based on a combination of surface irregularities and waviness using profiles such as those shown in Tables 5.13 and 5.14, or the chart shown in Table 5.15. Tables 5.13 and 5.14 are widely used in practice, but require judgment regarding the scale effects of *JRC*.

Less usual methods include measuring roughness using a mechanical profilometer or carpenter's comb (Tse & Cruden 1979) or conducting tilt and/or pullout tests on rock blocks (Barton & Bandis 1981).

The value of *JCS* may be assumed to be similar to the uniaxial compressive strength of rock, σ_c, if the defect rock walls are sound and not altered. If the rock walls are highly weathered and/or altered, the value of *JCS* may be smaller than $0.25\sigma_c$. The Schmidt hammer can be used to evaluate *JCS* using charts like the one shown in Table 5.16, or the correlation proposed by Deere and Miller (1966):

$$JCS = 6.9 \times 10^{(0.0087\rho R_n(L) + 0.16)} \qquad \text{(eqn 5.33)}$$

where *JCS* is in MPa units, ρ is the rock density in g/cm³ units and $R_{n(L)}$ is the rebound number of the L-type Schmidt hammer. Caution is suggested when using this correlation due to the large dispersion of values commonly found. There are several correlations between the uniaxial compressive strength of rock and the Schmidt hammer rebound number (see Zhang 2005). Alternatively, the ISRM empirical field estimates of σ_c shown in Table 2.3 can be used.

5.3.2.6 *Scale effects*

Although discussions about the effects of scale on the shear strength of defects as defined by the Mohr-Coulomb failure criterion (c_j and ϕ_j) are limited, the available data indicates that:

- laboratory tests frequently overestimate the shear strength of discontinuities, especially the cohesion;
- the results of several back analyses of structurally controlled instabilities indicate that the peak shear strength of clean structures with sound hard rock walls, at scales from 10–30 m and in a low confine-

Table 5.13: Defect roughness profiles and associated JRC values

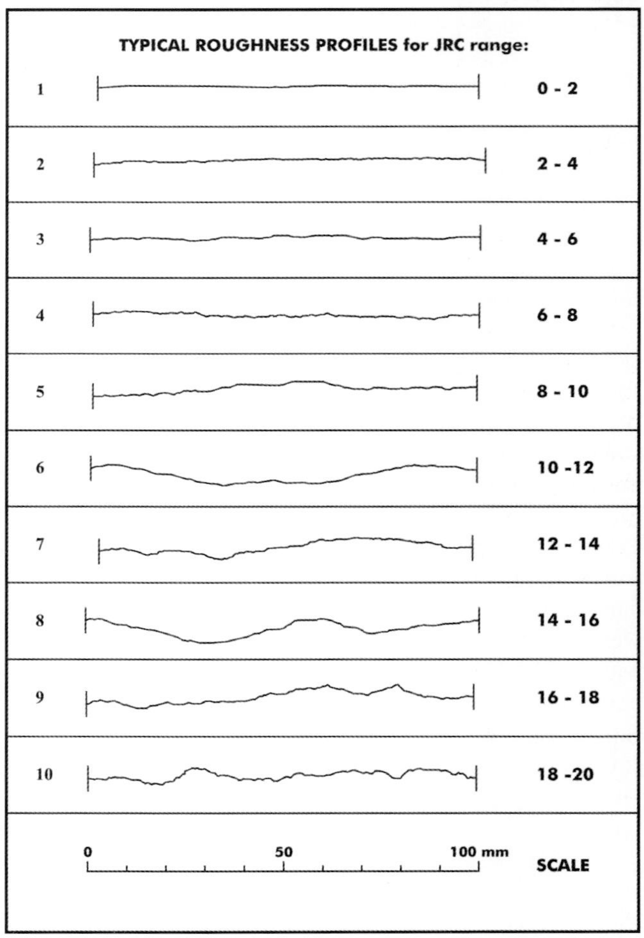

Source: Modified from Barton & Choubray (1977)

Table 5.14: ISRM-suggested characterisation of defect roughness

Class	Scale: Intermediate	Scale: Minor	Typical roughness profile	JRC20	JRC100
I	Stepped	Rough	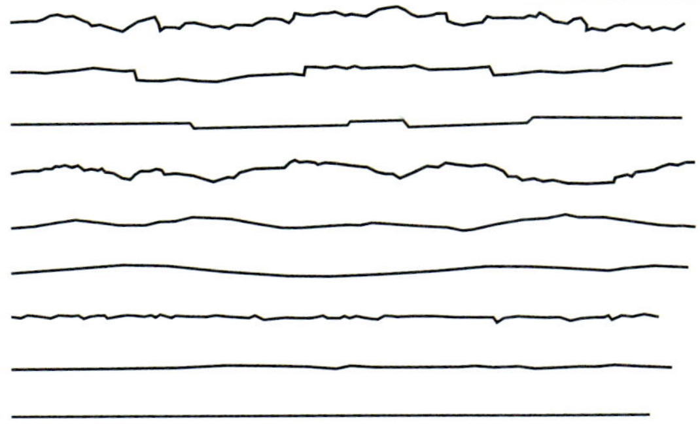	20	11
II		Smooth		14	9
III		Slickensided		11	8
IV	Undulating	Rough		14	9
V		Smooth		11	8
VI		Slickensided		7	6
VII	Planar	Rough		2.5	2.3
VIII		Smooth		1.5	0.9
IX		Slickensided		0.5	0.4

Notes
The length of the roughness profiles is intended to be in the range of 1–10 cm
The vertical and horizontal scales are identical
JRC_{20} and JRC_{100} correspond to joint roughness coefficient when the roughness profiles are 'scaled' to a length of 20 cm and 100 cm respectively
Source: Modified from Brown (1981) and Barton & Bandis (1990) by Flores & Karzulovic (2003)

Table 5.15: Estimating *JRC* from the maximum unevenness amplitude and the profile length

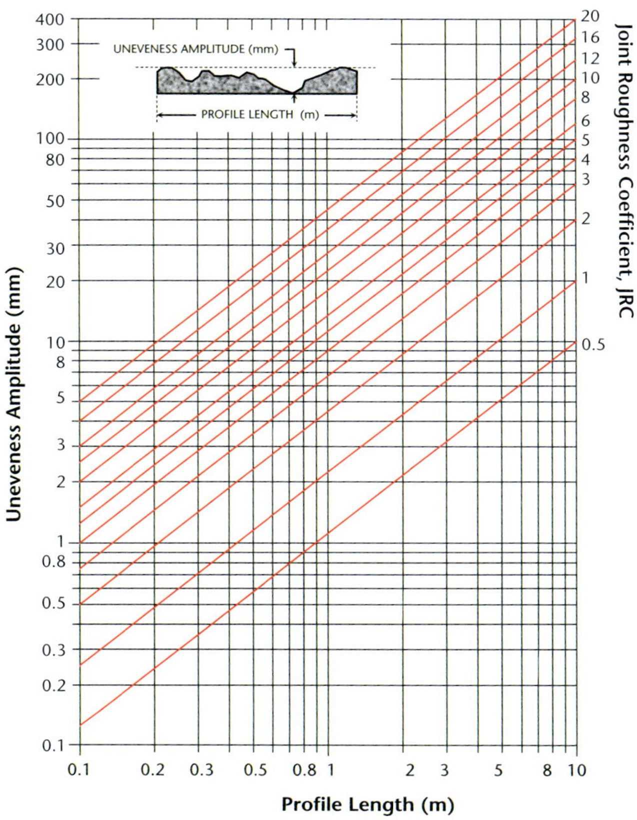

Source: Barton (1982)

Table 5.16: Estimating the uniaxial compressive strength, σ_c, of the defect rock wall from Schmidt hardness values

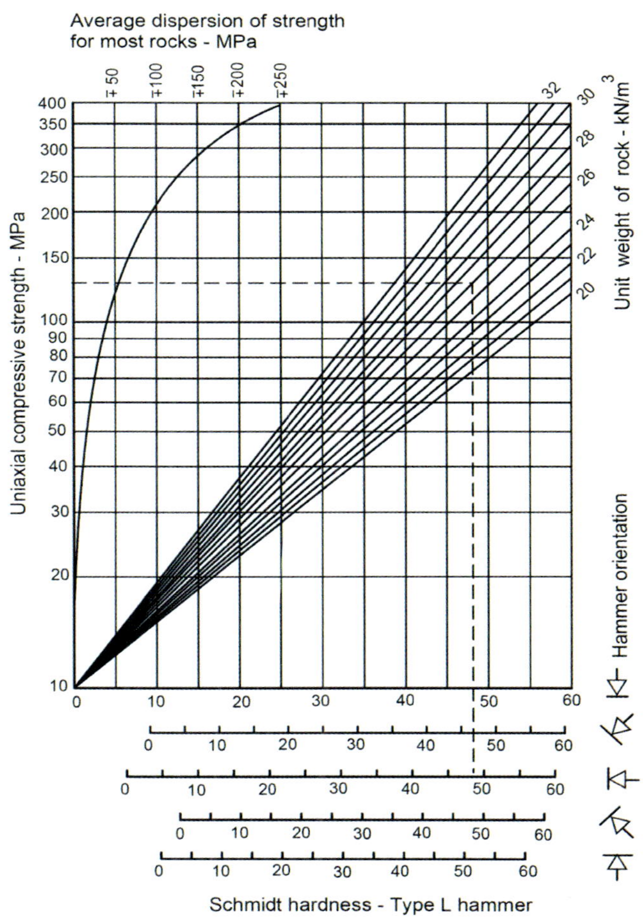

Source: Hoek (2002)

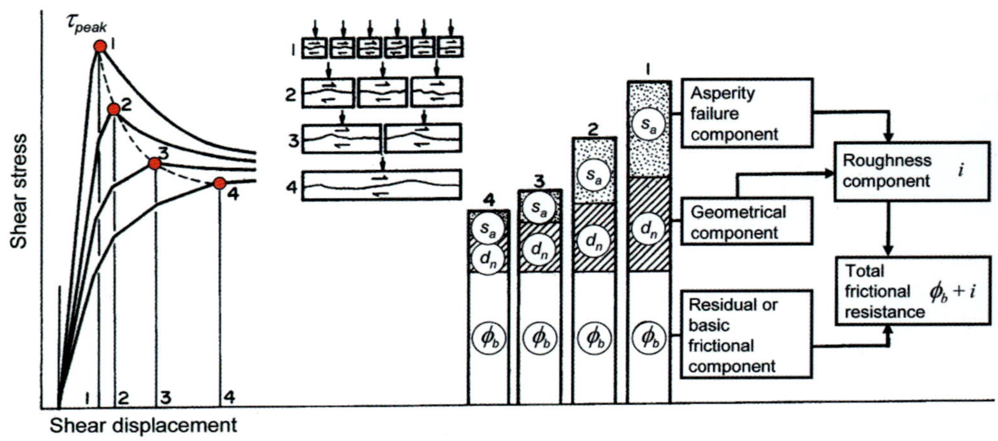

Figure 5.24: Summary of scale effects in the shear strength components of non-planar defects. ϕ_b is the basic friction angle, d_n is the peak dilation angle, s_a is the strength component from surface asperities, and i is the roughness angle
Source: Bandis et al. (1981)

ment condition (the predominant condition in the benches of an open pit mine) is defined by nil to very low values of cohesion and friction angles in the range of 45–60°;

- at low confinement and scales from 50–200 m, structures with centimetric clayey fillings have typical peak strengths characterised by cohesions ranging from 0–75 kPa and friction angles ranging from 18–25°;
- at low confinement and scales from 25–50 m, sealed structures with no clayey fillings have typical peak strengths characterised by cohesions ranging from 50–150 kPa and friction angles ranging from 25–35°.

Both *JRC* and *JCS* values are influenced by scale effects and decrease as the defect size increases. This is because small-scale roughness becomes less significant compared to the length of a longer defect and eventually large-scale undulations have more significance than small-scale roughness (Figure 5.24).

Bandis et al. (1981) studied these scale effects and found that increasing the size of the discontinuity produces the following effects:

- the shear displacement required to mobilise the peak shear strength increases;
- a reduction in the peak friction angle as a consequence of a decrease in peak dilation and an increase in asperity failure;
- a change from a brittle to a plastic mode of shear failure;
- a decrease of the residual strength.

To take into account the scale effect Barton and Bandis (1982) suggested reducing the values of *JRC* and *JCS* using the following empirical relations:

$$JRC_F = JRC_O\left(\frac{L_F}{L_O}\right)^{-0.02JRC_O} \quad \text{(eqn 5.34)}$$

$$JCS_F = JCS_O\left(\frac{L_F}{L_O}\right)^{-0.03JRC_O} \quad \text{(eqn 5.35)}$$

where JRC_F and JCS_F are the field values, JRC_O and JCS_O are the reference values (usually referred to a scale in the range 10 cm–1 m), L_F is the block size in the field and L_O is the length of reference (usually 10 cm–1 m).

These relationships must be used with caution because for long structures they may produce values that are too low. Ratios of $JCS_F/JCS_O < 0.3$ or $JRC_F/JRC_O < 0.5$ must be considered suspicious unless there are very good reasons to accept them.

The Barton-Bandis strength envelopes for discontinuities with different JRC values are shown in Figure 5.25, which also shows the upper limit for the peak friction angle resulting from this criterion.

From Table 5.14, the following values can be assumed as a first estimate for the joint roughness coefficient:

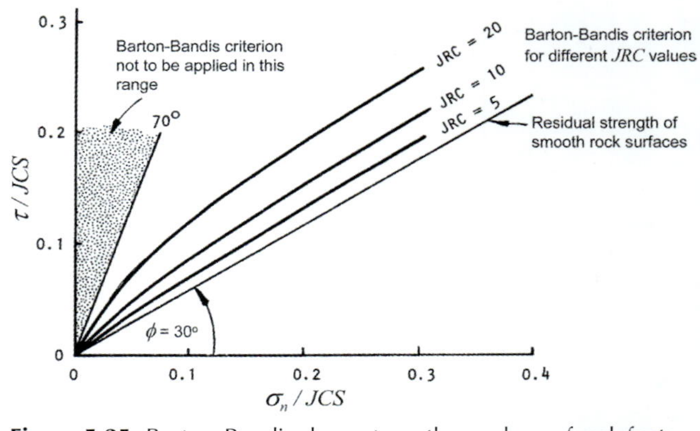

Figure 5.25: Barton-Bandis shear strength envelopes for defects with different JRC values
Source: Modified from Hoek & Bray (1981)

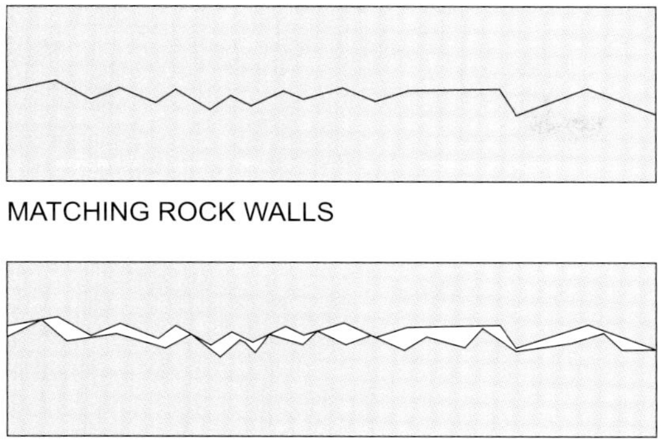

Figure 5.26: Examples of discontinuities with matching and mismatching rock walls
Source: Flores & Karzulovic (2003)

- rough undulating discontinuities: $JRC \approx 15$–20
- smooth undulating discontinuities: $JRC \approx 10$
- smooth planar discontinuities: $JRC \approx 2$

5.3.2.7 *Stress, strain and normal stiffness*

Numerical slope stability analyses require, in addition to the strength properties, the stress-strain characteristics of defects. Detailed discussions on the stress-strain behaviour of defects can be found in Goodman (1976), Bandis et al. (1983), Barton (1986), Bandis (1993) and Priest (1993).

The loading of a discontinuity induces normal and shear displacements whose magnitude depends on the stiffness of the structure, defined in terms of a normal stiffness, k_n, and a shear stiffness, k_s. These refer to the rate of change of normal (σ_n) and shear (τ) stresses with respect to normal (v_c) and shear (u_s) displacements (Bandis 1993):

$$\begin{Bmatrix} d\sigma_n \\ d\tau \end{Bmatrix} = \begin{bmatrix} k_n & 0 \\ 0 & k_s \end{bmatrix} \begin{Bmatrix} dv_c \\ du_s \end{Bmatrix} \quad \text{(eqn 5.36)}$$

where:

$$k_n = \left(\frac{\partial \sigma_n}{\partial v_c} \right)_{u_s} \quad \text{(eqn 5.37a)}$$

$$k_s = \left(\frac{\partial \tau}{\partial u_s} \right)_{v_c} \quad \text{(eqn 5.37b)}$$

Therefore, a discontinuity subjected to normal and shear stresses will suffer normal and shear displacements that depend on the following factors:

- the initial geometry of the discontinuity's rock walls;
- the matching between the rock walls, which defines the variation of the aperture and the effective contact area (Figure 5.26);
- the strength and deformability of the rock wall material;
- the thickness and mechanical properties of the filling material (if any);
- the initial values of the normal and shear stresses acting on the structure.

It is assumed that the defect cannot sustain tensile normal stresses and that there will be a limiting compressive normal stress beyond which the defect is mechanically indistinguishable from the surrounding rock (Figure 5.27).

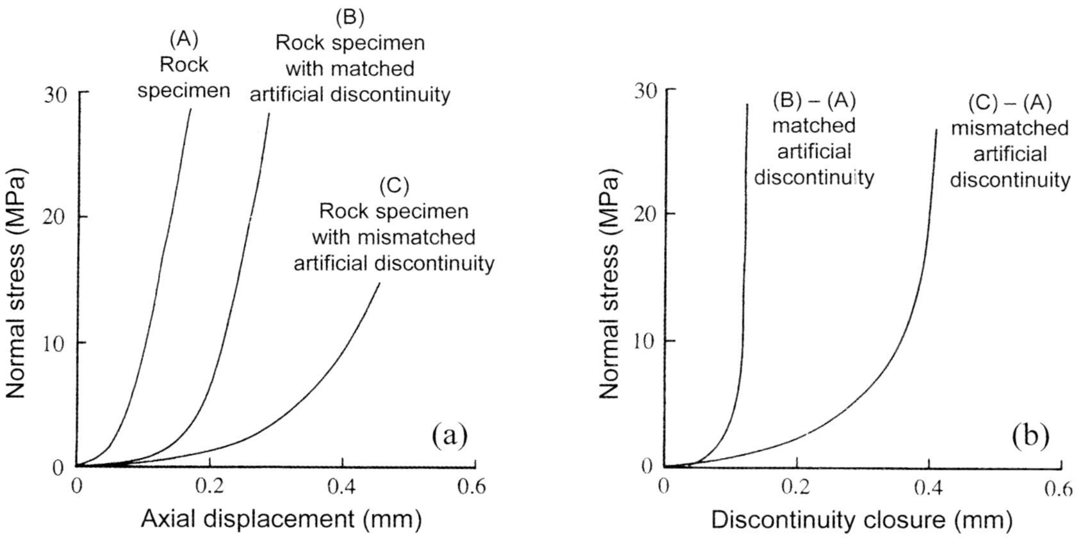

Figure 5.27: Determination of the normal stiffness of an artificial defect by means of uniaxial compression tests on specimens of granodiorite with and without a discontinuity. (a) Normal stress-total axial displacement curves. (b) Normal stress-discontinuity closure curves
Source: Goodman (1976)

The normal stiffness of a defect can be measured from a compression test with the load perpendicular to the discontinuity (Goodman 1976), or from a direct shear test if normal displacements are measured for different normal stresses (Figure 5.12). The following comments can be made.

1. Normal stiffness depends on the rock wall properties and geometry, the matching between rock walls, the filling thickness and properties (if any), the initial condition (before applying a normal stress increment), the magnitude of the normal stress increment and the number of loading cycles.
2. Generally, normal stiffness is larger if the rock wall and filling material (if any) are stronger and stiffer.
3. For a given set of conditions, normal stiffness is larger for defects with good matching than for mismatching ones.
4. Normal stiffness increases with the number of loading cycles. Apparently, the increment is larger in the case of stronger and stiffer rock walls.
5. The values quoted in the geotechnical literature indicate that normal stiffness ranges from 0.001–2000 GPa/m. It typically takes the following values:
   - → defects with soft infills: $k_n < 10$ GPa/m;
   - → clean defects in moderately strong rock: $k_n =$ 10–50 GPa/m;
   - → clean defects in strong rock: $k_n =$ 50–200 GPa/m.

The normal stiffness of a defect increases as the defect closes when σ_n increases, but there is a limit that is reached when the defect reaches its maximum closure, v_{cmax}. Assuming that the relationship between the effective normal stress, σ_n, and the defect closure, v_c, is hyperbolic (Goodman et al. 1968) it is possible to define the normal stiffness (Zhang 2005):

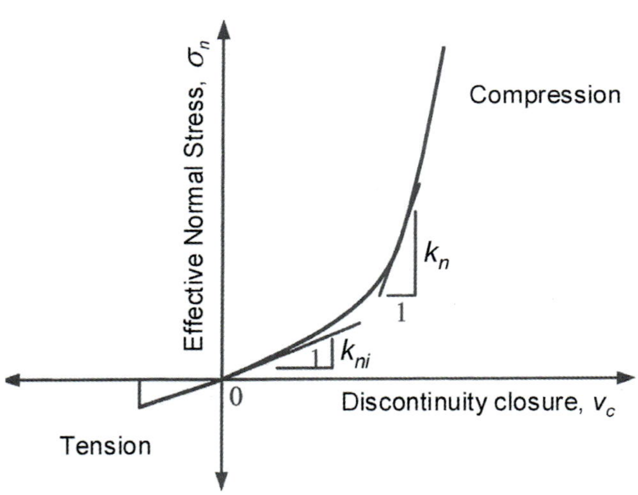

Figure 5.28: Definition of k_n and k_{ni} in an effective normal stress-discontinuity closure curve

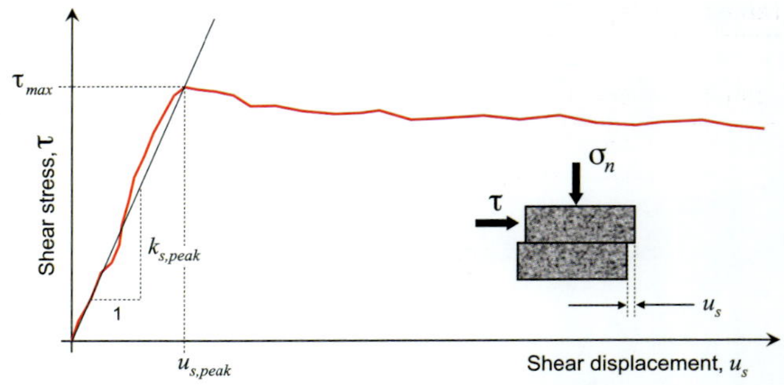

Figure 5.29: Determination of secant peak shear stiffness of a defect from a direct shear stress
Source: Goodman (1970)

$$k_n = k_{ni}\left(1 + \frac{\sigma_n}{k_{ni}v_{cmax}}\right)^2 \qquad \text{(eqn 5.38)}$$

where k_{ni} is the initial normal stiffness, defined as the initial tangent of the normal stress-discontinuity closure curve (Figure 5.29). As the defect's tensile strength is usually neglected, $k_n = 0$ if σ_n is tensile.

Hence, to determine the normal stiffness of a defect it is necessary to know the initial value of this stiffness and the defect's maximum closure. From experimental results, Bandis et al. (1983) suggested that k_{ni} for matching defects can be evaluated as:

$$k_{ni} \approx -7.15 + 1.75JRC + 0.02\left(\frac{JCS}{e_i}\right) \qquad \text{(eqn 5.39)}$$

where k_{ni} is in GPa/m units (or MPa/mm), *JRC* and *JCS* are coefficients of the Barton-Bandis failure criterion and e_i is the initial aperture of the discontinuity, which can be estimated as:

$$e_i \approx JRC\left(\frac{0.04\sigma_c}{JCS} - 0.02\right) \qquad \text{(eqn 5.40)}$$

where e_i is in mm, and σ_c and *JCS* are in MPa.

For the case of mismatching structures, Bandis et al. (1983) suggested the following relationship:

$$k_{ni,\,mm} = \frac{k_{ni}}{2.0 + 0.0004 \times JRC \times JCS \times \sigma_n} \qquad \text{(eqn 5.41)}$$

where $k_{ni,mm}$ is the initial tangent stiffness for mismatching defects. Regarding the scale effect on the normal stiffness, it can be implicitly considered by using 'scaled' values for *JRC* and *JCS*, and an 'adequate' value for e_i. Although these relationships have several limitations there are few practical tools to estimate k_n. Some reported values for the normal stiffness of discontinuities are listed in Tables 5.17 and 5.18.

Table 5.17: Reported values for normal stiffness for some rocks

Rock	Discontinuity	Load cycle	k_{ni} (GPa/m)	k_N (GPa/m)	Comments	Reference
SANDSTONE	Fresh to slightly weathered, good matching of rock walls	1	4–23		σ_{ni} = 1 kPa	Bandis et al. (1983)
		2	11–35			
		3	18–62			
	Moderately weathered, good matching of rock walls	1	4–26			
		2	9–27			
		3	15–45			
	Weathered, good matching of rock walls	1	2–5			
		2	9–14			
		3	11–20			
	Shear zone with clay gouge			1.7	Estimated from data in reference, assuming a 3 cm thickness	Wittke (1990)
	Bedding planes, good matching (*JRC* = 10–16)			13–24	Direct shear tests with σ_n ranging from 0.4–0.9 MPa	Rode et al. (1990)
	Bedding planes, good matching (*JRC* = 10–16)			7–12		
	Fresh fractures, good matching (*JRC* = 12–17)			17–25		
	Fresh fractures, poor matching (*JRC* = 12–17)			8–12		
LIMESTONE	Fresh to slightly weathered, good matching	1	8–31		σ_{ni} = 1 kPa	Bandis et al. (1983)
		2	54–134			
		3	72–160			
	Moderately weathered, good matching	1	5–70			
		2	26–91			
		3	53–168			
	Weathered, good matching	1	4–13			
		2	40–50			
		3	42–65			
	Joints in weathered limestone			0.5–1.0	σ_n = 5 MPa	Bandis (1993)
	Joints in fresh limestone			4–5		
QUARTZITE	Clean			15–30	σ_n = 10–20 MPa	Ludvig (1980)
	With clay gouge			10–25		
DOLERITE	Fresh, good matching	1	21–27		σ_{ni} = 1 kPa	Bandis et al. (1983)
		2	59–75			
		3	103–119			
	Weathered, good matching	1	8–13			
		2	24–92			
		3	37–130			
GRANITE	Clean joint (*JRC* = 1.9)	1		121	Estimated from ref.	Makurat et al. (1990)
	Clean joint (*JRC* = 3.8)	1		74	Biaxial tests σ_n : 25–30 MPa	
	Clean joint			352–635	Mes. Sist. Pac-ex.	Martín et al. (1990)
				50–110	σ_n: 8.6–9.3 MPa	
	Shear zone			2–224	Mes. Sist. Pac-ex. σ_n: 0.5–1.5 MPa	
				7–266	Mes. Sist. Pac-ex. σ_n : 18–20 MPa	

k_n = Normal stiffness
σ_n = Normal stress
k_{ni} = Initial normal stiffness
σ_{ni} = Initial normal stress
Pac-ex: Measured by the system Pac-ex, a special instrumentation system developed in the Underground Research Laboratory by Atomic Energy of Canada Ltd.
Source: Flores & Karzulovic (2003)

Table 5.18: Reported values for normal stiffness for some rocks

Rock	Discontinuity	Load cycle	k_{ni} (GPa/m)	k_N (GPa/m)	Comments	Reference
SILTSTONE	Fresh, good matching	1	14–26		σ_{ni} = 1 kPa	Bandis et al. (1983)
		2	22–64			
		3	22–70			
	Moderately weathered, good matching	1	10–11			
		2	20–22			
		3	20–26			
	Weathered, good matching	1	7–14			
		2	27–29			
		3	29–41			
QUARTZ MONZONITE	Clean			15.3	Triaxial testing (?)	Goodman & Dubois (1972)
PLASTER	Clean, artificial fractures			2.7–5.4	σ_n: 3.5–24 MPa	Barton (1972)
	Clean, artificial fractures			2.7		Karzulovic (1988)
SLATE	Fresh, good matching	1	24–47		σ_{ni} = 1 kPa	Bandis et al. (1983)
		2	98–344			
		3	185–424			
	Weathered	1	11–14			
		2	19–40			
		3	49–78			
RHYOLITE	Clean			16.4	Triaxial testing (?)	Goodman & Dubois (1972)
WEAK ROCK	With clay gouge			5–40	Increases with σ_n	Barton et al. (1981)
HARD ROCK	Soft clay filling			0.01–0.1	Typical range	Itasca (2004)
	Clean			37–93	Triaxial testing. Increases with number of loading cycles	Rosso (1976)
				8–99	Direct shear tests	
	Clean fracture			1620	Estimate for numerical analysis	Rutqvist et al. (1990)
	Good match, interlocked			> 100	Typical value	Itasca (2004)
	Fault with clay gouge			0.005	30–150 cm thick	Karzulovic (1988)
	Rough structure with a fill of rock powder			0.8	Mismatching	
GYPSUM	Fresh joints (*JRC* = 11)	1		3–11	σ_{ni} = 0.2 MPa	Rode et al. (1990)
	Fresh joints (*JRC* = 11)	> 1		10–13		

k_n = Normal stiffness
σ_n = Normal stress
k_{ni} = Initial normal stiffness
σ_{ni} = Initial normal stress
Pac-ex: Measured by the system Pac-ex, a special instrumentation system developed in the Underground Research Laboratory by Atomic Energy of Canada Ltd.
Source: Flores & Karzulovic (2003)

There are simple cases for which it is possible to compute the normal stiffness of the structures. If the Young's moduli of rock, E, and of rock mass, E_m, in the direction normal to the defects are known, and the rock mass contains only one set of defects with an average spacing s, then the normal stiffness of the structures can be computed as:

$$k_n = \frac{E_m E}{s(E - E_m)} \quad \text{(eqn 5.42)}$$

In the case of the defects with infills, if the defects are smooth or the infill thickness is much larger that the size of the asperities the normal stiffness can be computed as:

$$k_n = \frac{E_{\text{inf}}(1 - \nu_{\text{inf}})}{(1 + \nu_{\text{inf}})(1 - 2\nu_{\text{inf}})t} \quad \text{(eqn 5.43)}$$

where E_{inf} and ν_{inf} are the Young's modulus and Poisson's ratio of the infill and t is the infill thickness. This equation assumes that the infill cannot deform laterally, i.e. it is in an oedometric condition.

5.3.2.8 *Shear stiffness*

The shear stiffness of a discontinuity, k_s, can be measured from a direct shear test. The following comments can be made.

1 The shear stiffness depends on the rock wall properties and geometry, the matching between rock walls, the filling thickness and properties (if any), the magnitude of the normal stress increment and the length of the structure.
2 Generally, the shear stiffness is larger if the rock wall and filling material (if any) are stronger and stiffer.
3 For a given set of conditions, the shear stiffness is larger for structures with good matching than for structures with poor matching.
4 The shear stiffness values quoted in the geotechnical literature indicate that it ranges from 0.01–50 GPa/m. Typically it takes the following values:
   → defects with soft infills: $k_s < 1$ GPa/m
   → clean defects in moderately strong rock: $k_s <$ 10 GPa/m
   → clean defects in strong rock: $k_s < 50$ GPa/m

A secant peak shear stiffness can be evaluated from a direct shear tests as the ratio between the peak shear strength, τ_{max}, and the shear displacement required to reach this peak condition, $u_{s,peak}$ (Figure 5.29):

$$k_{s,\,peak} = \frac{t_{\max}}{u_{s,\,peak}} = \frac{\sigma_n \tan(\phi_j)}{u_{s,\,peak}} \quad \text{(eqn 5.44)}$$

It must be kept in mind that the peak shear stiffness of discontinuities is influenced by the scale effects affecting τ_{max} and $u_{s,peak}$ (Figure 5.30). Barton and Choubey (1977) found that the deformation $u_{s,peak}$ required to reach the peak shear stress, τ_{max}, typically is about 1% of the length of the discontinuity in the shear direction, L. Barton and Bandis (1982), from the analysis of observed displacements in direct shear tests (loading in shear) and earthquake slip magnitudes (unloading in shear), presented the following equation to estimate the shear displacement required to reach the peak shear strength of a discontinuity:

$$u_{s,\,peak} = \frac{L}{500}\left(\frac{JRC}{L}\right)^{0.33} \quad \text{(eqn 5.45)}$$

where L is the length (in m units) and JRC is the joint roughness coefficient of the defect.

Considering this and the Barton-Bandis criterion they presented the following expression to estimate the peak shear stiffness:

$$k_{s,\,peak} = \frac{\sigma_n \tan\left(\phi_b + JRC\log_{10}\left(\frac{JCS}{\sigma_n}\right)\right)}{\frac{L}{500}\left(\frac{JRC}{L}\right)^{0.33}} \quad \text{(eqn 5.46)}$$

where the values of JCS and JRC must be estimated for the length L (in m units). Regarding the use of equation 5.46, it must be pointed out that:

- applying this equation to structures with lengths from 0.1–10 m indicates that the slope of the $k_{s,peak} - L$ curve decreases as L increases;
- applying this equation to major geological faults results in quasi-residual values for the roughness coefficient ($JRC \approx 1$), and values of JCS equivalent to the uniaxial compressive strength of overconsolidated clays (in the range 1–10 MPa);
- this equation should not be applied to structures with clay infills, because if the infill thickness exceeds the maximum amplitude of the asperities the shear stiffness does not vary so much with the magnitude of the effective normal stress, and the scale effect is much less important.

The relation between shear stress, τ, and shear displacement, u_s, can be expressed as a hyperbolic function (Duncan & Chang 1970; Bandis et al. 1983; Priest 1993), making it possible to define the shear stiffness (Zhang 2005):

$$k_s = k_{si}\left(1 - \frac{R_f \tau}{\tau_f}\right)^2 \quad \text{(eqn 5.47)}$$

where k_{si} is the initial shear stiffness, defined as the initial tangent of the shear stress-shear displacement curve (Figure 5.31), τ is the shear stress at which k_s is evaluated, τ_f is the shear strength at failure and R_f is the failure ratio given by:

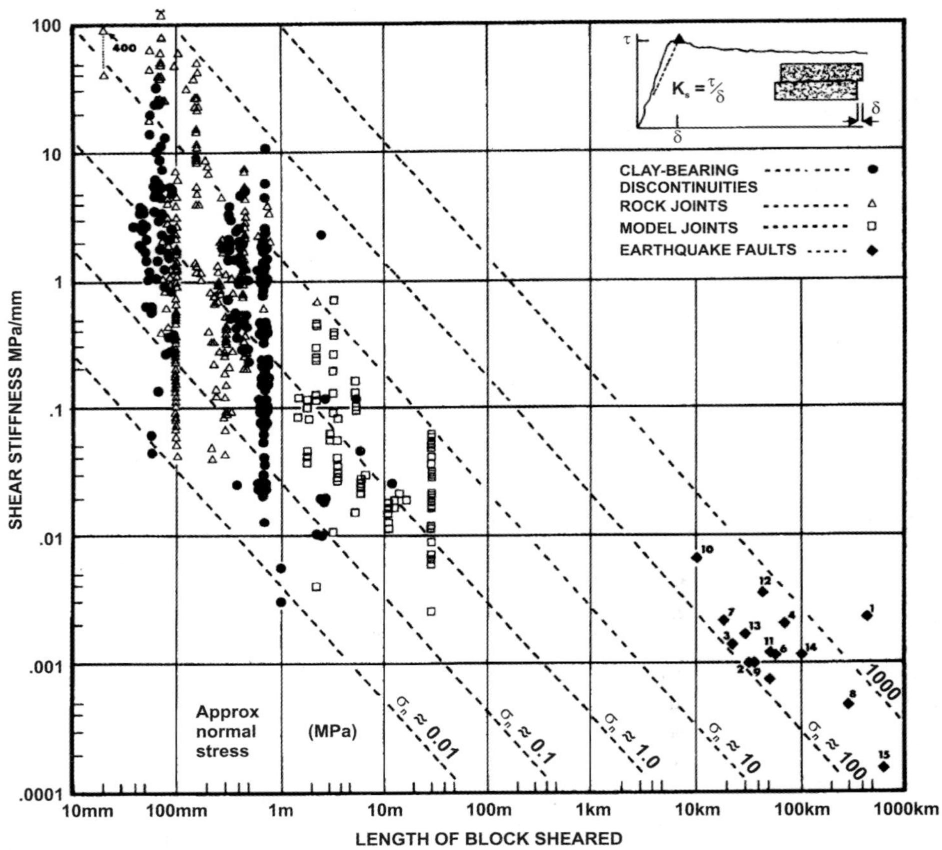

Figure 5.30: Experimental evidence for the scale effect on peak shear stiffness. The normal stress diagonals were tentatively extrapolated from tests at 100 mm size from the measured effects of scale on the *JRC, JCS* and $u_{s,peak}$ in the 100 mm to 1 m range
Source: Barton & Bandis (1982)

$$R_f = \frac{\tau_f}{\tau_{res}} \qquad \text{(eqn 5.48)}$$

where τ_{res} is the residual or ultimate shear strength at large shear displacements.

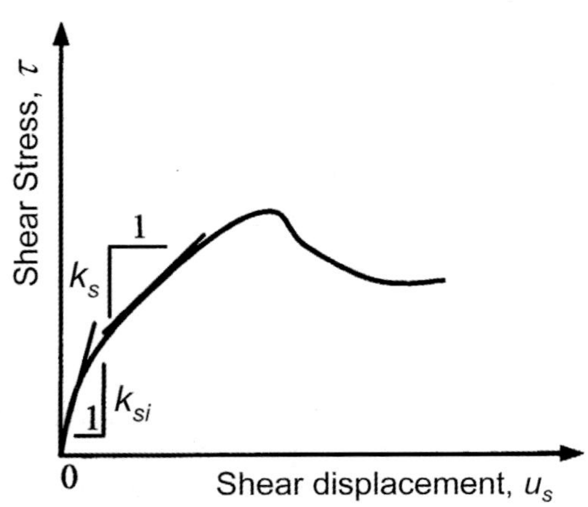

Figure 5.31: Definition of k_s and k_{si} in a shear stress-shear displacement curve

Hence, to determine the shear stiffness of a discontinuity at a shear stress τ it is necessary to know the initial value of this stiffness, the shear stress at failure and the failure ratio. Bandis et al. (1983) found that k_{si} increased with normal stress and could be estimated from:

$$k_{si} \approx k_j (\sigma_n)^{n_j} \qquad \text{(eqn 5.49)}$$

where k_j and n_j are empirical constants called the stiffness number and the stiffness exponent, respectively. Based on test results on defects in dolerite, limestone, sandstone and slate at σ_n ranging from 0.23–2.36 MPa and R_f ranging from 0.652–0.887, Bandis et al. (1983) found that n_j varied from 0.615–1.118 GPa/m, with an average of about 0.761. The stiffness number was found to vary with *JRC* and Bandis et al. (1983) suggested that for $JRC > 4.5$ it could be estimated as:

$$k_j \approx -17.19 + 3.86 \times JRC \qquad \text{(eqn 5.50)}$$

Although these relationships have several limitations, there are few practical tools to estimate k_s. Kulhawy (1975)

Table 5.19: Reported values for shear stiffness of some defects

Rock	Structure type	k_{si} (GPa/m)	$k_{s,yield}$ (GPa/m)	$k_{s,peak}$ (GPa/m)	Comments	Reference
AMPHIBOLITE	Schistosity plane			0.59	DST, σ_{ni} = 0.12 MPa	Kulhawy (1975)
SANDSTONE	Sandstone-basalt contact			0.11	DST, σ_{ni} = 0.13 MPa	
	Sandstone-chalk contact		0.3–2.1	0.1–0.2	DST, σ_{ni} = 0.1–1 MPa	
	Artificial fracture			29.8	DST, σ_{ni} = 0.26 MPa	
	Artificial rough fracture			1.3	DST, σ_{ni} = 2.4 MPa	
	Artificial clean fracture			5–38		Maki (1985)
	Fresh fracture, good matching	2.2–38		0.6–4.5	σ_n = 0.2–2.4 MPa	Bandis et al. (1983)
	Slightly weathered fracture, good matching	9–42		1.2–4.7	σ_n = 0.2-2.1 MPa	
	Moderately weathered fracture, good matching	1.2–6		0.5–1.7	σ_n = 0.2–2.0 MPa	
	Weathered fracture, good matching	2.1–7		0.6–1.4	σ_n = 0.5–2.0 MPa	
LIMESTONE	Clean smooth fractures		0.4–2.4	0.2–1.3	DST, σ_{ni} = 0.9–2.4 MPa	Kulhawy (1975)
	Artificial fracture			8.7	DST, σ_{ni} = 10.4 MPa	
	Clean artificial fracture			3–17		Maki (1985)
	Fresh to slightly weathered, good matching	8–51		1.7–7	σ_n = 0.2–1.8 MPa	Bandis et al. (1983)
	Moderately weathered, good matching	4–17		1.1–3.1	σ_n = 0.2–1.9 MPa	
	Weathered, good matching	1–11		0.7–1.9	σ_n = 0.2–1.5 MPa	
	Joint with large JCS		6.1	1.7–4.6	DST, σ_{ni} = 0.5 MPa	Kulhawy (1975)
	Rough bedding plane		0.2–13.8	1.2–2.6	DST, σ_{ni} = 1.5–4 MPa	
	Rough bedding plane		0.3–14.9	0.2–7.4	DST, σ_{ni} = 0.3–3.4 MPa	
	Moderately rough bedding plane		0.8–4.1	0.2–1.4	DST, σ_{ni} = 0.1–3.6 MPa	
	Mylonitised bedding plane		1.0–8.0	0.3–5.7	DST, σ_{ni} = 0.2–2.4 MPa	
	Chalk vein (0.2–20 mm)			2.3–23.6	DST, σ_{ni} = 0.5–1.5 MPa	
	Chalk vein (15–30 mm)		1.2–3.3	0.4–4.7	DST, σ_{ni} = 0.5–3 MPa	
	Chalk vein (0.2–2 mm), saturated		1.47	0.1–31.6	DST, σ_{ni} = 0.5–1.5 MPa	
	Chalk vein (1–3 mm), saturated		2.2–3.7	0.5–3.7	DST, σ_{ni} = 0.45–0.6 MPa	
	Chalk vein (1–50 mm), saturated		2.2–3.3	0.9–5.7	DST, σ_{ni} = 0.25–0.8 MPa	
	Shale layer		1.5–13.9	0.3–8.3	DST, σ_{ni} = 1.2–2.8 MPa	
	Shale layer (2–5 mm), wet			0.01–0.02	DST, σ_{ni} = 0.025 MPa	
	Fractured shale layer (2–5 mm)			0.01–0.02	DST, σ_{ni} = 0.02 MPa	
CHALK	Saturated joint		0.1–2.7	0.02–1.9	DST, σ_{ni} = 0.5–2.9 MPa	
	Sand filled fractures (1–2 mm)			2.34	DST, σ_{ni} = 0.98 MPa	
QUARTZITE	Clean fracture			5–9	σ_n = 10–15 MPa	Ludvig (1980)
	Fracture with clay gouge			2–4		

DST Direct shear tests
TT Triaxial tests
IST In situ tests
Source: Modified from Flores & Karzulovic (2003)

presented data on shear stiffness of defects evaluated both at the peak and yield points of the shear stress-shear displacement curves, $k_{s,peak}$ and $k_{s,yield}$. Some reported values for the shear stiffness of discontinuities are listed in Tables 5.19 and 5.20.

There are some simple cases for which it is possible to compute the shear stiffness of the structures. If the shear moduli of rock, G, and of rock mass, G_m, are known for shear in the direction parallel to the defects, and the rock mass contains only one set of discontinuities with an

Table 5.20: Reported values for shear stiffness of some defects

Rock	Structure type	k_{si} (GPa/m)	$k_{s,yield}$ (GPa/m)	$k_{s,peak}$ (GPa/m)	Comments	Reference
DOLERITE	Fresh to slightly weathered, good matching	8–19		1.8–5	σ_n = 0.2–2.1 MPa	Bandis et al. (1983)
	Weathered fracture, good matching	3.6–9		0.9–2.2	σ_n = 0.3–1.1 MPa	
SCHIST	Fracture		0.4–1.0	0.1–0.4	DST, σ_{ni} = 0.2–1.5 MPa	Kulhawy (1975)
GNEISS	Mylonitised plane (40–50 mm)		1.4–4.7	0.7–3.7	DST, σ_{ni} = 0.4–2.9 MPa	
	Foliation plane (?)		0.3–0.4	0.09–0.12	DST, σ_{ni} = 0.2–0.8 MPa	
GRANITE	Rough fracture (beam breakage)		1.3–1.6	1.0–1.6	DST, σ_{ni} = 1.1–1.4 MPa	
GREYWACKE	Bedding plane (5–8 mm)			0.23	DST, σ_{ni} = 1.24 MPa	
	Bedding plane			1.21	DST, σ_{ni} = 1.01 MPa	
	Sealed bedding plane			2.26	DST, σ_{ni} = 0.43 MPa	
SHALE	Clean artificial fracture			2–9		Maki (1985)
QUARTZ MONZONITE	Clean fracture			0.14	DST (?)	Goodman & Dubois (1972)
RHYOLITE	Clean fracture			0.44	DST (?)	
HARD PLASTER	Clean artificial fracture			0.003–0.04	σ_n = 0.2–11.2 MPa	Barton (1972)
	Clean artificial fracture			0.03		Karzulovic (1988)
SLATE	Fresh fracture, good matching			5–13	σ_n = 0.5–2.3 MPa	Bandis et al. (1983)
	Weathered fracture, good matching	2.8–8		0.6–1.3	σ_n = 0.4–1.5 MPa	
	Cleavage plane		0.9	0.8	DST, σ_{ni} = 4.4 MPa	Kulhawy (1975)
PORPHYRY	Joint		0.9–1.6	0.2–1.9	DST, σ_{ni} = 3.2–10.1 MPa	
HARD ROCK	Clean fracture			12–47	IST, σ_n = 0–6 MPa	Rosso (1976)
				20–93	TT, σ_n = 1–18 MPa	
				42–74	DST, σ_n = 3.5–10.5 MPa	
	Clean fracture			3	Estimation for numerical analysis	Rutqvist et al. (1990)
	Fault with clay gouge			0.12–0.23	DST, σ_n = 0.3–1.1 MPa	Kulhawy (1975)
	Fault with clay gouge, 30–150 cm thick			0.005		Karzulovic (1988)
	Rough structure filled with rock powder, mismatching			0.08		
WEAK ROCK	Structure with clay gouge			0.11–0.27	$\sigma_n \approx$ 5 MPa	Barton (1980)
				0.40–0.98	$\sigma_n \approx$ 20 MPa	

DST Direct shear tests
TT Triaxial Table I tests
IST In situ tests
Source: Modified from Flores & Karzulovic (2003)

average spacing s, then the shear stiffness of the structures can be computed as:

$$k_s = \frac{G_m G}{s(G - G_m)} \quad \text{(eqn 5.51)}$$

In the case of defects with infills, if the defects are smooth or the infill thickness is much larger that the size of the asperities, assuming that the behaviour is elastic a relationship between k_n and k_s can be derived (Duncan & Goodman 1968):

$$k_s = \frac{k_N}{2(1 + \nu_{fill})} \quad \text{(eqn 5.52)}$$

where ν_{fill} is the Poisson's ratio of the infill. Since Poisson's ratio for non-dilatant materials can range from 0–0.5, then k_s should be equal to $0.33k_n$–$0.5k_n$. However, Kulhawy (1975) presented data showing that this is not always the case, demonstrating that defects do not behave as elastic materials.

5.4 Rock mass classification

5.4.1 Introduction

The Mohr-Coulomb failure criterion is the backbone of all current limiting equilibrium and numerical methods of slope stability analyses, which creates a basic need to provide friction (Ø) and cohesion (c) values for the rock mass. However, triaxial testing of representative rock mass samples is difficult because of sample disturbance and equipment size limitations. Consequently, the preferred method has been to derive empirical values of friction and cohesion from rock mass rating schemes that have been calibrated from experience.

Rock mass rating schemes are based on subjective ratings of specific attributes of the rock mass in order to create discrete geotechnical zones or units. In this process, Bieniawski (1989) noted six specific objectives:

1. to identify the most significant parameters influencing the behaviour of a rock mass;
2. to divide a particular rock mass formation into group of similar behavior, i.e. rock masses classes of varying quality;
3. to provide a basis for understanding the characteristics of each rock mass class;
4. to relate the experience of rock conditions at one site to the conditions and experience encountered at others;
5. to derive quantitative data and guidelines for engineering design;
6. to provide a common basis for communication between engineers and geologists.

There are many different classification schemes, perhaps the oldest and best-known being that of Terzaghi, which was introduced for tunnel design in 1946 (Proctor & White 1946). Today, in open pit slope engineering the most used schemes are:

- Bieniawski's Rock Mass Rating (RMR) scheme (Bieniawski 1973, 1976, 1979, 1989), originally introduced for tunnelling and civil engineering applications;
- Laubscher's Rock Mass Rating (IRMR and MRMR) schemes (Laubscher 1977, 1990; Jakubec & Laubscher 2000, Laubscher & Jakubec 2001);
- Hoek and Brown's Geological Strength Index (GSI) (Hoek et al. 1995, 2002).

5.4.2 RMR, Bieniawski

5.4.2.1 Parameter ratings

The value of RMR determines the geotechnical quality of the rock mass on a scale that ranges from zero to 100 and considers the 5 classes presented in Table 5.21. The parameters and ratings used to determine the geotechnical quality of the rock mass are shown in Table 5.22. The table reflects changes to the ratings made by Bieniawski between 1976 and 1979, and restated in 1989. Because of these changes, it is important to indicate which version of the system is being used. The 1976 rating values are shown in Table 5.23. The 1979 changes for the RQD, joint spacing and joint condition rating values are shown in Tables 5.24, 5.25 and 5.26. Bieniawski's 1979 correlation between RQD and joint spacing is given in Figure 5.32.

Table 5.21: RMR calibrated against rock mass quality

RMR rating	Description
81–100	Very good rock
61–80	Good rock
41–60	Fair rock
40–21	Poor rock
<21	Very poor rock

5.4.2.2 Practical considerations

Regardless of which version is chosen, when using the Bieniawski system in open pit slope design applications a number of practical considerations must be kept in mind.

1. Groundwater parameter: the rock mass should be assumed to be completely dry and the groundwater rating set to 10 (1976) or 15 (1979). Any pore pressures in the rock mass should be accounted for in the stability analysis.
2. Joint orientation adjustment: joint orientations should be assumed to be very favourable and the adjustment factor set to zero. The effect of joints and other structural defects should be accounted for in the assessment of the rock mass strength (e.g. if using the Hoek-Brown strength criterion) and/or the stability analyses.
3. RQD parameter: RQD measures the total length of solid pieces of fresh, slightly weathered and moderately weathered core longer than 100 mm against the total

Table 5.22: Bieniawski RMR parameter ratings, 1976 and 1979

Parameter	Rating (1976)	Rating (1979)
UCS	0–15	0–15
RQD (drill core)	3–20	0–20
Joint spacing	5–30	5–20
Joint condition	0–25	0–30
Groundwater	0–10	0–15
Basic RMR	8–100	8–100
Joint orientation adjustment	0–60	0–60

Table 5.23: Bieniawski 1976 RMR parameter ratings

Parameter			Range of values						
1	Strength of intact rock material	Point-load strength index	>8 MPa	4–8 MPa	2–4 MPa	1–2 MPa	For this low range uniaxial compressive test is preferred		
		Uniaxial compressive strength	>200	100–200 MPa	50–100 MPa	25–50 MPa	10–25 MPa	3–10 MPa	1–3 MPa
	Rating		15	12	7	4	2	1	0
2	Drill core quality RQD		90–100%	75–90%	50–75%	25–50%	<25%		
	Rating		20	17	13	8	3		
3	Spacing of joints		>3 m	1–3 m	0.3–1 m	50–300 mm	<50 mm		
	Rating		30	25	20	10	5		
4	Condition of joints		Very rough surfaces Not continuous No separation Hard joint wall contact	Slightly rough surfaces Separation <1 mm Hard joint wall contact	Slightly rough surfaces Separation <1 mm Soft joint wall contact	Slickensided surfaces OR Gouge <5 mm thick Joints open 1–5 mm Continuous joints	Soft gouge >5 mm thick OR Joints open >5 mm Continuous joints		
	Rating		25	20	12	6	0		

length of the indicated core run, expressed as a percentage (section 2.4.9.2).

The use of RQD as a parameter in Bieniawski's RMR system presents particular problems. As devised by Don Deere and his colleagues at the University of Illinois in 1964/65 (Deere et al. 1967; Deere & Deere 1988), RQD is a modified core recovery percentage and an index of rock quality in that problematic rock that is highly weathered, soft, fractured, sheared and jointed is counted against the rock mass. Thus, it is simply a measurement of the percentage of 'good' rock recovered from an interval of a drill hole. As a parameter, it is poorly defined. It is highly subjective (different operators frequently report different values for the same interval of core) and inconsistent, often providing inaccurate and misleading results. Consequently, it must always be used with engineering judgment that takes proper account of the geological characteristics of the rock mass being classified.

Table 5.24: Bieniawski 1979 RQD parameter ratings

Rock mass quality	*RQD* (%)	Rating
VERY POOR geotechnical quality	<25	3
POOR geotechnical quality	25–50	8
FAIR geotechnical quality	50–75	13
GOOD geotechnical quality	75–90	17
EXCELLENT geotechnical quality	90–100	20

Table 5.25: Bieniawski 1979 joint spacing parameter ratings

Qualitative description of spacing	s (mm)	Rating
VERY CLOSE to EXTREMELY CLOSE	<60	5
CLOSE	60–200	8
MODERATE	200–600	10
WIDE	600–2000	15
VERY WIDE to EXTREMELY WIDE	>2000	20

Table 5.26: Bieniawski 1979 joint condition parameter ratings

Description of the condition of the structures	Rating
Continuous structures. Open structures (aperture >5 mm), or structures with soft gouge fillings (thickness >5 mm).	0
Continuous structures. Slickensided structures or open structures (aperture 1–5 mm), or structures with soft rouge fillings (thickness 1–5 mm).	10
Slightly rough structures. Structures with weathered and/or altered rock walls. Open structures (aperture <1 mm) or filled structures (thickness <1 mm).	20
Slightly rough structures. Structures with slightly weathered and/or slightly altered rock walls. Open structures (aperture <1 mm) or filled structures (thickness <1 mm).	25
Non-continuous structures. Very rough structures. Structures with unweathered and non-altered rock walls. Closed or sealed structures.	30

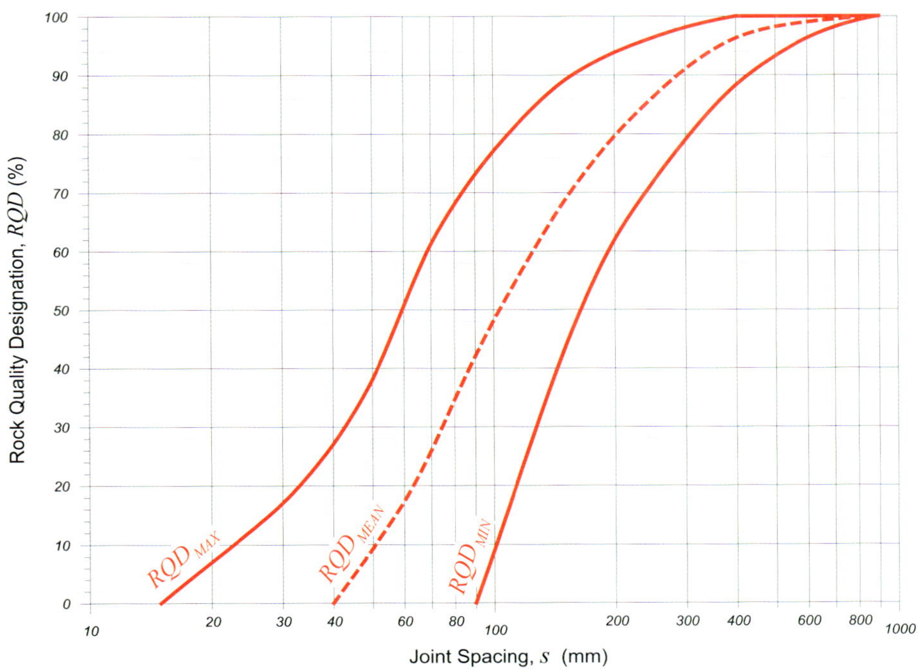

Figure 5.32: Bieniawski 1979 correlation between RQD and joint spacing

Sources of error that specifically undermine RQD's usefulness as a parameter include (Brown 2003; Hack 2002) the following.

- The RQD value is influenced by drilling equipment (single, double and triple-tube core barrels can all be used), drilling operators and core handling.
- The value of 100 mm of unbroken rock is an arbitrary and abrupt boundary, and small differences in joint spacing can produce large differences in the RQD value. For example, for a rock mass with a joint spacing of 90 mm perpendicular to the borehole RQD = 0%, while for a joint spacing of 110 mm RQD = 100%.
- RQD is biased by the orientation of the borehole (or scan line) with respect to the joint orientation. The apparent change in the joint spacing created by measuring from a different direction can produce a large difference in the RQD value (0–100%).

The ratings associated with the spacing between the structures assume that the rock mass presents three sets of structures. Laubscher (1977) suggested that if there are less than three joint sets the spacing of the structures could be increased by 30%.

Based on a correlation proposed by Priest and Hudson (1979), Bieniawski (1989) suggested that, if RQD or joint spacing data are lacking, the graph given in Figure 5.32 could be used to estimate the missing parameter. Given the bias that can be imposed on the RQD values by the orientation of a borehole or a scan line with respect to the joint orientation, this procedure must be used cautiously.

5.4.3 Laubscher IRMR and MRMR

Laubscher's In-situ Rock Mass Rating system (IRMR) and Mining Rock Mass Rating system (MRMR) were introduced by Laubscher as an extension of Bieniawski's RMR system for mining applications. The IRMR, so called to distinguish it from Bieniawski's RMR system, considers four basic parameters:

1. the intact rock strength (IRS), defined as the unconfined compressive strength (UCS) of the rock sample that can be directly tested;
2. the rock strength (RBS), defined as the strength of the rock blocks contained within the rock mass;
3. the blockiness of the rock mass, which is controlled by the number of joints sets and their spacings (JS);
4. the joint condition, defined in terms of a geotechnical description of the joints contained within the rock mass (JC).

The steps to determine IRMR and MRMR are illustrated in Figure 5.33. The IRMR value is established by adding the JS and JC values to the RBS value. Once the IRMR rating has been established, the MRMR value is determined by adjusting the IRMR value to account for the effects of weathering, joint orientation, mining-induced stresses, blasting and water.

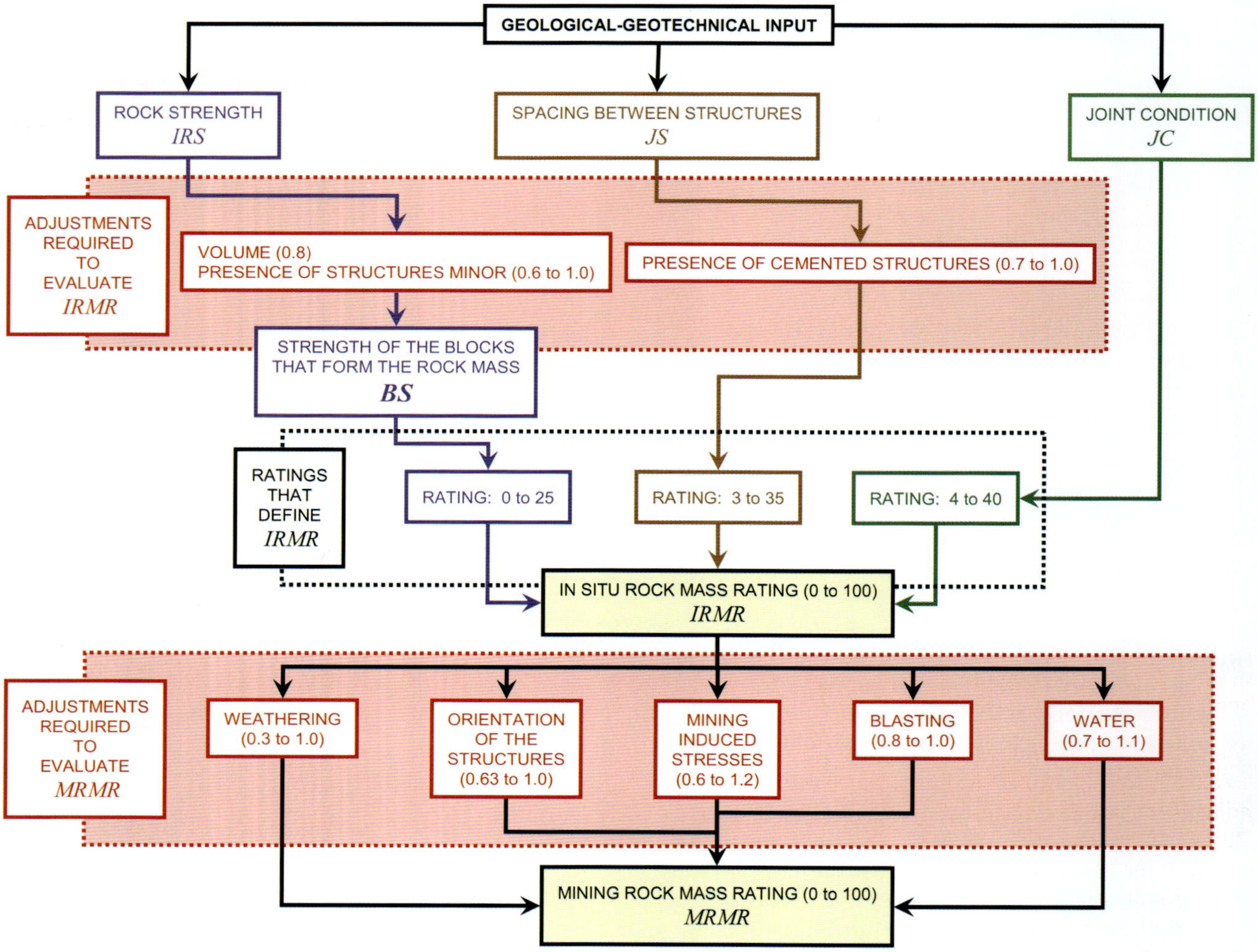

Figure 5.33: Procedures involved in evaluating IRMR and MRMR
Source: Modified from Laubscher & Jakubec (2001)

5.4.3.1 Intact rock strength

If the intact rock sample is homogenous, then it is considered that the IRS value is equal to the UCS value. If the sample is heterogenous, containing zones of weaker rock due to internal defects such as microfractures, foliation or weaker mineral clasts, then an equivalent value is determined considering the strength of both types of rock and their percentages in the sample, using the chart given in Figure 5.34.

As an example (Laubscher & Jakubec 2001), a strong rock sample (UCS = 150 MPa) contains zones of weak rock (UCS = 30 MPa) over 45% of its total volume. The relative strength of the weak rock is 20% of the strong rock ($30/150 \times 100$).

Using Figure 5.34, draw a horizontal line from point Y = 45 on the Y-axis until it intersects the 20% relative strength curve. Then draw a vertical line to the X-axis, which provides an equivalent IRS strength value of 37% of the stronger rock, i.e. 55 MPa.

5.4.3.2 Rock block strength

The rock block strength (BS in Figure 5.33) is the strength of the joint bound primary block of rock adjusted for sample size and any non-continuous fractures and veins within the block. The adjustment for sample size is such that the conversion from core or hand specimen to rock block is approximately 80% of the IRS value. If internal fractures and veins are present, a further adjustment is made based on the number of veins per metre and the Moh's hardness number of the vein infilling, using the chart given in Figure 5.35.

Only Moh's hardness values up to 5 are used in the procedure, as Laubscher and Jakubec (2001) considered that values greater than 5 are not likely to be significant. Open fractures and veins are allocated a value of 1.

As an example of the adjustment factor required for a block containing a number of gypsum veins (Laubscher & Jakubec 2001), a block with an IRS value of 100 MPa contains an average 8 veins of gypsum per metre. The

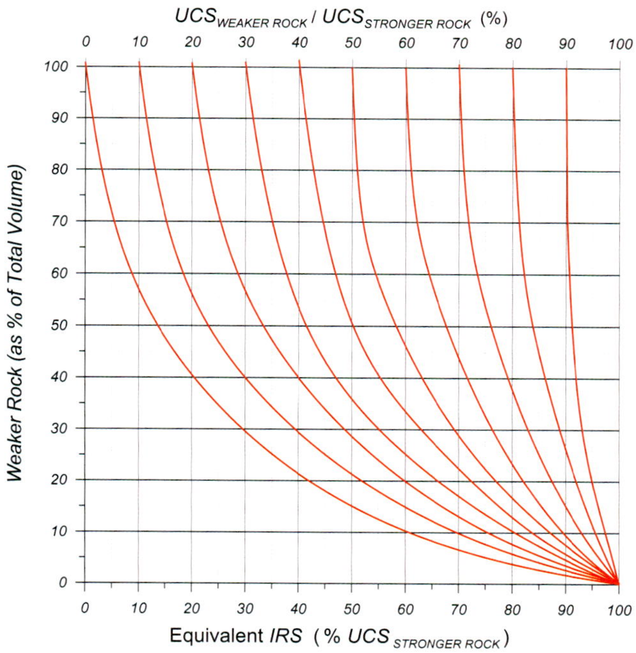

Figure 5.34: Evaluating an equivalent IRS value in the case of heterogenous rock samples of intact rock
Source: Laubscher & Jakubec (2001)

Moh's hardness of gypsum is 2. The ratio of the vein frequency to fill hardness is thus 4, which provides an adjustment factor on the Y-axis of Figure 5.35 of 0.75. The BS value is thus 60 MPa (0.8 × 0.75 × 100).

Once the BS value has been established, the corresponding rating is applied using the chart in Figure 5.36. For the example given above (BS = 60 MPa), the rating is 18.

5.4.3.3 *Joint spacing*

The rating for joint spacing (JS) is determined for open joints using the chart given in Figure 5.37. Laubscher and

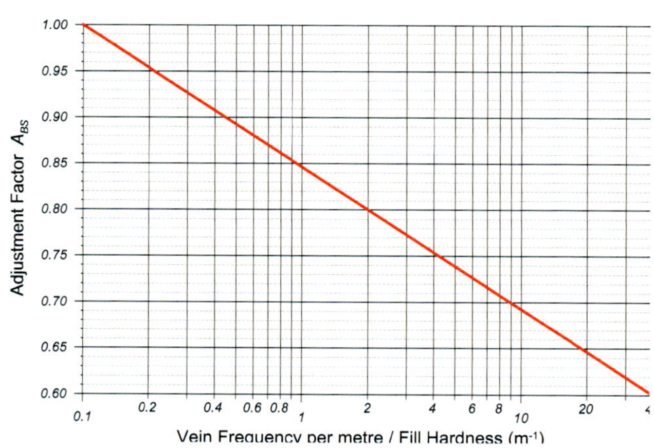

Figure 5.35: Adjustment factor for RBS as a function of the Moh's hardness of the fillings and the frequency of the veins within the rock block
Source: Laubscher & Jakubec (2001)

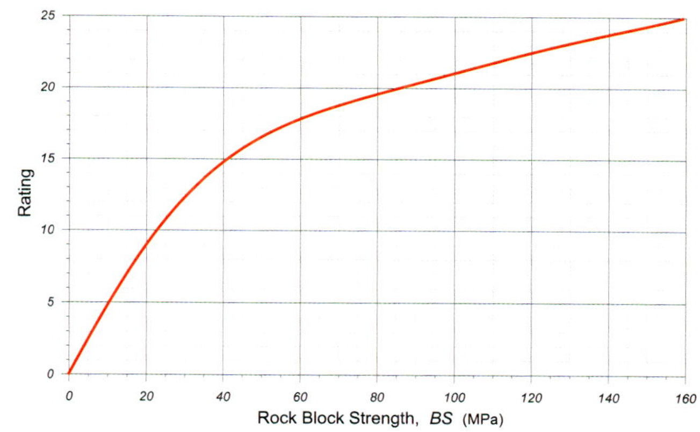

Figure 5.36: Rating values for BS
Source: Laubscher & Jakubec (2001)

Jakubec (2001) noted that where there are more than three joints sets, for simplicity they should be reduced to three. If cemented joints form a distinct set and the strength of the cement is less than the strength of the host rock, the rating for open joints is adjusted downwards using the chart given in Figure 5.38.

As an example (Laubscher & Jakubec 2001), if the rating for two open joints at a spacing of 0.5 m was 23, an additional cemented joint with a spacing of 0.85 m would have an adjustment factor of 90%. The final rating would thus be 21, which is equivalent to three open joint sets with an average spacing of 0.65 m.

5.4.3.4 *Joint condition*

If the rock mass contains only one set of structures the maximum rating of 40 is adjusted downward in line with relevant factors (see Table 5.27). As an example, if the joints in a single set are curved, stepped and smooth but do not have fillings and the walls are not altered, the adjusted JC rating would be 32 (0.90 × 0.90 × 40). If there

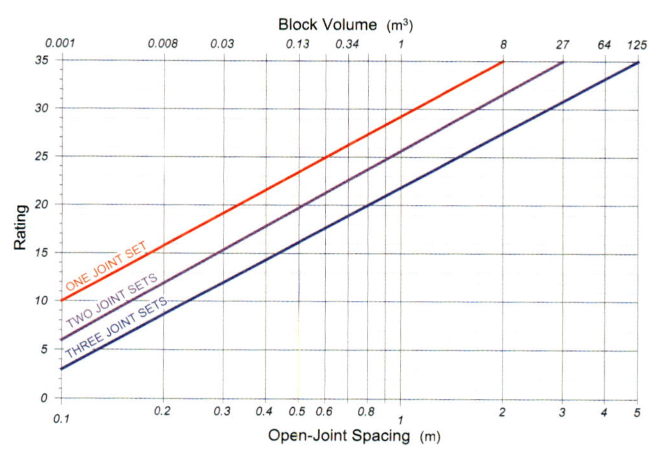

Figure 5.37: Rating for open joint spacing
Source: Laubscher & Jakubec (2001)

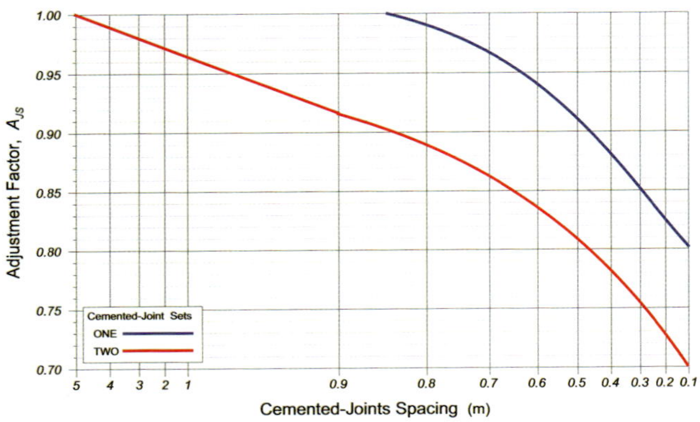

Figure 5.38: Adjustment factor for cemented joints where the strength of the cement is less than the strength of the host rock
Source: Modified from Laubscher &Jakubec (2001)

Table 5.27: Joint condition adjustments for a single joint set

Characteristics of the joints	Adjustment % of 40
A: Roughness at a large scale	
Wavy–multidirectional	1.00
Wavy–unidirectional	0.95
Curved	0.90
Straight/slight undulations	0.85
B: Roughness at a small scale (200 × 200 mm)	
Rough–stepped/irregular	0.95
Smooth–stepped	0.90
Slickensided–stepped	0.85
Rough–undulating	0.80
Smooth–undulating	0.75
Slickensided–undulating	0.70
Rough–planar	0.65
Smooth–planar	0.60
Slickensided–planar	0.55
C: Alteration of the rock walls	
The rock wall is altered and weaker than the filling	0.75
D: Gouge fillings	
Gouge thickness < amplitude asperities of the rock wall	0.60
Gouge thickness > amplitude asperities of the rock wall	0.30
E: Cemented structures/filled joints (infill weaker than rock wall)	
Hardness of the infill: 5	0.95
4	0.90
3	0.85
2	0.80
1	0.75

Source: Laubscher & Jakubec (2001)

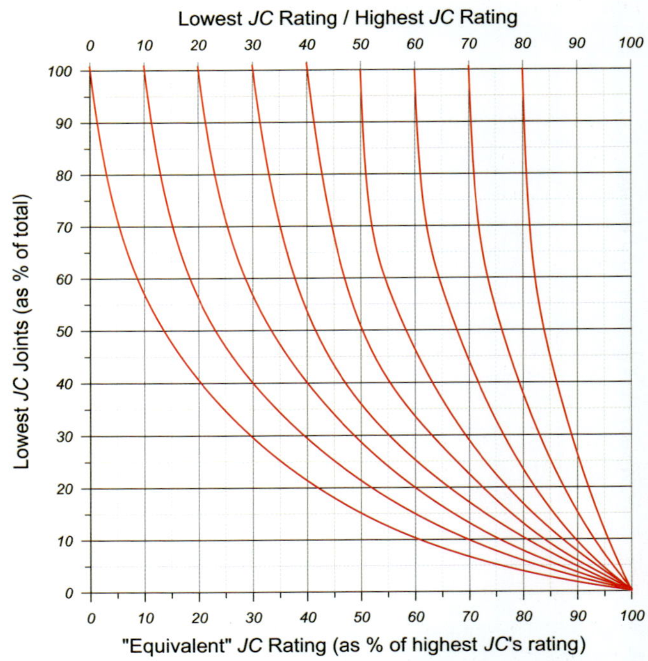

Figure 5.39: Estimating an equivalent rating for JC when the rock mass contains more than one joint set
Source: Laubscher & Jakubec (2001)

is more than one joint, the chart given in Figure 5.39 is used to determine an equivalent rating from the joint sets with the highest and lowest ratings.

As an example of determining an equivalent rating from the joint sets with the highest and lowest rating, assume that the best rating for several joint sets is 36 and the worst is 18, and the worst joints comprise 30% of the total number of joints. The relative rating of the worst to best joints is 50% of the best ones (18/36 × 100).

On Figure 5.39, draw a horizontal line from point Y = 30 on the Y-axis until it intersects the 50% lowest/highest relative rating curve. Then draw a vertical line to the X-axis, which provides an equivalent JC rating of 69% of the value for the best joints, i.e. 25.

5.4.3.5 Establishing MRMR from IRMR

To establish MRMR, the IRMR value is adjusted to account for the effects of weathering, joint orientation, mining-induced stresses, blasting and water. Tables outlining the adjustment factors for weathering, joint orientation, blasting and water are presented in Tables 5.28–5.31. Once the adjustment factors have been determined, the MRMR value is calculated as the product of the IRMR value and the adjustment factors.

Adjustment factors for mining-induced stresses are not tabulated by Laubscher and Jakubec. Mining-induced stresses are recognised by Laubscher and Jakubec (2001) as the redistribution of regional or mine-scale stresses as

Table 5.28: Adjustment factors for the effect of weathering

Degree of weathering	Time of exposure to weathering (years)				
	0.5	1	2	3	≥4
No weathered (fresh)	1.00	1.00	1.00	1.00	1.00
Slightly weathered	0.88	0.90	0.92	0.94	0.96
Moderately weathered	0.82	0.84	0.86	0.88	0.90
Highly weathered	0.70	0.72	0.74	0.76	0.78
Completely weathered	0.54	0.56	0.58	0.60	0.62
Residual soil (saprolite)	0.30	0.32	0.34	0.36	0.38

Source: Laubscher & Jakubec (2001)

a result of the geometry and orientation of an underground excavation. The adjustment factors are judged to range from as low as 0.60 to as high as 1.20, and their evaluation requires considerable experience of underground mining operations. The example given by Laubscher and Jakubec (2001) is for a caving operation where stresses at a large angle to structures will increase the stability of the rock mass and inhibit caving. In this case the allocated adjustment is 1.20. Conversely, stresses at a low angle will result in shear failure and have an adjustment factor of 0.70.

The example of mining-induced stresses emphasises that the MRMR system was primarily developed from the Bieniawski RMR system to cater for diverse mining situations, principally those underground. The fundamental difference noted by Laubscher (1990) was that the in situ rock mass rating (IRMR) needed to be adjusted according to the mining environment so that the final ratings (MRMR) could be used for mine design. Practical design applications of the MRMR system cited by Laubscher and Jakubec (2001) include the stability of open stopes, pillar design, the determination

Table 5.29: Adjustment factors for the effect of joint orientation

No. joints defining the block	No. block faces inclined from vertical	*JC* rating		
		0–15	16–30	31–40
3	3	0.70	0.80	0.95
	2	0.80	0.90	0.95
4	4	0.70	0.80	0.90
	3	0.75	0.80	0.95
	2	0.85	0.90	0.95
5	5	0.70	0.75	0.80
	4	0.75	0.80	0.85
	3	0.80	0.85	0.90
	2	0.85	0.90	0.95
	1	0.90	0.95	

Source: Laubscher & Jakubec (2001)

Table 5.30: Adjustment factors for the effect of blasting

Blasting technique	Adjustment factor, A_{BLAST}
Mechanical excavation/boring	1.00
Smooth-wall blasting	0.97
Good conventional blasting	0.94
Poor blasting	0.80

Source: Laubscher & Jakubec (2001)

of cavability, caving fragmentation and the extent of cave and failure zones.

It is important that the underground origins of the MRMR system are recognised and the appropriate judgments and interpretations made when it is applied to open pit mining situations. For example, as for Bieniawski RMR (Table 5.22), when dealing with pit slope design problems adjustments for joint orientation and groundwater should be unnecessary as both should be accounted for in the stability analyses. Similarly, mining-induced stress and their effect around a large open pit will be different from those underground. Adjustments for the effect of weathering and blasting may be, however, highly relevant in the open pit situation. Overall, the objective is for the engineering geologist, rock mechanics engineer and planning engineer to adjust the IRMR situation.

Finally, it is pointed out that the IRMR procedures and MRMR adjustments described above are the most recently publisheded (Laubscher & Jakubec 2001) and reflect changes since the system was first introduced (Laubscher 1977). As with Bieniawski RMR, it is important that the date of publication is stated if an earlier version of the procedure is being used.

5.4.4 Hoek-Brown GSI

The Hoek-Brown Geological Strength Index (GSI) concept was born in 1980 when it was used in the original Bieniawski RMR (Bieniawski 1974a, 1974b) format in support of the newly developed Hoek-Brown rock mass failure criterion (Hoek & Brown 1980b). Since then it has undergone numerous changes, principally between 1992 and 1995, with the name GSI officially emerging in 1995 (Hoek et al. 1995).

Table 5.31: Adjustment factors for the effect of water

Water condition	Adjustment factor, A_{WATER}
Moist	0.95–0.90
Water inflows 25–125 L/min, water pressures 1–5 MPa	0.90–0.80
Water inflows >125 L/min, water pressures >5 MPa	0.70–0.80

Source: Laubscher & Jakubec (2001)

The 1992 change (Hoek et al. 1992) is seminal, as it saw the use of RMR discontinued and the rock mass characterised in terms of:

- the block shapes and the degree of interlock;
- the surface condition of the intersecting defects.

The principal reason for moving from RMR to the new classification system was that it was judged to be a more adequate vehicle for relating the Hoek-Brown failure criterion to geological observations in the field (Hoek et al. 2002). It was also claimed to overcome an effective double-counting of the uniaxial compressive strength of the intact pieces of rock, which was included in both the RMR classification process and the Hoek-Brown strength computations.

The modified format proposed in 1992 was represented in 1993 (Hoek et al. 1993) without any changes except that the rock mass characterisation table was extended to include values for Young's modulus (E) and Poisson's ratio (ν) (Table 5.32).

Although many practitioners were comfortable with a system based more heavily on fundamental geological observations and less on the numbers provided by the RMR system, probably an equal number regretted that the modification had expunged the numerical accounting of RMR from the rock mass classification process. As a result, in 1995 a numerical system, known as the Geological Strength Index (GSI), was reintroduced and Table 5.32 was replaced (see Table 5.33). The tables are similar except for the addition of GSI to Table 5.33.

Table 5.32: Hoek-Brown rock mass classification system, 1993

MODIFIED HOEK-BROWN CRITERION

$$\sigma_1' = \sigma_3' + \sigma_c \left(m_b \frac{\sigma_3'}{\sigma_c} \right)^a$$

σ_1' = major principal effective stress at failure

σ_3' = minor principal effective stress at failure

σ_c = uniaxial compressive strength of *intact* pieces of rock

m_b and a are constants which depend on the composition, structure and surface conditions of the rock mass

STRUCTURE	SURFACE CONDITION	VERY GOOD Very rough, unweathered surfaces	GOOD Rough, slightly weathered, iron stained surfaces	FAIR Smooth, moderately weathered or altered surfaces	POOR Slickensided, highly weathered surfaces with compact coatings or fillings containing angular rock fragments	VERY POOR Slickensided, highly weathered surfaces with soft clay coatings or fillings
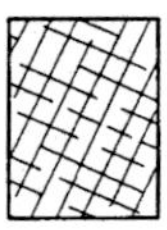 BLOCKY - very well interlocked undisturbed rock mass consisting of cubical blocks formed by three orthogonal discontinuity sets	m_b/m_i	0.7	0.5	0.3	0.1	
	a	0.3	0.35	0.4	0.45	
	E	50,000	40,000	25,000	5,000	
	ν	0.2	0.2	0.2	0.25	
 VERY BLOCKY - interlocked, partially disturbed rock mass with angular blocks formed by four or more discontinuity sets	m_b/m_i	0.3	0.2	0.1	0.04	
	a	0.4	0.45	0.5	0.5	
	E	25,000	15,000	5,000	1,000	
	ν	0.2	0.25	0.25	0.25	
 BLOCKY/SEAMY - folded and faulted with many intersecting discontinuities forming angular blocks	m_b/m_i		0.08	0.04	0.01	0.004
	a		0.5	0.5	0.55	0.6
	E		3,000	1,000	700	500
	ν		0.25	0.25	0.3	0.3
 CRUSHED - poorly interlocked, heavily broken rock mass with a mixture of angular and rounded blocks	m_b/m_i		0.03	0.015	0.003	0.001
	a		0.5	0.55	0.6	0.65
	E		1,000	800	500	300
	ν		0.25	0.3	0.3	0.3

Source: Hoek et al. (1992)

Table 5.33: Hoek-Brown rock mass classification system, 1995

GENERALIZED HOEK-BROWN CRITERION

$$\sigma_1' = \sigma_3' + \sigma_c \left(m_b \frac{\sigma_3'}{\sigma_c} + s \right)^a$$

σ_1' = major principal effective stress at failure

σ_3' = minor principal effective stress at failure

σ_c = uniaxial compressive strength of *intact* pieces of rock

m_b, s and a are constants which depend on the composition, structure and surface conditions of the rock mass

STRUCTURE	SURFACE CONDITION	VERY GOOD Very rough, unweathered surfaces	GOOD Rough, slightly weathered, iron stained surfaces	FAIR Smooth, moderately weathered or altered surfaces	POOR Slickensided, highly weathered surfaces with compact coatings or fillings containing angular rock fragments	VERY POOR Slickensided, highly weathered surfaces with soft clay coatings or fillings
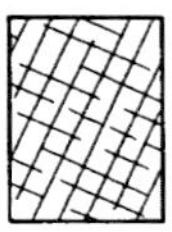 BLOCKY - very well interlocked undisturbed rock mass consisting of cubical blocks formed by three orthogonal discontinuity sets	m_b / m_i	0.60	0.40	0.26	0.16	0.08
	s	0.190	0.062	0.015	0.003	0.0004
	a	0.5	0.5	0.5	0.5	0.5
	E_m	75,000	40,000	20,000	9,000	3,000
	ν	0.2	0.2	0.25	0.25	0.25
	GSI	85	75	62	48	34
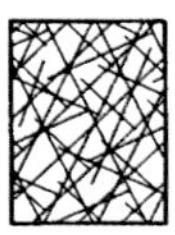 VERY BLOCKY - interlocked, partially disturbed rock mass with angular blocks formed by four or more discontinuity sets	m_b / m_i	0.40	0.29	0.16	0.11	0.07
	s	0.062	0.021	0.003	0.001	0
	a	0.5	0.5	0.5	0.5	0.53
	E_m	40,000	24,000	9,000	5,000	2,500
	ν	0.2	0.25	0.25	0.25	0.3
	GSI	75	65	48	38	25
 BLOCKY/SEAMY - folded and faulted with many intersecting discontinuities forming angular blocks	m_b / m_i	0.24	0.17	0.12	0.08	0.06
	s	0.012	0.004	0.001	0	0
	a	0.5	0.5	0.5	0.5	0.55
	E_m	18,000	10,000	6,000	3,000	2,000
	ν	0.25	0.25	0.25	0.3	0.3
	GSI	60	50	40	30	20
 CRUSHED - poorly interlocked, heavily broken rock mass with a mixture of angular and rounded blocks	m_b / m_i	0.17	0.12	0.08	0.06	0.04
	s	0.004	0.001	0	0	0
	a	0.5	0.5	0.5	0.55	0.6
	E_m	10,000	6,000	3,000	2,000	1,000
	ν	0.25	0.25	0.3	0.3	0.3
	GSI	50	40	30	20	10

Source: Hoek et al. (1995)

All the GSI values in Table 5.33 greater than 25 are exactly the same as those of the Bieniawski RMR_{1976} system. They can be determined visually from surface outcrops, using the chart, or numerically from drill core, using Bieniawski RMR_{1976}. If Bieniawski RMR_{1979} is used, the GSI value is RMR_{1979} - 5 (Table 5.32). If neither RMR nor GSI can be directly calculated, a suggested alternative is the empirical relationship between the Barton tunnelling index, Q (Barton et al. 1974), and Bieniawski's RMR_{1976} (Bieniawski 1979):

$$\text{Bieniawski } RMR_{1976} = 9 \ln Q + 44 \qquad \text{(eqn 5.53)}$$

This relationship must be used cautiously. The Q-index is used in tunnel design, not open pit mining. Furthermore, the correlation was developed from 111 tunnelling projects of which half (62) were from Scandinavia and a quarter (28) were from South Africa (Bieniawski 1979), so it is unlikely that it is unique for all geological environments and rock types.

The GSI cut-off value of 25 came about following the realisations that Bieniawski's RMR was difficult to apply to very poor quality rock masses and that the relationship between RMR and the Hoek-Brown strength criterion m and s parameters (section 5.5.2) was no longer linear when the RMR values were less than 25 (Hoek et al. 1995). The name 'geological strength index' was used to stress the importance of fundamental geological observations about the blockiness of the rock mass and

Table 5.34: Hoek-Brown rock mass classification system, 2000

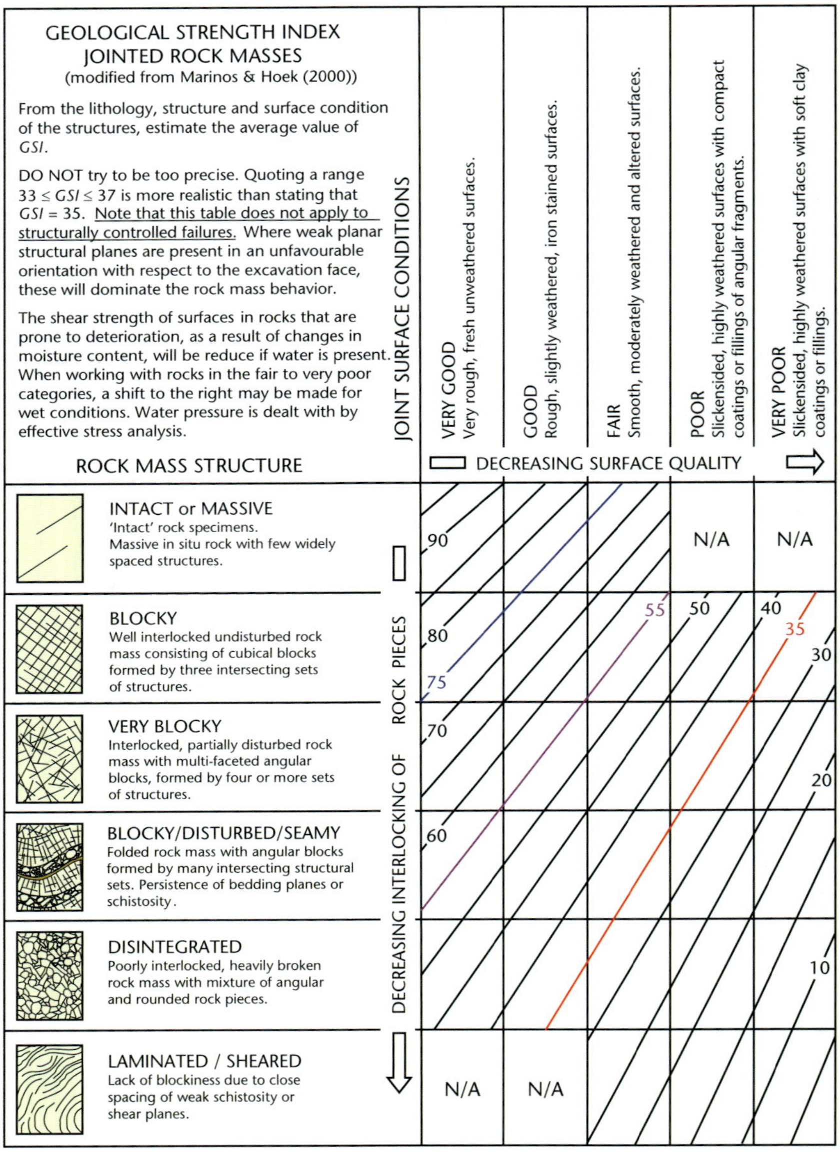

Source: Marinos & Hoek (2000)

the condition of the joint surfaces to the classification system.

Subsequent publications (Hoek & Brown 1997; Marinos & Hoek 2000) saw Table 5.33 modified and issued in the form shown in Table 5.34.

The principal changes between Table 5.33 and Table 5.34 are the presentation of only the GSI values across each box in the table and the introduction of the laminated/sheared rock mass structural classification.

Table 5.34 is now the GSI chart most used in practice. It has been extended to accommodate some of the most variable of rock masses and to project information gained from surface outcrops to depth (Hoek et al. 1998; Marinos & Hoek 2001; Marinos et al. 2005; Hoek et al. 2005).

Attempts have also been made to quantify GSI using joint frequency and orientation statistics (Sonmez & Ulusay 1999; Cai et al. 2004). The variety of these approaches emphasise the need to remember the assumptions that underpin the GSI classification system and the Hoek-Brown strength criterion it supports – that the rock mass is an isotropic clump of intact rock pieces separated by closely spaced joints for which there is no preferred failure direction. As noted in Table 5.34, it follows that the GSI system should not be used when a clearly defined, dominant structural system is evident in the rock mass. This is potentially the case for a number of the rock types nominated in some proposed extensions of the system, including bedded or fissile siltstone, mudstone, shale, flysch, schist and gneiss. These rock types should only be accommodated if they have been tectonically damaged and their structural preferences lost.

5.5 Rock mass strength

5.5.1 Introduction

Historically, the Mohr-Coulomb measures of friction (Ø) and cohesion (c) have been used to represent the strength of the rock mass. This practice was based on soil mechanics experience and methodology and assumed that the size of the rock particles in high, closely jointed rock masses were equivalent to an isotropic mass of soil particles. This assumption enabled rock slope design practitioners to adopt the Mohr-Coulomb measures of friction (Ø) and cohesion (c) and led to their embedment in the limit equilibrium stability analysis procedures that were introduced in the 1970s and 1980s. Subsequently, the use of Mohr-Coulomb strength parameters carried over into all the continuum and discontinuum numerical modelling tools that are now common in pit slope design.

It quickly became obvious that obtaining good triaxial measures of friction and cohesion for normal rock masses was not easy. The reasons were various, but usually included:

- the difficulty of performing tests on rock at a scale at the same order of magnitude as the real thing;
- the difficulty of getting good undisturbed samples from drill holes cored in rock which was already disturbed or damaged in some way;
- the scarcity of appropriate triaxial testing equipment and experienced operators;
- cost.

Initially, the preferred means of overcoming these difficulties was to derive empirical values of friction and cohesion from rock mass classification schemes that were calibrated from experience. The classic example of this practice is the calibration of friction and cohesion against RMR by Bieniawski, as shown in Table 5.35 (Bieniawski 1979, 1989).

Subsequently, many strength criteria were developed for rock (Franklin & Dusseault 1989; Sheorey 1997; Zhang 2005), but the best-known in mining engineering are the Laubscher and the Hoek-Brown rock mass strength criteria. A lesser-known but quite widely used system in open pit mines in North and South America is the CNI criterion developed by Call & Nicholas Inc. (Call et al. 2000).

The Laubscher, Hoek-Brown and CNI criteria are outlined below in sections 5.5.2, 5.5.3 and 5.5.4. They are followed by an outline of a method to account for the directional shear strength of a rock mass (section 5.5.5) and a newly developed synthetic rock mass model that may provide a means of honouring the strength of the rock mass without relying on Mohr-Coulomb, Hoek-Brown or other such constitutive models (section 5.5.6).

Table 5.35: RMR calibrated against rock mass quality and strength

RMR rating	Description	Ø°	Cohesion (kPa)
81–100	Very good rock	>45	>400
61–80	Good rock	35–45	300–400
41–60	Fair rock	25–35	200–300
40–21	Poor rock	15–25	100–200
<21	Very poor rock	<15	<100

5.5.2 Laubscher strength criteria

There are two Laubscher criteria, the rock mass strength criterion and the design rock mass strength criterion. Both are intended for use in underground mining.

5.5.2.1 Rock mass strength criterion

The rock mass strength (RMS) is derived from the IRS (section 5.4.3.1) and the IRMR (Figure 5.33 and section 5.4.3.5) according to the following procedure (Laubscher & Jakubec 2001).

The strength of the rock mass cannot be higher than the corrected average IRS of that zone. The IRS has been obtained from the testing of small specimens. However, test work on large specimens shows that large specimens have strengths 80% of the small specimen. As the rock mass is a 'large' specimen the IRS must be reduced to 80% of its value. Thus the strength of the rock mass would be IRS × 80% if it had no joints. The effect of the joints and their frictional properties is to reduce the strength of the rock mass.

In the IRMR classification ratings, a rating of 20 is given to all specimens with an IRS greater than 185 MPa because at those high values the IRS has little effect on the relative rock mass strength. On this basis the RMS must be calculated in a similar manner, i.e. that above 185 MPa the value of 200 MPa is used regardless of the IRS value.

Given these conditions, the following procedure is adopted to calculate the rock mass strength.

1. The IRS rating (B) is subtracted from the total rating (A) therefore the balance (i.e. RQD, joint spacing and condition) will be a function of the remaining possible rating of 80.
2. The IRS (C) is reduced to 80% of its value.
3. RMS = (A - B) × C × 100.

For example, if the total rating was 60 with an IRS of 100 MPa and a rating of 10:

$$\text{RMS} = 100\ \text{MPa} \times (60 - 10) \times 100 = 50\ \text{MPa}$$

5.5.2.2 *Design rock mass strength criterion*

The design rock mass strength (DRMS) is the unconfined rock mass strength in a specific underground mining environment. An underground mining operation exposes the rock surface and the concern is with the stability of the zone that surrounds the opening. The extent of this zone depends on the size of the opening and, except with mass failure, instability propagates from the rock surface. The size of the rock block generally defines the first zone of instability. Adjustments relevant to the specific mining environment are applied to the RMS to give the DRMS. As the DRMS is in MPa it can be related to the mining-induced stresses. Therefore, the adjustments used are those for weathering, joint orientation and blasting (Tables 5.28–5.30).

For example, for an RMS of 50 and weathering = 85%, orientation = 75% and blasting = 90%, then the total adjustments = 57%:

$$\text{DRMS} = 50 \times 57\% = 29 \text{ MPa}$$

5.5.3 Hoek-Brown strength criterion

The Hoek-Brown strength criterion was first published in 1980 (Hoek & Brown 1980a, 1980b) in the form:

$$\sigma_1' = \sigma_3' + \sigma_c(m * \sigma_3'/\sigma_c + s)^{1/2} \qquad \text{(eqn 5.54)}$$

where:

σ_1' = major principal effective stress at failure

σ_3' = minor effective principal stress at failure

σ_c = uniaxial compressive strength of the intact rock

m = dimensionless material constant for rock

s = dimensionless material constant for rock, ranging from 1 for intact rock with tensile strength to 0 for broken rock with zero tensile strength. c is 0 when the effective normal stress is 0.

In 1992 (Hoek et al. 1992) the criterion was modified to eliminate the tensile strength predicted by the original criterion:

$$\sigma_1' = \sigma_3' + \sigma_c(m_b * \sigma_3/\sigma_c)^a \qquad \text{(eqn 5.55)}$$

where m_b and a are constants for the broken rock. It was assumed that the jointed rock mass was undisturbed and only its inherent properties were considered. Values of m_b were estimated by substitution of the value for m_i into m_b/m_i (Table 2 in Table 5.36). Values of a were estimated directly from Table 3 within Table 5.36.

In 1995 (Hoek et al. 1995) the criterion was modified again. The generalised Hoek-Brown criterion was retained in the form of equation 5.55 but was replaced by the GSI. The introduction of GSI also saw the concept of 'disturbed' and 'undisturbed' rock being dropped. Originally, the 'disturbed' Hoek-Brown rock mass strength values were derived by reducing the RMR value by one row in the classification table, a somewhat arbitrary procedure. Instead, it was decided to let the users make their own judgments of how much to reduce the GSI value to account for the strength loss.

The values of m_b/m_i, s and a were set as follows:
For GSI > 25 (undisturbed rock masses):

$$m_b/m_i = e^{\left(\frac{GSI-100}{28}\right)} \qquad \text{(eqn 5.56)}$$

$$s = e^{\left(\frac{GSI-100}{9}\right)} \qquad \text{(eqn 5.57)}$$

$$a = 0.5 \qquad \text{(eqn 5.58)}$$

For GSI < 25 (undisturbed rock masses):

$$s = 0 \qquad \text{(eqn 5.59)}$$

$$a = 0.65 - \frac{GSI}{200} \qquad \text{(eqn 5.60)}$$

The third and final modification was made in 2002 (Hoek et al. 2002), when the values of m_b, a and s were restated:

$$m_b = m_i e^{\left(\frac{GSI-100}{28-14D}\right)} \qquad \text{(eqn 5.61)}$$

$$a = \frac{1}{2} + \frac{1}{6}(e^{-GSI/15} - e^{-20/3}) \qquad \text{(eqn 5.62)}$$

$$s = e^{\left(\frac{GSI-100}{9-3D}\right)} \qquad \text{(eqn 5.63)}$$

The introduction of the parameter D represents a re-evaluation of the 'undisturbed' versus 'disturbed' question that in the 1995 generalised equation had been left for the user to decide by making appropriate adjustments to the GSI value. It was reintroduced to represent the degree of disturbance to which the rock mass has been subjected by blast damage and stress relaxation, ranging from D = 0 for undisturbed rock to D = 1 for very disturbed rock masses (Table 5.37). The influence of the parameter can be large and its application requires experience and judgment. Hoek et al. (2002) gave an example using σ_{ci} = 50 MPa, m_i = 10 and GSI = 45. For D = 0 in a tunnel at a depth of 100 m the derived equivalent friction angle is 47° and the cohesion 580 Pa. For D = 1 in a highly disturbed slope 100 m high the equivalent friction angle is reduced to 28° and the cohesion to 350 kPa. Experience-based starting points for judging the extent of the blast-damaged zone resulting from open-pit mine production blasting are given by Hoek & Karzulovic (2000). Ultimately, the value selected for *D* should be validated through observation and measured performance.

The procedures for calculating the instantaneous effective friction angle and cohesion values for any particular normal stress are essentially the same as for the generalised 1995 criterion, although the process can be simplified by using the freeware RocLab program

Table 5.36: Modified Hoek-Brown failure criterion, 1992

Table 2. Values of constant m_i for intact rock, by rock group.

Grain size	Sedimentary Carbonate	Sedimentary Detrital	Sedimentary Chemical	Metamorphic Carbonate	Metamorphic Silicate	Igneous Felsic	Igneous Mafic	Igneous Mafic
Coarse	Dolomite 10.1	Conglomerate (20)		Marble 9.3	Gneiss 29.2	Granite 32.7	Gabbro 25.8	Norite 21.7
Medium	Chalk 7.2	Sandstone 18.8	Chert 19.3		Amphibolite 31.2		Dolerite 15.2	
Fine	Limestone 8.4	Siltstone 9.6	Gypstone 15.5		Quartzite 23.7	Rhyolite (20)	Andesite 18.9	Basalt (17)
Very Fine		Claystone 3.4	Anhydrite 13.2		Slate 11.4			

Values shown were derived from statistical analysis of triaxial test data for each rock type. Values in parenthesis have been estimated

Table 3. Estimation of m_b / m_i and a based on rock structure and surface condition

MODIFIED HOEK-BROWN FAILURE CRITERION

$$\sigma'_1 = \sigma'_3 + \sigma_c \left(m_b \frac{\sigma'_3}{\sigma_c} \right)^a$$

σ'_1 = major principal effective stress at failure

σ'_3 = minor principal effective stress at failure

σ_c = uniaxial compressive strength of *intact* pieces in the rock

m_b and a are constants which depend on the composition, structure and surface conditions of the rock mass

STRUCTURE	SURFACE CONDITION	VERY GOOD Unweathered, discontinuous, very tight aperture, very rough surface, no infilling	GOOD Slightly weathered, continuous, tight aperture, rough surface, iron staining to no infilling	FAIR Moderately weathered, continuous, extremely narrow, smooth surfaces, hard infilling	POOR Highly weathered, continuous, very narrow, polished/slickensided surfaces, hard infilling	VERY POOR Highly weathered, continuous, narrow, polished/slickensided surfaces, soft infilling
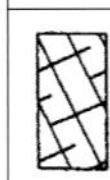 BLOCKY - well interlocked, undisturbed rock mass; large to very block size	m_b / m_i	0.7	0.5	0.3	0.1	
	a	0.3	0.35	0.4	0.45	
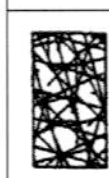 VERY BLOCKY - interlocked, partially disturbed rock mass; medium block sizes	m_b / m_i	0.3	0.2	0.1	0.04	
	a	0.4	0.45	0.5	0.5	
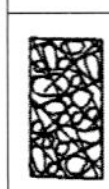 BLOCKY/SEAMY - folded and faulted, many interesting joints; small blocks	m_b / m_i		0.08	0.04	0.01	0.004
	a		0.5	0.5	0.55	0.6
 CRUSHED - poorly interlocked, highly broken rock mass; very small blocks	m_b / m_i		0.03	0.015	0.003	0.001
	a		0.5	0.55	0.6	0.65

Source: Hoek et al. (1992)

(Rocscience 2005a) using the appropriate values for GSI, σ_{ci} and m_i. Preferably, the value for m_i should be obtained from laboratory UCS (σ_{ci}) and triaxial tests of samples of the intact rock, which can be processed using the freeware RocData program (Rocscience 2004a). If this is not possible, m_i can be estimated from a tabulated list of examples in RocLab. Indicative values are given in Table 5.38.

Important points to remember when determining m_i from UCS and triaxial tests are:

- the confining pressure (σ_3) values used for the triaxial test should range from 0 to 50% of the UCS (σ_{ci}) strength. Confining pressures different from these limits can have a significant influence on the m_i values (Read et al. 2005);
- because of the inherent problems with UCS testing, e.g. sample splitting and the potential sensitivity of the sample to inclined imperfections, there tends to be quite a lot of scatter in the uniaxial data. If so, these data overwhelm the triaxial data in the fitting process. To overcome these problems the procedure should rely on the triaxial data. Where there are uniaxial data, the average of all uniaxial data points should be used. In this way the single uniaxial value will have the same

Table 5.37: Guidelines for estimating the disturbance factor, *D*

Appearance of Rock Mass	Description of Rock Mass	Suggested Value of D
	Excellent quality controlled blasting or excavation by tunnel boring machine results in minimal disturbance of the confined rock mass surrounding the tunnel.	$D = 0$
	Mechanical or hand excavation in poor quality rock masses (no blasting) results in minimal disturbance to the surrounding rock mass. Where squeezing problems result in significant floor heave, disturbance can be severe unless a temporary invert, as shown in photograph is placed.	$D = 0$ $D = 0.5$ No invert
	Very poor quality blasting in a hard rock tunnel results in severe local damage, extending 2 or 3 m, in the surrounding rock mass.	$D = 0.8$
	Small scale blasting in civil engineering slopes in modest rock mass damage, particularly if controlled blasting is used as shown on the left hand side of the photograph. However, stress relief results in some disturbance.	$D = 0.7$ Good blasting $D = 1.0$ Poor blasting
	Very large open pit mine slopes suffer significant disturbance due to heavy production blasting and also due to stress relief from overburden removal. In some softer rocks excavation can be carried out by ripping and dozing and the degree of damage to the slope is less.	$D = 1.0$ Production blasting $D = 0.7$ Mechanical excavation

Source: Hoek et al. (2002)

weight as the triaxial points, which were generally are limited to one value per confining stress (Hoek 2005, editor's pers comm).

Equations given in Hoek et al. (2002) for determining the Young's modulus of the rock mass (E_{rm}) using the GSI system for values of σ_{ci} either less than or greater than 100 MPa have been modified by Hoek and Diederichs (2006) into a single equation incorporating both GSI and D:

$$E_{rm}\,(MPa) = 100{,}000\left(\frac{1 - D/2}{1 + e^{(75 + 25D - GSI)/11}}\right)$$

(eqn 5.64)

5.5.4 CNI criterion

As an alternative to methods based on RMR assessments, Call & Nicholas Inc. (CNI) developed a criterion that relates the strength of the rock mass directly to the degree of fracturing present in the rock mass through a

Table 5.38: Indicative values for m_i for some rocks

<table>
<tr><th rowspan="3">Rock type</th><th rowspan="3">Class</th><th rowspan="3">Group</th><th colspan="12">Texture</th></tr>
<tr><th colspan="3">Coarse
(> 2 mm)</th><th colspan="3">Medium
(0.6–2 mm)</th><th colspan="3">Fine
(0.2–0.6 mm)</th><th colspan="3">Very fine
(< 0.2 mm)</th></tr>
<tr><th></th><th></th><th></th><th></th><th></th><th></th><th></th><th></th><th></th><th></th><th></th><th></th></tr>
<tr><td rowspan="9">SEDIMENTARY</td><td rowspan="4" colspan="2">Clastic</td><td colspan="3">← Conglomerates® →
(see Notes)</td><td colspan="6">← Sandstones® →
(15 ± 7)</td><td colspan="3">← Siltstones® →
7 ± 2</td></tr>
<tr><td colspan="3">← Breccias® →
(see Notes)</td><td></td><td></td><td colspan="4">← Greywackes® →
(16 ± 5)</td><td colspan="3">← Claystones® →
4 ± 2</td></tr>
<tr><td></td><td></td><td></td><td></td><td></td><td></td><td></td><td></td><td colspan="4">← Shales® →
(6 ± 2)</td></tr>
<tr><td></td><td></td><td></td><td></td><td></td><td></td><td></td><td colspan="5">← Marls® →
(7 ± 2)</td></tr>
<tr><td rowspan="5">Non-clastic</td><td rowspan="3">Carbonates</td><td></td><td></td><td></td><td colspan="4">← Crystalline limestone® →
(12 ± 3)</td><td></td><td colspan="4">← Micritic limestone® →
(9 ± 2)</td></tr>
<tr><td></td><td></td><td></td><td></td><td></td><td colspan="5">← Sparitic limestone® →
(10 ± 2)</td><td></td><td></td></tr>
<tr><td></td><td></td><td></td><td colspan="9">← Dolomites® →
(9 ± 3)</td></tr>
<tr><td>Evaporites</td><td></td><td></td><td></td><td></td><td></td><td colspan="4">← Gypsum® →
(8 ± 2)</td><td colspan="3">← Anhydrite® →
(12 ± 2)</td></tr>
<tr><td>Organic</td><td></td><td></td><td></td><td></td><td></td><td></td><td></td><td></td><td></td><td colspan="3">← Chalk® →
7 ± 2</td></tr>
<tr><td rowspan="8">METAMORPHIC</td><td rowspan="3" colspan="2">Non-foliated</td><td></td><td colspan="9">← Marble® →
9 ± 3</td><td></td><td></td></tr>
<tr><td></td><td></td><td colspan="3">← Hornfels® →
(19 ± 4)</td><td colspan="4">← Quartzites® →
20 ± 3</td><td></td><td></td><td></td></tr>
<tr><td></td><td></td><td></td><td colspan="4">← Meta-sandstones® →
(19 ± 3)</td><td></td><td></td><td></td><td></td><td></td></tr>
<tr><td rowspan="3" colspan="2">Lightly foliated</td><td></td><td colspan="6">← Gneisses® →
28 ± 5</td><td></td><td></td><td></td><td></td><td></td></tr>
<tr><td></td><td></td><td colspan="6">← Amphibolites® →
26 ± 6</td><td></td><td></td><td></td><td></td></tr>
<tr><td></td><td></td><td></td><td colspan="4">← Migmatites® →
(29 ± 3)</td><td></td><td></td><td></td><td></td><td></td></tr>
<tr><td rowspan="2" colspan="2">Foliated</td><td></td><td></td><td></td><td></td><td></td><td colspan="3">← Phyllites® →
(7 ± 3)</td><td colspan="4">← Slates® →
7 ± 4</td></tr>
<tr><td></td><td></td><td></td><td colspan="4">← Schists® →
12 ± 3</td><td></td><td></td><td></td><td></td><td></td></tr>
<tr><td rowspan="10">IGNEOUS</td><td rowspan="4">Intrusive</td><td rowspan="2">Light</td><td colspan="4">← Granites® →
32 ± 3</td><td></td><td colspan="4">← Diorites® →
25 ± 5</td><td></td><td></td><td></td></tr>
<tr><td></td><td></td><td colspan="5">← Granodiorites® →
(29 ± 3)</td><td></td><td></td><td></td><td></td><td></td></tr>
<tr><td rowspan="2">Dark</td><td colspan="3">← Norites® →
20 ± 5</td><td colspan="4">← Gabbros® →
27 ± 3</td><td></td><td></td><td></td><td></td><td></td></tr>
<tr><td></td><td></td><td></td><td></td><td></td><td colspan="4">← Dolerites® →
(16 ± 5)</td><td></td><td></td><td></td></tr>
<tr><td rowspan="2" colspan="2">Hypabyssal</td><td></td><td colspan="4">← Peridotites® →
(25 ± 5)</td><td colspan="4">← Diabases® →
(15 ± 5)</td><td></td><td></td><td></td></tr>
<tr><td colspan="9">← Porphyries® →
(20 ± 5)</td><td></td><td></td><td></td></tr>
<tr><td rowspan="4">Volcanics</td><td rowspan="2">Lavas</td><td></td><td colspan="4">← Rhyolites® →
(25 ± 5)</td><td colspan="4">← Basalts® →
(25 ± 5)</td><td colspan="3">← Obsidians® →
(19 ± 3)</td></tr>
<tr><td></td><td></td><td></td><td></td><td colspan="4">← Dacites® →
(25 ± 3)</td><td colspan="4">← Andesites® →
25 ± 5</td></tr>
<tr><td rowspan="2">Pyroclastics</td><td colspan="3">← Agglomerates® →
(19 ± 3)</td><td rowspan="2" colspan="6">← Tuffs® →
(13 ± 5)</td><td></td><td></td><td></td></tr>
<tr><td colspan="3">← Breccias® →
(19 ± 5)</td><td></td><td></td><td></td></tr>
</table>

Values in brackets are estimates; the others are from triaxial tests
Source: Karzulovic (2006)

combination of the intact rock strength and the natural fracture strength as a function of RQD (Call et al. 2000).

To determine the modulus of deformation for the rock mass, CNI noted that Bieniawski (1978) proposed the following relationship for the correlation between RQD and the ratio of the rock mass modulus to the intact rock modulus:

$$r = E_m / E_i \qquad \text{(eqn 5.65)}$$

where

E_m = rock mass deformation modulus
E_i = intact rock deformation modulus

and

$$r = \alpha e^{\beta(\%\text{RQD})} \qquad \text{(eqn 5.66)}$$

for

$\alpha = 0.225$
$\beta = 0.013$.

Deere and Miller (1966) demonstrated that the elastic modulus for intact rock can be related to the intact compressive strength, and defined a narrow range of observed ratios between elastic modulus and compressive strength for brittle and soft materials. Consequently, CNI judged it reasonable to expect that a similar relationship could exist between the rock mass modulus and the rock mass strength. Back analysis of slope failures by CNI indicated that the estimation of rock mass strength does follow Bieniawski's relationship for predicting deformation modulus. However, the strength properties were found to vary according to the square of the modulus ratio, r^2. For example, if the square of the modulus ratio r^2 is 0.3, the estimated rock mass strength is derived by compositing 30% of the intact rock strength with 70% of the natural fracture strength. The resulting equations for predicting the rock mass friction angle and cohesion are:

For RQD values of >50–60:

$$C_m = \gamma\left[r^2 c_i + (1 - r^2)c_j\right] \qquad \text{(eqn 5.67)}$$

$$\varnothing_m = \tan^{-1}\left[r^2 \tan\varnothing_i + (1 - r^2)\tan\varnothing_j\right] \qquad \text{(eqn 5.68)}$$

where

$\varnothing_m$ = rock mass friction angle
C_m = rock mass cohesion
c_i = intact rock friction angle
c_j = intact rock cohesion
$\varnothing_j$ = joint friction angle
c_j = joint cohesion

and

$\gamma = 0.5$ to 1.0
$\gamma = 0.5$, jointed medium to strong rock (>60 MPa)
$\gamma = 1.0$, massive weak to very weak rock (<15 MPa).

For simplicity, the CNI rock mass strength equations were presented for a linear Mohr-Coulomb failure envelope. However, CNI noted that the rock mass shear strength can be mapped to a power envelope by regression techniques using the calculated percentage of intact rock (r^2 . 100) and the power strength envelopes of both the intact rock and the fracture shear data.

The intact compressive strength exerts the primary control on the constant gamma (γ) in equation 5.67. However, the appropriate gamma (γ) value is also influenced by the degree of fracturing. In general, the gamma (γ) value increases as the intact compressive strength decreases. As the fracture intensity becomes greater, the gamma (γ) value lessens.

Applications of equations 5.67 and 5.68 by CNI indicated that for RQD values less than approximately 50, equation 5.68 tended to overpredict the rock mass friction angle. Consequently, the constants alpha (α) and beta (β) in equation 5.66 were revised to provide a better fit to back-calculated rock mass friction angles for lower RQD rock masses. For RQD values of less than 40 and up to 50, these constants are:

- $\alpha = 0.475$;
- $\beta = 0.007$.

The two relationships presented for predicting rock mass friction angle do not follow a smooth transition for RQD values between 40 and 60. Modifications to the equations in this RQD range are being investigated by CNI, which hopes to publish results in the future.

Because the relationships presented above for predicting rock mass strength are based on RQD, CNI noted that it is important to recognise that RQD can be an imprecise indicator of the degree of fracturing at RQD values below approximately 20 and above approximately 80. To overcome this deficiency, CNI believe a relationship based on fracture frequency would be preferable. However, existing mine databases typically lack these data or have very limited information. If more extensive databases for fracture frequency become available, CNI considers that these relationships can be readily converted and extended to find wider application in strongly fractured as well as massive rock units (Call et al. 2000).

5.5.5 Directional rock mass strength

The Hoek-Brown and CNI strength criteria assume that the rock mass comprises an isotropic clump of intact rock pieces separated by closely spaced joints for which there is no preferred failure direction, which is rarely the case. However, using the same concepts of the plane of weakness theory illustrated in Figure 5.40, it is possible to define directional shear strength for the jointed rock mass as follows.

1 Define the basic or isotropic rock mass shear strength by using the generalised Hoek-Brown criterion, and defining equivalent values for the cohesion and friction angle of the rock. This basic strength is the same in all

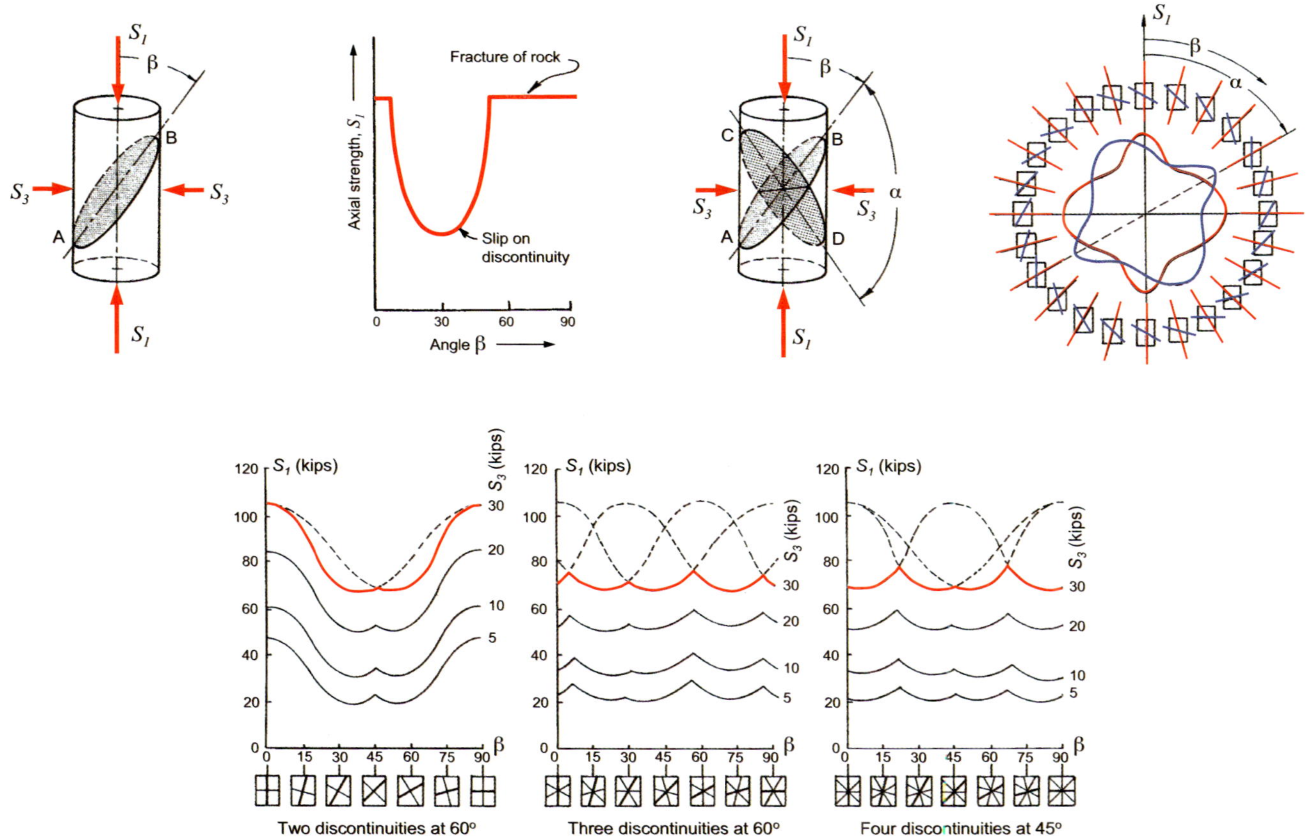

Figure 5.40: Effect of the pore pressure on (a) one, (b) two and several discontinuities with different orientations on the strength of a rock specimen. (a) Effect of a single discontinuity on the strength of a rock specimen. The plot on the right shows the variation of strength with the orientation of the discontinuity with respect to the direction of loading. (b) Effect of two discontinuities with different inclinations on the strength of a rock specimen. The polar plot on the right shows the variation of strength with the direction of loading with respect to the discontinuities. The perimeter of the blue area defines the directional strength of the rock specimen.
Source: Hoek & Brown (1980b)

directions and in a polar plot defines a circle (Figure 5.41).

2 If there are no discontinuity sets (faults and/or joints) parallel to the slope, then the shear strength of the rock mass can be assumed to be isotropic and corresponds to the basic strength defined in (1).

3 If there are one or more discontinuity sets parallel to the slope, the shear strength of the rock mass cannot be assumed isotropic, because the rock mass is weaker in the direction of these discontinuities. Hence, the rock mass shear strength is much smaller in the direction of these discontinuities and defines a butterfly-shaped curve in a polar plot (Figure 5.41). In the direction normal to the discontinuities the shear strength will be equal to the basic rock mass strength defined in (1), and in the direction parallel to the discontinuities it will be equal to the shear strength of the discontinuities. The shear strength of the discontinuities can be defined according to the following procedure:

→ if the discontinuities are persistent and can be assumed continuous for the slope being studied then the shear strength of the discontinuities can be assessed as described in section 5.3, and values for the cohesion and friction angle of the discontinuities can be defined;

→ if the discontinuities are non-persistent and their continuity is interrupted by rock bridges (see Figures 5.42 and 5.43) their shear strength will increase considerably. Unless the effect of rock bridges is accounted for, the shear strength of the discontinuities will be underestimated. Detailed discussions can be found in Jennings (1970, 1972), Einstein et al. (1983) and Wittke (1990). In a rock slope with non-persistent discontinuities a step-path failure surface will occur through a combination of discontinuities and rock bridges (Figure 5.43). An 'equivalent discontinuity' can be assigned to this step-path failure (Figure 5.43) and allocated 'equivalent' shear strength parameters.

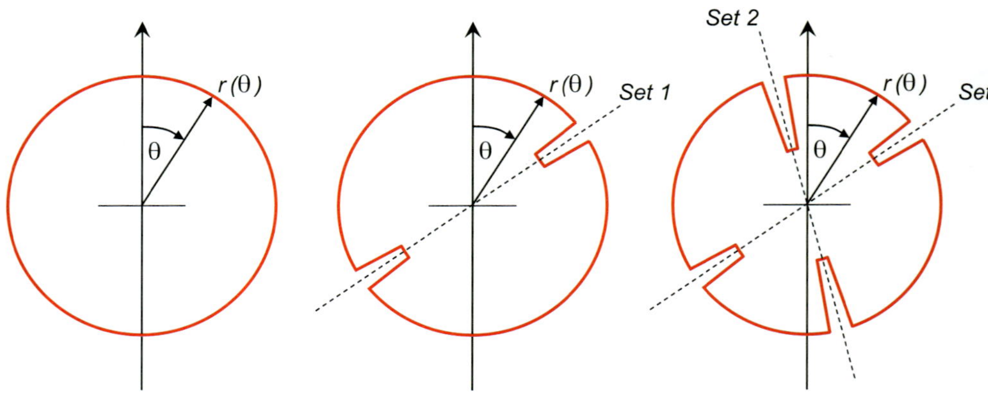

Figure 5.41: Polar plots illustrating the effect of discontinuity sets parallel to the slope in the shear strength of the rock mass. The magnitude of the shear strength for a given orientation θ is equal to the radial distance from the origin to the red curve

The definition of these equivalent shear strength parameters can be done using closed-form solutions (e.g. Jennings 1972). The simplest case is a planar rupture through coplanar joints and rock bridges, as shown in Figure 5.44.

In this case the equivalent strength parameters can be computed (Jennings 1972):

$$c_{eq} = (1 - k)c + kc_j \qquad \text{(eqn 5.69)}$$

$$\tan(\phi_{eq}) = (1 - k)\tan(\phi) + k\tan(\phi_j) \qquad \text{(eqn 5.70)}$$

where c_{eq} and ϕ_{eq} are the cohesion and friction angle of the equivalent discontinuity, c and ϕ are the cohesion and friction angle of the rock bridges, c_j and ϕ_j are the cohesion and friction angle of the discontinuities contained in the rock mass (joints) and k is the coefficient of continuity along the rupture plane given by:

$$k = \frac{\sum l_j}{\sum l_j + \sum l_r} \qquad \text{(eqn 5.71)}$$

where l_j and l_r are the lengths of the discontinuities and rock bridges. As discussed by Jennings (1972), these equations contain a number of important implied assumptions and become much more complex in the case of non-coplanar discontinuities and/or a rock mass with two discontinuity sets parallel to the slope orientation (Figure 5.43).

4 Once the shear strength of the discontinuities (persistent discontinuities) or equivalent discontinuities (non-persistent discontinuities containing rock bridges) have been defined, the directional strength of the rock mass can be defined as follows.

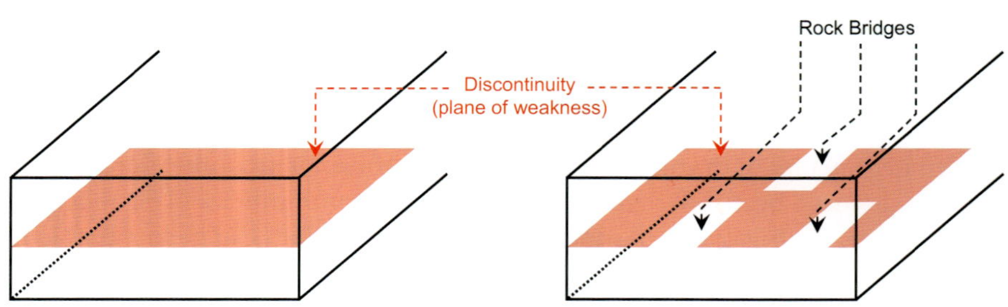

Figure 5.42: Simplified representation of the effect of rock bridges
Source: Modified from Wittke (1990)

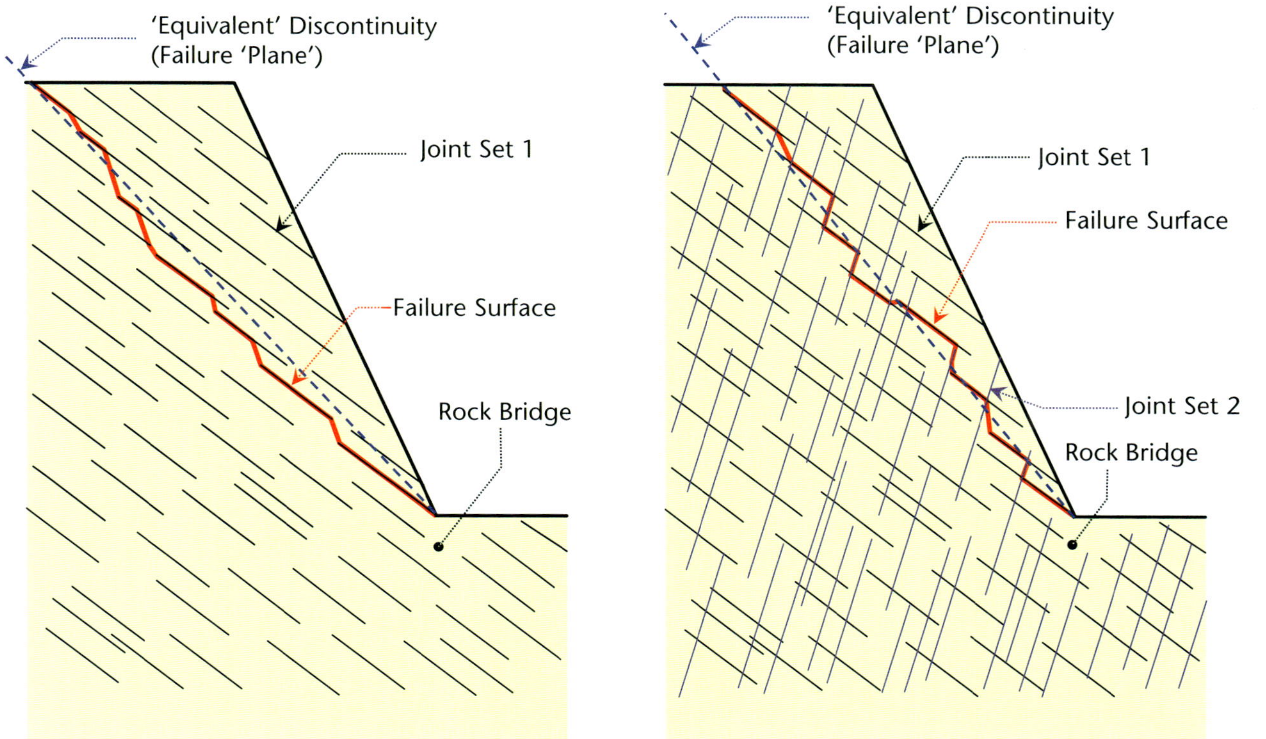

Figure 5.43: Step-path failure surface and 'equivalent' discontinuity for rock slopes containing one set (left side) and two sets (right side) of non-persistent discontinuities parallel to the slope
Source: Modified from Karzulovic (2006)

→ For each discontinuity set sub parallel to the slope orientation the most likely value of its apparent dip in the slope section, α_a, and the possible variation of this value, $\Delta\alpha_a$, must be determined. In most cases where good structural data are available $\Delta\alpha_a$ is about $\pm 5^{\circ}$, but where data are insufficient it can be much larger.

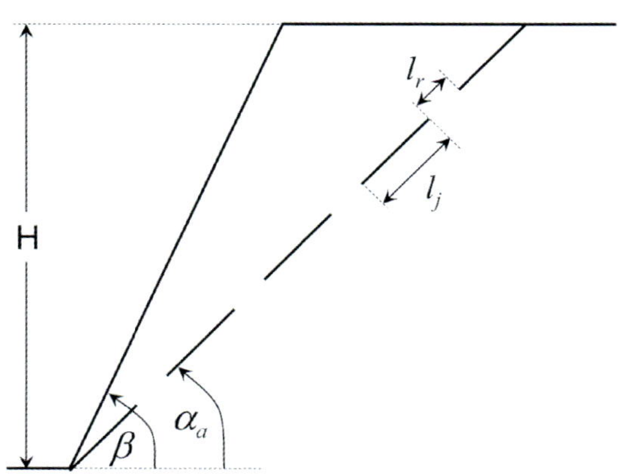

Figure 5.44: Planar rupture through coplanar joints and rock bridges in a rock slope with height H and inclination β. The rock bridges have a length l_r, and the discontinuities have a length l_j and an apparent dip α_a in the slope section

→ As shown in Figure 5.45, these values are used to define the directions where the strength corre-

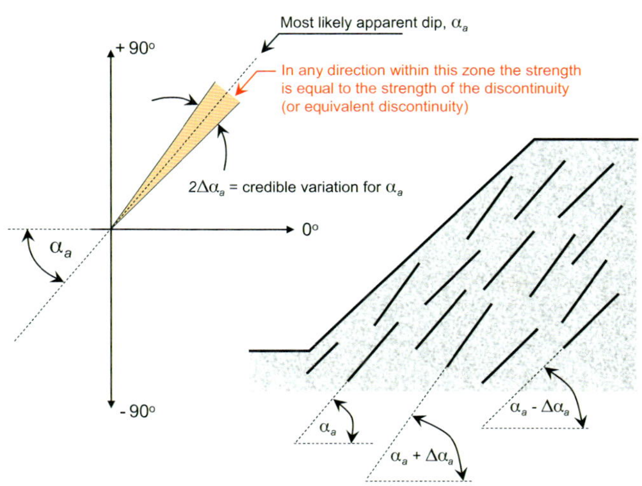

Figure 5.45: Definition of the set of directions where the strength of the rock mass is equal to the strength of the discontinuity (in the case of persistent discontinuities) or equivalent discontinuities (in the case of non-persistent discontinuities containing rock bridges), in terms of the most likely apparent dip of the discontinuities in the slope section, α_a, and its credible variation $\Delta\alpha_a$

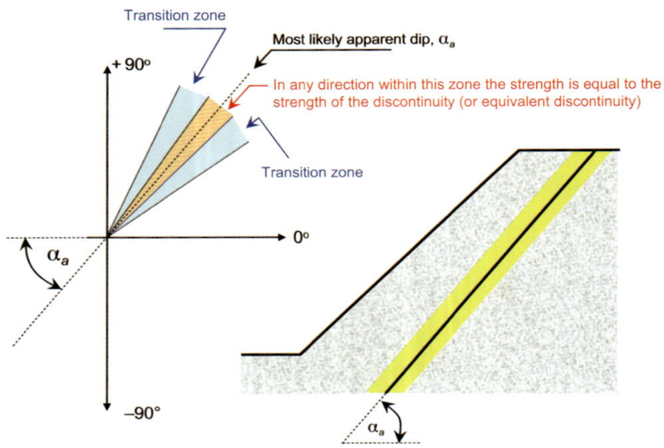

Figure 5.46: Definition of 'transition zones' to include the effect of an alteration zone associated with a discontinuity with a most likely apparent dip α_a.

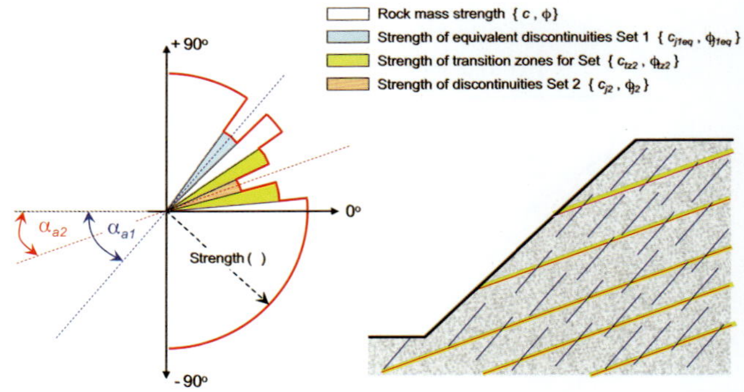

Figure 5.47: Definition of the directional strength of a rock mass containing two discontinuity sets. The discontinuities of Set 1 are non-persistent and include rock bridges, while the discontinuities of Set 2 are persistent and have an associated alteration zone

sponds to the strength of the discontinuities (if these are persistent) or equivalent discontinuities (if these are non-persistent and contain rock bridges).

- Some discontinuities such as faults may have an alteration zone associated with them, and the strength of this zone could be weaker than the strength of the rock. As shown in Figure 5.46, it is possible to include transition zones with a strength intermediate between those of the discontinuity and the rock mass:

$$c_{tz} = k_t c + (1 - k_t) c_j \qquad \text{(eqn 5.72)}$$

$$\tan(\phi_{tz}) = k_t \tan(\phi) + (1 - k_t) \tan(\phi_j) \qquad \text{(eqn 5.73)}$$

where c_{tz} and $\varnothing_{tz}$ are the cohesion and friction angle of the transition zone, c and $\varnothing$ are the cohesion and friction angle of the rock mass, c_j and $\varnothing_j$ are the cohesion and friction angle of the discontinuity, and k_t is a coefficient of transition that varies from 0 to 1 depending on the characteristics of the transition zone. For example, in the case of a transition zone with an intense sericitic alteration k_t would probably be 0.5–0.7, while if the sericitic alteration is slight to moderate k_t would probably range from 0.7 to 0.9. The size of the transition zone must be estimated considering the thickness of the alteration zone associated with the discontinuity, but typically values of about 10° are used to define the transition zone.

- These strengths are overlapped to define the directional strength of the rock mass, as illustrated in Figure 5.47 for the case of a rock mass containing two discontinuity sets. The discontinuities of Set 1 are non-persistent and include rock bridges while the discontinuities of Set 2 are persistent and have an associated alteration zone.

Once the rock mass strength has been defined, the slope stability analyses can be carried out. The importance of considering a directional strength for the rock mass with discontinuities subparallel to the slope is illustrated in Figure 5.48, for the case of 200 m rock slope with a 55° inclination, a 20 m deep tension crack and dry conditions.

In Figure 5.48 the examples, computed using the SLIDE software, assumed that the rock mass strength is defined by a cohesion of 400 kPa and a 35° friction angle, while the strength of the non-persistent joints with rock bridges is defined by a cohesion of 150 kPa and a 30° friction angle.

If there are no discontinuity sets subparallel to the slope the rock mass strength is isotropic and the slope has a factor of safety (*FoS*) equal to 1.29 (Case 1). If there is one discontinuity set subparallel to the slope, dipping 65° towards the pit, the rock mass strength is directional (i.e. weaker in the direction of the discontinuities) and the factor of safety decreases to 1.15 (Case 2). If the set dips 35° towards the pit the factor of safety decreases even more, to 0.99 (Case 3). If there are two sets subparallel to the slope, dipping 35° and 65° towards the pit, the rock mass strength is weaker in two directions and the factor of safety decreases to 0.88 (Case 4).

There is always variability in the length, spacing and orientation of discontinuities. Hence, in practice it may be preferable to use software such as STEPSIM for a probabilistic estimate for these equivalent shear strength parameters by considering the variability of parameters such as discontinuity persistency and strength. The STEPSIM 'step-path' routine was originally conceptualised as part of the pit slope design work performed at the Bougainville Copper Ltd mine, Papua New Guinea (Read & Lye 1983.) Baczynski (2000) described the latest version of this software, STEPSIM4,

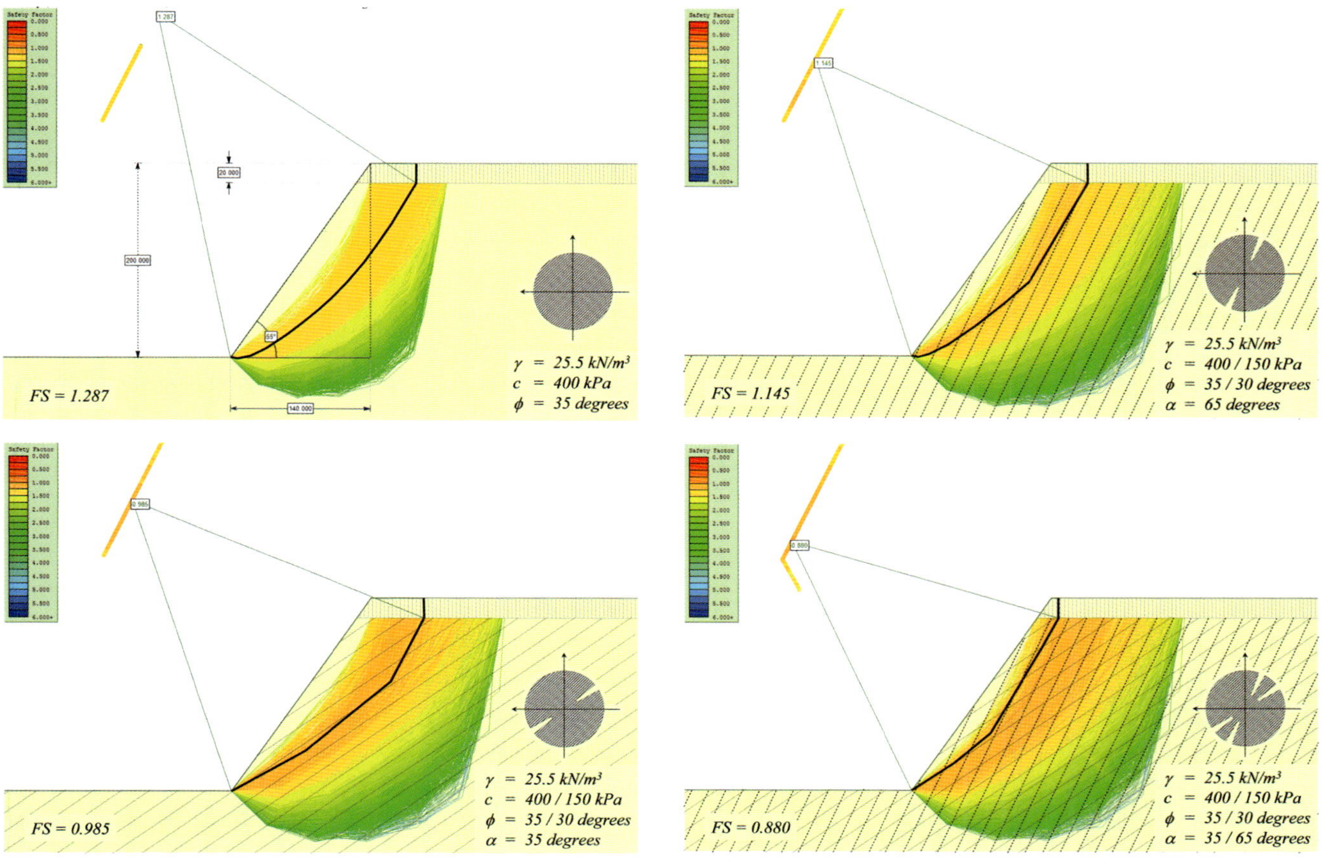

Figure 5.48: Factor of safety of a 200 m rock slope, with an inclination of 55°, for different conditions of rock mass strength

which envisages a potential rupture path through a rock slope as a series of adjacent cells (Figure 5.49).

Each cell is statistically associated with one or more of the following failure mechanisms: sliding along adversely oriented discontinuities (Set 1); stepping-up along another steeply dipping discontinuity set (Set 2); or shearing through a rock bridge.

The STEPSIM model assumes that the Set 1 and Set 2 discontinuities occur independently within the rock mass (Baczynski 2000). The computational procedures involve the following basic steps.

- The user defines the length of the failure path to be evaluated (e.g. 100 m, 250 m, 500 m). For each simulated failure path, the structural and strength characteristics of each cell are statistically assigned on the basis of the input parameters.
- A potential failure path starts at the toe of the slope. This position coincides with the first ground condition cell in the simulation process. Cell size should be statistically meaningful and, ideally, should mirror the size of the data windows used for structurally mapping. If this is impossible, an arbitrary cell size may be selected (e.g. 5 × 5 m or 10 × 10 m).
- Based on the input data for the probability of occurrence of the Set 1 and Set 2 discontinuities, the program uses a random number-generating technique to check whether one, both or none of the discontinuity sets should be simulated in the first cell. If neither of the sets occurs, then rock mass properties are assigned to the first cell.
- If one or both sets occur, the random number-generating Monte Carlo process is again used to systematically generate the respective discontinuities within the first cell. Based on the input statistical model for discontinuity type for the respective sets, a 'type' is assigned to the first structure. A similar process is used to assign orientation (apparent dip), length and shear strength parameters to the first discontinuity and to check whether the discontinuity terminates in rock or is cut-off by another discontinuity. If the first discontinuity is cut-off, then the second discontinuity starts at the end of the first one. If the first discontinuity is not cut-off, then an appropriate length rock bridge is simulated at its end. The second discontinuity starts at the end of this rock bridge. Depending on their size, such bridges may have either rock or rock mass shear strength assigned by Monte Carlo simula-

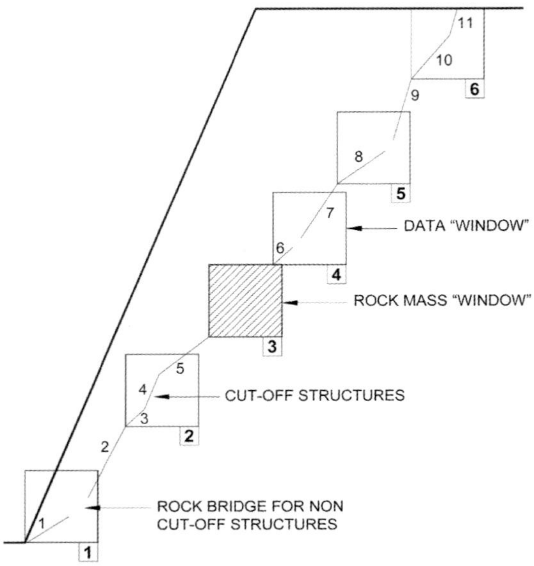

Figure 5.49: Conceptual STEPSIM4 model
Source: Baczynski (2000)

tion from the respective input statistical distributions for these parameters. If both Sets 1 and 2 occur in the first cell, the Monte Carlo process is used to decide whether the next discontinuity to be generated should be a Set 1 or a Set 2 member. This process is iterated until the last generated discontinuity or rock bridge terminates at the perimeter or just outside the current cell.

- The bottom left-hand corner of the second cell starts at the end of the last generated discontinuity or rock bridge. The above simulation process is repeated for the second cell.
- This process is repeated for successive cells until the target rupture path length has been simulated and the respective shear strength parameters and large-scale roughness characteristics are computed.
- The process is repeated for a large number of rupture paths (usually 2000–5000) and the ensuing statistical distribution of shear strengths is computed (i.e. a mean and standard deviation for the friction angle and cohesion).

5.5.6 Synthetic rock mass model

5.5.6.1 Introduction

As outlined in section 5.5.1, the Mohr-Coulomb measures of friction (Ø) and cohesion (c) that are used to represent the strength of the rock mass in the limiting equilibrium and continuum and discontinuum numerical modelling slope design tools are derived empirically from various rock mass classification schemes. Although this process is current practice it contains some basic flaws, which can be summarised as follows.

1 Mohr-Coulomb measures friction and cohesion at a point, which we transfer to a three-dimensional body of rock by assuming that the rock mass is isotropic, which is not the case in a jointed rock mass.
2 The empirical friction and cohesion values derived from the most popular classification schemes involve a number of idiosyncrasies and limitations. These include the inbuilt sources of error involved in some parameters used in these schemes (e.g. RQD, section 5.4.2 and GSI, section 5.4.4). As a result there is a high level of uncertainty in the realism of the adopted friction and cohesion values, which we attempt to overcome by calibrating them against existing slope movement data. This severely limits the chances of reliably predicting a future event.
3 We cannot simulate a brittle fracture that can propagate across the joint fabric within the intact pieces of rock (rock bridges) between the structural defects that cut through the rock mass as stress relaxation enables it to dilate and the pieces to separate and move. Instead, the empirical friction and cohesion values are applied as 'smeared' or 'average, non-directional' parameters across the rock bridges, which are assumed to behave as a continuum.
4 We cannot account for the effect of the degree of disturbance to which the rock mass has been subjected by stress relaxation on the strength of the rock mass (the 'D' factor in the Hoek-Brown strength criterion, Table 5.37).

These limitations lead to the recognition of two specific research needs (Read 2007):

- the need to construct an 'equivalent material' that honours the strength of the intact rock and joint fabric within the rock bridges that may occur along a candidate failure surface in a closely jointed rock mass;
- the need to be able to simulate the brittle fracture that can propagate across the joint fabric within the rock bridges as the rock mass deforms.

These needs and associated questions have formed one of the major research tasks of the Large Open Pit (LOP) Project. A number of approaches and numerical codes with the potential to construct an 'equivalent material' and model brittle fracture across the rock bridges were considered. The Itasca PFC code was selected as it uses a micro-mechanics based criterion to model brittle fracture. This offered the potential for stepping away from the Mohr-Coulomb and Hoek-Brown criteria, a feature that was consistent with the research objectives.

5.5.6.2 SRM model

In PFC the entire model is composed from the start as discrete elements bonded together (the bonded particle method [BPM], Potyondy & Cundall 2004), with the

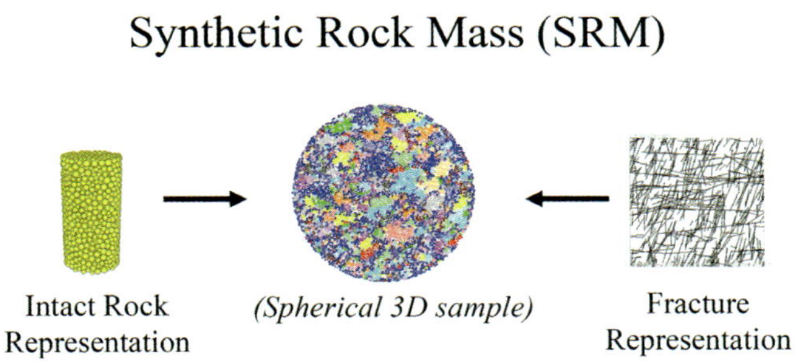

Figure 5.50: SRM model representation
Source: Courtesy Itasca Consulting Group, Inc.

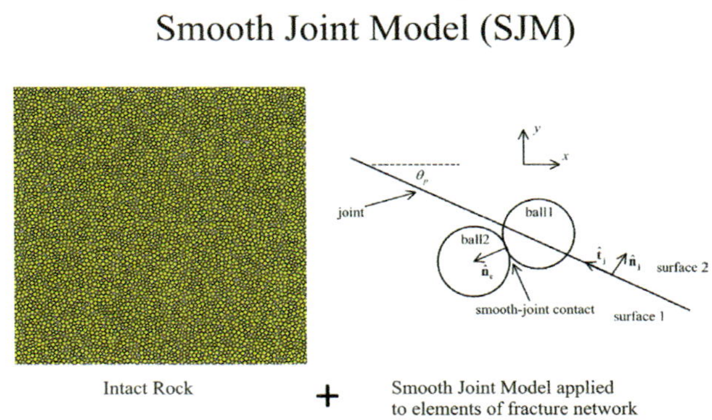

Figure 5.51: Smooth joint model representation
Source: Courtesy Itasca Consulting Group, Inc.

inputs (microproperties) restricted to stiffness and strength parameters for the particles and bonds. The initial state of such a bonded assembly of particles is taken as equivalent to an elastic continuum. The fracturing process consists of individual bonds breaking (micro-cracking) and coalescing to form macro-cracks. The PFC assembly is said to exhibit a rich constitutive behavior as an emergent property of the particle assembly without the use of supplied macro-mechanics constitutive models. Extensive tests on simulated laboratory samples have shown that the synthetic PFC material can be calibrated to produce quantitative fits to almost all measured physical parameters, including moduli, strengths and fracture toughness (Potyondy & Cundall 2004).

Development of the BPM method since 2004 in block caving studies by Itasca has shown that the BPM method can represent the strength of the intact rock and joint fabric within the rock bridges with an 'equivalent material' or synthetic rock mass (SRM) model (Pierce et al. 2007). In this model the intact rock is represented by an assemblage of bonded particles numerically calibrated using UCS, modulus and/or Poisson's ratio values to those measured for an intact sample (Figure 5.50). The joints are represented by a smooth joint model that allows associated particles to slide through, rather than over, one another and so represent joints that slide and open in the normal way (Figure 5.51).

Creating and testing the SRM sample illustrated in Figure 5.50 is essentially a three step process: creating the particle assembly that represents the intact rock in PFC3D; generating and importing the discrete fracture network (DFN) that represents the structural pattern of the rock mass into the particle assembly; and testing. Intermediate stages in preparing the sample involve using the DFN to estimate the average size of the rock bridges that will be modelled and calibrating the microproperties of the synthetic material (e.g. particle size distribution and packing, particle and bond stiffness, particle friction coefficients and bond strengths) to the measured properties of the physical material (e.g. Young's Modulus, UCS). When testing, a minimum of four tests, one tensile, one UCS and two triaxial tests at differing confining pressures have been found necessary to obtain an estimate of the strength envelope. Changes in the loading direction will also help determine the strength anisotropy of the rock mass. These activities require a working knowledge of one or other of the available DFN modelling packages (e.g. FracMan, JointStats or 3FLO, section 4.4.3), and PFC2D and its 3D equivalent PFC3D (Itasca 2008a, 2008b). To assist users, a Microsoft Excel workbook (the Virtual Lab Assistant or VLA) is available to help with the intact rock calibration process. The workbook can automatically retrieve test results and present them in a separate worksheet, and includes an option to record four predefined videos of each simulation.

From a slope stability point of view the SRM rock bridge is a potential break through. LOP Project research involving the above numerical tensile, UCS and triaxial tests on selected volumes of rock from sponsor mine sites has shown that the SRM does honour the strength of the intact material and the joint fabric within the rock bridges along a candidate failure surface in a closely jointed rock mass, and that it can provide a means of developing a strength envelope that does not rely on either Mohr-Coulomb and/or Hoek-Brown criteria. Similarly, the inverse of providing Hoek-Brown parameters and calibrating the Hoek-Brown strength envelope is possible. Of particular interest is the possibility of adjusting the calibrated Hoek-Brown and/or Mohr-Coulomb strength parameters to specific local conditions, including stress and slope orientation.

So far, the SRM tests have involved eight different rock types and have been performed at laboratory, bench and inter-ramp scales of 5 m, 10 m, 20 m, 40 m and 80 m. Initial outcomes of the research and the perceived benefits of using the SRM model in slope stability analyses are outlined in Chapter 7 (section 7.3.1) and Chapter 10 (section 10.3.3.4). Ongoing research outcomes will be brought into the public domain as they are reported and assessed.

6 HYDROGEOLOGICAL MODEL

Geoff Beale

6.1 Hydrogeology and slope engineering

6.1.1 Introduction

The presence of groundwater can affect open pit mine excavations in two ways.

1 It can change the effective stress and resulting pore pressures exerted on the rock mass into which the pit slopes have been excavated. Increased pore pressures will reduce the shear strength of the rock mass, increasing the likelihood of slope failures and potentially leading to slope flattening or other remedial measures to compensate for the reduced overall rock mass strength.

2 It can create saturated conditions and lead to standing water within the pit, which may result in:
   - → loss of access to all or parts of the working mine area;
   - → greater use of explosives, or the use of special explosives and increased explosive failures due to wet blast holes;
   - → increased equipment wear and inefficient loading;
   - → increased damage to tyres and inefficient hauling;
   - → unsafe working conditions.

The main purpose of this chapter is to discuss the first of these aspects – how the presence of groundwater and the resulting pore pressures may affect open pit slope design and performance.

Groundwater usually has a detrimental effect on slope stability. Fluid pressure acting within discontinuities and pore spaces in the rock mass reduces the effective stress, with a consequent reduction in shear strength. This is particularly evident in a weak deformable rock mass, where slope strain softening influenced by fluid pressure can ultimately lead to loss of peak shear strength. In steeper high-strength rock slopes, the potential for sudden brittle failure under small mining-induced strains is increased when the pore pressure is elevated.

This chapter includes:

- a discussion of how groundwater relates to pore pressure, and the relationship to total and effective stress;
- the main controls on pore pressure and its role in slope engineering;
- a distinction between general mine dewatering and pit slope depressurisation;
- a practical explanation of hydrogeology with respect to slope engineering, including the concepts of groundwater flow in fractures (section 6.2);
- how a conceptual hydrogeological model, which is the fourth and final component of the geotechnical model (Figure 6.1), is developed. Recharge, water table and piezometric surfaces, horizontal and vertical hydraulic gradients, discharge of water to the slope and the resulting pore pressure distribution are addressed in section 6.3;
- an outline of modelling for numerical hydrogeological models (section 6.4). Section 6.4.2 discusses the normal approach to mine scale numerical hydrogeological modelling. The approach to pit slope scale numerical modelling is outlined in section 6.4.3 and specific numerical modelling procedures for determining the pore pressures in pit slopes are discussed in section 6.4.4;
- methods that can be used to dissipate pore pressures in the pit slopes (section 6.5).
- a discussion of topics that need further research (section 6.6).

Definitions of the common terms that apply to groundwater in mine excavations are included in the Glossary.

6.1.2 Porosity and pore pressure

6.1.2.1 Porosity

Within most saturated porous formations such as sandstone, siltstone or shale, and within unconsolidated

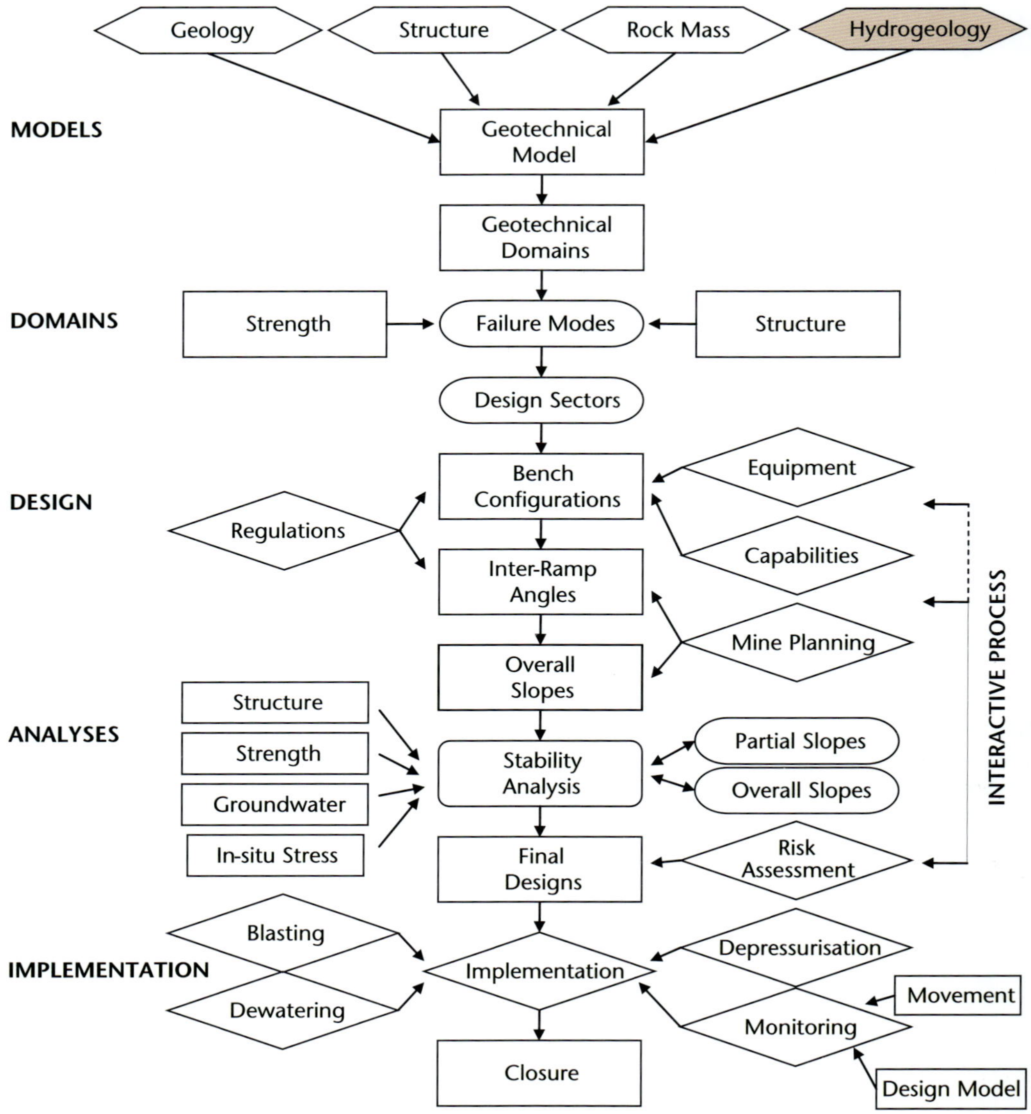

Figure 6.1: Slope design process

clastic sediments such as sand, silt and clay, most of the groundwater is contained within the primary interstitial pore spaces of the formation. Hard rock that is weathered or altered may also exhibit interstitial spaces between grains, particularly within zones of clay alteration or weathering. In addition, highly fractured and broken rock may exhibit similar hydrogeological properties to porous strata (commonly referred to as equivalent porous medium). Within porous strata, pore pressure is exerted on the entire rock mass.

The total porosity (n) of the rock mass in these settings is mostly controlled by the interstitial spaces between grains, which typically ranges from 10–30% of the total volume of the formation ($n = 0.1$ - 0.3) but may be up to 50% ($n = 0.5$) in poorly consolidated fine-grained materials. A cubic metre of the rock mass may therefore contain 100–300 L of groundwater. However, particularly for clay materials, the drainable porosity usually represents only a small proportion of the total porosity. Much of the groundwater may be held in place by surface tension and may not freely drain under gravity (Figure 6.2).

Within most saturated competent (hard rock) formations, including igneous, metamorphic, cemented clastic and carbonate settings, virtually all the groundwater is contained within fractures. Because there is no significant primary porosity, the pore pressure is exerted only on the fracture surfaces. However, in addition to the main faults and high-order fracture zones, the rock usually contains abundant low-order small-aperture fracture and joint sets distributed pervasively throughout the rock mass. These micro-fractures also contain groundwater and exhibit pore pressure.

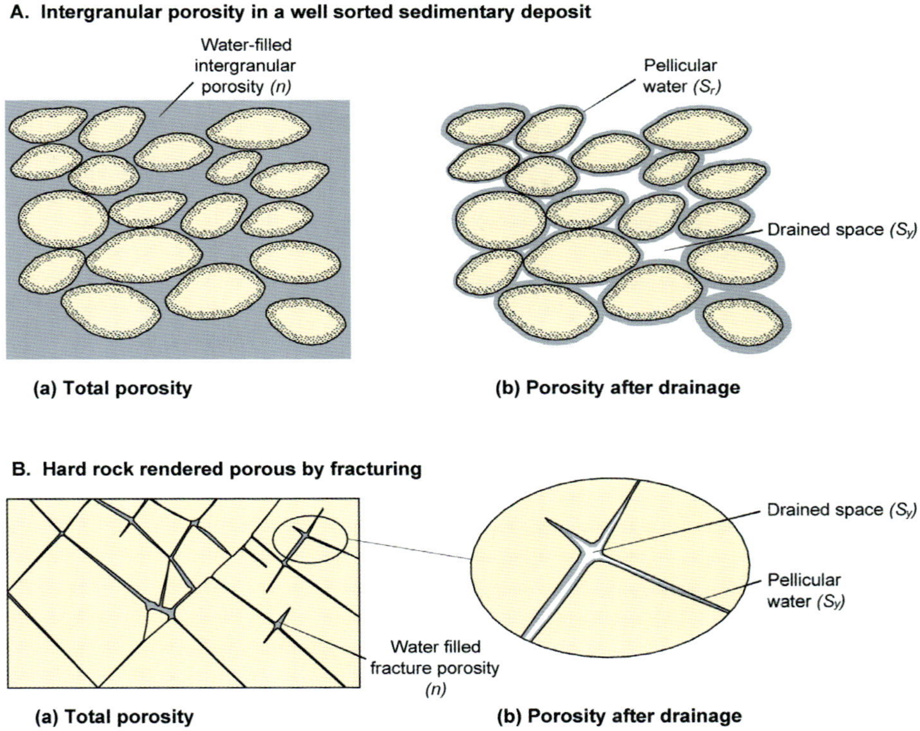

Figure 6.2: Illustration of porosity

The total porosity (n) of competent hard rock types is dependent on the frequency of open fractures and joints, and typically ranges from less than 0.1% to about 3% of the total volume of the formation (n <0.001 - 0.03). A cubic metre of the rock mass may therefore contain less than 1 L to as much as 30 L of groundwater. As with intergranular pore spaces, of the total groundwater contained within the fractures only a portion will drain out if the rock pore pressures are reduced by passive drainage or pumping. Typically, following drainage of the rock, most of the water in the small-aperture fractures is held in place by surface tension (with a slight negative pressure).

Most pit slopes are made up of a combination of consolidated (hard) and porous rock types. For example, the slopes of a porphyry copper pit may exhibit fracture-controlled porosity in the unaltered primary rock, porous-medium conditions in zones of argillic or sericitic alteration, and both types of porosity in the oxidised zone.

6.1.2.2 *Pore pressure in-pit slope engineering*

Pore pressure (u) is defined as the pressure of the groundwater occurring within the pore spaces of the rock or soil. The pore pressure may occur in the interstitial spaces between grains (porous strata) or in open fractures and joint sets (competent rock). Pore pressure is 0 at the water table, positive below the water table and negative above the water table.

Pore pressure is an integral parameter for any rock slope engineering assessment. It exerts the following geotechnical controls:

- it changes the effective stress of the rock mass within the slope;
- it may cause a change in volume of the slope material;
- it may cause a change in hydrostatic loading.

In most fractured rock settings, the first factor is by far the most important for slope performance. Because of the relatively low porosity of most fractured rock materials, the second and third factors are typically less important. However, the volumetric change can be important when draining silty or clayey rocks as settling may occur.

The relationship between the shear strength of a rock or soil mass and pore pressure is expressed in the Mohr-Coulomb failure law in combination with the effective stress concept developed by Terzaghi:

$$\tau = (\sigma_n - u)\tan \varnothing + c \qquad \text{(eqn 6.1)}$$

where:

τ = shear strength on a potential failure surface
u = fluid pressure (or pore pressure)
σ_n = total normal stress acting perpendicular to the potential failure surface
$\varnothing$ = angle of internal friction
c = cohesion available along the potential failure surface.

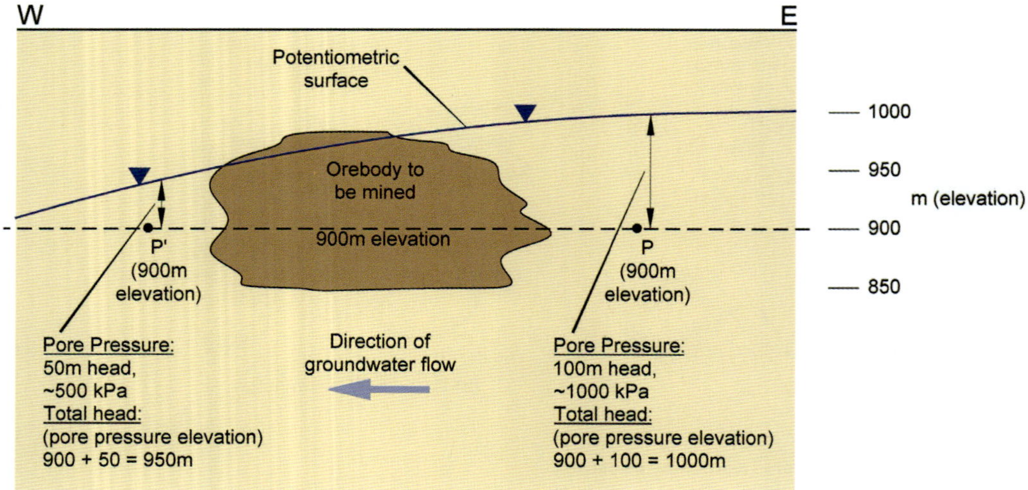

Figure 6.3: Lateral variation in pore pressure as a result of groundwater flow

The angle of internal friction and cohesion are strength properties of the material at any point on the potential failure surface.

Total normal stress (σ_n) is the pressure acting on the potential failure surface, caused by the weights of the overlying rock (lithostatic load) and the water (hydrostatic load). This total stress is counteracted partly by the granular or block components of the formation and partly by the fluid pressure within the pores (pore pressure).

The effective normal stress, usually expressed as σ_n', is defined as the difference between total stress and fluid pressure u. Thus, $\sigma_n' = \sigma_n - u$. The effective normal stress is the portion of the total stress actually applied to the grains or blocks of the formation. It therefore has a major control on the shear strength of the material. In accordance with the modified Mohr-Coulomb expression, effective normal stress mobilises a frictional component of the shear strength of the soil or rock mass.

If the pore pressure is decreased with no change in total stress it will lead to an increase in effective normal stress and hence an increase in shear strength on failure planes, with an improvement in slope stability. Therefore, slope depressurisation is a means of improving slope stability and achieving a more economical slope design.

6.1.2.3 *Storativity*

Porosity refers to the total void space of the rock mass. The term 'drainable porosity' or 'unconfined storage' refers to the part of the pore space that will release water by gravity when the water table is lowered within the rock unit (or fracture set) in question and the rock is drained (Figure 6.2). The term 'elastic storage' or 'confined storage' is applied when the potentiometric surface is lowered but remains above the top of the rock unit (or fracture set) in question. The distinction between a confined and unconfined groundwater setting is discussed in section 6.2.

Under confined conditions, the amount of water released per unit surface area of the rock mass per unit decline in head is referred to as the specific storage (S_s). Specific storage is a function of the elastic expansion of the rock mass and the compressibility of water, and is very small in comparison to unconfined storage.

When a confined groundwater system is pumped, the rock mass remains fully saturated until the potentiometric surface drops below the confining layer. The water is released from storage under conditions of decreasing head due to the compaction (low-strength materials, or soil) of the aquifer from increasing effective stress and the expansion of water from decreasing pressure.

6.1.2.4 *Controls on pore pressure*

Below the water table, pore pressure is determined by measuring the height of a column of water at a given point (depth and location) within the rock mass. In general, the deeper the point below the water table, the higher the pore pressure.

In any given setting, the pore pressure distribution will vary laterally following changes in the water table elevation. For rock masses characterised by interconnected granular or fractured pores with no impediments to groundwater flow, pore pressure can be represented as a single potentiometric surface (see Figure 6.3). At point P on the east side of the ore body, the potentiometric surface elevation is 1000 m. The pore pressure at point P is 100 m head (1000–900 m), which is equivalent to 1000 kPa. At point P' on the west side of the ore body, the potentiometric surface elevation is at 950 m. The pore pressure at point P' is 50 m head (950–900 m), which is equivalent to 500 kPa.

The distribution of pore pressure also varies vertically. Under static conditions, the pore pressure distribution is not directly related to the porosity. Provided the fracture sets are interconnected, at any given depth below the

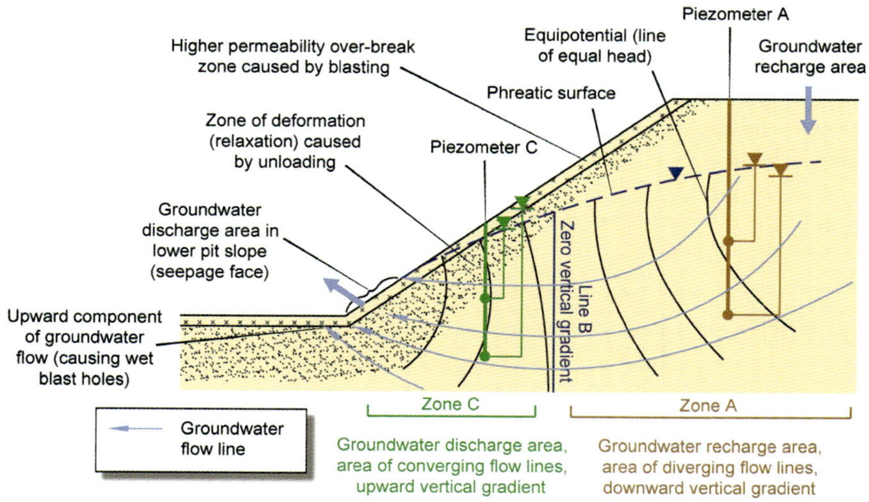

Figure 6.4: Simple groundwater flow net towards a pit slope

water table the same pore pressure will occur regardless of the fracture aperture and porosity. However, as soon as mining-induced flow or other changes occur, vertical components in the groundwater gradient will begin to develop (along with changes in the horizontal gradient). At any given location, the degree of vertical pore pressure variation will depend on the distribution of permeability below the potentiometric surface, and the elevation of groundwater recharge and discharge points. In Figure 6.4, a mine excavation has progressed below the elevation of the water table, creating a zone of discharge and causing the groundwater to flow towards the mining void from outside the crest of the pit slope.

Figure 6.5 illustrates the relationship of pore pressure and depth below the phreatic surface for the flow net shown in Figure 6.4.

As an open pit is progressively excavated deeper below the water table, the pore pressure behind the pit slope tends to reduce as groundwater drains through the pit walls into the excavation (see Figure 6.6). However, although the pore pressure has reduced due to seepage into the pit, an increase in the pore pressure gradient has occurred from the materials behind the pit slope to the newly exposed toe of the slope. The pore pressure gradient means that the pressure on the upstream side of every mineral grain or block is greater than the pressure on the downstream side. Thus, there is a force (the seepage force) attempting to move each grain or block in the direction of decreasing pressure (i.e. towards the slope). The increase in hydraulic gradient associated with the expansion of the pit has increased the seepage force and reduced the stability of the slope.

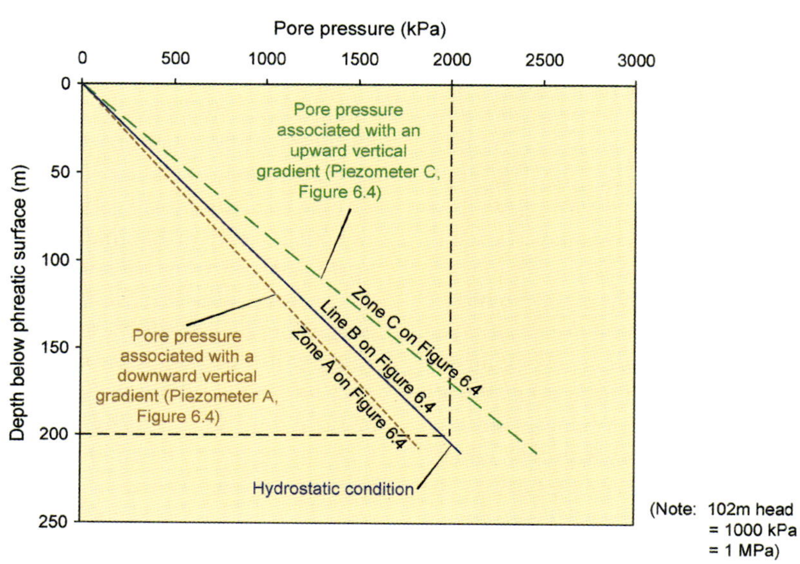

Figure 6.5: Variation in pore pressure with depth below the phreatic surface

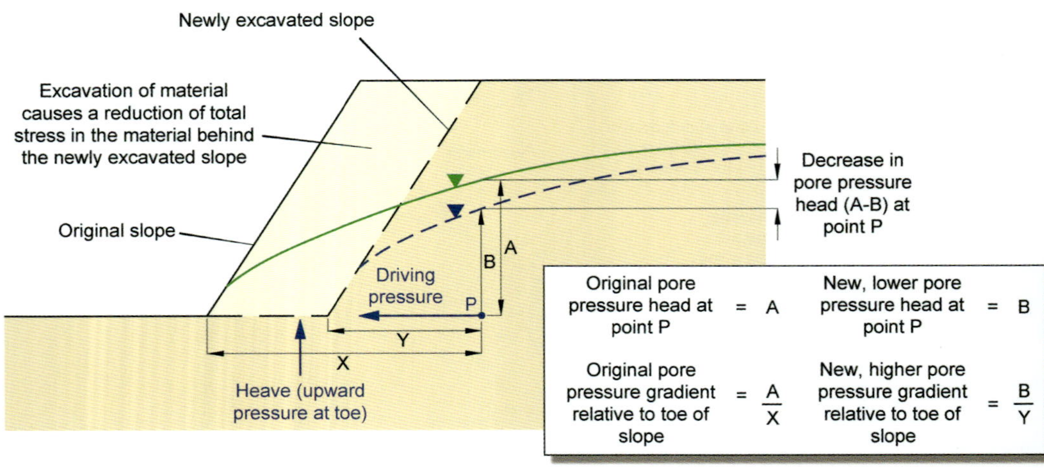

Figure 6.6: Increasing pore pressure gradient behind an advancing pit slope

Above the water table, the fractures and pore spaces are partially drained and the rock mass is not saturated. The water that remains is held in place within fractures and pore spaces by surface tension, so there is a negative pore pressure (suction) associated with these zones. In most instances, the magnitude of the negative pore pressure is too small to be significant in holding the rock mass together. Exceptions can occur in materials of very low permeability, where rebound following excavation increases the porosity of the rock but the permeability (hydraulic conductivity) is too low for the water pressures to respond (stabilise) immediately. This situation can result in significantly high negative pore pressures.

6.1.3 General mine dewatering and localised pore pressure control

All mines that are excavated below the water table need some form of dewatering. The scale of the dewatering effort depends on the following three factors:

1 the hydrogeological characteristics of the rock mass in which the excavation takes place;
2 the depth of the excavation below the water table;
3 the strength of the materials making up the pit slopes.

At some mines excavated below the water table, evaporation of minor groundwater seepage from the pit floor or pit walls in a strong and stable rock mass can take care of all dewatering requirements. At other mines, major pumping operations are necessary, using external wells to control groundwater inflow to the pit and to lower the pore pressure in the rocks making up the pit slopes. This is the case at the Goldstrike and Lone Tree mines in the Nevada Basin and Range in the USA and at the RWE Power's lignite mines in the Ruhr basin in Germany, which have pumped in excess of 2500 L/sec groundwater.

Mine dewatering and control of pore pressure in the pit slopes are interrelated. Five broad categories are recognised for open pit mines, based on their hydrogeological setting.

Category 1: Mines excavated below the water table within permeable rocks that are hydraulically interconnected

For this category, the general mine dewatering program can adequately reduce pressure in all pit slopes, requiring no additional localised measures to dissipate pore pressure.

Advanced lowering of the water table using wells causes gravity drainage of the pore spaces within the rock mass being excavated (Figure 6.7a). If the permeability of the rock mass is high and the rocks are hydraulically connected, the profile of the water table behind the pit slope will be relatively flat, indicating that the rock mass can drain easily.

An example of this category is the Cortez Pipeline mine in Nevada, where in excess of 1500 L/sec groundwater is pumped from peripheral and in-pit dewatering wells in a highly fractured limestone rock mass.

Category 2: Mines excavated below the water table with low-permeability rock in part of the walls

For this category, the rock mass may not drain fully as the water table is lowered. The rock mass to be excavated has a low permeability and effective advanced dewatering is not possible or requires a long time to be successful.

The permeability may be low in parts of the pit area and/or there may be poor interconnectivity of fractures. As a result, pore pressure within some parts of the rock mass will not dissipate. As the mine excavation is deepened, pore pressures within all or part of the slope may need to be controlled using localised measures (Figure 6.7b).

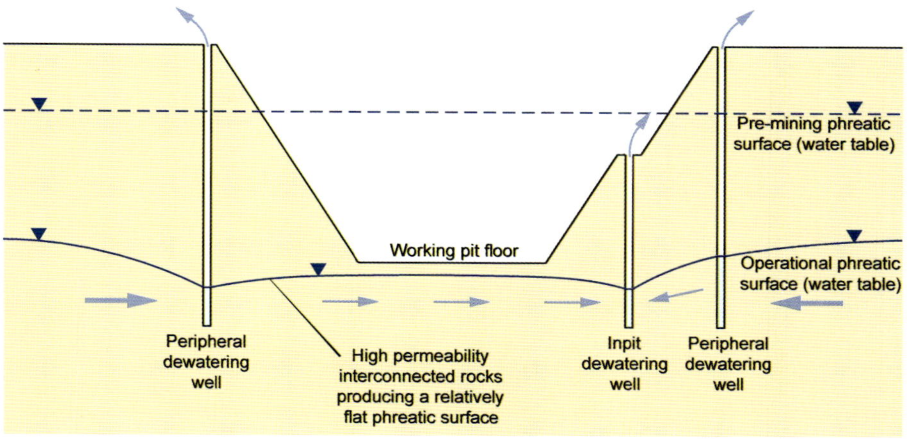

Figure 6.7a: Category 1: Mines excavated below the groundwater table in permeable and interconnected groundwater settings

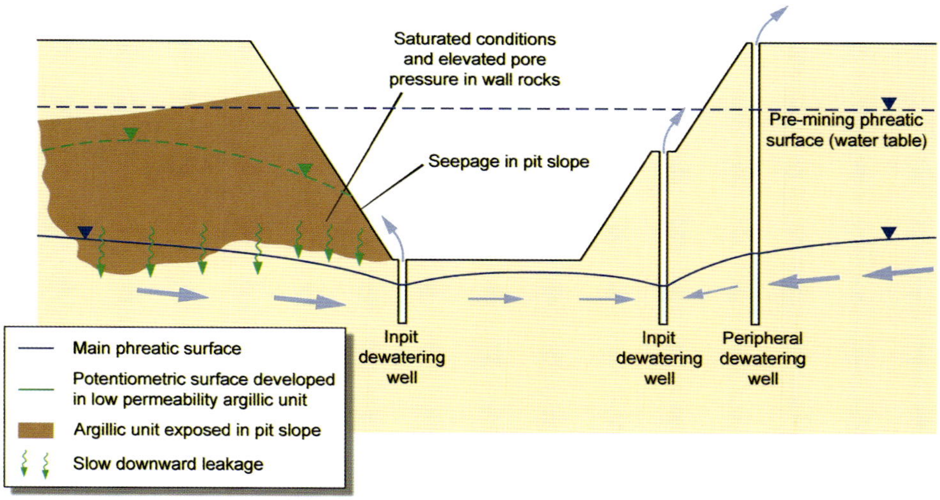

Figure 6.7b: Category 2: Mines with low-permeability rock in part of the pit walls

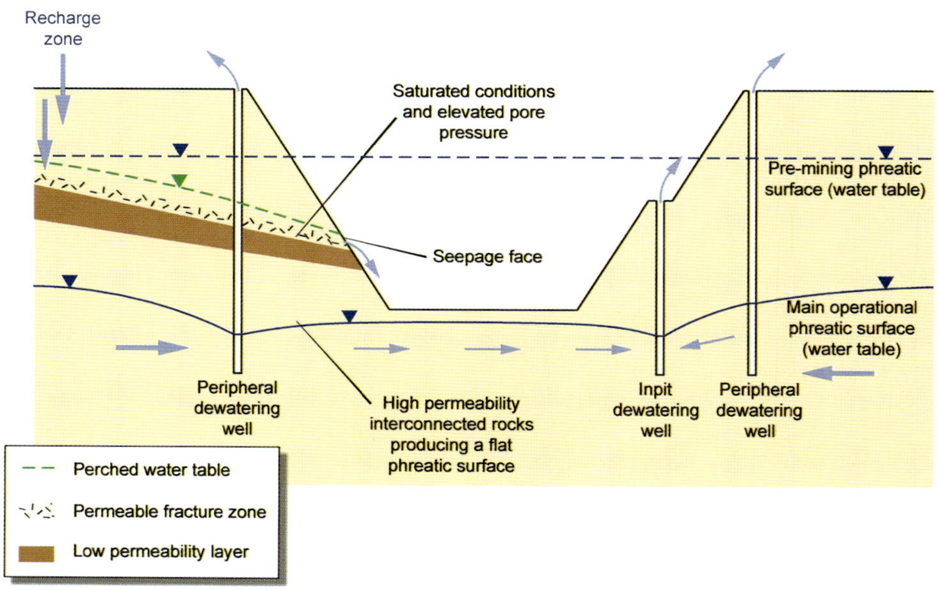

Figure 6.7c: Category 3: Mines with perched water tables

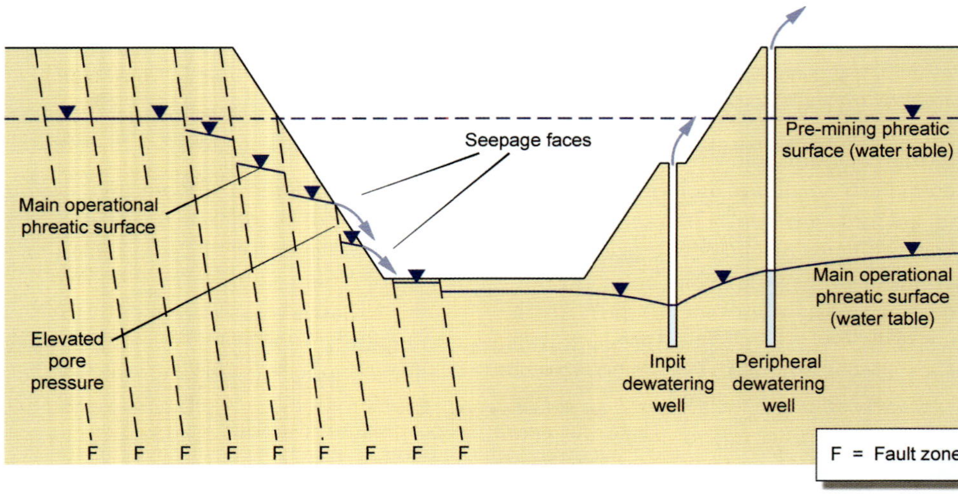

Figure 6.7d: Category 4 :Mines where structural compartments impede depressurisation of the pit slope

An example is the north-east and south-west walls of the Sleeper mine in Nevada, where in excess of 1300 L/sec was pumped from dewatering wells installed in alluvial gravels and permeable volcanic tuffs, but the drainage of clay-altered rocks in certain sectors of the pit wall was poor and required localised control.

Category 3: Mines excavated below the water table with perched groundwater zones

In other situations, perched groundwater zones may develop at higher elevations as the main water table is lowered. These perched zones may lead to residual pore pressures (Figure 6.7c). This was the case in the north wall of the Kori Kollo gold mine in Bolivia. The dewatering system at Kori Kollo pumped almost 1000 L/sec from wells, but a significant amount of seepage occurred to the pit faces during mining because of the perched groundwater conditions.

Category 4: Mines excavated below the water table in a fractured rock mass where subvertical geological structures form barriers to groundwater flow

For this category, the rock mass may not drain fully because geological structures act as impediments to horizontal groundwater flow, creating compartments of trapped water with unchanged (and therefore high) pore pressure. As a result, the general mine dewatering program does not dissipate pressure in all pit slopes. This was the case in the north-east wall of the Bajo de Alumbrera mine in Argentina.

As the excavation is extended and approaches the structural compartments, localised measures may become necessary to penetrate the structures and drain the water behind them. This situation is shown in Figure 6.7d.

Category 5: Mines excavated above the water table where seasonal precipitation or other recharge leads to perched groundwater in upper stratigraphic intervals

For this category, control of pore pressure may be required even though the open pit is entirely above the water table. Localised infiltration of precipitation can build up on less-permeable layers and form perched zones of groundwater, leading to locally elevated pore pressures in the pit slopes. This can also occur when there is artificial recharge from site facilities such as leaking pipelines or tailing areas close to the pit crest. This situation is shown in Figure 6.7e.

There are many examples of this category in tropical mine settings during early pit development, where infiltration of local rainfall can lead to permanent or transient pore pressures in the wall rocks. Another example may be where an open pit is excavated through paleochannel or active stream deposits.

6.1.4 Making the decision to depressurise

6.1.4.1 Quantifying the decision

Of the major factors that control pit slope stability, pore pressure is the one parameter that can often be readily modified. Other parameters such as lithology, structure and the inherent strength of geological materials (material strength friction and cohesion) cannot normally be changed.

The most obvious benefit of slope depressurisation is that it presents the opportunity to improve the overall slope performance and reduce stripping costs. The cartoon in Figure 6.8, which is based on the Hoek and Bray slope stability and groundwater condition charts presented in Figures 10.22 and 10.23 (Chapter 10), illustrates

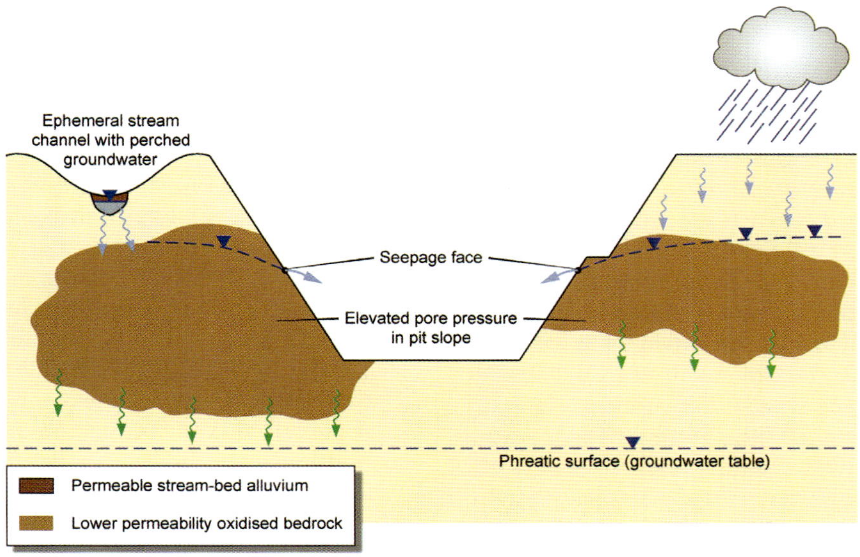

Figure 6.7e: Category 5: Mines occurring above the groundwater table

conceptually how the slope angle may be increased as a result of pore pressure reduction.

A limiting equilibrium analysis can be used to quantify the potential benefit of reducing pore pressure. As an example, consider an open pit gold mine with dimensions of approximately 2000 m along strike, 1000 m width and 400 m depth designed at an average overall slope angle of 38° and average stripping cost of US$1 per tonne. The effective stress concept described in section 6.1.2.2 can be used for calculating the balance between reducing the pore pressure and increasing the slope angle to achieve a constant minimum required safety factor. In the example, an increase in slope angle by 1° would reduce stripping requirements by about 90 million tonnes, with an

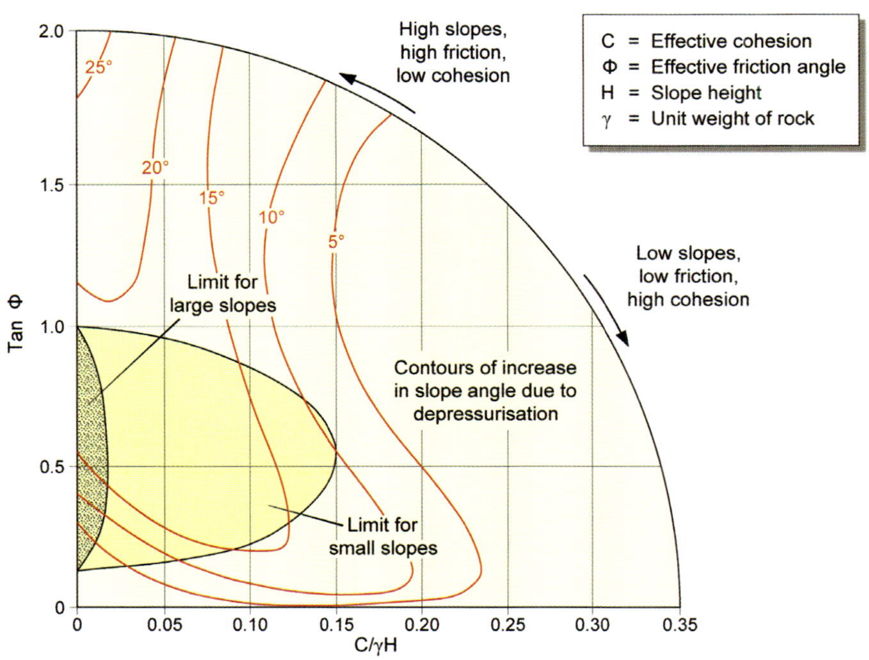

Figure 6.8: Conceptual slope angle increases with full depressurisation
Source: After Hoek & Bray (1977)

associated saving in stripping costs of about US$90 million.

The other side of the equation is harder to quantify: how would depressurisation work, and how much it would cost to achieve the target depressurisation levels? This chapter provides guidelines and alternative approaches for answering these questions.

6.1.4.2 General considerations

Natural drainage of the wall rocks begins to occur as soon as the slope is excavated below the water table. This drainage causes a reduction in pore pressure (see Figure 6.6). In some situations, this natural drainage (passive pore pressure control) is adequate for achieving the desired goals for slope stability. However, in many cases the rate of pore pressure dissipation achieved by passive drainage is too slow to keep up with the mining advance rate or to achieve the target depressurisation levels. Implementation of active pore pressure control (e.g. with pumping wells or horizontal drains, see section 6.5) can be used to accelerate the rate of pore pressure dissipation in the wall rocks.

The following are key considerations for deciding whether to implement an active mine dewatering and/or slope depressurisation program.

- Will slope depressurisation increase the effective stress of the materials enough to provide an economic benefit by allowing a reduction of stripping, resulting from an increase in the slope angle?
- Will slope depressurisation lead to a reduction of water on working benches, so as to increase the efficiency and reduce the cost of mine operations such as blasting, mucking or loading?
- Will slope depressurisation decrease the moisture content of the mined materials, so as to reduce the weight and cost of haulage?
- Can the slope depressurisation program achieve the desired results in the time available?
- Will the control of water reduce slope erosion and help prevent piping or squeezing of the softer materials within the slope?
- Will the overall reduction in water lead to a reduction in mobile equipment maintenance?
- Is the program necessary to achieve an increase in mine safety?

In some instances, the decision to implement a mine dewatering program can be made based on projected savings in operating or equipment maintenance costs alone. An example is the mine dewatering program for the Morenci mine Metcalf pit in Arizona (Table 6.1). The table illustrates only operating costs for a given mine design. In this example, the potential economic benefit of a depressurised and steeper wall was not considered in the cost–benefit analysis.

Table 6.1: Example of operating cost savings due to dewatering

Cost element	Benefit ($/yr)	Benefit ($/ton)
Savings in blasting costs	389 000	0.010
Reduced slope maintenance	960 000	0.024
Reduced operation of in-pit sumps	164 000	0.004
Savings in haulage costs	709 000	0.018
Savings in maintenance costs	800 000	0.020
Savings in power	(53 000)	(0.001)
Total cost benefit	2 969 000	0.074

Table refers to equipment operating costs only. Cost savings associated with improved slope performance are not included.

The Metcalf pit was successfully mined between 1998 and 2004. The relative absence of water significantly increased mine efficiency and allowed ore recovery down to the final planned bench.

6.1.4.3 Beneficial factors of a slope depressurisation program

A slope depressurisation program may be beneficial if:

- the slope height is large;
- depressurisation would significantly increase the effective stress of the slope materials;
- there are low-permeability materials or structural compartments that will not drain freely in response to normal mining and dewatering;
- it is practicable to install slope depressurisation measures, given the pit geometry and the nature of mine operations;
- it leads to greater mine safety.

6.1.4.4 Factors that reduce the need for a slope depressurisation program

The need for a slope depressurisation program is more questionable if:

- the effective stress is expected to show very little change as a result of depressurisation, e.g. in granite or other very strong rocks;
- passive drainage of the fractures and pore spaces occurs rapidly in response to excavation and sump operations, lowering the water table throughout the entire mining area;
- it is impracticable to achieve the required slope depressurisation. This may be the case in a high-rainfall tropical setting, where the pore pressures result from shallow water in the blast-damaged zone close to the pit wall and are rejuvenated with each rainfall event;
- the rocks have a very low permeability and will not depressurise significantly in the required time.

Judgment and experience are important in making a practical decision regarding whether to implement an active slope depressurisation program. Characterisation of the basic hydrogeology is required. Numerical modelling is often useful to help quantify the cost–benefit of the decision.

6.1.5 Developing a slope depressurisation program

A typical approach for implementing a slope depressurisation program is as follows.

Step 1: Collect hydrogeological data and develop an overall conceptual model for the mine site area. All mines require a general knowledge of hydrogeological conditions and need to collect data for the groundwater system. If not specifically required for dewatering or slope depressurisation, the information will be required for permitting and impact assessment. A conceptual groundwater model is often required by regulatory agencies.

Step 2: Determine the need for and scope of a pit slope depressurisation program. This step requires integration of mine planning and geotechnical information with the conceptual hydrogeological model. The cost–benefit of a slope depressurisation program is typically evaluated as follows:

- calculate slope angles and the associated safety factors assuming no depressurisation apart from passive drainage to the slope as mining progresses;
- analyse available data to determine the practicality and potential cost of slope depressurisation;
- calculate slope angles and the relevant safety factors assuming reduced pore-water pressures as a result of active depressurisation;
- evaluate the difference in slope design and stripping costs for an undrained, partially drained and fully drained slope;
- prepare a cost estimate for achieving the simulated depressurisation and compare with the total reduction in mining costs (cost–benefit analysis);
- evaluate the risk and contingency costs if the depressurisation measures do not perform as expected.

This is often an iterative process, and is carried out simultaneously with Steps 3 and 4.

Step 3: Carry out focused data collection specific to the area of the pit slope. The field methods described in Chapter 2 (section 2.5) of this book are typically the most appropriate. The monitoring network also requires updating to allow collection of data specific to the area of the pit slope.

Step 4: Prepare a conceptual hydrogeological model specific to the pit slope. Development of the conceptual model is described in section 6.3.

Step 5: Develop a numerical hydrogeological model of the pit slope, as required. The need for numerical modelling is discussed in section 6.4.

Step 6: Design and implement the required depressurisation measures to optimise the slope design, maximise safety and minimise stripping costs. Typical measures used for pit slope depressurisation are discussed in section 6.5.

Step 7: Carry out monitoring of pore pressures prior to and during excavation of the slope. The monitoring of groundwater pressures is outlined in Chapter 12 (section 12.2.2.5).

6.2 Background to groundwater hydraulics

6.2.1 Groundwater flow

6.2.1.1 Introduction

There are many readily available text books that describe groundwater hydraulics in detail. Those with a more practical focus include the following.

- Driscoll FD (1986). *Groundwater and wells*, 2nd edn. RG Designs, 1089 pp.
- Freeze RA & Cherry JA (1979). *Groundwater*. Prentice Hall, 604 pp.
- Price M (2003). *Agua subterránea*. Spanish translation of *Introducing groundwater*, translated and with additional material by JJ Carillo-Rivera & A Cardona. Editorial Limusa, Grupo Noriega, Mexico, 330 pp.

A brief summary of the principles of groundwater hydraulics in relation to slope depressurisation programs, and some of the properties commonly encountered in mine site settings, is provided in this section. Although the term hydraulic conductivity (which includes the density and dynamic viscosity of the fluid) is more correct, in this book permeability is used synonymously with hydraulic conductivity.

6.2.1.2 Darcy's law

Darcy's law is the basic equation governing the flow of groundwater through soil or rock. Darcy's law states that the volume rate of saturated flow (Q) of groundwater is directly proportional to the cross-sectional area (A) through which flow is occurring and the hydraulic gradient (i) (Figure 6.9). The hydraulic gradient is the difference in head between two points on the flow path divided by the distance (measured along the flow direction) between them. Thus, Darcy's law can be written:

$$Q = KiA \qquad \text{(eqn 6.2)}$$

The constant of proportionality, K, is the permeability. The hydraulic gradient (i) is determined by $\Delta h/L$.

The hydraulic head is the height to which the water rises above an elevation datum (typically mean sea level) and is a measure of the mechanical energy possessed by

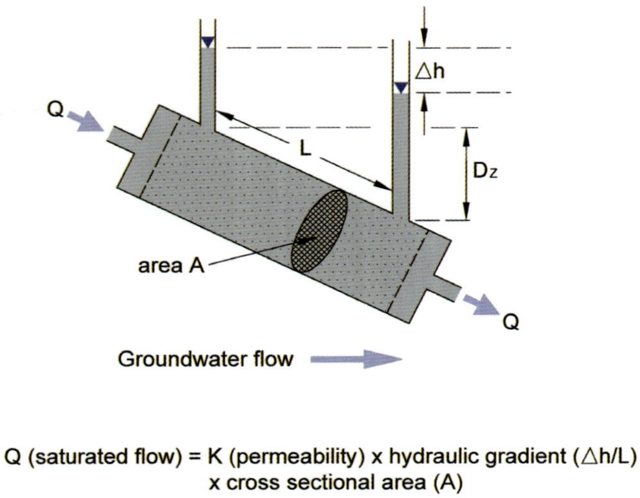

Figure 6.9: Darcy's law. (Note: The term 'permeability is used synonymously with hydraulic conductivity.)

the water. It represents the sum of the pressure head in the piezometer and the elevation of the measuring point in the piezometer. In Figure 6.10, the hydraulic head at point P is *h* and the pressure head (the head resulting from the pore pressure) is $(h-z)$.

At the down gradient point on Figure 6.10 (point P') the total head has reduced by an amount equivalent to Δh. The new head (h') is lower. However, the pore pressure at point P' is (h' - z'), which is higher.

6.2.1.3 Darcy's law in field situations

Consider a permeable unit of constant thickness, underlain and overlain by impermeable beds and containing water in which the pore pressure is everywhere greater than atmospheric pressure. When a borehole is drilled into the unit, the water rises above the top of the unit (Figure 6.11).

In Figure 6.11, the rate of flow through the permeable unit can be calculated using Darcy's law:

$$Q = kiA = K \times b \times w \times (\Delta h / L) \qquad \text{(eqn 6.3)}$$

The term transmissivity (T) is illustrated in Figure 6.12. Transmissivity is the rate at which water will move under a unit hydraulic gradient through a unit width of aquifer. Transmissivity is the product of permeability and saturated thickness (b) of the permeable unit (Kxb). Although commonly used in text books, the term is often difficult to apply in mine settings because of the highly variable nature of the hydraulic gradients and thicknesses of the permeable units. Thus, in mine settings it is usually more realistic to characterise the materials solely on the basis of their permeability.

In Figures 6.11 and 6.12, the potentiometric surface is above the surface of the permeable unit and the permeable unit is said to be confined. Figure 6.13 shows the situation where the potentiometric surface is within the permeable material. In this case, the potentiometric surface is the water table (phreatic surface) and the permeable unit is unconfined.

Figure 6.13 illustrates two important factors in an unconfined unit:

- the saturated thickness (b) is not constant;
- for flow to occur there must be a vertical component (gradient) – flow cannot be purely horizontal.

If there is a vertical groundwater flow component, there must also be a vertical hydraulic gradient and vertical difference in head (see Figure 6.14). A vertical groundwater flow component occurs whenever there is a vertical difference in head. Such conditions are common in porous medium formations around open pit mines, as groundwater moves progressively closer to the mine excavation.

In real-world situations, piezometers or observation wells do not lie conveniently along the direction of groundwater flow as they do in the illustrations. In most cases, measurements in several piezometers are used to derive the shape of the potentiometric surface, from which the magnitude and direction of groundwater flow is determined. At least three data points are needed to derive the slope of the potentiometric surface. In more complex settings, considerably more data points are required.

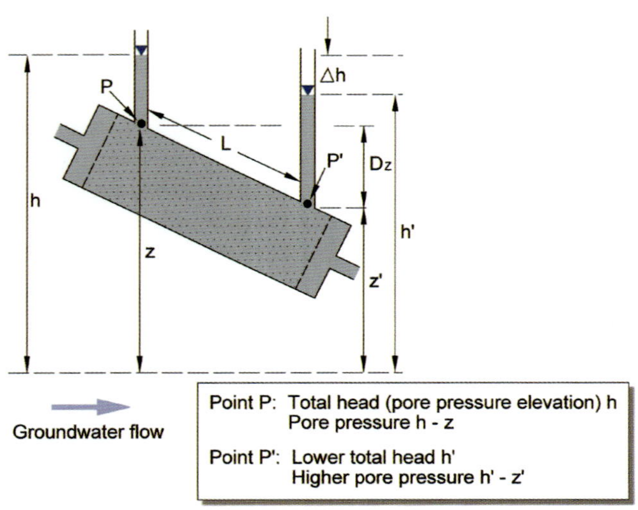

Figure 6.10: Piezometric head and head loss

6.2.1.4 Heterogeneity and anisotropy

The discussion so far has dealt with materials that are uniform or homogeneous (i.e. the properties are the same everywhere) and isotropic (i.e. the properties at any point do not vary with direction). However, most rocks and soils are heterogeneous and anisotropic to varying degrees.

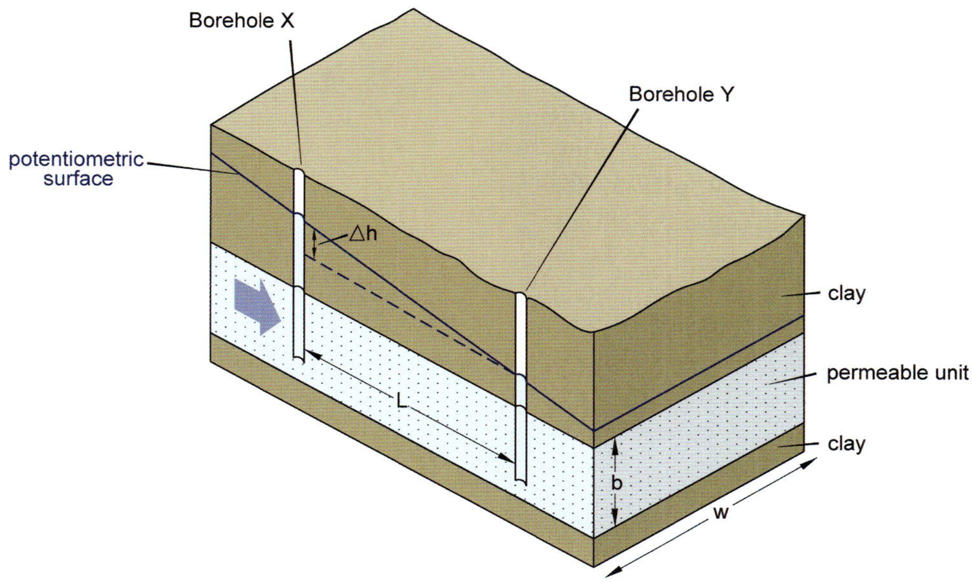

Figure 6.11: Field illustration of groundwater flow

The heterogeneity and anisotropy can arise in several ways and at different scales. In porous materials the grains or bedding are normally arranged in such a way that it is easier for water to flow in one direction than another. In Figure 6.15a the permeability in the direction parallel to the bedding is greater than that in the direction perpendicular to the bedding. The compaction or alignment of the grains themselves, or interbedded concretions or layers of finer silty materials, typically lead to a higher permeability in the horizontal plane (*Kh*) than in the vertical plane (*Kv*). An example would be argillaceous rocks in which the clay minerals tend to be aligned parallel to bedding or cleavage. In hard rock settings, there are often two or three main fracture or joint orientations and the alignment of these allows preferential flow directions to develop.

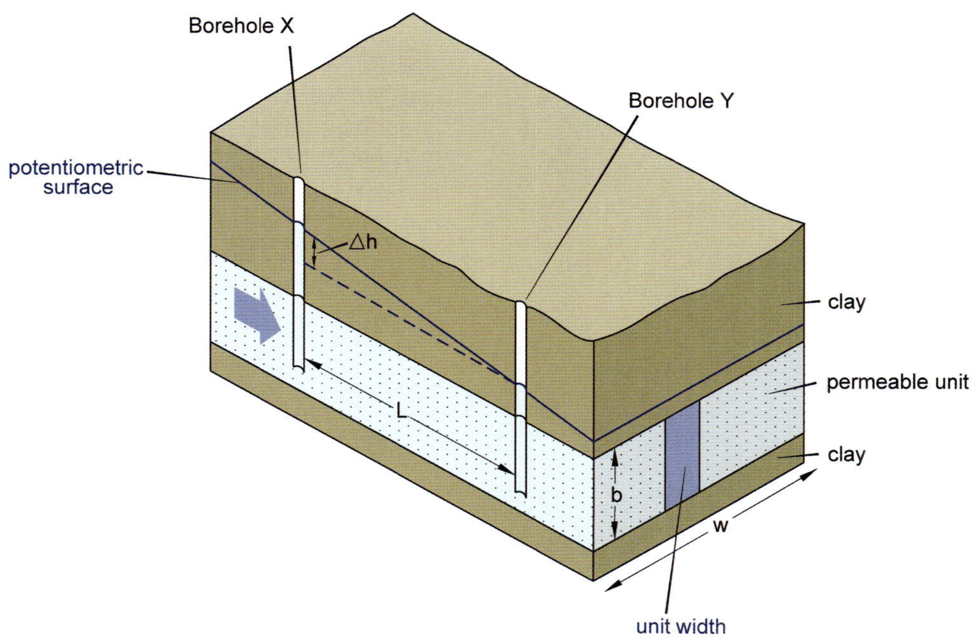

Figure 6.12: Transmissivity. (Note: the use of transmissivity is often more applicable to homogeneous overburden materials than in fractured rock, and is often difficult to apply around dynamic mining situations.)

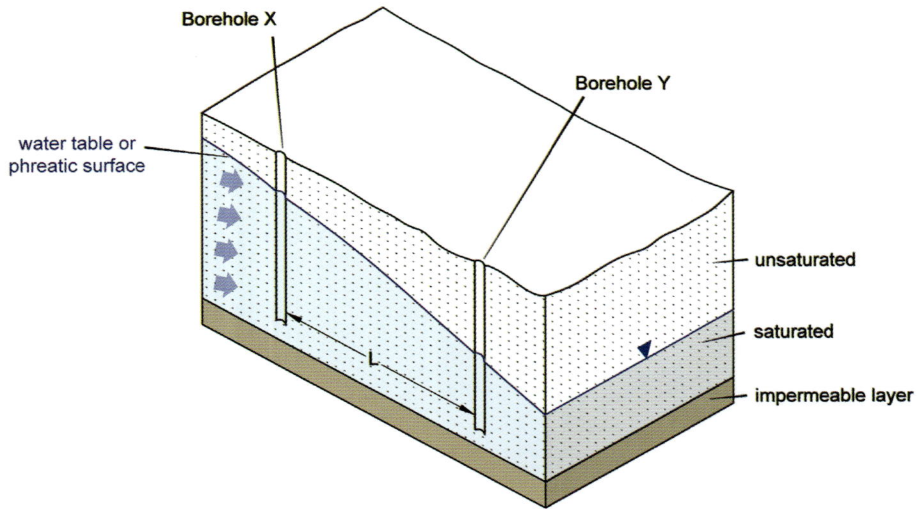

Figure 6.13: Unconfined groundwater flow

At a larger scale, adjacent rock layers in a bedded system usually have different properties (Figure 6.15b). Even if each individual bed is isotropic, within the whole system it will be easier for flow to take place parallel to the bedding than across it. Flow parallel to the beds will be dominated by the layer(s) of highest permeability, and flow across the beds will be dominated by the layer(s) of lowest permeability.

The ratio *Kh:Kv* can be less than one order of magnitude, e.g. in a beach sand or coarse gravel. In basin-deposited sequences, where interbedded finer-grained silt or clay layers occur, the effects of heterogeneity can lead to bulk anisotropy such that the average *Kh* can be one to three orders of magnitude, or more, greater than the average *Kv*.

6.2.2 Porous-medium (intergranular) groundwater settings

6.2.2.1 Introduction

Most of the commonly cited groundwater theory, and the examples shown in Figures 6.9 through 6.15, relate to saturated porous medium (intergranular) flow settings

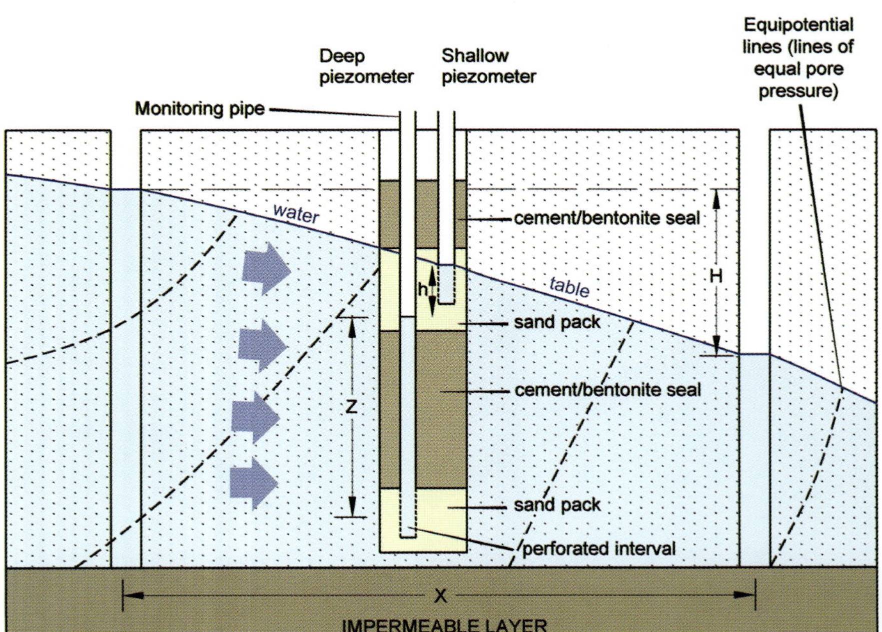

Figure 6.14: Lateral and vertical variation in head as a result of groundwater flow

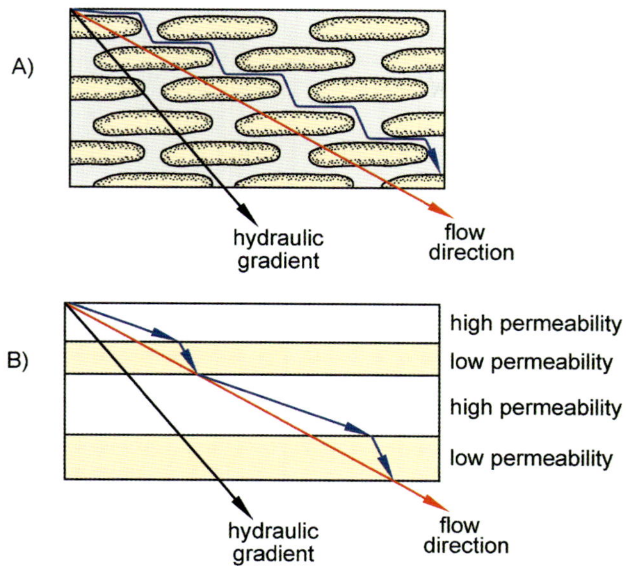

Figure 6.15: Anisotropy

such as unconsolidated overburden deposits, certain types of sandstone, or weathered bedrock and clay-altered bedrock.

Typically, characterisation and planning of groundwater control measures is more straightforward for intergranular settings. The flow system behaves in a more homogeneous manner, and is easier to extrapolate and interpret between installed piezometers and other data points. However, it is rare to encounter completely non-layered strata in which homogeneous unconfined conditions occur. Even an apparently homogenous sand unit will often have heterogeneities related to lateral depositional (facies) changes and vertical bedding changes.

6.2.2.2 *Unconsolidated deposits*

Figure 6.16 illustrates a typical overburden sequence found in many mine-site settings throughout the world. In the diagram, much of the groundwater flow occurs preferentially within discrete more-permeable gravel layers. As these layers begin to depressurise as a result of preferential flow to the pumping well, vertical groundwater leakage into the gravel layers begins to take place from the intervening less-permeable layers. This secondary leakage reduces the rate at which drawdown occurs in the permeable layers. This is termed a semi-confined ('leaky aquifer') setting.

6.2.2.3 *Sedimentary rocks*

Most poorly consolidated sedimentary rocks (sandstones and mudstones) exhibit predominantly porous-medium flow characteristics. Most competent (hard) sedimentary rocks exhibit fracture-flow, or characteristics of both fracture-flow and porous-medium flow.

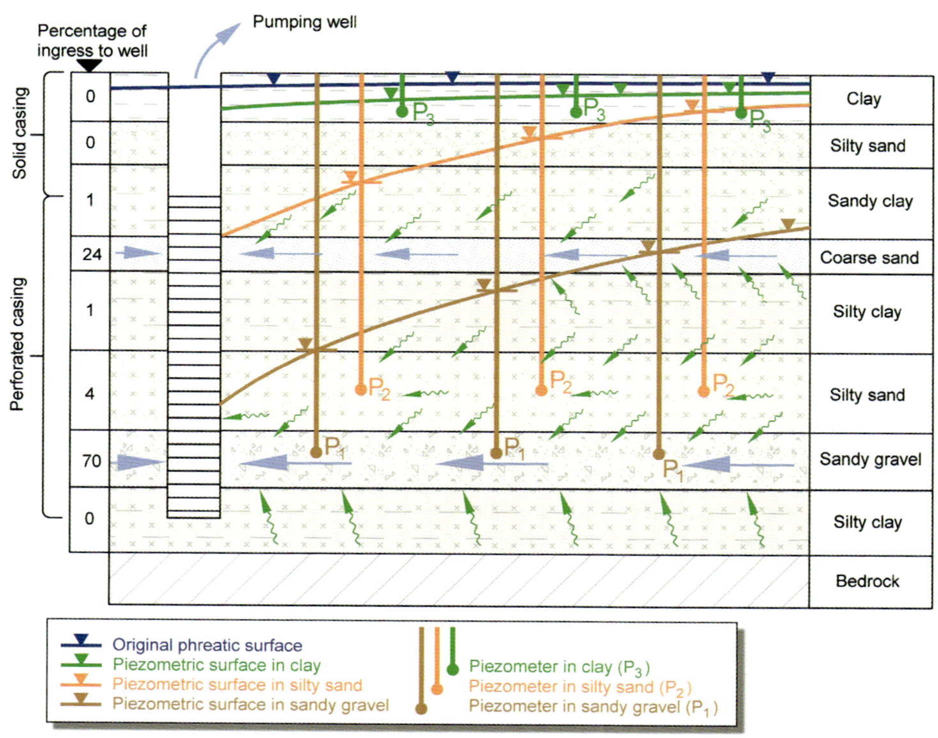

Figure 6.16: Typical overburden sequence found in mine settings

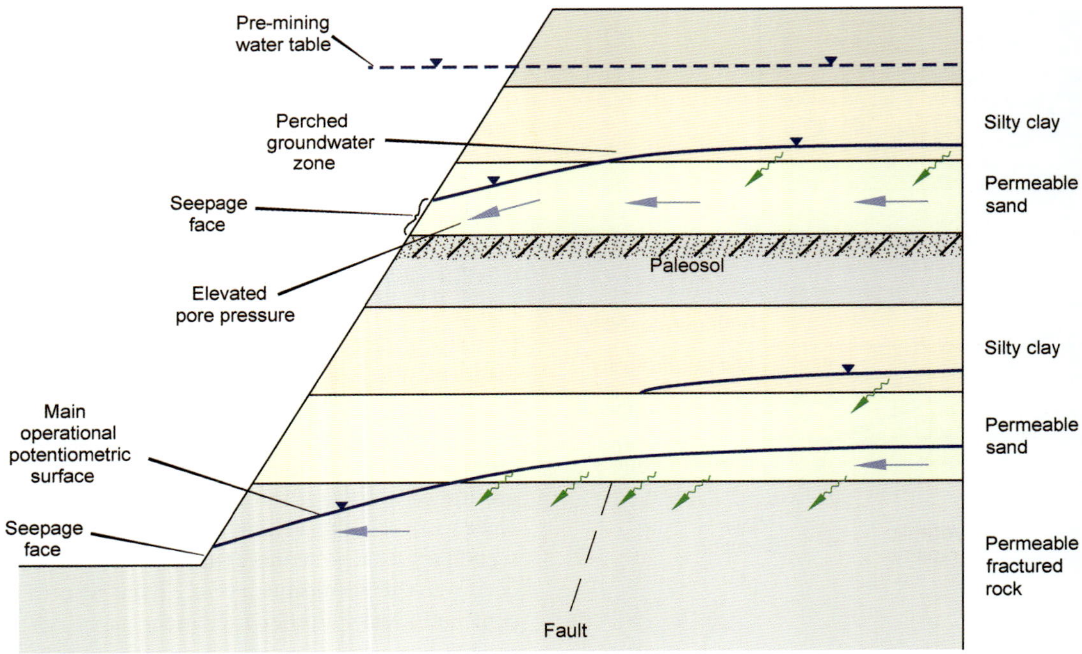

Figure 6.17: Perched zone development in a pit slope cut through a layered sequence

Bedding is often an important factor controlling groundwater flow in sedimentary rocks. Strong vertical hydraulic gradients can develop around mine excavations as a result of resistance to vertical flow caused by bedding. Perched groundwater zones can also develop as the main water table is lowered. The perched zones may be controlled by low-permeability inter-beds or paleosols between the main units (see Figure 6.17).

6.2.2.4 Clay alteration and weathering

In zones of weathering or clay-altered bedrock, the original rock fabric often becomes destroyed and alteration of the rock mass creates zones of interstitial porosity.

Because of their low permeability, zones of clay alteration associated with faults or within volcanic sequences can be some of the most difficult units for pit slope depressurisation and may require the largest amount of lead time. These zones may also be associated with a lower material strength, increasing the need for an active slope depressurisation program implemented in advance of each mine cut. Groundwater flow in these zones is often highly anisotropic. Preferential flow can occur along certain relict structural features, while other relict structures may act as barriers to flow.

Within tropical areas, zones of weathered bedrock or saprolite may also be difficult to depressurise. However, the tropical weathering process is typically associated with a gradual unloading and expansion of the original rock mass over geological time. The process may mean that permeability values in tropical areas are somewhat higher than in zones of clay alteration associated with deeper mineralisation. Furthermore, subvertical features such as quartz dykes may weather to a sand and provide preferential flow conduits.

Clay-altered leached caps associated with porphyry copper deposits may also exhibit somewhat higher porosity values because of the leached nature of the rock mass.

Consideration may need to be given to water-sensitive materials such as swelling clays or kimberlites. For these materials, in addition to reducing pore pressure it is also necessary to minimise contact with water so that the materials do not suffer a loss of strength when they swell, weather and relax and new fracture surfaces develop.

6.2.3 Fracture-flow groundwater settings

6.2.3.1 General

In hard rock settings, most of the groundwater movement occurs in joints and fractures. Volcanic and intrusive rocks typically exhibit only fracture-flow, except in porous clay-alteration and weathered zones as described above. Well-cemented sedimentary rocks and limestones are primarily a fracture-flow medium but may also contain interstitial pore space that provides secondary drainage as the fracture zones become depressurised. This condition is termed a dual-porosity system.

In most hard rock lithologies, some degree of pervasive fracturing typically occurs throughout the entire rock mass. As a result, pore pressure is distributed throughout the rock, not just around the major geological structures.

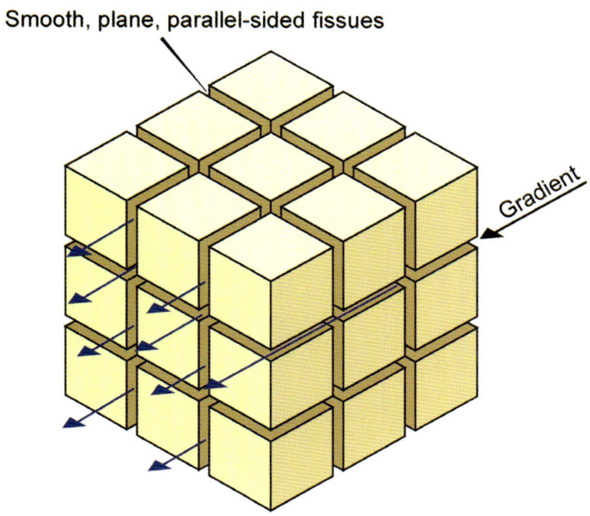

Figure 6.18: Fracture flow under uniform conditions (rock mass containing three idealised, mutually orthogonal fracture sets)

Under static conditions, pore pressure occurs in the low-order joints and micro-fractures, no matter how small the fracture aperture or opening.

6.2.3.2 Fracture flow

A fracture can be thought of as an extreme example of a highly permeable layer. Theoretical studies of fracture flow are usually based on the assumption that the fracture can be treated as an opening bounded by smooth, plane, parallel plates with uniform aperture (Figure 6.18).

The values of porosity and hydraulic conductivity in a homogenous fractured system are shown in Figure 6.19. This figure shows the relationship between fracture spacing, fracture aperture, porosity and permeability for a fracture system consisting of three ideal joint sets which are mutually orthogonal and smooth-walled. For 'real' fractures, roughness decreases the permeability but does not affect porosity (except where alteration occurs along the fracture surfaces). This means that realistic values for porosity tend to be higher for a given permeability than predicted by Figure 6.19.

Continuous fracture zones typically have a higher water-carrying capacity than an equivalent porous medium (Figure 6.20). Open fractures tend to have high permeability but most fractured lithologies tend to exhibit low porosity. Most fractured rocks are highly heterogeneous and anisotropic. However, the anisotropy almost always has a pattern that is determined by the orientation of the dominant fracture sets.

An important concept in groundwater interpretation is that the magnitude of flow (flux) is always controlled by the least permeable (or most resistive) part of the flow system (i.e. the least permeable material along the groundwater path). For fracture-flow settings, the magnitude of flow is normally controlled by the presence of discrete barriers, which may be related to bedding or fault structures and may be subhorizontal or subvertical. The most fractured and permeable zones along the flow path usually exert little influence on the total magnitude of flow, because the amount of groundwater available to feed the permeable zones is governed by the water's ability

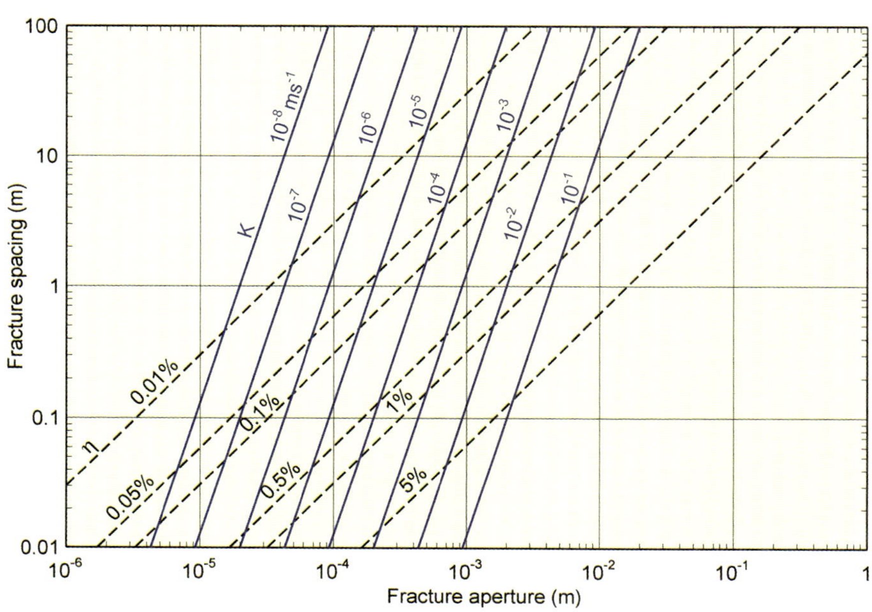

Figure 6.19: Values of porosity (%) and permeability (m/sec) for fracture flow under the uniform conditions of Figure 6.18. Permeability values assume fractures are filled with pure water at 10°C
Source: After Snow (1968)

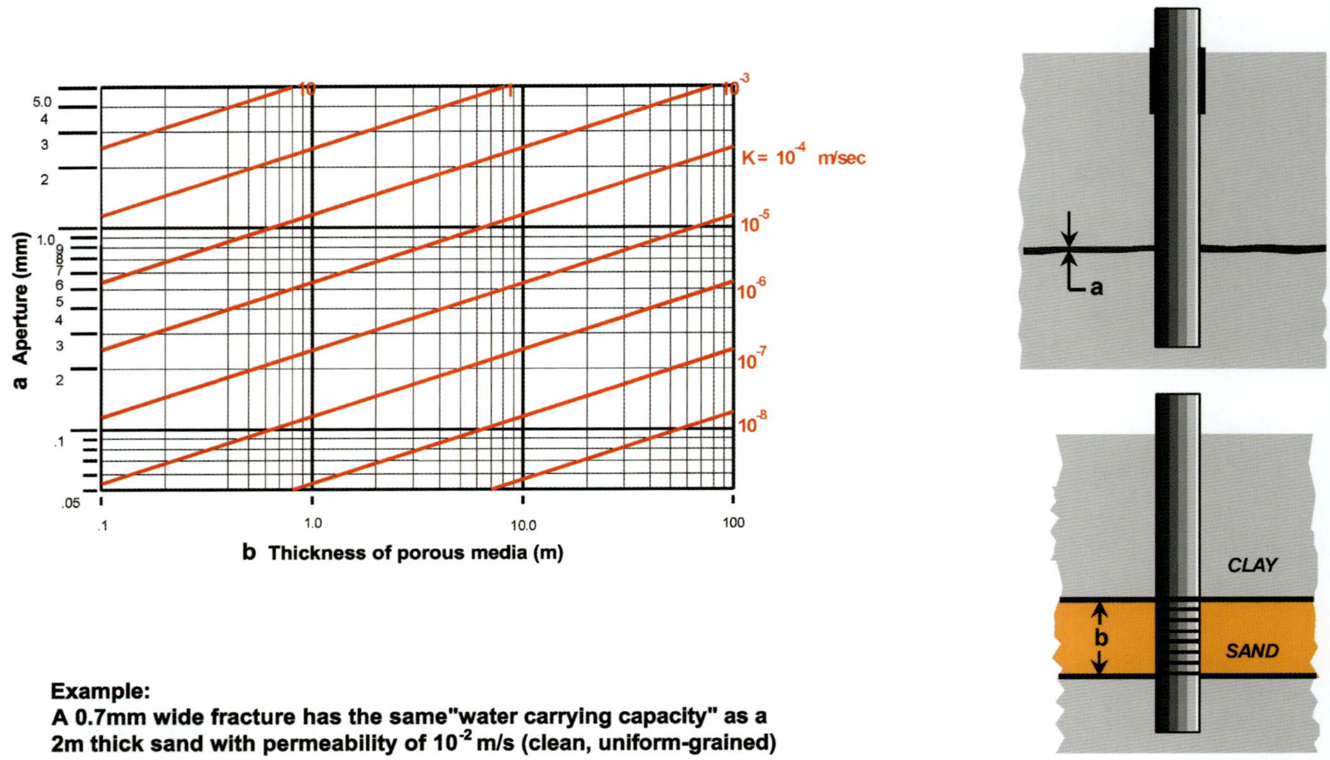

Figure 6.20: Equivalence of fracture flow to porous medium flow

to cross less-permeable structures or bedding planes (Figure 6.21).

6.2.3.3 *Lateral flow barriers and groundwater compartmentalisation*

The presence of high-angle geological structures is often the most important factor influencing groundwater flow within and around mine sites. Geological structures can result in enhanced permeability and groundwater flow along the direction of their strike, and in subparallel fractures within the adjacent rock mass. However, the same structures tend to create barriers to flow across their strike because they may offset the brittle units where fracturing and groundwater flow is occurring, and may create a zone of disturbance (often with clay gouge) of low permeability that further disrupts groundwater flow.

In Figure 6.22 the hydraulic gradient is from north to south, but most flow vectors occur oblique to the overall hydraulic gradient as a result of the structural orientation. If the north-west–south-east and north-east–south-west trending structures are contemporaneous, as is common, the structures offset each other, giving rise to the discontinuous flow system shown in the cross-section. The presence of the two structural orientations causes the groundwater system to become compartmentalised. Within each compartment the fracturing and groundwater flow system is relatively continuous, so the water table is relatively flat. Large hydraulic gradients can develop across each boundary, causing groundwater levels to be 'stair-stepped' and discontinuous.

Compartmentalisation is common in fracture-flow groundwater settings. It can minimise the spread of drawdown away from dewatering centres and thus minimise the amount of water that has to be pumped. If the principal structures are widespread the compartments can be up to a square kilometer. In southern Africa, regional dykes can create compartments in fractured karstic dolomite encompassing the whole mine or larger areas. If the structures are more closely spaced, the compartments can be much smaller.

In certain situations, the flow system can be more continuous along a particular structural direction. For example, a test dewatering well was installed to target a principal fault zone. An array of six piezometers was placed radially around the pumping well at distances of 10–300 m. The well was pumped for 14 days at an average rate of about 35 L/sec. No response was observed in any of the piezometers, because none of them had penetrated the principal fault zone. However, an open exploration drill hole located in the principal fault zone about 3000 m to the north of the pumping well responded strongly, demonstrating very discrete and rapid flow along the structural zone.

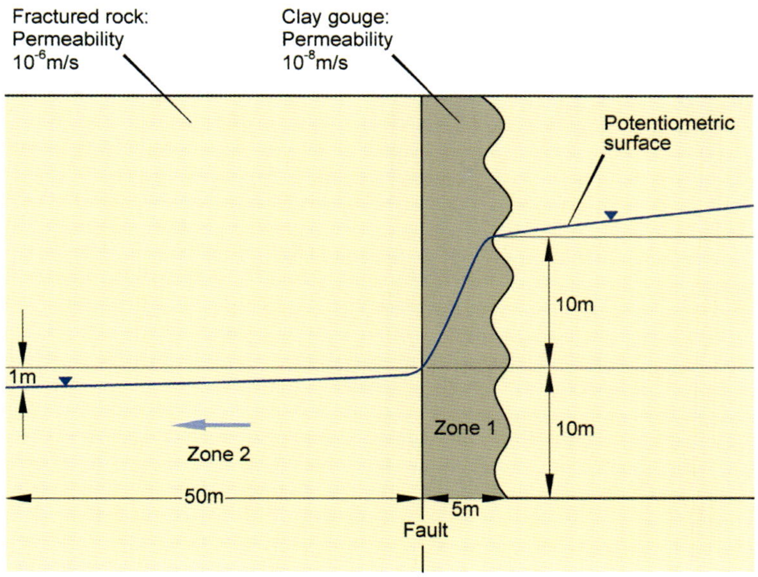

Zone 1: Unit flow (Q) = 2×10^{-8} m/sec ($10^{-8} \times (^{10}/_{5}) \times 1$) : Volume of flow controlled by permeability of clay gouge (barrier)

Zone 2: Unit flow (Q) = 2×10^{-8} m/sec ($10^{-6} \times (^{1}/_{50}) \times 1$) : Volume of flow controlled by amount of water passing upstream barrier (clay gouge)

Figure 6.21: Magnitude of groundwater flow being controlled by the lowest-permeability zone

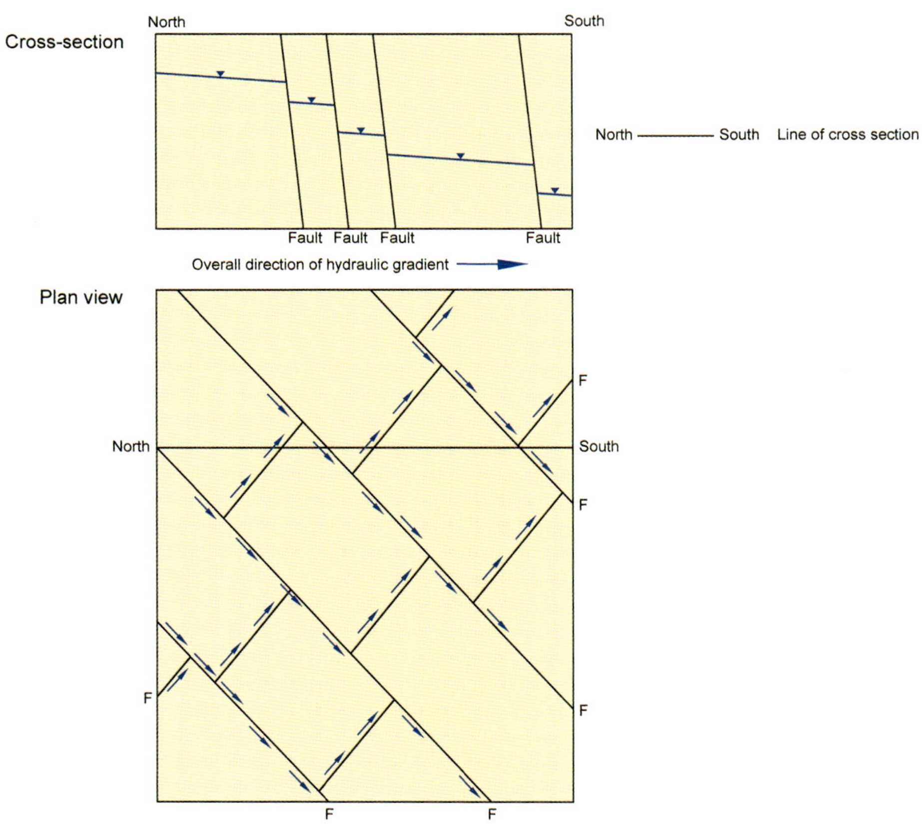

Figure 6.22: Influence of high-angle faults on groundwater flow

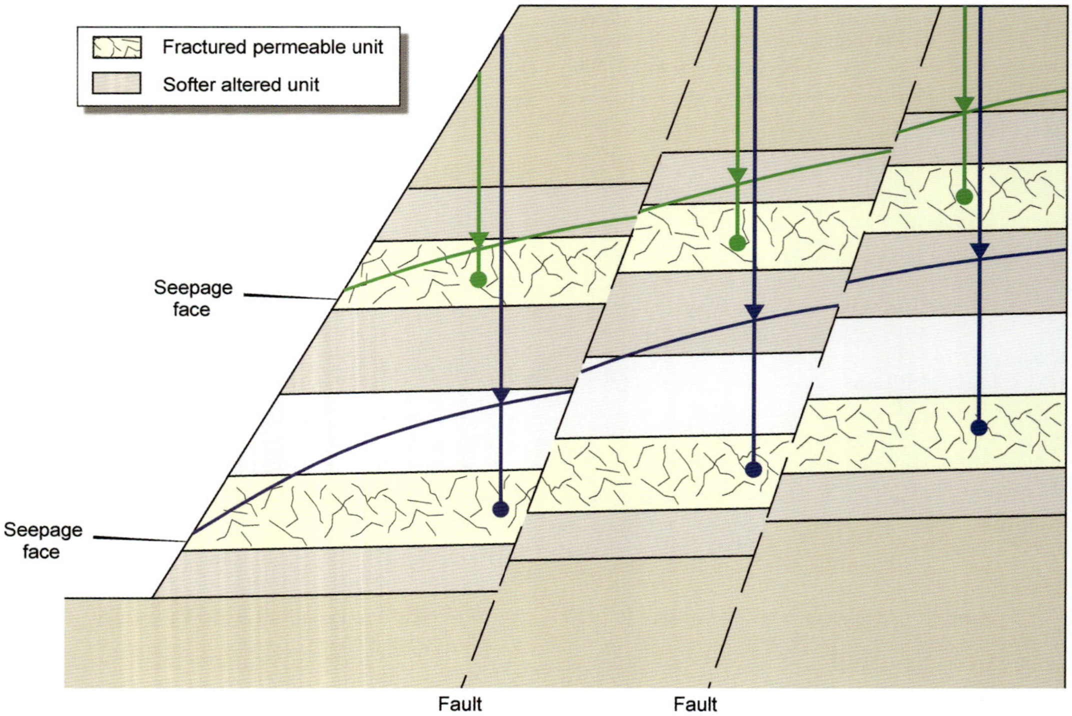

Figure 6.23: Vertical and lateral compartmentalisation

6.2.3.4 Vertical flow barriers

Vertical compartmentalisation is also common in fracture-flow settings. Figure 6.23 shows a cross-section through a layered volcanic sequence. Much of the groundwater flow occurs within discrete layers or interflow horizons, where the rock is more brittle, easily fractured and more permeable. The intervening layers create vertical barriers to flow and allow vertical hydraulic gradients to develop.

6.2.3.5 Influence of vertical jointing

Subhorizontal layering and the presence of bedding planes or paleosols tend to create barriers to vertical flow. However, there are a few situations where continuous vertical jointing can increase the fracture connectivity within or between layers. Continuous vertical joints spanning large vertical intervals are not common in mining situations. When examples do occur they are typically found in limestone (or marble), in some sandstones (or quartzite) and very occasionally in some volcanic sequences (e.g. some basalts).

6.2.3.6 An example mine-site setting

The preceding illustrations demonstrate that, for most fractured rock settings, it is the continuity of the main fracture zones and the presence or absence of bounding structures that form the main influence on the groundwater flow system, rather than the in situ permeability of the rocks themselves. For example, Figure 6.24 shows groundwater flow within the dynamic setting of a pit slope. The figure shows:

- residual high pore pressure and a minor seepage face in the near-surface overburden clays;
- dewatering of the underlying permeable gravel overburden using alluvial interceptor wells;
- the effect of structural boundaries and compartmentalisation within the bedrock;
- clay-altered zones associated with steeply dipping structures within the bedrock, causing elevated pore pressures behind the pit walls.

Comparing Figure 6.24 with Figure 6.9, the application of Darcy's law to an actual mine setting can be seen as follows:

- Darcy's law is applicable in the gravel overburden and the flow rate towards the mine within the gravel may be reasonably estimated. Although the underlying structures continue upward and offset the overburden materials, the gravel units are too recent to be fault-affected and the groundwater flow system remains continuous;
- for fractured bedrock, the application of Darcy's law requires more interpretation because of the complex and discontinuous nature of the flow system. It is relatively straightforward to characterise the in situ

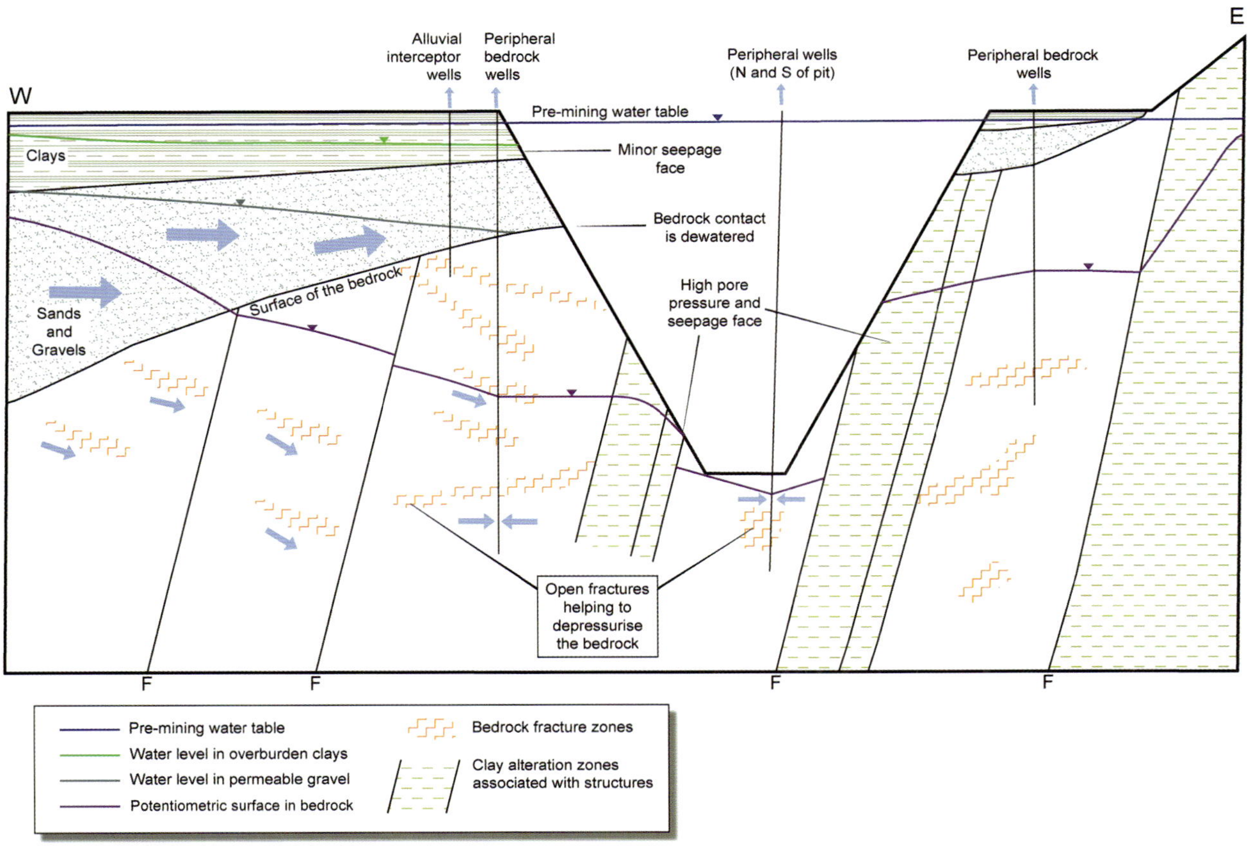

Figure 6.24: Slope depressurisation from an active open pit
Source: Sleeper mine, Nevada, USA

permeability of the fractured rock between the structures. However, the flow rate through the section is controlled by the permeability across the structures, which it is not practicable to measure in the field and requires interpretation and estimation. In the bedrock, the magnitude of groundwater flow is controlled by water availability and not by in situ permeability. Therefore, the application of Darcy's law is much more subjective.

6.2.4 Influences on fracturing and groundwater

The most common factors controlling groundwater flow around open pit mines are as follows.

1 **Depth below topography.** As a general rule, the aperture of open fractures (and therefore the porosity and permeability) can be expected to decrease with increasing depth below ground surface. Around open pits, the natural fracture pattern is further influenced by the stress field and zone of relaxation, ZOR (deformation), created by the mine void.
2 **Geological structure.** This is the major contributor to the distribution and alignment of fractures in most mine settings. Typically, two or three prominent fracture orientations can be recognised, depending on the structural setting. Fracture sets can be classified as first-order, second-order and third-order, as required (Figure 6.25). In some cases the main fault zones may control the first-order fractures, which typically have the largest apertures (in excess of 10 mm). Lower-order fracture apertures can be as small as a few micrometres.
3 **Lithology.** Certain lithological types are brittle and sustain extensive open fracturing (e.g. quartzite, rhyolite). Other lithological types are more ductile and open fractures are less common, or fractures become healed and closed. Hence, subtle variations in lithology can be equally or even more important than the structures themselves for maintaining open fractures in the rock mass, and therefore for controlling groundwater flow patterns.
4 **Alteration** can change the fabric of the rock mass and therefore cause significant changes to its hydraulic properties. Silicic or potassic alteration may increase

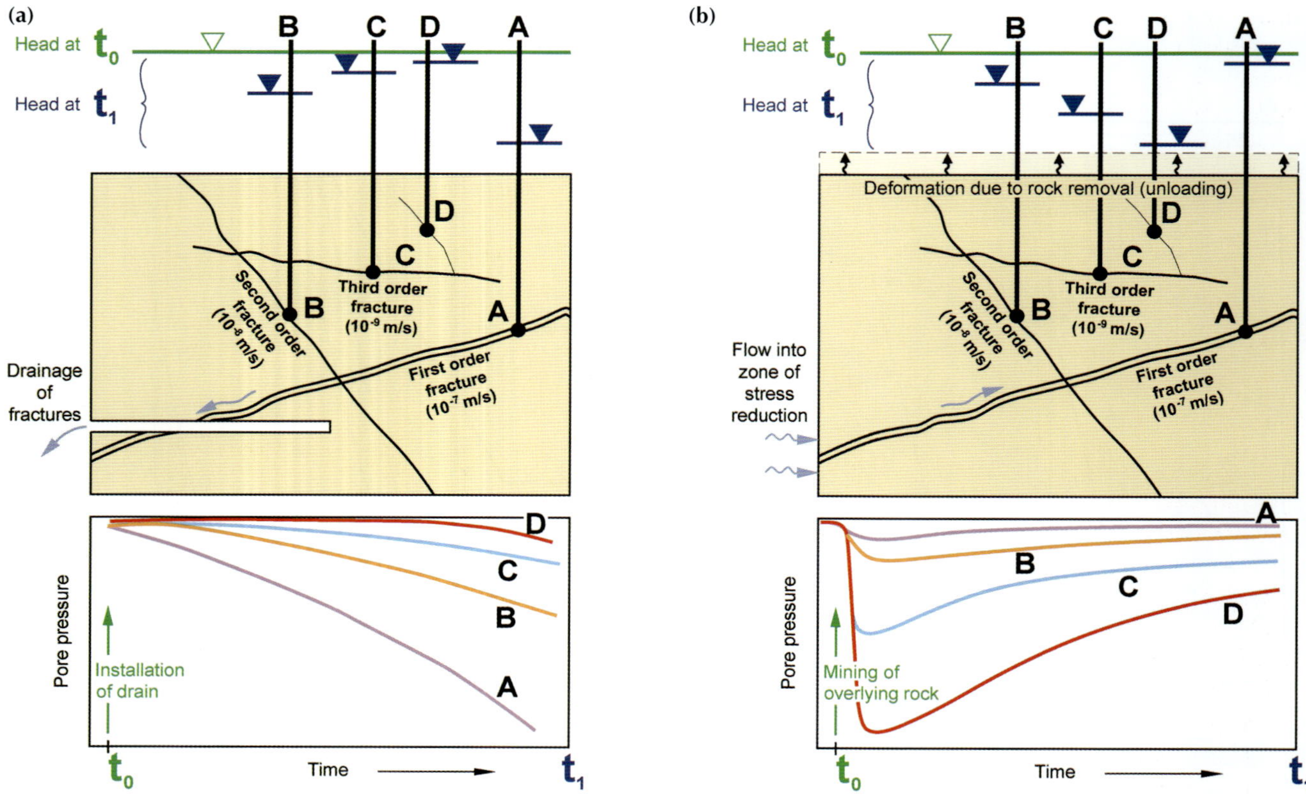

Figure 6.25: (a) Fracture-controlled (dual-porosity) drainage response due to groundwater flow; (b) unloading response in a fractured medium

the rock's tendency to sustain open fracturing. In many porphyry settings, zones of potassic alteration provide greater groundwater ingress to dewatering wells or horizontal drains. Conversely, sericitic or argillic alteration tends to cause fractures to heal, greatly reducing their potential to transmit groundwater. Clay-rich minerals can be self-sealing and, although water-bearing on exposure, become closed over days or months. Argillisation may change the nature of groundwater movement from fracture-flow to intergranular (porous-medium) flow.

5 **Mineralisation** may alter the fabric of the rock mass. For example, in several of the porphyry copper deposits leaching has caused healing of fractures, and groundwater movement in the highly leached zones can be a combination of fracture and intergranular flow. For many leached zones, the rock mass itself is more permeable but the absence of open fractures causes a reduction in the overall groundwater flow potential. In kimberlites, the ore body can become 'unbonded' from the country rock, providing open fracturing between the kimberlite and country rock.

6 **Weathering** alters the physical nature of the rock. In many tropical areas, most of the fracturing in the upper weathered regolith or saprolite zone has become healed and intergranular flow dominates. Often, there is a transition zone near the base of the weathered zone, where some of the relict fractures remain partially open and a combination of intergranular and fracture-controlled flow may occur.

7 **Rock mass unloading** occurs as a result of the stress field created by the mining operation and often causes the opening or widening of fractures in the zone of relaxation around the mine excavation (see section 6.2.5). Open fracturing, and higher permeability and porosity, can be related to the zone of rock deformation. This zone may extend for a considerable distance behind the pit slope.

8 **The blast-damaged zone** represents the area where rock properties and conditions are altered because of the excavation processes. The extent of the blast damage depends on the type of rock, the pattern of blast holes and the charges used. In smaller mines with 6 m high catch benches, or in an operation where the mining process is purely mechanical, the damaged zone may be less than 5 m behind the pit wall. In larger open pits with >15 m high benches, the blast-damaged zone may extend more than 40 m behind the wall. It is generally accepted that blast damage can increase permeability by up to three orders of

magnitude, but the increase may be higher in unaltered brittle rock types. As a result, there is a tendency for groundwater to move preferentially as unseen seepage at shallow levels behind the pit slope. The porosity of the rock is also greatly increased within this rind. The over-break zone may also be present in the base of the pit and can drain pit wall seepage water into sumps.

6.2.5 Mechanisms controlling pore pressure reduction

6.2.5.1 General

A reduction in pore pressure within a pit slope may occur as a result of three mechanisms:

1 groundwater flow away from the zone in question (to a seepage face or to a well or drain as a function of mining);
2 increase in the total porosity, caused by deformation and expansion of the rock, as a result of stripping the overlying materials (lithostatic unloading and relaxation);
3 increase in total porosity caused by expansion of the rock mass as a result of drainage and removal of water from the overlying rock (hydrostatic unloading).

6.2.5.2 Pore pressure reduction from groundwater flow

In most mine site settings, changes in pore pressure usually occur as a result of groundwater flow. In many hard rock environments, the first-order fracture sets are related to the main (primary) fault zones and the rock's overall permeability is mostly controlled by the degree of interconnection of the first-order fractures. Most groundwater movement therefore occurs within the first-order fractures. Pressure changes can develop quickly in the first-order fractures in response to flow of water towards seepage faces, pumping wells or horizontal drains.

As the head in the first-order fractures begins to decrease, flow along the second-order less-permeable fracture sets begins to occur towards the first-order fractures. This dual-permeability response is illustrated in Figure 6.25a. Similarly, as flow occurs in the second-order fractures and their pressure reduces, flow and drainage from the third-order fractures will occur, and so on until all the drainable water is removed by gravity and the rock becomes depressurised. Depending on the permeability and continuity of the various fracture sets, this process may occur over several days or several years.

The relationship between groundwater flow, time and space can be expressed as:

$$\Delta h(x, t) = \Delta h_o \, erfc\left(\frac{4Kt}{S_s}\right)^{-0.5} \qquad \text{(eqn 6.4)}$$

where

$\Delta h(x,t)$ = change in head (drawdown) at a point distance x from the dewatering point at time t
Δh_o = change in head at the dewatering point at t = 0
K = permeability (hydraulic conductivity)
S_s = storativity (specific storage)
t = time
x = distance.

The variables of permeability and storativity are controlled by the nature of the rock mass. Within reasonable limits they do not change, apart from within zones of deformation behind the pit slope (see section 6.2.5.3). The ratio is often referred to as hydraulic diffusivity and equation 6.4 is often referred to as the hydraulic diffusivity equation. Time is an important factor in terms of when a dewatering system is implemented and how long it takes to achieve the target pore pressure at a particular point. If the permeability is low the dewatering volume will be small but the time required for dewatering and the number of dewatering points will be large.

Because of the simplicity of the one-dimensional equation relative to real-world problems, it is almost never used by itself. Groundwater flow models use the same relationship in solving more complex problems, and include other factors such as recharge and boundaries. Nonetheless, the equation can be used simplistically to estimate how much drawdown might be achievable in a certain amount of time. It can easily be programmed in Excel.

6.2.5.3 Pore pressure reduction from lithostatic unloading

When lithostatic unloading occurs, the removal of the overlying rock by mining results in a decrease in total stress. This causes the formation to deform and expand slightly, leading to an expansion of the pore space and a drop in pore pressure within the zone of relaxation.

In response to the reduction in hydraulic head, water from the surrounding material flows into the depressurised region and the pore pressure partly rebounds to give a new (lower) equilibrium condition (see Figure 6.25b). Significant reductions in pore pressure resulting from active push-backs have been observed in surface piezometers at several mine sites.

The theory of hydromechanical (HM) coupling can be used to help understand the physical interaction between hydraulic and mechanical processes in a pit slope. These processes act under dynamic conditions to control fracture aperture, permeability and pore pressure. The alteration of the fracture aperture and pore pressure due to the decrease (or increase) of the effective stress (load) is known as direct HM coupling. In relation to pore pressure in a pit slope, direct HM coupling is solid-to-fluid, when a change in stress produces a change in fluid pressure. Direct HM coupling can also be fluid-to-solid, when a change in fluid pressure results in a change in the volume of a porous or fractured medium, for example in hydro-fracturing applications.

An 'undrained' HM response occurs when an applied load is reduced quickly by rapid mining and/or the permeability is low, so that the water contained within the rock mass has insufficient time to equalise. The resulting expansion in pore/fracture space may result in a significant decrease in pore pressure. A 'drained' response occurs when the applied load is reduced slowly and/or the permeability is higher, and water can flow towards and equalise the pressure in the area of deformed rock. In this case, the pore pressure may show no decrease.

In many crystalline rock settings, there is a correlation between depth (and stress) and permeability. The most pronounced depth-dependency occurs in the upper 100–300 m of bedrock and, according to Rutqvist and Stephansson (2003), can be explained by the non-linear normal stress-aperture relationship of single extension joints. Snow (1968) described a fracture with an average hydraulic aperture of 200 μm near the ground surface whose aperture decreased to 50 μm at a depth of 60 m. In the blast-damaged zone (or over-break zone), the alteration of the physical properties of the rock may be much more pronounced.

Huffman (2002) concluded that permeability variations at low stress levels are more sensitive than at high stress levels, a phenomenon attributed to non-linear-fracture normal-stress-displacement relationships. Liu and Elsworth (1998) described how alterations in overburden characteristics due to mining activity induce large undrained changes in pore fluid pressures recorded in the underlying mining zone.

Under normal circumstances, where the permeability is relatively high (e.g. above 10^{-8} m/sec) the redistribution of fluid pressure in the rock mass following a decrease in total stress takes place fairly quickly. The pressure reduction caused by unloading is equalised instantaneously (or very quickly). Only where the surrounding formation or rock mass has a very low permeability (e.g. below 10^{-8} m/sec) is the pressure drop caused by lithostatic unloading normally detectable. In addition, the increase in the fracture aperture caused by the unloading can often create an increase in permeability, which also acts to dampen the pore pressure reduction caused by the rock expansion.

An example where pore pressures are influenced by lithostatic unloading is the east wall of the Chuquicamata pit in Chile. A significant downward pore pressure gradient occurs throughout the east wall of the pit and beneath most of the pit floor area, where downward vertical pressure gradients of greater than 1.5:1 have been observed. The downward pressure gradient has been explained in terms of changes in fracture porosity resulting from deformation, as follows:

- at shallower depths in the slope (<150 m), some deformation and unloading already exists so the fracture porosity is comparatively high (e.g. 0.001). Further mining-induced deformation causes further increases in porosity so the pore pressure falls;
- at greater depths within the slope (>300 m), the pre-existing deformation is smaller and the fracture porosity is lower (less than 0.001). Mining-induced deformation causes the porosity to increase. The absolute porosity increase is much less than it would be shallower in the slope, but because the average fracture porosity typically decreases with depth, rock expansion and an increase in void space in response to unloading is proportionally greater at depth. Hence, the magnitude of pore pressure dissipation in response to material unloading is more pronounced with depth. Furthermore, the permeability values lower in the slope (10^{-11} m/sec) are significantly lower than those at shallower depths (10^{-9} m/sec) so the rate at which water can flow towards and equalise the pressure is much lower.

Figure 6.26 shows lithostatic unloading from a diatreme unit in the north-east highwall of the Cerro Yanacocha Sur pit in Peru. No flow to the toe of the slope could occur and no other drainage measures were installed. Pore pressures in the diatreme show a general downward trend as a result of mining the overburden above the slope. Between each phase of overburden removal a recovery (rebound) in pore pressure can be observed due to groundwater flow into the depressurised area.

The development of the blast-damaged (over-break) zone is also important for pore pressure control and for helping to calibrate pore pressure models (section 6.4). The zone is characterised by an area of reduced fluid pressure extending in every direction away from the zone. However, the shape and extent of the zone depends on many factors, including blasting procedures and rock properties, which vary considerably. Therefore, uniform pore pressures related to the over-break zone would not be expected.

Figure 6.27 shows a laminated shale sequence at the Jwaneng diamond mine in Botswana. The photograph shows strongly developed bedding, with well-developed orthogonal subvertical joint sets cross-cutting the bedding planes. During deformation due to unloading, an increase in aperture of the joint sets is thought to be an important factor for reducing pore pressure in the zone of deformation (relaxation) behind the slope. The observed

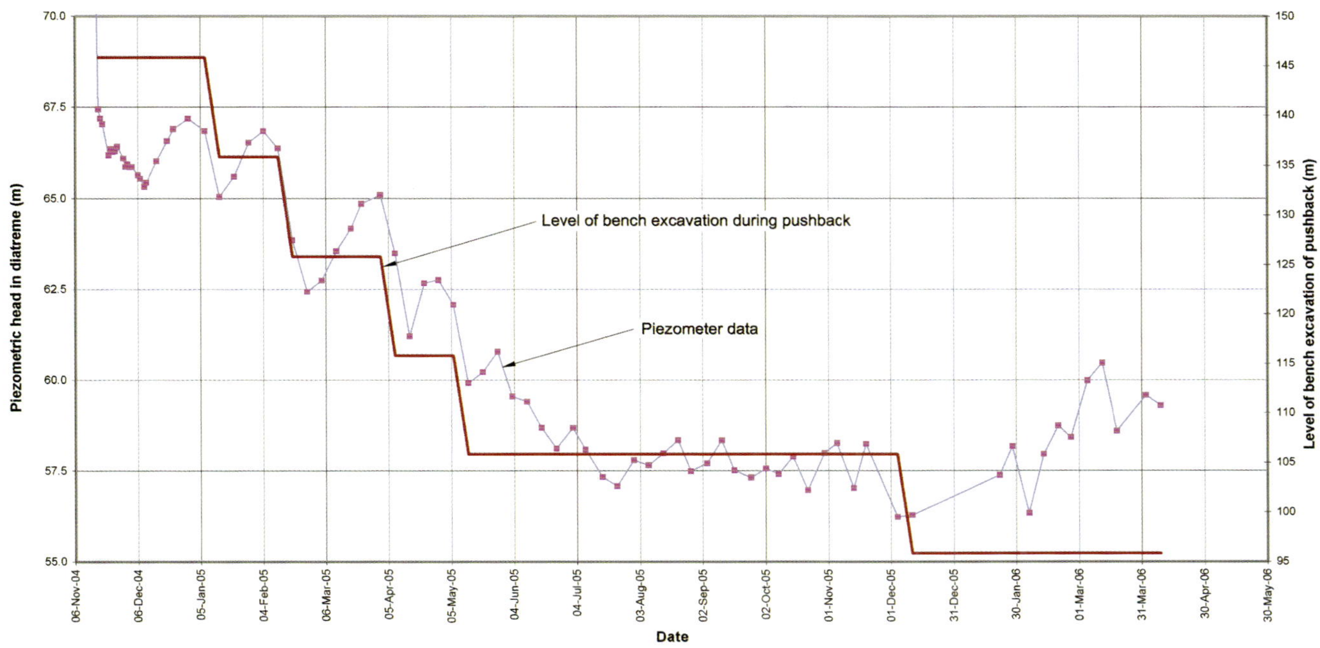

Figure 6.26: Pore pressure response in diatreme material, Yanacocha Sur pit, north wall
Source: Courtesy of Minera Yanacocha SRL

Figure 6.27: Shale sequence at the Jwaneng diamond mine, Botswana

fracturing associated with the bedding itself is mostly the result of blast damage, so bedding is thought to have less influence in controlling how the shales depressurise deeper behind the slope. However, the blast-induced fracturing along the bedding planes is important for bringing minor seepage in the overbreak zone to the face.

6.2.5.4 *Pore pressure reduction from hydrostatic unloading*

In some instances, hydrostatic unloading can lead to deformation (expansion) of the underlying rock, creating a zone of relaxation. However, in most mine settings the potential for this to cause significant changes in pore pressure is low. It might occur, for example, where a significant thickness of permeable and porous overburden (e.g. a thick gravel unit) becomes dewatered. The weight of water removed from the gravel may be large enough to cause expansion of the underlying rock.

In most normal dynamic hard rock mining situations, the effect of groundwater flow greatly exceeds the effect of lithostatic or hydrostatic unloading. Lithostatic and hydrostatic unloading are more important when the initial permeability is low. Both Wang and Ellsworth (1999) and Rutqvist and Tsang (2002) attributed this to the opening of horizontal fractures that were closed until the point of lithostatic unloading. However, if there is a large difference in permeability between the first-order and lower-order fracture sets a transient situation may occur where the discrete first-order fractures show significant depressurisation but the lower-order fracture sets retain residual pressure for some time.

6.2.5.5 *Piping of slope materials*

Flow resulting from rapid transient pressure changes in softer materials may cause piping. Typical geotechnical models do not include piping, at least not directly. Piping can be a significant cause of slope instability in poorly consolidated fine granular materials.

6.3 Developing a conceptual hydrogeological model of pit slopes

6.3.1 Integrating the pit slope model into the regional model

For most pit slope depressurisation studies, it is necessary to integrate the specific program for the pit slopes into the total hydrogeological or dewatering program for the mine site. The specific groundwater conditions around the pit slopes are influenced by the regional setting. An example is Barrick Goldstrike in Nevada, where the carbonate formations within the pit are interconnected for tens of kilometres away from the pit area and it is necessary to view the hydrogeological model within a broad regional setting.

The first step in developing a conceptual hydrogeological model of the pit slopes is to determine the regional hydrogeological model.

Only in certain circumstances can the pit slope depressurisation program be carried out in isolation. Examples include:

- pit slopes above the water table where the pore pressure concerns are the result of localised infiltration of precipitation or runoff;
- mine sites in pervasively low-permeability environments where there is minimal potential for regional scale groundwater flow to influence conditions in the slope.

At the Chuquicamata pit in Chile, the permeability of all units is very low and the hydrogeological response in the mine area is independent of conditions away from the pit. The total groundwater flux in the mine area is less than 10 L/sec, mostly derived from artificial recharge at the pit rim. In this situation the conceptual model needs to cover only the immediate area of the pit.

As a very general rule of thumb, hydrogeological units of permeability greater than about 10^{-6} m/sec to 10^{-7} m/sec may drain as a result of site-wide dewatering whereas units within the permeability range of 10^{-7} m/sec to 10^{-9} m/sec may require localised drainage measures to enhance the rate of pressure reduction. Units with a permeability below 10^{-9} m/sec may become more dominated by unloading effects, depending on the elastic properties of the material. This guideline is very general and will depend on many factors, particularly the time available to achieve the desired pore pressure profile.

6.3.2 Conceptual mine scale hydrogeological model

The conceptual mine scale hydrogeological model usually includes the entire area of the groundwater system that may be influenced by the open pit mining operation. It may include:

- definition of the area of hydrogeological influence of the mine site;
- definition of the hydraulic boundaries within the site;
- definition of the porosity and the amount of water within each hydrostratigraphical unit at the mine site.

The conceptual hydrogeological model for an open pit mine therefore usually includes the following items.

1 A detailed description of the overburden and bedrock geology and hydrogeology of the mine site, including the relationship between groundwater, lithology and geological structure. This may include:

- → regional groundwater-level elevations plotted onto a geological base map showing surface lithology and structure;
- → a number of regional geological cross-sections showing the geology and the available water level data;
- → where alluvial or thick overburden is present, maps showing the elevation of the bedrock surface beneath the alluvium/overburden, the base of the permeable horizon within the alluvium/overburden, water levels within the alluvium/overburden and the saturated thickness of the alluvium/overburden. If the mine is already in production, the position of waste rock, tailing areas and other mine facilities should be added to the maps and cross-sections.

2 A description of the regional groundwater flow system, groundwater elevations, lithological controls on groundwater elevations, and structural control on groundwater elevations. This usually includes definition of:
- → surface topography, location of water bodies and surface watersheds;
- → the principal groundwater flow paths based on the geology, hydrochemistry and water level information;
- → where areas of deep alluvium, paleochannels or other coarser alluvial materials may occur;
- → less-permeable layers within the system which may impair vertical flow and create large vertical hydraulic gradients when stress is applied to the groundwater system, e.g. fine-grained layers within unconsolidated alluvium, weathering and development of a low-permeability layer at the bedrock surface, palaeosols that indicate breaks in the stratigraphic sequence and create low-permeability weathering horizons, and general layering within the sedimentary or volcanic sequence;
- → the geological structures that may create boundaries to horizontal flow across their strike and enhance flow along their direction of strike.

3 Identification of the principal hydrostratigraphical units forming the regional groundwater flow system and estimates of their horizontal and vertical permeability, porosity and storage characteristics, together with:
- → definition of which units will transmit most of the groundwater and which units will act as barriers to flow;
- → definition of how the structural geology may influence the direction and magnitude of the groundwater flux in each hydrostratigraphical units on a site-scale.

4 Evaluation of the groundwater recharge areas and the magnitude of groundwater recharge, including groundwater/surface water interaction. The assessment may include definition of areas of recharge to the groundwater system that may be caused by:
- → infiltration of precipitation or snowmelt;
- → leakage from rivers, lakes or other surface water bodies;
- → losses from alluvial underflow systems beneath river valleys;
- → infiltration from mine facilities, including old workings and the plant.

The assessment should also include a definition of which sources may provide constant recharge towards the mining operation as groundwater heads are lowered in and around the mine area.

5 Evaluation of groundwater elevations, groundwater flow directions and groundwater flux, including calculation of the magnitude and direction of the groundwater flux in each stratigraphic unit and how these may change as a result of the lowering of groundwater heads in and around the mining operation.

6 Evaluation of the natural groundwater discharge areas and how excavation of the open pit may alter the pattern of discharge. Natural discharge features include evapotranspiration, springs and seeps at the surface. Groundwater discharge may also occur to rivers, lakes or other surface water bodies. Other discharge features may exist, including historical pumping wells, and there may be discharges to historical or existing excavations. Upon excavation of the pit, additional discharge sources may include dewatering wells, horizontal drains, evaporation from within the pit walls or floor, or seepage faces and inflow to the pit walls or floor. Change or removal of vegetation may affect evapotranspiration and alter discharge.

Evaluation of pre-existing and/or natural changes to the groundwater system is also important in helping to establish pre-mining environmental impacts.

6.3.3 Detailed hydrogeological model of pit slopes

The main purpose of this section is to describe the evaluation and control of pore pressures primarily for increasing the effective stress of the material and the stability of the pit slopes. Typically, the spatial coverage of the conceptual pit slope model will be the geotechnical domain used in the slope stability analysis.

A major factor in formulating the pit slope model is how the pit slope is expected to be influenced by the mine scale flow system in two important ways – the potential for continuing recharge from the mine scale flow system to the local hydrogeological units in the pit walls, and the potential for drainage of the local hydrogeological units and dissipation of pore pressures in the pit walls as a result of site-scale dewatering.

Therefore, the following are usually included in a detailed model of the pit slopes:

- the hydrogeological units that occur in the pit slope, typically defined by the main geological controls (lithology, alteration, mineralisation and structure). It often makes an integrated analysis easier if the hydrogeological units are coincident with the geotechnical units, although this is not always possible;
- the nature of groundwater flow in each unit (fracture-flow, porous-medium flow or a combination);
- the horizontal and vertical permeability of each unit. Data for horizontal permeability are usually obtained by a slug test, packer test or pump test program. Data on vertical permeability cannot usually be obtained by simple field testing but can be estimated by previous experience in similar hydrogeological units, or by evaluating the response of the hydrogeological system via monitoring wells and piezometers;
- the specific yield and specific retention (drainable and non-drainable porosity) of each unit. Data can be obtained by laboratory testing and possibly downhole geophysical logging, but usually the model relies on piezometer observations of how the unit responds to drainage stress;
- the pressure head in each unit, lateral gradients in pore pressure, vertical gradients in pore pressure and the total pore pressure profile that develops. A map showing the depth of the potentiometric surface below the pit slope is a valuable tool. Data are typically obtained from piezometers, augmented by water level measurements in geological or geotechnical drill holes and possibly by visible seepage on the pit slopes;
- the main structural alignments and how the structures influence the direction and magnitude of the groundwater flow in each unit, including the way that structural orientations create anisotropy in the flow system. Definition of the structural controls is typically obtained by:
  - → detailed knowledge of the structural geology;
  - → continuity or discontinuity of groundwater heads along or across the main structural orientations;
  - → the response patterns of the piezometers to pit wall seepage or pumping stress;
  - → packer testing of core holes to obtain permeability values and shut-in pressures for specific structures.

The detailed model must be developed to identify or differentiate between:

- units that show free drainage and pore pressure dissipation in response to seepage to the mine excavation;
- units that show free drainage and pore pressure dissipation as a result of dewatering on a site scale;
- units that show pore pressure dissipation in response to rock mass unloading;
- units that show a drainage response to localised enhanced measures at the pit slope (section 6.5);
- units that are likely to be difficult to drain, so that elevated pore pressures may have to be included as part of the final slope design.

For each unit, it will be necessary to predict:

- the current pore pressure profile and rate of change;
- the rate of future pore pressure dissipation that can be achieved;
- the level of expenditure that is appropriate to enhance the rate of pore pressure dissipation and allow slope stability or slope angles to be increased, which is also a function of acceptable risk.

A typical detailed hydrogeological model interpreted along a 2D geotechnical section in preparation for numerical pore pressure modelling is shown in Figure 6.28.

6.4 Numerical hydrogeological models

6.4.1 Introduction

Consistent with the requirements set out in section 6.3 for the development of a conceptual hydrogeological model for pit slopes, there are four 'scales' of numerical hydrogeological modelling:

1. regional scale;
2. mine scale;
3. pit slope scale;
4. models to address specific hydrogeological issues (e.g. infiltration from site facilities, investigations of groundwater chemistry).

Regional groundwater models are often used to help assess the overall spread of drawdown and potential impacts of mine dewatering. They are frequently used to support environmental documents for permitting.

Where appropriate, mine scale models may be used to help design the overall dewatering system for the mine. In some cases, the mine scale model may be independent of the regional scale model. However, in many cases the regional scale and mine scale models are combined.

Section 6.4.2 discusses the normal approach for mine scale numerical hydrogeological modelling. The approach to pit slope scale numerical modelling is discussed in section 6.4.3 and specific numerical modelling procedures for determining the pore pressures in pit slopes are discussed in section 6.4.4.

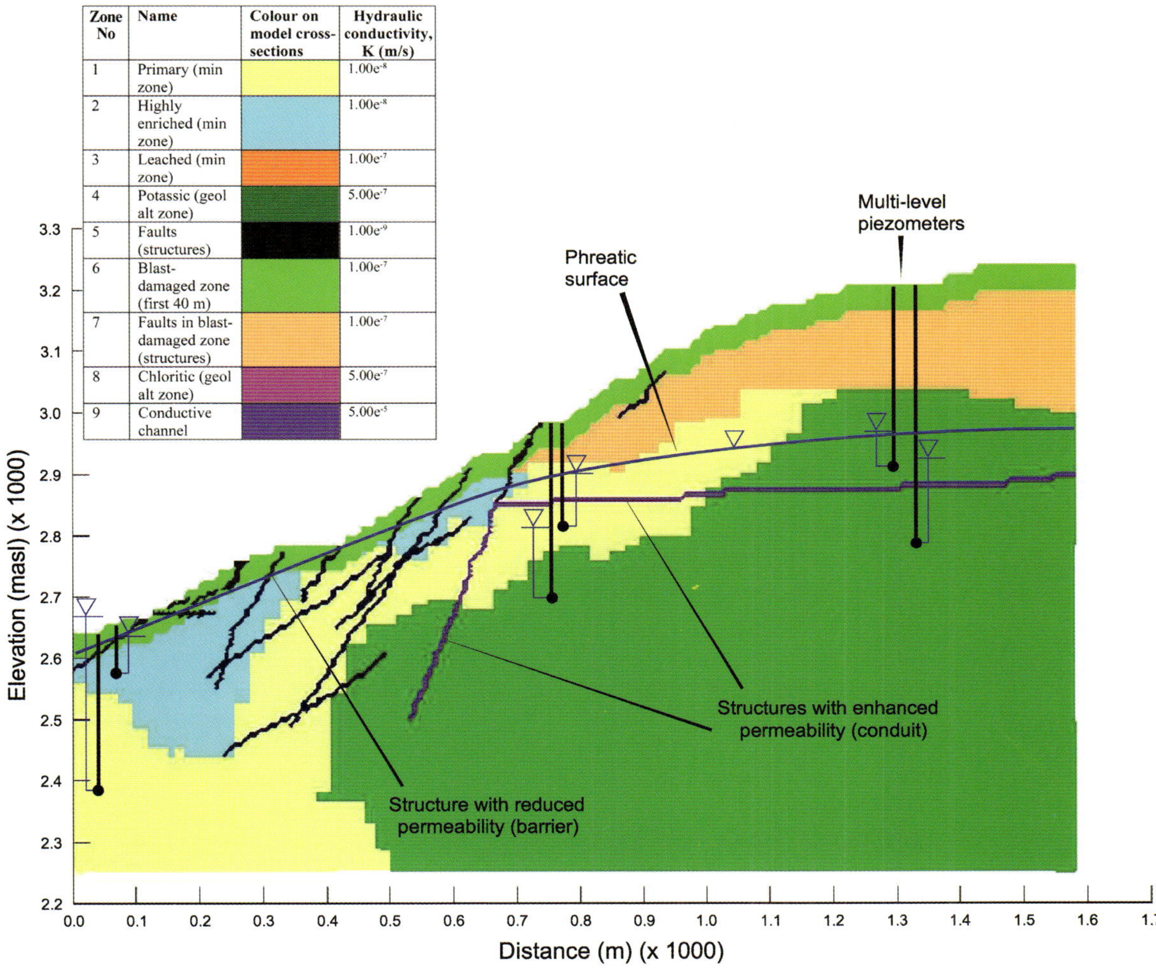

Figure 6.28: A detailed hydrogeological model used for numerical 2D pore pressure modelling
Source: Courtesy Minera Escondida Limited

6.4.2 Numerical hydrogeological models for mine scale dewatering applications

6.4.2.1 General

The typical applications of site-wide numerical models for planning dewatering systems are:

1. to help predict the required water level drawdowns and pumping rate and design the mine dewatering system and/or the discharge system for the pumped water;
2. to help investigate the sensitivity of alternative mine plans to dewatering requirements and sequencing;
3. to help investigate the rate of drawdown within and surrounding the mine site area;
4. to determine the time required to reduce heads in site-wide groundwater units and plan implementation of dewatering systems;
5. to help investigate potential impacts on the environment.

6.4.2.2 Requirement and applicability of a mine scale numerical model

Typically, a mine scale numerical groundwater flow model is appropriate in situations where:

- there is potential for widespread regional groundwater flow to provide sustained recharge to the mine dewatering system;
- the groundwater system is relatively homogeneous and the structural geological setting is of relatively low complexity;
- conceptualisation of the groundwater system is relatively straightforward;
- data and experience exist or can be obtained to support a conceptual model and to calibrate the numerical model;
- a tool is desired to predict potential hydrogeological impacts, to investigate sensitivity of the site-scale

hydrogeological system to mining operations and to design mitigation systems;
- regulatory authorities (and sometimes funding agencies) require a site-scale groundwater model.

Table 6.2 shows a typical modelling sequence that may be applied to evaluate dewatering and slope depressurisation requirements for new projects. In the early phases of project development, in addition to the use of analytical solutions, a simple possibly axisymmetric model can be set up and used to develop the initial predictions.

Table 6.2 shows that environmental issues will mostly be addressed at Level 4. In many parts of the world, however, the environmental issues and potential impacts of the mining and dewatering operations on water resources and aquatic habitat must be addressed much earlier in the program, certainly by the pre-feasibility stage, as part of the permitting process. Many regulatory agencies demand that the levels of confidence in predictions of environmental impacts be much higher at the pre-feasibility and feasibility level than those shown in Table 6.2. This usually means that the hydrogeological investigation, including the field investigation and the predictive numerical groundwater flow modelling, must be much more extensive and hence more costly in the earlier stages than would normally be required for just planning the engineering aspects of dewatering and depressurisation.

For assessing slope depressurisation, it is important that the modelling be focused on what is achievable in terms of pore pressure reduction in a given time frame and what can be gained in terms of increasing slope angles or increasing factors of safety.

A site-scale numerical groundwater flow model may not be required in situations where:

- there is little potential for regional groundwater flow to influence the mine dewatering system, and vice versa;
- the site-scale groundwater flux is relatively small;
- there are insufficient data to prepare a conceptual model on which to base a numerical model;
- the site is geologically complex such that effective calibration and operation of a model would be unable to provide more reliable predictions without substantial additional data acquisition;
- hydrogeological issues can be evaluated to the required level of detail with more simplistic calculations, and dewatering rates and potential impacts can be reliably predicted using empirical data or by analytical methods.

6.4.2.3 Available numerical codes

There is a wide range of groundwater flow numerical modelling software codes obtainable from academic, government and private distributors. Listing them all is beyond the scope of this book, but there are a few major categories of models with commonly used examples:

- fully saturated flow (used for confined and unconfined aquifers);
- variably saturated flow (used for unconfined aquifers or evaluation of infiltration processes);
- multiphase flow (used when evaluation of flow of multiple phases such as air and water is important, e.g. for leaching systems).

In general, factors that should be considered in a groundwater model for a mining operation include:

- whether the code is 2D or 3D;
- the numerical methodology (e.g. finite difference, finite element, boundary integral) used by the code;
- the code's ability to simulate mining-specific features such as time-variable excavation of a pit, seepage faces, multiple faults of irregular orientation, dewatering wells and subhorizontal drainholes, shafts, drifts and drainage galleries;
- simulation of the phreatic surface and re-wetting of nodes or cells (i.e. the simulation of saturated conditions after desaturation has occurred);
- pre-processors and post processors, including graphical output.

For mine dewatering applications, probably the most widely used codes are *MODFLOW* (developed by the US Geological Survey), *MODFLOW-SURFACT* (an enhanced version of *MODFLOW* modified by HydroGeologic), *FEFLOW* (developed by WASY), *Seep/W* (developed by GeoSlope) and *MINEDW* (developed by HCItasca; not currently commercially available except to mining companies). The principal attributes of these codes are summarised in Table 6.3.

FEFLOW has the ability to simulate mass- and heat-transport, attributes primarily applicable to environmental issues, and *MODFLOW* and *MODFLOW-SURFACT* can be coupled to mass transport codes such as *MT3D* (sspa.com/software/mt3d99), but these considerations are outside the scope of this text. There are numerous other codes with special attributes that might apply to other mining-related issues (e.g. seepage from tailings), but they are generally less applicable to solving the general mine dewatering and slope depressurising problems that are the focus of this chapter.

In general, model grids are easier to set up with finite difference codes. However, it is easier to replicate the geometry of the hydrogeological setting of most mines (with their complex boundaries between geological units and faults with numerous orientations) using the non-geometrically constrained finite element method rather than with the finite difference method. In the

Table 6.2: Typical modelling sequence to evaluate dewatering and slope depressurisation requirements for new projects

Stage	Type	Application (see Table 8.1)	Input	Calibration	Predictions	Target level of data confidence (see Table 8.1)
Preliminary	Analytical	Levels 1 and 2 Conceptual and pre-feasibility studies	Simplified (homogeneous and isotropic) geology Estimate of hydraulic properties Cylindrical or conical, time-variable mine plan	None or minimal	Preliminary estimates of inflow over time	>20%
	Axisymmetric (numerical)		Layered geology with lateral-to-vertical anisotropy Estimate of hydraulic parameters Cylindrical or conical time variable mine plan	Minimal	Preliminary estimates of inflow over time and pore pressures within highwalls Scoping analysis	30–50%
Intermediate	Numerical	Level 3 Feasibility studies Preliminary design of dewatering systems	Fully 3D representation of geology Field-derived values of hydraulic properties Actual shape vs time of mine plan Data on recharge, pumping and surface water Some measured water levels and flows (for calibration)	Some	Amounts of water to be managed and effects of various active dewatering schemes Changes in water levels and impacts on water resources Uncertainty analysis	40–65%
Comprehensive		Level 4 Detailed design of dewatering systems Environmental impact assessments	Fully 3D representation of geology including structures Field-derived values of hydraulic properties and good understanding of range of values Detailed mine plan with applicable geotechnical input (e.g. location and timing of mine excavation) Data on recharge, pumping and surface water Extensive water level and flow data (for calibration)	Intensive	Detailed design and optimisation (location and timing) of dewatering system Changes in water levels and impacts on water resources Distribution of pore pressure within highwalls and effects of various dewatering schemes Uncertainty analysis Inflow to a pit lake and water level recovery	60–75%

Table 6.3: Commonly used codes for mine-related groundwater models

Code	Source	Dimensions	Method	Additional information
MODFLOW	USGS	3D	FD	water.usgs.gov/nrp/gwsoftware/modflow.html
MODFLOW-SURFACT	HydroGeologic	3D	FD	modhms.com/software/modsurfact
FEFLOW	WASY	3D	FE	wasy.de/english/produkte/feflow/index
Seep/W	GeoSlope	2D/3D	FE	geoslope.com/products/seepw2007
MINEDW	HCItasca	3D	FE	hcitascacg.com/mining_hydro

2D = 2-dimensional (for vertical slices such as slopes)
3D = fully 3-dimensional
FD = finite difference
FE = finite element

finite difference codes described above, the grids must be orthogonal and the discretisation (size of the mesh) must be continuous to the boundaries of the model.

All the widely used codes use the continuum approach, relying on the assumption that a fractured rock mass behaves as an equivalent porous medium. The equivalent porous-medium assumption suggests that, if there is a reasonably large density of interconnected fractures, the secondary porosity and permeability created by the fracturing will cause the rock to behave hydraulically as the pores in a porous medium. Therefore, it is possible to measure the bulk properties of the medium, including the effect of structural deformation on the fluid flow properties, without having to map and characterise each fracture. Although this assumption may not hold true on a micro-scale (section 6.2) it has proven adequate to simulate groundwater flow on a bulk scale.

A number of discrete fracture network (DFN) codes are available to help map and characterise individual fractures. Many mine sites collect a large amount of data through exploration and geotechnical drilling and logging, which may be used to map structural features. These have been applied in cases such as nuclear waste repositories where the dimensions are relatively small and extremely large and detailed databases on fractures exist. However, such codes have not yet been successfully applied to the scale of an open pit mine.

6.4.2.4 *Application of hydrogeological models to mine dewatering and slope design*

The overall ability of the model to predict groundwater conditions accurately regardless of the code used is governed by:

- the understanding of the geological and hydrogeological setting;
- the availability of data and ability to construct a representative conceptual hydrogeological model;
- the knowledge and experience of the modeller;
- the ability to calibrate the model to historic and current conditions prior to making predictions.

One of the most important advances is the increasing understanding by mine managers and operators of the applicability of groundwater flow models to mine dewatering problems, as well as the limitations of modelling. Many managers and operators have tended to move away from large all-encompassing models. Rather, there has been a tendency to focus modelling efforts on a particular application, such as the investigation of a specific aspect of the mine dewatering system (e.g. the prediction of pore pressure in a specific sector of the pit slope).

For successful prediction and design of dewatering systems, it is often appropriate to use these steps.

1 Prepare a conceptual hydrogeological model in as much detail as possible, identifying the key hydrogeological controls for mine dewatering, as well as the uncertainties in the conceptual model.
2 Estimate the required dewatering rate for a given mine plan using analytical methods based on the available geological and hydrogeological information.
3 Identify the key issues for the mine hydrogeology and clearly define the objectives of numerical modelling.
4 Construct the numerical model to focus on the key issues identified, and provide support and refinement for the analytical estimates.
5 Conduct a practical, experience-based review of the model construction and results to evaluate the model's applicability to the simulation of the 'real world', particularly the practical ability to achieve dewatering targets and the lead time required.

It is essential that the numerical model represents the conceptual hydrogeological model and that it can be modified by and calibrated to actual field data (e.g. water levels, pumping trials and pilot tests). Calibration is the process of modifying the magnitude and distribution of model properties within a set range of values to produce a truer representation of reality, which is a key step before proceeding to the predictive phase. All models are simplifications of reality, and all models have some degree of uncertainty. Model calibration assumes that the conceptual model is appropriate and that the model uncertainty lies in the parameters used in the model. Hydraulic properties are known to vary and can have a range of values for each type of fractured rock or porous medium.

Calibration generally utilises three approaches.

- The hydraulic properties (permeability, specific storage, specific yield) of the various hydrogeological units can be assigned values based on measured data or professional judgment. Then the values can be modified within a range of possible values to improve correlation with observed water levels and pore pressures.
- Geological and hydrogeological interpretation of the distribution of types of materials and their hydraulic behaviour can be incorporated into the model as zones of different property values. Then the property values and spatial distribution of the zones can be modified within the possible range to improve prediction.
- Geostatistical or other statistical methods can be applied to predict the distribution and magnitude of property values (i.e. a different value is applied to various zones) based on a limited measured data set.

The process of calibration was automated and formally became inverse modelling in the late 1980s, starting with the work of Carrera and Neuman (1986).

Recent developments in the calibration process (e.g. the combined Modflow-UCODE and the combined *FEFLOW-PEST*) are increasingly used by model operators. However, although these codes add value where limited data are available, the uncertainty in the conceptual model itself must be carefully evaluated prior to any calibration process.

In almost all instances it is beneficial to make a first-order estimate of the dewatering rate prior to construction of a numerical model. This is often best done using a water balance approach, which sums the groundwater storage removal from individual hydrogeological units, mine scale or regional scale groundwater flow towards the mine area in individual hydrogeological units, and ongoing groundwater recharge from precipitation or surface water bodies. Simple radial flow equations or axisymmetric models may also be used to predict groundwater flow towards the mine area, but care needs to be taken to ensure that the hydrogeological properties of the various units, and hydrogeological boundary conditions, are not oversimplified in the assumptions to the point of having major effects on the model's predictions. An experience-based evaluation is invariably useful, drawing on operating experience of dewatering systems around the world in analogous hydrogeological settings.

The use of an analytical or experience-based approach may help the mine operator as well as the modeller to focus on which factors will be important for controlling the dewatering rate, what the uncertainties are and whether there is enough information to address the uncertainties. If the overriding assumptions are the same, a simple analytical estimate can be as valid as a sophisticated numerical model.

With the increasing distribution of user-friendly and easy-to-use software, one of the challenges for the mining industry is to ensure modelling efforts stay focused on practical issues and that models are constructed at an appropriate level for:

- the type of project;
- the support data available;
- planning and operational decisions;
- the experience level of the mine operator.

Peer review is recommended. It is also important that the conceptual model be challenged regularly and updated when necessary.

6.4.3 Pit slope scale numerical modelling

A suggested approach for pit slope scale numerical modelling is as follows.

Step 1: Model development

- A conceptual hydrogeological model of the pit slope is developed using a series of hydro-geotechnical cross-sections and base maps that describe sources of water, geology, structure, groundwater units, historical pore pressures, boundary conditions, current mine water balance and future mine plans.
- A numerical model grid is constructed and discretised to show the geometry of the main hydrogeological units and structures in the slope system. It is usually built with the same domains as the slope geotechnical model.
- Field data are used to assign representative hydraulic parameter values to each hydrogeological unit.
- Structural data are fed into the model and key controlling structures are discretised in the model domain.

Step 2: Input of mine planning data

- The model is initially developed using the current profile of the pit slope.
- Future mine cuts are discretised into the model in space and time so that portions of the mesh can be removed from the model domain, and material properties and boundary conditions can be updated in accordance with the mine plan.
- The current and future pit slopes are heavily discretised so that vertical pore pressure gradients and any depressurisation installations (e.g. horizontal drain arrays) can be accurately simulated and moved to new locations in the model as mining progresses.

Step 3: Simulation of hydraulic property changes due to unloading and deformation

- Deformation contours for the slope materials are obtained from the geotechnical model or estimated using geotechnical parameters.
- The model grid is further discretised to allow deformation zones to be input to the model as zones of increased/changed permeability and porosity due to the mining activities (Figure 6.29).
- Changes in material properties are put into the model to simulate the effect on pore pressures for each future time step.
- The overbreak zone of the pit wall is defined and put into the model for each successive pushback as a zone of higher permeability and porosity. When calibrating the model, it is important to appreciate that seepage may move 'unseen' downslope within the more permeable overbreak zone without a surface expression and may, therefore, be difficult to account for.

Step 4: Model calibration

- Appropriate boundary conditions are assigned, such as a prescribed head to represent regional or site scale flow and possibly a recharge boundary on the crest area. For a telescoped (or window) model, the boundary conditions for the pore pressure model are extracted from the larger site scale model.

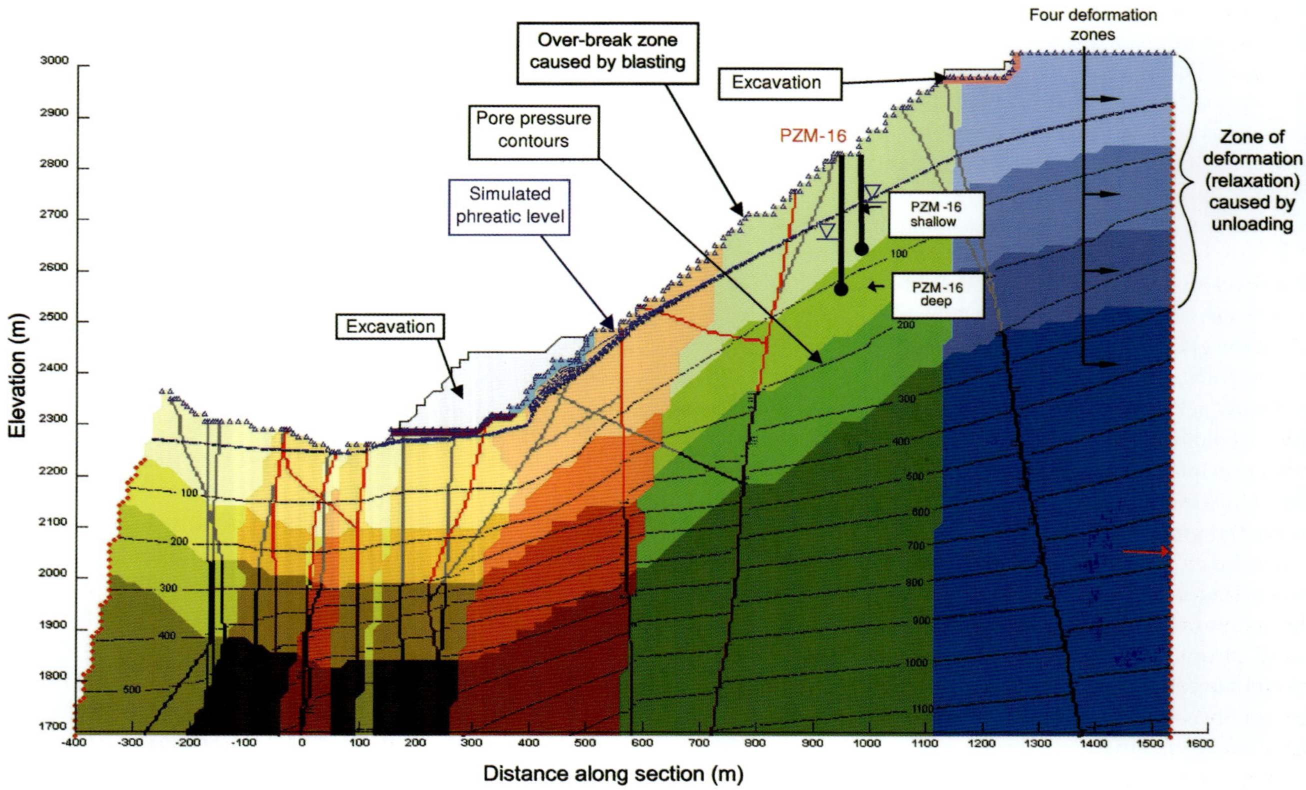

Figure 6.29: A 2D groundwater flow model with pore pressure contours
Source: Courtesy Codelco Norté

- The model is calibrated to historical data from in vertically discretised slope piezometers (preferably including some point pressures from vibrating-wire installations). Where data allow, a transient mode calibration is developed to match the model with historical depressurisation responses.
- In zones of low permeability, calibration to observed historical unloading responses in piezometers is incorporated into the calibration.
- Calibration should be confirmed by comparison of transient pressure responses to depressurisation activities (e.g. production well pumping or horizontal drain flows).

Step 5: Predictive simulations

- Once a satisfactory calibration is attained, the model is run in predictive mode for specified future time periods, dictated by the mine plan.
- Elements or cells are removed from the model to simulate advance of the pit slope (Figure 6.29).
- Deformation zones are varied with each time step, based on the elements or cell removed and the predicted deformation of the materials beneath the newly mined slope.
- Predicted pore pressure profiles are generated at selected future time intervals and at selected stages of slope development.
- Predicted pore pressures and water flows are compared with field data and historical depressurisation experience as a reality-based verification of results (i.e. model validation).
- Simulation of changes due to recharge can be added, e.g. monthly changes in head influenced by seasonal events.

Step 6: Interaction with the geotechnical model

- The pressure distributions are transferred from the groundwater flow model and used for input into the slope stability model.
- The geotechnical model can then be used to determine slope performance and to assess critical slope sectors where pore pressure is of most concern.
- This information can be fed back into the groundwater flow model to assess the robustness of the pore pressure predictions in the critical sectors and whether the desired level of pressure dissipation can be achieved, given the available lead time.
- The groundwater and slope stability models are interactively rerun to examine changes in slope

pressure and factors of safety for a range of active depressurisation options.

- An optimised pore pressure profile is input to the geotechnical model as the basis for the final slope design.

6.4.4 Numerical modelling for pit slope pore pressures

6.4.4.1 Requirements for specific pit slope pore pressure models

If the slope materials are permeable and freely draining, a site scale groundwater flow model can be used to simulate pore pressures and drainage of the slope. In this case, the geological units that form the slope are likely to be an important part of the overall hydrogeological setting. There are several examples of a site scale model being adequate to predict pit slope pore pressures, such as the Cortez Pipeline mine in Nevada and the Alumbrera mine in Argentina.

Pore pressures within the pit slopes tend to be of greater concern if the slope materials are of low permeability or if they contain isolated or perched groundwater as a result of complex geology, structures and/or alteration. If so, groundwater may not drain freely in response to pumping the surrounding permeable units. Pore pressures within the slope often vary within a very small area. Frequently, this level of detail is difficult to fit within the mine scale conceptual hydrogeological model. As a result, the value of applying a mine scale model may be limited – development of a separate pore pressure model for the pit slopes is increasingly common for large open pit planning.

A specific pit slope pore pressure model should be considered in the following circumstances:

- if the materials in the slope are of low permeability and/or variable, remain isolated and do not drain in response to general mine dewatering;
- if the geological structure and/or hydrogeology of the slope materials creates compartments from which the groundwater does not freely drain;
- if there is high precipitation to provide continual recharge to sustain pore pressure in the slope materials;
- if pore pressures have the potential to significantly decrease the effective stress of the materials in the slope.

A specific pit slope pore pressure model may not be necessary in situations where:

- the materials in the slope are above the water table and contain no perched groundwater and the mine site is located in a dry climate;
- the materials within the slope are permeable and homogeneous and drain readily in response to site-wide mine dewatering efforts;
- the rock properties and strength of the material in the slope are such that it is judged that pore pressures have little influence on the effective stress of the material (e.g. fresh granite) and that pore pressures and seepage may not necessarily be a major concern for the slope design.

Monitoring data and the conceptual model of the groundwater system in the area of the pit slopes are used to constrain the boundary conditions for the groundwater flow model. If a larger mine scale model is available, telescoping (or window modelling) may also be used to define the boundary conditions for the groundwater flow model. The telescoping method uses boundary conditions for the pit slope model that are transferred at a common ring of nodes or cells in the larger mine scale model. In this way, it may be possible to capture the detailed required for the pit slope model without incurring the long running time of a larger-scale model.

To increase the ease and accuracy of data transfer between the two models, the mesh in each should be designed to share certain features. For example, at the intersection of the 2D pit slope model and the mine scale 3D model the mesh can be made to correspond exactly. This reduces the potential for introducing numerical error when transferring data from one model to the other. Using this type of approach, it is vital to ensure that the pit slope model is updated to cover changes in the larger model. Changes in the larger model that seem relatively small on a site scale could be very significant in the pit slope model.

6.4.4.2 Goals for modelling slope depressurisation

The typical goals of a numerical groundwater flow model to determine pore pressure in the walls of the pit are:

- to improve the understanding of the current pore pressure distribution and to help identify whether pore pressures within the materials behind the slope may have a significant influence on the slope design and performance as the mine operation is advanced;
- to provide a numerical simulation of the pore pressure profile that can be directly input to the geotechnical model;
- to analyse the potential effects and benefits of alternative depressurisation systems for the pit slope and to determine the most cost-effective way to reduce elevated pore pressures behind the slope, given the lead time available for pore pressure dissipation;
- to guide location of pore pressure monitoring instrumentation.

In some pit slopes, the magnitude of the vertical hydraulic gradient is similar to or greater than the lateral component. Thus, a realistic representation of the vertical gradient of the pore pressures within a slope is critical input to a geotechnical model. However, even the most robust models tend to suffer from a lack of real high-resolution hydraulic data related to both observations and hydraulic properties. Consequently, interpretation and judgment is required when evaluating the results of such models.

6.4.4.3 Modelling in 2D

The overall direction of groundwater flow in pit slopes is frequently subperpendicular to the slope. For effective pore pressure modelling, it is essential to simulate the vertical pore pressure distribution that will develop within the slope. It is unlikely that pore pressures will increase linearly with depth.

As a result, it is often more effective to use a vertical 2D cross-section (or slice) approach. Because most geotechnical modelling for planning and operational settings is carried out in 2D, using a 2D model for pore pressure simulations is often appropriate for evaluation of slope stability. In general, a 2D approach may offer the following benefits:

- complex vertical pore pressure gradients can be simulated with a greater level of detail. Calibration of the model is a simpler process because it only requires data points specific to the slope sector of concern. Calibration is easier than a 3D model because the model domain is smaller and more tightly controlled;
- the orientation of the model is subparallel to the typical groundwater flow lines within the pit wall;
- where appropriate, the hydrogeological model domain can be made coincident with the 2D geotechnical model domain;
- hydrogeological data collection can be focused around the key geotechnical sections. The model can be quickly set up, often using the information already available from the geotechnical model. Cross-sections of lithology, alteration and mineralisation can be used to conceptualise the hydrogeological units in the 2D model. In certain applications, it may also be applicable to make the hydrogeological units consistent with the geotechnical units;
- the model is easily managed and running times are rapid compared to 3D models. This allows the model to rapidly simulate alternative slope depressurisation methods. It allows the model to be used in 'what if' mode to address uncertainty and to see if the required level of pore pressure dissipation is achievable in the available lead time.

An example output from a 2D groundwater flow model developed using *Seep/W* to determine pore pressure in the walls of the pit is shown in Figure 6.29.

The main disadvantage of working in 2D is that it implicitly assumes the cross-section is representative of the surrounding conditions. It is important to appreciate the potential for a groundwater flow component oblique to the slope and to understand the assumptions of the model. Groundwater flow directions (vectors) oblique to the slope may result from structural controls or lithological or alteration contacts oriented subparallel to the slope, as illustrated in Figure 6.30.

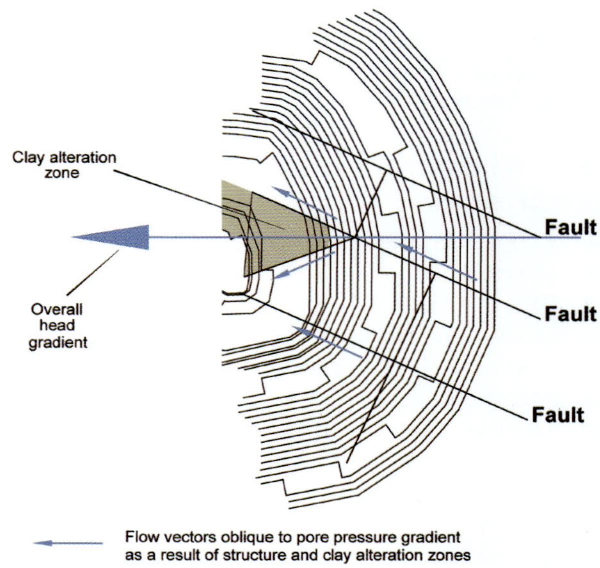

Figure 6.30: Flow vectors oblique to the pore pressure gradient

Incorporating variable structures into a 2D model oriented perpendicularly to the slope requires significant interpretation. Enhanced permeability occurs along the strike of many structures, but the same structure may form a barrier to flow across its strike. In these situations the exact model input details will require careful consideration of the local site conditions and the nature of the structures.

Another drawback of working in 2D is the difficulty of simulating horizontal drains or vertical wells that are off the section line of the model, and the difficulty in using the model to simulate different drain or well spacings. When working purely in 2D it is often better to calculate the interference effects of multiple drains or wells outside the model (see Figure 6.31). An example output from a 2D groundwater flow model that includes drains drilled from a tunnel is shown in Figure 6.32.

6.4.4.4 Slice models

A method that avoids the complexity of using a 3D model but overcomes the difficulty of inputting drain spacing is to use a slice model (fence diagram) to extend the hydrogeological units and model grid in the third dimension (see Figure 6.33). The central axis of the slice model is the 2D hydrogeological section. The model grid is often symmetrical around the 2D axis, but this is not always necessary. Figure 6.33 also shows how existing

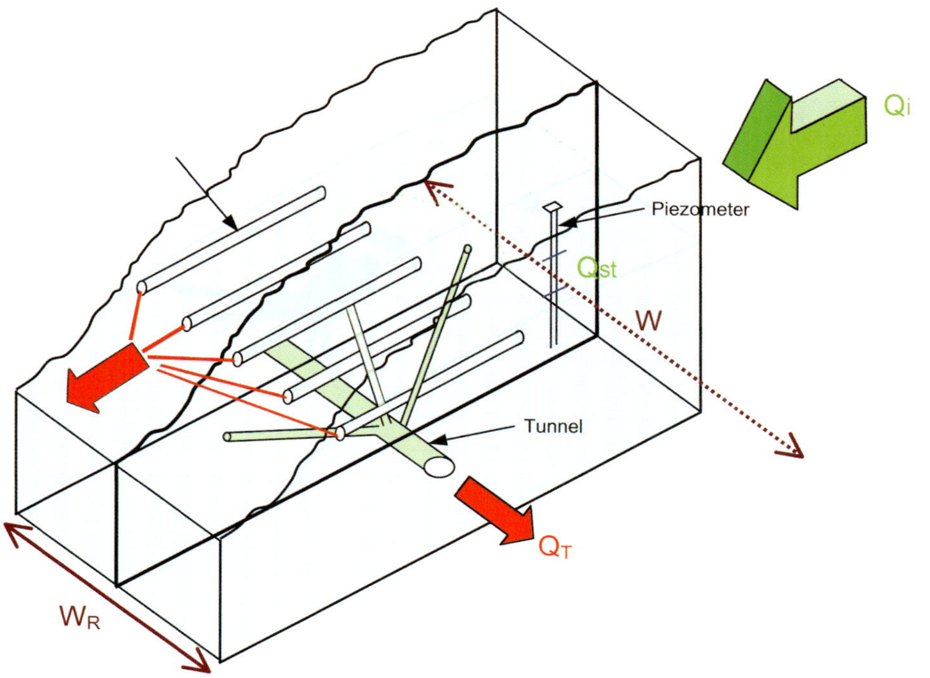

Figure 6.31: Flow balance components to estimate representative width for flow calculation in a 3D model

underground workings occurring off-section have been input to one part of the model domain. The model set-up allows the effect of different drain and well spacings surrounding the section to be directly input and investigated by the model. Figure 6.34 shows an output from a slice groundwater flow model developed in *FEFLOW* to account for the effect of drains installed from an underground gallery. The illustrated pressure profiles

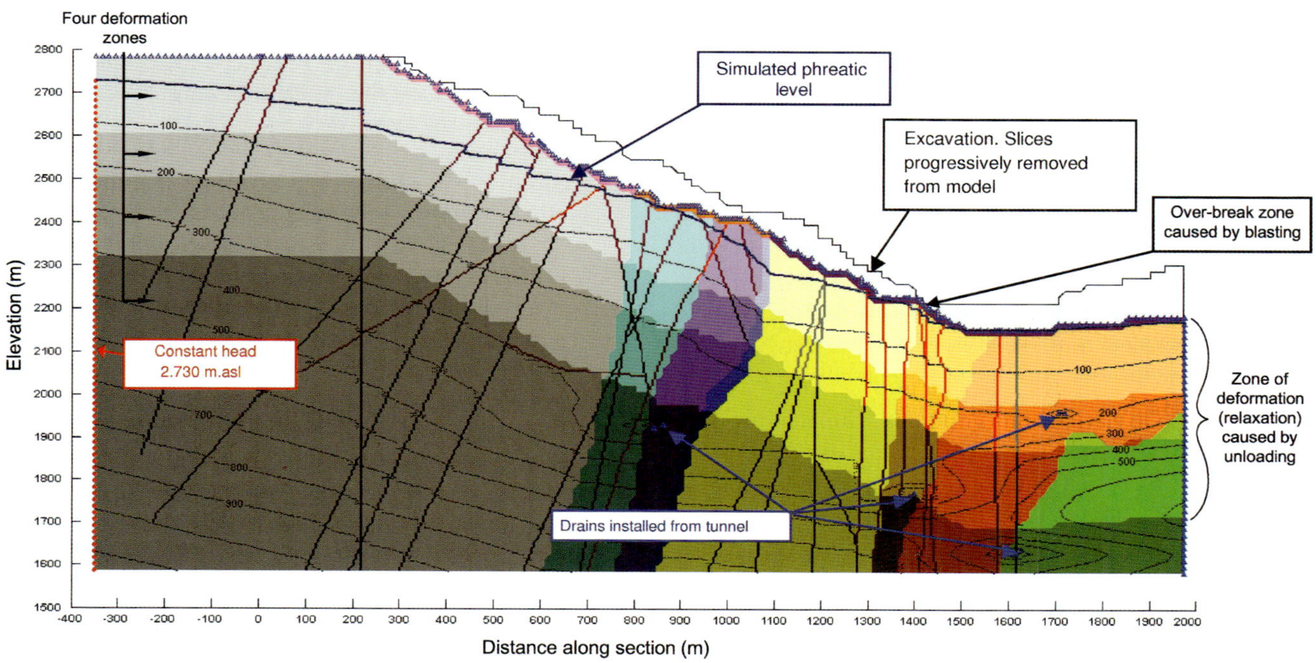

Figure 6.32: A 2D groundwater flow model incorporating underground drains
Source: Courtesy Codelco Norté

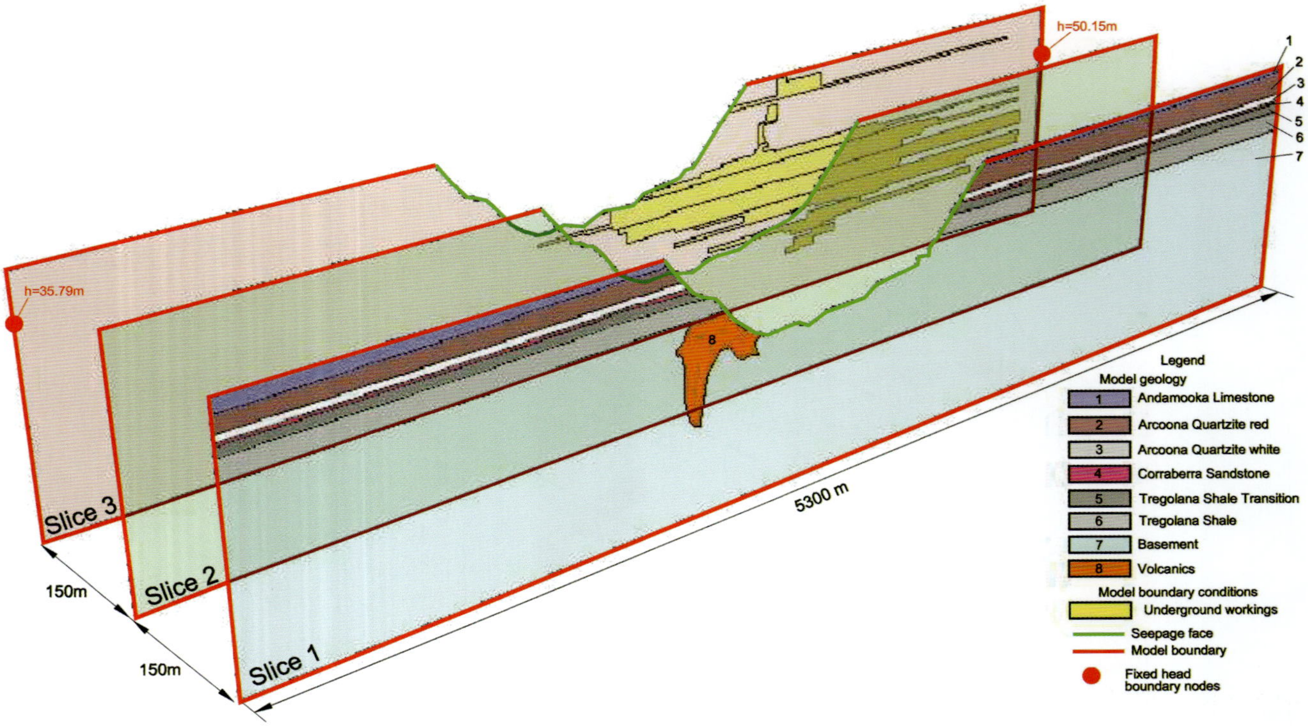

Figure 6.33: A slice model set-up
Source: Courtesy Olympic Dam Expansion project

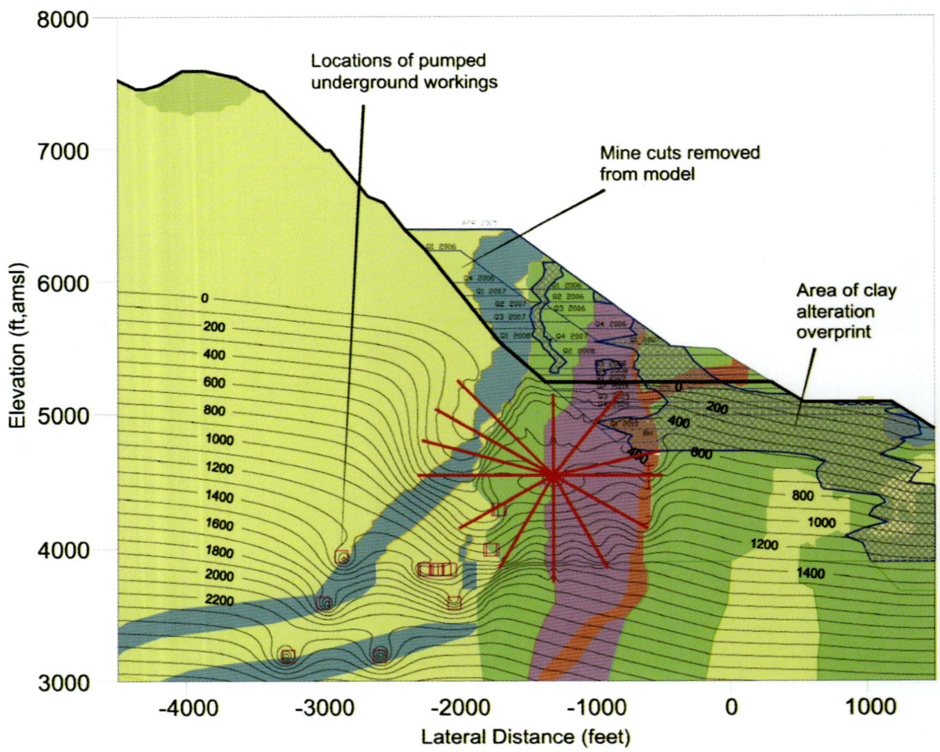

Figure 6.34: Simulation of drains in a slice model
Source: Courtesy Kennecott Utah Copper

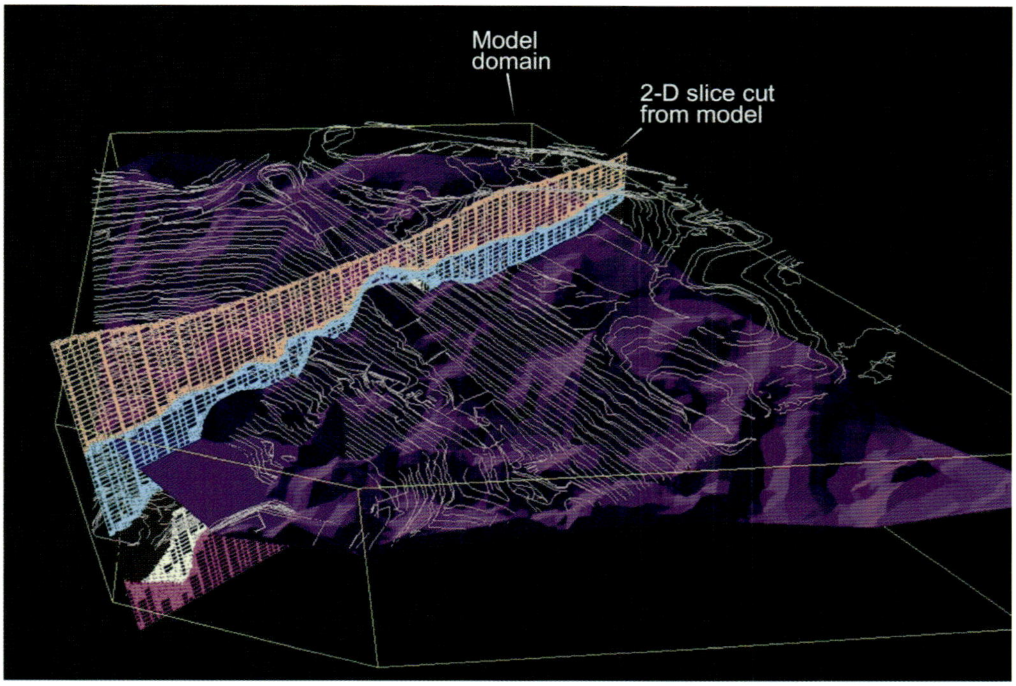

Figure 6.35: A 3D block model construction

can be converted to *ASCII* files (in terms of total head or simple pressure) and used as direct input to the geotechnical model. The effects of different drain spacings and orientations can be directly simulated in terms of their influence on the pore pressure and slope performance. When evaluating the results, it should be noted that the slice model may misrepresent an extended portion of a curving pit wall. As for any model output, interpretation and judgment of the results is required.

6.4.4.5 3D models

If the nature of the geological materials, the structural orientation in the slope or the geometry of the pit makes 2D or slice modelling unrealistic, a local scale 3D block model of the slope (sometimes referred to as a sector or wedge model) can be constructed. A 3D groundwater flow model may also be appropriate when a 3D geotechnical model is being used for that sector of slope. Construction of a 3D model may be less onerous in situations where a mine block model (e.g. *MineSight)* can be transformed into a hydrogeological model, with different rock types and structures having distinct hydraulic characteristics.

An example of a 3D block model to simulate pore pressure is shown in Figure 6.35. In this case, the eastern boundary of the model (around the crest of the slope) was a permeable rhyolite. Heads at the eastern boundary were fixed in accordance with observed groundwater levels in the rhyolite. Future heads at the eastern boundary were progressively stepped down according to future year-by-year predicted dewatering levels in the rhyolite.

When considering 3D modelling of a slope sector, the following guidelines may be helpful.

- To provide realistic simulations, the complexity of the model should reflect the conceptual understanding of hydrogeological conditions within the slope and should be commensurate with the data available. Unless there is real justification, the model should not be overly complex.
- Sufficient vertical discretisation will need to be built into the model to simulate vertical pressure gradients and multiple levels of horizontal drains. To accurately replicate the known vertical pore pressure gradients, the model in Figure 6.35 required 32 layers.

6.4.5 Coupling pore pressure and geotechnical models

For many pit slopes developed in rocks with moderate permeability, the pore pressure distribution is principally controlled by groundwater flow based on natural aquifer properties. In slopes dominated by low-permeability rocks, changes in hydraulic properties (fracture aperture and interconnection) as a result of mining-induced deformation response become important.

Fracture-controlled permeability at depth is less sensitive to disturbance than permeability near the surface. Typically, the materials adjacent to the excavated surface have been found to be the most sensitive to changes in stress, and are therefore where the greatest variations in hydraulic and mechanical properties can be expected. Experience has shown that properties such as stress-dependent permeability are most pronounced in intact rocks with macro-fractures. The sensitivity of these responses depends on hydraulic properties (fracture permeability and interconnectivity) and mechanical properties (fracture-normal stiffness/shear strength) of the fractures.

As discussed in Section 6.2.5, strain of the rock mass and fluid pressure are related and a change in one affects the other (thus the term 'coupled'). The coupled response of fluid extraction and pore pressure change (dissipation) is controlled by the specific storage term S_s (section 6.1.2.3), in the equation:

$$S_s = \rho g(\alpha + n\beta) \qquad \text{(eqn 6.5)}$$

where

S_s = specific storage
ρ = density of water
g = acceleration due to gravitation
α = compressibility of the aquifer
n = porosity
β = compressibility of water.

As the rock mass deforms in response to excavation, there is a corresponding increase of pore space and permeability. In soil mechanics, it is reasonable to assume that liquids are incompressible because the bulk modulus of water is high compared with that of the soil mass. This relationship does not apply to rock mechanics, however, where the rock mass can have a significantly higher compressibility and both the rock and fluid compressibility must be accounted for.

When evaluating active pit slopes, conditions of existing stressed ground and pore pressure must be determined and accounted for in the analysis of future conditions. Any model must have a transient calibration based on the historical mine excavation. Historical conditions must be simulated over a number of time steps, from the start of the excavation to current conditions. Results can be used to assess how the pore pressure profile has changed because of the excavation and whether changes in pore pressure have caused the state of stress to become greater than the rock strength at any time during mine development.

In the uncoupled approach, the effective stress distribution is determined by subtracting pore pressure distribution (attributed only to flow through the excavation) from the calculated total stress. Thus, conditions of stress and pore pressure do not interact (are 'uncoupled'). Transient conditions of mining-induced pore pressure changes due to rock mass deformation following excavation cannot be evaluated by an uncoupled analysis.

The coupled approach enables the evaluation of rapid excavation on the pore pressure profile and stability of the slopes. Under conditions of rapid excavation, coupled analysis can be used to show how transient pore pressures may change as a result of rock deformation.

Normally, most current models adopt the semi-coupled or iterative approach described in section 6.4.4. Deformation contours for the slope material are extracted from the geotechnical model. Discretised deformation zones are then input to the pore pressure model. The commonly used numerical groundwater codes accommodate change by varying the hydraulic parameters of the slope materials over successive time steps, based on predicted future deformation of the materials as the slope is mined.

The model output in Figure 6.29 shows four depth zones within each of the main modelled hydrogeological units. To allow the model to simulate historical pore pressure reductions due to rock mass deformation, hydraulic parameters were increased over successive model time steps to allow calibration of piezometer monitoring data. To allow the model to simulate future pore pressure reductions, the hydraulic parameters were increased in the predictive model time steps using estimates based on predicted future rates of deformation.

Other factors may influence the model's ability to accurately predict future conditions. For example, it was shown by Carrera and Neuman (1986) that permeability may decrease near tunnels and similar underground openings. There are also documented cases of decreasing permeability near the bottom of pit slopes due to apparent closing of fractures.

There is currently no fully coupled code that has been used for mining applications. However, coupled geomechanical modelling is widely used in the oil industry and developments are currently underway to adapt these codes for use in mining.

6.5 Implementing a slope depressurisation program

6.5.1 General mine dewatering

Most open pit mine dewatering systems use some form of vertical pumping wells. These may either be outside the crest of the pit wall or within the pit. In a relatively homogenous groundwater system the wells can be used to lower the groundwater flow system below the working pit floor, as shown in Figure 6.7a.

However, many ore bodies are associated with more complex groundwater settings and may include permeable alluvium or overburden deposits at shallow levels within the slope. In this case, wells need to be targeted to specific groundwater units or into individual groundwater compartments, as illustrated in Figure 6.24. In that example, interceptor wells are used to prevent groundwater in the permeable alluvial materials from reaching the slope. The wells pump a relatively high volume. However, because they intercept water at a relatively shallow depth their installation and operating costs are relatively low. Without these wells intercepting the shallow recharge water, it would not be possible to depressurise the slope materials below. The objective of pumping is to lower water levels, not to produce large volumes of water.

Groundwater cut-off systems such as slurry walls, grouting of permeable fracture systems or freeze walls are occasionally used in open pit mine dewatering applications. In an open pit setting, their primary function is to reduce the permeability of a particular formation or zone along a defined flow path, with the goal of reducing the amount of groundwater reaching the pit. If correctly installed, they act as flow barriers so the piezometric head will build up on their upstream side and be reduced on the downstream side. The use of polymers is also being investigated to reduce the permeability of fractures and hence reduce the magnitude of groundwater flow. While these measures may be applied to an overall mine water management program, they do not specifically apply to slope depressurisation and are therefore not discussed in detail here.

6.5.2 Specific programs for control of pit slope pressures

6.5.2.1 Methods of slope depressurisation

As noted in section 6.1.4, pore pressure is the only major parameter in slope stability that can readily be modified. Methods for reducing pore pressure in a pit slope can be divided into four categories:

1 natural seepage – allowing pressures to dissipate as a result of seepage to the slope, with no enhanced dewatering/depressurisation measures (passive drainage);
2 enhanced gravity drainage – installation of gravity-flowing drains from the pit slope. These may be horizontal, vertical or inclined (active drainage using gravity flow);
3 pumped drainage – installation of localised pumping wells or well points, targeting specific units within the slope (active drainage with pumping);
4 drainage tunnels – use of an underground drainage gallery or tunnel installed behind the slope (active drainage that may use a combination of gravity and pumping).

As the size of open pits and depths of excavation increase, control of groundwater and pore pressure in the pit walls plays a greater part in slope design. As the heights of pit slopes increase, the cost–benefit of depressurising the slope materials becomes greater.

Selecting the preferred category involves a detailed understanding of the geology, the likely pressure gradients within the slope and the cost–benefit of achieving the anticipated reduction in pressure. In general, there is a cost increase from Category 1 to Category 4. Obviously, the higher the cost of the pressure dissipation methods, the greater the level of understanding required to optimise the design. In reality, most pressure dissipation systems use a combination of methods, which may be installed progressively.

6.5.2.2 Seepage to the slope: Category 1

In some instances, seepage from the slope may itself provide enough pressure dissipation to achieve the desired slope performance goals without any additional active measures. This method is applicable where:

- the materials in the slope have high strength properties and pore pressure is not a main factor in the slope-stability assessment;
- the materials are more permeable and homogenous and the potentiometric surface has a low vertical and lateral gradient;
- it may not be practicable to install dewatering measures because of the geometry and/or accessibility of the slope or because of the extreme low permeability of the materials, meaning that a sufficient level of depressurisation may not be achievable given the available lead time.

The seepage water can be collected and managed using a series of sumps located below prominent zones of seepage.

At the Mag pit at the Pinson mine in Nevada, it was necessary to excavate through about 25 m of saturated low-permeability alluvium in the east wall. The alluvium extended a considerable distance into the pit and without pre-drainage it was not possible to run heavy equipment over the bench being mined. Figure 6.36 shows how a 6 m deep trench was cut at the toe of the slope prior to mining each new bench. Pumping from the trench allowed the water level in the alluvium to be lowered to allow mining. The trench also increased the drainage rate of the alluvial material in the slope. Figure 6.37 shows a photograph of the operations.

6.5.2.3 Installation of gravity drains from the pit slope: Category 2

Horizontal drains

Horizontal drains are common in open pits throughout the world to relieve pore water pressure behind the pit slopes. There are many construction methods, but a typical construction involves holes with diameters of 100–150 mm, with 25–50 mm diameter slotted pipe installed in the drain. The drains may be installed using a diamond drill by coring methods, but are more usually installed using conventional tricone drilling methods with

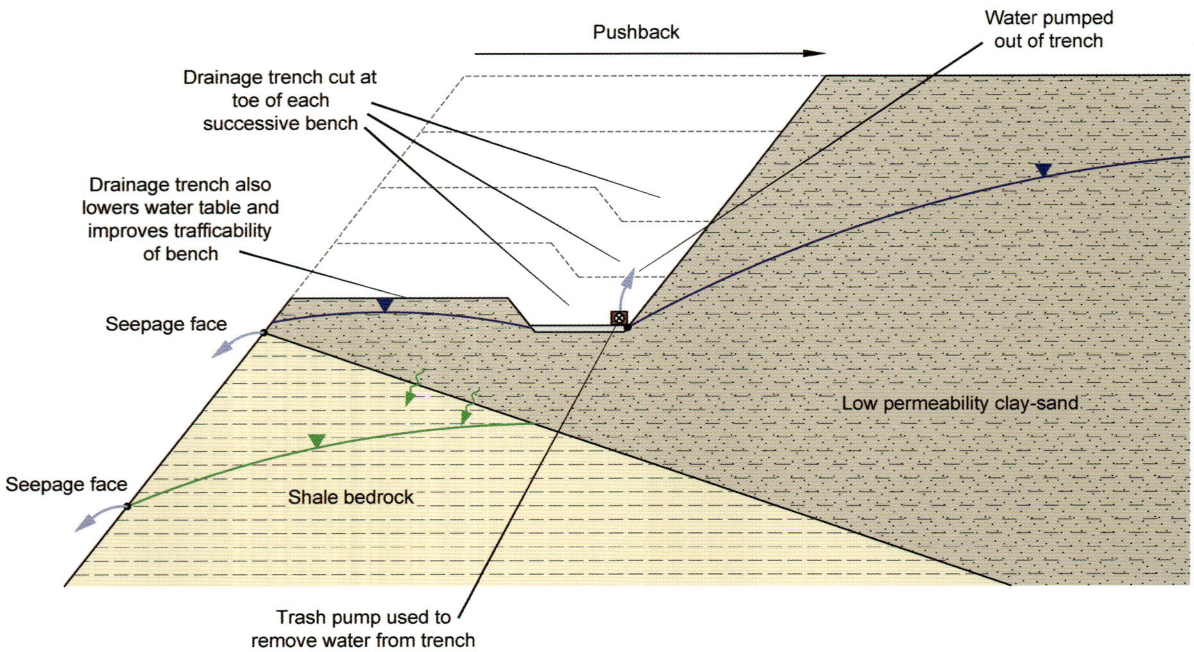

Figure 6.36: An alluvial toe trench
Source: Courtesy Pinson Mining Company

Figure 6.37: An alluvial toe trench construction (see Figure 6.36 also)
Source: Courtesy Pinson Mining Company

air or water flush. Many drilling companies operate purpose-built horizontal drain drills such as the one illustrated in Figure 6.38, which can install drains in rock to depths of up to 450 m behind the slope.

Figure 6.39a shows a typical simple design for a horizontal drain. Surface (collar) casing is frequently installed to depths ranging from 2 m to about 10 m. In some cases it may be advisable to install the collar casing deep enough to penetrate the overbreak zone in the slope, to minimise the risk of water seeping from the completed drain into the overbreak zone. It is not always necessary to install pipe in the completed hole but it is often advisable, to minimise the potential for blockage due to hole collapse.

As shown in Figure 6.39b, a packer can be used where it is necessary to limit the amount of water flowing in the annulus and prevent it from recharging depressurised fractures at a shallower depth in the hole. In these cases, a packer is set around the casing above the target water-bearing zone. The packer forces the water to enter the screened interval, and therefore prevents it from flowing along the annular space around the casing and back into the formation at a shallower depth in the hole.

Prior to starting a horizontal drain program, it is important to determine the objectives of the drains and develop specific targets for the program. The design of the drain program can then be optimised. The two sets of drains in Figure 6.40 have different objectives, as follows.

- A few long drains are needed to dewater permeable fractures that contain compartmentalised water in unaltered rock more than 250 m behind the slope. For these drains, intersection of permeable fractures by the drains is more important than the actual drain spacing. Significant variability in drain yields is to be expected because of the nature of the fracture-controlled groundwater system. Drains which do not hit permeable fractures will have lower yields; those which do hit permeable fractures may have high yields.
- A greater number of short drains are needed to depressurise the poorly-permeable altered material closer to the slope. More consistent drain yields are to be expected because of the porous-medium nature of the flow system. The drain spacing is important and depends on the permeability of the material. Yields are likely to be low (<0.2 L/sec) but consistent.

It may also be advantageous to install the drains so that they intersect the maximum number of open joints and fractures. Thus, the majority of the initial drain holes should be as orthogonal as possible to the strike of the main water-bearing structures. Once a number of drains have been installed in differing directions, the results can

Figure 6.38: Track mounted drill for horizontal drain construction (set back from bench face, with safety berm)

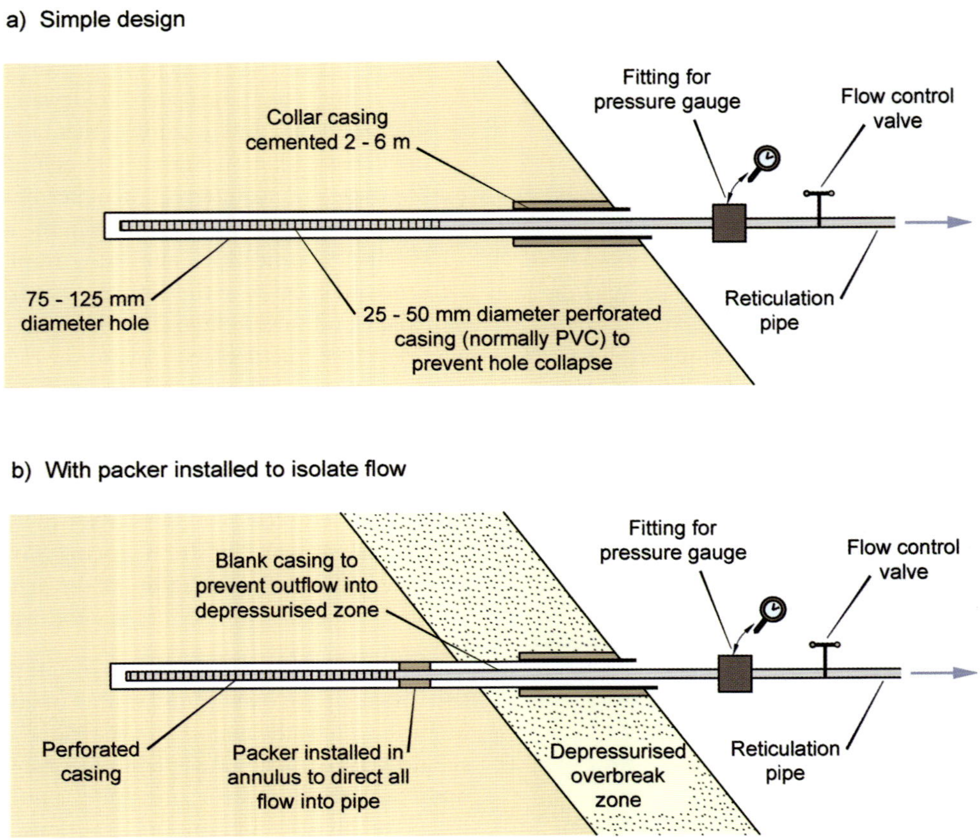

Figure 6.39: A horizontal drain design

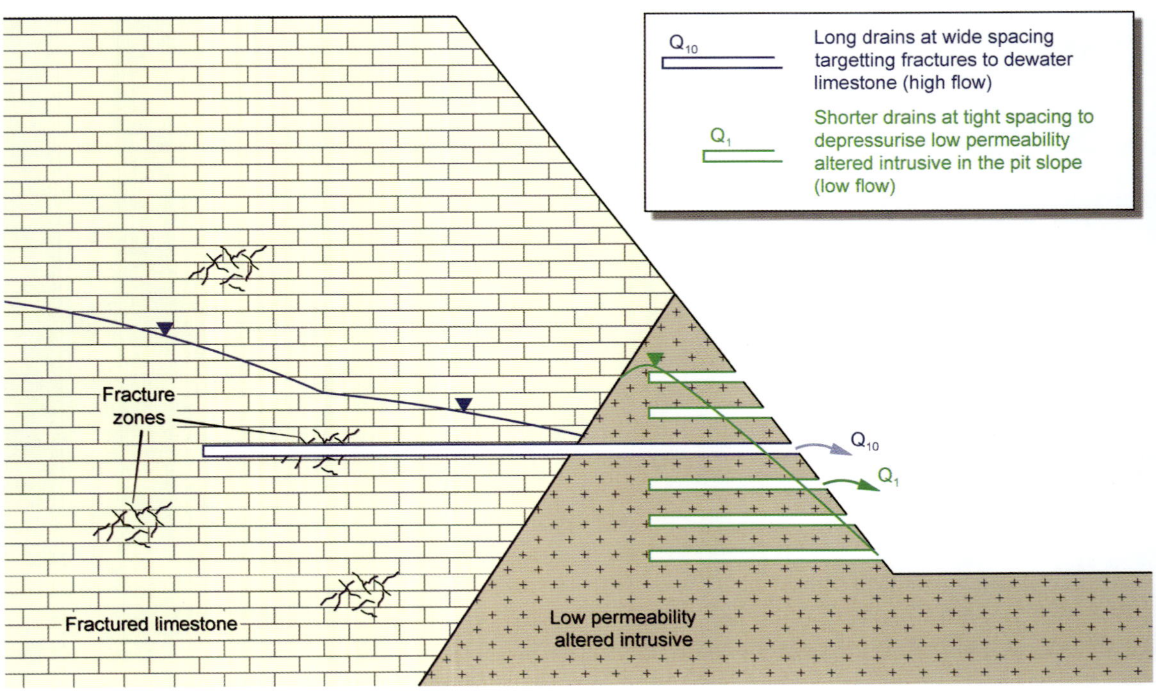

Figure 6.40: Cross-section showing two different objectives of horizontal drains

be evaluated to determine if there is a relationship between the drain hole direction and the flow rate.

Typical advantages of horizontal drains are:

- they flow by gravity and do not require pumping;
- they can be used to target zones of elevated pore pressure behind the slope;
- they can often be conveniently installed from the pit slope from haul roads or special benches as mining progresses;
- they are the most efficient way to dissipate pore pressures in a rock mass compartmentalised by steeply dipping structures.

Disadvantages of horizontal drains are:

- they are generally inefficient, because the first portion of the drain is often drilled through unsaturated material. As the water table declines due to drainage, their efficiency decreases further;
- they can only be installed after the slope has been cut. They cannot lower the potentiometric surface below their collar elevation. Therefore, they cannot be used to achieve advanced depressurisation prior to excavation;
- they require continual maintenance to collect and pipe out-flowing water to prevent infiltration into the catch bench below;
- if uncontrolled, outflow water may become a nuisance to operations;
- they can become cut and lost during a subsequent pushback of the slope. If this occurs while they are flowing they may feed water into the newly cut slope.

To tap more-permeable fracture-controlled systems (see Figure 6.41) fewer drains are typically required, so the logistics of drilling from within the pit are easier. For clay-alteration zones of low permeability, where more drains at a closer spacing are required, installation from the pit can be difficult to achieve and manage.

In settings where the topography is steep, horizontal drains can be drilled beneath the pit floor from down-slope to lower the water pressure before mining (Figure 6.41).

Vertical drains

Gravity-flowing vertical drains can be considered where a large vertical hydraulic gradient is developed within the slope or where groundwater is perched above a less-permeable unit that is underlain by more-permeable units at depth. The use of vertical drains is illustrated in Figure 6.42. The purpose is to allow water to enter the drains from the upper zone. The water flows down the drains and into the depressurised more-permeable zone below.

Angled drains

For horizontal drain installations from the pit wall, some operators may prefer to install the drains slightly upward or downward (e.g. ±10°), depending on the type of formation.

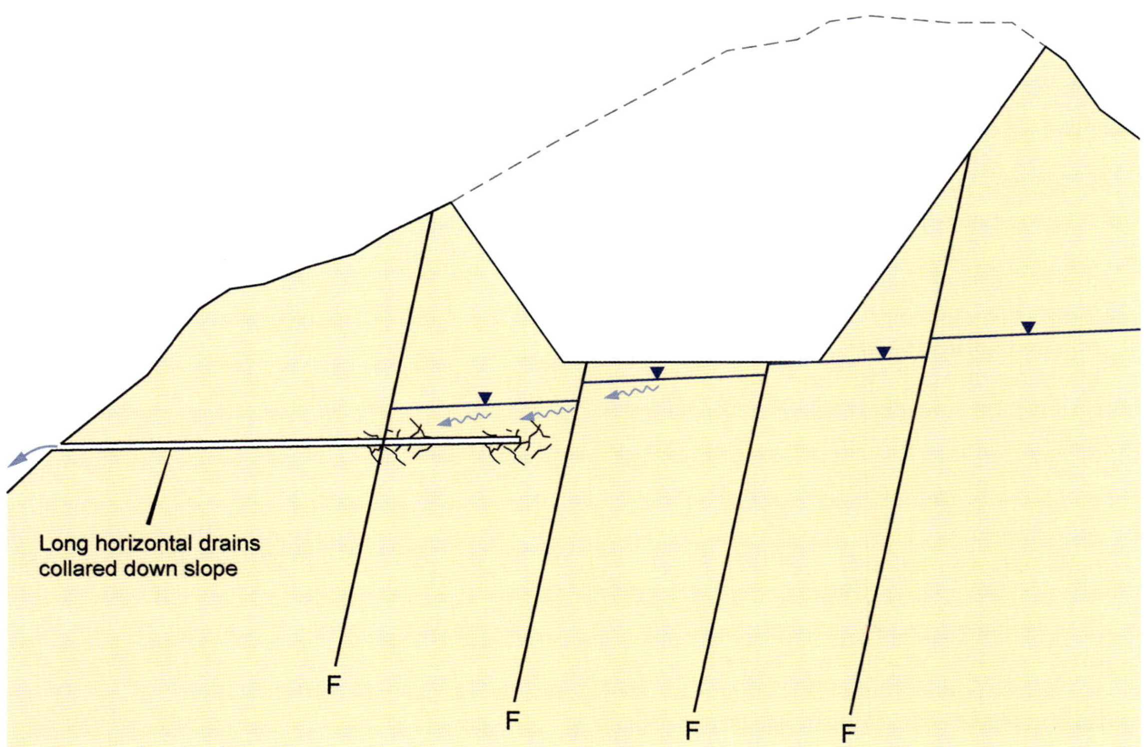

Figure 6.41: Horizontal drains used to underdrain pit

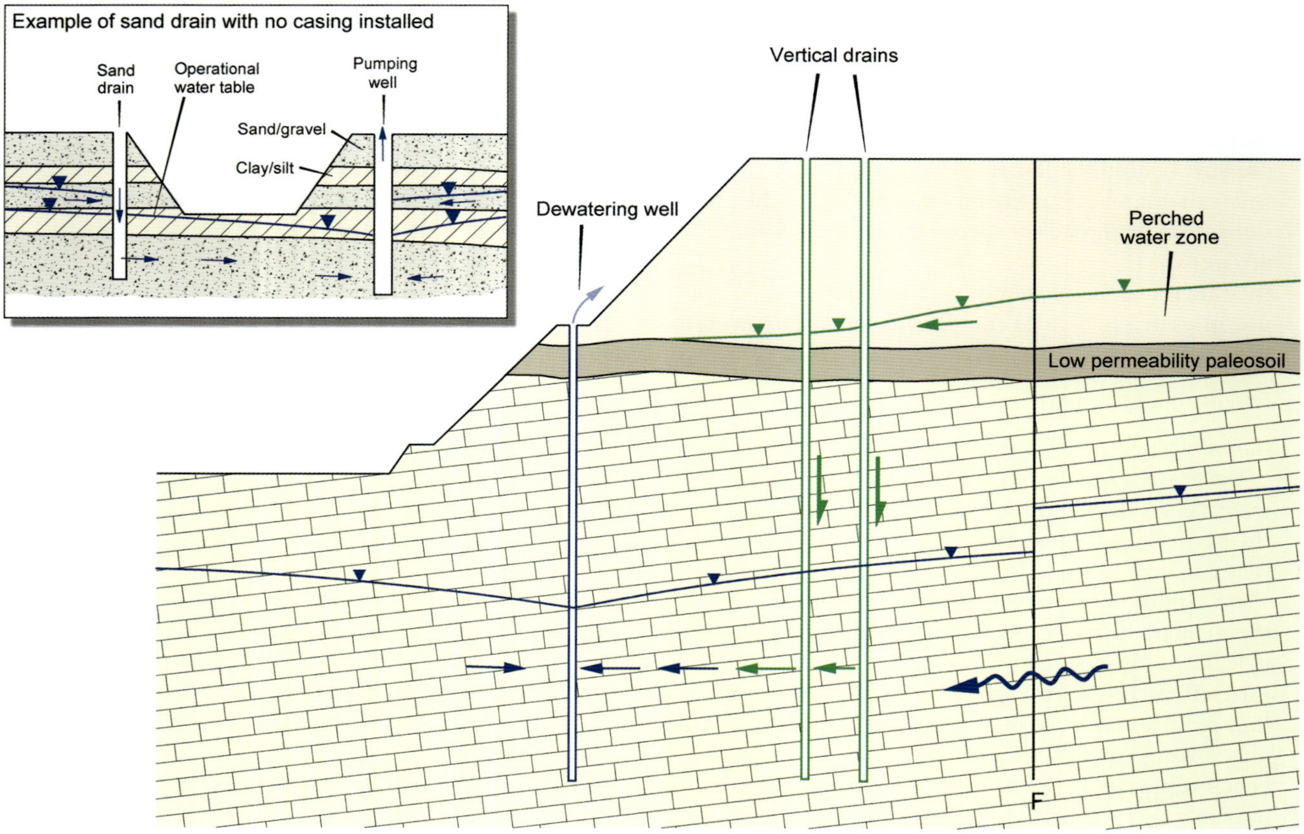

Figure 6.42: Use of vertical drains

Drains drilled slightly upward may have the advantage of reducing the effects of collapse and blockage. Drains drilled downward may have the advantages of keeping the well bore saturated and of reducing the potential for oxidation and chemical precipitation to lessen the drain's efficiency. Regardless of the angle, the flow from the drain is controlled by the head difference between the formation and the drain collar, not the drain angle itself.

In certain geological settings, drains may be installed at steep angles to increase the probability of encountering permeable zones or to provide better cross-connection of perched groundwater zones. If drains with a steeper angle are to be considered, the target zone for the drains should be defined first.

6.5.2.4 Localised pumping wells: Category 3

There are many applications for vertical pumping wells for mine dewatering and slope depressurisation. Examples where pumping wells could be specifically applied to slope depressurisation are shown in Figure 6.43. In the lower part of the slope, low-yielding wells are used to depressurise compartmentalised fracture zones ahead of a planned pushback where depressurisation ahead of mining could not be achieved using horizontal drains. In the upper part of the slope, shallow wells are used to achieve final dewatering of a shallow fracture zone. Figure 6.44 shows a line of low-yielding wells used to pump water from fractured argillites and siltstones to depressurise a low-permeability alluvium horizon.

Low-yield slope depressurisation wells can often be installed at relatively low cost using a reverse circulation (RC) drill. Holes can be drilled at diameters of 150–200 mm to allow installation of 100–150 mm diameter casing and screen, depending on the anticipated yield and the pumping head. Maintenance requirements are often small for low-yielding wells, and small-diameter well pumps (40 HP or less) require a minimal amount of maintenance. The wells typically pump to an in-pit sump, or can pump out of the pit if conditions are suitable. However, pumping directly out of the pit can increase the power requirements and hence the diameter of the pump, which in turn may lead to larger-diameter, higher-cost well installations. In some cases it has been possible to pump the water from low-yielding slope depressurisation wells into larger dewatering wells and remove it using the main dewatering well pumps.

To decrease the cost of multiple well installations, airlift pumping for wells or horizontal drains may be

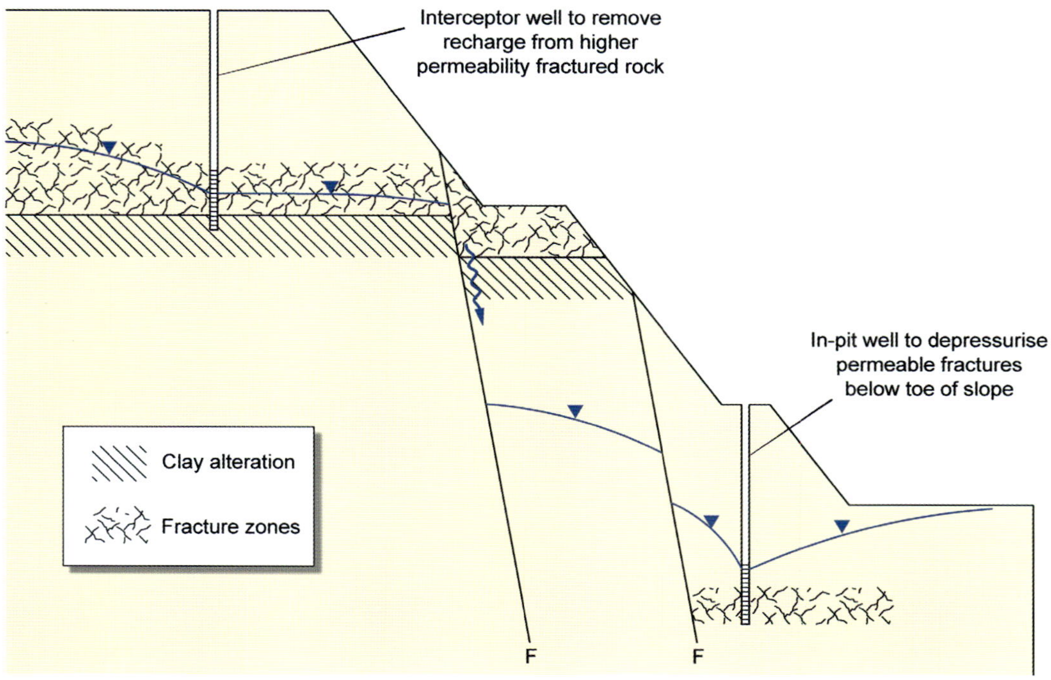

Figure 6.43: Pumping well installations for pit slope depressurisation

suitable. In the case of the low-yielding angled drains shown in Figure 6.45, 15–25 mm diameter airlines were installed in each drain and a single small compressor was used to remove the water from all drains in the array. Typical flow rates from individual drains were 0.02–0.2 L/sec and 30 drains were used in the array. For this application, because of the geometry of the slope and access constraints, angle drains and a low-cost airlift

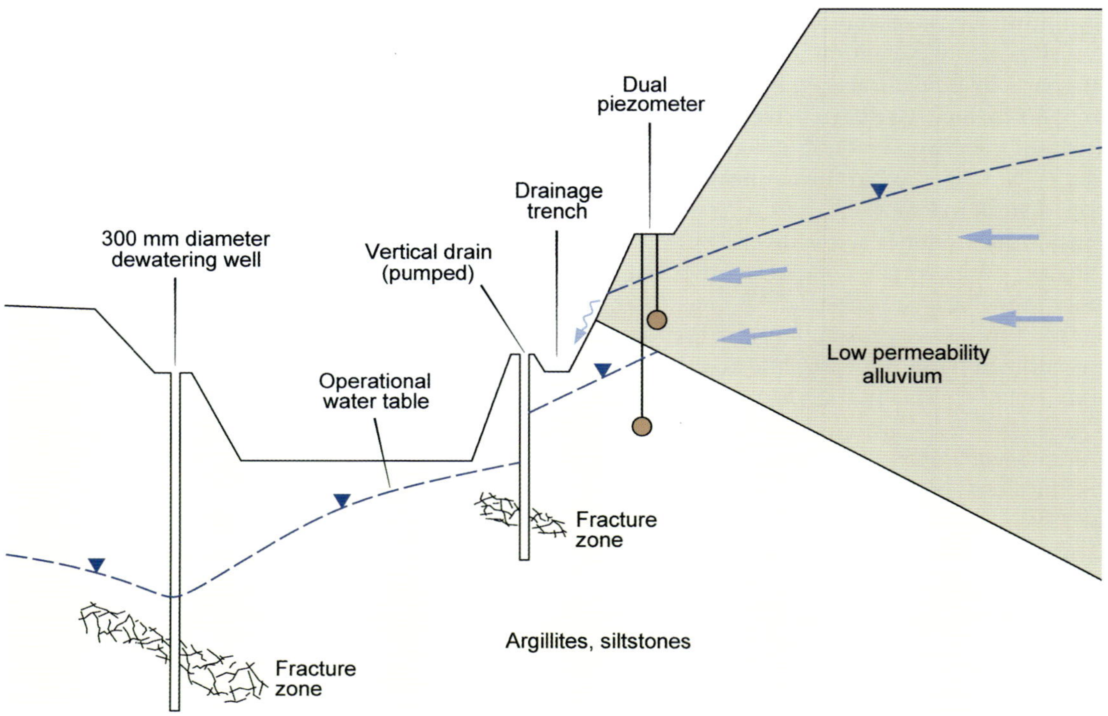

Figure 6.44: Low-yielding ejector well (pumped drain)

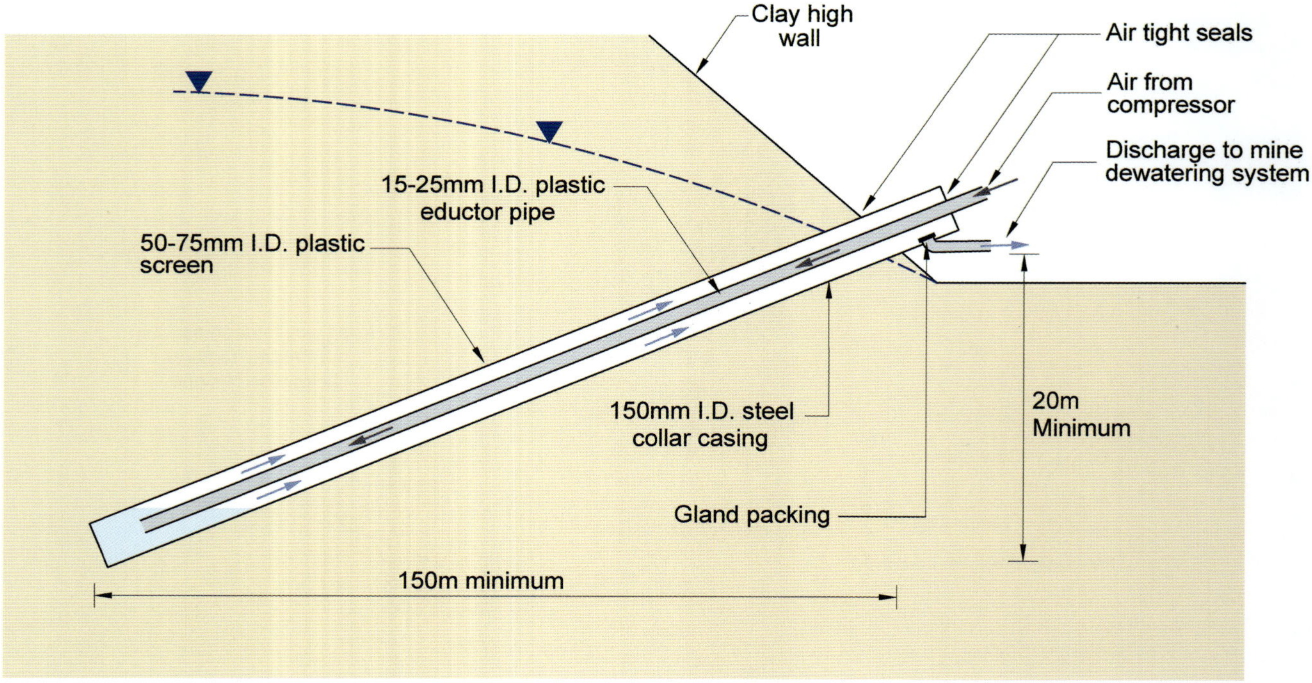

Figure 6.45: Airlift educator arrangement for angled drains

pumping system were more efficient than horizontal gravity drains.

6.5.2.5 *Underground drainage tunnels: Category 4*

Drainage tunnels are being increasingly considered by large open pit mining projects, both for dewatering and to achieve pit slope depressurisation. In mine settings where the topography is favourable, gravity tunnels installed from downslope have been driven below the workings to drain the ore body. The Sutro tunnel in the Comstock Lode, Nevada, the Carlton tunnel in the Cripple Creek district in Colorado and the Amole tunnel at the Grasberg Mine, Indonesia, are historic examples from the 19th and 20th centuries. More recent examples are the gravity tunnel driven beneath the Boinas and El Valle pits for the Rio Narcea project in Spain, the San Pablo tunnel at the El Indio mine in Chile and the Socavon tunnel beneath CODELCO's Mina Sur pit in Chile.

Each of these involve gravity draining. The tunnel was driven below the lowest ore zone from a suitable portal location downslope, and relied on good vertical connection by gravity drainage through the ore body to achieve the required dewatering. In some cases, drain holes were drilled upward or laterally from the tunnel to improve the vertical hydraulic connection within the ore body, or to allow the drainage of fault-controlled hydrogeological compartments. Figure 6.46 shows the layout of a gravity drainage tunnel. For some operations, underground workings beneath the active pit may form a drainage gallery and allow underdrainage of the water within the pit (e.g. the Nchanga mine in Zambia and the Bougainville Copper mine in Papua New Guinea).

For most gravity tunnels, the goal has been to underdrain the ore body. However, several of the larger open pit mines have considered the use of drainage tunnels specifically for depressurising the pit slopes. The main purpose of these drainage tunnels is to provide access for drilling drain holes that do most of the actual depressurising. Drain holes drilled from a tunnel behind the slope are more efficient because they penetrate rock which is saturated, whereas drains drilled from within the pit are typically drilled dry for the first part of their length (Figure 6.47).

A drainage tunnel was installed in the north-east wall of the Escondida mine in Chile in 2001 for slope depressurisation. Expansion of the tunnel is continuing (as of 2008). A drainage tunnel was installed beneath the south wall of the Chuquicamata pit in Chile in 1997. More recently, tunnels were driven behind the east wall of the Chuquicamata pit to allow depressurisation as part of a slope-steepening project. At both Escondida and Chuquicamata the tunnels were installed from a portal within the pit. Their primary goal was to provide access for drilling depressurisation drain holes from behind the slope. Figure 6.48 shows the alignment of the tunnel, the drain holes and piezometers installed in the north-east wall at Escondida.

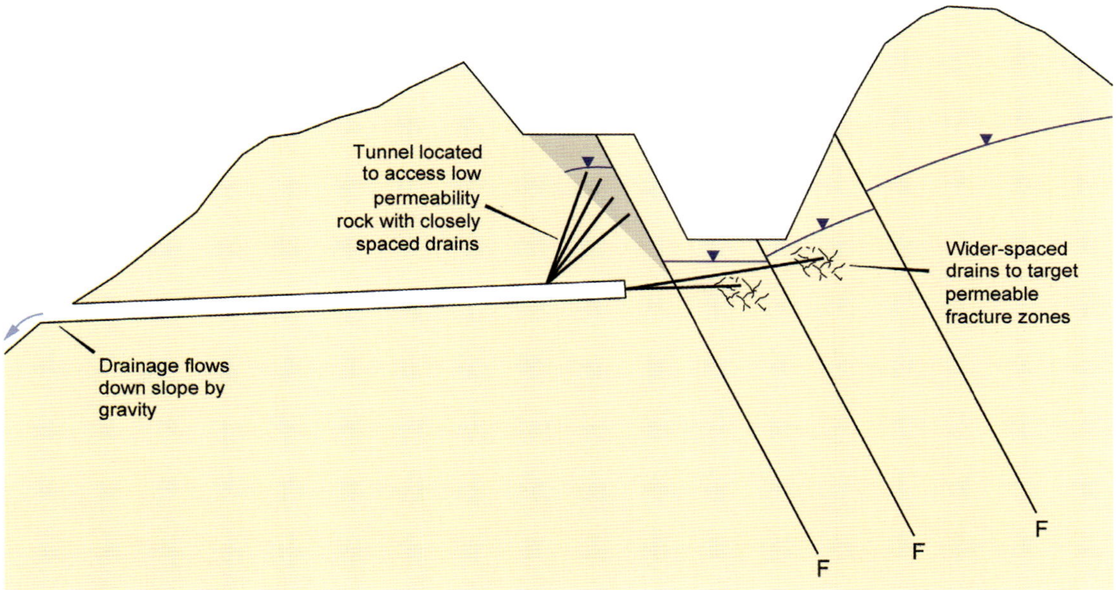

Figure 6.46: A gravity drainage tunnel

Drilling drains from a tunnel behind the slope offers the following advantages:

- all drain collars and collection pipes are outside the active open pit mining operation;
- the drains are not affected by subsequent slope pushbacks;
- the drains are under a higher driving hydraulic head and are more efficient than drains drilled from within the pit;
- the drains are saturated throughout their entire length, leading to greater efficiency.

Figure 6.49 shows the measured pore pressures following installation of the drainage tunnel and drain holes at Escondida. The level of the phreatic surface was not significantly reduced, but the area surrounding the tunnel was depressurised by more than 1 MPa. This depressurisation in a critical area behind the slope significantly reduced the amount of movement in the slope above.

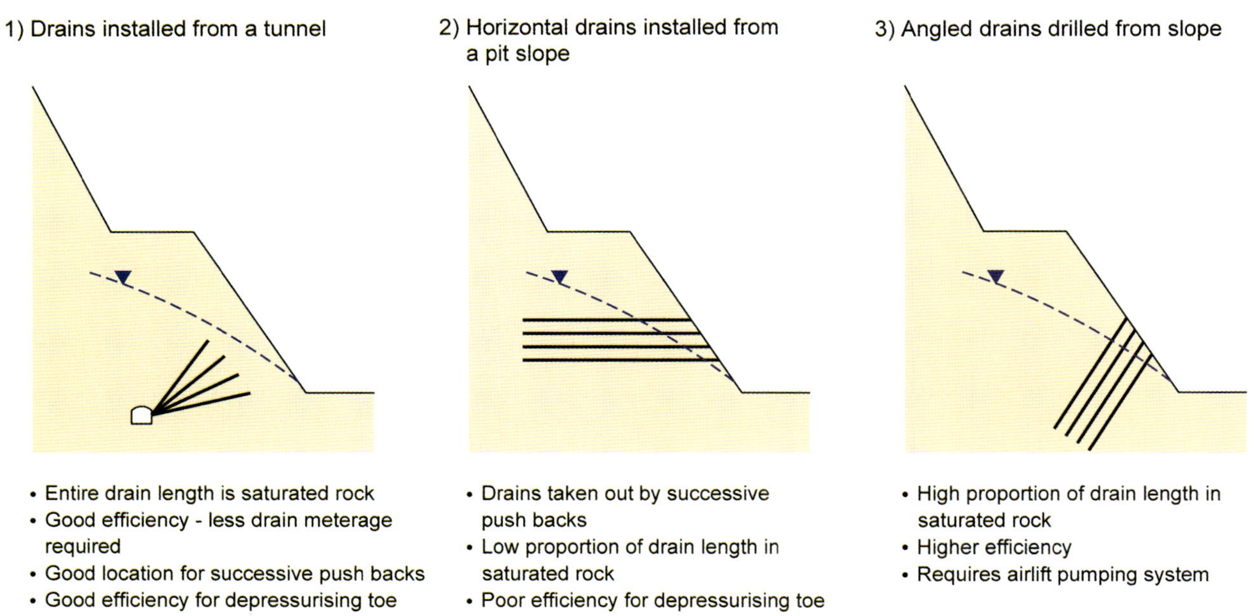

Figure 6.47: Comparison of drains drilled from pit slope with drains drilled from tunnel

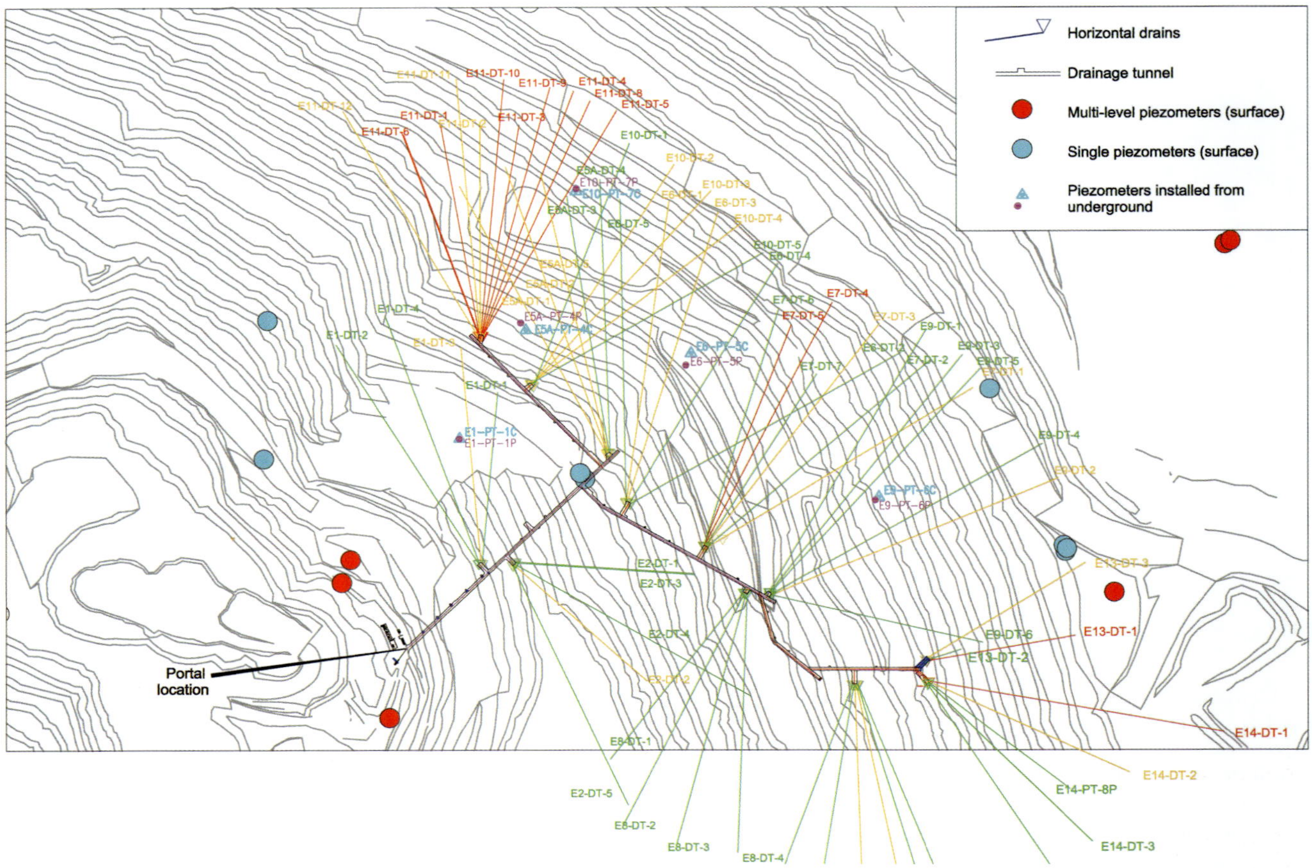

Figure 6.48: Pit wall drainage tunnel (Escondida Mine, Chile)

Figure 6.49 illustrates the difference between 'drainage' (the removal of water from the pores and/or structures in the rock mass to create unsaturated conditions in previously saturated materials) and 'depressurisation' (the reduction of pressure in the still fully saturated material). Although the pore pressure targets were achieved, most of the material within the slope above the tunnel has remained saturated as a result of minor recharge moving down the near-surface zone of deformation and increased permeability.

In most cases the primary purpose of the tunnel is simply to provide access for drilling the depressurising drain holes. Figure 6.50 is a schematic diagram of a drain hole drilled from a tunnel. Design of the tunnel location and its alignment should depend on consideration of the conceptual hydrogeological model of the slope, the location of the zones requiring slope depressurisation and the lead time required to achieve the depressurisation goals. For example, if the goal of the tunnel is to allow drain drilling into permeable compartmentalised fracture zones, fewer drains will be needed. The tunnel alignment can be further from the target zone and the required amount of depressurisation can be achieved relatively quickly. However, if the goal of the tunnel is to allow closely spaced drain drilling into low-permeability clay zones, more drains will be required. The tunnel alignment must therefore be closer to the pit slope and planning and installation must allow greater lead time for depressurisation.

Installation of exploration tunnels or underground mining operations beneath the open pit or behind the pit slope also provide an opportunity to enhance slope depressurisation or mine dewatering. The tunnel installed in the early 1990s at the Cove pit in Nevada was used to enhance pre-dewatering of the ore body ahead of the advancing floor of the open pit. A similar pre-dewatering effect was recently achieved as part of the exploration tunnel driven at the Round Mountain mine in Nevada. Such underground developments around the open pit provide the opportunity to install localised slope depressurisation measures from drill stations within the tunnel.

When selecting the layout of a drainage tunnel for a large open pit operation, the inclination of the tunnel and its ability to achieve gravity drainage in a timely manner are important. It may not always be possible to achieve optimal inclination for a drainage tunnel installed behind a pit slope, especially if deepening of the tunnel may be required to depressurise the toe of the slope. If

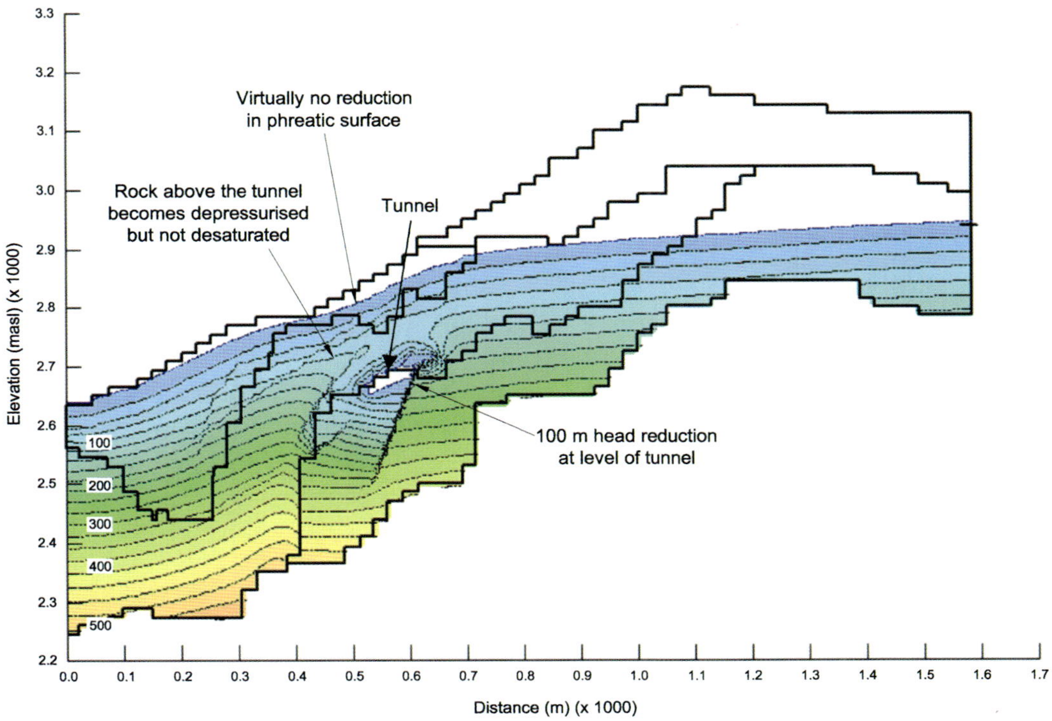

Figure 6.49: Pore pressures around an active drainage tunnel
Source: Courtesy Minera Escondid Limitada

relatively large amounts of water have to be pumped from the drainage tunnel, the potential effect of interruptions to the power supply and/or maintenance requirements for the pipelines and pumps must be carefully assessed.

Another consideration related to tunnel construction is mine closure and the possible need to install permanent bulkheads to seal the tunnel when mining finishes. This may be less relevant for tunnels installed behind the walls

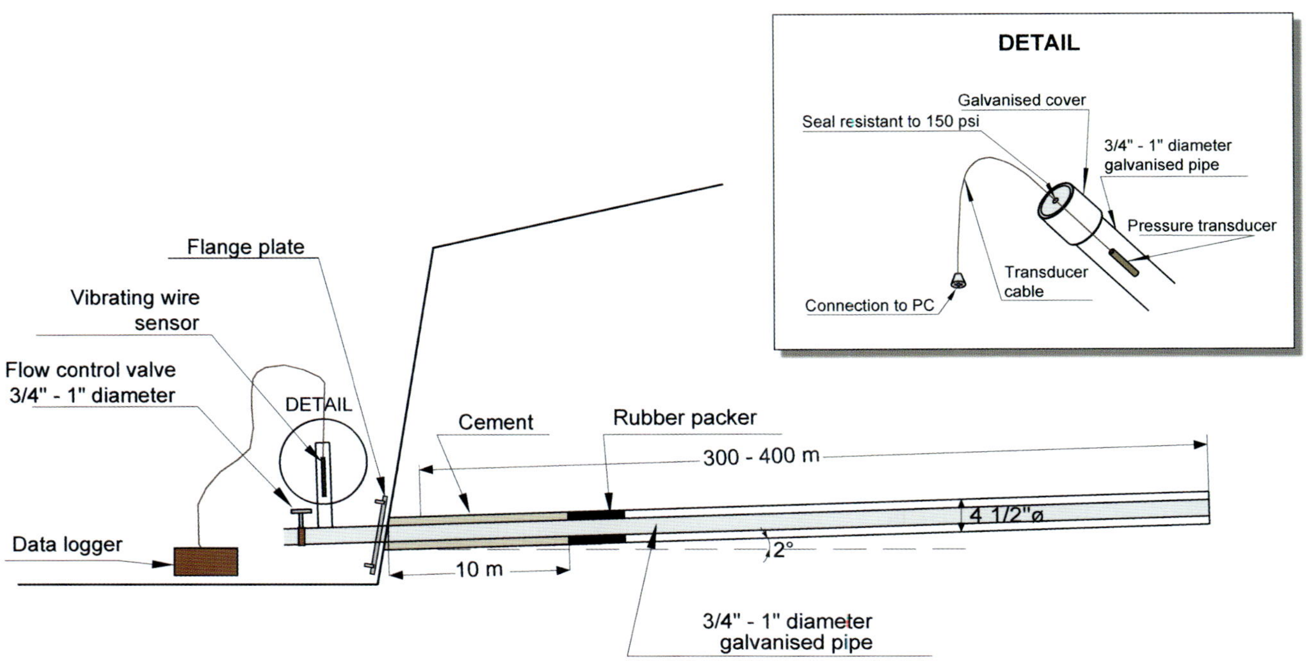

Figure 6.50: A drain installed from a tunnel

from within the pit, but is nonetheless an important factor to consider when designing a drainage tunnel and depressurisation system.

6.5.3 Selecting a slope depressurisation method

The following four factors must be considered as part of any slope depressurisation design:

1. the scale of the operation and the potential cost–benefit of depressurising the slope;
2. the conceptual hydrogeological model developed for the slope, how it relates to the mine dewatering system and the requirement to cut off recharge to successfully achieve slope depressurisation;
3. the target pore pressures and the time available to achieve them, given the slope design;
4. access considerations for the installation of drainage measures within the pit, including the rate of pit advancement and the time necessary to achieve the required depressurisation of the slope materials for a given dewatering/depressurisation system.

The ability to integrate the required pore pressure controls with mine planning and operations is a critical factor for design and a major consideration in what can practicably be achieved. It is normally possible to provide the required access by sequencing the drainage measures or by making minor adjustments to the mine plan. Slope depressurisation often results in the opportunity to steepen slope angles, reducing stripping and waste rock handling and increasing productivity. Consequently, minor adjustments to the mine plan to accommodate slope drainage measures can have a large payback.

A number of operational considerations can be useful when planning the location of in-pit measures:

- possible installation of horizontal drains concurrently with mining, i.e. after mucking the bench but before production drilling and blasting of the underlying bench;
- creation of slightly wider catch benches at certain elevations to provide access for drilling and monitoring;
- installation of drainage measures on catch benches close to the point where the benches intersect the haul ramps.

A significant operational advantage of a drainage tunnel is that the drains and other slope depressurisation measures can be installed and operated from within the tunnel, without interfering with mining operations (once the portal is established). An obvious potential downside of a tunnel is the up-front cost, but the overall cost of a tunnel is often competitive considering the costs of installing and maintaining a large number of drains from within the pit and the associated in-pit nuisance factor. However, potential poor ground conditions in a tunnel and cost overruns need to be carefully considered in any possible tunnel application. As for all depressurisation options, the overall cost must be viewed in terms of the potential benefit of achieving steeper slope angles.

Several mines are considering the transition from open pit to underground operations, particularly block caves. Some may be able to use the underground mineral exploration tunnel for installing drain holes for depressurising the pit walls and pre-dewatering the proposed underground workings. This has the advantages of shared benefits and costs.

6.5.4 Use of blasting to open up drainage pathways

It is sometimes possible to use controlled blasting to increase the permeability of tight ground and open up drainage pathways (see Figure 6.51). In that case, a low-permeability clay gouge zone was impairing drainage of rocks on the footwall side of the structure which was, in turn, causing elevated pore pressures in the pit slope above the bench.

Three lines of 45 m deep blast holes were installed across the tight zone. The lines of holes were about 20 m apart and the aim was to create three drainage pathways from the saturated rocks in the footwall into the dewatered rocks on the hanging wall side of the structure. The blasts were progressed from the hanging wall to the footwall with long delays. The resulting drainage caused pore pressures in the pit wall above the bench to reduce, as shown by the inset in Figure 6.51.

At the Robinson mine in Nevada, development of new drop cuts has historically been difficult due to water in a low-permeability rock mass. A procedure was established of drilling 30 m deep blast holes to open up fractures below the pit floor to act as sumps, then to install temporary well casing and submersible pumping equipment to drain localised areas ahead of mining.

6.5.5 Water management and control

6.5.5.1 *In-pit water management*

It is important that water produced by in-pit dewatering and depressurisation is removed from the vicinity of the slope as soon as possible, as any localised infiltration back to the slope materials may defeat the purpose. In particular, it is necessary to keep water away from materials that are sensitive to re-wetting (e.g. argillic materials and kimberlite), so that when these materials relax and new fracture surfaces develop their strength is not reduced when they react with water. Management options for in-pit water may include:

- piping by gravity to a central sump;
- using lined channels to convey the water to a central sump;

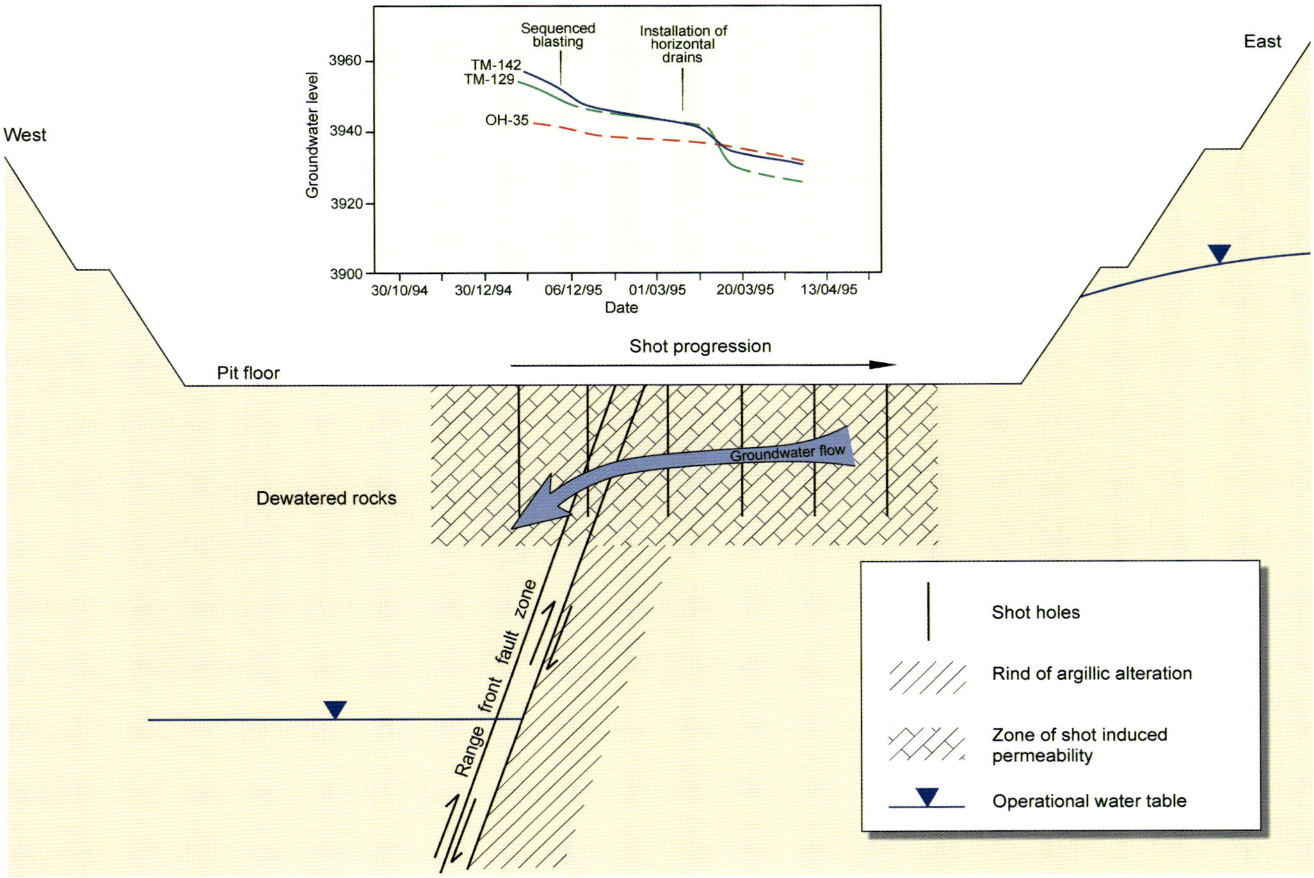

Figure 6.51: Blasting to enhance permeability across a structure

- integration with the in-pit surface water management sump system (in wetter climatic regions);
- piping by gravity into higher-volume dewatering wells (if the incremental pumping rate is relatively low).

6.5.5.2 Surface water control

Three important goals for surface water control are:

- minimising the potential for surface water to enter tension cracks above the crest or within the slope;
- minimising the extent to which infiltration into the overbreak zone can create high transient pore pressures in the near surface materials (even if the rocks deeper in the slope are depressurised) – the transient pressures can sometimes lead to bench scale failures;
- minimising the extent to which erosion and back-cutting into the slope occurs due to surface water erosion.

In wet and seasonal climates, slope damage due to surface water can be difficult to control and there is often no easy solution. Figure 6.52 shows surface water rilling and back-cutting into a cemented gravel unit exposed in an upper slope. If left uncontrolled, coalescing of the back-cuts may lead to bench scale failures and a general deterioration of the slope, potentially creating large-scale failures.

Surface water control measures for large open pits usually involve:

- diversion of water around the crest of the slope. If the topography is suitable, construction of gravity diversion ditches around the crest of the pit is often beneficial for minimising the amount of water reaching the slope. Diversion ditches should be lined if there is a risk of infiltration into the materials above the crest. It is often beneficial to place permanent ditches far enough away from the crest to minimise the potential for water to enter tension cracks and for diversion damage due to the development of tension cracks. In areas of steeper topography, diversion ditches may be placed on high catch benches in the upper slope rather than at the pit crest. In drier climates, simple runoff control berms around the pit crest may be adequate for preventing water from entering the slope;
- collection of runoff water on catch benches. Installation of diversion ditches around prominent catch

Figure 6.52: Bench scale erosion and failures due to runoff on pit wall
Source: Courtesy of Minera Yanachocha SRL

benches or along the inside of haul roads is commonly used to remove surface water runoff from the slope. If topography allows, the ditches may drain by gravity to a low point in the high wall and out of the pit. Otherwise, collection sumps and the ability to store runoff from high-intensity storm events must be included in the mine design. Ideally, the diversion ditches should be designed to shed water from the slopes as rapidly as possible and minimise the potential for ponding and infiltration into the slope materials. In drier climates, ditches around the inside of the haul roads may be all that is required. In-pit trenches close to water-sensitive material, such as kimberlite, should be lined.

It is also necessary to develop a plan for managing water that enters the pit from horizontal drain holes. Ideally, this water would be collected by pipes that are routed directly to sumps, with no contact with the wall rocks or catch benches. However, many operations allow the drain water to flow directly onto the working benches. This may be acceptable where the water can be routed directly into collection ditches (e.g. drains drilled along the inside of a haul ramp). However, the flow is often allowed to infiltrate into the catch benches below, where it may join the invisible water moving down the slope within the overbreak zone and may help to sustain pore pressures in the wall rocks below. Maintaining access to catch benches to manage sustained horizontal drain flows can be difficult and there is often no easy operational solution.

Normally, in-pit seepage and runoff is routed to the pit floor or other storage areas as rapidly as possible to minimise the potential for infiltration into the slope. It is often necessary to provide one or more runoff water storage areas inside the pit. Typically, the main storage area is the pit floor. Pump design for water removal involves a balance between higher capacity (and higher cost) pumping equipment and the length of time that water is allowed to remain inside the pit. A rule of thumb for a large pit is to design a surface water pumping system that can remove the runoff from a 1 in 50 year runoff event in 30 days. However, this will depend on many factors including the intensity of the runoff events, the flexibility of the mine plan to provide alternate dig faces away from the pit floor and the frequency of use of the pumping equipment. In some tropical areas, lower levels of the pit are closed during the wet season and all operations are carried out on upper benches above the water.

For mines located in regions with a seasonal climate, it is often necessary to implement a program of inspection, cleaning and maintenance of diversion ditches prior to the onset of the wet season. In tropical areas, where the upper slope materials are weathered, removal of sediment from ditches and sumps needs to be part of the maintenance plan. In colder climates, winter maintenance and snow/ice removal from diversion ditches often needs to be carried out to help ensure the diversion systems have maximum capacity during the peak springtime runoff period.

Figure 6.53 shows a slope designed with 0.5% outward grade on the catch benches, allowing installation of surface water interception trenches every second bench to prevent erosion. In this case, the mean annual rainfall was relatively high (1500 mm/yr) and the relatively long dry season allowed sufficient time for adequate trench maintenance.

6.6 Areas for future research

6.6.1 Introduction

This chapter has assessed the current methods of characterising, modelling and managing pore pressures within large open pits. Many of the difficulties in characterising and understanding pore pressure distributions relate to the anisotropy and heterogeneity in structurally complex fractured rock environments. While the understanding of hard rock mining hydrogeology has increased significantly over the past 15 years, there is still a tendency to oversimplify approaches for assessing pore pressure distributions. As large vertical pore pressure gradients develop in high walls of open pits, if the output of geotechnical models is to be relied upon to provide an optimised slope design it is increasingly necessary to adequately characterise the vertical pressure gradients and include them in the geotechnical models.

Given the approaches common at many mine sites throughout the world, the following are important areas for research and improvement.

1. Better characterisation of the dynamic relationship between more-permeable high-order fractures (e.g. large-aperture fractures occurring in fault zones) and less-permeable low-order fractures created by joints (e.g. small-aperture joints throughout the rock mass). Due to the different hydraulic behaviour of the various fractures, at any given moment the magnitude of pore pressure and the rate of pore pressure change may vary widely on a local scale (tens of metres) within a pit slope.
2. Creation of a standardised format for inputting pore pressure fields into geotechnical models, and development of procedures for the coupling of pore pressure and geotechnical models. A primary consideration is the scale at which pore pressures are important (i.e. whether in the larger-aperture fractures associated with the main structures or in the lower-order fractures created by joints that pervade the rock mass).
3. Better characterisation and modelling of transient pore pressures, ranging from a seasonal timescale related to recharge in wet climates to a millisecond timescale related to blasting.
4. Provision of a practical approach for coupled modelling in terms of data collection, numerical analysis and field application. It will be necessary to ensure that coupled modelling is developed at a level commensurate with available data and knowledge, so that it can provide solutions that offer reliable support for slope designs. For example, there is little value in developing models that require a density of data that is practically or economically impossible to achieve. There is equally little point in collecting large amounts of data that are unlikely to be used within the timescale of the operation.

6.6.2 Relative pore pressure behaviour between high-order and low-order fractures

The more-permeable high-order fractures in discrete fault zones are likely to respond to changes in stress differently from the less-permeable low-order fractures associated with the joints that are more widely distributed throughout the rock mass. For example, the more-permeable fractures associated with fault zones may depressurise rapidly in response to an array of horizontal drains that cut the faults, whereas the intervening low-order fractures will respond much more slowly. Conversely, low-order poorly connected fractures may depressurise quickly as a response to unloading and deformation of the slope, whereas the permeable and interconnected fractures associated with fault zones may hardly depressurise at all. Such differences may occur not only on the overall scale of the pit slope but also on a local scale, such as tens of metres.

In such situations, use of a simple water table (phreatic surface) in a geotechnical analysis may lead to an unrealistic representation of pore pressure in the wall rocks. Following mining and unloading of the slope, the poorly connected lower-order fracture sets may have a lower pore pressure than the more-permeable first- and second-order fracture sets, in which the pressure has equalised because of groundwater flow. The poorly permeable fracture sets may show an undrained hydromechanical response, while the permeable fractures may show a drained response. Thus, the situation may develop where the bulk of the rock mass has a reduced transient pore pressure relative to the main structures.

There are no known monitoring data for large open pits that can verify whether lower-order fractures may have a lower pore pressure following unloading than the more-permeable first- and second-order fracture sets in which the pressure has equalised because of groundwater flow. Thus, future research could involve instrumenting a pit slope using closely spaced (e.g. 5–10 m) horizontal and/

Figure 6.53: Surface water control on an active pit wall
Source: Courtesy Minera Yanachocha SRL

or vertical vibrating-wire piezometer arrays. The use of a closely spaced piezometer interval would allow different rates of unloading pressure response to be correlated with different values of rock mass permeability. Such a test should occur in an area where:

- future mining and unloading is rapid;
- the permeability of the rock mass is low;
- the overall porosity is less than 0.001;
- the rock is a brittle nature that will allow clean fracturing;
- it would be possible to measure the rock mass permeability over a 1–2 m interval around each piezometer by packer testing prior to piezometer installation.

The instrumentation and associated testing would be best carried out from underground (e.g. in a drainage tunnel) so that it is out of the way of the mining operation. The drilling depth of the holes would be less for underground mining, so that high resolution of the packer testing and instrument installation would be easier.

The data from such a test could be included in a geotechnical code that would allow pore pressures to be input independently in the main structural zones and in the rock mass. The sensitivity of observed differences in pore pressure on the factor of safety against failure could then be investigated.

6.6.3 Standardising the interaction between pore pressure and geotechnical models

Analytical and numerical design work for pit slopes has historically used a simple water table or phreatic surface as input to the geotechnical model. While this may be appropriate for smaller slopes in more homogenous material, it does not allow vertical pore pressure gradients to be applied. The assumption of a simple water table and hydrostatic conditions beneath it can lead to an overestimate of the pore pressure in pit slope zones that have a downward hydraulic gradient, and an underestimate of the pore pressure in pit slope zones that have an upward hydraulic gradient. In some circumstances, the input of a simple phreatic surface may cause the pore pressure to be under- or overestimated in the geotechnical evaluation by a factor of 2, which may have a significant effect on the probability of failure assumed for the slope design.

Specific pore pressure models have been increasingly used in geotechnical models. The pore pressure models can produce a grid that considers both lateral and vertical pressure gradients. Bingham Canyon (USA), Diavik (Canada), Antamina (Peru), Escondida (Chile), Chuquicamata (Chile), Olympic Dam (Australia), Grasberg and Batu Hijau (Indonesia), Letlhakane and Orapa (Botswana) and Venetia (South Africa) are large open pit mine developments that used groundwater flow models to develop pore pressure fields for input to the geotechnical models. At some operations, the pore pressure models and geotechnical models were run interactively to examine changes in the safety factor for a range of slope depressurisation options.

The output of the pore pressure model is provided as *x, y, h* or *P* (for 2D models) or *x, y, z, h* or *P* (for 3D models) in *DXF* format, for a defined time step, for transfer and input into the geotechnical model. Often, it is more straightforward for hydrogeologists if the output of the pore pressure is given in pore pressure elevation (total head in metres elevation) than in MPa, psi, or other units of pressure that are more familiar to geotechnical engineers. It would be beneficial if a standardised format was developed for inputting pore pressure into geotechnical codes, so that a corresponding standard output format could be developed for pore pressure models. A standard input format would also ensure that the significant structural features affecting the hydrogeology, and the zones where pore pressures are potentially liable to change rapidly, correspond with the appropriate structural zones in the geotechnical model.

An integral part of this is a better determination of the scale at which pore pressures become important. Most pore pressure and geotechnical models include key structural zones as discrete features within the model domain. However, because the output of most current pore pressure models is in simple *x, y* (or *x, y,* z) format, there is no consideration of the relative importance of pore pressures within larger-aperture fractures associated with the main structures compared to pore pressures in lower-order fractures that pervade the rock mass.

6.6.4 Investigation of transient pore pressures

In many areas of higher rainfall, or where seasonal groundwater recharge occurs close to the crest of the slope (e.g. from waste dumps), transient seasonal pore pressures may develop within the overbreak zone and the underlying zone of deformation. Such transient pore pressures can sometimes lead to significant seasonal changes in slope performance, such as in the east wall of the Bingham Canyon pit in the USA, where seasonally infiltrating water builds up pore pressure above low-permeability intrusive sills that cut more-permeable limestone beds.

Research may be warranted to document and interpret slope failures that have resulted from seasonal pore pressure increases. The research may focus on:

- the magnitude of seasonal pore pressure changes in different environments;
- resulting changes to the effective stress of the slope materials;
- whether it would be possible to minimise recharge to the slope;
- the cost–benefit of installing measures to reduce the effect of transient pore pressure.

A further area of uncertainty relates to transient pressure responses during blasting and their potential to create slope instability. Blasting in saturated slopes can have the effect of causing a sudden increase in fluid pressure. At such times, slope stability is potentially vulnerable. In time, depressurisation occurs naturally and slope stability is improved. If the slope is depressurised prior to excavation, this potentially unstable transitory state can be avoided. This transitory stress condition is generally overlooked in most standard uncoupled analyses.

Some mines collect data on how blasting affects pore pressures. Sealed vibrating-wire piezometers have been installed close to blast patterns and research would initially draw on this existing information. A more comprehensive monitoring program could be implemented on a number of test slopes within different geological and hydrogeological settings, using closely spaced vibrating-wire piezometers installed in the toe of the slope and recording at millisecond intervals. The influence of blasting at various distances from the instrumented area would then be documented. Sudden transient pore pressure increases related to blasting may trigger initial instability along structural zones of weakness, which lead to larger-scale slope instability.

6.6.5 Coupled pore pressure and geotechnical modelling

Only a limited amount of integrated pore pressure and geotechnical modelling has been applied to large pit slopes. True coupled modelling would consider the changes in the hydraulic parameters that result from unloading the slope and deformation of the materials behind the slope. Deformation of rock due to unloading typically causes an increase in permeability, porosity and fracture interconnection, which in turn reduces pore pressures and improves slope performance.

Figure 6.54 shows an example from the east wall of the Chuquicamata pit where pore pressures have been reduced to low levels by a combination rock mass unloading and drain holes drilled from with tunnels behind the slope. In this case, the use of a groundwater flow code alone,

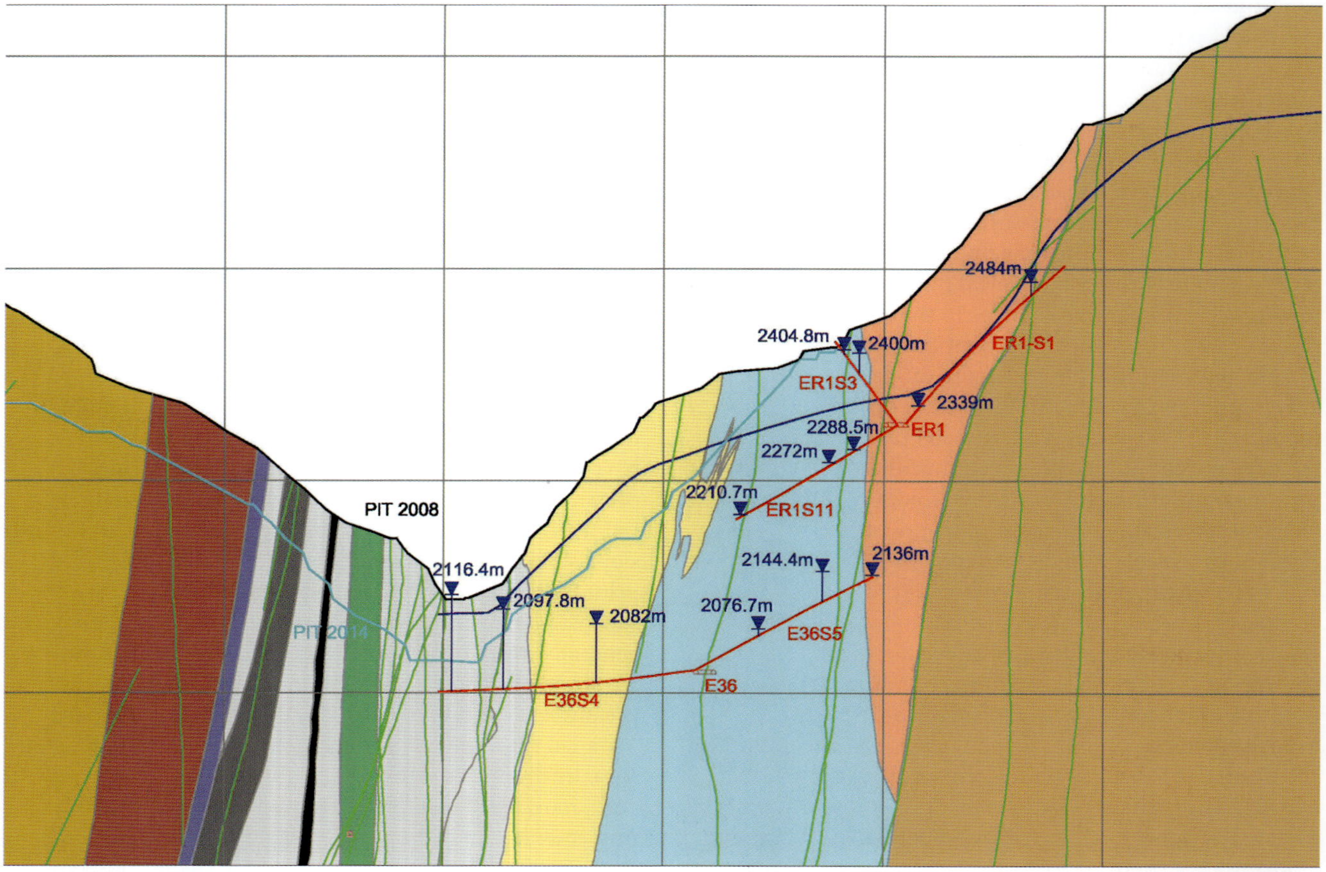

Figure 6.54: Example where pit slope pore pressures have shown a large response to rock mass unloading (Chuquicamata east wall)
Source: Courtesy of Codelco Norte

without changing the hydraulic properties in response to deformation of the rock mass, would have lead to a significant over-estimate of the actual pore pressures in the model output.

Several mines have used the output from deformation models as the basis for changing hydraulic parameters in successive time steps in pore pressure models (e.g. Bingham Canyon, USA, and Chuquicamata, Chile). The geotechnical and groundwater flow models have been run interactively, with the output of one model being used to define the input of the other. Although the models used exactly the same grid (and in some cases consistent hydrogeotechnical units) they were run independently.

A fully coupled modelling approach is complex and thus would be difficult to implement and interpret. The main uncertainties stem from an inability to adequately characterise heterogeneous rock masses and to include all critical material features. It may be possible to predict general changes, but detailed local-scale modelling is much more difficult. Any practical advances in modelling which will produce meaningful results must be consistent with advances in the quality and detail of field data collection (Chapter 2, section 2.5).

Most coupled modelling has been undertaken by the oil and gas, nuclear waste disposal and geothermal energy industries. In the mining industry, one area where coupled modelling may have a practical application is simulating a change in hydraulic parameters that may result from slope failure. The failure may cause an increase in permeability and porosity, leading to a reduction in pore pressure. This in turn may lead to increased stability of the affected part of the slope. Such an interactive process between material deformation and hydraulic parameters can only be simulated using a coupled modelling approach.

In addition to modelling, the relationship between deformation and changes in permeability and porosity for a range of lithological and alteration types should be evaluated. This would be an essential input parameter for any coupled model. Potential tools for this modelling include *Fracod 2D and 3FLO*.

Fracod-2D is a 2D boundary element model that allows the simulation of fracture initiation and propagation. It can handle tensile and shear failures and makes use of the displacement discontinuity method. The input data include the geometry of the model domain and the geometry of pre-existing fractures, boundary conditions, far-field stresses and elastic properties of the rock mass, fracture toughness, fracture stiffness of pre-existing and created fractures, fracture friction and cohesion. However, boundary element models typically require the domain to be uniform (homogeneous).

3FLO is a software applied to the 3D simulation of flow and transport in porous and fractured media. The code can generate a 3D DFN based on the orientation and spatial distribution of the structures that intersect the rock mass (Chapter 4, section 4.4.3). Once the fracture network has been built, *3FLO* can compute the steady or transient flow in the network alone or the network coupled with porous media. Flow problems can be solved in models with permeability contrasts of 10^7.

As with any model application, to provide reliable solutions that can be used as the basis for improved slope design and steeper wall angles, it will be important to have adequate data to further understand actual field conditions in the slope as a result of movement and deformation. Joint instrumentation of active slopes with extensometers, TDRs and vibrating wire piezometers in the same holes would be the first step in helping to build an empirical understanding on which to calibrate a coupled model.

7 GEOTECHNICAL MODEL

Alan Guest and John Read

7.1 Introduction

The introduction to Chapter 2 of this book noted that the geotechnical model is the cornerstone of open pit design – it must be in place before the steps of setting up the geotechnical domains, allocating the design sectors and preparing the slope designs can commence. Chapters 3, 4, 5 and 6 outlined the procedures that should be followed when preparing each of the geotechnical model's four components – the geological, structural, rock mass and hydrogeological models. Chapter 7 outlines the iterative processes used to bring these components into the geotechnical model so that geotechnical domains and design sectors can be fixed and employed in the slope design process (Figure 7.1).

Standard procedures for linking each component and constructing the model are outlined in section 7.2. Different approaches to how the data held in the model are processed and made ready for use in the design analyses are discussed in section 7.3. The aim of section 7.3 is to highlight and provide guidance on the slope design issues for which clarification is frequently sought, including scale, the merits of the differing rock mass classification systems and the issues associated with deriving and applying the generalized Hoek-Brown strength criterion in open pit slope designs, and pore pressure considerations.

7.2 Constructing the geotechnical model

7.2.1 Required output

The information required by each component of the geotechnical model is summarized in Figure 7.2. When brought together, the information from these components should provide the following representative design values for each geotechnical domain and design sector:

- material type(s), including alteration variants (type and/or degree);
- orientation, spatial distribution and shear strength values for the major structures, including the shear strength of the individual faults, bedding planes and any laminated structures associated with metamorphic rocks such as slate, phyllite and schist that are continuous along strike and down dip within each domain;
- orientation, spatial distribution and shear strength values for the rock fabric within each domain, including the strength of micro-bedding, minor faults, joints, schistosity and cleavage;
- rock mass strength values, including the point load (Is_{50}), uniaxial and triaxial strength test values for the intact rock, the rock mass classification information and the estimated shear strength values of the rock mass within each domain. If laminated features such as bedding or foliation have imposed a recognizable anisotropy to the rock mass, the strength of the rock mass along and across these anisotropic features must be evaluated;
- elastic moduli values for the rock mass in each domain, for use in the numerical slope stability analyses;
- pore water pressure data derived from regional, mine and pit slope scale groundwater flow models that have been calibrated with pore pressures observed in vertically discretized slope piezometers during mining. The purpose of the calibrated models is to predict the pore pressure distributions in each domain for input into the slope stability analyses and estimates of the need for artificial depressurization of the slopes.

Single parameter values should be retained for deterministic analyses purposes, with discrete and/or continuous distributions (see Appendix 2) retained for probabilistic analyses.

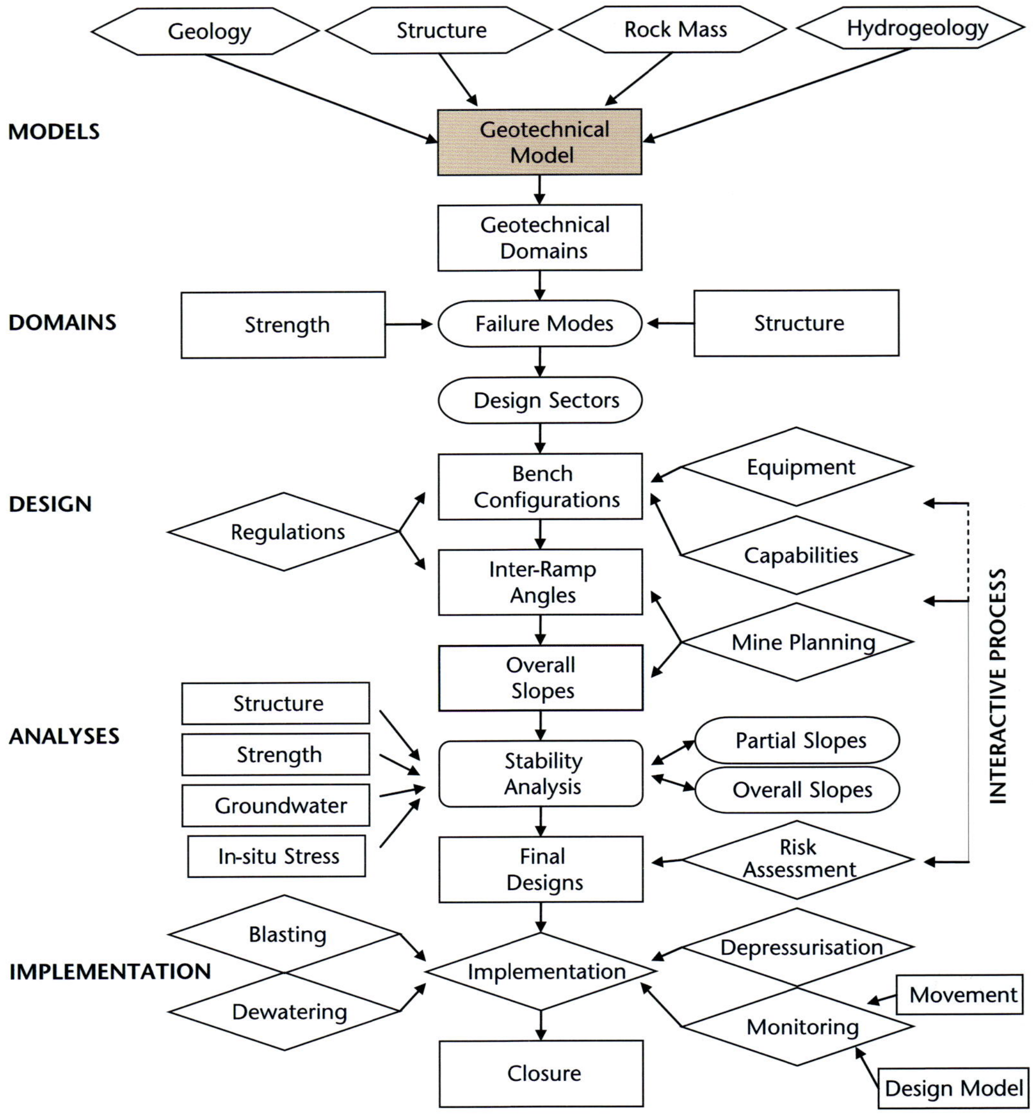

Figure 7.1: Slope design process

7.2.2 Model development

The construction of the geotechnical model is an evolving process through the various development levels of an open pit mine. In many projects sufficient data to compile a detailed model would only be available at the feasibility or construction stages (Levels 3 and 4, Table 1.2). At earlier stages, such as scoping and pre-feasibility (Levels 1 and 2, Table 1.2), a geotechnical model containing much less detail may only be possible.

Early stage models would still need to address the four main components of geology, structures, rock mass characteristics and hydrogeology, but may achieve this through estimation of geotechnical domains for which only general characteristics may be available. In this case, reliance is often placed on distributing parameters by means of the geological model, such as a combination of lithology and alteration.

7.2.3 Building the model

Building the geotechnical model is a step-by-step process of bringing successive layers of individual or combinations of individual data sets into a 3D solid model using a modelling systems such as Vulcan™, DataMine™, MineSite™ or Surpac™. To illustrate the process, a simple but typical example is outlined below.

The geological model describes the regional and mine site geology and is fundamental to the slope design process. Hence, the starting-point in any geotechnical model is an overlay that shows the rock type boundaries.

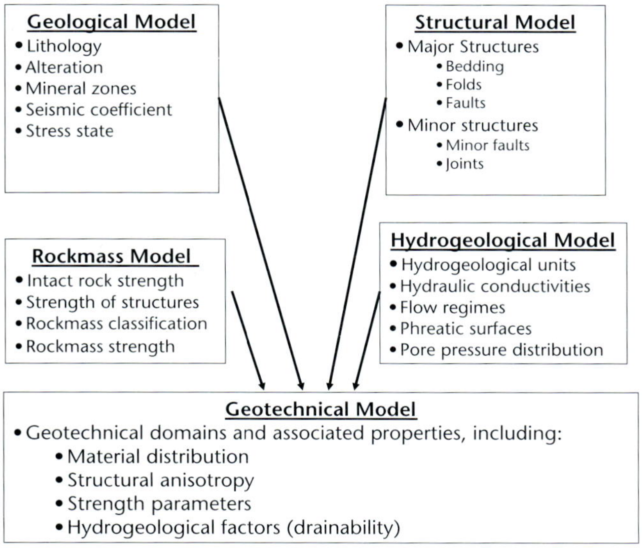

Figure 7.2: Component information and output from the geotechnical model

This is illustrated in Figure 7.3, which represents Layer 1 in the model, viewed for simplicity in 2D. In the figure, country rock (Unit A) is intruded by Units B and C and all three are cut by a series of dykes, represented by Unit D.

Additionally, a weak alteration zone has been mapped that associates with the dykes that form Unit D. The alteration has lowered the unconfined compressive strength of the intact country rock and the zone has been brought into the model as Layer 2 (see Figure 7.4).

The third layer is drawn from the structural model and is represented in Figure 7.5. In the example, which overlies the rock type boundaries presented in Layer 1, four mapped major faults are aligned with the dykes of Unit D.

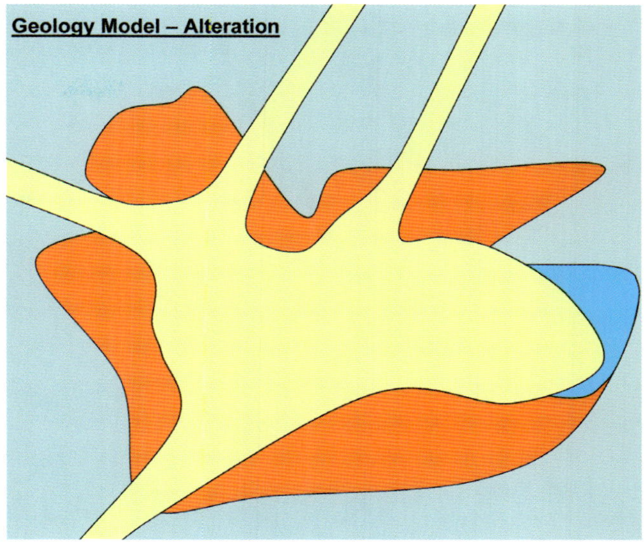

Figure 7.4: Layer 2, alteration zone

These faults form the boundaries to five structural domains, each of which has a distinctively different structural fabric, represented by the five stereonets in Figure 7.5.

With the relevant geological and structural model data accounted for, the next step is including the required data sets from the rock mass model. Separate layers are created for the strength of the intact rock (Layer 4), fracture frequency (Layer 5) and joint condition (Layer 6). Layer 4 is illustrated in Figure 7.6, which shows the effect of the alteration zone introduced by the dykes (Layer 2, Figure 7.4) on the uniaxial compressive strength of the country rock (Unit A) and the two intrusive stocks (Units B and C). Layer 5 is illustrated in Figure 7.7, which plots the available fracture frequency data against the background

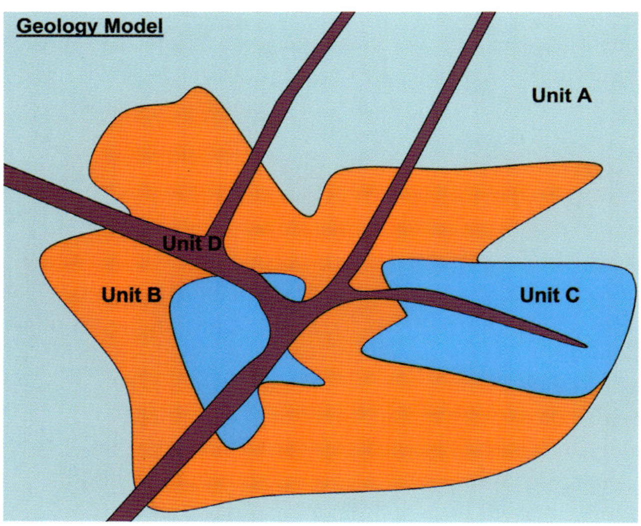

Figure 7.3: Layer 1, rock type boundaries

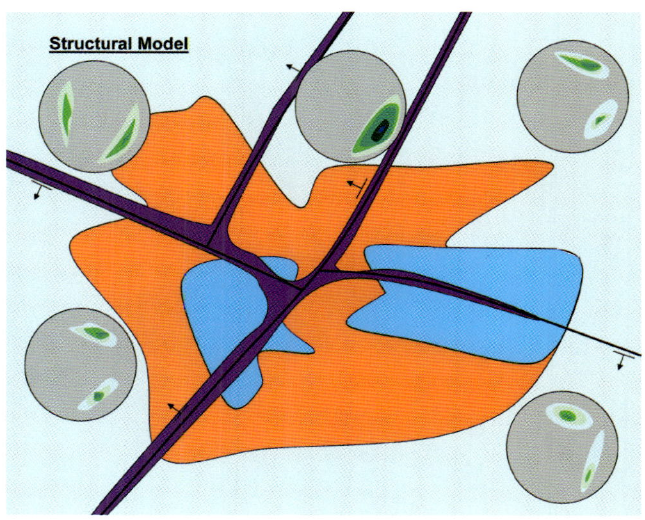

Figure 7.5: Layer 3, structural data

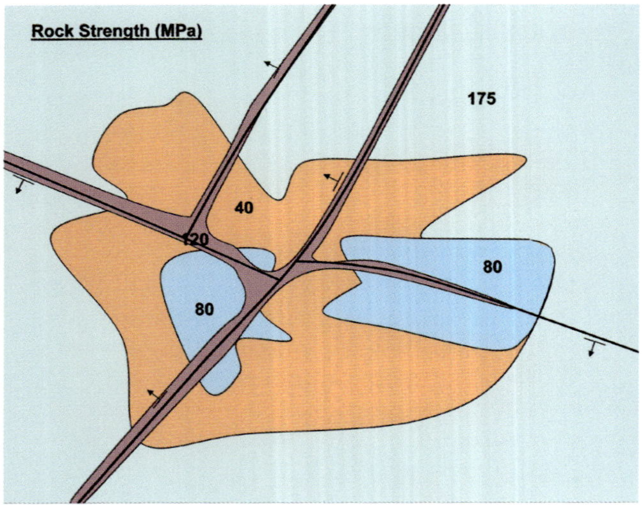

Figure 7.6: Layer 4, intact rock strength

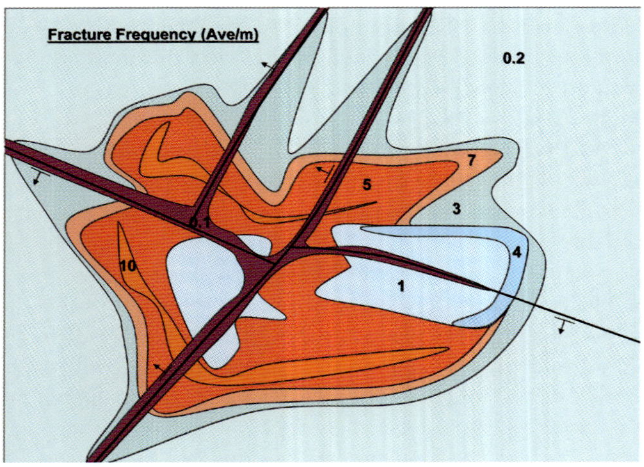

Figure 7.7: Layer 5, fracture frequency

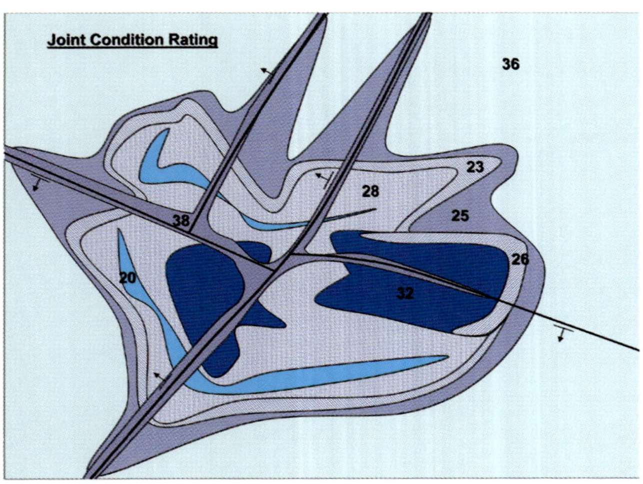

Figure 7.8: Layer 6, joint condition

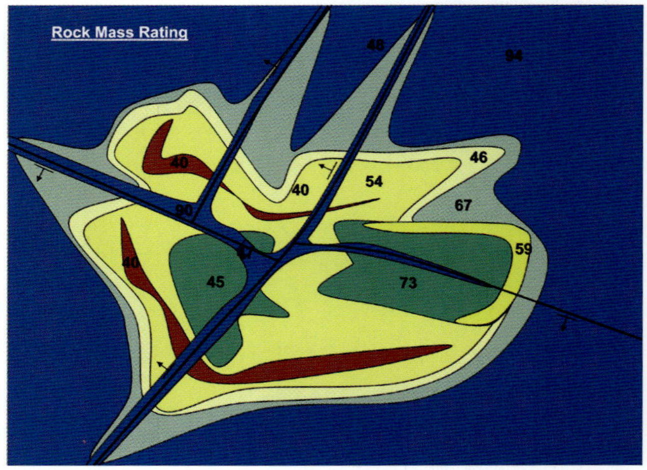

Figure 7.9: Layer 7, rock mass rating

of the unconfined compressive strength zones of Layer 4. Layer 6 is illustrated in Figure 7.8 and is constructed against the background of the fracture frequency data plotted in Layer 5.

With Layers 4, 5 and 6 completed, it is possible to merge the intact rock strength, fracture frequency and joint condition data in a composite rock mass rating layer (Layer 7), as represented in Figure 7.9.

The final step in the process is to bring into the model the information contained in the hydrogeological model. A simple example is illustrated in Figure 7.10, which presents a layer (Layer 8) of six units based on the hydraulic conductivity of the different fresh and altered rock types and the major faults

The geotechnical model is completed by bringing the individual units together, as illustrated in the upper left quadrant of Figure 7.11. For clarity, the individual geotechnical units are usually numbered, as shown in the remainder of the figure. The upper right and lower left quadrants have similar units, due to the similarity of the structural domains in the two quadrants. The lower right quadrant is different because of the influence of a different structural domain (Figure 7.4).

It is stressed that the example illustrated by Figures 7.3–7.11 is a simplified 2D explanation of the construction of the 3D model. As such, it is not a rigid guideline for constructing and bringing the layers of the geotechnical model together. It is stressed that no two sites will be the same – differing data sets and levels of complexity levels will be encountered and should be allowed for. A key element is not to overload the system with superfluous data that will not be required in the stability analyses, so critical evaluation of the model is important. Where possible, the geotechnical domains should be simplified before implementation into the analysis for pit slope design.

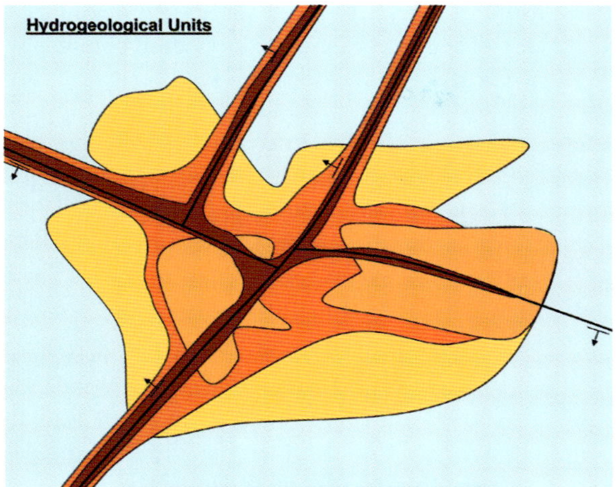

Hydrogeological Unit	Hydraulic Conductivity K (cm/sec)	Colour Code
Unit A	2×10^{-6}	
Unit B	1×10^{-5}	
Unit C	1×10^{-4}	
Unit D	3×10^{-6}	
Alteration	5×10^{-7}	
Faults	2×10^{-2}	

Figure 7.10: Layer 8, hydrogeological units

7.2.4 Block modelling approach

Fitting geotechnical parameters such as UCS, RMR and RQD to 3D block models is often suggested as an alternative means of bringing geotechnical information into the stability analyses. This may be achieved by simply overlaying the data from the geotechnical model on the geological model in a deterministic manner; Figure 7.12 is a generic example of this approach. However, care is required with this approach since, in situations where the information is scattered or widely dispersed, there is a

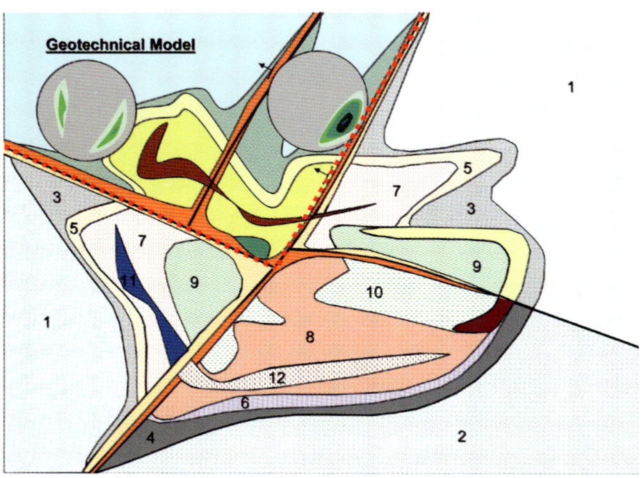

Figure 7.11: Completed geotechnical model

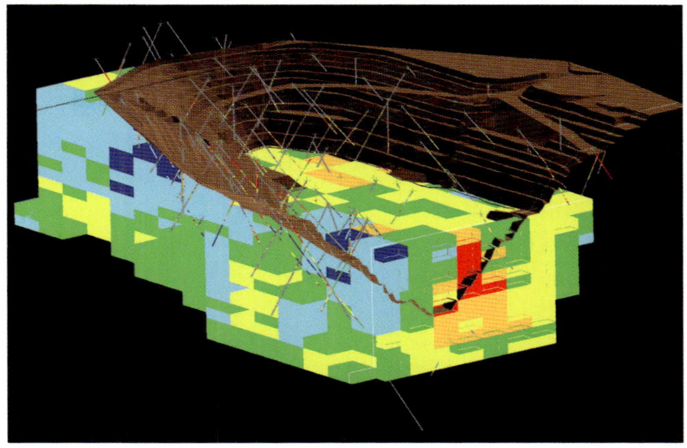

Figure 7.12: Block model of geotechnical parameters
Source: Courtesy BHP Billiton, Nickel West

temptation to average or smear the results over the area of interest, which can result in highly misleading output. For example, the average of very good quality rock containing a number of through-going zones of weakness is likely to average out to a good quality rockmass, which is not at all representative of the actual conditions. The emphasis should be on creating geotechnical domains that accurately describe parameters of significance, within which each of these parameters are consistent, followed by describing the variability of these values within each of the chosen areas.

Attempts have also been made to take a more statistical approach by Kriging the various geotechnical parameters. Figure 7.13 provides an actual example of

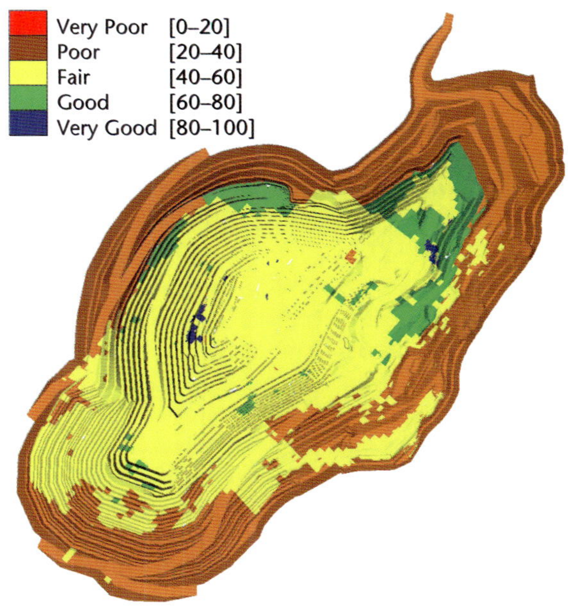

Figure 7.13: Kriged RMR values
Source: Courtesy AngloGold Ashanti

kriged RMR values draped on the pit slopes at a mine in Western Australia.

Although popular with the statistically minded, caution needs to be exercised with this approach. The kriging technique and variograms are important for ore body block modelling, where the grade information is usually orderly and closely spaced. Statistically, the process is less suited to geotechnical applications, where the information is often scattered and/or widely dispersed. It should be remembered that an RMR value is calculated from a range of overlapping data sets, some with well defined domains and others with poorly defined variability. Therefore, kriging of these data may not produce a meaningful result.

Once the geotechnical domains have been described in three dimensions, it is often useful to load this information into a block model as a means of better utilising the geotechnical information within the designing process, an example of which is that given in Figure 7.12.

7.3 Applying the geotechnical model

Building a geotechnical model is one matter, but applying the information it contains to the slope design is another, and there are always questions. The most frequently asked questions invariably concern:

- the scale or relationship between the size of the slope being analysed and the strength of the rock mass and its defects;
- which rock mass classification system should be used and why;
- how the generalized Hoek-Brown strength criterion should be used in open pit slope designs.

These three questions and the need to develop good mine scale groundwater flow and pore pressure distribution models are addressed below.

7.3.1 Scale effects

The issue of size must be addressed when assessing the shear strength of the defects that cut through the rock mass and the shear strength of the rock mass itself. Rock mass strengths determined by conventional means are based on a range of small scale laboratory tests and combined with medium scale field measurements and point estimates. Therefore, potential scale effects must always be a consideration in deciding how appropriate it might be to use a value for determination of large scale strength. As an example, consider the difficulty in determining an appropriate intact rock UCS value when laboratory test results exhibit considerable variability and a comparison with core logging suggests that it was only possible to test the stronger more competent sections. These issues will be addressed in more detail below.

7.3.1.1 Defects

The effects of scale on the shear strength of the defects that cut through the rock mass are outlined in Chapter 5, section 5.3. As noted there, hard data on the topic are limited. However, there are some important points that should be re-emphasised.

First, experience has shown that:

- at low confinement and at scales of 10–30 m (i.e. bench scale), the peak shear strength of clean structures with sound hard rock walls is defined by nil to very low values of cohesion and friction angles in the range of 35–55°, depending on the roughness of the natural fractures;
- at low confinement and scales of 25–50 m (i.e. multibench scale), sealed structures with no clayey fillings have typical peak strengths characterized by cohesions ranging from 50–150 kPa and friction angles of 25–35°;
- at low confinement and scales of 50–200 m (i.e. inter-ramp scale), structures with 10+ mm thick clayey fillings have typical peak strengths characterized by cohesions ranging from 0–75 kPa and friction angles of 18–25°.

Second, when reliable laboratory and/or field back-analysis data are not available, the usual fallback is the Barton–Bandis criterion (section 5.3.2.5 and Equation 5.30). To take scale effects into account, Barton and Bandis (1982) suggested empirical relationships (Equations 5.34 and 5.35) to reduce the values of *JRC* and *JCS*. These relationships and the Barton-Bandis criterion itself must be used with caution. Specifically, it must always be remembered that the criterion was established only for defects of geological origin, meaning defects formed as a consequence of brittle failure (Barton 1971, 1973). Defects are excluded from the criterion if they were modified by processes such as the passage of mineralizing solutions, which left behind a variety of infillings ranging from soft to weak to hard and strong such as clay, talc, gypsum, pyrite and quartz on the defect faces, or by tectonic events, for example faulting and plastic deformation such as foliation, slaty cleavage and gneissosity. Although the criterion has the advantage of explicitly including the effects of surface roughness through the parameter *JRC* and the magnitude of the normal stress through the ratio *(JCS/sn)*, the net effect of the exclusions make it difficult to apply the Barton-Bandis criterion to many of the geological environments found in pit slope engineering.

The limitations of the Barton-Bandis criterion set up a preference for direct shear testing of field samples.

Figure 7.14: Laboratory test sample compared to field scale situation

However, obtaining good, representative samples is always difficult. This issue, combined with the difficulty of performing laboratory tests that do not overestimate the shear strength of defects, especially the cohesion, leads to a bottom line that encourages the sharing and application of experience gained in operating mines.

7.3.1.2 Rock mass

The terms 'intact rock', 'rock mass' and 'scale effects' are widely used in rock slope engineering to describe the fact that mechanical properties are measured by laboratory testing of small rock specimens, and these properties should be scaled to a field scale in order to include the effect of defects such as joints and other geological structures contained in the rock mass. Given the standard wisdom that the specimen diameter should be at least 10 times the size of the largest grain, the situation becomes a little unreal. This is clearly illustrated in Figure 7.14, which shows a standard 50 mm diameter core sample with micro-defects compared to a blockily jointed rock mass at bench scale (it also shows a gentleman who, although wearing a hard hat, is standing in a potentially hazardous location, a situation that would not be allowed in many mines today).

The heart of the problem is that geological structures have different sizes, and the ones to be included in the rock mass will depend on the height of the slope and the volume considered. For example, joints could be included as an integral part of the fabric of the rockmass bridges when analysing the stability of an overall slope, but considered explicitly as discontinuous structures for bench stability analyses. Therefore, the blockiness of the rock mass depends on the relative size of its blocks compared to the size of the slope being analysed. The same rock mass could behave as very blocky for an overall slope, blocky for an inter-ramp scale and almost massive at bench scale (see Figure 7.15).

Until now, when considering pit slope design the accepted solution is usually to consider joints explicitly as discontinuous structures for bench and inter-ramp scale analyses and as part of the fabric of the rock mass at the

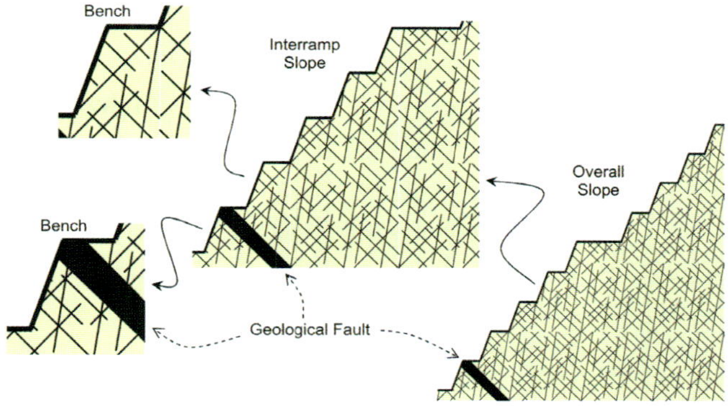

Figure 7.15: The blockiness of the rock mass depends on the volume considered

overall slope scale (section 10.1). Usually, the strength of the rock mass is described by the Hoek-Brown criterion (section 5.3.3). The practitioner decides on its applicability according to the perceived scale and degree of anisotropy of the rock mass, based on the criteria represented in the well-known Hoek-Brown diagram (Figure 7.16).

However, the Hoek-Brown criterion does not solve the scale effect. Sjöberg (1999) highlighted the importance of scale in the analyses (Figure 7.17) but until now a usable scaling function has remained elusive. However, a breakthrough has been achieved. Studies now in progress have shown that the synthetic rock mass model (section 5.5.6) can provide a strength envelope that honours the strength of the intact material and the joint fabric at different scales. Initially, PFC2D biaxial tests were performed on simulated 20 m, 50 m and

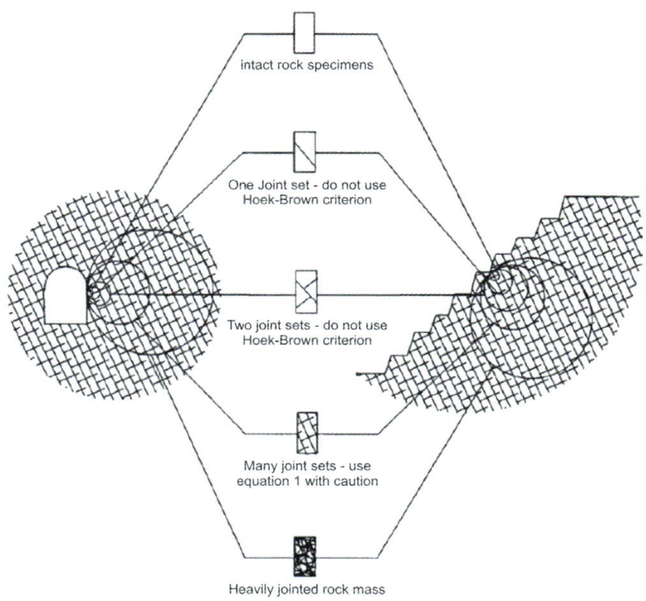

Figure 7.16: Transition from intact to heavily jointed rock mass with increasing sample size
Source: Hoek & Brown (1997)

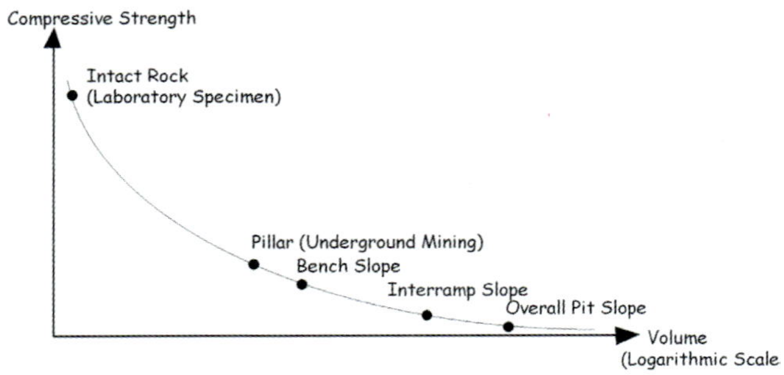

Figure 7.17: Scale effect of rock mass strength
Source: Sjöberg (1999)

100 m diameter SRM samples (Figure 7.18), which provided the distinctly different stress-strain curves shown in Figure 7.19.

The promising results from these initial tests were carried forward into a series of 3D tests on different sized samples using intact rock and structural information from different LOP project sponsor mine sites. Figure 7.20 shows a set of scaled carbonatite test samples from Palabora. The intact strength of the carbonatite was obtained from routine laboratory testing and the structural fabric from underground and surface mapping. Simulated laboratory scale and 20 m, 40 m and 80 m cubes of carbonatite were tested. The results (Figure 7.21) all show a distinct size effect whereby the smaller samples are stronger and stiffer than the larger samples, which reflects the conceptual relationship shown in Figure 7.17. The scaled SRM results were also compared with Hoek-Brown carbonatite strengths derived from GSI estimates. The results (Figure 7.22) show that the Hoek-Brown strengths may either be higher or lower than the strengths observed for large samples, depending on the GSI value used in the estimate. Although there was variability, these relationships repeated themselves in all of the different rock types tested that have been tested so far.

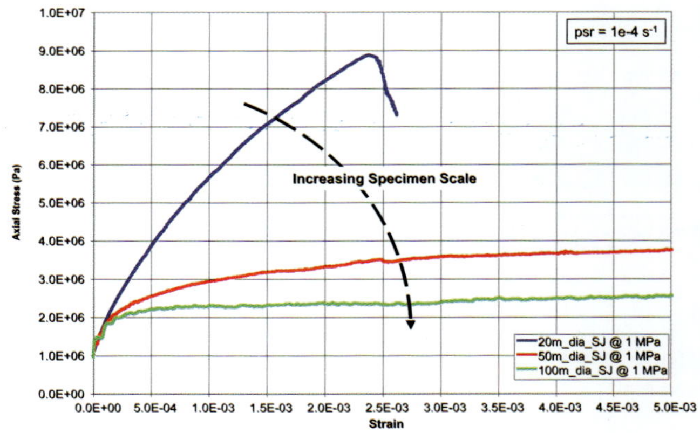

Figure 7.19: Results of tests performed on 2D biaxial test samples shown in Figure 7.18
Source: Courtesy Itasca Consulting Group Inc.

Equally important outcomes of the tests were studies of the effect of different loading directions on the samples. As shown in Figures 7.23 and 7.24, different loading directions generated different stress/strain responses, which reflected the differing orientations of the joint fabric in the sample to the loading direction. In each figure x = East–West, y = North–South, and z = Vertical.

The conclusion drawn from these tests is that the SRM approach is able to supply information that is missing from empirical strength estimates. It uses all of the information that is available from the field and offers different ways (of increasing complexity) of using the results from the SRM element tests in stability analyses.

1. The effect of scale (e.g. small slope versus large slope) can be introduced into simple limit equilibrium analyses by using the SRM derived GSI/Hoek-Brown

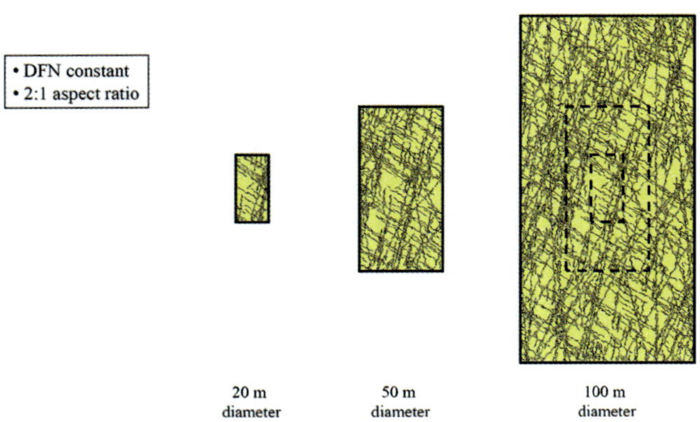

Figure 7.18: Different sized PFC2D biaxial test samples
Source: Courtesy Itasca Consulting Group Inc.

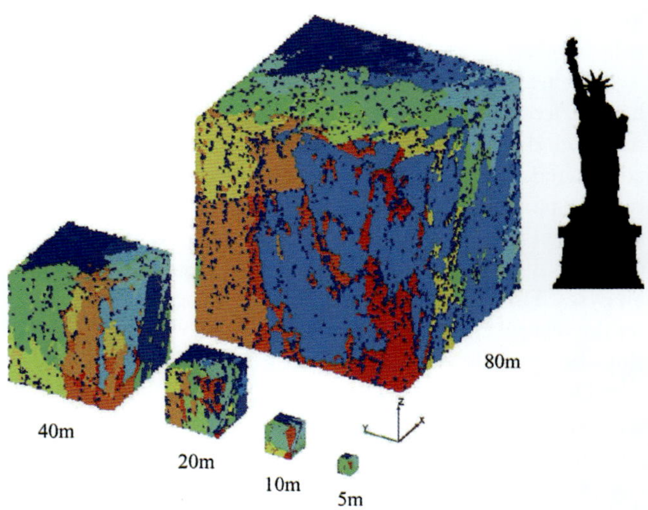

Figure 7.20: Different sized cube test samples of carbonatite
Source: Courtesy Itasca Consulting Group Inc.

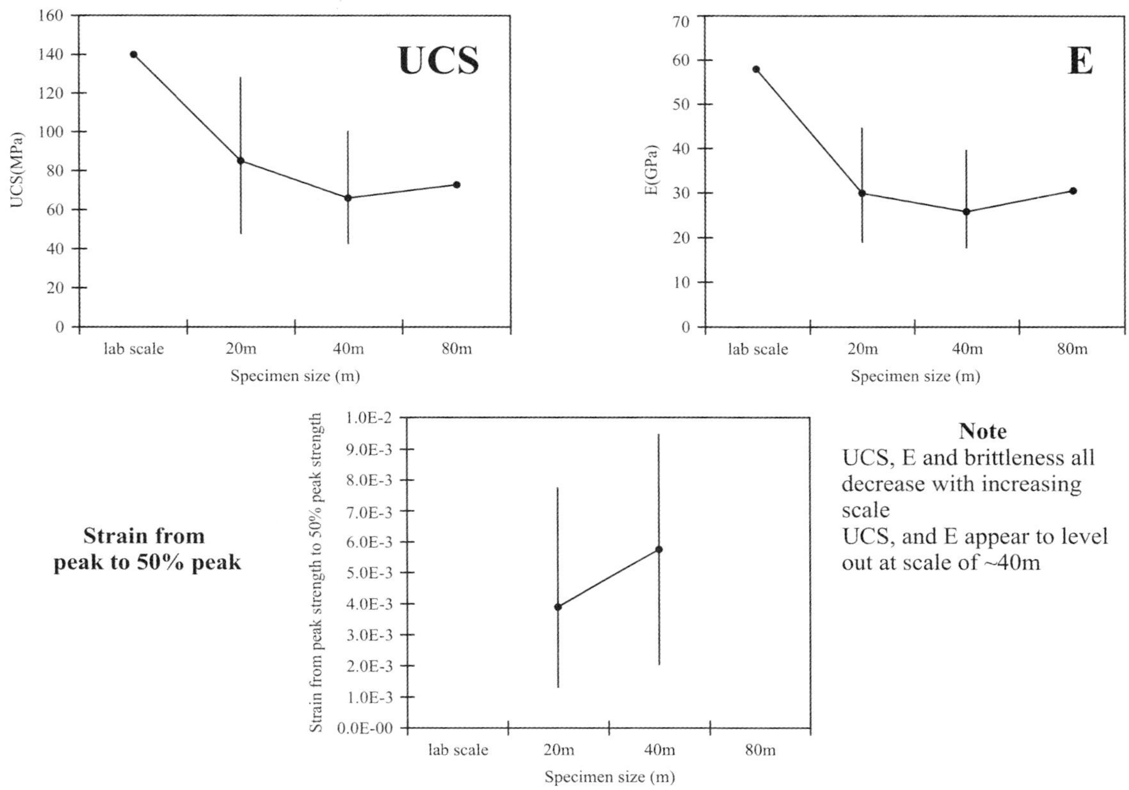

Figure 7.21: Carbonatite test results, showing diminishing strength and E values with increasing sample size
Source: Courtesy Itasca Consulting Group Inc.

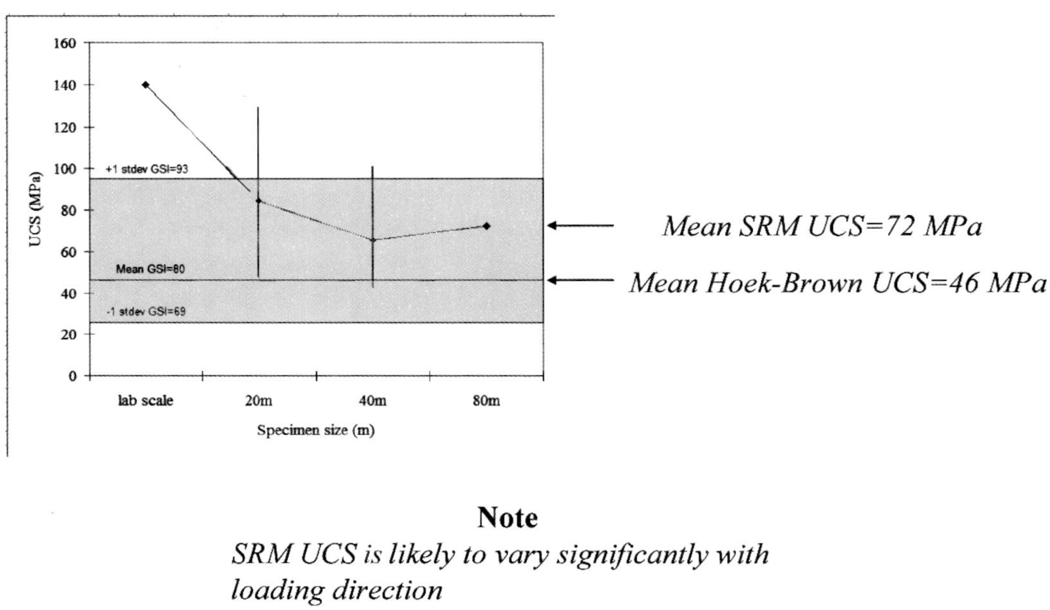

Figure 7.22: Comparison of SRM and Hoek-Brown carbonatite strength values
Source: Courtesy Itasca Consulting Group Inc.

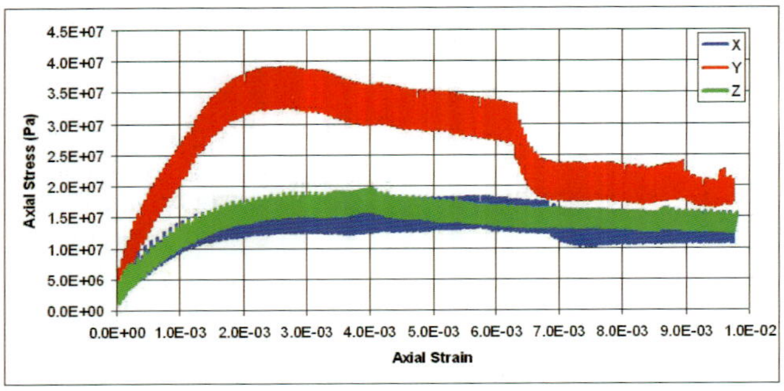

Figure 7.23: Carbonatite 40 m x 80 m sample, UCS stress-strain response
Source: Courtesy Itasca Consulting Group, Inc.

values in the analyses instead of the empirically derived values.

2. The SRM results can be used in 'standard' numerical analyses (e.g. finite element and DEM analyses) by exporting SRM derived strength envelopes, deformation moduli, softening rates and the effects of anisotropy into the models.
3. The SRM approach can be used to assess the inherent variability of rock mass properties by generating and testing different samples of the same unit. This variability can then be introduced into more advanced numerical analysis tools that have recently emerged (Jefferies et al. 2008)
4. The SRM approach can be used directly in a full slope simulation, although considerable computing capacity is required at this time to perform the simulations in 3D. Steps now being taken to improve the resolution and speed of full 3D simulations are outlined in section 10.3.4.5.

7.3.2 Classification systems

As noted in section 5.4.1, in open pit mining the most used classification schemes are the Bieniawski RMR model, the Laubscher IRMR and MRMR models and the Hoek-Brown GSI model, with the MRMR and GSI models frequently

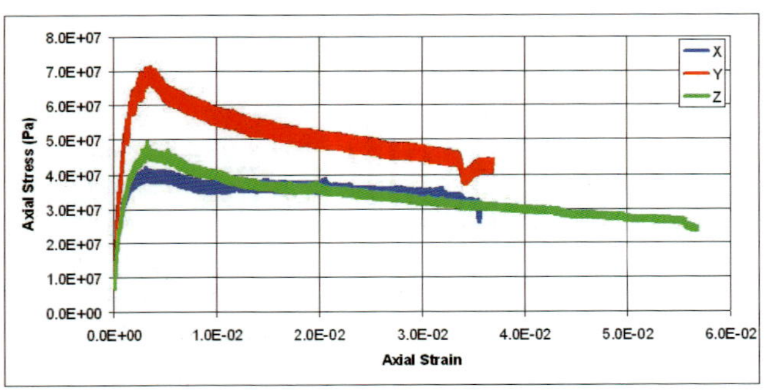

Figure 7.24: Carbonatite 40 m x 80 m sample, triaxial stress-strain response
Source: Courtesy Itasca Consulting Group, Inc.

used interchangeably when estimating rock mass strength using the Hoek-Brown criterion.

To avoid misuse or misapplication of the Laubscher MRMR and Hoek-Brown GSI models in the Hoek-Brown criterion, basic differences between the models must be understood by users. Two points are made.

1 The Hoek-Brown GSI model is an RQD-based model that originates from Bieniawski's 1976 RMR classification scheme. The GSI values in Table 5.34 greater than 25 are exactly the same as those of the Bieniawski RMR_{1976} scheme. When using the model, the following procedures must be adopted:
   - → surface mapping – the GSI values must be obtained from Table 5.34;
   - → drill hole logging – the GSI values must be obtained via Bieniawski RMR_{1976}. If Bieniawski RMR_{1979} is used, the GSI value is (RMR_{1979} – 5);
   - → the rockmass should be considered to be drained.

2 The Laubscher MRMR model is a fracture frequency-based model. This is because it was developed primarily for underground applications, which is why it also contains an adjustment factor for mining-induced stresses. It also contains adjustment factors for weathering, the orientation of structures, blasting and water (Figure 5.33). If the MRMR values will be used instead of the GSI values in the Hoek-Brown rock mass strength calculations, it is suggested that:
   - → because the values will be used in an open pit not an underground environment, adjustments should not be made for mining stress;
   - → adjustment should not be made for water – as with GSI, any pore pressures in the rock mass water should be accounted for in the stability analyses;
   - → no attempt should be made to convert the fracture frequency values to RQD values in order to switch from MRMR to GSI, or vice versa. Priest and Hudson (1979) and Bieniawski (1989) suggested conversion factors, but their use is not recommended because of the directional bias associated with RQD and the empiricism of the suggested correlations. Both of these can introduce even more errors and uncertainties into procedures that are already empirical and likely to contain high levels of uncertainty (Chapter 8, section 8.5.1);
   - → if an adjustment has been made for blasting, emphasize that this has been done to avoid double-counting when dealing with the disturbance factor *D* in the Hoek-Brown strength criterion.

7.3.3 Hoek-Brown rock mass strength criterion

Although it seems likely that the SRM model will provide a strength envelope that honours the strength of the intact material and the joint fabric in a rock mass at different scales, until it has been fully tested and verified by experience it is likely that the Hoek-Brown criterion will remain the strength criterion of choice.

When using the Hoek-Brown criterion, users must understand its origins and that it is an empirical not a constitutive relationship for an ostensibly homogenous and isotropic rock mass (Figure 7.16). They must take account of the origins of the values they are using. Two elements are particularly important in this regard.

1 Users must check the veracity of the σ_c, m_i and GSI values they are using. The questions that must be asked are:
   - → whether the σ_c values properly represent the uniaxial compressive strength of the intact rock;
   - → whether the m_i values were obtained from laboratory triaxial tests on samples of intact rock or whether they were indicative values drawn from supplementary tables (e.g. Table 5.24);
   - → whether the *GSI/MRMR* values were derived from field mapping or drill hole logs, or a combination of both.

   Unless these questions can be answered in full, it is extremely difficult for anyone to assess the reliability of the data relative to the target levels of data confidence that are expressed in Table 8.1.

2 Users must understand the implications of the disturbance factor, *D*, in their deliberations. They must check to see whether the GSI values originate from MRMR values and, if so, whether they have been adjusted for blasting. Further, they must have a clear understanding of the likely depth of the blast-affected zone in the pit walls. As noted in section 5.5.3, the influence of the parameter can be large and its application requires experience and judgment. The largest value of D ($D = 1$) effectively reduces the cohesion of the rock mass by a factor of 2, which is a particularly severe reduction (or punishment) of the rock mass strength.

7.3.4 Pore pressure considerations

Pore pressures control the effective stress of the rock mass in the pit walls. Acting within the jointed rock mass, increased pore pressures reduce the effective stress which, in turn, leads to a reduction in the shear strength of the rock mass (section 6.1.2.2, equation 6.1). Hence, the fundamental assumption underlying all stability analyses in jointed rock slopes with water present is that the effective stress principle applies at all scales of analysis, from large-scale overall slopes to inter-ramp slopes and benches. It is also recognised that the pore pressures within the slope are usually the only element of a slope design that can readily be modified by artificial intervention.

The methods used to incorporate pore pressures in limiting equilibrium and numerical slope design analyses are detailed in Chapter 10 (sections 10.3.3.2 and 10.3.4.3). Depending on experience and available analytical tools, practitioners will, however, follow a variety of approaches in setting up these analyses, ranging from the simplest to the most complex. In most cases, they will fall into one of three groups: a 'dry' slope approach, a 'wet' slope approach, and a 'hybrid' approach.

- **'Dry' Slope Approach.** In this approach the slopes are assumed to be 'dry', although what 'dry' means is not always well defined. It may mean that water flow and seepage may appear on the slope face, as long as no significant pressures develop, it may mean that no water pressure will be present between the slope face and any candidate failure surface, or it may mean that no water should appear on the face of the slope. In all cases, the 'dry' slope requirement shifts the responsibility to the hydrogeologists, who must provide depressurisation measures (e.g. wells and/or horizontal drains) to ensure a 'dry' slope. It must also be confirmed that the 'dry' (depressurised) condition can be achieved in the time available and then maintained.
- **'Wet' Slope Approach.** This is probably the most common approach. It assumes that the rock mass below the phreatic surface is fully saturated and that pore pressures act on all fractures regardless of their scale and/or connectivity. Essentially, the slope is assumed to be a gravel. However, in almost all cases the jointed rock mass is represented as an equivalent continuum. Flow analyses usually are performed to determine the steady-state distribution of the pore pressures in the slope. The resultant pressure distributions ignore possible pore pressure reductions due to mining-induced slope deformation (lithostatic unloading and relaxation, section 6.2.5). From the point of view of stability analysis, the concept of a phreatic surface is not useful in a jointed rock mass with poor connectivity. There are simply water pressures and water content, which may exist in separate regions. A continuous boundary between saturated and unsaturated parts of the slope may not exist.
- **'Hybrid' Approach.** This approach attempts to acknowledge the different water pressure regimes that can exist within a slope, which is represented as a system of rock blocks separated by explicit fractures. The explicit fractures typically have pore pressures specified. The rock blocks typically behave as a continuum that implicitly include minor fractures (or fabric) and may or may not specify pore pressures. The inherent assumption is that explicit fractures often have high permeabilities and connectivity such that the pore pressure within them is not affected by slope deformation. The 'hybrid' approach offers the most

flexibility as different pore pressures can be specified separately in different components of the rock slope.

Theoretically, back-analyses of slope failures should be capable of identifying the 'correct' approach. However, when back analysing failures, all three of the approaches described have met with varying degrees of success. This is mostly because there are uncertainties in both the initial rock mass strength and pore pressure distributions, and there are many combinations of strength and pore pressure that can reproduce slope failures. Hence, no single approach has been or can be accepted universally.

The Large Open Pit (LOP) Project research has shown that the synthetic rock mass model approach (section 5.5.6) may provide a means of significantly reducing uncertainty with respect to the rock mass strength. This means that attention can now be focussed on trying to understand what pore pressures should be applied to the various components of the rock mass that makes up the slope.

The starting point in this process is the erection of good mine and pit slope scale groundwater flow models (sections 6.3 and 6.4). Unfortunately, at a disturbingly large number of mines a good groundwater flow model and an understanding of the distribution of the pore pressures in the rock mass behind the pit walls is a rarity. The usual excuses for the lack of a good groundwater flow model are the lead time required and the capital cost of obtaining the data needed to build the model. It is recognised that cost is an issue. However, the lack of good model to support the slope design will almost certainly result in a conservative design, so early characterisation of the regional and mine scale hydrogeological regime is considered to be of paramount importance.

In Tables 6.2 and 8.1, it is suggested that regional groundwater surveys should be performed during the conceptual (Level 1) project studies and that mine scale airlift, pumping and packer testing to establish initial hydrogeological parameters should at least be commenced during the pre-feasibility (Level 2) stage of the project. And sensible "piggy backing" of the data collection program on mineral exploration and resource drilling programs at this stage of the program (Section 2.5.1.2) can go a long way towards reducing the cost of obtaining the data. By the time the project feasibility (Level 3) studies start, piezometer installation and targeted pumping and airlift testing based on the information collected during the pre-feasibility studies are an absolute requirement.

Other factors which have to be considered when setting up the groundwater flow model include:

- the interconnection between the explicit first order fractures and the less permeable second and third-order fractures and the fabric within the intervening rock blocks (see Figure 6.24), and the effect of these structures on the flow of water through the rockmass. The ability and time taken to remove all the drainable water by gravity and depressurise the rock mass will depend on the permeability and connectivity of these structures;
- the effect of lithostatic unloading (i.e. mining) on pore pressures and groundwater flow, particularly in low permeability rocks. Currently this is not well documented and is not considered in either of the 'wet' slope or 'hybrid' approaches to stability analysis outlined above. Groundwater flow analyses typically ignore the potential role of slope deformation in changing pore pressures within fractures and/or changing the permeability of the fractures. For example, some practitioners have speculated that small-scale fractures (C and D, Figure 6.25) experience volumetric increase during slope unloading such that the pore pressures within them essentially drop to zero. Furthermore, these fractures have low permeability and are connected so poorly that pore pressure is not likely to re-establish in the short term (say, a year or so).

The industry needs guidance on these issues, a need which has been recognised and has been taken up by the LOP Research Project. Further to the research needs outlined in section 6.6, field tests of fracture flow and instrumentation designed to record pore pressure fluctuations and deformation during mining are being instigated at sponsor mine sites in a research program that is designed to achieve the following objectives.

1 Develop an understanding of the flow process in rock masses at different scales, particularly those with poor or limited connectivity.
2 Develop a numerical model that realistically couples fluid flow, pressure distribution and rock deformation.
3 Extend and apply the understanding to an assessment of the effects of pore pressure on the stability of fractured rock slopes.
4 Develop and document a methodology that will allow the industry to assess the effects of groundwater on the stability of their slopes.
5 Validate that methodology against existing conditions at different sites.

The outcomes of this research will be brought into the public domain as it is reported and assessed.

8 DATA UNCERTAINTY

John Read

8.1 Introduction

To this point , the chapters of this book have focused on data collection and preparation of the individual components of the geotechnical model (Figure 8.1). The next step, one of the most important in the slope design process, is to determine and report the uncertainties in the collected data at levels that are commensurate with each stage of project development.

Determining and reporting the uncertainties in each component of the geotechnical model requires an understanding of the causes of data uncertainty, its potential impact on the reliability of the pit slopes, how it is quantified and how it is reported to corporate mine management and the investment community. This chapter will provide a basic framework for each of these topics. Section 8.2 addresses the causes of data uncertainty, section 8.3 examines the impact of data uncertainty and section 8.4 describes the tools that are most frequently used to quantify data uncertainty. Section 8.5, the most important part of the chapter, addresses the pressing need for a geotechnical reporting system that matches the uncertainties in each component of the geotechnical model with each stage of project development. A summary of the essential concepts of probability and statistics is given in Appendix 2.

8.2 Causes of data uncertainty

In open pit mining, data uncertainty stems from the recurrent difficulties geologists, engineering geologists and geotechnical engineers face to correctly predict the inherently variable properties and characteristics of natural materials. There is a voluminous literature dealing with this variability, well beyond the scope of this book. However, the relevant types of uncertainty can be placed into three groups: geological uncertainty, parameter uncertainty and model uncertainty.

1 Geological uncertainty embraces the unpredictability associated with the identification, geometry of and relationships between the different lithologies and structures that constitute the geological and structural models. It encompasses, for example, uncertainties arising from features such as incorrectly delineated lithological boundaries and major faults, as well as unforeseen geological conditions.

2 Parameter uncertainty represents the unpredictability of the parameters used to account for the various attributes of the geotechnical model. Typically, it includes uncertainties associated with the values adopted for rock mass and hydrogeological model parameters such as the friction angle, cohesion, deformation moduli and pore pressures.

3 Model uncertainty accounts for the unpredictability that surrounds the selection process and the different types of analyses used to formulate the slope design and estimate the reliability of the pit walls. Examples include the various two-dimensional methods of limit equilibrium stability analysis and the more recently developed three-dimensional numerical stress and displacement analyses now used in pit slope design. Model uncertainty exists if there is a possibility of obtaining an incorrect result even if exact values are available for all the model parameters.

This chapter is specifically concerned with geological and parameter uncertainty.

8.3 Impact of data uncertainty

Geological and parameter uncertainty lead directly to unreliability and possible poor performance of the pit slopes. An international review of poorly performing mine projects (IPA Inc. 2006) has shown that that four key drivers of underperformance are:

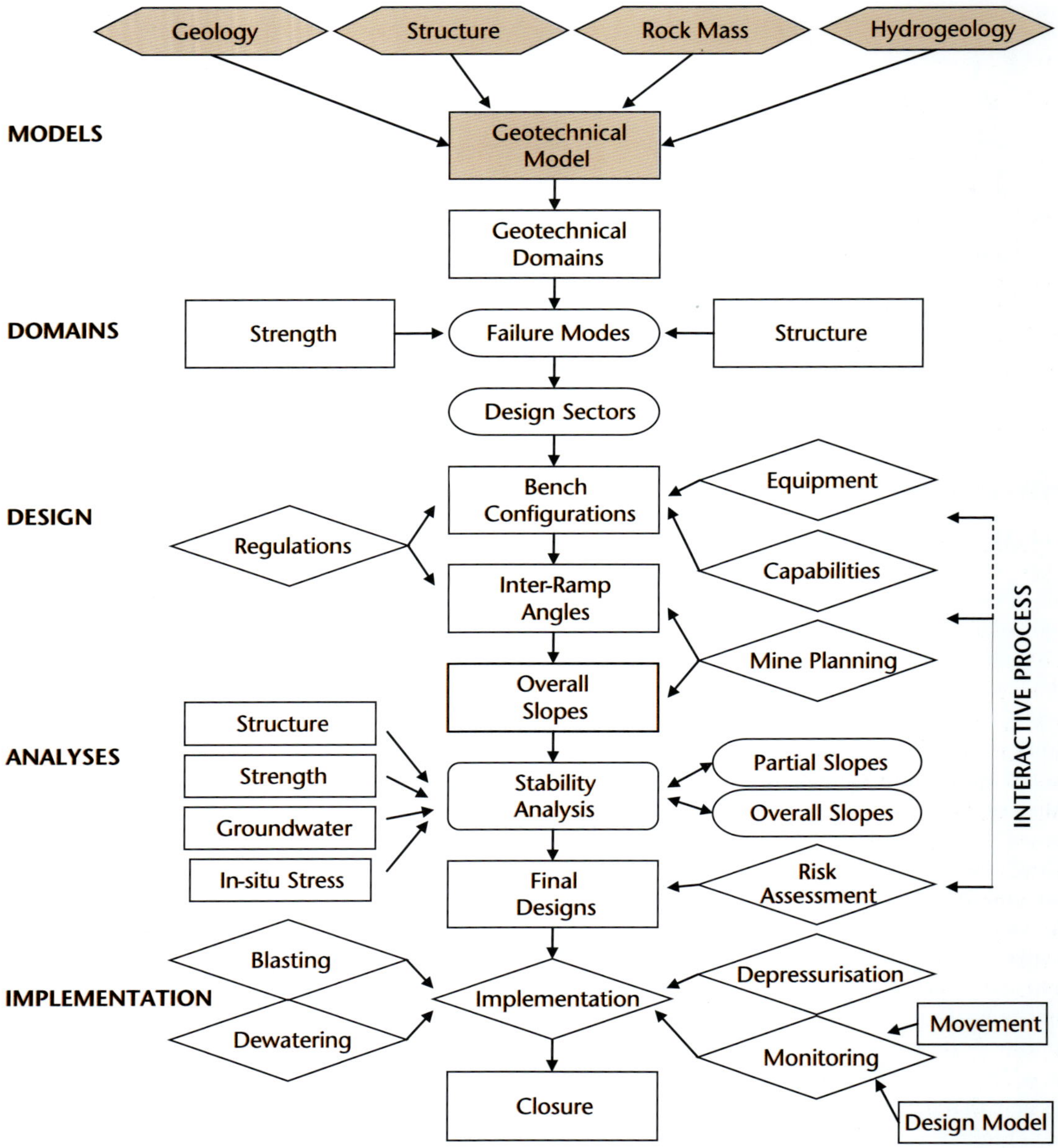

Figure 8.1: Slope design process

1 misunderstanding of grade variability;
2 inadequate metallurgical testing and poor characterisation of ore/waste;
3 inadequate drilling to define orebody and overburden to support interpretation of geological structure, and to support geotechnical and geohydrological interpretation;
4 inadequate drilling to support detailed mine planning, grade control and scheduling.

These findings support considerable anecdotal evidence that a number of large open pit mining projects commenced operating without a complete understanding of the geotechnical model used to develop the slope designs. In effect, the level of certainty in the locations of features such as lithological boundaries and major faults, and the values of geotechnical parameters, has not been commensurate with the needs of a detailed design. All too often, operating level investment decisions have been made using geotechnical data that are more appropriate to a conceptual or pre-feasibility level of investigation. This imbalance has adversely affected the reliability of the slope designs and hence the operational and economic viability of the projects. It also has demonstrated the need to develop standards for the reporting of geotechnical information that are commensurate with each stage of project development (Table 1.2).

8.4 Quantifying data uncertainty

8.4.1 Overview

In our daily lives we cope with uncertainty intuitively by using previous experience to rank and guide our choice. In open pit mining, we evaluate and update the uncertainties in the geological, structural, rock mass and hydrogeological parameters within each geotechnical domain and design sector using relative frequency concepts and probability distributions aided, if necessary, by subjective assessments of how the data was collected.

The boundaries between the geotechnical domains and design sectors, however, are positional, which makes it difficult if not impracticable to derive probability distributions from measured values that reflect their locations. The alternative is to gather boundary data using subjective assessments prepared by competent geologists, engineering geologists and geotechnical engineers, acting individually or as members of a review panel.

In the past, subjective assessment in geotechnical engineering has usually been in the form of that hardy perennial 'engineering judgment' (Read 1994). However, many aspects of the process by which individuals making a judgment accept responsibility for their judgments raise questions of credibility and defensibility. Consequently, as more sophisticated slope design and risk assessment procedures are introduced into open pit mining, more rigorous techniques of quantifying the measure of the confidence in the outcome will need to be adopted by the mining industry.

The next parts of this section outline essential tools that geotechnical professionals can use to help quantify their uncertainty about each part of the geotechnical model. The section focuses on subjective assessment and relative frequency concepts, which can be used to evaluate the reliability of the structural and rock mass parameters within each domain.

8.4.2 Subjective assessment

It is not always possible to estimate representative probability distributions from measured values. Nor is it always possible to precisely model natural phenomena such as progressive slope failure, or explicitly form value judgments on questions such as mine rehabilitation. In these instances we tend to rely on subjective assessment, usually in the form of 'engineering judgment'. Engineering judgment (the 'competent person' or 'expert opinion' approach) may be sufficient for the requirements of conventional projects. However, it is easily challenged, particularly by those who disagree with the outcome. Similarly, consensus may be difficult to achieve, especially if the proceedings are dominated by a strong and possibly biased individual.

Although they have not been widely used in the mining industry, subjective assessment methods have often been used to help overcome these challenges and disagreements in formal assessments of the reliability of underground nuclear waste storage facilities. The best-known methods probably are:

- Bayesian probability;
- calibrated assessment;
- Delphi panels;
- probability encoding.

Bayesian probability (Harr 1996) provides an organised system for using new information to update prior knowledge, indicating how opinions held before an experiment should be modified by the results. It is a good approach when the fundamental mechanism is understood and the data comprise a representative sample of the value being assessed. Geostatistical estimation of ore reserves is one example, evaluating concrete strengths is another. However, the method relies on objectively derived subsidiary probabilities. Thus, a truly subjective Bayesian assessment must still be based on another model of subjective assessment.

The calibrated assessment approach adjusts individual assessments to reflect the assessor's known biases. Thus, two sets of assessment are required: assessments of the values in question, and an assessment of the assessors. The assessors can be assessed by their peers or through a set of questionnaires that quantify their biases with respect to known conditions.

In the Delphi approach the individuals in a defined group of experts are each given the same set of background information and requested to perform assessments in writing. These assessments are provided anonymously to each of the other experts, who are encouraged to adjust their assessments in light of their peer's assessments. The iterations are continued until the results stabilise. In situations where consensus cannot be achieved, the group average may be used.

Probability encoding is similar to the calibrated assessment approach except that an encoding analyst works with each expert to obtain a more accurate assessment instead of simply correcting the expert's assessments based on predetermined calibration factors. The method tacitly assumes that the experts are incompetent in quantitatively assessing their own uncertainty and uses the encoding analyst to bridge the gap. The limitations of the method are that it depends on the credibility of the analysts and there is no mechanism for achieving consensus.

A number of texts outlining the concepts and principles that underpin subjective assessments are available in the public domain, but they are particularly well outlined in *Degrees of belief: subjective probability and engineering judgement* by Steven Vick (2002).

8.4.3 Relative frequency concepts

Statistics and probability concepts are widely used in geotechnical engineering. Typically, the emphasis is on direct measurement and organising the data in a structured manner as a means of examining variability within a range of values, or distinguishing between populations within or across different domains. Using the same concepts to assess levels of confidence in the data is an accepted but less common practice and frequently requires specialist knowledge. Even so, direct measurement to determine probabilities is a standard technique and all geotechnical practitioners should be familiar with the statistical measures of central tendency and scatter, notably the expected value ($E[x]$), the standard deviation ($\sigma[x]$) and the coefficient of variation ($V(x)$). In the absence of more definitive information the coefficient of variation can be used to assess uncertainty and provide reasonable values for the parameters in calculations. The coefficient of variation is defined as:

$$V(x) = \frac{\sigma[x]}{E[x]} \times 100\ (\%) \qquad \text{(eqn 8.1)}$$

Generally, coefficients of about 10% are considered low and values greater than 30% are high. If the expected value of a parameter is unknown, one can be estimated and the uncertainty quantified with an appropriate coefficient of variation.

In addition to the simple concept of the coefficient of variation, geotechnical practitioners should also be aware of the fact that, because the expected value is obtained from the probability distribution function of a random variable, the individual outcomes may have quite different probabilities of occurring. They should also have a working knowledge of cumulative distribution functions, which provide the means of progressively estimating the likelihood that the occurrence of a given phenomenon will equal or exceed a given set of values, and at least the binomial, uniform, normal and lognormal distributions.

To help quantify the level of confidence without the added need for specialist assistance, statistics and probability routines that assess data uncertainty and determine confidence limits for specified data and/or attributes in each part of the geotechnical model have been developed by the LOP project for use within the JointStats data management system (Brown 2003). As outlined in Chapter 2, the original JointStats database accepted standard structural data from face mapping or borehole scanlines, organised the data hierarchically then sorted and statistically analysed the data according to the standard structural attributes of orientation, length, spacing and persistence. These capabilities have now been enhanced to include quantitative measures of rock mass parameters, enabling data uncertainty to be assessed and confidence limits determined for the structural and the rock mass parameters from within any geotechnical domain.

Detailed information about the concepts of uncertainty is available in a substantial number of basic and advanced texts. A well-known introduction to the basic concepts and applications of statistics and probability in engineering is *Reliability-based design in civil engineering* by Milton E. Harr (1996). Useful parts of this text are reproduced in Appendix 2. These parts include the axioms of probability (A2.1), commonly used probability distributions including the binomial (A2.2.1), Poisson (A2.2.2), normal (A2.2.3), uniform (A2.2.4), exponential (A2.2.5), lognormal (A2.2.6) and beta distributions (A2.2.7), information and distributions (A2.3) and confidence limits (A2.4).

8.5 Reporting data uncertainty

8.5.1 Geotechnical reporting system

As outlined in section 8.3, there is a demonstrated need for a system that reports the confidence in the geotechnical information used in slope designs to everyone involved in the project at levels commensurate with each stage of project development. 'Everyone' includes project development staff, mine operators, corporate mine management and the investment community.

It is proposed that the target levels of geotechnical effort outlined in Table 1.2 be matched by target levels of confidence in the data, as outlined in Table 8.1. These levels are subjective, but are intended to provide guidelines to the level of certainty required at each stage of project development. They also provide an indication of the level of expenditure that may be required.

Descriptive guidelines for estimating the level of confidence in the data are outlined below. As noted in Chapter 1, to maintain consistency with the codes already used in different countries for reporting exploration results, mineral resource and ore reserves, similarly to the reporting framework proposed by Haile (2004), the terminology used in the guidelines to describe the different levels of uncertainty is equivalent to the 'inferred', 'indicated' and 'measured' levels of confidence used by JORC (2004) to define the level of confidence in mineral resources and ore reserves, as illustrated in Figure 8.2.

The boundary levels of low, reasonable and high confidence are not explicitly stated by the JORC code. There is, however, anecdotal evidence that confidence levels of ±25% for Indicated Mineral Resources and ±10% to ±15% for Measured Mineral Resources are used by the industry. These boundaries are also consistent with the target levels of data confidence suggested for Levels 2, 3 and 4 in Table 8.1.

Table 8.1: Suggested levels of geotechnical effort and target levels of data confidence by project stage

	PROJECT STAGE				
Project level status	**Conceptual**	**Pre-feasibility**	**Feasibility**	**Design and Construction**	**Operations**
Geotechnical level status	Level 1	Level 2	Level 3	Level 4	Level 5
Geological model	Regional literature; advanced exploration mapping and core logging; database established; initial country rock model	Mine scale outcrop mapping and core logging, enhancement of geological database; initial 3D geological model	Infill drilling and mapping, further enhancement of geological database and 3D model	Targeted drilling and mapping; refinement of geological database and 3D model	Ongoing pit mapping and drilling; further refinement of geological database and 3D model
Structural model (major features)	Aerial photos and initial ground proofing	Mine scale outcrop mapping; targeted oriented drilling; initial structural model	Trench mapping; infill oriented drilling; 3D structural model	Refined interpretation of 3D structural model	Structural mapping on all pit benches; further refinement of 3D model
Structural model (fabric)	Regional outcrop mapping	Mine scale outcrop mapping; targeted oriented drilling; database established initial stereographic assessment of fabric data; initial structural domains established	Infill trench mapping and oriented drilling; enhancement of database; advanced stereographic assessment of fabric data; confirmation of structural domains	Refined interpretation of fabric data and structural domains	Structural mapping on all pit benches; further refinement of fabric data and structural domains
Hydrogeological model	Regional groundwater survey	Mine scale airlift, pumping and packer testing to establish initial hydrogeological parameters; initial hydrogeological database and model established	Targeted pumping and airlift testing; piezometer installation; enhancement of hydrogeological database and 3D model; initial assessment of depressurisation and dewatering requirements	Installation of piezometers and & dewatering wells; refinement of hydrogeological database, 3D model, depressurisation and dewatering requirements	Ongoing management of piezometer and dewatering well network; continued refinement of hydrogeological database and 3D model
Intact rock strength	Literature values supplemented by index tests on core from geological drilling	Index and laboratory testing on samples selected from targeted mine scale drilling; database established; initial assessment of lithological domains	Targeted drilling and detailed sampling and laboratory testing; enhancement of database; detailed assessment and establishment of geotechnical units for 3D geotechnical model	Infill drilling, sampling and laboratory testing; refinement of database and 3D geotechnical model	Ongoing maintenance of database and 3D geotechnical model
Strength of structural defects	Literature values supplemented by index tests on core from geological drilling	Laboratory direct shear tests of saw cut and defect samples selected from targeted mine scale drill holes and outcrops; database established; assessment of defect strength within initial structural domains	Targeted sampling and laboratory testing; enhancement of database; detailed assessment and establishment of defect strengths within structural domains	Selected sampling and laboratory testing and refinement of database	Ongoing maintenance of database
Geotechnical characterisation	Pertinent regional information; geotechnical assessment of advanced exploration data	Assessment and compilation of initial mine scale geotechnical data; preparation of initial geotechnical database and 3D model	Ongoing assessment and compilation of all new mine scale geotechnical data; enhancement of geotechnical database and 3D model	Refinement of geotechnical database and 3D model	Ongoing maintenance of geotechnical database and 3D model
Target levels of data confidence					
Geology	>50%	50–70%	65–85%	80–90%	>90%
Structural	>20%	40–50%	45–70%	60–75%	>75%
Hydrogeological	>20%	30–50%	40–65%	60–75%	>75%
Rock mass	>30%	40–65%	60–75%	70–80%	>80%
Geotechnical	>30%	40–60%	50–75%	65–85%	>80%

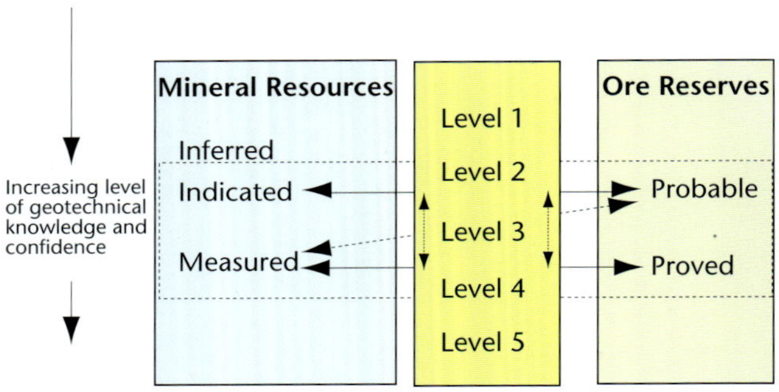

Figure 8.2: Geotechnical levels of confidence relative to the JORC code

8.5.1.1. Conceptual stage (Level 1)

At the conceptual stage it is considered that the reliability of the geotechnical model will have been estimated at a low level of confidence defined as 'Level 1'. The model will have been entirely inferred from existing reports and interpretations based on available regional data from mines in similar geological environments. These preliminary data may be supplemented by aerial photographic interpretations of the regional lithology and structure and any outcrop mapping performed during exploratory project surveys. Overall, the information will be sufficient only to provide indicative slope designs and plan pre-feasibility stage investigations.

At this stage of the project the data assessments have been almost entirely performed subjectively.

8.5.1.2 Pre-feasibility stage (Level 2)

At the pre-feasibility stage it is considered that the reliability of the geotechnical model will have been estimated at a low level of confidence defined as 'Level 2'. The model will have been inferred from interpretations based on the information provided during the conceptual stage of development augmented by data from outcrops, exposures in road cuttings and river banks, trenches, pits, underground workings and oriented drill holes at the proposed mine site. All these data may be limited or variably distributed and/or of uncertain quality. Any sampling, field testing and laboratory testing procedures must be sufficient to satisfy designated international standards for site investigation and laboratory testing (e.g. ISRM, ASTM). The information will be sufficient to form working plans and Level 2 pre-feasibility slope design studies.

At this stage of the project the data assessments have still largely been performed subjectively, but they have been supplemented by quantitative assessments as measurable data became increasingly available.

8.5.1.3 Feasibility stage (Level 3)

At the feasibility level it is considered that the reliability of the geotechnical model will have been estimated at a reasonable level of confidence defined as 'Level 3'. For the chosen option, the interpretations will have been based on the results of the mine site feasibility investigations. Sampling locations will have been spaced closely enough to sustain 3D interpretations of the geotechnical domain boundaries to the limits of mining based on boundary intersections and the continuity of the structural fabric, rock mass properties and hydrogeological parameters within each domain. Some structural analyses will have been performed, utilising estimates of joint frequencies, lengths and conditions. All major features and joint sets should have been identified. Testing (small sample) for the physical properties of the in situ rock and joint surfaces will have been carried out. Similarly, groundwater data will be based on targeted pumping and airlift testing, and piezometer installations. All sampling, field testing and laboratory testing procedures must be sufficient to satisfy designated international standards for site investigation and laboratory testing (e.g. ISRM, ASTM). At the completion of the investigations variations may occur and alternative interpretations may be possible, but in the view of a competent person these would be unlikely to affect the potential economic viability of the project.

At Level 3, project features such as structural and lithological domain boundaries, especially those at depth, have mostly been assessed subjectively. However, there will have been a significant increase in the availability of measurable data, enabling the uncertainty in the values assigned to the structural, rock mass and hydrogeological parameters within each domain to be assessed quantitatively.

8.5.1.4 Design and construction stage (Level 4)

At the design level it is proposed that the reliability of the geotechnical model will have been estimated at a high level of confidence defined as 'Level 4'. The work will be performance-based to confirm the results obtained during the feasibility investigations. It will include detailed mapping, observation of initial slope behaviour, the possible installation of trial slopes, observation of groundwater behaviour and confirmation of pumping parameters, field testing and laboratory testing. All sampling, field testing and laboratory testing procedures must be sufficient to satisfy designated international standards for site investigation and laboratory testing (e.g. ISRM, ASTM). The data will be sufficient to confirm the results of the Level 3 feasibility slope design.

At Level 4, the uncertainty in the values assigned to the structural, rock mass and hydrogeological parameters within each domain have mostly been assessed quantitatively. With the increased amount of outcrop and subsurface information, it will have become possible to apply quantitative assessments to geological boundaries that were previously assessed subjectively.

8.5.1.5 Operations stage (Level 5)

Designated as 'Level 5', the operations stage commences with mining. It is marked by the ongoing maintenance and refinement of the geotechnical database and the ongoing comparison of the expected mining conditions with reality. At this advanced stage of the project the majority of the data assessments have been performed quantitatively.

It is suggested that the quantity, distribution and quality of data and the levels of confidence attached to the data at each project stage in Table 8.1 should be ratified by a geotechnically competent person and/or reviewer. It is also suggested that, as proposed in Chapter 1, the basic criteria for a competent person be an appropriate graduate degree in engineering or a related earth science, a minimum of 10 years post-graduate experience in pit slope geotechnical design and implementation, and an appropriate professional registration.

8.5.2 Assessment criteria checklist

When assessing the levels of confidences in the boundaries of the geotechnical domains and design sectors, there are key items that must be checked:

- the nature of the information used to set the domain boundaries. Was the geological and other information qualitative or quantitative? What was the spacing and distribution of the data relative to the complexity of the deposit, especially at depth below surface to the limits of mining? Were core and other field samples logged to a level of detail sufficient to support the interpretation? What assumptions were made when preparing the interpretation?
- the effect, if any, of alternative interpretations of the data;
- the results of any audits or reviews of the data and interpretations;
- the nature and scale of planned further work.

When assessing the levels of confidence in the structural, hydrogeological and rock mass parameters within each geotechnical domain and design sector, particular attention must be paid to the following items:

- the integrity of the database (e.g. what quality control procedures were adopted);
- the nature and quality of sampling (e.g. disturbed, undisturbed);
- field sampling techniques (e.g. chip, diatube, hand-trimmed cube, moisture loss protection);
- drilling techniques (e.g. auger, core, core diameter, triple-tube, orientation of core);
- drilling bias, especially with respect to the orientation of the borehole relative to any major structures;
- drill sample recovery;
- core logging techniques (e.g. qualitative, quantitative, level of detail);
- sample bias, especially with respect to the possibility of only the stronger materials remaining intact following core recovery and handling;
- sample preparation (e.g. hand-trimmed, cut, sawn);
- laboratory testing (e.g. nature, quality and appropriateness of test procedures used);
- location of data points (e.g. nature and accuracy of surveys used to locate field sample points and borehole collars);
- nature and scale of planned further sampling and laboratory testing work.

8.6 Summary and conclusions

The principal objectives of Chapter 8 were to:

- provide an understanding of the causes of data uncertainty and its potential impact on the reliability of pit slopes;
- highlight the need for uniform industry standards to report the uncertainties in the geotechnical data used in slope design;
- present a geotechnical reporting system that defines levels of confidence in the data that are commensurate with each stage of project development.

A further consideration was that the system needed to be consistent with the codes already used in different countries for reporting mineral resource and ore reserves (e.g. JORC 2004).

In developing the system, five levels of confidence have been defined.

1. Level 1, with a low level of confidence at the conceptual development stage.
2. Level 2, with a low level of confidence at the pre-feasibility development stage.
3. Level 3, with a reasonable level of confidence at the feasibility development stage.
4. Level 4, with a high level of confidence at the design and construction stage.
5. Level 5, with an increasingly high level of confidence as mining proceeds.

Target levels of confidence for each level were presented in Table 8.1 and a checklist of assessment criteria outlined in section 8.5.2.

A key driver of the need to develop the system has been that too often operating level investment decisions have been made using geotechnical data that is more appropriate to a conceptual or pre-feasibility level of investigation. For example, the project may have advanced to the design and construct stage (Level 4), but the level of confidence as judged by items such as the number of drill holes and laboratory tests may still be at Level 2.

The key benefit of the system is that it provides a quantitative measure that can be used by corporate mine management and the investment community to assess their level of exposure to risk. The costs of moving from Level 1 to Levels 2, 3, 4 and 5 can be estimated and incorporated in a project risk assessment. For example, the risk of moving from design into construction when confidence in the data is at Level 2 is likely to be unacceptable. On the other hand, if confidence is at Level 3 corporate management may consider the risk acceptable for development purposes. Either way, the system provides a yardstick that can be understood by everyone.

The next major initiative is to introduce the system into the industry and the investment community at all levels of management. This will require two steps. The first will be for executive mine management and geotechnical practitioners to agree on the definitions and requirements of each level of confidence. The second will be for these parties to agree on the definition of a 'geotechnically competent person' that is proposed in Chapter 1.

9 ACCEPTANCE CRITERIA

Johan Wesseloo and John Read

9.1 Introduction

The data collected (Chapters 2–7) and the reliability assigned to them at each level of project development (Chapter 8) must now be applied to the iterative design and analysis components of the slope design process outlined in Figure 9.1. Before the final designs can be accepted, they must be aligned with the slope failure criteria specified by the owner.

In open pit mining slope failure is not easily defined. Whereas in some engineering systems failure occurs immediately and is not reversible (e.g. the buckling of a structural column or the failure of a dam), in an open pit mine slope failure may take place gradually so that determining the stage at which the pit wall ceases to perform adequately may be highly subjective.

Inherently, the owners and managers of any open pit mine expect that the system will be optimised to meet the essential needs of safety, ore recovery, financial return, and the environment (section 1.2). Accordingly, the requirement for the pit slope designs involves walls that will be stable for the required life of the open pit, which may extend into closure. At the very least, any instability must be manageable at every scale of the walls, from the individual benches to the overall slopes. The owner's acceptance criteria, which form the basis of a slope design, must reflect these requirements in terms of the corporate risk profile.

Traditionally, assessments of the performance of open pit mine slopes have been made on the basis of the allowable Factor of Safety (FoS), which is the ratio of the nominal capacity (C) and demand (D) of the system. Over the years other acceptance criteria have been introduced, including the probability of failure (PoF), the consequences of slope displacement on mine operations, and risk. This chapter examines the principles of each criterion.

The FoS is addressed in section 9.2 and the PoF in section 9.3. Section 9.3 also outlines a procedure that can combine FoS and PoF with the physical consequences of slope instability as a means of assessing their effect on the integrity of the slopes at bench, inter-ramp and overall scale. Section 9.4 outlines how the probability and the consequences of slope failure are brought together in acceptance criteria based on risk. A summary of typical acceptance criteria values is provided in section 9.5.

9.2 Factor of safety

9.2.1 FoS as a design criterion

The FoS is a deterministic measure of the ratio between the resisting forces (capacity) and driving forces (demand) of the system in its considered environment:

$$\text{FoS} = \frac{C}{D} \qquad \text{(eqn 9.1)}$$

The FoS is the most basic design acceptance criterion in engineering. In geomechanics it came to prominence in the middle of the 20th century when geotechnical engineering was developed as an independent engineering discipline. In 1940, Taylor defined it as the ratio of the average shear strength of the material constituting the slope and the average shear stress developed along the potential failure surface, or the factor by which the shear strength would have to be divided to give the condition of incipient failure.

In concept, limiting equilibrium is achieved when the FoS has a value of 1.0. In reality, uncertainty about the likely performance of the system over a specified period under the proposed operating conditions usually results in the setting of a prescribed minimum design acceptance value for the FoS, learned from experience based on factors such as the analytical method used in the design

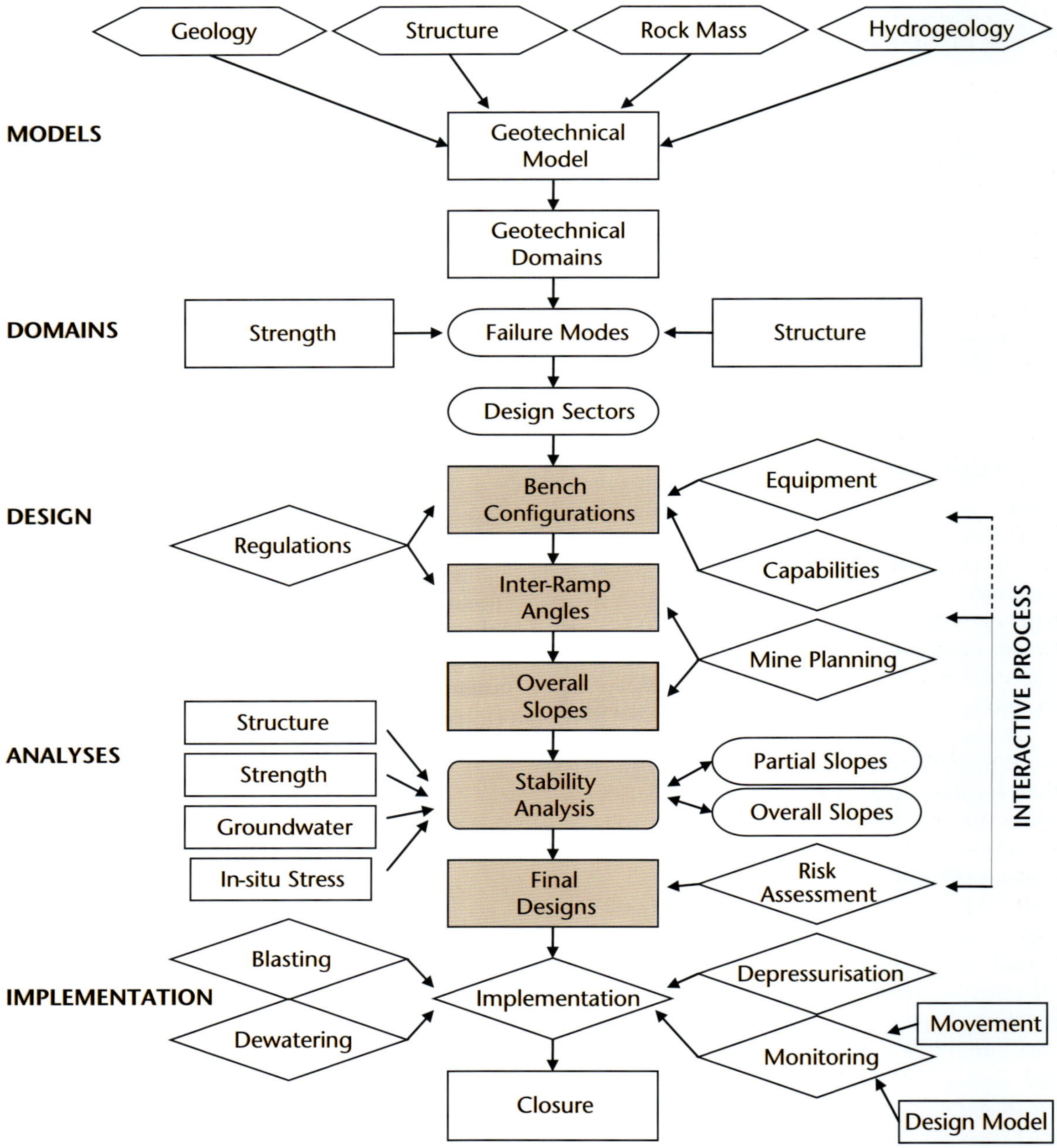

Figure 9.1: Slope design process

calculations, the degree of confidence in the input parameters, and the consequences of failure.

In limit equilibrium analyses, the FoS is calculated for a slope with the underlying assumption that all the material along a potential failure surface has the same FoS. Hence, the calculated FoS relates to a single ultimate strength for all the materials in the slope. Progressive failure mechanisms and strain softening are not accounted for in the calculations. If they are to be addressed, then finite element or finite difference codes and the shear strength reduction technique must be used (section 10.3.4.3)

The degree of confidence in the capacity function (*C*) depends on the variability in the rock mass shear strength parameters, testing errors, mining procedures, inspection procedures and so on. Similarly, the demand function (*D*) includes factors such as the gravitational load of the rock mass, earthquake accelerations, stress history, the location of the water table and equipment loadings. Common to both are the assumed formulae and equations used to scale the parameters.

Attempts to reduce the effect of the variability and uncertainty in the capacity and demand functions have mainly focused on creating a ratio of single-valued expected or characteristic values, with a the central factor of safety (CFoS), defined as:

$$\text{CFoS} = \frac{E[C]}{E[D]} \qquad \text{(eqn 9.2)}$$

where

$E[C]$ = expected value of the capacity
$E[D]$ = expected value of the demand.

In equation 9.2, the CFoS is considered to represent a single-valued measure that theoretically should have a result equivalent to that obtained from a full stochastic analysis.

Early attempts to set a single-valued capacity function ($E[C]$) stem from the US Army Corp of Engineers slope stability manual (1970), which specified that design strengths be chosen such that two-thirds of the test values are greater than the design strength selected. A more recent process is the characteristic value approach, which stems from Eurocode7 and suggests that a credible range for the characteristic strength lies between the 90th percentile (that is, 90% of the domain, by volume, when tested will display a measured strength greater than that used for analysis of stability) and something a little less than the mean.

Mostly, however, the uncertainty in the value of the conventional FoS is accounted for by the traditional method of setting a prescribed minimum design acceptance value based largely on experience.

9.2.2 Tolerable factors of safety

Few authors have published recommended design acceptance levels for the FoS. This leads to a question: how did we determine the FoS? Typical values have been set by observation and trial-and-error experience over time, taking into account issues such as the reliability of the data, the types of analyses utilised and the simplifying assumptions made. An example of tolerable FoS values established with these methods is given in Table 9.1.

Table 9.2 outlines acceptable design FoS values recommended in the literature for civil engineering applications. For normal operating conditions and long-term stability, the FoS may vary from 1.25 to 2, depending on the author, while for short-term slopes the recommended values vary between 1.3 and 1.5. The required FoS for severe loading conditions varies from 1.25 to 1.3. This may be a lower value since it caters for a condition that is unlikely to happen and that, when it does happen, lasts for only a short time.

The applicability of the FoSs used for civil engineering slopes to open pit mine slopes can be debated due to the different operating environments. However, the values most frequently used in both disciplines are very similar, ranging from 1.2 for non-critical slopes to 1.5 for slopes containing critical access ramps or infrastructure such as in-pit crushers. It should be noted that these levels are for static analyses. If pseudo-static analyses are performed to account for seismic effects, the FoSs should be adjusted in accordance with the recommendations provided in Chapter 10, section 10.3. Typical static and pseudo-static values used in mining are summarised in Table 9.9.

Table 9.1: Examples of acceptable FoS values (Priest & Brown 1983)

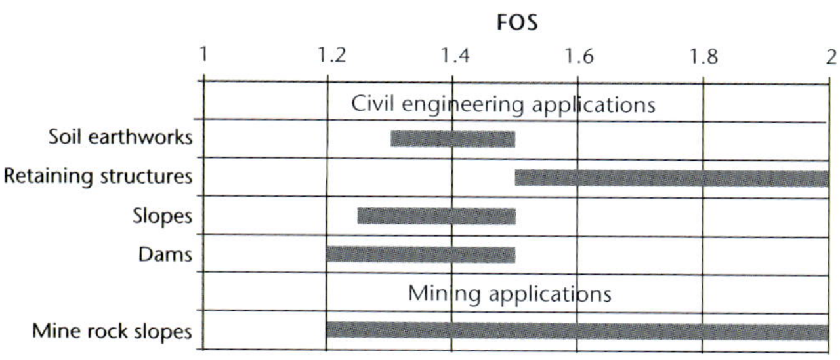

Source: Priest & Brown (1983)

9.3 Probability of failure

9.3.1 PoF as a design criterion

The PoF has become increasingly used as an acceptance criterion during the past 35 years, albeit with varying degrees of enthusiasm and scepticism. During his 1982 Terzaghi Lecture, Whitman (1983) was of the opinion that probability theory was regarded with doubt or even suspicion by the majority of geotechnical engineers. Attitudes have changed and use of the PoF as a design criterion has strengthened.

There are two options, both of which take into account the variability in the capacity (C) and demand (D) functions.

1 Option 1 – recognising the FoS as a random variable and seeking the probability of it being equal to or less than 1:

$$\text{PoF} = P[\text{FoS} \leq 1] \qquad \text{(eqn 9.3a)}$$

2 Option 2 – seeking the probability that the demand (D) exceeds the capacity (C):

$$PoF = P[C - D \leq 0] \qquad \text{(eqn 9.3b)}$$

Option 1 is used most often, but using either option has three particular attractions.

- It enables the variabilities in the capacity (C) and demand (D) functions to be taken into account and helps establish the level of confidence in the design. The reliability of a structure is its probability of success. Thus, if the estimated PoF of a slope is 20%, its reliability is 80% (Equation A2.3), which reflects the level of confidence required for the design and construction (Level 3) stage of project development (Table 8.1).
- It scales linearly, i.e. a PoF of 10% is twice as great as a PoF of 5%.
- It is an essential parameter in the calculation of risk, where risk (R)is defined as (section 9.5):

$$R = \text{PoF} \times (\text{consequences of failure}) \qquad \text{(eqn 9.4)}$$

Table 9.2: Acceptable FoS values, civil engineering applications

Material type	Conditions	Acceptance level (static)	Reference
Soil earthworks	Normal loads and service conditions	1.5	Meyerhof (1984)
	Maximum loads and worst environmental conditions	1.3	
Earth retaining structures and excavations	Normal loads and service conditions	2	
	Maximum loads and worst environmental conditions	1.5	
Slopes	Cohesionless soils	1.3	
	Cohesive soils	1.5	
	Based on field vane tests corrected for strain rate and anisotropic effects	1.3	Bjerrum (1973)
		1.25	Bowles (1979)
	Highest value for serious consequence of failure or high uncertainty	1.25–1.5	Gedney & Weber (1978)
		1.5	Hansen (1967)
		1.3–1.5	Meyerhof (1970)
		1.3–1.4	Sowers (1979)
	Lower values for temporary loading	1.5 1.25–1.3	Terzaghi (1943)
	Permanent or sustained conditions	1.5	US Navy Department (1962)
	Temporary	1.25	SAICE COP (1989)
	Permanent	1.5	SAICE COP (1989)
Dams	End of construction, no reservoir loading, pore pressure at end of construction estimates with undissipated pore pressure in foundations	1.3	Hoek (1991)
	Full reservoir, steady state seepage with undissipated pore pressure in foundation	1.3	
	Full reservoir with steady state flow and dissipated pore pressure	1.5	
	Flood level with steady state flow	1.2	
	Rapid drawdown pore pressure in dam with no reservoir loading	1.3	

9.3.2 Acceptable levels of PoF

As with the FoS criterion, few recommendations exist in the literature for acceptable PoFs for design. Notable contributions are those of Priest and Brown (1983), Kirsten (1983), SRK Consulting (2006) and Sullivan (2006).

The design FoSs and PoFs suggested by Priest and Brown (1983) are presented in Tables 9.3 and 9.4. In Table 9.3, Priest and Brown use three slope categories based on the consequence of failure and suggest design values for the FoS and PoF for:

- the probability of the FoS being less than 1.0 (P[FoS ≤ 1.0]);
- the PoF being less than 1.5 (P[FoS ≤ 1.5]).

If one of these criteria is not met, the slope is deemed to be potentially unstable, as described in Table 9.4.

Current industry experience suggests that the acceptance levels suggested by Priest and Brown in Tables 9.3 and 9.4 are conservative.

Kirsten (1983) suggested the use of Table 9.5, which is based on a literature study and several back-analyses of soil slopes and earth and rockfill dams. It incorporates the service life, public liability and type of monitoring applied. The table also provides guidance for interpreting the PoF level in terms of the frequency of failed slopes, including unstable movements. Although this may sometimes be helpful, it should be used with caution as it was based on a frequency-of-event interpretation of the PoF not a degree-of-belief, subjectively assessed PoF (Vick 2003), and therefore implicitly assumes the PoF to be a property of the slope and not of the design.

Table 9.6 is a simple but effective system that has been used successfully by SRK Consulting for several diamond mines in southern Africa. In general terms, there appears to be a reasonable correlation between this system, that presented by Kirsten (1983) and that presented by Swan and Sepulveda (2001).

Swan and Sepulveda (2000) developed Table 9.7 to describe the acceptance criteria for the design of the slopes at the Ujina open pit, Chile. The process combines FoSs and PoFs with the physical consequences of slope instability and their effect on the integrity of the slopes at bench, inter-ramp and overall (global) scale. In financial terms, the physical consequences can include the costs of sterilising ore, clean-up of the ramps and benches, remedial stripping and down-time. Because of

Table 9.3: FoS and PoF guidelines

Consequence of failure	Examples	Acceptable values		
		Mean FoS	Minimum P[FoS < 1.0]	Maximum P[FoS < 1.5]
Not serious	Individual benches; small (< 50 m), temporary slopes, not adjacent to haulage roads	1.3	10%	20%
Moderately serious	Any slope of a permanent or semi-permanent nature	1.6	1%	10%
Very serious	Medium-sized (50–100 m) and high slopes (<150 m) carrying major haulage roads or underlying permanent mine installations	2.0	0.30%	5%

Source: Priest & Brown (1983)

its success in practice, the process developed for the Ujina mine has been introduced at a number of related open pit mining operations in Chile, with suitable local variations.

It should be noted that the values given in Table 9.7, which include a number of FoS levels that some practitioners would consider relatively high, are specific to the Ujina mine, but the process can be utilised at any mine. Its significant merit is that conceptually it matches mine management's design expectations with actual slope performance at bench, inter-ramp and overall scale, and links those expectations to the slope performance and the capacity of the mining equipment being used at the mine site.

Significant features of the system noted by Swan and Sepulveda (2000) include:

- bench-scale failures are inevitable and permissible provided the acceptable contained volumes of material on benches are unlikely to be exceeded. In general, the larger the volume, the smaller the acceptable PoF. Also, benches located immediately above and below ramps and those in the final wall must have lower tolerances of failure;
- acceptance of inter-ramp instabilities depends on the amount of ramp loss and the overall volume affected. The minimum permissible values can be determined in terms of FoS and a maximum limit to the PoF. Final wall inter-ramp slopes must have an operational life in excess of those for the purpose of expansion;
- overall instability must consider the possibility of the loss of ramps in the affected sector(s), given the likelihood that the volumes will be substantially greater than those affecting inter-ramp failures. Acceptance is defined in terms of a minimum permissible value for the FoS. Additionally, because of the uncertainties likely to be associated with the geotechnical model, a maximum limit is also defined for the probability of a permissible failure. Other factors that must be considered are that overall slopes only reach their maximum at the completion of the final wall pushback, and whether important infrastructure is located within or on the surface close to the perimeter of the pit.

Table 9.4: Interpretation of Priest & Brown (1983) FoS and PoF guidelines

Performance of slope with respect to Table 9.3	Interpretation
Satisfies all three criteria	Stable slope
Exceeds minimum mean FoS but violates one or both probabilistic criteria	Operation of slope presents risk that may or may not be acceptable; level of risk can be reduced by comprehensive monitoring program
Falls below minimum mean FoS but satisfies both probabilistic criteria	Marginal slope: minor modifications of slope geometry required to raise mean FoS to satisfactory level
Falls below minimum mean FoS and violates one or both probabilistic criteria	Unstable slope: major modifications of slope geometry required; rock improvement and slope monitoring may be necessary

Overall, the system provides an unequivocal statement of what is expected of the slopes and a direct communication channel between executive mine management, mine design and mine operations. It leads to the development of a unique set of acceptance criteria that suit the site specific failure mechanisms, model uncertainty and owner's risk profile, which recommends it strongly as a process for developing design acceptance criteria.

9.4 Risk model

9.4.1 Introduction

The acceptance criteria outlined above calibrate the performance of the pit slopes but do not quantify the risks that may be associated with slope failure.

In slope design, the risks (R) associated with slope failure are defined and quantified as:

$$R = \text{PoF} \times (\text{consequences of failure}) \quad \text{(eqn 9.5)}$$

Broadly, the consequences of slope failure can be categorised in the following six ways.

Table 9.5: PoF design acceptance guidelines

	Design criteria			Aspects of natural situation	
PoF (%)	**Serviceable life**	**Public liability**	**Minimum surveillance required**	**Frequency of slope failures**	**Frequency of unstable movements**
50–100	None	Public access forbidden	Serves no purpose	Slope failures generally evident	Abundant evidence of creeping valley sides
20–50	Very very short-term	Public access forcibly prevented	Continuous monitoring with intensive sophisticated instruments	Significant number of unstable slopes	Clear evidence of creeping valley sides
10–20	Very short-term	Public access actively prevented	Continuous monitoring with sophisticated instruments	Significant instability evident	Some evidence of slow creeping valley sides
5–10	Short-term	Public access prevented	Continuous monitoring with simple instruments	Odd unstable slope evident	Some evidence of very slow creeping valley sides
1.5–5	Medium-term	Public access discouraged	Conscious superficial monitoring	No ready evidence of unstable slopes	Extremely slow creeping valley sides
0.5–1.5	Long-term	Public access allowed	Incidental superficial monitoring	No unstable slopes evident	No unstable movements evidence
<0.5	Very long-term	Public access free	No monitoring required	Stable slopes	No movements

Source: Kirsten (1983)

1 Fatalities or injuries to personnel, including the costs of industrial and legal action.
2 Damage to equipment and infrastructure, including the costs of replacing equipment and infrastructure.
3 Economic impacts on production, including the costs of:
   - → removing failed rock material to the extent that mining can safely continue;
   - → slope remediation – the slope may have to be cut back to prevent secondary failures due to steeper upper slopes, or slope support systems may be required;
   - → haul road repair and re-access – the haul road and ramp may be damaged and re-access to the mine may need to be considered;
   - → equipment re-deployment – the cost of equipment being isolated by the failure and the cost of moving equipment to other parts of the mine unaffected by the failure where it can be used productively should be considered;
   - → unrecoverable ore – the loss of a ramp or part of an inter-ramp slope may lead to sterilising sections of the orebody, at least on a temporary basis.
4 Force majeur (a major economic impact), which should normally equate to failure of an overall slope or loss of medium- to long-term access to ore such that contracts cannot be fulfilled.
5 Industrial action, i.e. loss of worker confidence.
6 Public relations, such as stakeholder resistance due to social views and/or environmental impacts arising from the failure. Increased regulatory supervision.

Table 9.6: Acceptable PoFs, mining rock slopes

Category	Description	Acceptable PoF
1	Critical slopes where failure may affect continuous operation and pit safety	<5%
2	Slopes where failure have a significant impact on costs and safety	<15%
3	Slopes where failure has no impact on costs and where minimal safety hazards exist	<30%

Source: SRK Consulting (2006)

Traditionally, the consequences of slope failure have been taken into account using cost–benefit analyses. During the last decade, a risk model process has been proposed as a means of providing the range of real consequences from potential failures in order to give management the opportunity and responsibility to define the risk appropriate for their mining business. Both methods are outlined below.

9.4.2 Cost–benefit analysis

Cost–benefit analyses that compare the financial effects of slope failures or other modifications to the base case slope design have long been an essential requirement of the mine planning cycle. Usually, the analysis calculates the effect relative to the base case of steepening the pit walls, including waste stripping savings and the costs of instability. The benefits and costs are determined for each

Table 9.7: Acceptance criteria, FoS, PoF and category of slope instability

		Characteristics of instability		Acceptability Criterion		
Slope type	Case	Loss of ramp berm (%)	Material affected (ktons/m)	FoS	PoF (%)	Comments
Bench	Expansion, not adjacent to a ramp	<25	<0.5/<1.0			Berms should have a nominal width to contain unravelling wedges whose probability of occurrence is >30%; controlled blasting will be used to minimise induced damage and presplitting on the final wall slopes
		25–50	<1.0/<2.0		<45	
		>50	>1.0/>2.0		<35	
	Expansion, adjacent to a ramp	<25	<0.5/<1.0			
		25–50	<1.0/<2.0		<40	
		>50	>1.0/>2.0		<30	
	Final wall, not adjacent to a ramp	<25	<0.5/<1.0			
		25–50	<1.0/<2.0		<35	
		>50	>1.0/>2.0		<25	
	Final wall, adjacent to a ramp	<25	<0.5/<1.0			
		25–50	<1.0/<2.0		<30	
		>50	>1.0/>2.0		<20	
Inter-ramp	Expansion	<25	<5	>1.20	<30	Stability analysis must include explicit effect of rock mass structures; two independent access ramps will be made to the pit bottom; measures will be implemented for slope drainage
			>5	>1.25	<25	
		25–50	<5	>1.25	<25	
			5–10	>1.30	<22	
			>10	>1.35	<20	
		>50	<10	>1.30	<22	
			10–20	>1.35	<20	
			>20	>1.45	<18	
	Final wall	<25	<5	>1.20	<25	
			>5	>1.25	<20	
		25–50	<5	>1.30	<22	
			5–10	>1.35	<20	
			>10	>1.45	<18	
		>50	<10	>1.35	<20	
			10–20	>1.40	<18	
			>20	>1.50	<15	
Global	Expansion		<25	>1.30	<15	Stability analysis must include mass structures; all mine infrastructure lie outside pit perimeter limits
			25–50	>1.40	<12	
			>50	>1.50	<10	
	Final wall		<25	>1.30	<12	
			25–50	>1.45	<10	
			>50	>1.60	<8	

Source: Swan & Sepulveda (2000)

year of the prospective mine life and discounted to the present (CANMET 1977). The effects of instability can be included in the analysis stochastically (Ryan & Pryor 2000). Other factors that must be included in the analysis are tax rates, royalties and capital expenditure.

Detailed probabilistic stability analyses and a full financial analysis are required at the detailed mine design stage and should be repeated for alternative layouts before the optimum wall design and final layout are decided. This type of cost–benefit analysis for design has been applied at several large open pits, including the Bingham Canyon mine in the USA.

9.4.3 Risk model process

The objective of the risk model is to provide a basis for management decision by:

- defining the risks in terms of safety and economics;
- quantifying risk levels for different slope configurations;
- quantifying the economic value added for increased levels of risk.

The model suggests that stability is not the end objective, but rather that safety is not to be compromised as the economic impact of the chosen slope angles is optimised. A corollary to this objective is that slope failures are acceptable on the condition that they can be safely managed without compromising the business plan. The process is outlined schematically in Figure 9.2. It involves four steps.

1 The first is a fault tree analysis to determine the slope PoF; this is the PoF shown in the first column in Figure 9.2 and is termed the 'top fault'. The process is a geotechnical function that utilises all the information and measures of uncertainty in the geotechnical model. It commences with a conventional stochastic stability analysis to determine the level of stability of the slope for the given input parameters. This analysis represents the 'normal' condition for the fault tree. This is followed by analyses that take into account the uncertainties in the geotechnical model such as changes in geological boundaries (lithologies and through-going structures), rock mass strength, groundwater levels and rock strengths, together with mining-related issues such as overdigging or blasting.

The approach allows different levels of uncertainty to be included in the overall assessment of the design's reliability. Uncertainties in the data are accounted for as outlined in Chapter 8 and the stability analyses are performed as outlined in Chapter 10. The stability analyses can be at bench, inter-ramp or overall slope scale.

2 An event tree analysis to determine the risks that may be associated with a slope failure. The probabilities used in the event tree are knowledge-based probabilities (Vick 2003) as distinct from the

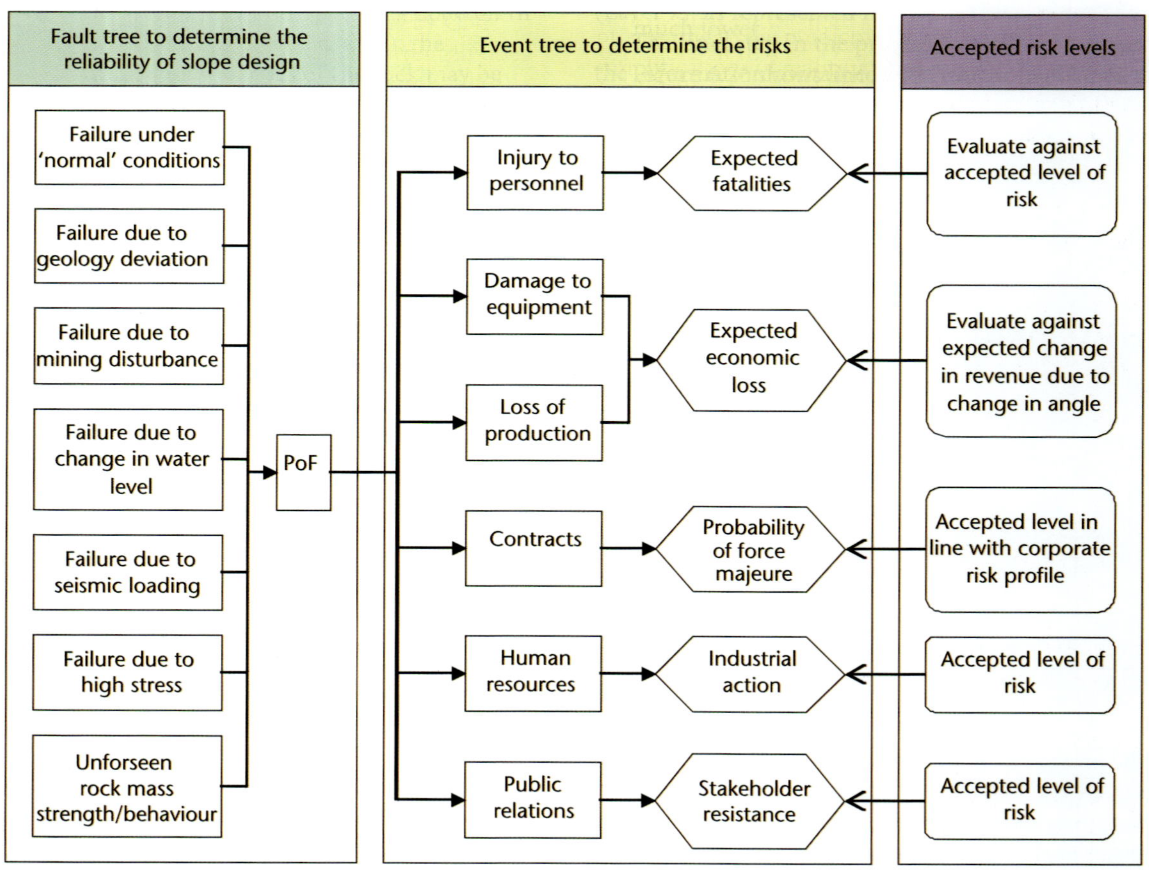

Figure 9.2: Risk/consequence model process
Source: Steffen et al. (2006)

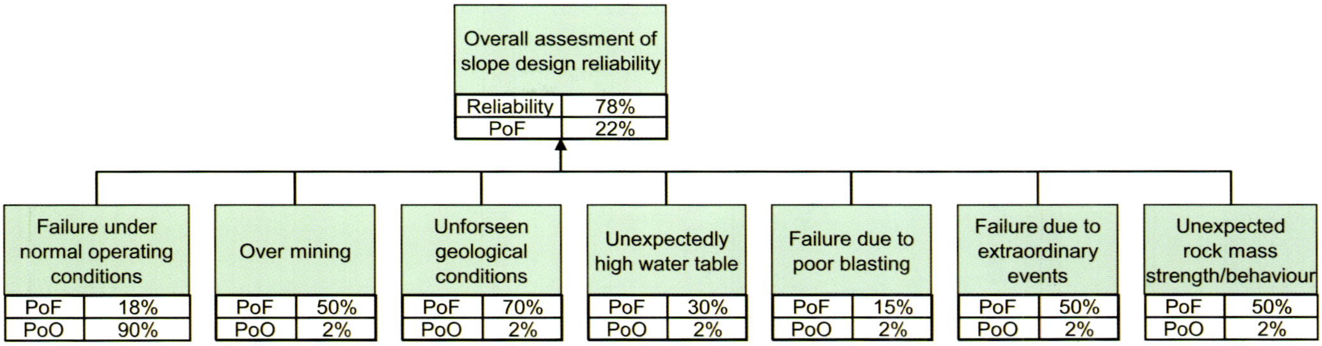

Figure 9.3: PoF–PoO fault tree analysis
Source: Steffen et al. (2006)

frequency-based probabilities used to estimate the top fault PoF and are determined subjectively (Section 8.4.2) with input from experienced, site-based personnel.

3 Carriage of the top fault value into the event tree, where the risk of a defined incident (e.g. fatality, economic loss) is evaluated. This part of the analysis is known as the risk/consequence analysis. It can be performed independently to determine the appropriate slope design reliability needed for the desired level of confidence in achieving the mine plan, or to ensure the desired safety level at the mine.

4 A comparison of the outcome of the top fault/event tree analysis against the acceptance criteria (risk levels) decreed by management.

9.4.3.1 *Fault tree analyses*

Two methods of estimating the PoF or top fault have been propagated. The first method involves modifying the estimated PoF under 'normal' conditions by other contributing factors, such as the uncertainties in the geological boundaries or pore pressures, using the single-pass PoF–Probability of Occurrence (PoO) approach illustrated in Figure 9.3.

The PoF for the normal operating condition is first determined (PoF = 0.18 in Figure 9.3) and then combined with its estimated PoO (PoO = 0.90 in Figure 9.3); the resultant value is 0.162. The combined values of the PoF and PoO for each contributing factor are then determined and progressively added to the combined initial value, to provide the overall assessment. In the example given in Figure 9.3, this results in a final PoF of 0.22 (22%), which is a reliability of 78%.

The second method involves modifying the FoS under 'normal' conditions by the other contributing factors using a response surface approach (Calderon & Tapia 2006; Jefferies 2006), as follows.

1 The FoS is determined for the normal operating condition. Triangular distributions, providing upper and lower limits and the best estimate (mode), are used for each parameter in the model.

2 A response surface estimation is performed to determine the effect of the uncertainty in each parameter on the analysis outcome. The response surface is formed by varying each of the uncertain parameters in turn while the others are kept fixed at their best estimate value.

3 Two 'step-out' cases are used for each parameter, using the lowest ('-' case) and highest ('+' case) values from each triangular distribution. The resultant is expressed as:

$$\beta = \text{FoS}_{\text{step-out}}/\text{FoS}_{\text{best case}}$$

4 A second-order polynomial is fitted to the three β values associated with each parameter. This includes the values for the lowest and highest cases and a central value that does not need to be computed as it is always unity.

5 The process is continued for each assumption, with the resultant equations for the respective β values defining the response surface.

6 The overall FoS for any given scenario of realised values for each parameter (x_n)is given as:

$$\text{FoS} = \text{FoS}_{\text{best case}} * \beta 1(x_1) * \beta 2(x_2) * \ldots \beta n(x_n)$$

A schematic response surface is shown in Figure 9.4. Figure 9.5 gives an example outcome involving six different parameters: geological strength index (GSI); unconfined compressive strength (UCS); major fault (VIF) direction; major fault (VIF) strength; groundwater; and deviation in the slope angle. The process can be repeated sufficiently to provide a probability distribution of the FoS values, from which the slope's PoF can be determined.

9.4.3.2 *Event tree analyses*

The event tree represents the slope management strategies and processes at the mine. Since the value of the process is dependent on site ownership, it is essential that the available site knowledge is included in the analysis. This can be done subjectively (section 8.4.2) within a workshop

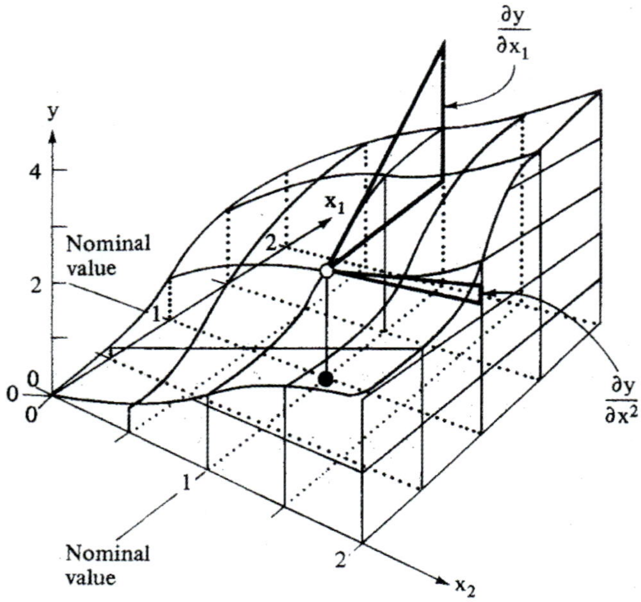

Figure 9.4: Schematic of a response surface defined by $y = f(x_1, x_2)$
Source: Jefferies (2006) after Morgan & Henrion (1990)

environment at the mine site, when decisions on the likelihood of the success or failure of different components of the mine's slope management system can be made by experienced mine staff. A simplified event tree representing the economic consequences of slope failures is presented in Figure 9.6.

The estimates of component failures can be presented simply as triangular distributions obtained from the best estimate and lowest and highest credible estimates. Note that event trees are generally not sensitive to small changes in the assigned probabilities, but they are sensitive to changes in the tree structure. The more independent questions that can be raised, the more accurate and repeatable the end result will be. However, overcomplicated structures can lead to overlaps and misunderstandings.

9.4.3.3 *Risk/consequence analyses*

The primary use of the risk/consequence analysis is to evaluate the effect of an incident on the operation and compare the outcome to the risk levels established by executive management. It can also be used to compare different slope configurations and mine planning scenarios on a common basis. Figure 9.7 illustrates a comparison based on the probability of achieving different values of NPV. Although the optimistic mine plan may be able to unlock more wealth, the probability of achieving this is low. The probability of achieving the lower NPVs is higher for the conservative mine plan than for the optimistic mine plan. Assuming that the risk to personnel

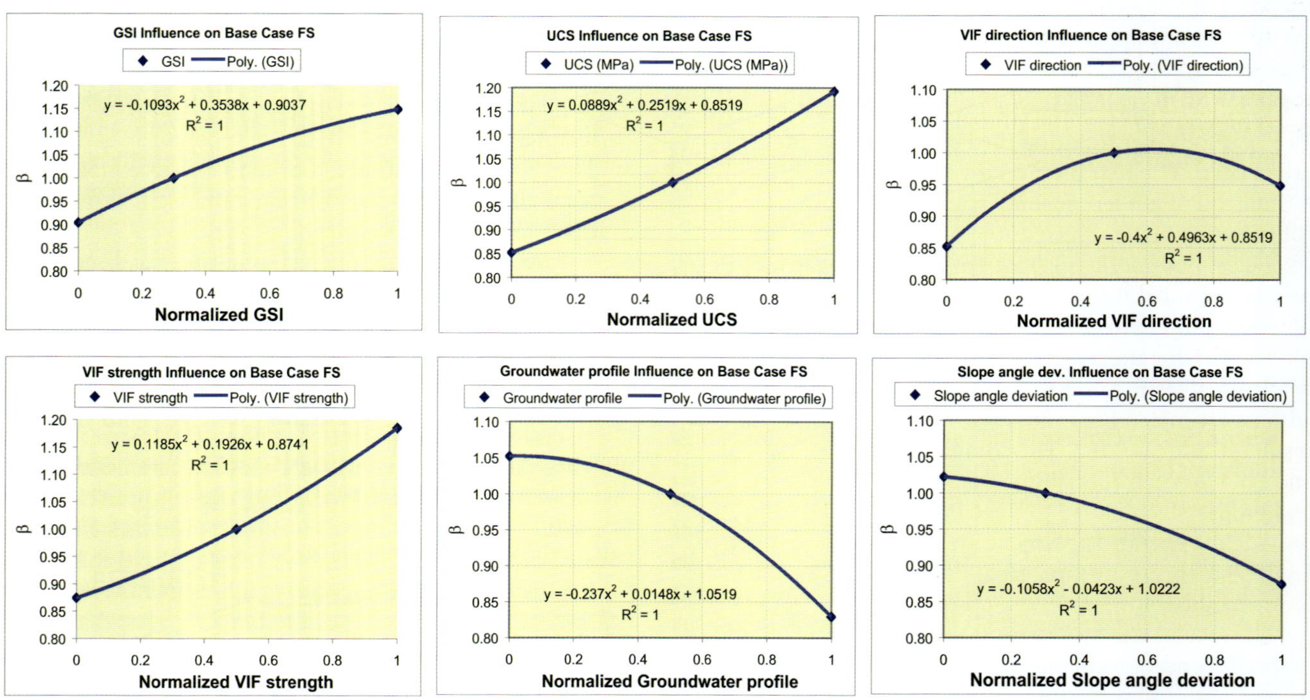

Figure 9.5: Example outcome of response surface analysis showing triangular distribution used as input for each parameter and the resultant equations for the respective β values defining the response surface
Source: Calderon & Tapia (2006)

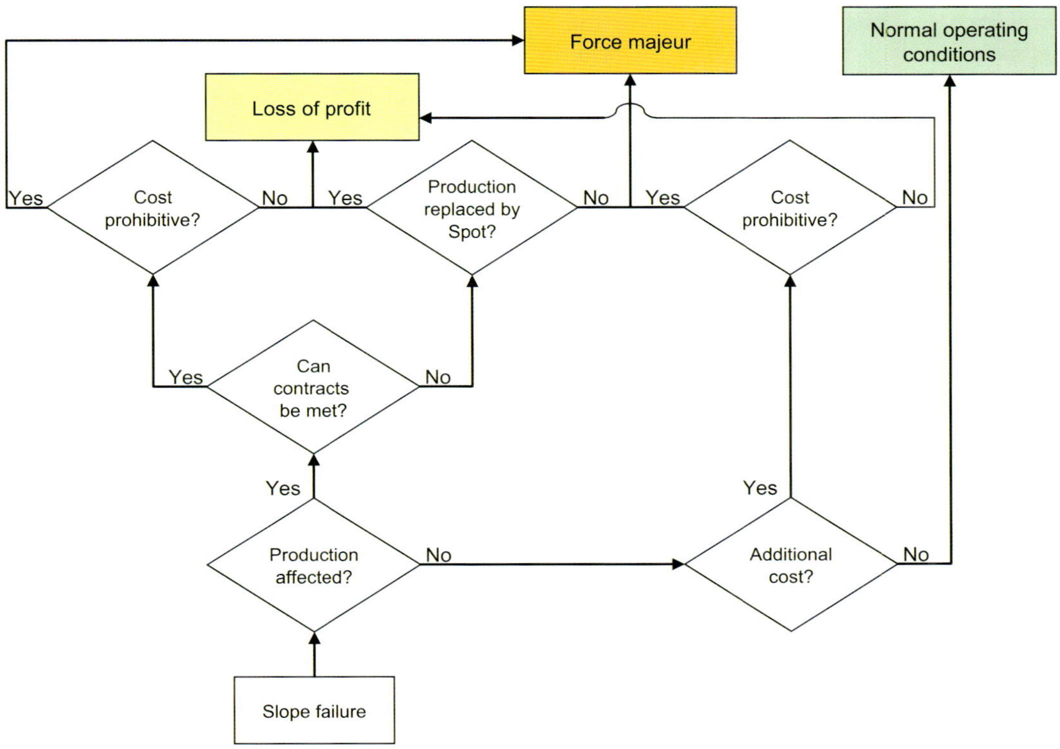

Figure 9.6: Event tree for evaluating the economic consequences of slope failure
Source: Steffen et al. (2006)

for both alternatives is at acceptable levels, the decision whether to accept the conservative or optimistic mine plan is purely a management decision, weighing up the economic risk character of the alternatives within the corporate risk profile.

Figure 9.8 shows a comparison between different slope designs for a mine in South Africa, based on NPV and the risk of fatalities. A substantial increase is shown in the expected NPV with an increase in the stack (inter-ramp) angle up to 65°. The increase in the NPV was, however, expected to be marginal with an increase in the slope angle to greater than 65°. The increase in risk to personnel, however, exceeded the mine's chosen limit for the risk of a fatality at an angle of greater than 65°. The management at this mine decided that the 65° stack angle option offered a good compromise between maximising profit without exposing the workforce to unacceptable risk levels.

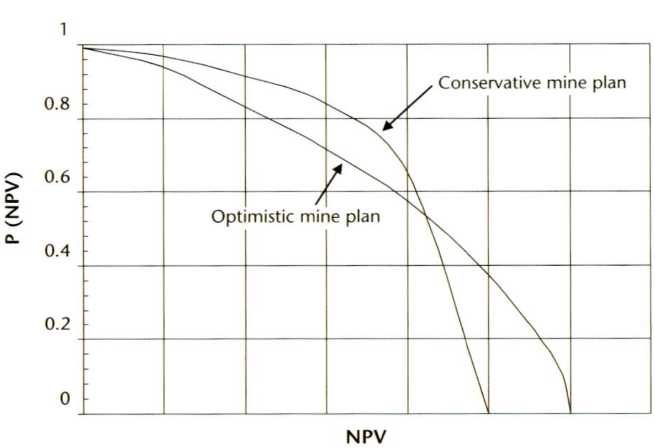

Figure 9.7: Comparison of conservative and optimistic mine plans on the basis of probability of achieving the expected NPV
Source: Steffen et al. (2006)

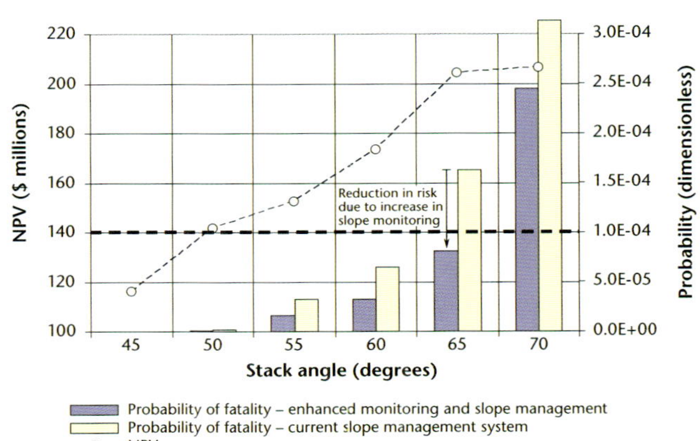

Figure 9.8: Comparison between different designs based on NPV profit and risk of fatalities for a mine in South Africa. The thick dashed horizontal line defines the 'acceptable fatality rate'
Source: Steffen et al. (2006)

9.4.4 Formulating acceptance criteria

The level of risk that may be accepted by a mining company is an executive management decision. It is likely to be governed by a complex mixture of company culture and attitude to risk, legislative requirements, economics and societal views. The task of risk management and the application of risk theory in the decision-making process during the life of an open pit mine is addressed in Chapter 13, and this book does not offer any hard-and-fast rules about acceptance criteria and risk. It offers general comments about three of the major elements of risk featured in Figure 9.2, i.e. economic loss, force majeur and fatalities and injuries.

9.4.4.1 Fatalities and injuries

The acceptable risk of a fatality from a slope failure is the most sensitive of all the risks, with most companies holding 'zero harm' accident policies. While this is certainly a worthy aim, accident statistics, which provide a means of quantifying and evaluating risk on a comparative basis, show that it is not a reality. This raises the question, what is the reality?

Figure 9.9 presents some of the statistics reported in the literature for the risks associated with many common activities such as drinking water, staying at home or partaking in sport as the probability of a fatality/person/year. Figure 9.10 collates fatalities from various countries, including the USA.

In Figure 9.9, a distinction is made between voluntary and involuntary risk. Involuntary risks are those to which the average person is exposed without choice, which includes many diseases and general accidents. For voluntary risk, only the select few who choose to take part in certain activities are exposed. Examples of voluntary risk are extreme sports such as skydiving, dangerous employment such as an astronaut, and health-threatening habits such as cigarette smoking and alcohol or drug abuse.

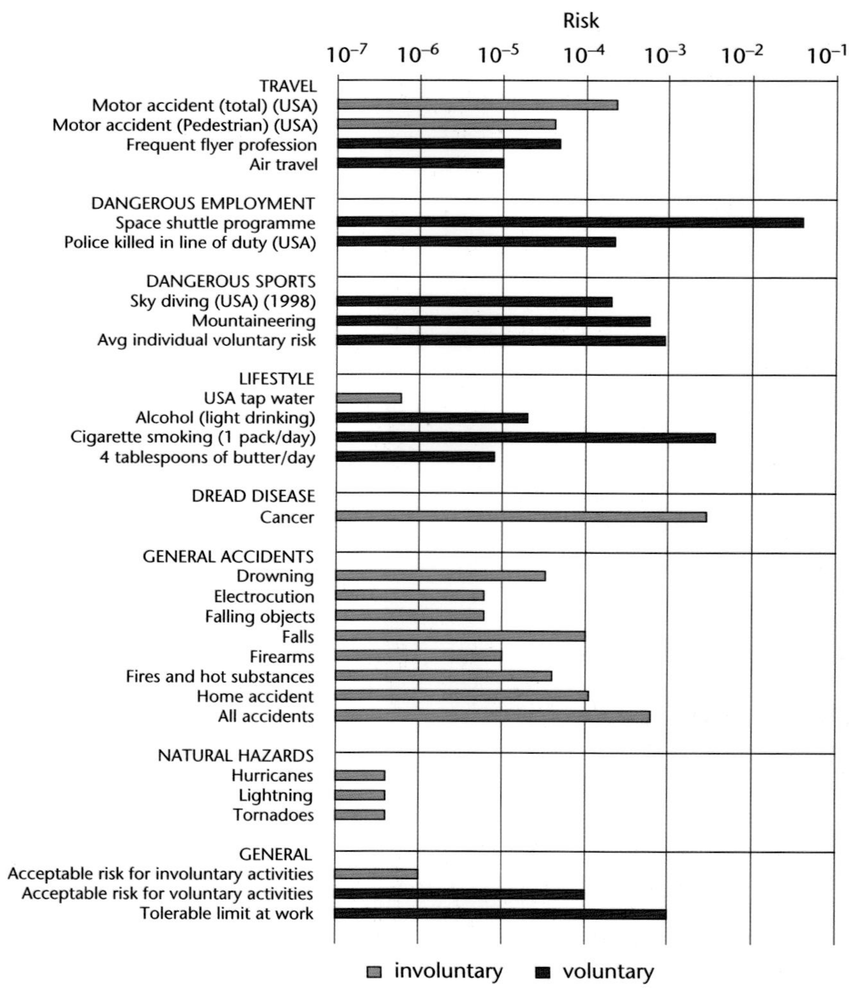

Figure 9.9: Comparative fatality statistics
Source: Steffen et al. (2006)

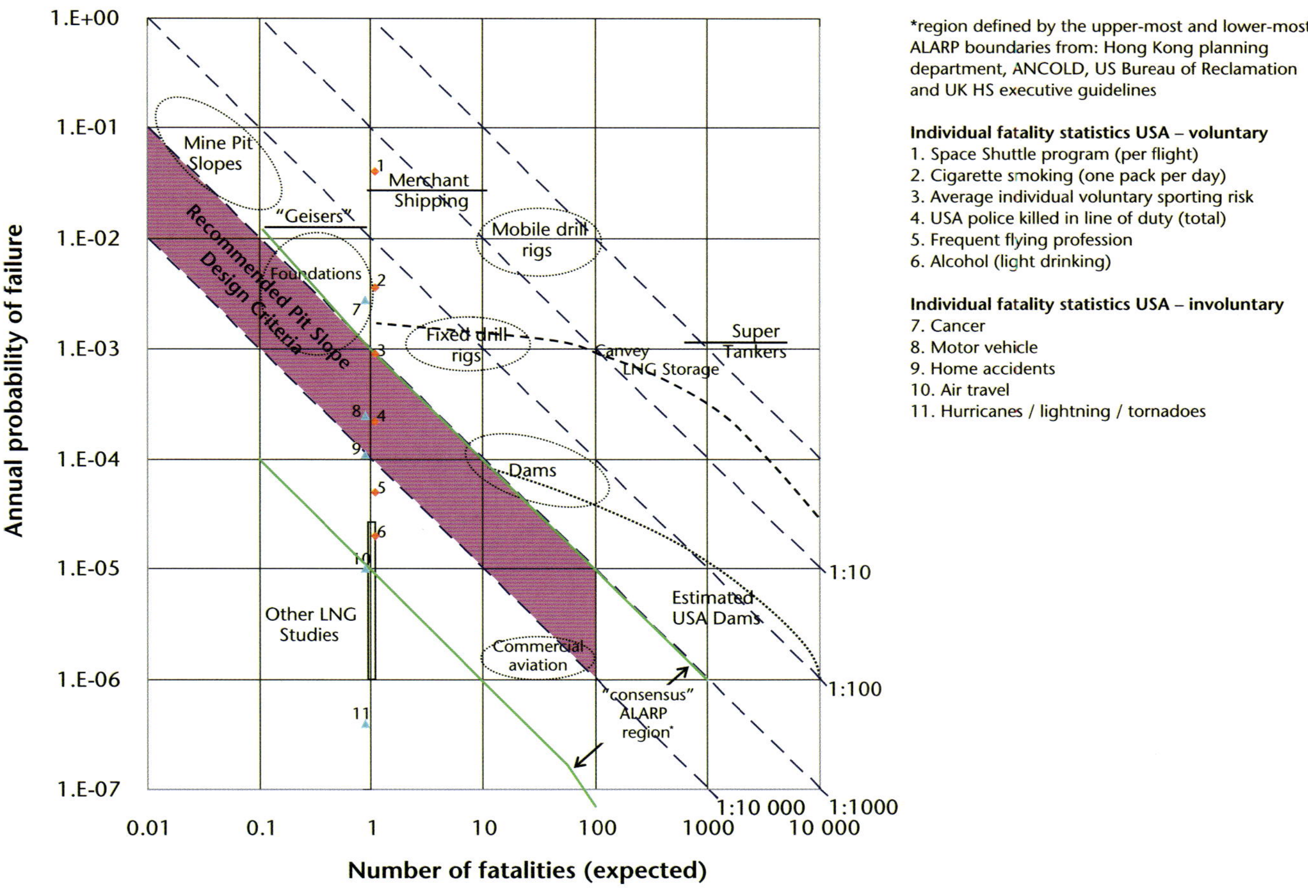

Figure 9.10: Comparison of risk acceptability criteria with statistics
Source: Steffen et al. (2006)

For open pit mining and other industrial professions, there is often a difference in opinion on whether employees' exposure to risk in the workplace should be regarded as voluntary or involuntary. Social risk acceptance studies have shown that people will accept risk if they perceive the benefit to outweigh the risk. It has been suggested that industrial risk can be regarded as voluntary if and only if the employee has been empowered to consciously accept the risks in order to obtain the reward.

In Figure 9.10, the diagonal dashed lines are constant risk lines as defined by the UK Health and Safety Executive guidelines. Risk values between 1 and 1:10 are intolerable, between 1:10 and 1:100 may be justifiable, and below 1:10 000 are negligible and of no concern. The upper limit of the ALARP (as low as reasonably possible) region is defined by a constant risk of 1:1000.

It is interesting to note that although the risk posed by mobile drill rigs at sea falls within the 'intolerable' zone of the UK Health and Safety Executive guidelines, there has been very little public outcry. Presumably, working on an oil rig is perceived as well-paid and the benefit of oil production is considered very important. In comparison, the individual risk posed by household accidents (9) falls almost on the 'negligible risk' line of the UK Health and Safety Executive guidelines.

Data from Baecher (1983) (see Figure 9.10) show that the historical risks associated with open pit mine failures and dam failures are similar (between 1:100 and 1:1000). This is consistent with the annual probability of mine slope failures being about 1000 times greater than that of dam failures, while the expected fatalities per failure is about 1000 times lower.

With due consideration of the data and literature presented above and in Chapter 13, it is suggested that open pit mines slopes should be designed to a fatality risk level between 1:1000 and 1:10 000, within the upper level of the ALARP region shown in Figure 9.10. Designing mine slopes to the same risk level as that prescribed for dams and other civil engineering structures may seem conservative or even unrealistic. It must be realised, however, that, contrary to civil engineering structures, this

Table 9.8: Slope angles related to levels of effort and target levels of confidence

Level of effort	Target level of confidence (%)	Slope angle
1	>30	**Conceptual**, representing slope angles based on experience in similar rocks using data inferred entirely from reports and interpretations based on available regional data gathered from mines in similar geological environments
2	40–60	**Pre-feasibility**, representing slope angles based on experience in similar rocks, but using data inferred from interpretations based on the information provided during the conceptual stage of development, augmented by data obtained from outcrops, exposures in road cuttings and river banks, trenches, pits, workings and drill holes at the proposed mine site
3	50–75	**Feasibility**, representing slope angles based on data gathered during mine site feasibility investigations. Sampling locations will have been spaced closely enough to sustain 3D interpretations of the geotechnical domain boundaries to the limits of mining based on boundary intersections and the continuity of the structural fabric, rock mass properties and hydrogeological parameters within each domain
4	65–85	**Design and construction**, representing slope angles based on data gathered to confirm and update the results obtained during the feasibility investigations. The data will include detailed mapping, observation of slope behaviour, the possible installation of trial slopes, observation of groundwater behaviour and confirmation of pumping parameters, field testing and laboratory testing

risk level will not be achieved by designing a more conservative slope but by properly managing the slope to avoid compromising the business plan.

9.4.4.2 *Economic loss and force majeur*

Typically, a major economic impact is expressed as a percentage impact on the NPV or on revenue. The criterion for economic risk cannot be determined in isolation from the risk–reward relationship for a mine, but is typically found to be optimum at the level of 5% of NPV. The same is true for determining a suitable criterion for force majeur. Typical values for a force majeur criterion are found to be 1% or less. However, under no circumstances should the design increase the risk to life beyond the accepted criteria for a fatality.

9.4.5 Slope angles and levels of confidence

To obtain a balanced design, it is essential that the measures outlined in Chapter 8 (Table 8.1) to report the confidence level in the geotechnical information that is used in the slope designs for the pits that define the reserves are carried through into the acceptance criteria. It is therefore recommended that the levels of effort and data confidence suggested in Table 8.1 and illustrated in Figure 8.2 be used with the criteria outlined in sections 9.2, 9.3 and 9.4 as the standard by which the design is judged and the slope angles are set. This is discussed further below and summarised in Table 9.8.

9.4.5.1 *Level 1: Conceptual slope angle*

A Level 1 conceptual slope angle corresponds to the application of typical slope angles based on experience in similar rocks. The design basis will have been entirely inferred from reports and interpretations based on available regional data gathered from mines in similar geological environments. These preliminary data may be supplemented by aerial photographic interpretations of the regional lithology and structure and any outcrop mapping performed during exploratory project surveys. Overall, the information will be sufficient only for providing indicative slope designs and the planning of pre-feasibility stage investigations.

9.4.5.2 *Level 2: Pre-feasibility slope angle*

A Level 2 pre-feasibility slope angle will correspond to the application of typical slope angles based on experience in similar rocks, but with quantification based on a preliminary rock mass classification and a reasonable inference of the geological and groundwater conditions within the affected rock mass. The data used are inferred from interpretations based on the information provided during the conceptual stage of development, augmented by data obtained from outcrops, exposures in road cuttings and river banks, trenches, pits, workings and drill holes at the proposed mine site. All these data may be limited or variably distributed and/or of uncertain quality. Any sampling, field testing and laboratory testing procedures must be sufficient to satisfy designated international standards for site investigation and laboratory testing (e.g. ISRM, ASTM). The information will be sufficient to allow simplified design models to be developed and sensitivity analyses to be carried out.

9.4.5.3 *Level 3: Feasibility slope angle*

A Level 3 feasibility slope angle corresponds to a design based on a geological model that allows a reasonable assumption on the continuity of stratigraphic and lithological units. The data have been based on the results of mine site feasibility investigations. Sampling locations have been spaced closely enough to sustain 3D

Table 9.9: Typical FoS and PoF acceptance criteria values

Slope scale	Consequences of failure[b]	Acceptance criteria[a] FoS (min) (static)	FoS (min) (dynamic)	PoF (max) P[FoS ≤ 1]
Bench	Low–high	1.1	NA	25–50%
Inter-ramp	Low	1.15–1.2	1.0	25%
	Medium	1.2	1.0	20%
	High	1.2–1.3	1.1	10%
Overall	Low	1.2–1.3	1.0	15–20%
	Medium	1.3	1.05	5–10%
	High	1.3–1.5	1.1	≤5%

a: Needs to meet all acceptance criteria
b: Semi-quantitatively evaluated (see Figure 13.9)

interpretations of the geotechnical domain boundaries to the limits of mining, based on boundary intersections and the continuity of the structural fabric, rock mass properties and hydrogeological parameters within each domain. Some structural analyses have been performed, utilising estimates of joint frequencies, lengths and conditions. All major features and joint sets should have been identified. Testing (small sample) for the physical properties of the in situ rock and joint surfaces will have been carried out. Similarly, groundwater data will be based on targeted pumping and airlift testing, and piezometer installations. All sampling, field testing and laboratory testing procedures must be sufficient to satisfy designated international standards for site investigation and laboratory testing (e.g. ISRM, ASTM). The information will be sufficient to allow full design models to be developed and sensitivity analyses to be carried out.

9.4.5.4 Level 4: Design and construction slope angle

A Level 4 design and construction slope angle requires that the reliability of the geotechnical model have been estimated at a high level of confidence. The work will be performance-based to confirm and update the results obtained during the feasibility investigations. It will include detailed mapping, observation of slope behaviour, the possible installation of trial slopes, observation of groundwater behaviour and confirmation of pumping parameters, field testing and laboratory testing. All sampling, field testing and laboratory testing procedures must be sufficient to satisfy designated international standards for site investigation and laboratory testing (e.g. ISRM, ASTM). The data will be sufficient to confirm the results of the Level 3 feasibility slope design.

9.5 Summary

Acceptance criteria are a means by which mine operators, mine owners and the investment community can establish the required performance of pit slopes with respect to safety, ore recovery and financial return.

Historically, the most used criterion has been the factor of safety (FoS), a deterministic measure of the ratio between the resisting and driving forces in the system. A state of balance or limiting equilibrium occurs when the FoS has a value of 1.0. As the FoS is a deterministic measure, the uncertainty in its value is usually accounted for by setting a prescribed minimum design acceptance value, derived from experience.

During the last 35 years an additional measure, the probability of failure (PoF), has become increasingly used as an acceptance criterion. In this option, the accepted practice has been to recognise the FoS as a random variable and seek the probability of it being equal to or less than 1. The measure is considered to have three main attractions:

- it enables the uncertainties in the capacity and demand functions in the system to be taken into account;
- it scales linearly;
- it is an essential parameter in the calculation of risk.

There are a limited number of published recommended design acceptance levels for the FoS or PoF. Those that have been published mostly relate to civil engineering not to mining, a fact which is evident in most of the tables presented in sections 9.2 and 9.3. To clarify this situation, Table 9.9 summarises the values of the FoS and PoF that are typically used as acceptance criteria in the mining industry. The measures of low, medium and high consequence of failure are based on the generic procedures used to develop the semi-quantitative risk matrix illustrated in Figure 13.9.

It should also be noted that, in addition to the FoS and PoF, numerical models can be used to estimate other acceptance criteria such as displacements. However, in practice such criteria can be difficult to establish since they depend on a thorough understanding of the failure mode. For example, toppling can accept several metres of annual movement without causing alarm. In contrast, more brittle failures such as daylighting wedges can accept very little displacement.

Although the acceptance criteria outlined in Table 9.9 calibrate the performance of pit slopes, they do not quantify the risks associated with slope failure. The risk model process outlined in section 9.4 was introduced to help overcome this shortcoming, albeit subjectively. The process is complex and remains to be widely adopted across the industry, but it provides a means of defining mining risks in terms of safety and economics. It also enables risk to be quantified for different levels of data confidence and varying slope configurations, and provides a means of quantifying the economic value added to the operation for increased levels of risk. Importantly, it also helps to implement the measures outlined in section 8.5.1 (Table 8.1), which are aimed at reporting the degree of confidence in the slope designs at each stage of project development (see Table 9.8).

Acceptance criteria form a corporate guideline that must be defined by management in consultation with the other stakeholders and take into account regulatory requirements. While the framing of the criteria is a management responsibility, input from the geotechnical and other specialist areas (e.g. risk, hydrogeology) is an essential part of the process.

10 SLOPE DESIGN METHODS

Loren Lorig, Peter Stacey and John Read

10.1 Introduction

The purpose of this chapter is to outline the essential steps in the formulation of pit slope design criteria. An integral part of this process involves slope stability analyses of the rock slopes in an open pit mine using the geological, structural, material property and hydrogeological information that has been brought together in the geotechnical model. The fundamental objective of the slope design process is to enable a safe and economic design for the pit walls at the bench, inter-ramp and overall slope scales. The process is outlined in Chapter 1 and re-summarised in Figure 10.1.

10.1.1 Design steps

The formulation of slope design criteria fundamentally involves analysis of the predicted failure modes that could affect the slope at bench, inter-ramp and overall scales. The process starts with the division of the geotechnical model for the proposed pit area into geotechnical domains with similar geological, structural and material property characteristics. The characteristics of each domain can be used to formulate the basic design approach. This essentially involves evaluating the critical factors that will determine the potential failure mode(s) at the respective scales (bench, inter-ramp, overall) against which the slope elements will be designed.

When assessing potential failure mechanisms, a fundamental consideration for any rock mass is that in stronger rocks structure is likely to be the primary control, whereas in weaker rocks strength can be the controlling factor, even down to the bench scale, as summarised in Figure 10.2.

For each domain, potential failure modes are assessed and designs are based on the required acceptance levels against instability as defined by company policy, industrial standards or regulatory requirements.

Where structure is expected to be a controlling factor, the slope orientation may exert an influence on the design criteria. In this case a further subdivision of a domain into design sectors is normally required, based upon kinematic considerations related to the potential for undercutting structures (planar) or combinations (wedges), or toppling on controlling features. The sectorisation can reflect controls at all levels, from bench scale, where fabric provides the main control for bench face angles, up to the overall slope scale, where a particular major structure may be expected to influence a range of slope orientations within a domain.

In the case of weak rocks, where the rock mass strength is expected to be the controlling factor for slope stability, the design process should commence with analyses to establish the overall and inter-ramp slope angles. Angles meeting the acceptance criteria should then be translated down in scale into bench face configurations.

Combinations of benches provide the inter-ramp slope, which may simply represent the height between access ramps in the pit. However, in larger pits with higher slopes, the slope designer may choose to provide more flexibility or stability by incorporating wider geotechnical berms (risk management berms) at prescribed height intervals on the wall. This approach is often used for the pre-mining design stages, when data are limited. It is also frequently used to ensure access onto the slope for surface water control, cleanup and the installation of dewatering wells or monitoring installations.

The inter-ramp angles are normally provided to mine planners as the basic slope design criteria. Only when ramps have been added does the overall slope angle become apparent. Thus, for initial mine design and evaluation work, an overall slope angle involving the inter-ramp angle, flattened by 2–3° to account for ramps, may be used for Whittle cone analyses and similar studies. This is discussed further in Chapter 11 (section 11.2).

Other factors in the slope designs could include:

- excavation equipment (controls operating bench height);

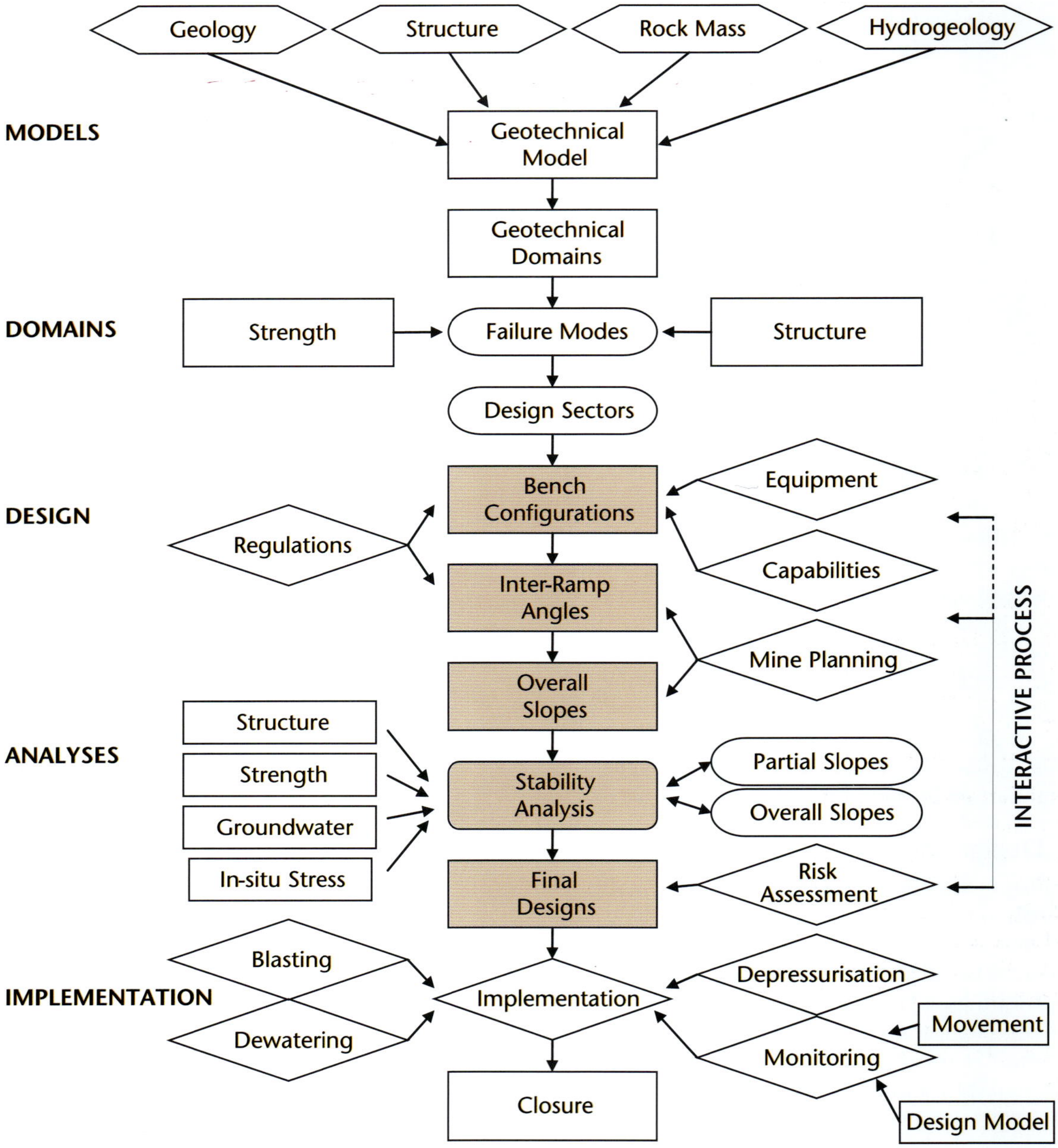

Figure 10.1: Slope design process

- equipment and operator capabilities;
- surface water control requirements (bench width);
- mine planning constraints (ore control and resulting mining height);
- regulatory restrictions (e.g. minimum bench width).

Safety considerations (high ravelling potential) may also be a factor that prevents stacking of benches.

10.1.2 Design analyses

The formulation of the slope design criteria for each element of a pit wall involves performing stability analyses to the required acceptance level (factor of safety or probability of failure). The type of analysis is largely dictated by the anticipated failure mode, the scale of the slope, available data and the level of the project. The process is often iterative, involving interaction with the mine planners.

The main types of analyses include:

- kinematic analysis, which is based on stereographic projections and is mainly applied to bench designs;
- limit equilibrium analysis applied to:
  - → structurally controlled failures in bench and inter-ramp design,

	Rock Mass Strength		
	Weak	**Moderate**	**Strong**
Slope Element			
Bench Face	Strength (structure)	Structure	Structure
IRA	Strength	Structure (strength)	Structure
Overall	Strength	Structure (strength)	Structure (strength)
Design approach	General Overall ↑↓ IRA ↑↓ Bench	By Sector Bench ↓ IRA ↓ Overall	By Sector Bench ↓ IRA ↓ Overall
	Major structures may impact Overall (and IRA)		

Notes

WEAK ROCKS
- Less susceptible to wall orientation unless major structures present
- Start by assessing Overall slope
- Fit bench configuration to Overall and/or IRA
- Bench height or angle may be controlled by strength
- Multiple benching (stacking) unlikely
- Water pressures likely to play major role

MODERATE TO STRONG ROCKS
- Sectorizallon required
- Structural control of BFAs
- Catchment design based upon anticipated failure quantity: minimum may be regulated
- Bench height controlled by equipment.
- Multiple benching (stacking) may be possible, especially in strong rock

Figure 10.2: Slope design controls by rock strength

→ inter-ramp and overall slopes where the stability is controlled by rock mass strength, with or without structural anisotropy;

- numerical analysis using finite element and distinct element methods for the assessment and/or design of the inter-ramp and overall slopes.

The stability analyses may form the basis of a risk assessment that incorporates mitigating factors to achieve acceptable levels of risk in terms of safety and economics.

The design methods in each situation are outlined below in two sections: kinematic analyses, which deal with the structurally controlled bench and inter-ramp failures; and rock mass analyses, which deal with the inter-ramp and overall slope failures controlled by the strength of the rock mass or combinations of the rock mass and major structures.

10.2 Kinematic analyses

10.2.1 Benches

The principal function of the benches is to provide a safe environment for personnel and equipment that must work near the slope face. Accordingly, they must satisfy needs for:

- reliability, which requires stable bench faces and bench crests. The primary variables controlling the stability of the bench faces and crests are the joint geometry and the shear strength of the joints;
- safety, which requires bench widths sufficient to arrest and mitigate the danger of rockfalls and contain any spillage from the benches above;
- long-term access along the benches for operators involved in activities such as slope monitoring and clean-up of rockfall and spillage.

The components of a bench are:

- the bench height, which is the vertical height between catch benches. Where benches are stacked this will be a multiple of the operating bench height;
- the bench width;
- the bench face angle.

The relationship is illustrated in Figure 10.3. A summary of the typical approach to analyses at a bench scale is illustrated in Figure 10.4.

10.2.1.1 Bench height

Bench heights in the range of 10–18 m are common in most large open pit mines. Fifteen metres is perhaps the most common, but the final decision is usually made by matching the height with the capacity of the excavating equipment (e.g. rope or hydraulic shovels) that will be

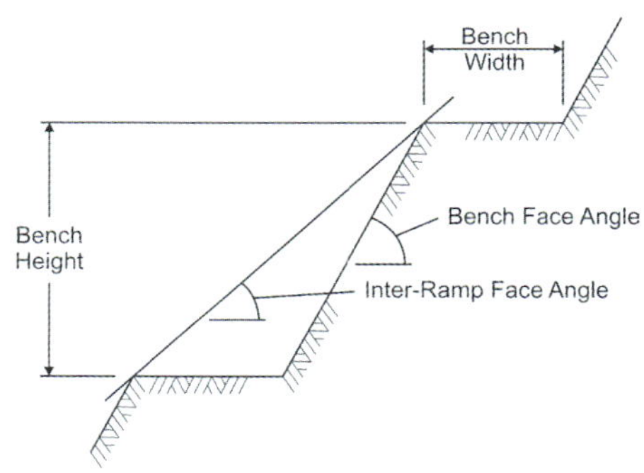

Figure 10.3: Components of bench configurations

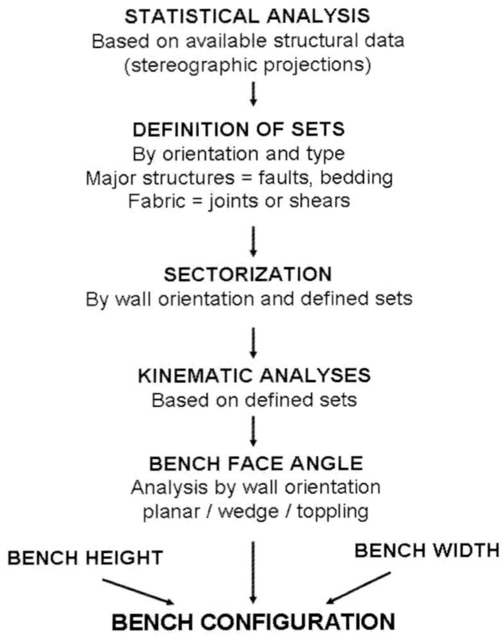

Figure 10.4: Bench face angle design process for moderate to strong rocks

used at the mine. Stacking of benches to steepen the inter-ramp slopes is common where rock strength and operating procedures permit. Although it was probably introduced to improve productivity, with improved drilling and blasting techniques double- and even triple-benching is increasingly being used as a means of improving safety, enabling wider, more reliable benches on the steepened inter-ramp slope.

10.2.1.2 Bench width

The benches must be wide enough to arrest potentially hazardous rockfalls and contain any spillage from the benches above. They must also allow long-term access to features such as slope movement and groundwater monitoring stations.

Most often the 'minimum' or design bench width is based on a mixture of company policy and operating experience; for bench heights of 15 m it is rarely less than 7 m. In some localities (e.g. British Columbia, Canada), it is codified together with bench height in relation to the capacity of the excavating equipment.

When considering rockfall, many practioners in North and South America use the modified Ritchie criterion developed by Call & Nicholas Inc. as a guide. The original criterion was published by Ritchie (1963) following an evaluation of highway shoulders for catching rockfalls from natural and excavated slopes. Ritchie's investigation was limited to a small number of geometries and therefore required extrapolation to typical mine geometries. The empirical relationship developed by Call & Nicholas Inc. defines the preferred catch bench width (Ryan & Pryor, 2000):

$$\text{bench width (m)} = 0.2 \times \text{bench height} + 4.5 \text{ m} \qquad \text{(eqn 10.1)}$$

In the case of retention capacity, it is neither practicable nor economical to design the benches to contain the spillage from every potential sliding plane or wedge failure in the benches above. Instead, the limit is set to the failed volume that that can reasonably be contained on the bench. Usually, this is 70–85% of the potential failed volumes. When calculating the failed volumes, structures included in the wedge analyses should strike at angles greater than 20° to the strike of the bench face. Planar failure structures should strike at angles of less than 20° to the strike of the bench face.

After rockfalls and spillage have been considered, the final step in the design process is to assess the likelihood of achieving the design bench width. Even with good blasting and excavation control, experience shows that a specific bench width can rarely be achieved with 100% reliability. The width that can actually be achieved is controlled by the amount of backbreak that occurs along the crest as the bench is excavated. The amount of backbreak is controlled by the same features that control the effective bench face angle (Figure 10.5). As noted above, there are no unique acceptance criteria, but catch bench reliabilities of around 80% (i.e. only 20% of the benches are less than the selected design width) are usually required.

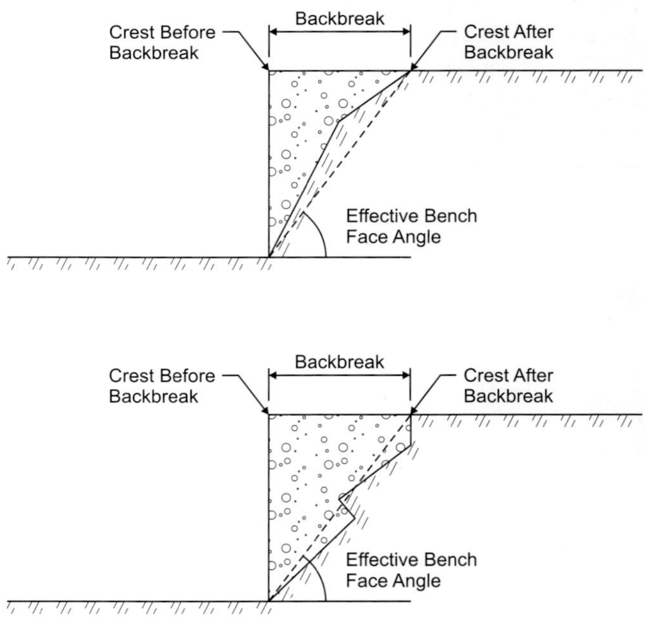

Figure 10.5: Definition of backbreak and effective bench face angle
Source: Ryan & Pryor (2000)

10.2.1.3 *Bench face angle*

The amount of backbreak and the effective bench face angle (Figure 10.5) is controlled by the joints and faults that intersect the bench.

The stability of these structures is controlled by a number of factors, including:

- their orientation relative to the direction of the bench face;
- the amount of undercutting that takes place as the bench is excavated;
- blast damage and reduced shear strength;
- blast-induced fractures on the bench face or in the vicinity of the bench crest.

The typical analytical approach used to establish bench face angles in moderate to strong rocks is presented in Figure 10.4. For weak rocks, the impact of both the rock mass strength and any relict structures must be considered (Figure 10.2).

The principal design task is to determine a face angle that undercuts as few of the structurally formed planes, wedges or blocks as possible. In this process, all the planes or wedges that could daylight in a defined window around the direction of the proposed slope are extracted from the structural fabric database and analysed in a break-back or cumulative frequency analysis to determine their stability and their probability of being flatter than the desired face angle (e.g. Read & Lye 1983; Ryan & Pryor 2000).

Sliding plane, block and wedge failures can be assessed using stereographic projection methods (John 1968; Phillips 1968; Ragan 1985; Lisle & Leyshon 2004; Wyllie & Mah 2004) or limit equilibrium techniques. For limit equilibrium solutions, the undercut planes, blocks and wedges are free bodies and the induced forces are due to weight and shearing resistance only. In these circumstances there is an equal amount of unknowns and equations, which means that the factor of safety can be estimated using deterministic and stochastic procedures based on the general limit equilibrium solution for plane failure:

$$\text{FoS} = \frac{\tan\varphi}{\tan\theta} + \frac{2c\sin\beta}{\gamma H\sin(\beta-\theta)} \qquad \text{(eqn 10.2)}$$

where

FoS = factor of safety
ϕ = angle of frictional
c = cohesion
γ = unit weight of material
H = bench height
β = effective bench face angle
θ = joint dip angle.

Because of blast damage and the relatively low stresses involved, the cohesion of the structures is often ignored in bench scale analyses.

Limit equilibrium sliding plane and wedge solutions for bench design are contained in a number of commercially available software packages, the best-known of which are probably SBlock (SRK 2006), Swedge (Rocscience 2006) and RocPlane (Rocscience 2004b). Proprietary break-back analysis software is also used by a number of consultants. Most of these programs are limited in that they can work at bench scale only. For the LOP project, CSIRO developed the software package Siromodel (Poropat & Elmouttie 2006), which uses DFN techniques (section 4.4.3) to create a 3D structural model of any section of the pit and perform break-back analyses at single bench and inter-ramp scale. The software will accept scan line or digitally collected structural data and a DXF file-defined or a user-defined pit geometry. For user-defined pit models, the software allows the users to simulate mining the pit to see if changes in pit geometry will affect the stability of the benches and structurally controlled inter-ramp slopes.

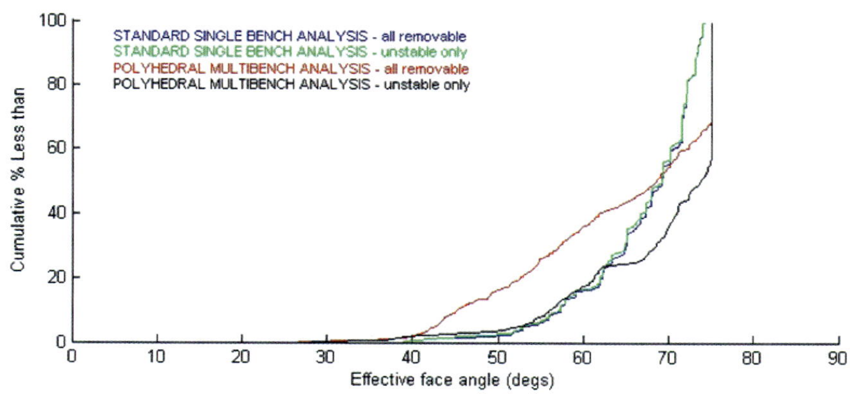

Figure 10.6: Screen snapshot of Siromodel results for a standard single-bench effective face angle analysis (blue and green curves) and a polyhedral multi-bench inter-ramp slope angle analysis (red and black curves)
Source: Courtesy CSIRO

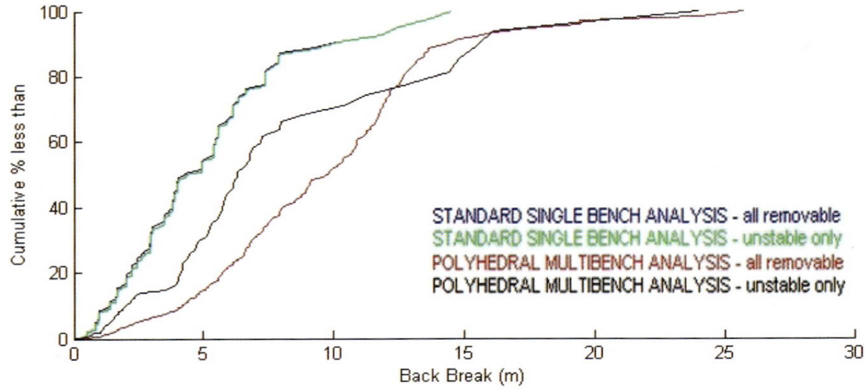

Figure 10.7: Screen snapshot of Siromodel results for a standard single-bench break-back analysis (blue and green curves) and a polyhedral multi-bench inter-ramp break-back analysis (red and black curves)
Source: Courtesy CSIRO

In the Siromodel analyses, at bench scale all possible block and wedge combinations are evaluated to produce estimated cumulative frequency curves for the effective face angle and the amount of break-back for the selected bench. At the inter-ramp scale the analysis looks for potentially unstable block and wedge combinations (polyhedra) across all the benches that are stacked on the geometry being assessed.

The results of the bench scale break-back analysis are usually shown as cumulative frequency curves such as the blue and green curves in Figure 10.6 (see also Chapter 12, Figures 12.6 and 12.7). In this figure, which was performed using Siromodel for Bench 3 from a stack of five benches on a target inter-ramp slope, the blue curve represents all potentially unstable blocks and wedges and the green curve only those blocks and wedges that have a FoS of less than 1.0. The matching break-back cumulative frequency curves for Bench 3 are shown in blue and green in Figure 10.7. The comparative effective face angle and break-back curves for the inter-ramp slope above Bench 3 are shown in red and black on Figures 10.6 and 10.7. The ability to view the results of the single-bench and multi-bench inter-ramp break-back analyses together enables the predicted face angle and break-back values to be double-checked before the respective designs are finalised.

10.2.1.4 Bench toppling

There are two types of toppling failure: block toppling and flexural toppling. Block toppling (Figure 10.8) occurs where individual columns formed by closely spaced joints dip into the bench at angles of 65–85° and a second set of more widely spaced orthogonal joints are undercut at the toe of the bench. Toppling is promulgated if the centroids of the individual blocks lie on the pit side of the toe of the block or as the blocks at the toe are pushed forward by loads from the overturning blocks at the top. The action is often enhanced if the bench face angles are steeper than 50°.

Flexural toppling (Figure 10.9) differs from block toppling in that the inward-dipping columns are more continuous and maintain face-to-face contact as they bend over in flexure. Usually, at a bench scale flexural toppling is associated with thinly bedded and/or slightly metamorphosed rocks such as shale and phyllite rather than jointed sedimentary or igneous rocks. Characteristically, it exhibits interlayer shearing, obsequent scarps at the crest and tension cracks behind the crest that decrease in width at depth.

The analysis and prediction of toppling is a less than exact science. In civil engineering, limit equilibrium methods have been used to analyse block toppling, mostly with a view to calculating the loads required to anchor individual blocks (Wyllie & Mah 2004). Because of the bending moment, limit equilibrium methods cannot be used to analyse toppling. Techniques used for toppling include base friction models (Goodman 1976),

Figure 10.8: Bench scale block toppling on joints in granodiorite

Figure 10.9: Bench scale flexural toppling on cleavage in sericitic phyllite

centrifuges (Adhikary et al. 1997) and numerical models including the Itasca codes FLAC (section 10.3.4.1) and UDEC (section 10.3.4.2). However, numerical approaches require an understanding of the mechanics involved in the process.

As illustrated in Figure 10.10, bending of a layered rock mass introduces non-uniform stresses into the layers. Several numerical models, including UDEC, average stresses over the element. This means that the numerical models that use this simple formulation must have more than one element across the thickness of the rock layer in order to represent the non-uniformity in stress.

Since the bending stiffness of the intact rock layers is typically neglected in continuum formulations and the bending-induced stress non-uniformity vanishes upon averaging, the conventional theories (e.g. the ubiquitous joint model) lack the capability to accurately model the behaviour of layered rocks subjected to bending. The only alternative in continuum models is to incorporate moment stresses in the model formulation, which will not vanish upon continuum formulation.

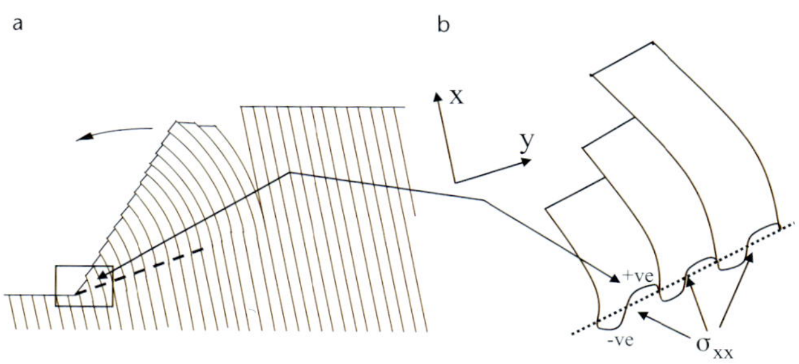

Figure 10.10: Bending-induced stress non-uniformity in a layered rock mass. (a) Flexural toppling of a layered rock slope. (b) Bending-induced tensile and compressive stresses in rock layers

Relative joint displacement in a layered rock with bending stiffness results in asymmetric macroscopic shear stresses, which are physically equilibrated by bending moments. Hence, in describing the behaviour of layered rocks, in addition to the conventional force equilibrium equations, in continuum formulations it is necessary to incorporate an extra equilibrium equation balancing the differences in shear stresses by bending moments. In such conditions, bending moments and associated rotations become additional independent variables. Appropriate models are those based on micro-polar theories which incorporate these independent variables in their formulation.

Such equivalent continuum models can be formulated successfully on the basis of Cosserat theory (Cosserat & Cosserat 1907) and were followed in Mühlhaus (1993), Dawson and Cundall (1996) and Adhikary and Dyskin (1998), where the rock layers were assumed to be elastic. In Adhikary and Dyskin (1998), provision was made for plastic deformation along the joints only. Recently, CSIRO developed a fully elasto-plastic 2D plane strain Cosserat model, COSFLOW, in which both joints and intact rock (rock layers) are allowed to undergo plastic deformation (Adhikary & Guo 2002). The yield of both the rock matrix and the joints is defined by the Mohr-Coulomb criterion with tension cut-off. Recent developments indicate that it may be possible to simulate the toppling process using the particle flow code PFC2D and its derivative PFC3D (section 10.3.4.10).

The efficiency of numerical codes depends on the layer thickness relative to the slope height. Modelling high slopes with thick layers is the numerical equivalent to studying bench scale slopes with thin layers. At bench scale the best indicators are the inward-dipping orientation of the closely jointed or laminated structures as observed in surface outcrops and drill core. Experience has shown that prevention depends on good excavation control as the bench is mined. Undercutting at the toe must be prevented and transitory pore pressures must not be allowed to develop in any tension cracks formed behind the crest. Once started, toppling is often difficult to stop. The only effective remedy is to carefully recut the bench at a flatter angle.

10.2.2 Inter-ramp slopes

Combinations of benches form the inter-ramp slope. Plane and wedge failures traversing these benches can interfere with the integrity of the inter-ramp slope and any intervening access ramps (Figure 10.11). In the extreme, they can disrupt the entire slope (Figure 10.12).

Inter-ramp scale plane and wedge failures can be formed by singular or multiple combinations of faults and/or persistent joint sets that are long enough to define failure geometries ranging from two or three benches to full slope height. Kinematically, the methods used to design the slopes are the same as for benches – only the scale is different. Additionally, because of the greater heights inherent in inter-ramp slopes there is the possibility of more complex failure modes that may include rock mass failure (e.g. non-daylighting wedges). These more complex failure modes typically require analysis by numerical methods.

Figure 10.11: Wedge failure on an inter-ramp slope

Figure 10.12: Wedge failure disrupting the entire slope

10.2.2.1 Inter-ramp slope height

There are no unique criteria governing the height of the inter-ramp slopes, which may simply be represented as the height between the access ramps in the pit. In large pits with high slopes the geotechnical engineers and mine planners may choose to provide more flexibility for pit access by introducing wider benches or ramps at prescribed intervals on the wall. Wider benches may also be introduced to decouple high slopes to provide additional stability or a safer working environment on the slopes below. Typical maximum heights are 100–200 m between the wider benches or ramps.

10.2.2.2 Inter-ramp slope angle

Although kinematically the methods used to design the inter-ramp slopes may be the same as for the benches, the increased slope heights and the potential consequences of large plane and wedge failures through a number of the benches on the slope add a new dimension to the design process. Spillage onto ramps, the loss of ramps (Figure 10.11) and overspillage onto the slopes below the ramps (Figure 10.12) are examples of the types of hazards to be anticipated and avoided.

When searching for these types of instability, the following procedures are critical and must be followed in each design sector to help determine the optimum inter-ramp slope angle.

1 The definition of the orientation, spacing, length, location and/or probability of occurrence of the faults and persistent joints that are long enough to define plane and wedge failure geometries within the inter-ramp slope. Although they may be relatively widely spaced, major regional and mine scale faults are likely to be continuous along strike and down dip, and thus are liable to have the most impact. However, if it is

closely spaced, bench scale joint fabric also has the potential to link in step-path fashion to form a more continuous failure surface or one side of a wedge. In all cases, the level of confidence in the data should be linked with the stage of project development and specified using the criteria suggested in section 8.5.1 and Table 8.1.

2 The determination of the shear strength of the faults and joints. Persistent joints and faults, particularly the major regional and mine scale faults, are likely to be planar and/or slickensided and filled with crushed and sheared material (fault gouge). The nature of all these features must be thoroughly described in engineering terms and the shear strength of the faults and joints determined following the procedures outlined in section 5.3. Cohesion is often considered in inter-ramp analyses.

3 The performance of stability analyses using the information gained from Steps 1 and 2 to determine the likelihood of planes and wedges failing and undercutting the benches and ramps on the inter-ramp slope. The output should include the volume and tonnage of material generated by the failure. The analyses may be performed to determine the effect of individual planes and wedges at particular locations or on a sector-wide scale to estimate, for example, the number of failures and the total failure volumes and tonnages expected for each inter-ramp slope angle. Joints and faults included in the wedge analyses must strike at an angle of more than 20° to the azimuth of the design sector. Joints and faults that strike at angles of less than 20° to the azimuth of the design sector should be included in the plane failure analyses. Limit equilibrium sliding plane and wedge solutions suitable for the analyses are available in software packages such as SBlock (2006), Swedge (Rocscience 2006), RocPlane (Rocscience 2004b) and Siromodel (Poropat & Elmouttie 2006).

10.2.2.3 Inter-ramp toppling

Flexural toppling can develop very large obsequent scarps (Figure 10.13) and can easily propagate from the benches into the inter-ramp and overall slopes (Figure 10.14).

As outlined in section 10.2.1.4, predicting the development of flexural toppling from its start point at a bench level to almost the full slope height and assigning the slope a factor of safety or a probability of failure is possible but difficult, even with astute modelling. However, experience has shown that although such performance criteria are difficult to predict, toppling slopes can be managed by matching the rate of deformation to the rate of mining and the slope angle. A key factor in this regard is that flexural toppling tends to be non-catastrophic as long as the slope is not so steep that the rubble generated by the degradation of the benches falls. This tends to restrict the inter-ramp slope angles to approximately 40°. Although visually extreme, the obsequent scarp pictured in Figure 10.13, which is located below the haul road towards the centre of the photo in Figure 10.14, was at no stage unsafe, at least in the view of the person walking along it. It developed over three to four years as the pushback was mined out. Although the final slope is ragged and the form of the benches has been lost, the slope did not collapse and most of the ore at the toe was mined out successfully in a carefully controlled operation. The keys to success are good geological modelling, recognition of the likelihood of toppling, the selection of conservative slope angles, careful preparation of the benches, monitoring of the deformation with mining, and reconciliation of the rate of deformation with the rate of mining.

Figure 10.13: Large obsequent scarp associated with flexural toppling in sericitic phyllite

Figure 10.14: Flexural toppling developed across inter-ramp slopes in altered phyllitic rocks

Besides directly undercut features, the potential for failures controlled by non-daylighting structural configuration, e.g. non-daylighting wedges, should also be considered.

10.3 Rock mass analyses

10.3.1 Overview

Rock mass stability analyses examine the potential for inter-ramp and overall slope failures where the failure mechanisms are controlled by the strength of the rock mass. They are an essential final step in the design process, to check that there are no fatal flaws in the inter-ramp slopes selected by the kinematic analysis process and that the rock mass can sustain the proposed design over the full height of the slope.

Analysis of rock mass controlled failures commenced in the 1950s and 1960s and was based on soil mechanics experience and methodology. The analyses incorporated fundamental assumptions of scale and discontinuity density, for example that the size of rock particles in high, closely jointed rock masses were considered equivalent to an isotropic mass of soil particles. This assumption enabled slope design practitioners to adopt the Mohr-Coulomb measures of friction (Ø) and cohesion (c) to represent the strength of the rock mass. This led to the direct use of the emerging limit equilibrium slope failure analyses such as Bishop, Janbu, Morgenstern and Price and Spencer in slope design, and the embedment of Mohr-Coulomb strength parameters in the limit equilibrium stability chart procedures introduced in the 1970s and 1980s. The use of Mohr-Coulomb strength parameters was also carried over into all the continuum and discontinuum numerical modelling tools now common in pit slope design.

Limit equilibrium and numerical modelling slope design tools that use the Mohr-Coulomb criterion to represent the strength of the rock mass are all used today at some stage of project development. Regrettably, even a cursory examination of the literature reveals that it is overflowing with articles addressing the perceived advantages, idiosyncrasies and limitations of these tools, a situation that is often more confusing than helpful. This section will attempt to cut through this excess of information and concisely review the background to each method, how it is used, when it should be used and, equally importantly, when it should not be used. It will also examine current research trends and developments aimed at closing the critical gaps in our understanding of rock mass failure in large open pits.

10.3.2 Empirical methods

At an early stage of project development when data are limited and the geotechnical model has not been fully developed, empirical charts can be extremely useful for establishing preliminary slope designs, provided that their limitations are recognized.

10.3.2.1 Slope angle versus slope height charts

A number of authors have published slope angle versus slope height charts. Well-known examples include charts published by Hoek (1970) and Sjöberg (1999). Hoek reported the investigation of many slopes and included a range of slope angles, shown in Figure 10.15. Similar charts appear in Hoek and Bray (1981) and Wyllie and Mah (2004).

Hoek's approach of simply comparing slope angle with slope height from actual successful and failed slopes was used in a more recent study by Sjöberg (2000), which also classified the slopes by the characteristic rock-hardness rating. Sjöberg's data for the two cases are plotted in Figure 10.16, using the notation that open symbols represent successful slopes and solid symbols represent failed slopes.

Hoek's correlation of slope angle versus slope height to an approximate FoS trend is also indicated in Figure 10.16. As with Hoek's findings, some slopes appear stable when a slope angle versus slope height classification would suggest failure, while others failed where stability might have been expected. Sjöberg's update of Hoek's work suggests a wider range of uncertainty.

The question of uncertainty raises difficulties with viewing slope angle versus slope height charts as

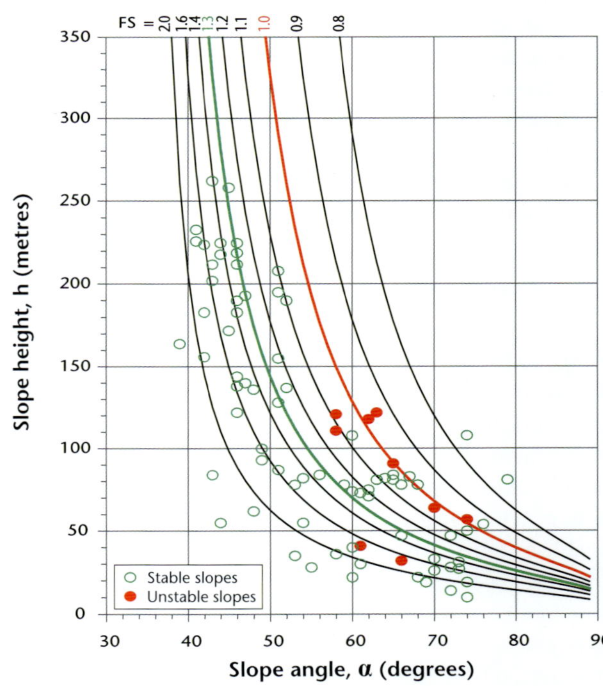

Figure 10.15: Rock slope versus slope height, distinguishing between failures and non-failures

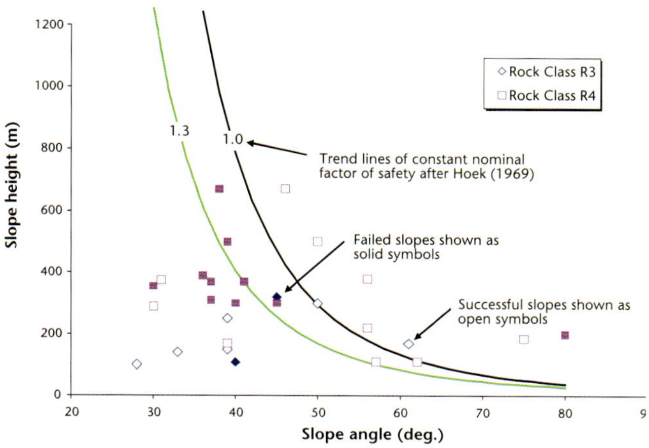

Figure 10.16: Rock slope success and failure designated by rock strength
Source: data from Sjoberg (2000)

anything other than interesting information. Basically, it is a case of comparing apples with oranges and thus has significant limitations. First, even when some measurement of rock mass strength is introduced the approach does not recognise differences or biases in the characterisation data between slopes. Second, the failure mechanism, be it structurally controlled or rock mass controlled, is not accounted for. Provided these limitations are recognised, however, the approach can provide a reality check on any outcome. For example, Figure 10.15 shows that 150–350 m high slopes at angles of 70° are unlikely to be stable, which does provide useful information to those who are unfamiliar with the world of slope stability geomechanics.

10.3.2.2 Empirical design charts

Design charts provide design slope angle and slope height guidelines by combining the experience gained from the known performance of slopes at various mine sites with the information provided by a classification scheme. One of the best-known and most widely used charts is that published by Haines and Terbrugge (1991), which is based on the Laubscher MRMR rock mass rating scheme (section 5.4.3). The chart is shown in Figure 10.17.

The limitations of using design charts are their experiential and semi-quantitative nature. These limitations are clearly recognised in the Haines and Terbrugge chart, which indicates where it believes slope angles and slope heights can be determined solely on the basis of the MRMR rating of the rock mass, where the estimate is marginal and where additional analyses are required.

Design charts are useful tools for estimating preliminary slope designs at the conceptual and pre-feasibility stages of project development. However, they should not be used at the feasibility or design stage of a project unless it can demonstrated with reasonable to high levels of certainty, that the failure mechanisms being studied involve only rock mass failure. In this

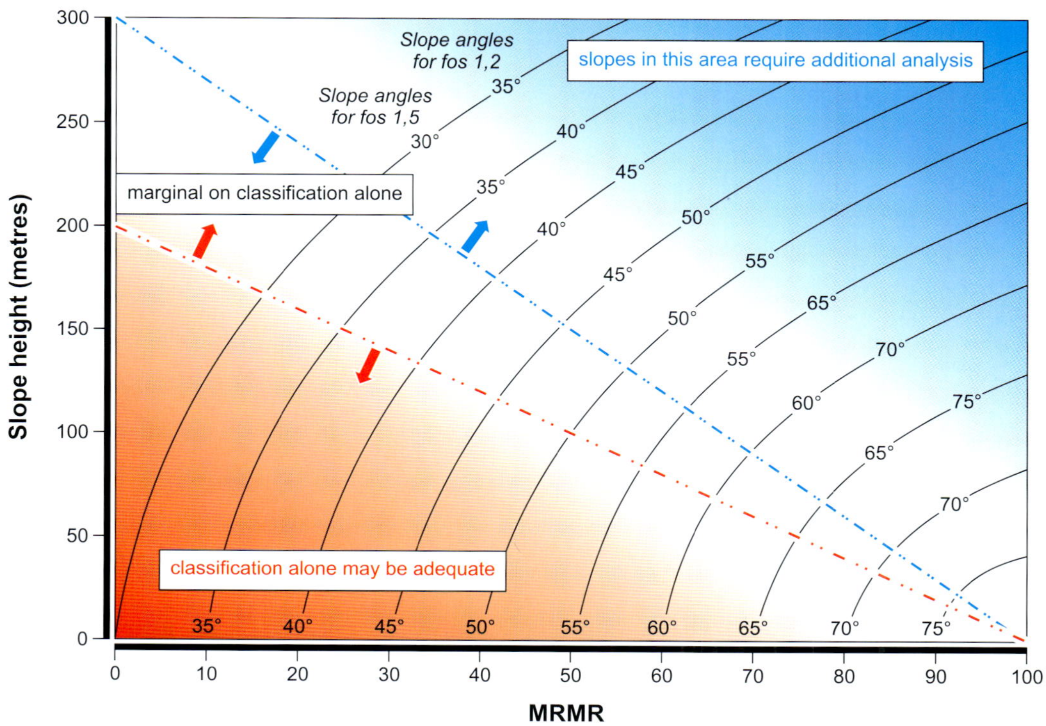

Figure 10.17: Haines & Terbrugge chart for determining slope angle and slope height

context they can be a useful tool for checking that the rock mass can sustain the slope heights proposed for inter-ramp and overall slopes selected on the basis of kinematic planar and wedge analyses.

10.3.3 Limit equilibrium methods

Limit equilibrium methods use representative geometry, material and/or joint shear strength, material unit weights, groundwater and external loading/support conditions to determine slope safety factors based on a set of simplifying mechanical assumptions that are outlined below. A practical, complete and accessible description of limit equilibrium methods is in Duncan and Wright (2005).

10.3.3.1 Method of slices

The limit equilibrium methods used to determine the stability of sliding planes, blocks and wedges are solved for a single free body and do not depend on the distribution of the effective normal stresses along the failure surface. However, if the mobilised Mohr-Coulomb strength of the rock mass is to be calculated, the distribution of the effective normal stresses along the candidate failure surface must be known. Solutions for this condition are usually based on the 2D method of slices, which divides the body into *n* slices above the candidate surface. This surface is often assumed to be circular, but may take any shape as the method of slices can readily accommodate complex slope and candidate failure surface geometries, variable rock mass conditions and external boundary loads. However, the problem is statically indeterminate as the solution has more unknowns than equations.

The forces acting on an individual slice in the method of slices are illustrated in Figure 10.18. The associated equations and unknowns are summarised in Table 10.1.

The most widely known methods of analysis based on the method of slices and the static equilibrium conditions satisfied by each are summarised in Table 10.2. Additional details of these and other methods can be obtained from a number of different sources: comprehensive descriptions and examples are provided in Abramson et al. (1996) and Duncan and Wright (2005).

The ordinary method of slices (OMS) (Fellenius 1927, 1936) is the earliest and simplest method and perhaps the only one that can be solved without a computer. However, it neglects all interslice forces and does not satisfy force equilibrium for the slide mass or the individual slices. Bishop's simplified method (Bishop 1955) and Janbu's simplified method (1954, 1957, 1973) are also only partial equilibrium solutions. Both assume zero interslice forces, reducing the number of unknowns to (4n – 1), leaving the solutions overdetermined. Bishop leaves horizontal force equilibrium unsatisfied for one slice and Janbu does not completely satisfy moment equilibrium; Janbu presents

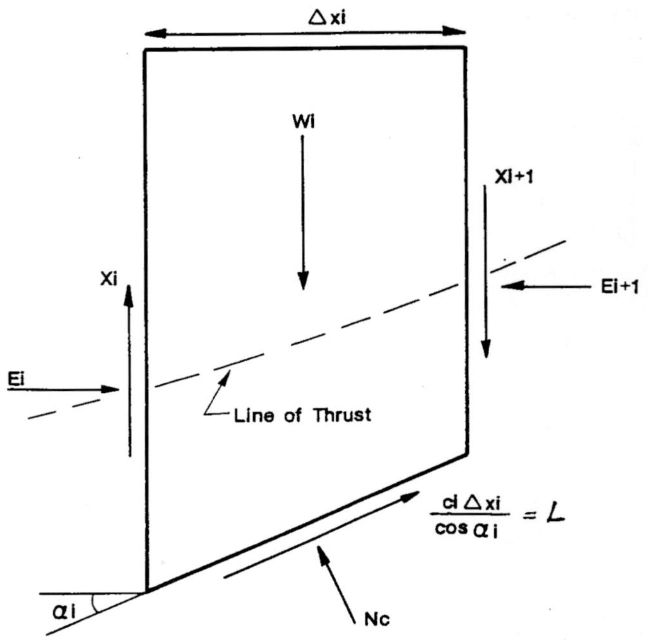

Figure 10.18: Forces acting on an individual slice in the method of slices

the correction factor, f_o, to account for this deficiency. Bishop and Janbu also outline more rigorous methods that allow them to better satisfy equilibrium. Those methods similarly suggest that the position of the line of thrust is an additional unknown. In the subsequent analyses, equilibrium is said to be rigorously satisfied if the assumption selects the correct thrust line.

The US Army Corps of Engineers method considers the inclination of the interslice forces to be horizontal to the

Table 10.1: Equations and unknowns associated with the method of slices

Equations	Condition
n	Moment equilibrium for slice (∑M = 0)
2n	Horizontal and vertical equilibrium for slice (∑Fh & ∑Fv = 0)
n	Mohr-Coulomb equation
4n	Total number of equations
Unknowns	**Variable**
1	Factor of safety, FoS
n	Normal force, N
n	Position of N on sliding plane
n	Shear force, T
n-1	Horizontal interslice forces, Ei and Ei + 1
n-1	Vertical interslice forces, Xi and Xi + 1
n-1	Line of thrust, position of Ei, Ei + 1
6n-2	Total number of unknowns

Table 10.2: Static equilibrium conditions satisfied by the method of slices

Method	Force equilibrium x	Force equilibrium y	Moment equilibrium
Fellenius OMS	No	No	Yes
Bishop's simplified	Yes	No	Yes
Janbu's simplified	Yes	Yes	No
US Corps of Engineers	Yes	Yes	No
Lowe & Karafiath	Yes	Yes	No
Morgenstern & Price	Yes	Yes	Yes
Spencer	Yes	Yes	Yes
Sarma	Yes	Yes	Yes

ground surface or equal to the average slope between the left and right end points of the failure surface. As with Bishop and Janbu, this produces an overdetermined solution which does not satisfy moment equilibrium for all slices. The Lowe and Karafiath method is another partial equilibrium solution that also fails to satisfy moment equilibrium. It assumes that the interslice forces are inclined at an angle that is equal to the average of the ground surface and the slice base angles, leaving (4n – 1) unknowns and an overdetermined solution.

Morgenstern and Price (1965), Spencer (1967, 1973) and Sarma (1973) all satisfy force and moment equilibrium. The Sarma solution is different from the others in that it uses the method of slices to calculate the magnitude of a horizontal seismic coefficient (section 10.3.3.5) needed to bring the slide mass into a state of limiting equilibrium. The procedure develops a relation between the seismic coefficient and a presumed FoS, with the static FoS corresponding to the case of a zero seismic coefficient. If Sarma's method is used, it should be noted that the candidate failure surface corresponding to the static FoS may be different from the surface determined using more conventional approaches.

The method of slices solutions that completely satisfy equilibrium have been shown to provide similar values for the FoS (Fredlund & Krahn 1977; Duncan & Wright 1980). Well-known slope stability programs that incorporate most of the available options include SLOPE-W (Krahn 2004), SLIDE (Rocscience 2004b) and XSTABL (Sharma 1992). In the case of hard rock slope stability, the preferred solution techniques are those of Spencer, Morgenstern and Price and Janbu, because they can model irregular failure surfaces.

Overall, limit equilibrium methods are popular because they are relatively fast and easy to use. Their main disadvantage is the assumption that the failure is of a rigid body (i.e. deformations within the sliding body are ignored completely). Out-of-slice forces are also ignored. Judging from observed failure modes in large-scale slopes, these are oversimplifications. Furthermore, to some extent the failure surface must be known. Care is also required when attempting to simulate the effect of anisotropy. Many programs allow anisotropy within rock mass units to be specified. However, the anisotropy is not involved unless the failure surface corresponds to the direction of anisotropy. Methods for defining the directional strength of a jointed rock mass are outlined in section 5.5.4. Situations where limit equilibrium models cannot be used include toppling, block flow failures and crushing failures at the slope toe.

10.3.3.2 Incorporating water pressures

Two methods are common for specifying the distribution of pore pressure within slopes. The most rigorous method is to perform a complete flow analysis and use the resultant pore pressures in the stability analyses. A less rigorous but more common method is to specify a phreatic or a piezometric surface. In this case, the correct method for calculating the pore pressure forces on an end slice or an intermediate slice in any analysis, based on the method of slices, is illustrated in Figure 10.19.

When calculating the pore pressure forces it is important not to confuse phreatic and piezometric surfaces. Phreatic surfaces represent the free groundwater level within the slope. In most slopes the groundwater level will be inclined, indicating groundwater flow. Such conditions require the pore pressure calculations to account for seepage losses. This requires determination of the equipotential line passing through the centre of the slice base.

If the equipotential line is assumed to be a straight line, the inclination of the phreatic surface and the magnitude of the vertical distance between the phreatic surface and the slice base may be used to estimate the pore pressure head. The pore pressure, u, will then be (Figure 10.20):

$$u = h_w \cos^2\alpha \qquad \text{(eqn 10.3)}$$

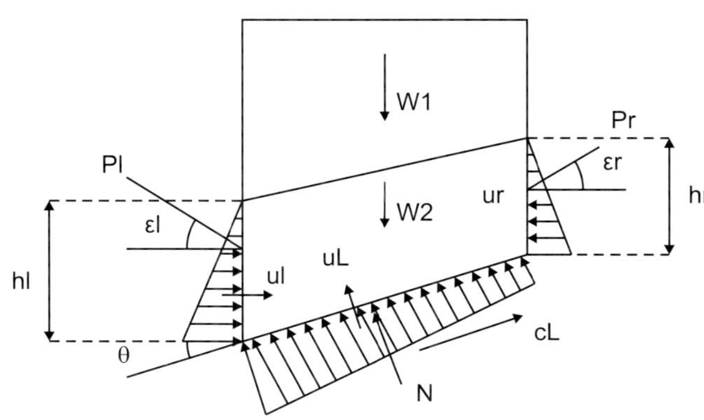

ul = 0.5γwhl² : uL = 0.5γwLhl+hr : ur = 0.5γwhr²

Figure 10.19: Introduction of pore pressures in stability analyses based on the method of slices

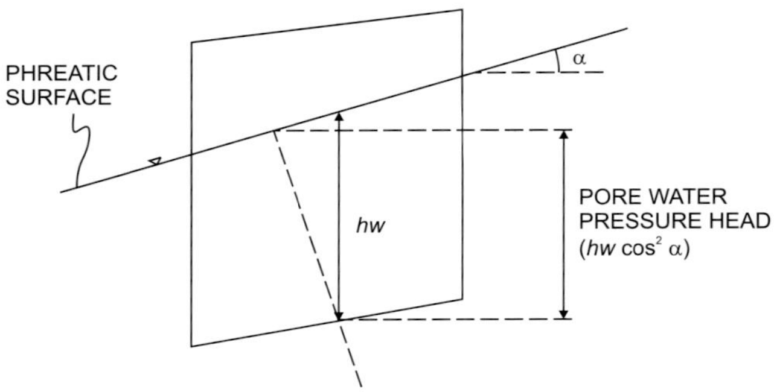

Figure 10.20: Pore pressure head measured from the phreatic surface

The piezometric surface represents the actual pressure head relative to a surface within the slope. In 2D this surface will correspond to a line such as a candidate failure surface or a rock type boundary. If the piezometric head is known (measured from a piezometer in the slope), the pore pressures should be calculated according to the vertical distance between the base of the slice and the piezometric surface (Figure 10.21):

$$u = h_w \qquad \text{(eqn 10.4)}$$

In situations where the phreatic and piezometric surfaces have been confused, if the inclination of the phreatic surface is small (e.g. <5°) the results of the analyses will be only slightly affected. However, for larger angles the calculated differences will be significantly greater, with the phreatic surfaces always generating lower pore pressures than the piezometric surface. It is important to note that the introduction of water pressure into a limit equilibrium solution assumes that water pressure acts on all rock surfaces at all scales. In essence, the rock mass acts as a soil or gravel. This notion is confirmed by observing that the water pressure is referred to by the soil mechanics term 'pore pressure'.

10.3.3.3 Design charts

The first slope stability design charts were published by Taylor (1937, 1948). They were developed in terms of a total stress approach for simple, dry, homogenous soil slopes using the statically determinant friction circle method of analysis. Classical examples of the use of the charts are provided in Abramson et al. (1996).

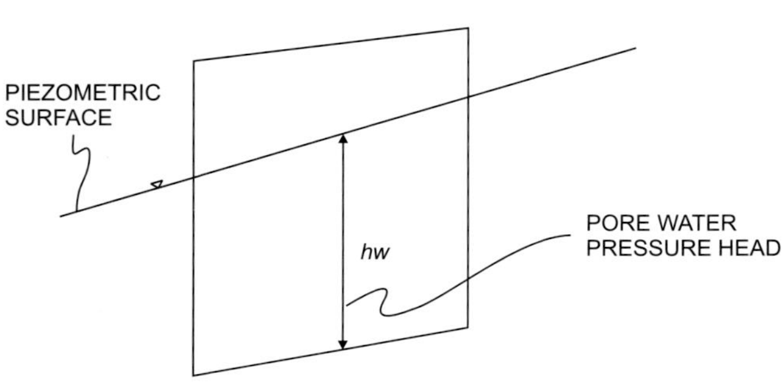

Figure 10.21: Pore pressure head measured from the piezometric surface

A number of charts for soil slopes have been published since 1948, including those developed by Bishop and Morgenstern (1960), Gibson and Morgenstern (1962), Spencer (1967) and Janbu (1968). However, the first and most widely used stability charts for slopes in closely jointed rock masses are those developed by Hoek and Bray (1981). An example is given in Figure 10.22.

The use of the stability charts requires that slope conditions meet the following assumptions:

- the material forming the slope is homogenous, with uniform shear strength properties along the candidate failure surface;
- the shear strength of the material (τ) is characterised by the Mohr-Coulomb equation where $\tau = c + \sigma \tan \emptyset$;
- failure occurs on a circular or rotational slide surface that passes through the toe of the slope;
- a vertical tension crack occurs in the upper surface or the face of the slope;
- the location of the tension crack and the slide surface are such that the slope FoS is a minimum for the slope geometry and groundwater conditions considered.

It should also be noted that the charts are optimised for a rock mass density of 18.9 kN/m³. Higher densities may provide higher FoSs than indicated by the charts and lower

CIRCULAR FAILURE CHART NUMBER 3

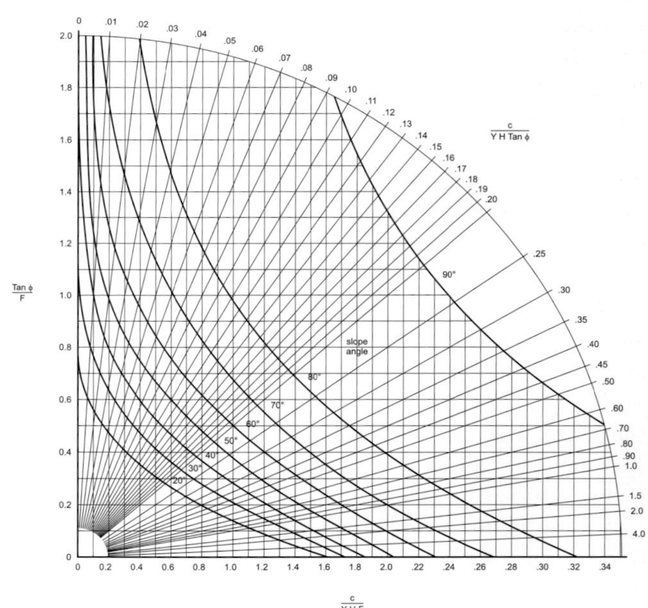

Figure 10.22: Hoek and Bray slope failure chart no.3, corresponding to the groundwater conditions shown for chart no. 3 in Figure 10.23

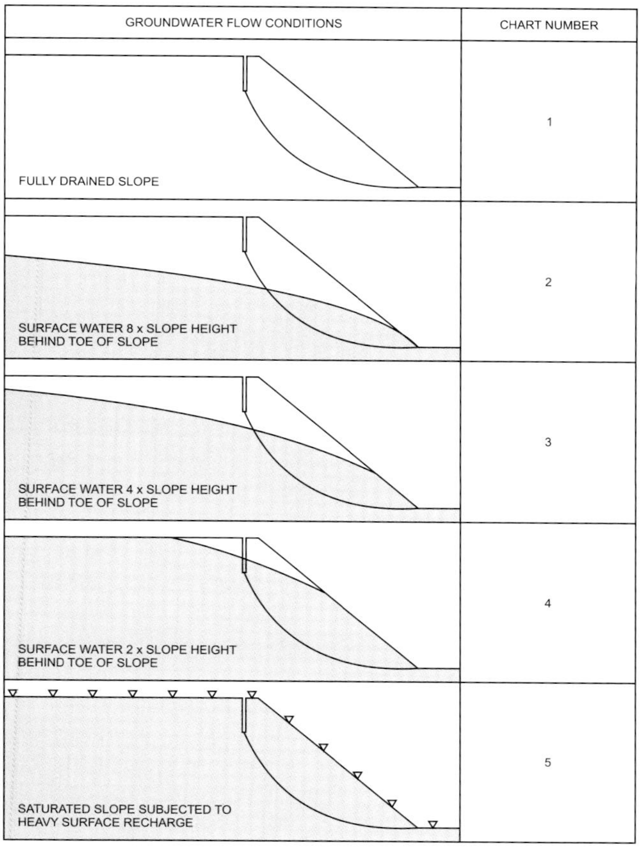

Figure 10.23: Groundwater conditions for use with Hoek and Bray slope failure charts

densities may give lower FoSs. Many users are not aware of this point and do not realise that it may be inappropriate to apply the charts to slopes in which the material density is significantly different from 18.9 kN/m^3, which includes most rock slopes.

10.3.3.4 3D methods

Two-dimensional programs examine the stability of a unit-width slice of the slope and ignore any shear stresses on the sides of the slice. While 2D procedures are a reliable method of analysis, circumstances can arise for which 3D analysis is required to define the slide surface and slope geometry more precisely.

Most of the general-purpose 3D slope stability analysis procedures are based on a method of columns (or prisms). The columns method is the 3D equivalent of the slices method in two dimensions. In the columns method, the rock mass is subdivided into an approximately square cross-section in plan view. However, a considerable number of assumptions must be made to satisfy the six equations of static equilibrium and achieve a statically determinate solution. Several methods employ simplifying approaches rather than fully satisfying the six equations. Unfortunately, the effects of these assumptions may be as large as the 3D effects themselves, which can generate a considerable amount of uncertainty in the results. Hence, 3D procedures should be used cautiously, especially when they are used as a basis for acceptance in cases where 2D analyses may indicate unacceptably low FoSs (Duncan & Wright 2005).

Programs that solve 3D problems using limit equilibrium methods include CLARA (Hungr 1987; Hungr et al. 1989), CLARA-W (Hungr 2002), 3D–SLOPE (Lam & Fredlund 1993) and EMU–3D (Chen et al. 2001).

10.3.3.5 Seismic analysis

There is considerable debate about the need for seismic analyses for open pit slopes. There are few, if any, recorded instances in which earthquakes have been shown to produce significant slope instabilities in hard rock conditions, a statement supported by evidence from a number of mines in highly active seismic zones. These include the Bougainville and Ok Tedi mines in Papua New Guinea, and a number of mines in Chile and Peru.

Earthquakes have produced small shallow slides and rockfalls in open pits but none on a scale sufficient to disrupt mining operations. The Bougainville mine, which is located within 60 km of the Pacific plate subduction zone (Figure 3.14) and experiences a level of seismic activity three times that of California, was subjected to the design earthquake (magnitude 7.6) in July 1975. Although there were a number of spectacular tailings liquefaction failures, small face failures on the waste dumps and some overburden failures on ridge crests around the pit, there were no incidents on the pit slopes.

Because of the high level of seismicity in the Bougainville Island region, in 1981 Bougainville Copper Ltd commissioned a consultant to collate the record of open pit behaviour under earthquake loading in Chile, selected because of its mining history and its tectonic similarity to Bougainville. The west coast of Chile is underlain by the Peru-Chile subduction zone (Figures 3.16 and 3.17). The basic conclusion of the report (Hoek & Soto 1981) was that the Chilean open pit mines exhibited a high degree of stability, with the 1969 East Wall (in pit crusher) slide at the Chuquicamata mine noted as a possible exception (still debated). At that time the Chuquicamata mine slopes were about 550 m high and had successfully withstood an earthquake of magnitude 7.0 at a hypocentral distance of 30 km, which produced a maximum horizontal acceleration of the order of 0.25 g. At the Penoso iron ore mine near Vallenar, which is situated near the coast midway between Santiago and Antofagasta, slopes up to 320 m high had withstood an earthquake of magnitude 7.0 at a hypocentral distance of 20 km and a ground acceleration of 0.31 g.

Most recently (14 November 2007), the northern regions of Chile experienced a magnitude (Mw) 7.7 earthquake (Figure 3.16). Although there was considerable damage to property and loss of life, rock slopes at the many open pit mines in the region were not damaged by the main earthquake or two major (Mw 6.8 and Mw 6.2) events that occurred the next day. It was noted that slope monitoring prisms and piezometers reacted to the earthquake, but no incidents of slope instability were reported.

These incidents contrast experiences with natural slopes, where earthquakes have produced numerous landslides, large and small. The process responsible for earthquake induced landslides in natural rock slopes is generally considered to be topographic amplification, which is an increase in shaking associated with ridges and topographic changes. It is believed to be a function of slope geometry and seismic wavelength resulting from at least two interacting processes: focusing and de-focusing of seismic waves from the free surfaces of hills and canyons; and excitation of whole topographic edifices, which occurs when the wavelength of the incoming seismic wave is similar to the width of the topographic feature (Murphy 2008).

Ashford and Sitar (1997) noted that maximum topographic amplification, which may result in a two to five-fold increase in the peak ground acceleration (Faccioli et al. 2002), occurs when the wave propagates downwards into the slope. Meunier et al. (2008) demonstrated that S waves are significantly more important to topographic amplification than P waves and Ashford et al. (1997) noted that the effect of topographic amplification decreases significantly within even one slope height distance behind the slope crest. Given these descriptors, it does not seem unreasonable to suggest that topographic amplification and consequent slope failure does not occur in large open pit slopes because the slopes are outside the range of geometries that would experience topographic amplification or, if amplification does occur, the slope geometries are such that the amplification is too weak to promote slope failure. Another possibility, of course, is that the rock mass is simply too strong. Whichever is true, in some jurisdictions it may be that executive management and/or the regulatory authorities will raise the familiar maxim 'absence of evidence is not necessarily evidence of absence' and declare that an analysis is necessary.

If it is decided that earthquake effects should be considered, designers can use limit equilibrium models that consider seismic loading pseudo-statically by specifying a horizontal static acceleration, similar to gravity, which is meant to be representative of the design earthquake. There are at least three important assumptions inherent in the pseudo-static approach:

- earthquakes can be modelled as a static force acting on the mass of a potential slide;
- no dynamic water pressures are generated;
- materials show no significant loss of strength as a result of cyclic loading.

Selection of the design earthquake is usually left to experts in seismology. Choice of an appropriate horizontal acceleration (or seismic coefficient) is the main difficulty with the pseudo-static approach. It is an approach that has no physical basis, but relies on a fictitious parameter (the seismic coefficient) which cannot be derived using logical or physical principles. There is no simple, universally accepted way to determine an appropriate seismic coefficient since earthquakes involve different durations and frequency contents. In almost all cases the horizontal acceleration is less than or equal to half the maximum acceleration of the design earthquake (Pyke 1997). The horizontal acceleration acts to produce an inertial force out of the slope, therefore the determination of the safety factor using the limit equilibrium method proceeds as usual. Since earthquakes are not static, the analysis with constant horizontal acceleration is usually considered to be conservative. If the seismic coefficients in Figure 10.24 are used, resultant FoSs greater than 1.0 (Pyke 1997) to 1.15 (Seed 1979) usually indicate that seismic displacements will be acceptably small.

Appropriate selection of the seismic coefficient is avoided in the Newmark analysis (Newmark 1965). The Newmark analysis computes the displacement of a single block subject to seismic motion. The portion of the design acceleration record above the critical acceleration, a_c, is

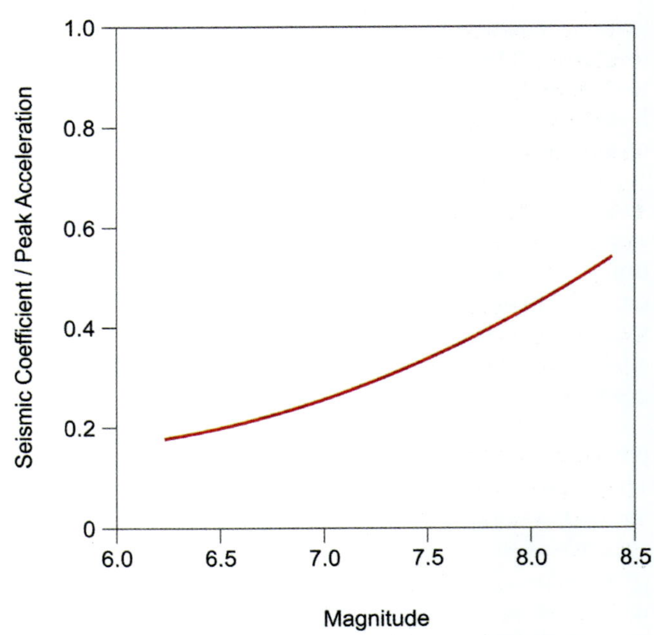

Figure 10.24: Recommendations for seismic coefficient based on earthquake magnitude and peak acceleration
Source: Pyke (1997)

integrated twice to obtain displacement. The critical acceleration is defined as:

$$a_c = (FoS - 1)g\sin\beta \qquad \text{(eqn 10.5)}$$

where

FoS = static safety factor
β = slope angle
g = gravitational acceleration.

The result of a Newmark analysis is an estimation of block displacement, which can be compared to the roughness scale of sliding discontinuities in order to assess whether the discontinuities will displace sufficiently to pass from peak to residual strength. If residual shear strength results, the slope stability should be analysed using the residual shear strength. Jibson and Jibson (2005) give details of a US Geological Survey open file report containing Java programs intended to facilitate rigorous and simplified Newmark sliding-block analyses and a simplified model of decoupled analysis (http://earthquake.usgs.gov/resources/software/slope_perf.php).

10.3.4 Numerical methods

Although limit equilibrium methods of analysis such as the method of slices are simple to use and have been well adapted to slope stability problems in jointed rock masses, they cannot represent deformation and/or displacement of the failing rock mass. This deficiency has largely been filled by numerical methods of analysis, which can model many of the complex conditions found in rock slopes such as nonlinear stress–strain behaviour, anisotropy and changes in geometry. They can be used to help explain the observed physical behaviour of the rock mass and to evaluate different geological models, failure mechanisms and slope design options.

Numerical models divide the rock mass into elements. Each element is assigned an idealised stress–strain relation and properties that describe how the material behaves. The elements may be connected in a continuum model or separated by discontinuities in a discontinuum model. Discontinuum models allow slip and separation at explicitly located surfaces within the model. The essential features of these models, together with some advanced hybrid continuum–discontinuum models, are outlined below.

10.3.4.1 Continuum models

Continuum codes assume material is continuous throughout the body. In large rock slopes much of the rock mass must therefore be represented by an equivalent continuum in which the effect of discontinuities is to reduce the intact-rock elastic properties and strength to those of the rock mass. As mentioned above, numerical models divide the rock mass into elements. Each element is assigned a material model and material properties. The material models are stress–strain relations that describe how the material behaves. The simplest model is a linear elastic model that uses only the elastic properties (Young's modulus and Poisson's ratio) of the material. Linear elastic/perfectly plastic stress–strain relations are the most common rock mass material models. These models typically use Mohr-Coulomb strength parameters to limit the shear stress that an element can sustain. The tensile strength is limited by the specified tensile strength which, in many analyses, is taken to be 10% of the rock mass cohesion. Using this model, the rock mass behaves in an isotropic manner.

Although the models described typically use Mohr-Coulomb strength parameters to limit the shear stress that an element can sustain, in practice the most common failure criterion for rock masses is the Hoek-Brown failure criterion (section 5.5.2). It has been used indirectly in numerical analyses by finding equivalent Mohr-Coulomb shear strength parameters that provide a failure surface tangent to the Hoek-Brown failure criterion for specific confining stresses or ranges of confining stresses. The tangent Mohr-Coulomb parameters are used in traditional Mohr-Coulomb type constitutive relations, and the parameters may or may not be updated during analyses. There has been little direct use of the Hoek-Brown failure criterion in numerical models. Numerical models solve for displacements as well as stresses and can continue the solution after failure has occurred in some locations. In particular, plasticity constitutive relations require a flow rule that supplies a relation between the components of strain rate at failure. There have been several attempts to develop a full constitutive model from the Hoek-Brown criterion (e.g. Pan & Hudson 1988; Carter et al. 1993; Shah 1992). These formulations assume that the flow rule has a fixed relation to the failure criterion and that the flow rule is isotropic, whereas the Hoek-Brown criterion is not. Cundall et al. (2003) proposed a scheme that does not use a fixed form of the flow rule, but a form that depends on the stress level and, possibly, some measure of damage.

Major structural features such as mine scale faults are represented as interfaces between major regions of continuum behaviour. Persistent closely spaced joints in continuum models are represented implicitly by a ubiquitous joint model, which limits the shear strength according to a Mohr-Coulomb criterion in a direction corresponding to that of the structure.

Finite element and finite difference continuum codes widely used by slope design practitioners include PHASE^2 (Rocscience 2005b), FLAC (Itasca 2005), FLAC3D (Itasca 2006) and ABAQUS.

10.3.4.2 Discontinuum models

Discontinuum codes start with a method designed specifically to model faults and joints (discontinua) and treat continuum behaviour as a special case. These codes are generally referred to as discrete element codes. Two widely used discrete element codes for slope stability studies are UDEC (Itasca 2004) and its 3D equivalent, 3DEC (Itasca 2003).

A discrete element code embodies an efficient algorithm for detecting and classifying contacts. It will maintain a data structure and memory allocation scheme that can handle hundreds or thousands of discontinuities. The discontinuities divide the problem domain into blocks. Blocks within discrete element codes may be rigid or deformable, with continuum behaviour being assumed within deformable blocks. Selection of the structural geometry that defines the shape and extent of these blocks is a crucial step in discontinuum analyses. Typically, only a very small percentage of the faults and joints can be included, in order to create models of reasonable size for practical analysis. The structural geometry data must be filtered to select only the faults and joints most critical to the mechanical response. This is done by identifying the structures that are most susceptible to slip and/or separation for the prescribed loading condition. This may involve determining whether sufficient kinematic freedom is provided, especially in the case of toppling, and calibrating the analysis by comparing observed behaviour to model response.

As an example of this form of analysis, Sainsbury et al. (2007) reported on the back-analysis of a complex non-daylighting wedge failure mechanism at the Cadia Hill open pit using 3DEC. A three-dimensional discontinuum analysis was required to simulate the failure, which resulted from a combination of rock mass failure and slip along a geological structure.

A termination criterion, stipulating whether the structure terminates in the rock mass or against other faults or joints, is also required. This criterion is fundamental to providing the rock mass with strength derived from rock bridges and other natural rock mass features that are not considered when all the structures in the rock mass have infinite persistence. Chapter 4 (section 4.4.3) discusses discrete fracture network modelling tools that can be used to help visualise the structured rock mass and set the criterion, including JointStats (Brown 2007), FracMan (Golder Associates Inc. 2006), 3FLO (Billaux et al. 2006) and SIMBLOC (Hamdi & du Mouza 2004).

The development and use of discontinuum codes in slope stability analyses represented a major step forward in modelling the effect of structures in closely jointed rock mass. Overall, however, the need to limit the number of faults and joints in order to create models of a size that can actually be handled by these codes represents a critical shortcoming in our ability to understand and adequately model the failure mechanisms that may develop in closely jointed rock masses. The advanced codes outlined in the next section represent an improvement but not a complete solution to the difficulty; further research and development is required. The objectives and outcomes of such research and development, currently underway in the LOP project, is outlined in section 10.3.4.5.

10.3.4.3 Modelling considerations

Element size

To adequately capture stress and strain gradients within the slope, it is necessary to use relatively fine discretisations. By experience, it has found that at least 30 lower-order elements (elements with constant or uniform stress) are required over the slope height of interest. Figure 10.25 shows results when only 10 elements per zone height are used. The figure also shows that higher-order elements, or elements employing mixed discretisation, show reasonably accurate results when only 10 elements are used in the slope heights. Finite element programs using higher-order elements probably require fewer zones than the constant-strain/constant-stress elements common in finite difference codes. If flexural toppling is involved, a minimum of four zones across the rock column is usually required.

Initial conditions

Initial conditions are those conditions that exist prior to mining. The initial conditions of importance at mine sites are the groundwater conditions and the in situ stresses. Groundwater conditions are discussed in Chapter 6. This section focuses on the in situ stresses, which traditionally have been ignored in slope analyses. There are several possible reasons for this.

- Limit equilibrium analyses, which are used widely for stability assessments, cannot include the effect of stresses in their analyses. Nevertheless, they are thought to provide reasonable estimates of stability in many cases, particularly where structure is absent, such as in soil slopes.
- Most slope failures are gravity-driven, and the effects of in situ stress are thought to be minimal.
- In situ stresses in rock masses are not measured routinely for pit slopes, and their effects are largely unknown.

Open pit design practice as presented in this chapter assumes that the effect of in situ stress is an issue only when the stresses induced in pit wall slopes are substantial enough to approach or exceed the rock mass strength. This could lead to rock mass damage, producing an enlarged zone of weakened rock which may subsequently fail, mainly in shear under gravity loading. Appendix 3 provides additional perspective through discussion of the origins and characteristics of in situ stress as well as typical stress fields surrounding open pit mines. The

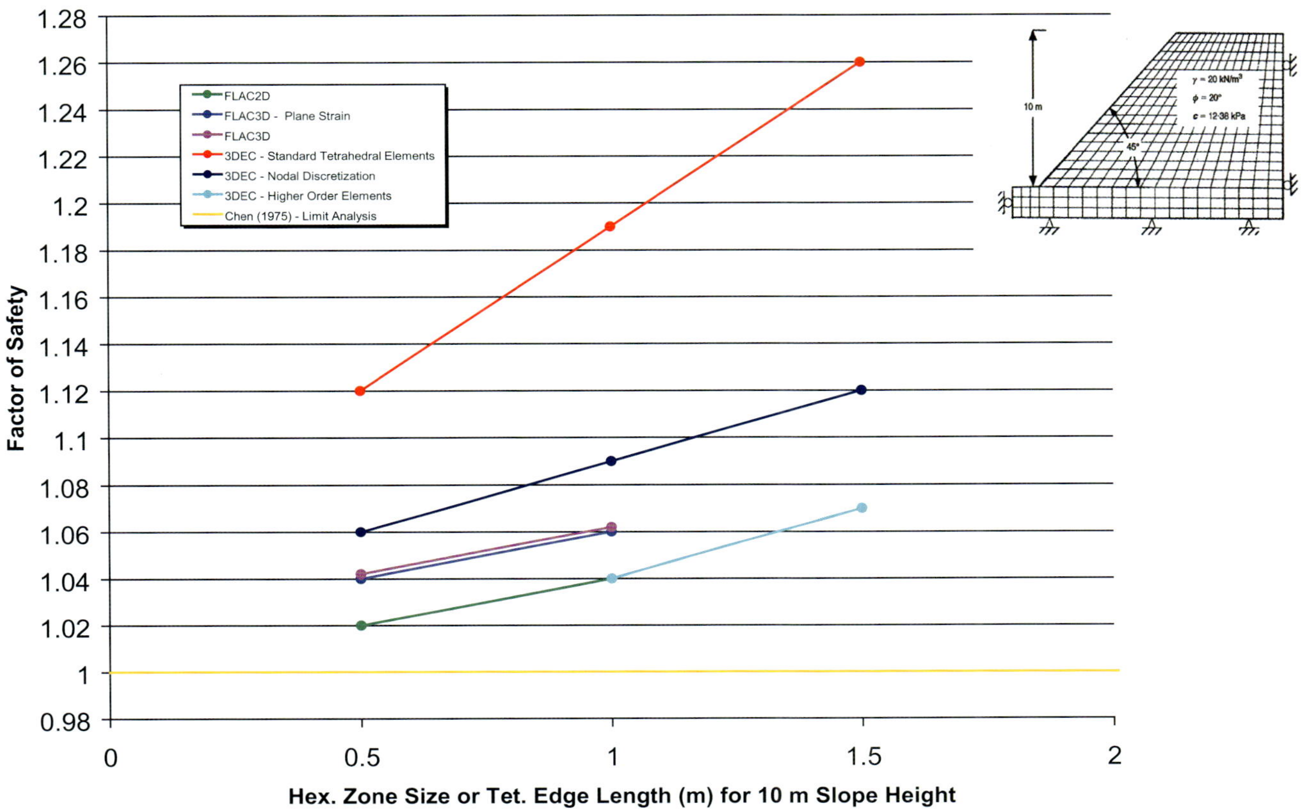

Figure 10.25: Comparison of calculated FoSs for the Chen (2007) solution for different element types. (Inset: numerical mesh for homogeneous embankment.)

appendix concludes with a procedure to be followed when it is possible that the in situ stresses may have a significant impact on the stability of mine excavations. The procedure includes a recommendation to use simple numerical tools to evaluate the influence of pit geometry and regional stresses on the induced stresses and displacements around the pit.

One advantage of numerical models lies in their ability to include pre-mining initial stress states in the stability analyses and to evaluate their importance through constitutive relations that simulate the rock behaviour under mining induced stress conditions. Current constitutive relations appear adequate to simulate typical shear failure modes, including strain softening. However, as open pits deepen there is a possibility to encounter previously unexperienced behaviours and failure modes, including brittle rock failure. The topic of brittle rock failure is one of considerable research and debate. Few, if any, numerical models currently contain an adequate constitutive relation to permit the correct simulation of brittle failure propagation. What seems clear from experience is that many brittle fracture processes are self-stabilizing and, looking at the stress fields in which they propagate, it seems that the choking off of the propagation process takes place when the stress field into which the failure is propagating is such that the principal stress ratio (σ_3/σ_1) increases over a short distance. This could be the case in the toe of a steep slope in massive rock. On the other hand, when a pillar is formed as in a traditional room and pillar operation or by two excavations in close proximity, the principal stress ratio increases more gradually or perhaps not at all and this allows for the unstable propagation of the failure. Based on this logic it seems that there could certainly be local surface spalling when the compressive stress on the slope surface exceeds the spalling limit (about 40% of the UCS) of the massive rock. However, propagation of this spalling into the slope is unlikely unless there is a very weak persistent discontinuity parallel to and behind the wall or existing excavations are located behind the slope face.

In general it is difficult to say what effect the initial stress state will have on any particular problem, because behaviour depends on factors such as the orientation of the major structures, rock mass strength and water conditions. Notwithstanding these comments, some observations on the effects of in situ stress on stability can be made.

- The larger the initial horizontal stresses, the larger the horizontal elastic displacements, although this is not much help since elastic displacements are not particularly important in slope stability studies.
- For slopes involving toppling behaviour, initial horizontal stresses in the plane of analysis that are less than the vertical stresses tend to decrease stability slightly and reduce the depth of significant shearing with respect to a hydrostatic stress state. This observation may seem counterintuitive as smaller horizontal stresses would be expected to increase stability. The explanation lies in the fact that the lower horizontal stresses actually slightly decrease normal stress on potential shearing surfaces and/or joints within the slope. This observation was confirmed in a UDEC analysis of a slope in Peru (Carvalho et al. 2002), where in situ horizontal stresses lower than the vertical stress led to deeper levels of joint shearing in toppling structures compared to cases involving horizontal stresses that were equal to or greater than the vertical stress.
- It is important to note that the regional topography may limit the possible stress states, particularly at elevations above regional valley floors. As a result, three-dimensional models can be very useful in addressing some regional stress issues.

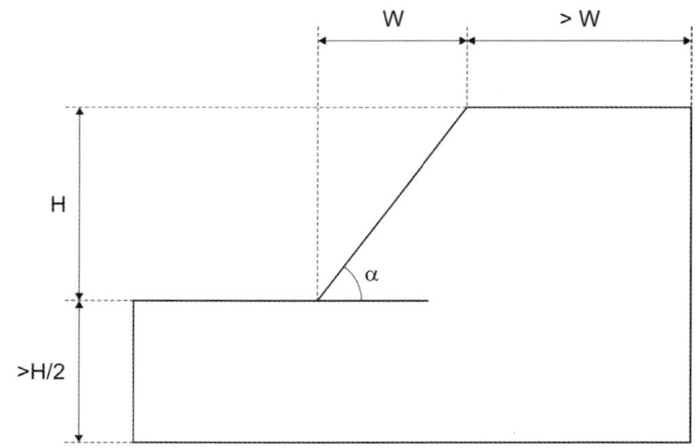

Figure 10.26: Typical recommendations for the locations of artificial far field boundaries in slope stability analyses

Boundary conditions

Boundaries can be real or artificial. Real boundaries in slope stability problems correspond to the natural or excavated ground surface that is usually stress-free. Artificial boundaries do not exist in reality. All problems in geomechanics, including slope stability problems, require that the infinite extent of a real problem domain be truncated artificially to include only the immediate area of interest. Figure 10.26 shows typical recommendations for locations of the artificial far field boundaries in slope stability problems. Artificial boundaries can be prescribed-displacement or prescribed-stress. Prescribed-displacement boundaries inhibit displacement in the vertical or horizontal direction, or both. Prescribed-displacement boundaries represent the condition at the base of the model and toe of the slope.

Displacement at the base of the model is always fixed in both the vertical and horizontal directions to inhibit rotation of the model and sliding along the base. Two assumptions are made regarding the displacement boundaries near the toe of any slope. One is that the displacements near the toe are inhibited only in the horizontal direction. This is the mechanically correct condition for a problem that is perfectly symmetric with respect to the plane or axis representing the toe boundary. Strictly speaking, this condition only occurs in slopes of infinite length, which are modelled in two dimensions and assume plane strain, or in slopes that are axially symmetric, in which the pit is a perfect cone. In reality, these conditions are rarely satisfied. Therefore, some models are extended laterally to avoid the need to specify any boundary condition at the toe of the slope. It is important to note that difficulties with the boundary condition near the slope toe are usually a result of the 2D assumptions. This difficulty seldom exists in 3D models.

The far field boundary location and condition must be specified in any numerical model for slope stability analyses. The general notion is to select the far field location so that it does not significantly influence the results. If this criterion is met, it is not important whether the boundary is prescribed-displacement or prescribed-stress. In most slope stability studies, a prescribed-displacement boundary is used. In some cases, a prescribed-stress boundary has been used without significantly differing from the results of a prescribed-displacement boundary. The magnitude of the horizontal stress for the prescribed-stress boundary must match the assumptions regarding the initial stress state for the model to be in equilibrium. However, following any change in the model, such as an excavation increment, the prescribed-stress boundary causes the far-field boundary to displace toward the excavation while maintaining its original stress value. For this reason, a prescribed-stress boundary is also referred to as a following-stress or constant-stress boundary, because the stress does not change and it follows the displacement of the boundary. Following stresses usually occur where slopes are cut into areas where the topography rises behind the slope. Even where slopes are excavated into an inclined topography, the stresses would flow around the excavation to some extent, depending on the effective width of the excavation perpendicular to the downhill topographic direction.

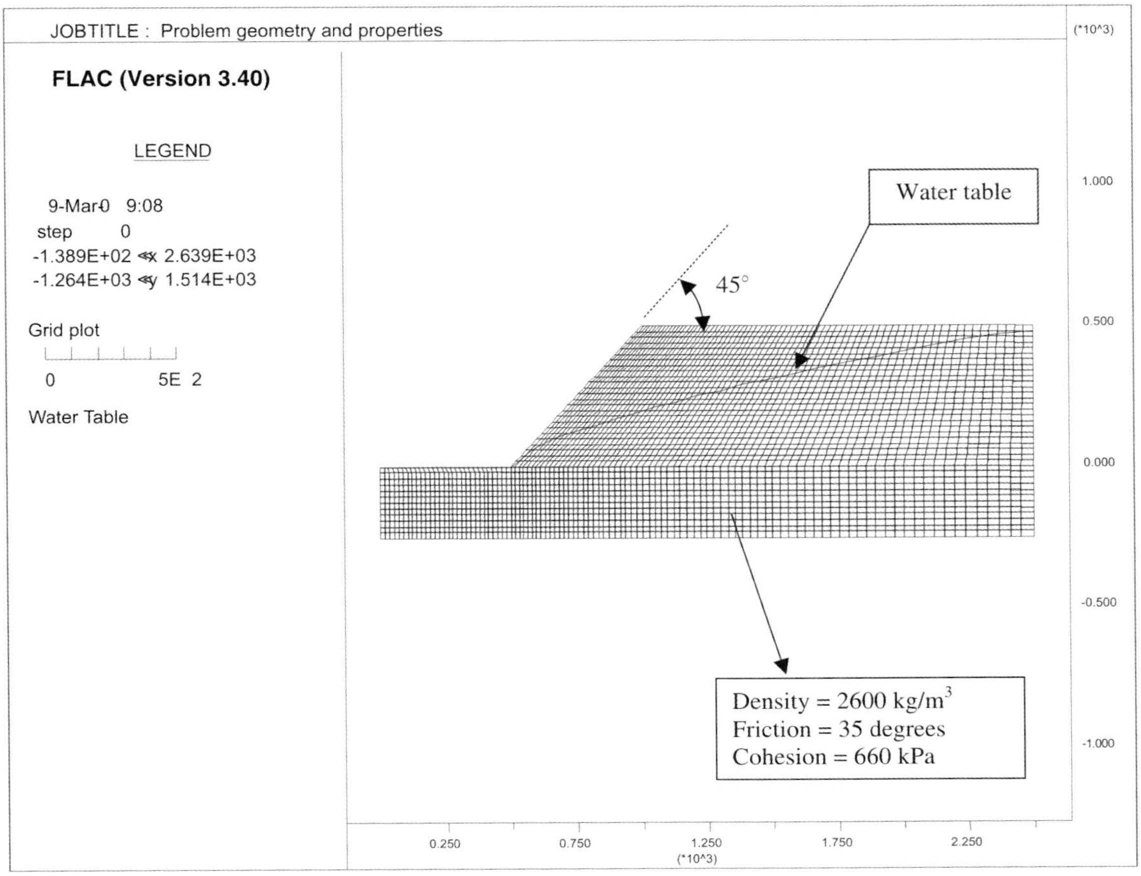

Figure 10.27: Problem geometry and conditions used in examining two different methods to specify water pressure in a slope
Source: Courtesy Itasca Consulting Group, Inc.

The effects of boundary conditions on analysis results can be summarised as follows:

- a fixed boundary causes both stresses and displacements to be underestimated, whereas a stress boundary does the opposite;
- the two types of boundary condition bracket the true solution; conducting tests with smaller models then averaging the results may lead to a reasonable estimate of the true solution.

One final point to be kept in mind is that all open pit slope stability problems are 3D in reality. This means that the stresses acting in and around the pit are free to flow beneath and around the sides of the pit. It is therefore likely that, unless there are faults of very low strength parallel to the analysis plane, a constant-stress or following-stress boundary will overpredict the stresses acting horizontally.

Incorporating water pressures

As outlined in section 10.3.3.2, the most rigorous method of specifying the distribution of pore pressure within the slope is to perform a complete flow analysis and use the resultant pore pressures in the stability analyses. The less rigorous but more common method is to specify a phreatic surface.

The error resulting from specifying a phreatic surface without doing a flow analysis can be evaluated using the results of two identical problems. In one case, a flow analysis was performed to determine the pore pressures. In the second case, the pressures were determined using a piezometric surface taken from the flow analysis. The material properties and geometry for both cases are shown in Figure 10.27. The right-hand boundary was extended to allow the far field phreatic surface to coincide with the ground surface at a horizontal distance of 2 km behind the toe. Hydraulic conductivity within the model was assumed to be homogeneous and isotropic. The error caused by specifying the water table can be seen in Figure 10.28. The largest errors (up to 45%) are found just below the toe, while errors in pore pressure values behind the slope are generally less than 5%. The errors near the phreatic surface are insignificant as they result from the relatively small pore pressures just below the phreatic surface, where small errors in small values result in large relative errors.

For a phreatic surface at the ground surface at a distance of 2 km a FoS of 1.1 is predicted using

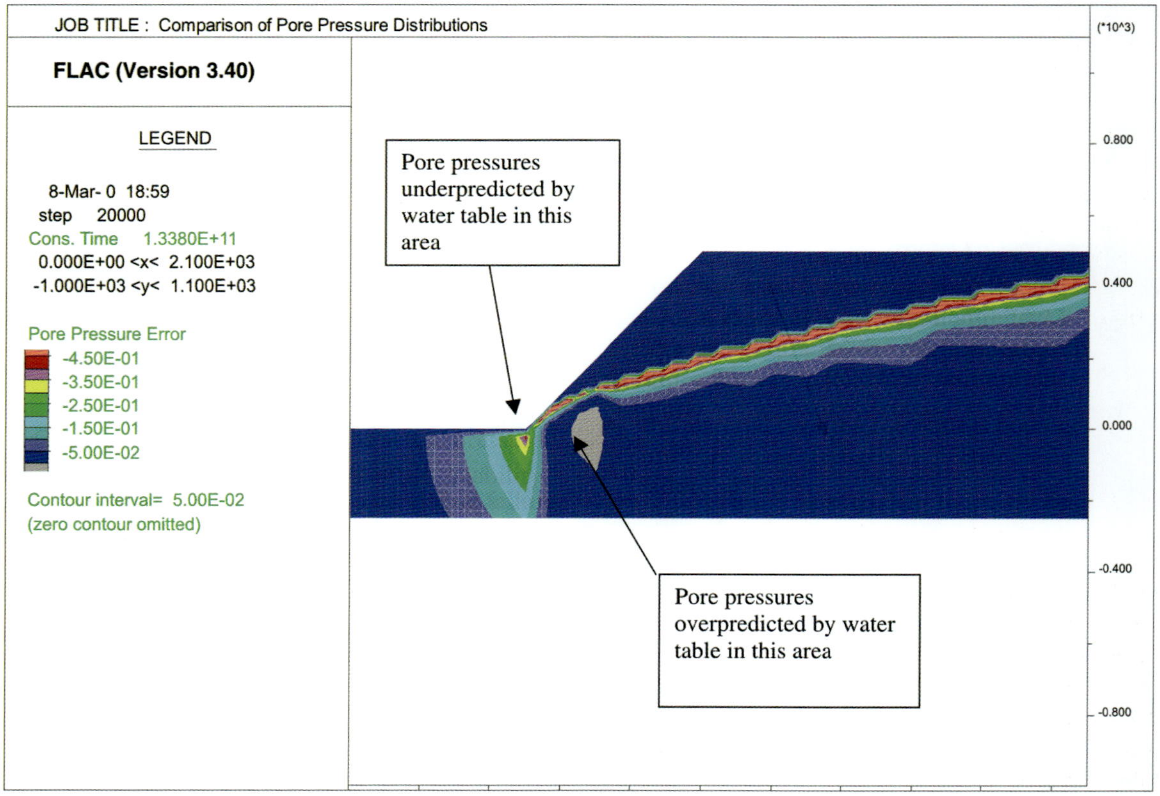

Figure 10.28: Error caused by specifying the position of the water table rather than performing a flow analysis
Source: Courtesy Itasca Consulting Group, Inc.

circular-failure chart no. 3 (Figure 10.22). The FoS determined by FLAC is approximately 1.15 for both cases. The FLAC analyses give similar safety factors because the distribution of pore pressures in the area behind the slope where failure occurs is very similar. The conclusion is that there is no significant difference in predicted stability between a complete flow analysis and simply specifying a phreatic surface. However, it is not clear if this conclusion can be extrapolated to other cases involving, for example, anisotropic flow.

Excavation sequence

Simulating excavations in numerical models poses no conceptual difficulties. However, the amount of effort required to construct a model directly depends on the number of excavation stages simulated. Therefore, most practical analyses seek to reduce that number. The more accurate solutions are obtained using the largest number of excavation steps, because the real load path for any element in the slope will be followed closely. In theory, it is impossible to prove that the final solution is independent of the load path followed. However, for many slopes, stability seems to depend mostly on slope conditions, such as geometry and pore pressure distribution at the time of analysis, and very little on the load path taken to get there.

A reasonable approach to the number of excavation stages has evolved in which only one, two or three excavation stages are modelled. Two calculation steps are taken for each stage. In the first step, the model is run elastically to remove any inertial effects caused by sudden removal of a large amount of material. In the second step, the model is run allowing plastic behaviour to develop. Reasonable solutions to a large number of slope stability problems have been obtained with this approach. However, the elastic solutions may involve stresses well outside the failure envelope for structures and/or rock masses. The return path to admissible stresses may not be realistic in all cases. Alternatively, the stabilising effects of overlying material which is excavated can be represented by equivalent forces that are gradually reduced to zero in order to simulate excavation. This gradual reduction approach is preferred because it provides a more realistic simulation involving admissible stresses during all phases of the excavation procedure.

Determining the FoS

For slopes, the FoS often is defined as the ratio of actual shear strength to minimum shear strength required to prevent failure. A logical way to compute the FoS with a finite element or finite difference program is to reduce the shear strength until collapse occurs. The FoS is then the ratio of the rock's actual strength to the reduced shear strength at failure. This shear-strength reduction technique was first used with finite elements by Zienkiewicz et al. (1975) to compute the FoS of a slope composed of multiple materials.

To perform slope stability analysis with the shear-strength reduction technique, simulations are run for a series of increasing trial FoS (f) values. Actual shear-strength properties, cohesion (c) and friction angle (ϕ) are reduced for each trial according to the following equations:

$$c_{\text{trial}} = \left(\frac{1}{f}\right)c \qquad \text{(eqn 10.6)}$$

$$\phi_{\text{trial}} = \arctan\left(\frac{1}{f}\right)\tan\phi \qquad \text{(eqn 10.7)}$$

If multiple materials and/or joints are present, the reduction is made simultaneously for all materials. The trial factor of safety is increased gradually until the slope fails. At failure, the factor of safety equals the trial factor of safety (i.e. f = factor of safety). Dawson et al. (1999) show that the shear-strength reduction factors of safety are generally within a few percent of limit analysis solutions when an associated flow rule, in which the friction angle and dilation angle are equal, is used.

The shear-strength reduction technique has two main advantages over slope stability analyses done with limit equilibrium methods. First, the critical slide surface is found automatically and it is not necessary to specify the shape of the slide surface (circular, log spiral, piecewise linear) in advance. In general, the failure surface geometry for slopes is more complex than simple circles or segmented surfaces. Second, numerical methods automatically satisfy translational and rotational equilibrium, whereas not all limit equilibrium methods do. Consequently, the shear-strength reduction technique usually determines a FoS equal to or slightly less than that determined with limit equilibrium methods.

Seismic analysis

Seismic analyses with numerical models can be performed in two ways:

- pseudo-static analysis using a seismic coefficient derived from earthquake records;
- time-domain analysis using applied earthquake records.

The first method involves pseudo-static analysis with an applied horizontal acceleration as described in section 10.3.3.5. Time-domain analyses compute stresses and displacements using earthquake records as input. In these analyses the time is real and the stresses and displacements are computed at discrete points in time (every millisecond or so). Time-domain analysis using numerical models is not a trivial task. The user must consider appropriate damping, boundary conditions and element sizes to propagate waves. Interested readers are referred to the user manuals for the specific model. The advantages of using numerical models to perform dynamic analysis (Glass 2000) are that:

- numerical methods incorporate realistic earthquake motions;
- numerical models allow the use of realistic properties for soil and rock slopes;
- numerical models compute the displacement time history of the slope, allowing assessment of the displacement impact on the ultimate behaviour of the slope.

The disadvantage of using time-domain analysis for dynamic analysis is that factors of safety are not determined. Rather, displacements can be calculated for different magnitude earthquakes and the probability of exceeding a limiting displacement can be related to the probability that an earthquake of sufficient magnitude to cause the limiting displacement will occur during the mine's life.

10.3.4.4 Advanced numerical models

Two advanced numerical codes have been leading the search to improve the way slope failures in jointed rock masses are modelled. The codes are ELFEN (Rockfield 2001) and PFC2D (Itasca 2004a), and its 3D equivalent PFC3D (Itasca 2005b).

ELFEN is a hybrid 2D/3D numerical modelling package that incorporates finite element and discrete element analysis. It was developed for the dynamic modelling of impact loading on brittle materials such as ceramics, but has been increasingly used in rock mechanics. A feature of ELFEN is its ability to allow fractures to develop according to a failure criterion specified through macro-mechanics constitutive models employing Mohr-Coulomb, Drucker Pager or Rankine failure criteria. At some point in the analysis the adopted constitutive model predicts the formation of a failure band within a single element or between elements. The load-carrying capacity across such localised bands decreases to zero as damage increases until the energy needed to form a discrete fracture is released. At this point the topology of the mesh is updated, leading initially to fracture propagation within a continuum and eventually resulting in the formation of a discrete element as a rock fragment (Figure 10.29).

Motion of these discrete elements and further fracturing of the remaining continuum and previously created discrete elements is simulated. The evolution is continued until the system comes to equilibrium or up to the time of interest. Klerck (2000) and Crook et al. (2003), when modelling drill hole breakout, showed that by

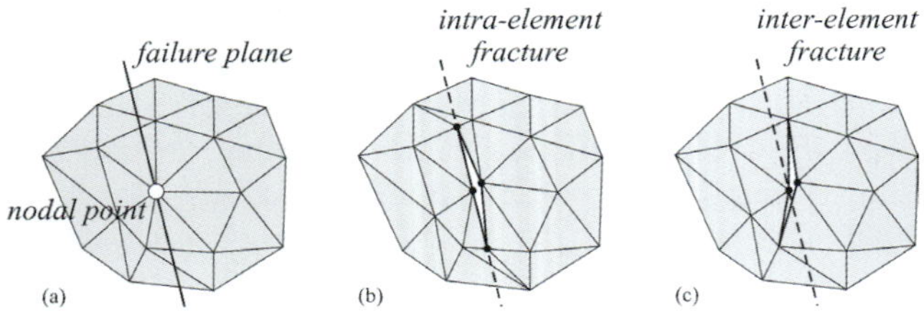

Figure 10.29: ELFEN crack insertion procedure. (a) An initial configuration. (b) Crack development through an element. (c) Crack development along an element boundary
Source: After Yu (1999)

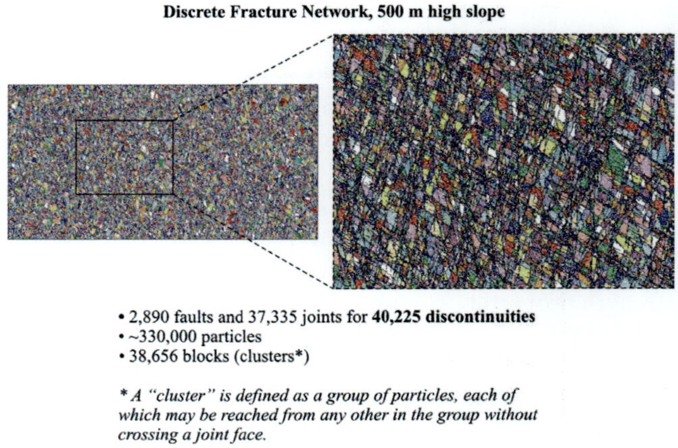

Figure 10.30: DFN network for a 500 m high, 1000 m wide slope intersected by eight persistent fault sets and two joint sets
Source: Courtesy Itasca Consulting Group, Inc.

augmenting the standard Mohr-Coulomb yield function with a Rankine tensile cutoff, thereby coupling tensile and shear damage, they could effectively model brittle, tensile axial-splitting fractures and more ductile shear features. By applying such techniques, it becomes possible to model the behaviour of a continuum and a discontinuum, and the transformation of the rock mass from a continuum to a discontinuum. Pre-existing faults and joints can be introduced into the model in 2D and 3D using DFN codes such JointStats or FracMan.

PFC2D and PFC3D are distinct element codes that represent rock as an assembly of rigid bonded particles which have deformable contacts that can break. The assembly can be used to simulate the progressive yield of a jointed rock mass where failure involves the yielding of faults and joints and the fracture of intact rock. In this model a macro-mechanics based failure criterion is not required as the mechanical behaviour of the rock is governed by the emergent growth and eventual coalescence of microcracks into macroscopic fractures when load is applied (Potyondy & Cundall 2004).

Improvements to the bonded particle method have shown that in 2D it is possible to make a complete slope model with a realistic representation of the in situ fracture network and then simulate the progressive failure of the slope, where failure involves sliding along the major structures and fracture across the intact rock bridges or blocks of rock left between these structures (Cundall 2007).

Figure 10.30 shows a portion of the discrete fracture network generated by 3FLO (Billaux et al. 2005) for a PFC2D simulation of a 500 m high, 1000 m wide slope involving eight faults with trace lengths from 74 m to 532 m spaced 15–140 m apart, and two joint sets with a trace length of 16.6 m spaced 5 m apart. The slope is later carved from this PFC2D model. The example is based on the upper section of the West Wall at the Chuquicamata mine in northern Chile and was prepared as part of the LOP project research program. The total network represents over 40 000 discontinuities separated by almost 39 000 blocks of rock or rock bridges (Cundall 2007).

The strength of the intact rock and joint fabric within the rock bridges between the major structures is represented by an equivalent material or synthetic rock mass (SRM) that combines the response of the intact rock and the joint fabric into the spherical 3D SRM unit (section 5.5.6). The complete SRM model for the slope in Figure 10.30 is shown in Figure 10.31.

Analysis of the slope illustrated in Figure 10.31 showed that rock fracture did not occur but there was considerable yielding and dislocation of the smaller blocks to depths of 130 m (Figure 10.32) and toppling on the major structures (Figure 10.33).

Although rock fracture did not occur, the toppling and dislocation of the smaller blocks of rock to depths of about 130 m closely resembles the observed slope behaviour. It has been suggested that the lack of rock fracture is possibly a 2D artefact in that the intersection of the given fault and joint sets created many discrete blocks or closed areas in 2D. Geometrically this is artificial, because in 3D the discontinuity traces are much less likely to form closed volumes. In 3D simulations of block caving where these limitations are not present it has been found that rock bridge fracture is widespread and a seemingly essential component in determining the behaviour of the rock mass (Cundall 2007).

10.3.4.5 Research targets

The ability of the SRM model to construct an equivalent material that honours the strength of the intact rock and joint fabric within the rock bridges along a candidate failure surface in a closely jointed rock mass (sections 5.5.6 and 7.3.1.2) is a significant development. In particular, it provides a means of:

- establishing a constitutive material model (strength envelope) that is not reliant on Mohr-Coulomb or Hoek-Brown criteria;
- establishing a strength envelope from which the Hoek-Brown parameters can be derived, i.e. it provides a means of calibrating the Hoek-Brown strength envelope.

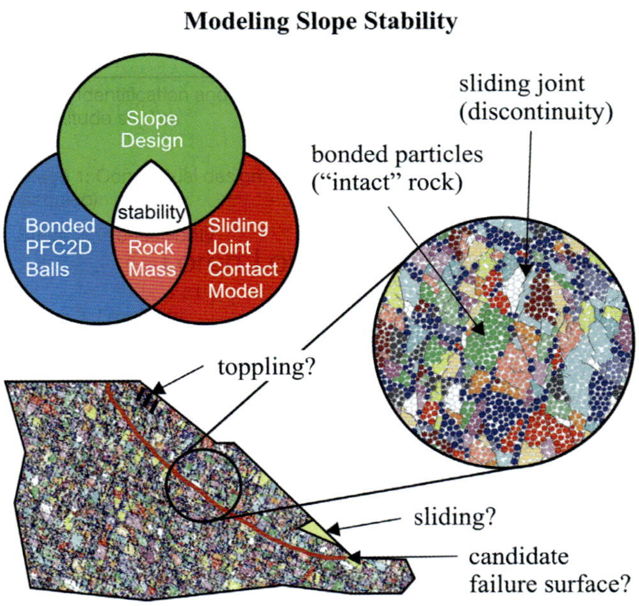

Figure 10.31: Fully bonded PFC2D model for high slope in closely jointed rock
Source: Courtesy Itasca Consulting Group, Inc.

Current hardware limitations make it difficult to perform 3D SRM simulations at the same resolution as the 2D example given in Figure 10.30. However, research targeted at improving the resolution and speed of full 3D simulations have shown that a special purpose code, based on a simplified lattice (using nodes and springs rather than the balls and contacts of PFC3D), will run 10 times faster than PFC3D and will be able to handle much larger models. Accordingly a new 3D code, *Slope Model*, is being written for the LOP project. *Slope Model* embodies the SRM concept so that a DFN may be imported, with failure involving both movement on the faults and joints that

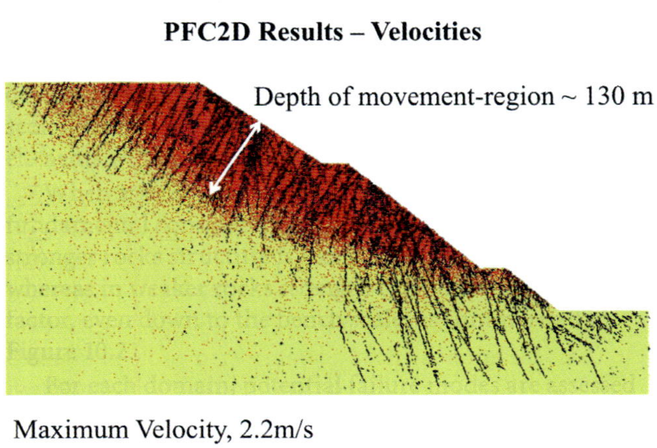

Figure 10.32: PFC2D model showing velocity vectors (red) overlying shear joints, with yielding and dislocation of smaller blocks to depths of 130 m
Source: Courtesy Itasca Consulting Group, Inc.

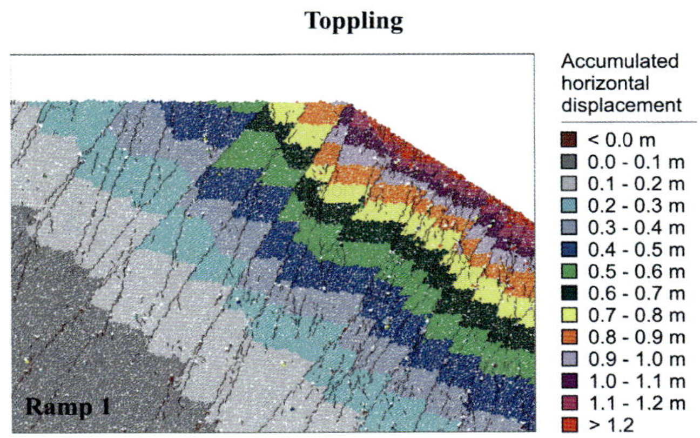

Figure 10.33: PFC2D model showing toppling on major structures
Source: Courtesy Itasca Consulting Group, Inc.

intersect the rock mass and breakage through the intervening rock bridges. The extra capacity should be sufficient to enable direct modelling of significant portions of a real slope in three dimensions (e.g. a potentially unstable region in one part of the slope). In support of the work on topics suggested in Chapter 6 for hydrogeological research (sections 6.6.2. 6.6.3 and 6.6.5), *Slope Model* will also have the ability to couple fluid flow, pore pressure distribution and rock deformation. The hydrogeological computations will have three parts: an initial static model to assess slopes with fractures into which pore pressures can be imported (effective stress); a flow-only version (quasi-static flow within joint segments); and a fully coupled version to model transient flow within an heterogeneous environment, including transient evolution of pore pressures as the slope is excavated. To illustrate an application of a preliminary version of *Slope Model*, Figure 10.34 shows a 3D slope with several benches and a simple DFN. The colours correspond to magnitudes of displacement, and two discontinuous joint sets are denoted by black dots (at the locations where penny-shaped joint segments intersect lattice springs). Note that the displacements in this example are due to gravity being imposed on an existing slope; another option in *Slope Model* allows the simulation of excavation in a pre-existing stress field that is in gravitational equilibrium.

For the same model (with two discontinuous joint families generated from stochastic parameter sets), Figure 10.35 shows groundwater flow vectors that develop a short time after imposing a step pressure increment at the left-hand boundary. Note that the flow is contained within joint segments, and that there is a moving fluid front that eventually will reach across the entire model, assuming that there is enough connectivity of joint segments. These examples were generated with a preliminary version of *Slope Model*, but it is anticipated that a beta version the code will be tested at LOP project sponsors' sites towards the end of 2009.

Table 10.3: Methods of slope stability analysis for each stage of project development

Method of stability analysis	Level 1: Conceptual	Level 2: Pre-feasibility	Level 3: Feasibility	Level 4: Design and Construction	Level 5: Operations
Empirical	¤	¤			
Limit equilibrium		¤	¤	¤	¤
Continuum and discontinuum numerical models			¤	¤	¤
Advanced numerical models			?	?¤	?¤

Research is also being direct at utilising an open-source code, YADE_OPEN DEM, in the SRM-based slope stability and hydrogeological studies. The YADE framework (Kozicki and Donzé 2008a, 2008b) uses object-oriented programming techniques to provide a flexible platform that is capable of handling different algorithms within a single package without the restrictions that often accompany commercial software. With YADE, new algorithms can be added, existing algorithms can be re-used, exchanged or extended, and different methods of simulation (e.g. discrete element, finite element and lattice geometrical methods) can be coupled all within the same framework. To reduce the peripheral work load, common low level functions (e.g. data input/output, mesh generation, visualisation of results) are provided through plug-ins and libraries. The outcome of this research will be

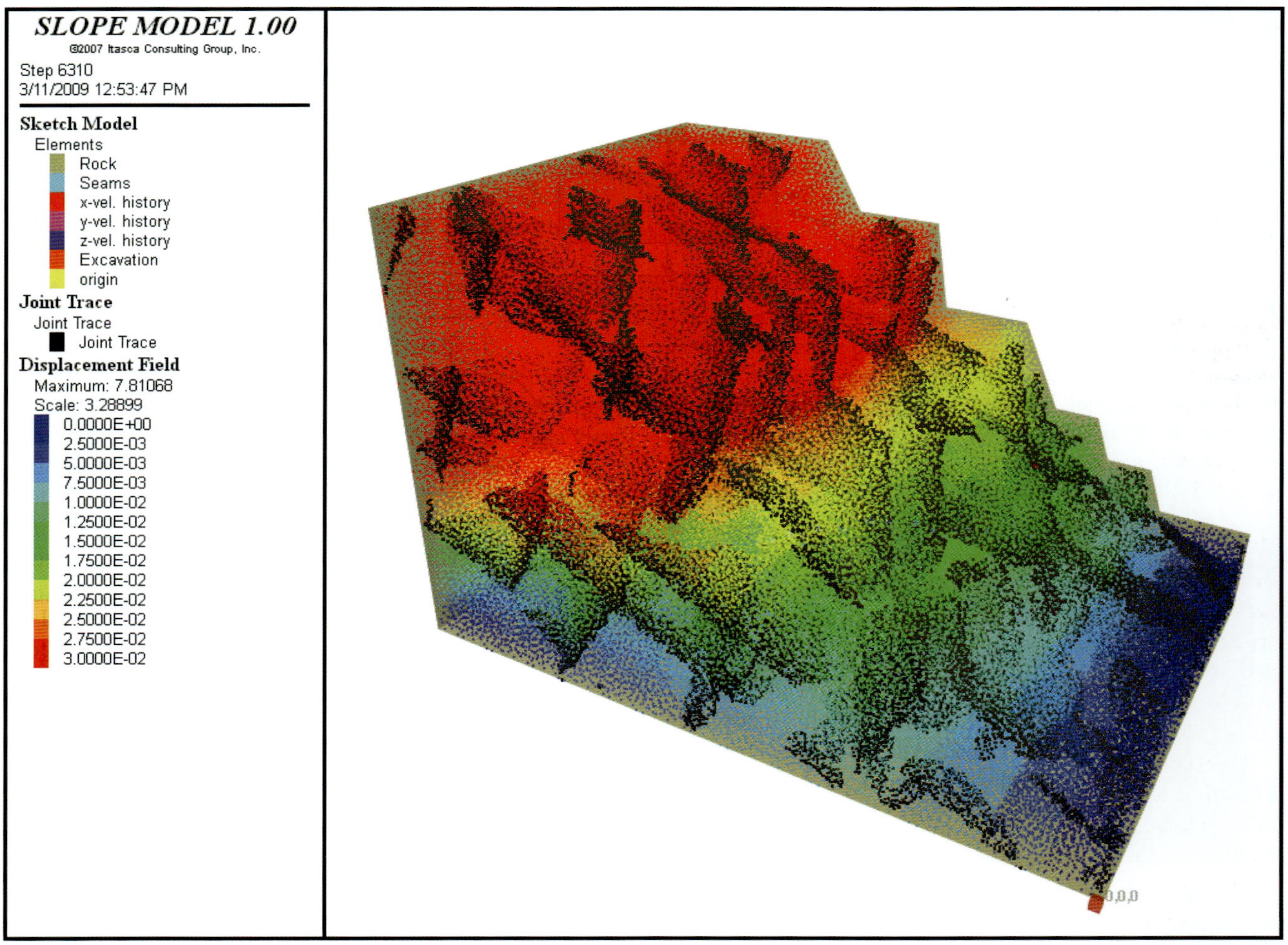

Figure 10.34: An example of a small benched slope created with a preliminary version of *Slope Model*. Colour contours denote displacement magnitudes and black dots indicate segments of two discontinuous joint sets.
Source: Courtesy Itasca Consulting Group, Inc.

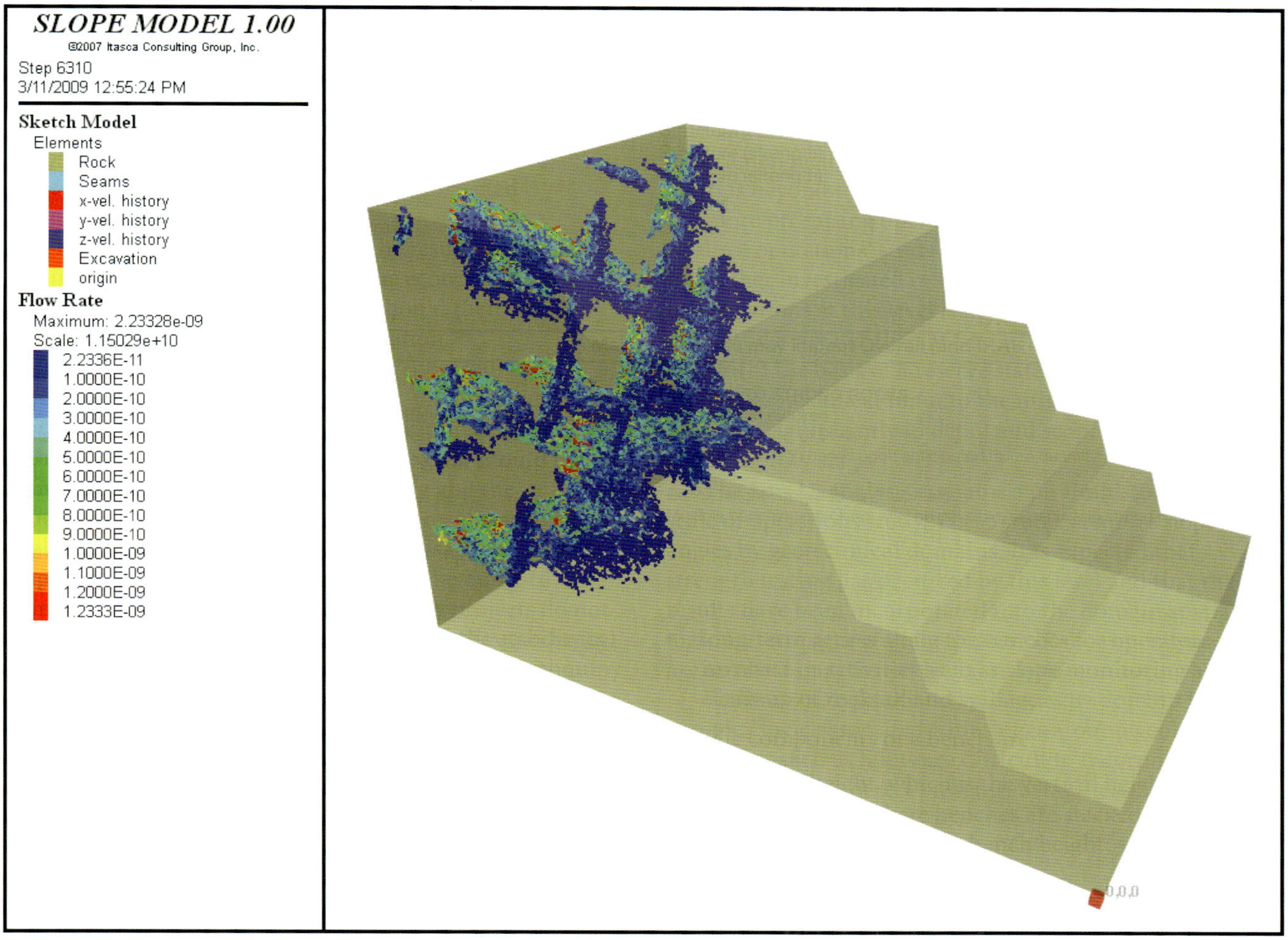

Figure 10.35: Transient fluid flow vectors calculated by *Slope Model* after a short time has elapsed following a step pressure increment applied to the left-hand boundary. Colours correspond to flow magnitude. Note the discontinuous nature of flow, as fluid migrates between intersecting joint segements.
Source: Courtesy Itasca Consulting Group, Inc.

brought into the public domain as it is assessed and reported.

10.3.5 Summary recommendations

Section 10.3 provides the background to the principal methods of rock mass stability analysis, when they are used and how they are used. This final section suggests the different methods of stability appropriate in each stage of project development, as outlined in Tables 1.2 and 8.1.

Consistent with the largely subjective nature of the available data, empirical methods are regarded as suitable only for indicative slope designs at the Conceptual (Level 1) and Pre-feasibility stages (Level 2) of project development (Table 10.3). As a next step, limit equilibrium methods can be introduced at Pre-feasibility (Level 2) and used at all levels up to and including Mine Operations (Level 5).

Before numerical methods of analysis are brought into the slope design task, two questions need to be asked.

1 Will the analyses add value to the design studies? There are many circumstances (e.g. shallow pits with a short mine life) where limit equilibrium analyses will provide perfectly adequate design information and value will not be added by more sophisticated numerical methods of analysis. The trend is to use numerical methods of analysis for even the most simple tasks, but this is often driven by a desire rather than a need to use these analyses to solve design problems. It is suggested that numerical methods of analysis should not be contemplated until it is perceived that deformation and/or displacement of the rock mass may detract from the required performance of the slope.
2 Is the level of certainty in the input data commensurate with the sophisticated nature of the method of analysis? It is suggested that continuum and discontinuum methods of numerical analysis should not be introduced until the level of

data certainty has reached at least Level 3 (Feasibility), and preferably Level 4 (Design and Construction). Level 4 data would be required for the advanced group of numerical models. At Level 3, although testing for the physical properties of the in situ rock and joint surfaces and targeted hydrogeological testing will have been carried out, the data assessments have largely been performed subjectively, especially for data from depth. By Level 4 there has been a significant increase in the availability of measurable data, allowing quantitative assessment of the structural, rock mass and hydrogeological parameters that are likely to be needed for the analyses.

11 DESIGN IMPLEMENTATION

Peter Williams, John Floyd, Gideon Chitombo and Trevor Maton

11.1 Introduction

The slope designs (Figure 11.1) are brought into the mine design through mine planning, which is usually an iterative process between the slope designer and the mine planner.

The subsequent operational implementation of the slope designs in accordance with the mine plan typically includes requirements for controlled blasting, excavation control, scaling, and occasionally slope support to ensure the designs are achieved safely and economically. From this perspective, however, even during times of high product prices, mine operators remain under pressure to minimise mining costs. To address these constraints, mining equipment of ever-increasing size is being introduced. As expected, there are some disadvantages to this trend. For example, where large equipment is used in minimum-width pushbacks to reduce the instantaneous stripping ratio, although the advance rates can be high and are therefore not necessarily conducive to the concurrent use of measures such as controlled blasting, careful scaling, support or drain hole installation that may be required to improve stability.

The resulting conflicts between the interests of production and those of slope stability are often exacerbated by the fact that the stabilisation techniques such as controlled blasting and slope support actually increase the operating costs on which the operations manager is frequently judged, even though there is an overall increase in profit. Therefore, meeting the objective of achieving slope designs that are practicable in terms of the operating constraints in the specific pit requires interaction and compromise between the geotechnical engineers, the mine planners and the operating staff as the design criteria are formulated.

The purpose of this chapter is to outline the needs and interaction of these different operating constraints. Section 11.2 addresses the mine planning aspects of slope design and section 11.3 outlines the steps required to achieve good wall control through the application of controlled blasting techniques. Section 11.4 examines excavation and scaling techniques, and section 11.5 outlines different methods of applying artificial slope support.

11.2 Mine planning aspects of slope design

11.2.1 Introduction

The required inputs and deliverables to the mine planning process change with the nature of the deposit and the stage of the project development. Similarly, the form and use of geotechnical information used by a mine planner changes with the stage of the project. In general, geotechnical inputs to the mine planning process start with high-level assumptions when projects are at early-stage analysis. More complex inputs are required for late-stage analysis and operations.

When determining the level of geotechnical input necessary for a given stage of project analysis, the question of material or financial impact needs to be addressed. The use of geotechnical information and the accuracy required at each stage can vary considerably depending on the characteristics of a given deposit. An example is the early-stage analysis for a large shallow copper deposit compared to a deep low-grade gold deposit. Final wall slope angles in the shallow mine do not represent a material financial consideration to the project viability, whereas the achievable slope angle for the deep gold deposit may be the most critical parameter in determining project financial viability.

11.2.2 Open pit design philosophy

The open pit slope design philosophy implemented at the mine must be well defined. Usually a philosophy of 'slope management' rather than one of ensuring slope stability

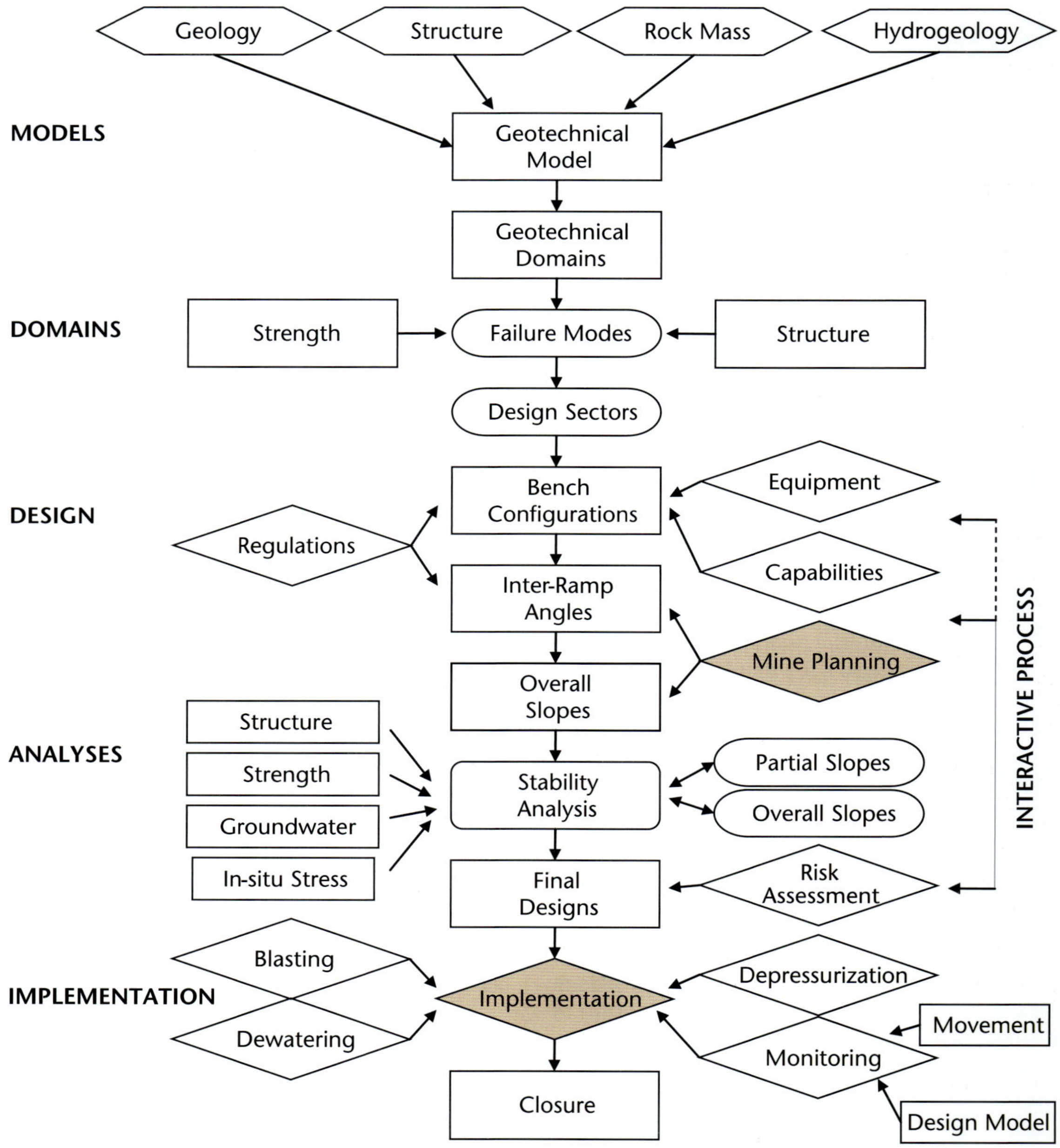

Figure 11.1: Slope design process

is followed. That is, there is expectation of manageable slope instability at acceptable levels of risk rather than a design that focuses on achieving stable pit walls (section 9.4.3).

The term 'slope management' is not new to the open pit mining industry. In reality it is how most slopes are developed in many, if not all, large open pit metalliferous mines. Simply put, if slopes were designed for complete stability the majority of mineral deposits would be uneconomic and would remain undeveloped. The term 'slope management' can be difficult to define, or is easily misunderstood. In consequence, mine owners may be incurring unexpected levels of risk, or the slopes may be managed without the owners understanding the processes taking place. For this reason, it is essential that a transparent process allows all those involved to understand the design and operating philosophies.

Often this process takes the form of a document referred to as a Ground Control Management Plan (section 12.3), which must be understood and approved by the most senior individual on the property and therefore implicitly by all stakeholders. It is important that senior personnel ensure that the designs, plans and implementation are such that the philosophy is being honoured and achieved. If the level of risk is considered unacceptable, the philosophy may need to be modified to reflect this, perhaps requiring changes to the slope design criteria. Alternatively, a higher level of risk may be considered acceptable and the management plan altered to reflect, for example, more aggressive slope design criteria.

It is essential that those making final decisions regarding slope design criteria are completely familiar with the deposit, the slope designs and the philosophies being adopted. For the slope management approach to be successful there must also be a well-understood and accepted open pit slope design philosophy, a team of highly skilled and professional site-based employees and a requirement to maintain a slope management process encompassing all aspects of continuous improvement. If these conditions cannot be met, risk levels may increase significantly and misleading slope designs can translate into unachievable mine planning decisions. This can be particularly dangerous if a slope steepening program is adopted but later found to be unworkable, since this can lead to a vicious cycle of sub-optimal planning in order to regain production-related losses.

It is important that all stakeholders in the mine design process recognise the changes that have taken place in open pit mining relative to the slope design process. Over the past three decades there has been a move towards the development of larger open pits, mined at higher mining rates as larger earthmoving equipment becomes available. The slope design methods outlined in Chapter 10 have essentially remained the same over this period – they are tools that must be applied with engineering judgement and take experience into account.

11.2.3 Open pit design process

The development of an ultimate pit design is often based on a series of interim pit phases that reflect successive cutbacks. This phased pit development allows the final pit wall design criteria to be based on the operational performance of each preceding phase. Lessons learned from any slope instability are applied to the design of subsequent mining phases. The understanding of geotechnical issues is progressively improved and the associated risk is reduced through the implementation of appropriate engineering designs and operating practices. To this end, observations and results from past and current experience should be applied to future mine planning. Specifically, this includes:

- the definition of geotechnical domains established on the basis of consistent structure and rock mass properties, as a basis for the formulation of respective slope designs (Figure 11.1);
- ongoing documentation of the excavation and operational performance of the slopes to judge the validity of the geotechnical model and its underlying assumptions. These practices are outlined in Chapter 12 and allow assessment of the consistency of operating practices. The reconciliation process should involve all departments associated with design, planning and execution of pit walls;
- monitoring and technical evaluation of slope instability (see Chapter 12) in order to understand its relationship to the slope design.

Implicit in the design approach is the commitment to continuous improvement, and ongoing review of the design process and operating considerations.

The pit design and execution process outlined in Figure 11.1 should follow accepted engineering and operating principles in line with the philosophy being adopted, as summarised in Figure 11.2. This design process flowchart is similar to others commonly used in the open pit mining industry. The front end is a conventional geotechnical program that would be implemented for any large open pit development; the remainder highlights the process that integrates the geotechnical program with mine planning, operating practices, economic evaluation, production requirements and acceptable levels of risk.

The conventional geotechnical program forms a critical part of the process, providing support to designs based on past and current experience. The joint approach is used predicatively to reduce the geotechnical risk associated with subsequent mining phases, by increasing the reliability of the geotechnical models. To ensure success, the process must be able to withstand rigorous audit and relies on a highly skilled and professional site-based team of employees, supported by adequate levels of peer review.

In Figure 11.2, a requirement to short-cut the design process means that complete geotechnical analysis of changing mine plans may not always or immediately be implemented. This is sometimes necessitated by the development of unexpected slope instability issues. This is not to say that various geotechnical considerations are not taken into account, but due to time constraints (decisions may need to be finalised within days or weeks) changes must be made and implemented immediately. As a result, there may not be time for rigorous geotechnical analysis; on occasion these may be conducted after the event. Again, this highlights the need to have an experienced and knowledgeable geotechnical team in which an operator has complete confidence. The ultimate aim of a geotechnical program is to ensure safety and minimise the likelihood of unexpected slope stability related events that negatively affect the operation.

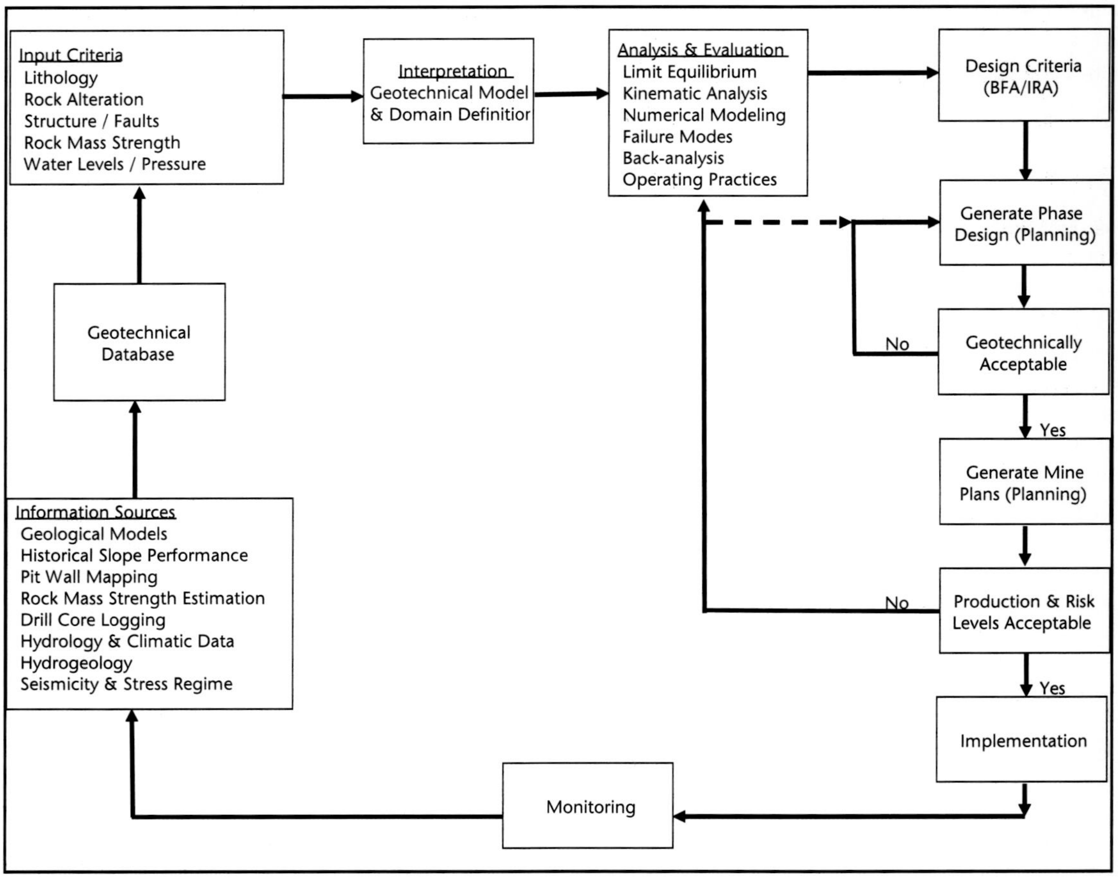

Figure 11.2: Typical open pit slope design process flowchart

11.2.4 Application of slope design criteria in mine design

The process to decide if a mining project should be built or rejected requires a series of studies (e.g. technical, economical, environmental, social and commercial) to increase knowledge of critical aspects of the project and to minimise investment risks. Since these studies are expensive and take time, the mining projects are usually developed in several stages related to the available level of information, the required accuracy of the engineering estimate and the level of planned costs. This approach provides enough information to make a decision without committing the total investment.

Although the definition of mining study stages varies, the terms 'conceptual', 'pre-feasibility', 'feasibility', 'design and construction' and 'operations' that are outlined in Table 1.2 (in Chapter 1) and Table 8.1 (in Chapter 8) are well accepted in the mining industry. In mine planning terms, Table 11.1 outlines the objectives and purposes of each stage in relation to the mining project.

The graph shown in Figure 11.3 illustrates the relation between capital and operating cost variation and project stage development through the stages of project development.

The following sections outline the geotechnical engineering input to each level from target identification to operations, consistent with the target levels of data confidence and slope angles outlined in sections 8.5.1 and 9.4.5. Mine closure is addressed in Chapter 14.

11.2.4.1 Target identification stage

The target identification stage confirms that the mineralisation is of sufficient consistency and grade to warrant additional exploration drilling, geophysics and other exploration level investigation. Target identification focuses on understanding the rock or mineral package so that subsequent investigations can concentrate on certain geological structures or mineralisation traces.

Engineering or geotechnical analysis at this stage is high-level with little direct data. Business or economic analysis at this stage is not recommended as a guiding force. Applying engineering, geotechnical and economics criteria too early will hinder the exploration process by making assumptions on start-up capital infrastructure,

Table 11.1: Project stages sequence and objectives

Stages	Objective
Target identification and order of magnitude studies	Confirm that the identified mineralisation is sufficient in consistency and grade to warrant additional exploration. Provide a first financial estimate, albeit high-level base on comparative analysis.
Stage 1: Conceptual design (scoping)	The objective of this stage is to look for fatal flaws and to refine investigation with new data. The major drivers of the project should be identified if possible.
Stage 2: Pre-feasibility	The objective of this stage is to ensure that a robust business case exists before proceeding to Stage 3. Stage 2 does not look to optimise the project, capital or operating plan. The stage will test major operating options to establish a robust case. The major drivers of the project need to be identified.
Stage 3: Feasibility	The objective of this stage is to determine the best operating configuration and to refine it for capital and operating costs and revenues. This stage will refine data and assumptions to enable the corporation to book reserves.
Stage 4: Design and construction	The objective of this stage is to construct the project on time and within budget and to transfer skills to operations staff from construction staff. Pre-stripping would fall into this stage. Developing the necessary action plans by discipline or function in enough detail to assure a smooth implementation from commencement of construction until early operations.
Stage 5: Operations	The objective of this stage is to execute mine production plans to the most efficient level possible including health, safety and loss prevention while maximising return on investment. For the geotechnical discipline, this stage involves using the pit phase implementation as a test of what to expect when the final walls are achieved, based on present material interpretation and current procedures and practices for loading and blasting.
Closure	This stage is focused on the physical and chemical stability of materials and effluent products of mining activity to ensure care of human health and environment. Develop a sustainable usage of land post mining activity and mitigate the social effects in communities when operations are ceased (Chapter 14).

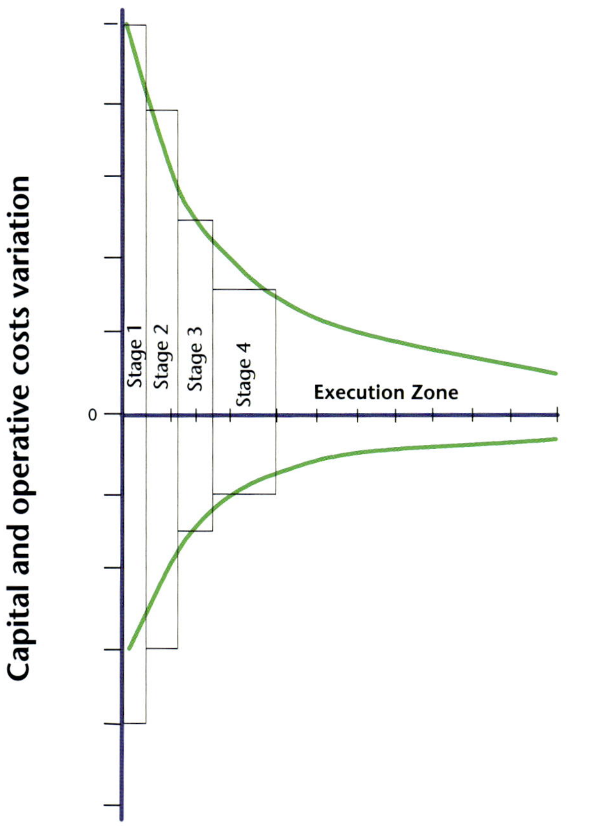

Figure 11.3: Capital and operating cost variation and project stage development

plant and mobile fleet capital and operating costs. The objective at this point is to understand the basics of the mineralisation.

A geological model is not usually created at this stage, so if base economics are applied they are more in the form of stripping ratios than slope angles. If a geological and geostatistical model is present, slope angles can be applied to a pit limit exercise. The slope angles would be based on comparable mineralisation nearby or on previous experience with the given target mineralisation types.

It is unlikely that geotechnical or engineering test data would be available at this point.

11.2.4.2 Level 1: Conceptual design

This level of study can be subdivided into two parts, although they are frequently combined or overlap. The sub-divisions are:

- order of magnitude studies;
- scoping studies.

Order of magnitude studies

The purpose of this stage is to determine, within an order of magnitude, if the style of mineralisation has potentially positive economics. Additional exploration drilling has occurred since the target identification stage, with some research on district logistics and the cost of doing business in the area. An order of magnitude look at the business potential is developed.

From a geotechnical perspective, this stage is similar to the target identification stage.

Engineering or geotechnical analysis at this stage is high-level with little direct data. High-level economics are applied to various assumptions concerning the size of the mineralisation. The results are judged on an order of magnitude basis. Positive economics are not necessary at this stage, but economics are considered with respect to the kind of mineralisation grade and depth needed to enhance economic results.

A geological model is usually available and slope angles are assigned based on the available geological domains. A pit limit exercise would be carried out. The slope angles would be based on comparable projects nearby or on previous experience with the given mineralisation types. The slope angles must consider groundwater if it was present in the exploration drilling.

High-level functional sensitivities could be defined at this stage, but data may not warrant this exercise, so activities may be delayed until the next stage of the project. Sensitivities to slope angles, major process options or mining methods would be assessed.

Scoping studies

The objective of scoping studies is to look for fatal flaws. Good district mineralisation has been established in previous stages, but obvious fatal flaws have not typically been investigated. Fatal flaws can occur in forms such as technical, commercial, social or political. Rarely would geotechnical issues be considered a fatal flaw, although geotechnical aspects may contribute to large variability in project economics.

Scoping analyses may take a few months of investigation for a small team of mineral professionals. Geotechnical investigation is at a high level and may encompass core inspection with site geologists, with a view to developing a simple geotechnical model. The geotechnical model will allow mine engineers to run pit limit routines to assign slope angles by geotechnical domain. Geotechnical domains are often lithological or based on alterations.

It is important that mine engineering communicates to geotechnical engineering the project sensitivity to overall slope angles. In the case of a deposit that is mineralised from the surface with a short payback period and a somewhat longer production life, it is likely that the assumed design is not overly sensitive to slope angles during this level of analysis. A decision to commit capital (project go-ahead) at this level is more important than the precise period of net revenue.

Relative slope angles for a deep ore deposit with multiple years of stripping and a long payback period may require more geotechnical investigation. In these cases it may make sense to run a small geotechnical drilling program to help determine pit wall angles during the pre-stripping phase. This kind of deposit also signals that, in Level 1, data collection should start in areas such as hydrology since the presence of groundwater may flatten slope angles or indicate a need for depressurisation, which would involve a financial impact.

In general, the mine engineering objectives at this level is two-fold:

1 to test the relative sensitivity of project economics to major technical, commercial and social variables;
2 to see if any of these variables represents a fatal flaw.

The mine engineer will determine the critical drivers that dictate the financial results of a potential operation. That investigation will set the focus of future investigations at this level and future stages. Programs such as level of drilling, testing or additional research can be planned and justified on each driver's level of impact on the go-ahead decision.

At this level, there is very little geotechnical activity. Geotechnical domains are created using core log or chip samples based on lithology or alteration. Structural knowledge of the deposit is relatively sparse due to a lack of geotechnical drill data (e.g. oriented core) and is typically based only on surface mapping. Initial slope recommendations are usually made with little or no geotechnical testing. Mine planners incorporate the domain-based slope angles into pit limit optimisation programs.

11.2.4.3 Level 2: Pre-feasibility

The objective of Level 2 is to ensure that a robust business case exists before proceeding to Level 3, the feasibility stage. Level 2 does not look to optimise the project, capital or operating plan. The stage will quickly test major operating options and seek to establish a robust business case. The definition of robust is frequently specific to corporate culture and is beyond the scope of this book.

In general the geotechnical data available and the slope design approach are similar to that of the conceptual design stage (Level 1). Unless the project is very sensitive to slope angle, similar inputs are expected from the geotechnical engineer. The geotechnical engineer will review new core drilling with the geologist and refine the resulting geotechnical domains used in pit limit and pit phase analysis. The geotechnical engineer should design, set up and start data collection in this stage so that the appropriate data and required level of confidence will be available in later stages. The project sensitivity to overall slope angles will determine the content of the proposed geotechnical program. The geotechnical engineer should understand the final proposed mining method to determine if additional data collection programs are needed to achieve the mine plan.

Hydrological programs should mesh with geotechnical programs. The geotechnical engineer should be familiar enough with the hydrological program and results to give mine engineers slope angles adjusted to the likely hydrological conditions predicted during mine operation. A close relationship between mine engineering, geotechnical and hydrological groups is expected throughout all stages of a project.

The project owner may require the results of Level 2 to qualify the deposit as a resource if the economics look promising. Geotechnical programs should then consider requirements of the 'resource' designation.

Typically, a detailed geotechnical report will be required in Level 2 studies. The report should review past reports for the property, current exploration drilling, hydrological data and core logging. The data should be itemised, summarising the methodology and the conclusions about the slope angles. The recommend slope angles should be labelled on specific areas of a pit surface map. Recommended future work should be outlined, citing the reason for specified work. Many projects do not progress beyond pre-feasibility, so the level of geotechnical engineering at this stage should be limited. However, many rejected projects are resurrected numerous times, so even a limited geotechnical report will speed future analysis.

The level of detail in slope angle guidance will increase in Level 2. Typically, slope angles will be specified by geotechnical domain together with bench configurations, as shown in Figure 11.4.

11.2.4.4 Level 3: Feasibility

The objective of a feasibility study (Level 3) is to find the best capital facility layout and operating plan (defining 'best' is beyond the scope of this text). Typically, a few of the more promising operating options are investigated in a 'scale of operations' study to determine which option will be further investigated in Level 3. The cost of investigations at this level is usually one to two orders of magnitudes greater then at Level 2. Level 3 investigations may form the basis of a reserve statement.

The geotechnical input to the mine planning process increases significantly in detail and scope from the pre-feasibility stage.

The Level 3 analysis contains a pit limit study and a scale of operations study. These studies require slope angles by geotechnical domains. It is expected that strength testing has occurred by this level and that the results have been incorporated into the domain spatial logic and strength characterisation. Bench configurations are not necessary for pit limit analysis.

Pit limit analysis can accept slopes by domain but more complex logic associated with slopes around geological structures such as faults are difficult to incorporate and can lead to an incorrect pit size. Further testing of the pit

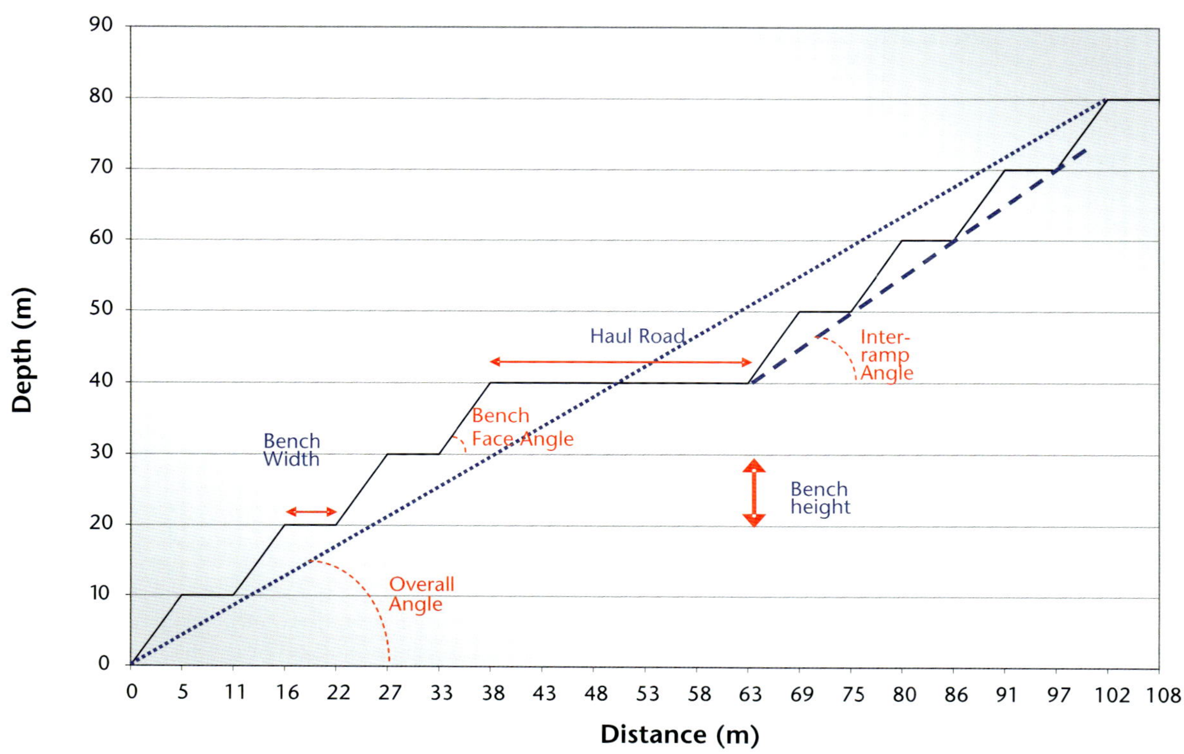

Figure 11.4: Inter-ramp and bench geometries used in open pit evaluations

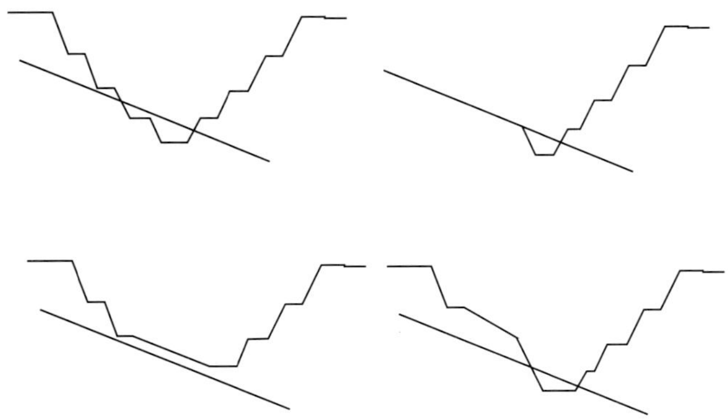

Figure 11.5: Alternative configurations for dealing with a fault

limit should occur, and is even more critical when geological structures such as faults are present. Testing should include drawing the pit in different configurations, using the pit limit data as a guide. Each pit configuration should be compared for relative economics and other deterministic measures.

Geotechnical professionals should help pit limit software developers with the complexity of slope logic around major structures. Figure 11.5 shows a typical cross-section of a pit with alternative outcomes. Assuming that the general country rock has the same slope angle, dealing with the fault may involve staying in front of it, taking the risk of under-cutting the fault low in the pit or following the fault down, or a combination of these.

After the pit limit and scale of operations studies, a more formal final pit and phase design is developed. In formal pit design, the geotechnical group must supply bench configurations specific to areas in the pit. Given that no production data are available, the bench configuration guidance is subject to change, therefore it should not be overly complex. Complex bench configuration will probably not be required until an operation is in the mid to late stage of excavation.

Formal pit and phase design will require the following inputs from the geotechnical group:

- domain modelling and importation into the mine planning package;
- structural modelling and importation into the mine planning package;
- bench configuration by domain;
- inter-ramp slopes by domain;
- geological structures in pit areas;
- design reconciliation for angles and structure.

The geotechnical engineer should check the final and phase designs for compliance with the slope design criteria and whether stability issues result from wall placement.

Dealing with major structures and resulting wall placements can be an iterative exercise between the mine planner and the geotechnical engineer. Typically, the mine planner draws up the pit and the geotechnical engineer checks for pit wall–geological structure intersection or proximity, and for any other features that could impact on the stability of the wall (e.g. alteration zones). Geological structures can be put into general mine planning packages in the form of strings, surfaces or solids (Chapter 4, section 4.4.1). The inclusion of such geological structures will help the mine planner draft the pit design. Model blocks may be given rock codes to identify whether the blocks are touching or outside a geological structure. This will allow geotechnical engineers to apply specific recommendations that may define step-outs, changes in simple/double benches configurations or local flatter slopes. Continuous and close coordination between engineering and geotechnical disciplines is especially important in this process.

The next step for mine engineering is to produce a production plan, which will form the basis of a life-of-mine (LOM) operating plan. Process, Human Resources, Geotechnical, Land and all other site departments base their plans on the mine engineer's production plan. Site-wide cash flows and resulting financials are also based on the production plan. The geotechnical group guides the mine planners in the following areas during development of the production plan:

- vertical rate of advance;
- phase lag;
- blasting practice;
- limit excavation practice;
- face cleanup practice;
- surface and water diversion (upstream and in-pit);
- groundwater impact on pit operations.

The geotechnical group also works with mine planning and mine operations in the areas of facility planning and cost estimation from Stage 3 to closure. A geotechnical program should estimate the following:

- strength and rock mass characterisation;
- domain modelling and importation into the mine planning package;
- structural modelling and importation into the mine planning package;
- surface and groundwater analysis, quantity, infrastructure and routing;
- geotechnical program content and cost from Stage 3 to closure (staff, equipment, consumables) by period.

Inter-ramp and bench configurations are typically entered into the generalised mine planning package using zones within the geological or the geotechnical model if available. These models should extend at least to the limit of any potential pit. Modelled domains and pit design zones are transferred to mine planning packages, labelling each block in the resource model with the code that identifies a particular pit design zone or a domain. Slopes and other bench geometrical characteristics associated with the code are input to generate the economic and geotechnical boundary (Lerchs-Grossman cone) that will guide the mine planner through the pit design process.

Continuing close coordination between the geotechnical engineers and mine planners is particularly important at this level.

11.2.4.5 Level 4: Design and construction

The purpose of this stage is to develop the necessary action plans by discipline or function in enough detail to ensure a smooth implementation during the first years of the project. Classically, this level involves 'detailed engineering', which is the refinement of a concept design into design drawings, component costing and construction project scheduling. In simple terms – what to build, how to build it, when to build it, what it will cost and how to manage the total process. As such, Stage 4 may form the basis of the owner's commitment to construction funds.

The geotechnical group needs to be part of the Level 4 analysis. Its involvement includes:

- external work flow (determination of data and process links with other departments from construction to closure);
- internal work flow determination (beyond the scope of this book);
- period-based action plans (beyond the scope of this book);
- project schedule (beyond the scope of this book);
- staffing schedule and skill level determination (beyond the scope of this book).

Ongoing data collection

The design stage may take months and requires ongoing data collection. Exploration drilling could continue and may result in changes to the mine designs and plans. It is advisable to involve geotechnical engineers in the data collection phase as gathering data will enhance the characterisation of specific geotechnical domains and add confidence to models. It will also be important to reconcile new mine designs with updated geotechnical guidance.

External work flow

The geotechnical group needs to plan its interaction with other departments through the stages of project development, the degree of interaction varying as the operation evolves from construction to a mature operation. This interaction ideally results in the early identification of issues that may affect slope design and excavation plans, resulting in a safer operation and potentially improved operation economics.

Interdepartmental facilities that require geotechnical input include construction of foundations for any major structure, surface and groundwater facilities, waste dumps, temporary and permanent access roads, leach pads, dykes, in-pit haul roads, top soil storage areas and water reservoirs.

Construction

Pre-stripping may occur during the construction phase. Data are still limited but functional group interaction is increasing. Data collection and operations procedures are the focus.

Pits, dumps and pads may be in early development, but this is a good time to test and refine field procedures, construction, blasting and excavation methods. The final walls in the pit are normally not excavated at this stage, so it is a good time to test blast patterns for back break and the potential to damage walls. Waste dump foundation construction and monitoring, along with leach pad foundations, are areas where the geotechnical group must be involved.

Ongoing data collection to refine knowledge of geological structures and material strengths should be done in conjunction with mine planning. The mine engineers will need to use the best available geotechnical knowledge to update short- and long-term pit designs and access ramps.

The evaluation of overall performance in phase walls by geotechnical engineers is an essential input for mine planners to adjust their designs so that, by the time a pit is mature, early difficulties have been resolved and discrepancies between early characterisation of geotechnical domains and reality can be mitigated. By the time final walls are starting to develop there should be optimised pit slope criteria, based on significant hard data and experimentation.

Mine planning involves short- and long-term waste dump and leach pad production plans. Loading rates for these facilities depends on the geotechnical response to loading. Mine planners usually want to haul waste rock the least distance possible, so dump heights and rapid loading will be requested and debated with the geotechnical group.

Access ramp design can change quickly in the short term. The geotechnical group needs to map features (e.g. structures, alteration zones) that may influence the placement of access ramps. The mapped features must be input into the mine design software database for use by the mine engineer.

The geotechnical and mine planning groups are also involved in developing a pit wall management manual or ground control manual, a joint effort between mine operations, planning, geotechnical, hydrology and other groups who are active in and around pits.

11.2.4.6 Level 5: Operations

The objective of this stage is to progress mine production plans to the most efficient level possible, including health, safety and loss prevention, while maximising the return on investment.

The geotechnical group is involved in the following areas: mine design, mine production planning and daily production.

Mine design

Mine design is an ongoing process. In the early stages of development, mine design may change frequently due to exploration drilling of the ore body. In mature operations, mine designs change as the nature of the geotechnical limitations of pit wall segments and the influence of groundwater and storm water on the pit walls is better understood. Metal prices may change mine designs but they should not change slope angles, just the wall placements.

Large or deep open pits may change as a result of testing wall height versus slope angle.

Mine operations, planning and geotechnical groups typically work together on the following items when the pit is mature:

- face angle and catch bench configuration by domain;
- inter-ramp and overall slope angle by domain;
- review of final and phase designs, including ramps and wide benches for geotechnical compliance of slope angles, bench configurations, lag height and structural intercepts;
- predictive programs to ensure that slope angle guidance for the last layback is sound;
- predictive changes in slope angles due to changes in operating practice.

Systematic geotechnical data collection can be used to build relationships between operational practices, the size or type of equipment, effective bench face angles and catch bench performance by geotechnical domain in several zones of the pit.

The systematic collection and analysis of data regarding bench scale failures and the mechanics of failure can also lead to better catch bench and face angle configurations or multi-benching from single bench configurations. From a mine design and production point of view, multi-benching is usually more desirable as less equipment, time and expense are involved in reaching the final wall limit.

Mine production planning

Well-founded production planning requires input from many departments. The geotechnical group usually contributes to the production plan guidance in the following areas:

- height lag or maximum distance between phases;
- vertical advance rate;
- dewatering or depressurisation necessary for stable slope angles;
- dewatering rate and infrastructure necessary to achieve the vertical mining rate;
- blast reconciliation regarding blast damage;
- excavation reconciliation for over- or under-digging of toe, face and crest;
- face scaling reconciliation to determine if the operating techniques are working;
- operating practice guidance (blasting, bench face and catch configuration);
- slope monitoring program;
- how to manage wall instability;
- how to manage wall failure;
- how to operate under or around rockfall hazards, instability or failure;
- safe operating distance under normal, high, unstable or failed walls;
- planning access to the pit for testing, data collection, performance monitoring or other geotechnical initiatives;
- reconciliation of wall performance to design.

Daily production planning

Daily production activities require procedures to ensure a safe and efficient work area. Tight coordination is required between Mine Operations, Mine Engineering and the Geotechnical and Hydrological groups. The geotechnical group contributes in the following areas:

- mitigation of unstable areas (volume, risk and approach);
- mitigation of failed areas (volume, risk and approach);
- blast monitoring for damage to walls, catch bench and bench faces;
- excavation damage monitoring;
- pit slope monitoring;
- monitoring of groundwater drawdown rate versus mining vertical advance rate and impacts to slope angles if they are out of sync;
- structural monitoring and analysis of pit and foundation surfaces;
- strength monitoring and analysis for walls in pits, dumps and dams;
- strength monitoring in waste dump and leach pad foundations;
- daily, weekly and monthly maps of failed or unstable areas such as rock fall hazards, engineering controls (bunds), one-ways, special dig permits and no-dig areas;
- procedure manual for the above;
- training shift supervisors in early detection of failures.

11.2.4.7 Closure

Mine closure planning may start years before mining commences. As outlined in Chapter 14, designing for mine

Figure 11.6: Backfilled pit showing reclaimed area

closure is a very important part of providing sustainability for the local community or future land tenants and may be considered during the original feasibility study.

The following areas have geotechnical aspects and are usually part of a closure plan:

- pit, waste dump, leach pad and road slope stabilisation;
- surface water control and diversion or ditch stability;
- pit backfill stability (see Figure 11.6);
- pit dewatering;
- pit lakes formation and stability;
- revegetation stability (see Figure 11.7).

Closure design criteria can be dictated by local or federal regulations, company standards or standards set by lending companies helping finance the initial capital, and vary considerably. In all situations the geotechnical engineer should pay special attention the class of facility being designed (temporary or permanent), and whether a temporary facility has the potential to become permanent.

Figure 11.7: Concurrent reclamation of waste rock dump

11.2.5 Summary and conclusions

There should be an alignment between the geotechnical and the mine planning field through all stages of project development, but some issues remain. Standardisation of slope angle inputs to the mine planning software should be undertaken or checked carefully for each project. Of greater concern are the depth, complexity and marginal nature of many deposits now being developed as deep and complex open pits.

The changing economic factors that drive the life of mine operations imply a continued review of field practices and open pit designs. This aspect, associated with changes in the understanding of material characteristics and a general lack of engineering resources through the entire industry, demands increases to the level of coordination in all disciplines. It also requires appropriate systems to ensure knowledge is transmitted from designer to mine operator.

The next generation of deposits and the resulting operations will increase in complexity and depth. These deposits will represent a major challenge for design engineers and mine operators to ensure acceptable financial results, environmental compliance and occupational safety. Data collection, analysis and evaluation methods must adapt to the changing landscape. Slope management should therefore form the basis of the open pit design philosophy and process.

11.3 Controlled blasting

11.3.1 Introduction

Large open pit mines depend on economies of scale to meet their business targets. Consequently, the trend in large open pit mining has been towards high-energy blasting to increase in-pit comminution and excavation performance, coupled with large equipment that is capable of high levels of productivity. Table 11.2 provides examples of increased energy concentrations and the corresponding outcomes documented from mine to mill comminution studies (Scott et al. 1998; Kanchibotla et al. 1998).

This increasing scale of energy concentration and production rate can threaten the integrity of the pit walls.

Table 11.2: Recent documented mine to mill demonstrations

Mining operation	Powder factor increases	SAG mill throughput
Gold mine (Papua New Guinea)	0.60–1.25 kg/m^3	Increase of 15%
Copper/gold mine (Australia)	0.91–1.21 kg/m^3	Increase of 10%
Gold mine (Australia)	0.55–1.20 kg/m^3	Increase of 10%

Consequently, discipline and effective controlled blasting strategies, driven by technical issues that are not significantly compromised by short-term production requirements, are necessary to ensure that the walls are not damaged by blasting.

It is essential to monitor technical developments to ensure that the best available approaches are applied to large open pit operations. Table 11.3 lists some established and emerging technologies that might assist in engineering blasting operations for large open pits. A number of controversial issues are also given. These are issues on which blasting experts disagree and for which more quantitative solutions are required.

The goal of an effective wall control blasting strategy is to produce a well-fragmented, loose muck pile as well as an on-design and undamaged slope. The upper photo in Figure 11.8 illustrates a notably unsuccessful attempt to achieve this goal. The lower photo shows a successful outcome.

Figure 11.8: Conditions of final walls. Damage control techniques not applied (upper), damage control techniques applied (lower)
Source: Photos courtesy of A. Karzulovic

Table 11.3: Established and emerging technologies and controversial issues

Established technologies	Emerging technologies	Controversial issues
High-precision GPS Drill control Fleet management systems Image analysis systems Slope monitoring with radar Variable explosive density Environmental monitoring Computer design of blasts Wired electronic detonators Ore body modelling Drilling technologies	Wireless electronic detonators Low-density explosive products SMART trucks Remote structural mapping (digital photogrammetry) Strata recognition Excavation monitoring Movement sensors Numerical modelling of explosive/rock interaction Rock mass modelling (joint simulation) Rock mass characterisation Blastability Grindability	Designing based on powder factor Size of limit blasts Choked vs free face blasting of limit blasts Shock dependence on confinement Controlling diggability Pre-split design (explosive distribution) Pre-split design (timing) Blasting in saturated ground (dynamic water) Use of gadgets (stemming plugs) Berm width design and protection Using information in blast design Blast damage for destressing purposes

To achieve success, several site-specific conditions must be evaluated, including:

- the intended slope design;
- geology, especially structure and hardness;
- water conditions;
- vibration characteristics;
- pattern shape;
- available free faces.

Once these conditions have been defined, a controlled blast design can be developed that takes site factors into account. A suggested approach that allows engineers to react to changing conditions is outlined in Figure 11.9. It is based on a combination of models, measurements and experience, and provides a sound basis for the derivation and refinement of blast designs required for future large open pit mines.

The methodology is considered appropriate for all kinds of blast designs in open pit mines, including production and perimeter or limit blasts. The goal, however, is controlled limit blast design through the efficient synchronisation of explosive energy distribution, energy confinement and energy level, as illustrated in Figure 11.10. Practical means of achieving this goal are discussed below, including blast damage mechanisms, geological considerations, controlled blasting techniques, delay configurations, site evaluation, design development, design implementation, excavation, performance monitoring and design refinement.

11.3.2 Design terminology

Different terms are used by the mining industry to describe controlled blast designs. For clarification, the terms used in this text are shown in Figure 11.11.

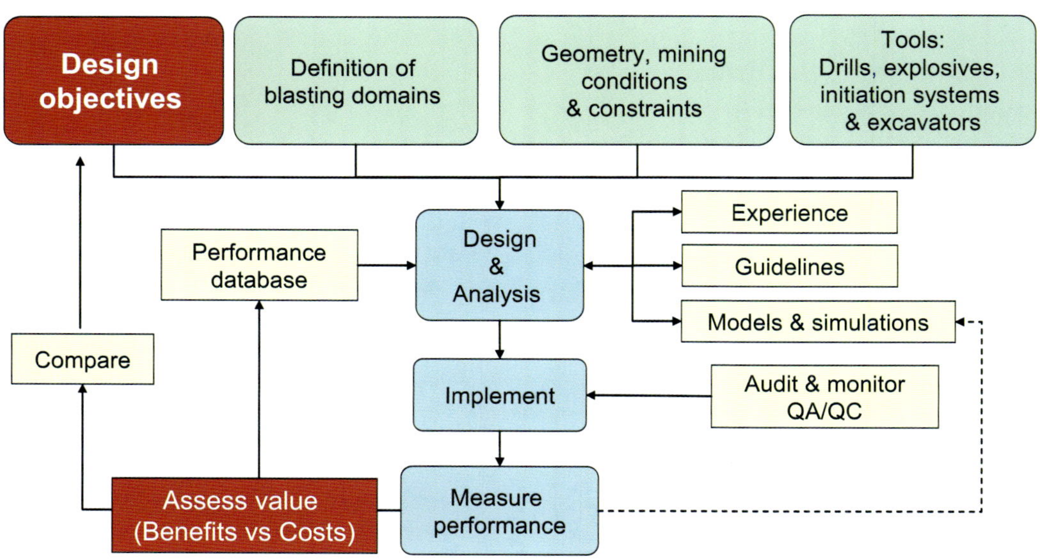

Figure 11.9: Engineering design and optimisation methodology for open pits

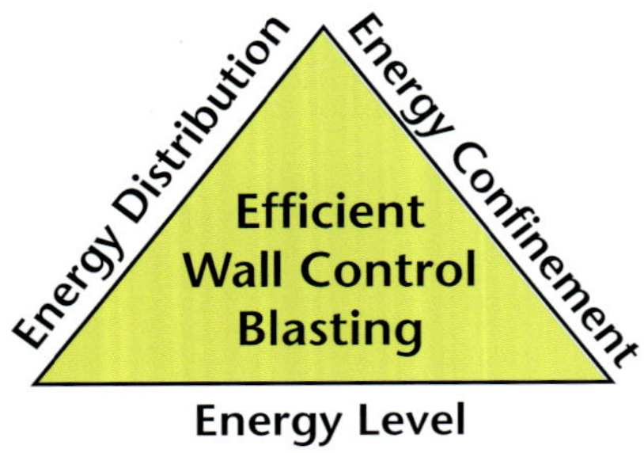

Figure 11.10: Explosive energy efficiency triangle

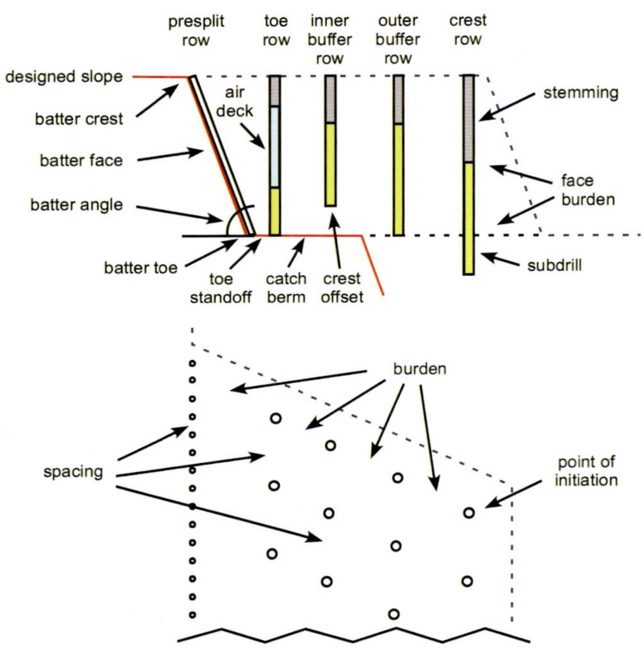

Figure 11.11: Design terminology

11.3.3 Blast damage mechanisms

The blast damage mechanisms that cause reduction in slope stability can be categorised into near and far field effects.

In the near field (less than 50 m from the charge), shock, crack extension, gas-related displacement and tensile failure are the major contributors to rock mass damage.

When an explosive detonates (Figure 11.12), the quasi-solid, relatively cool explosives are quickly converted into a high-temperature, high-pressure gas. A 10 m long explosives column is consumed in 2 milliseconds, expands approximately 1000 times its volume and generates theoretical blast hole pressures in excess of 2 GPa. Immediately around the blast hole the high-detonation pressures propagate a shock wave into the rock mass. The pressure of this initial shock wave is much greater than the strength of the rock and, as a result, a zone of 2–3 charge diameters is crushed in compression (Figure 11.13).

This crushing and expansion of the blast hole reduces the pressure to the point where the shock wave is reduced to a strain pulse. As the strain pulse propagates through the rock mass at a rate equal to the P-wave or seismic velocity it compresses the rock radially, which results in tangential tension or hoop stress. If the tangential tension is greater than the tensile strength of the rock, fractures are created (usually for 20–30 charge diameters) radially in all directions (Figure 11.14).

When the strain pulse reaches a rock/air interface (e.g. an open joint), the pulse is reflected back in tension; if the tension is greater than the tensile strength of the rock, spalling occurs. Tensile fracturing and spalling relieves the stress within the rock mass enough that no new fractures from shock are created.

One way to quantify the shock/strain pulse applied to the slope is to measure the blast's peak particle velocity millimetres/second (mmps) in the vicinity of the designed slope. The peak particle velocity required to damage rock depends on the strength and structure of the rock mass. Typical damage thresholds are:

- at 50–100 mmps, loose structures fall;
- at 125–500 mmps there is damage to weak rock;
- at 375–1000 mmps there is damage to hard fresh rock.

The next phase of blast-induced rock damage involves the expansion and penetration of the high-pressure gases into the rock mass.

Figure 11.12: Explosive detonation

This phase starts when the explosive detonates but it takes place at a much slower rate than shock- and strain-related damage. Initially, the gas pressure produces quasi-static stress in the rock mass around the blast hole. This stress can be enough to create new cracks, but the primary benefit in terms of fragmentation occurs as the high-pressure gases wedge into existing fractures and cause them to expand. As the fractures expand, they intersect with other cracks and produce a significant portion of the fragmentation. The envelope of damage created during tensile failure and crack extension is generally thought to be 20–30 charge diameters depending on the strength of the rock, the explosives used and the degree of confinement (Figure 11.15).

When pre-split techniques (section 11.2.5) are correctly employed and the geology is favourable, the slope damage caused by tensile failure and radial cracking can be significantly reduced. Typical compressive (crushed zone), tensile and radial cracking related damage are shown in Figure 11.16.

The last phase of the rock fragmentation process occurs as gas pressure bends and breaks the rock mass (flexural rupture) toward the path of least resistance. This can occur in any direction and can lead to excessive overbreak if relief away from the slope is not provided. In addition, if weak, adversely oriented discontinuities exist in the batter face, the expansion of gases can cause block heave beyond the designed slope (Figure 11.17).

As the gases expand they apply a load to the slope that is released as the blasted rock mass swells. If the initial load placed on the slope is excessive, due to confinement, tensile failure can occur well beyond the designed limit (Figure 11.18).

It should be noted that, while pre-split techniques can reduce the damage caused by tensile fracturing and crack extension, they have limited use in controlling block heave and release of load damage. Gas expansion damage can only be reduced by providing relief for the explosive energy to displace material away from the wall, thus limiting the amount of pressure applied to the wall.

In the far field (greater than 50 m from the blast), the damage that results from blasting is mainly shear strength reduction and ravelling of loose material that eventually fills the catch berms. This type of damage can be controlled by minimising peak particle velocities and, if possible, maintaining dominant vibration frequency above the natural resonance of the slope (section 11.3.8).

11.3.4 Influence of geology on blast-induced damage

The geological characteristics of the region adjacent to the blast and further up the slope dictate the potential for blast-induced damage. As with slope design, it is important to consider the nature of the rock mass when developing

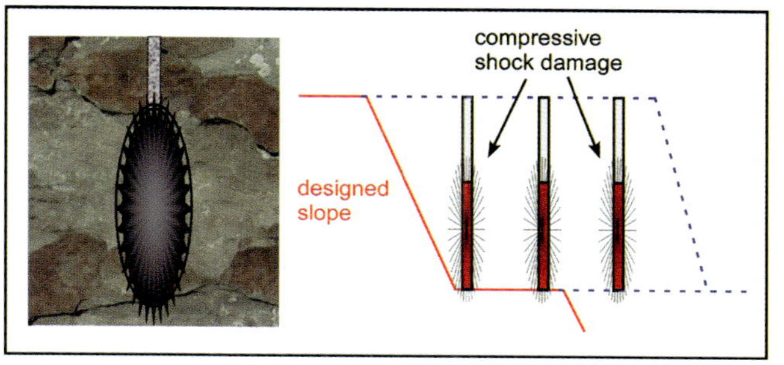

Figure 11.13: Compressive rock damage immediately around explosive

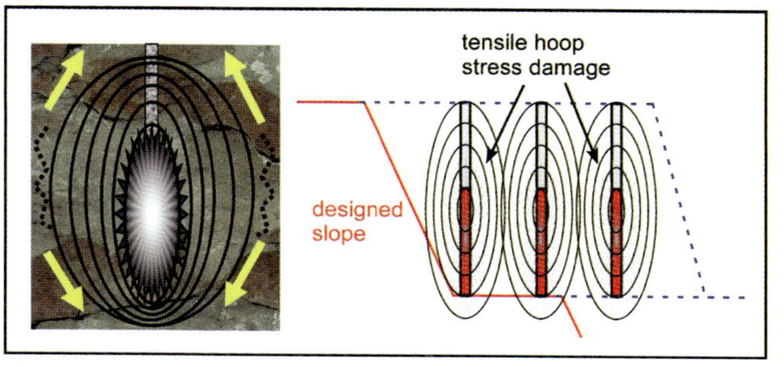

Figure 11.14: Strain-induced tensile failure

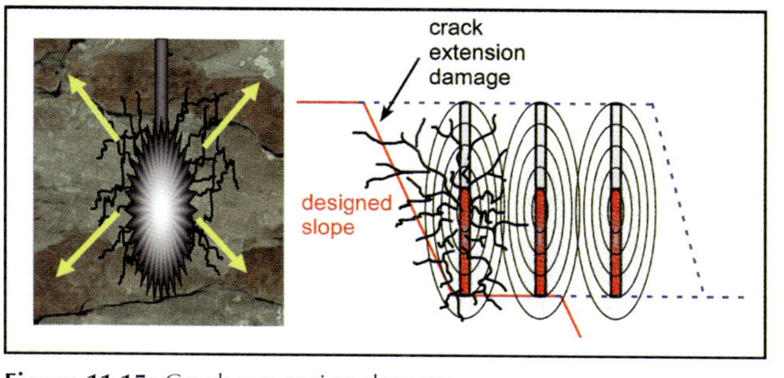

Figure 11.15: Crack expansion damage

Figure 11.16: Near field blast-induced damage

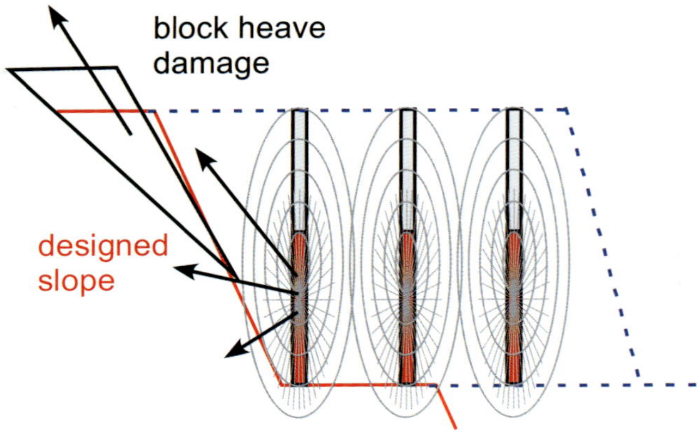

Figure 11.17: Block heave slope damage

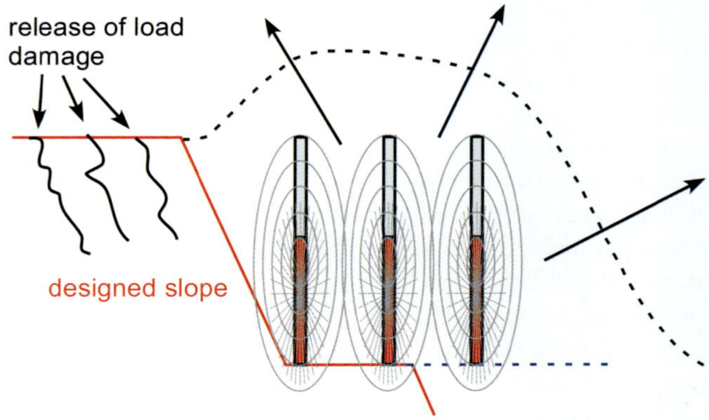

Figure 11.18: Release of load damage

efficient wall control blast designs. Typically, tests are performed on core samples from the slope regions to determine the nature of the rock mass. These tests include rock strength and structural evaluations.

Dynamic rock strength or strength under stress can be divided into three main categories:

- compressive strength;
- tensile strength;
- shear strength of discontinuities.

The tensile strength is approximately one tenth of the compressive strength, which helps explain why more blast damage occurs in tension than in compression. By determining the relationship between the lateral deformation and longitudinal deformation of the core (Poisson's ratio), it is possible to determine how prone the rock will be to propagate a crack during pre-splitting. The lower the ratio, the easier the individual rock blocks will split (not considering structural orientation). Young's modulus (modulus of elasticity) measures the rock's ability to withstand deformation. The higher the value, the harder the rock will be to break. Some dynamic rock strengths are shown in Table 11.4.

While the strength of the rock is certainly important, it is virtually meaningless unless the structure of the rock mass is also taken into consideration.

Massive rock (Figure 11.19) has few defects and makes it relatively easy to achieve slope design parameters. Pre-split blast designs are usually successful and potentially very cost-effective, as shown in Figure 11.19.

When the rock has layers (bedded) it is described in terms of:

- layer thickness;
- bond between layers (open, tight, fill material);
- strength of bond;
- orientation to the batter face (e.g. horizontal, vertical, dipping into pit, dipping into slope).

In sub-horizontal strata, steep batter faces can be achieved, but the thickness of the bedding planes at the top may cause overhangs and poor excavator productivity if

Table 11.4: Typical rock strength characteristics

Rock type	Density (g/cc)	Compressive strength (mPa)	Tensile strength (mPa)	Young's modulus (GPa)	Poisson's ratio	Longitudinal velocity (mps)
Basalt	2.9	149	11	62	0.27	5229
Dolomite	2.5	55	3	28	0.32	4024
Gneiss	2.8	224	14	81	0.22	5732
Granite	2.7	186	9	43	0.33	4844
Limestone	2.7	159	5	55	0.25	5000
Marble	3.1	251	15	106	0.28	6705
Sandstone	2.5	134	1	7	-	3933
Sandstone	1.8	11	0	6	0.3	2095
Schist	2.9	166	9	77	0.20	5482
Slate	2.6	85	6	66	0.17	5168
Taconite	2.9	251	17	93	0.25	6140

Figure 11.19: Relatively massive rock

the explosive energy of the blast is not well-distributed (Figure 11.20). Trim or pre-split blasting techniques are most appropriate for this type of rock mass.

Thin, horizontally bedded and well-cemented rock masses as shown in Figure 11.21 can produce batter faces of 70–80° with conventional trim or cushion blast techniques. When the bedding is dipping into the pit, it is often difficult (and inadvisable) to achieve a batter face steeper than the dip angle (Figure 11.22).

Near-vertical bedding can cause overhangs and very irregular faces, depending on the thickness of the bedding planes (Figure 11.23). This type of rock mass is best suited for trim blasting techniques. If the strike of the structure is perpendicular to the face (Figure 11.24), the overhang problem will be non-existent and the face will be much more consistent.

Joints or cracks within the rock mass can also dictate the slope design and the most appropriate controlled blasting technique. Jointing can be described in terms of:

- block size;
- spacing;
- persistence;

Figure 11.20: Horizontally bedded rock mass with relatively thick layers at crest

Figure 11.21: Thinly bedded silicified shale

Figure 11.22: Bedding dipping into the pit, strike parallel to crest

Figure 11.23: Near-vertical bedding dipping into slope, strike parallel to crest

Figure 11.24: Dipping thinly bedded structure, strike perpendicular to crest

- crack characteristics (e.g. aperture width, roughness, fill type and strength);
- orientation to batter face.

Blocky jointing (Figure 11.25) requires excellent explosive energy distribution to control overbreak and maintain excavator productivity. Pre-split techniques may be appropriate if the joints are tight. Otherwise, trim blast designs with good horizontal relief are most effective.

The fracture frequency shown in Figure 11.26 usually adversely affects the establishment of pre-split blasting failure planes. If the designed batter face angle is less than 60°, it is recommended that buffer blasting be used to define the batter face. Batter face angles greater than 70° in such material typically require trim blasting.

Wedge failures (Figure 11.27) can occur when dominant joint sets intersect each other and the surface on the catch berm. This kind of failure is very difficult to prevent, especially when the cracks are open or filled with weak material. The failure mode is block heave and can only be reduced by limiting the load placed on the slope by the blast (pre-splitting will not prevent this type of damage). This type of rock mass requires excellent horizontal relief away from the wall. In many cases,

Figure 11.25: Blocky jointing

Figure 11.26: Highly jointed, dipping into slope

smaller blast holes are used to disperse and limit the amount of charge that fires at one time. In extreme cases, the production blast adjacent to the trim blast will cause block heaving beyond the design crest. This often requires that the width of the trim blast be increased to move the production blast farther from the slope.

Water-saturated slopes transmit shock energy more efficiently than dry rock masses. As a result, the vibration and pressure levels do not attenuate as quickly and the damage envelope is likely to be greater. The keys to minimising slope damage in these environments are to reduce the charge weight that fires at a time and to provide good relief away from the wall.

Major contacts between rock types (Figure 11.28) can also lead to slope instability, especially when the contact is adversely dipping into the pit. It is important to control both near and far field damage in this type of rock. Low-frequency blast vibrations can weaken the contact strength and may cause premature slope failure.

11.3.5 Controlled blasting techniques

Several techniques are used to reduce blast-induced slope damage. They include:

Figure 11.27: Wedge failure caused by intersecting joints

Figure 11.28: Major contacts

- buffer blasting;
- trim blasting;
- pre- or mid-split blasting;
- post-split blasting;
- line drilling.

Each technique has advantages and disadvantages that limit its use to specific rock conditions or design requirements. The choice of technique depends on the slope design and characteristics of the rock mass. As stated previously, all techniques benefit from good horizontal relief away from the slope. It is therefore recommended that relief be carefully considered during the design process.

11.3.5.1 Buffer blasting

Buffer (or cushion) blasting is typically used for weaker rocks and involves modification of the production blast designs to reduce wall damage. The common modifications are:

- a free face is created for horizontal relief;
- the pattern width is often reduced to three to six rows deep;
- the delay sequence is modified to control vibration levels and displacement;
- subdrilling is reduced or eliminated above the catch berm.

The last row of holes is placed in front of the designed batter face; this is known as the 'toe standoff' distance. The toe standoff distance is critical to achieving the designed batter face angle without damaging the slope. The example in Figure 11.29 is for a relatively weak rock mass and the excavator defines the batter face by free digging essentially unshot material.

Buffer blast designs are also appropriate when the designed batter face angle is less than 60°, as in Figure 11.30.

In harder or more structurally complex rock masses, the optimum toe standoff distance is difficult to determine. If the last row of holes is close enough to define the toe of the batter, a portion of the crest is usually lost. Moving the last row farther away to protect the crest makes it difficult to achieve the designed toe. In both cases, the carry capacity of the catch berm is reduced and the required capacity is not achieved. As a result, trim blasts are more common in average to hard, structurally complex rock masses.

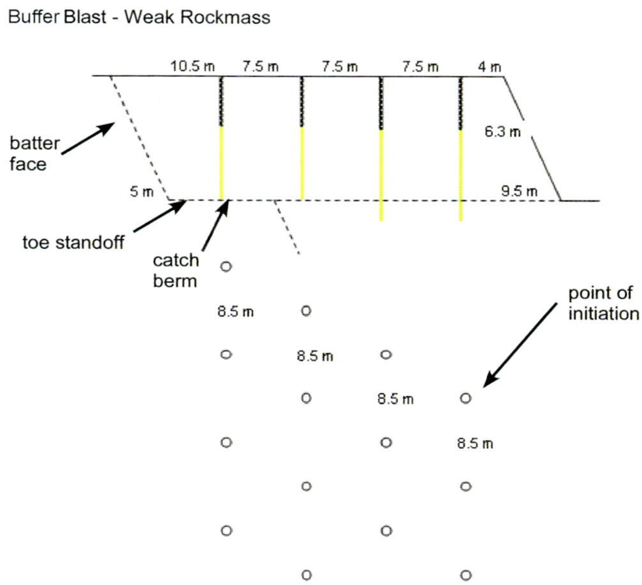

Figure 11.29: Buffer blast design

11.3.5.2 Trim blasting

Trim blasting is the most commonly used controlled blasting technique. It can provide good results in a wide variety of geological conditions as long as the designs are refined in a logical manner. The relationship between

Figure 11.30: Rock mass well-suited for buffer blasting

explosive energy distribution, confinement and level can be enhanced with the use of air decks, pattern modifications and/or reduced hole diameters. In some cases, a pre-shear row is included in the trim blast design (section 11.3.5.3).

Trim blasts are typically three to five rows deep and are shot to a free face that has a consistent burden. In adverse geology, extra rows may have to be added to the blast to protect the slope from damage caused by the production blast.

Batter face angles of 60–75° are fairly typical for trim blast designs. However, in many cases the angle achieved is directly controlled by the structure of the rock mass.

Typical trim blast design modifications (using the initial trim blast as a starting point) for harder or more structurally challenging rock masses are shown in Figure 11.31.

The purpose of the toe row is to define the toe of the batter, not the crest. The burden, spacing and charge are reduced accordingly. The burden is greater than the spacing to promote breakage between the toe holes. The spacing is initially set at half the normal spacing to make it easier to tie-in.

Air decks are commonly used in the toe and inner buffer rows to reduce blast hole pressures and to increase fragmentation in the top portion of the bench. It should be noted that bench cratering typically occurs at the bottom of the stemming column when air decks are used.

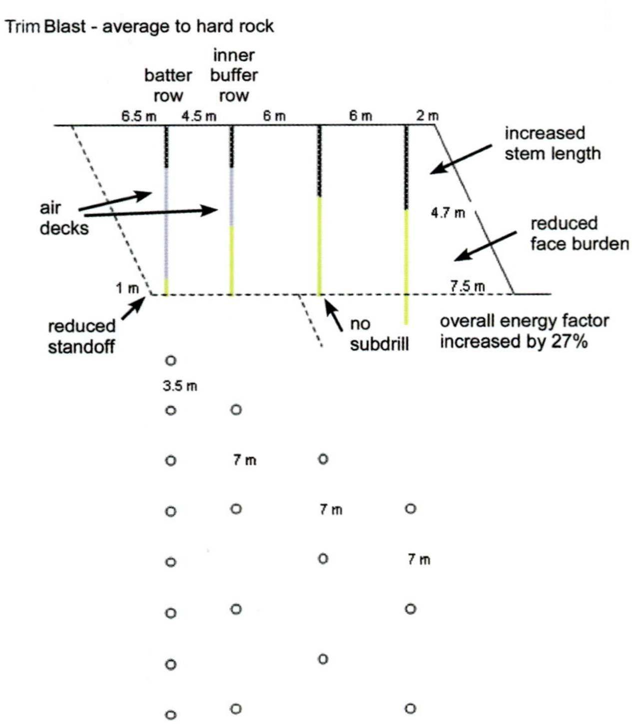

Figure 11.31: Trim blast modifications

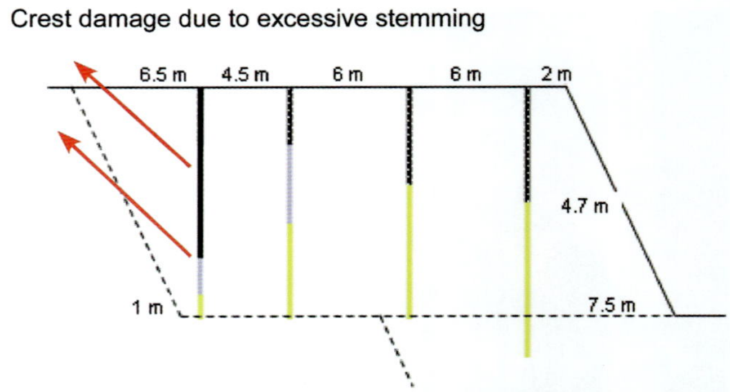

Figure 11.32: Damage caused by overconfined toe row

If the stem length is too long, the charge will be overconfined and excessive crest damage can occur (Figure 11.32).

The toe row can be partially stemmed without an air deck if oversize fragmentation from the top portion of the bench is not a concern. The length of stemming will be defined by the amount of confinement required to define the toe without damaging the crest (Figure 11.33).

The inner buffer row is designed to define the crest, so its charge and standoff from the batter face requires careful determination. No subdrilling is used above and immediately adjacent to the catch berm.

To improve the horizontal explosive energy relief, the face burden is reduced. Additional stemming is placed in the front row to confine the explosive energy long enough to move the toe out.

Energy factors are increased by 27% (an overall average of 740 kJ/t in this example) to compensate for the harder material. If there is good horizontal relief away from the wall, the energy factor commonly used for production blasting does not need to be reduced when implementing controlled blasting techniques.

Examples of the type of geological conditions that are well-suited for trim blasting are shown in Figure 11.34. Two examples of successful trim blast designs are shown in Figures 11.35 and 11.36. General guidelines for trim blast design are shown in Table 11.5. It should be stressed that

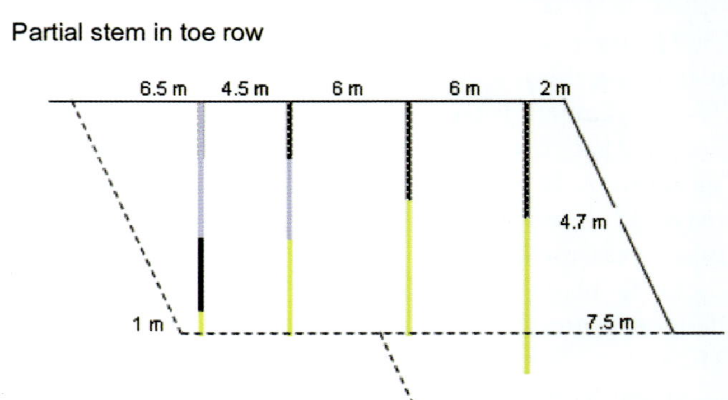

Figure 11.33: Toe stemming when top oversize is not an issue

Figure 11.34: Typical trim blast geological conditions

Table 11.5: Initial trim blast guidelines

Blast hole diameter (mm)	Charge (kg/m)	Burden (m)	Spacing (m)a	Offset from toe (m)
76	0.6	1.7	1.4	0.30
89	0.75	1.9	1.5	0.36
102	0.9	2.0	1.7	0.41
114	1.2	2.4	1.9	0.46
127	1.4	2.6	2.0	0.51
153	2.0	3.1	2.4	0.61
165	2.3	3.3	2.6	0.66
200	3.4	4.1	3.1	0.80
229	4.6	5.0	3.5	0.92
270	6.0	5.5	4.1	1.08
311	7.8	6.1	4.8	1.24

a: Spacing may need to be half the inner buffer row spacing to help pattern tie-in

these guidelines are for initial design development. It is unlikely that they will provide optimum performance without modification.

11.3.5.3 Pre- or mid-split blasting

Pre- or mid-split blasting involves drilling a row of closely spaced holes along the designed dig limit. These holes are loaded with decoupled charges to split the gap between holes in tension without causing compressional damage to the slope (Figure 11.37). The purpose of the crack is to minimise the damage caused by tension cracking and crack extension. It will not prevent block heave or damage in adverse geology if the blast is overconfined.

Pre-split differs from mid-split blasting only in the way the holes are initiated. In pre-split blasting, all the holes are fired prior (milliseconds or days) to the first hole in the adjacent pattern.

Ideally, the pre-split will be fired before the holes in the adjacent blast are drilled. However, this is not always possible so the pre-split is fired next to loaded blast holes. If the time between the detonation of the pre-split and adjacent holes is too great, the performance of the explosive in the adjacent holes can be adversely affected. As a result, a mid-split (shot in the middle of the timing sequence) is fired a short time (100 ms) before the detonation of the adjacent holes. For the purposes of this discussion, the term pre-split will be used interchangeably for pre- or mid-split blasting since all other design parameters remain the same.

As with other controlled blasting techniques, the characteristics of the geology control the pre-split performance (Figure 11.38). Favourable pre-splitting conditions include:

- massive rock;
- tight joints;
- dominant joint orientation more than 30° off strike of the designed face;
- absence of weak structures that form wedges of daylight on the batter face and catch berm.

Initial pre-split designs call for the blast holes to be spaced approximately 14 hole diameters apart and that the total charge (kg) in the blast hole be approximately half the surface area between blast holes (bench height × spacing/2). Table 11.6 provides initial pre-split guidelines.

Table 11.6: Initial pre-split guidelines

Blast hole diameter (mm)	Charge (kg/m)	Spacing (m)	Minimum decoupled charge diameter (mm)	Maximum decoupled charge diameter (mm)
76	0.5	1.1	22	25
89	0.6	1.2	22	29
102	0.7	1.4	25	32
114	0.8	1.6	32	38
127	0.9	1.8	32	44
153	1.1	2.1	38	51
165	1.2	2.3	44	51
200	1.4	2.8	51	64
229	1.6	3.2	64	76
270	1.9	3.8	68	89
311	2.2	4.4	78	103

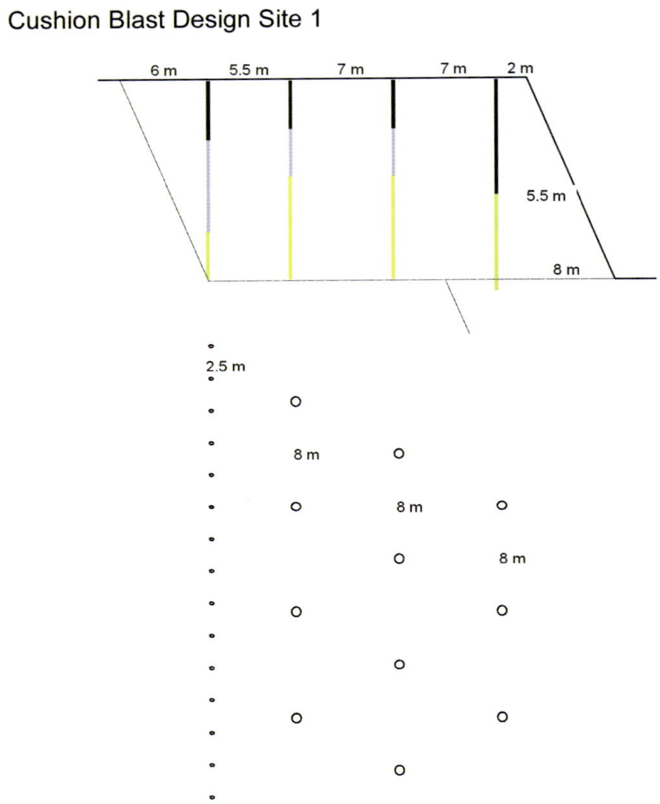

Design Data	Toe or Presplit Row	Inner Buffer Row	Outer Buffer Row	Crest or Modified Production Row	
Hole Diameter (mm):	127	280	280	280	
Hole Depth (m):	15.0	15.0	15.0	16.0	
Hole Angle (deg):	0	0	0	0	
Burden (m):	5.5	7.0	7.0	5.0	
Spacing (m):	2.5	8.0	8.0	8.0	
Airdeck (m):	7	4	4	0	
Subdrill or Standoff (m):	0	0	0	1	
Stemming (m):	4.5	3.7	3.7	8.7	
Total Charge Weight (kg):	35	375	375	350	Average
Powder Factor (kg/tonne):	0.04	0.18	0.18	0.23	0.14
Energy Factor (kcal/tonne):	36	146	146	191	120

Figure 11.35: Trim blast design, Site 1

Cushion Blast Design Site 2

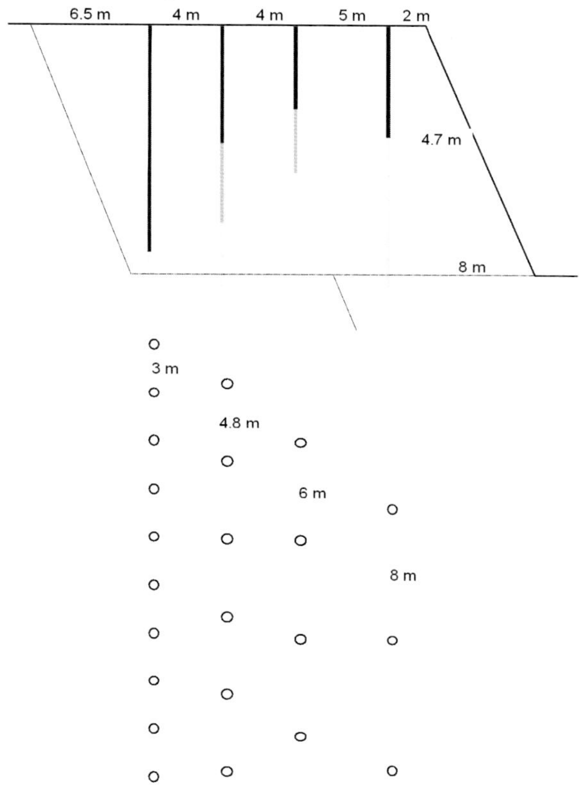

Design Data	Toe or Presplit Row	Inner Buffer Row	Outer Buffer Row	Crest or Modified Production Row	
Hole Diameter (mm):	270	270	270	270	
Hole Depth (m):	15.0	16.0	15.0	16.0	
Hole Angle (deg):	0	0	0	0	
Burden (m):	4.0	4.0	5.0	5.0	
Spacing (m):	3.0	4.8	6.0	8.0	
Airdeck (m):	0	5	4	0	
Subdrill or Standoff (m):	0	1	0	1	
Stemming (m):	13.8	7.0	5.0	6.6	
Total Charge Weight (kg):	80	260	390	600	Average
Powder Factor (kg/tonne):	0.09	0.36	0.34	0.39	0.27
Energy Factor (kj/tonne):	74	291	280	323	219

Figure 11.36: Trim blast design, Site 2

Figure 11.37: Pre-split crack between drill holes

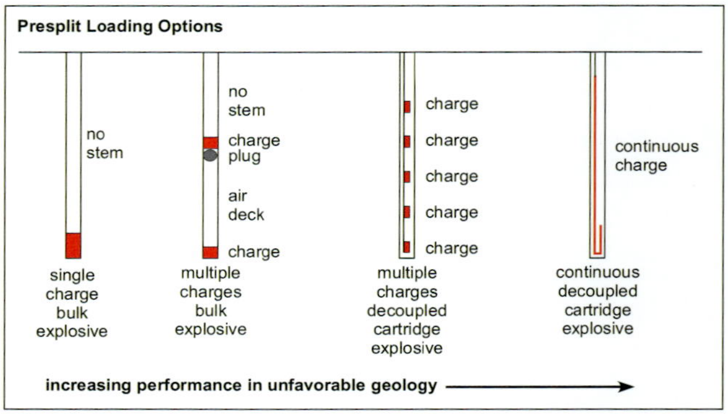

Figure 11.39: Pre-split loading options

While these guidelines are appropriate for initial design development, it will be necessary to fine-tune the spacing and charge weight based on site geological conditions. Adverse structure typically requires the spacing to be reduced while massive structures allow the spacing to be increased.

The method used to place the charge in the blast holes influences the cost and the design's ability to overcome adverse rock structure.

Bulk charges are typically ten times less expensive than continuous cartridge explosives and work well in weak massive rock. Continuous decoupled charges provide excellent energy distribution that can more effectively overcome the influence of adverse geology (Figure 11.39).

When continuous decoupled charges are used it is important to achieve high enough blast hole pressures without exceeding the compressive strength of the rock. Typically, the charge diameter required for pre-splitting ranges from 0.25 to 0.33 times the hole diameter (Table 11.6), depending on the rock type and structure. In most cases, it is inadvisable to extend decoupled charges closer than eight hole diameters from the surface. In weaker rock, it may be necessary to have a distance of as much as 15 hole diameters from the surface to protect the crest.

Stemming the pre-split hole will cause cratering of the top of the bench (Figure 11.40). Therefore, pre-split holes are typically left unstemmed unless air overpressure must be controlled. If noise control is required, the minimum amount of stemming needed to muffle the sound should be used.

It is recommended that the pre-split row be drilled 10–20° from the vertical for most geological structures. This positions the crest further away from the adjacent buffer row, which helps to reduce damage.

Initial pre-split designs should be evaluated in non-critical areas to allow refinement of the design. Pre-splitting is the most expensive controlled blasting technique, so its performance must be continuously evaluated to maintain cost-effectiveness.

A typical pre-split design is shown in Figure 11.41. The blast adjacent to the pre-split should be designed using trim blast guidelines. Actual pre-split designs are shown in Figures 11.42 and 11.43. Again, the key factor in controlling overbreak is the standoff of the toe row from the pre-split. In some cases, the use of pre-splitting is not recommended due to narrow bench widths or highly fractured rock. When pre-splitting is used with narrow bench widths and is shot next to loaded holes, the detonation of the pre-split can shift and cut-off the

Figure 11.38: Favourable geology for pre-splitting

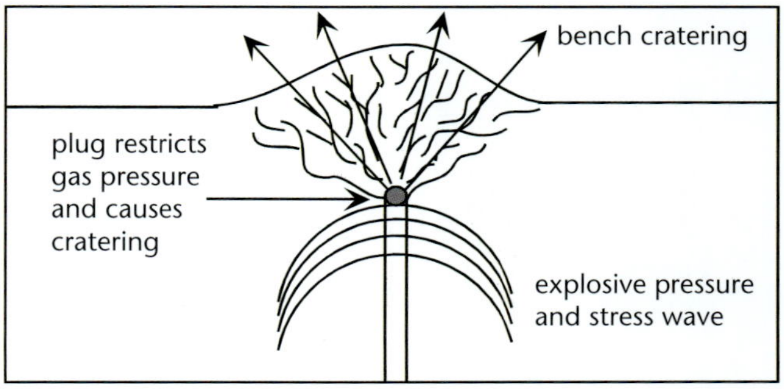

Figure 11.40: Crest damage caused by stemming pre-split holes

Pre-split Blast - Average to hard rockmass

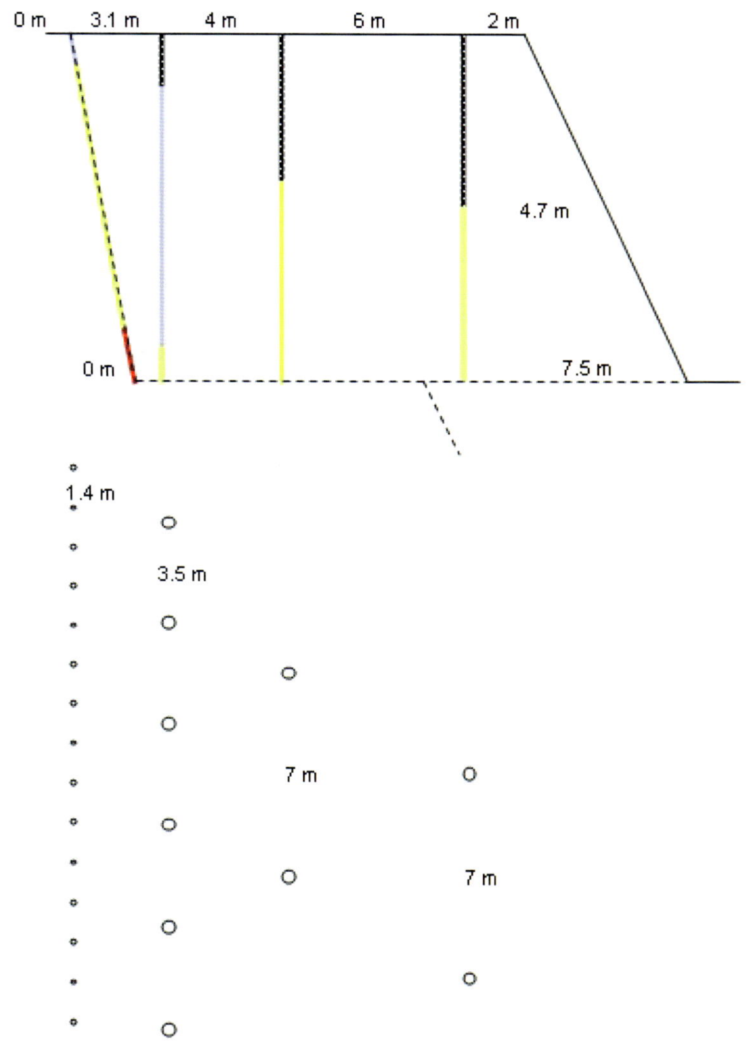

Design Data	Toe or Presplit Row	Inner Buffer Row	Outer Buffer Row	Crest Row	
Hole Diameter (mm):	102	251	251	251	
Hole Depth (m):	12.2	12.0	12.0	12.0	
Hole Angle (deg):	10	0	0	0	
Burden (m):	3.1	4.0	6.0	4.8	
Spacing (m):	1.4	3.5	7.0	7.0	
Airdeck (m):	0	0	0	0	
Subdrill or Standoff (m):	0	0	0	0.0	
Stemming (m):	0.0	1.8	5.0	6.0	
Total Charge Weight (kg):	8.4	50	285	245	**Average**
Powder Factor (kg/tonne):	0.07	0.14	0.26	0.28	0.24
Energy Factor (kjoule/tonne):	246	506	961	1044	757
Comments: typical modified production blast with free face					

Figure 11.41: Pre-split design

Pre-split Design Site 3

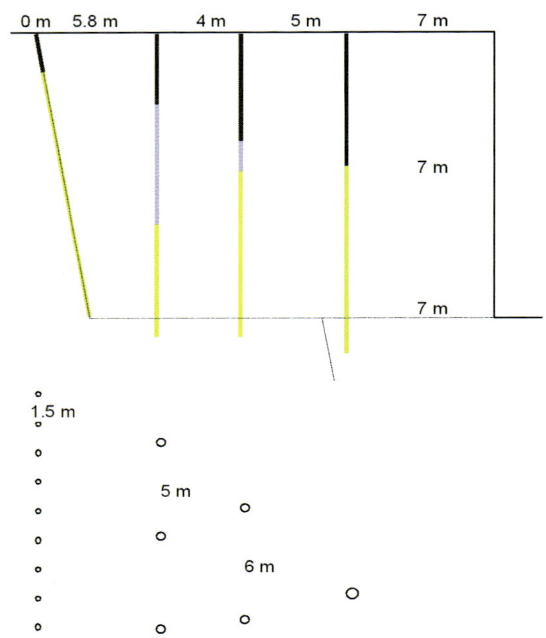

Estimated Ground Vibration	
Attenuation Constant (K):	80
Distance Constant (b):	1.50
Charge Weight (kg):	100
Distance Away (ft):	20
Estimated PPV (ips):	**28.28**

Displacement Analysis	
Estimated Face Velocity	
Top of Charge (m/s):	6
Bottom of Charge (m/s):	6
Estimated Face Displacement *	
Displacement Correction:	100%
Top of Charge (m):	9
Bottom of Charge (m):	3

* **Note:**
The estimated displacement does not consider the "rolling" of the material once it hits the ground. This rolling can extend the muckpile well beyond the estimated displacement.

Design Data	Toe or Presplit Row	Inner Buffer Row	Outer Buffer Row	Crest or Modified Production Row	
Hole Diameter (mm):	165	270	270	311	
Hole Depth (m):	15.2	16.0	16.0	17.0	
Hole Angle (deg):	10	0	0	0	
Burden (m):	5.8	4.0	5.0	7.0	
Spacing (m):	1.5	5.0	6.0	9.0	
Airdeck (m):	0	6	2	0	
Subdrill or Standoff (m):	0	1	1	2	
Stemming (m):	2.0	3.8	5.6	7.0	
Total Charge Weight (kg):	17	380	570	760	Average
Powder Factor (kg/tonne):	0.05	0.50	0.50	0.32	0.32
Energy Factor (kcal/tonne):	45	409	409	244	259
Comments: T1 trim					

Figure 11.42: Pre-split design, Site 3

Pre-split Design Site 4

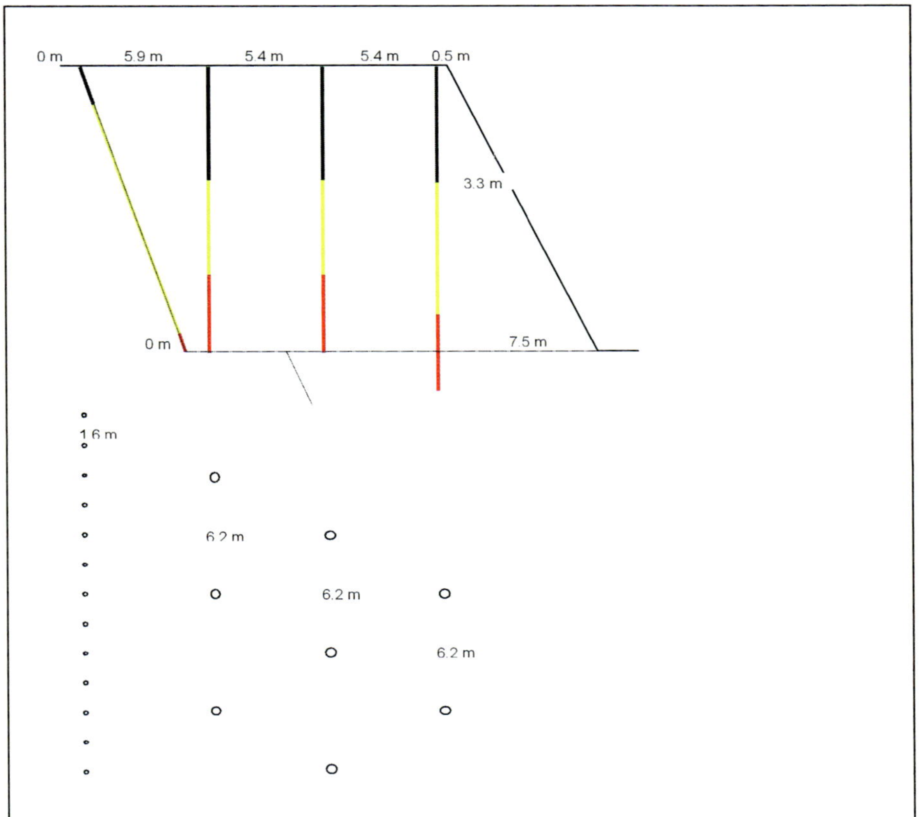

Design Data	Toe or Presplit Row	Inner Buffer Row	Outer Buffer Row	Crest Row	Average
Hole Diameter (mm):	102	229	229	229	
Hole Depth (m):	15.8	15.0	15.0	17.0	
Hole Angle (deg):	18	0	0	0	
Burden (m):	5.9	5.4	5.4	4.0	
Spacing (m):	1.6	6.2	6.2	6.2	
Airdeck (m):	0	0	0	0	
Subdrill or Standoff (m):	0	0	0	2.0	
Stemming (m):	2.0	6.0	6.0	6.0	
Total Charge Weight (kg):	20.4	375	375	443	
Powder Factor (kg/tonne):	0.05	0.28	0.28	0.44	0.30
Energy Factor (kcal/tonne):	176	950	950	1532	842

Figure 11.43: Pre-split design, Site 4

Table 11.7: Initial post-split guidelines

Blast hole diameter (mm)	Charge (kg/m)	Spacing (m)
76	0.8	1.2
89	0.9	1.4
102	1.0	1.6
114	1.1	1.8
127	1.3	2.0
153	1.5	2.4
165	1.6	2.5
200	2.0	3.1
229	2.3	3.5
270	2.7	4.2
311	3.1	4.8

adjacent holes. In these conditions the use of post-splitting is warranted.

In Figure 11.42, note the crest damage caused by the subdrill from the bench above.

11.3.5.4 Post-split blasting

Post-split blasting utilises a closely spaced, lightly charged row of blast holes placed along the designed batter face. The row of holes is shot after the adjacent blast. In highly fractured rock, post-split holes have more relief and typically cause less damage to the slope.

The blast adjacent to the post-split should be designed using the trim blast guidelines above. Since the post-split row is shot last, there is an increased risk of column cut-off due to block heave from the adjacent hole. In adverse geology that is prone to block heaving (e.g. daylighting structures), it is recommended that the post-split be fired no more than 50 ms later than adjacent holes.

11.3.5.5 Line drilling

Line drilling consists of a line of unloaded holes drilled along the final limit. When the material between the holes is placed under tension from the adjacent blast, a plane of breakage occurs (depending on the strength of the rock mass and spacing of the holes). This breakage plane helps guide the excavation of the slope. In weak material, the hole spacing is typically around 12 hole diameters. Hard massive rock requires the spacing to be reduced to three to six hole diameters.

Initially, the buffer row should be placed 50–75% of the normal burden away from the line drill row. If the ground is saturated the burden must be increased to prevent overbreak.

Line drilling is usually most cost-effective in weakly cemented alluvium (Figures 11.44, 11.45 and 11.46).

Figure 11.44: Line drilling weak alluvium

11.3.6 Delay configuration

Once a controlled blast design has been developed, it is important that the timing configuration provides relief and promotes horizontal displacement away from the wall. Correct timing designs can make a good blast design perform better. However, timing alone cannot improve the performance of a bad design.

Trim and cushion blasts should be laid out in a staggered pattern and shot to two free faces. The point of initiation should be on the corner at the point of maximum relief (Figure 11.47). The beginning and end of the blast should be angled to reduce confinement along the wall. To reduce overbreak, the direction of displacement should be at a low angle (20–40°) to the desired crest (Figure 11.48), not perpendicular to the wall (Figure 11.49).

When blasting to only one free face, a flat chevron or V configuration should be used (Figure 11.50). If a deep V pattern is used, excessive backbreak usually occurs at the point of the V (Figure 11.51). The orientation of the dominant joint structure to the azimuth of the crest and initiation direction also influences the amount of overbreak produced (Figures 11.52 and 11.53).

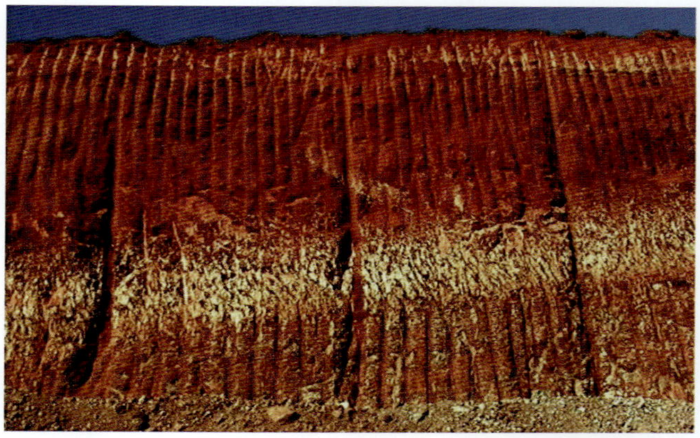

Figure 11.45: Line drilling to guide excavation

Line Drill Design

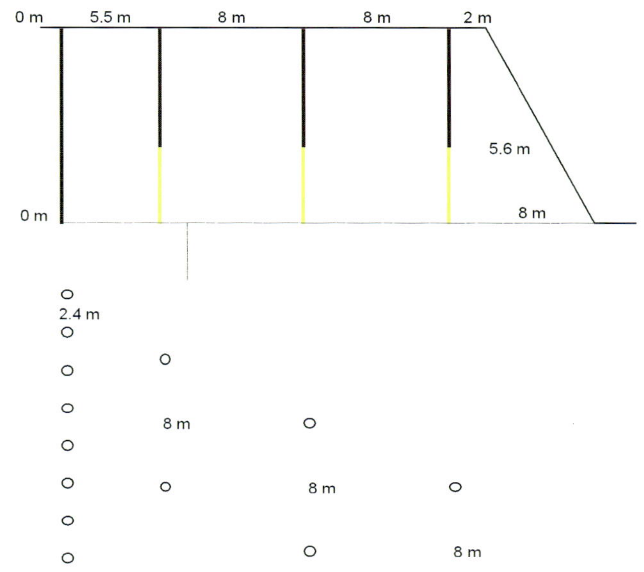

Design Data	Toe or Presplit Row	Inner Buffer Row	Outer Buffer Row	Crest Row	
Hole Diameter (mm):	251	251	251	251	
Hole Depth (m):	12.2	12.2	12.2	12.2	
Hole Angle (deg):	0	0	0	0	
Burden (m):	5.5	8.0	8.0	5.0	
Spacing (m):	2.4	8.0	8.0	8.0	
Airdeck (m):	0	0	0	0	
Subdrill or Standoff (m):	0	0	0	0.0	
Stemming (m):	12.2	7.3	7.3	7.3	
Total Charge Weight (kg):	0	204	204	204	Average
Powder Factor (kg/tonne):	0.00	0.12	0.12	0.19	0.13
Energy Factor (kcal/tonne):	0	106	106	169	96
Comments: soft rock- excellent results					

Figure 11.46: Line drill design, Site 5

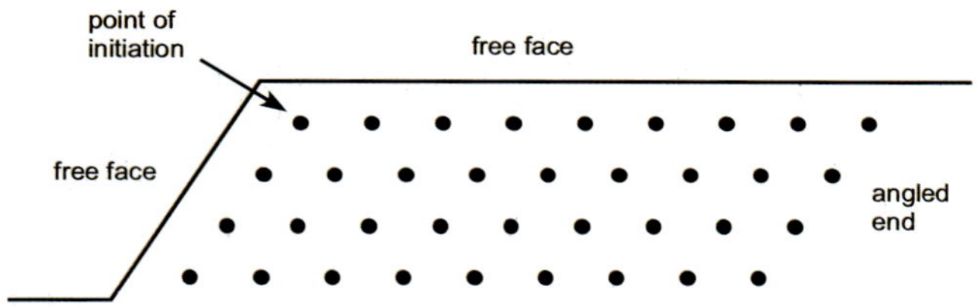

Figure 11.47: Point of initiation

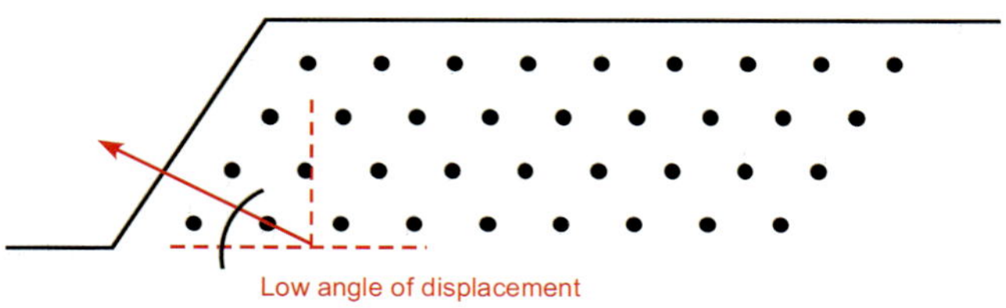

Figure 11.48: Preferred angle of displacement

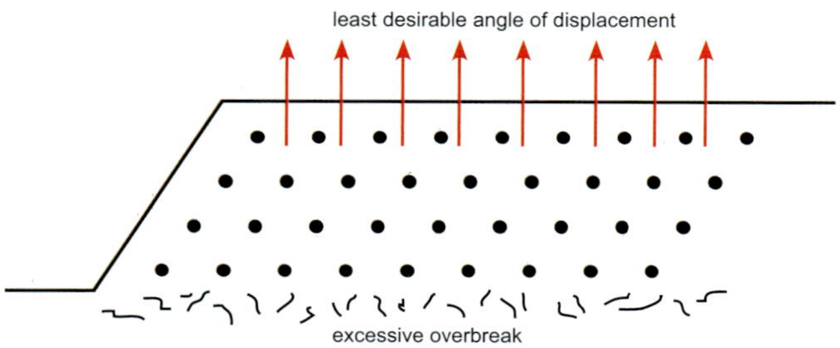

Figure 11.49: Adverse angle of displacement

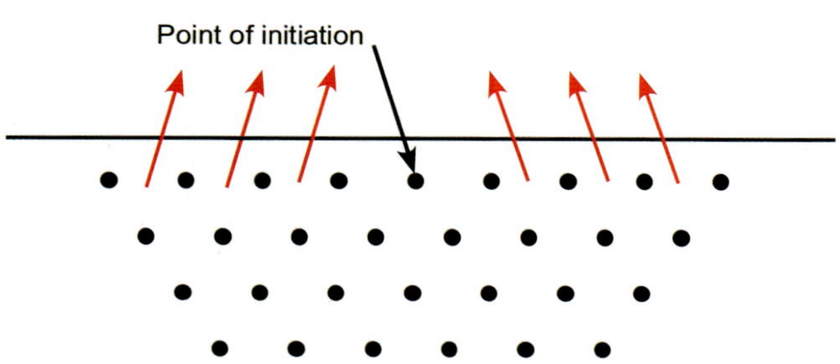

Figure 11.50: Flat V displacement

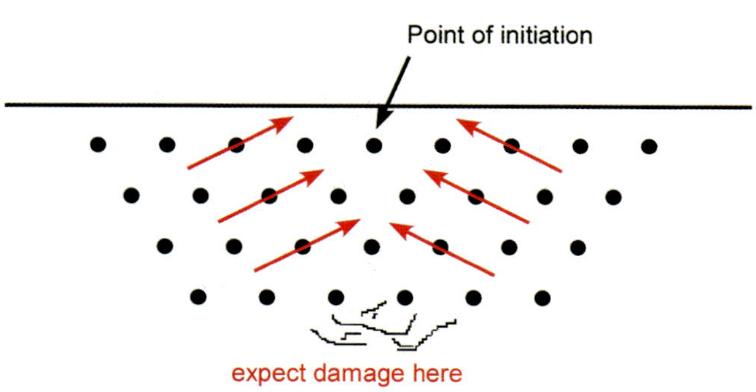

Figure 11.51: Deep V damage

Pre-split holes are fired instantaneously whenever vibration control is not an issue. If the slope is sensitive to vibration, groups of holes (5–10 per group) should be fired with around 25 ms between groups. In very sensitive walls, only one hole can be shot at a time. In this case, it may make sense to use post-splitting instead of pre-splitting techniques.

The detonation of pre-split holes should take place before the adjacent holes are drilled, but this is not always practical from a production point of view. To reduce the impact on production the trim blast and the next pre-split should be shot together (Figure 11.54).

Mid-splitting may cause surface cut-offs or explosive column shifting if too much time passes between the firing of the mid-split and the adjacent buffer row. It is recommended that the mid-split be sequenced with adjacent holes to reduce the possible adverse effects (Figure 11.55).

Frequency control timing configurations should be used when blasting adjacent to vibration sensitive slopes (section 11.3.7).

11.3.7 Design implementation

The process of design and implementation of efficient wall control blasting requires communication and coordination between the following groups at the mine:

- long-range planning;
- geotechnical engineering;
- geology/ore control;
- short-range planning;
- drill and blast engineering;
- explosive supplier;
- drill crew;
- blast crew;
- production.

A careful review of the existing slope designs and field conditions must be made before the blast design is developed. The review process should include evaluation of:

- rock structure;
- rock hardness;
- vibration sensitivity of slope;
- designed batter angle;
- designed berm width;
- past blast performance;
- fragmentation;
- bench requirements;
- drill requirements;
- explosive requirements;
- labour requirements.

After the review, the initial designs are based on the site-specific information. If the design calls for placement of a row above a designed crest it should be vertically offset from the crest to reduce the potential for damage (see Figure 11.56). The template

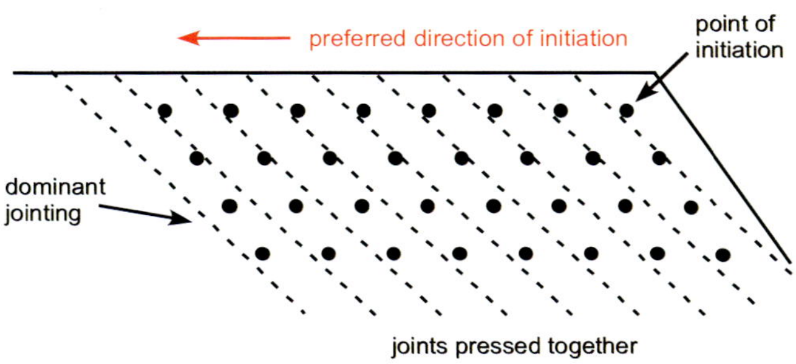

Figure 11.52: Direction of initiation that limits wall damage

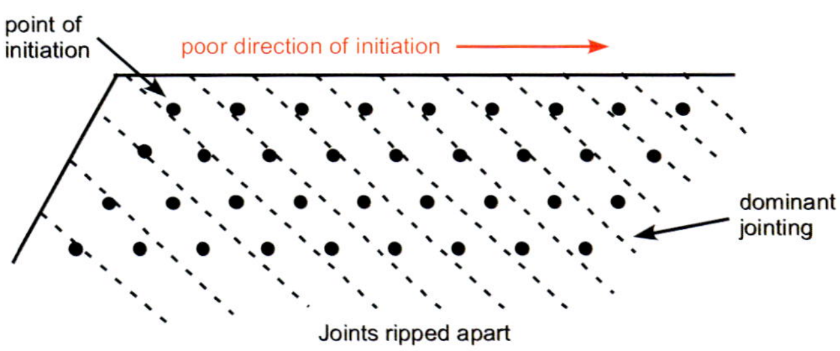

Figure 11.53: Direction of initiation that increases wall damage

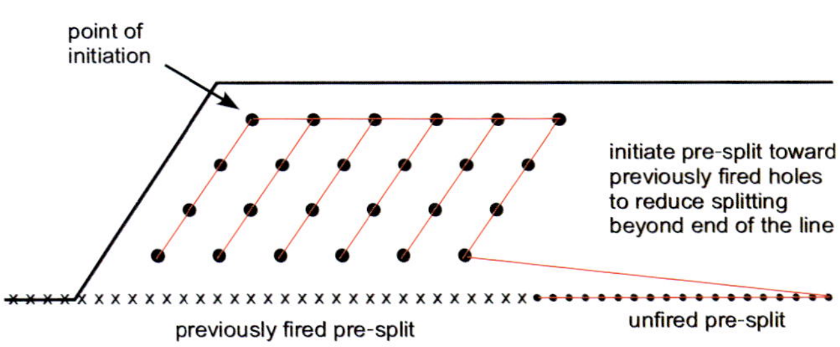

Figure 11.54: Detonation of trim blast with the next pre-split row

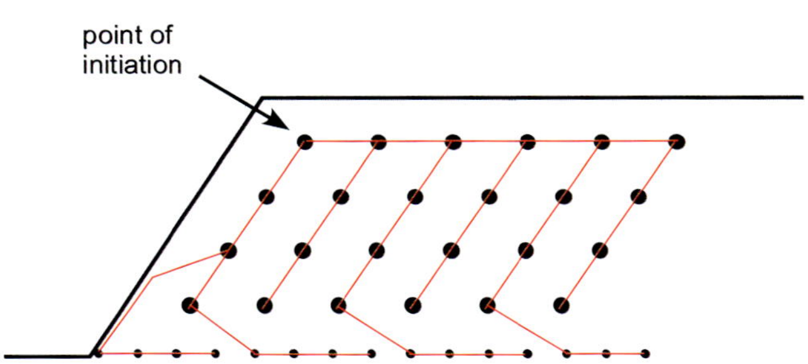

Figure 11.55: Mid-split sequencing

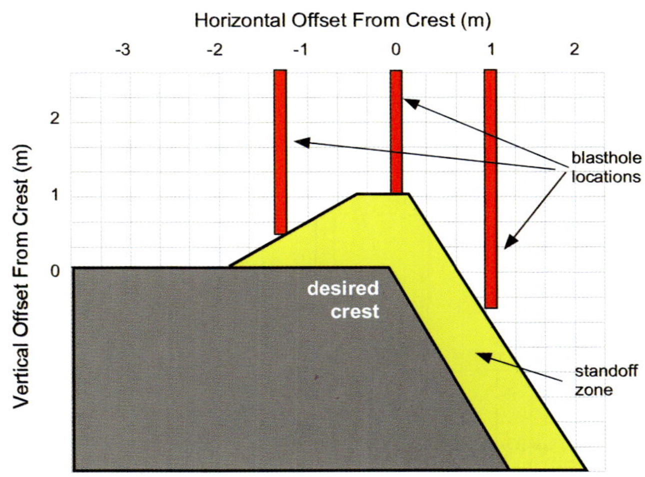

Figure 11.56: Drill offsets to protect crest

represents typical conditions and must be adapted to site-specific conditions.

Performance goals should be established for each design, along with the performance indicators that will be tracked to quantify blast efficiency. Typical performance indicators include:

- face preparation;
- drilling accuracy;
- loading accuracy;
- muckpile displacement;
- overbreak;
- vibration levels;
- excavator productivity;
- cost.

It is recommended that a database be used to track blast implementation and performance information.

When blasting to a free face it is imperative that the faces be prepared in a consistent manner. Excavator operators and supervisors must be aware of the need for straight, steep and clean faces. A series of drop and keep-out stakes should be placed along the design crest to guide face excavation (Figure 11.57).

Blast hole patterns should be based on actual field conditions, taking into consideration the existing face and adjacent wall locations. Patterns should be precisely located using accepted survey techniques.

It is important that the blast holes are drilled and logged correctly. The following drill-related procedures are considered best practice:

- instruct drillers about the need for accurately drilled blastholes (depth, angle, azimuth);
- monitor drill penetration;
- after the hole is drilled and the drill has pulled off the collar, measure the hole and place a stake in the

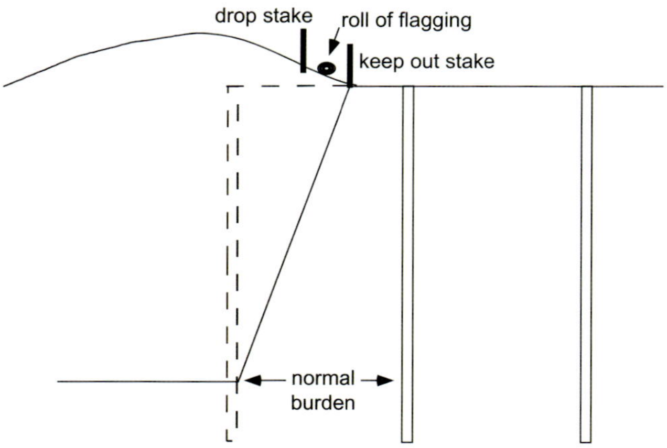

Figure 11.57: Drop and keep-out stakes

cuttings with the following information: driller's name, date drilled, depth drilled, relative hardness, amount of broken material in collar of pre-split holes and any unusual conditions (e.g. voids);

- survey the hole locations to monitor drill accuracy and record crest and toe burdens for each face hole;
- enter location deviation into the database;
- determine the offset of the bottom of the holes from the desired crest on the next bench below;
- mark the bench with location of the crest of the bench below (when GPS-assisted drill navigation is not used).

The amount of broken material in the collar region of pre-split holes will indicate how much crest damage is created by blast holes from the bench above.

During loading of the blasts the following steps should be taken:

- measure and record hole depths;
- redrill or backfill holes as required;
- record the charging of each hole;
- record the stem length of each hole;
- record the timing configuration used.

11.3.8 Performance monitoring and analysis

The performance of each controlled blast should be monitored and analysed to make sure the design is refined to meet changing slope conditions. The performance should be evaluated in terms of damage in the near and far fields as well as the vibration levels produced.

In the near field, one of the best methods to study overbreak is with standard speed video cameras that focus on specific regions of the blast (Figure 11.58). Figure 11.59 illustrates the recommended remote video camera setup to study face displacement and overbreak.

Motion analysis targets (empty primer boxes painted orange, or cones) are placed on the desired crest and towards the wall at known increments (approx. 3 m). A

Figure 11.58: Typical video analysis information

detonation indicator (shock tube or detonating cord) is placed in the blast hole immediately adjacent to the target set, to define when the hole fired. The cameras should be set up to view the back and face of the shot. In some cases it may be possible to see both areas of the blast, but it is generally better to zoom in on specific areas for detailed

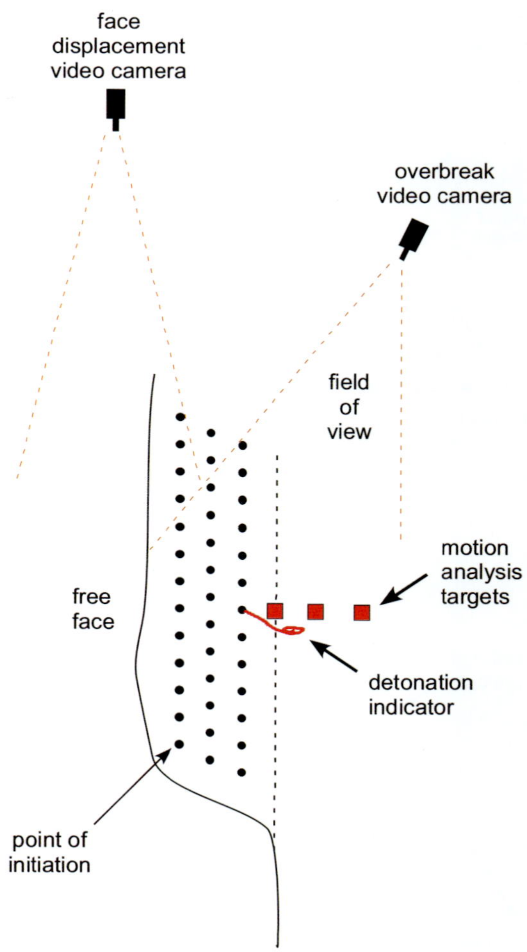

Figure 11.59: Remote video camera setup

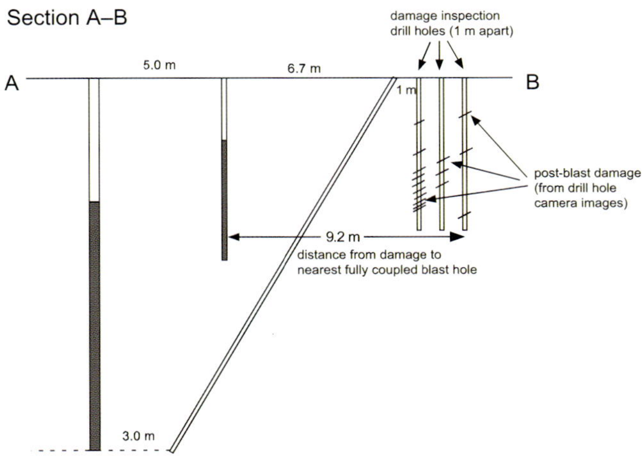

Figure 11.60: Subsurface damage inspection holes

analysis. It is not unusual for the video operator to set the camera up, turn it on during the final clearing of the blast and retreat to a safe location. This allows the blast to be framed in close without using the digital zoom. After the blast, the video should be downloaded onto a computer for storage and review.

Drill hole cameras (Figure 11.60) and optical televiewers (section 2.4.9.5) can be used to determine the extent of subsurface damage behind the crest. Typically, a series of holes is drilled behind the designed crest. Images are taken before and after the blast to identify subsurface damage.

Blast vibrations should be monitored in the near and far field (Figures 11.61 and 11.62). A reference seismograph should be placed a set distance (approx. 50 m to avoid over-ranging geophone-based instruments) from each production and controlled blast. Far field units are placed at distances of 50–500 m from the blast at points of concern. Placing the blast next to slope prisms can help establish the link between vibration levels and slope response. The locations of the seismographs should be surveyed to determine the slope distance to the nearest blast hole. The following information should be recorded to assist future analysis:

- date of shot;
- time of blast;
- pattern ID;
- location;
- geotechnical zone;
- blast type, i.e. production or controlled;
- maximum instantaneous charge;
- seismograph ID;
- slope distance from nearest blast hole;
- scaled distance;
- radial PPV (mmps);
- vertical PPV (mmps);

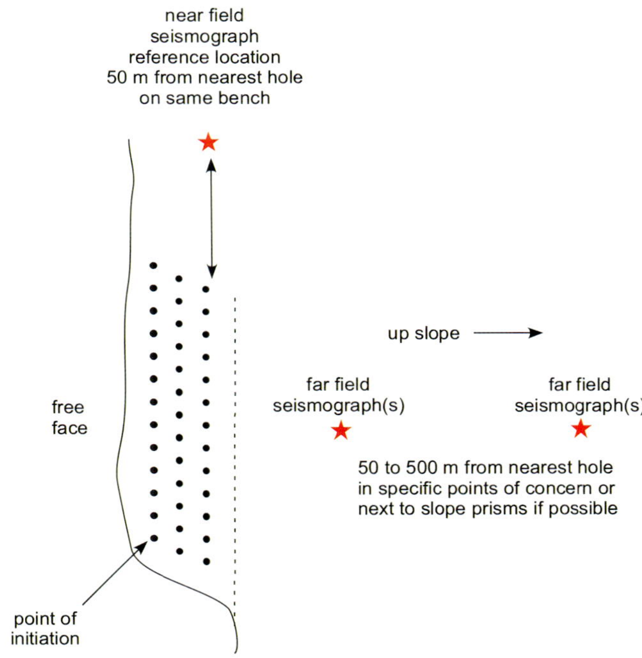

Figure 11.61: Near and far field seismograph locations

- transverse PPV (mmps);
- maximum PPV (mmps);
- vector sum PPV (mmps);
- dominant radial frequency (Hz);
- dominant vertical frequency (Hz);
- dominant transverse frequency (Hz);
- peak radial displacement (m);
- peak vertical displacement (m);
- peak transverse displacement (m);
- peak displacement (m);
- peak radial acceleration (g);
- peak vertical acceleration (g);
- peak transverse acceleration (g);
- peak acceleration (g);
- slope response;
- blast hole diameter (mm);
- burden (m);
- spacing (m);
- semming (m);
- subdrill (m);
- relative confinement (free faces);
- number of holes per shot;
- total charge weight;
- delay interval between holes;
- delay interval between rows;
- blast duration.

This information will be used to perform linear regression analysis of the vibration data to help establish the relationship between blast design and slope response.

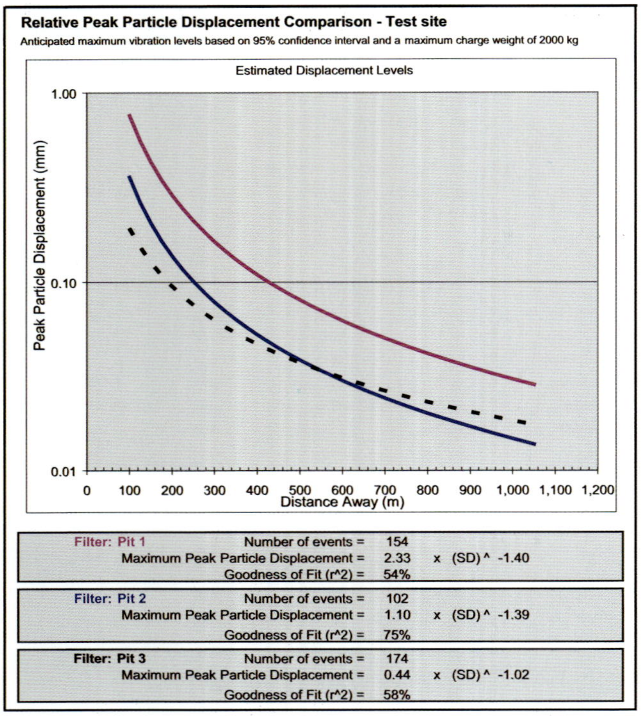

Figure 11.62: Relative peak particle displacement analysis

Blast vibration control is an important part of the optimisation of blast designs for slope protection.

The dominant frequency content of the vibrations generated by production and wall control blasting should not match the resonant frequency of the pit slope. If possible, delay sequences should minimise the production of unwanted frequencies. This generally requires the use of specific electronic delay times (not generic pyrotechnic delay times) and a good understanding of the site's vibration characteristics, gained by conducting single-charge tests and measuring the frequency content and attenuation rate of blast vibrations.

The test procedure is as follows:

- detonate a single blast hole with a similar charge weight and confinement conditions of typical blast holes:
- monitor the vibration levels produced up the slope at locations of concern (tension failures etc.) with a line of seismographs placed 95–500 m from the charge;
- use the minimum charge weight, related to the distance away (see Table 11.8);
- record the charge weight and distance from the hole(s) to the monitoring locations;
- if the recorded levels are below 8 mmps consider shooting a larger charge to improve the quality of the vibration signature;
- if two or more holes are needed for the proper amplitude, they should be detonated instantaneously (preferably with detonating cord and no in-hole delays or electronic delays).

A typical vibration characterisation test setup is shown in Figure 11.63. Note that it is suggested that at least two monitors be used to evaluate the slope's blast vibration attenuation characteristics. These units should be placed in a straight line from the charge and located in the near and far field.

This information is processed to reveal the most appropriate delay intervals for frequency control.

11.3.8.1 Post blast inspection

Once it is safe to re-enter the blast site, the overbreak should be reviewed and recorded on the timing diagram of the blast report. This will allow the timing configurations to be critically evaluated.

Table 11.8: Single hole charge recommendations

Minimum charge weight for single hole tests	
Distance away (m)	Minimum charge (kg)
50	10
100	40
150	90
200	159
250	249
300	359
400	638
500	996
600	1435
700	1953

Basis is a scaled distance of 35

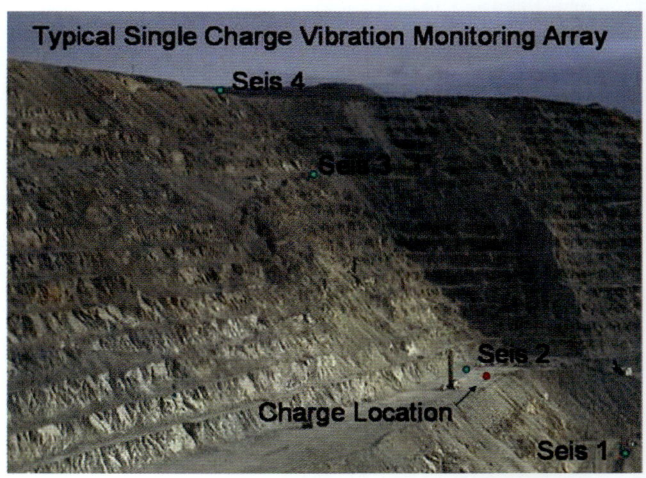

Figure 11.63: Vibration monitoring array

11.3.8.2 Post excavation inspection and batter quantification

When the excavator reaches the final limit, the designed toe and crest should be achieved and no blast-induced damage should be visible on the face. After excavation is completed the face should be inspected and analysed for excessive overbreak. The damage should be classified into the following categories to help guide design refinement. Different degrees of visible damage are illustrated in Figures 11.64 to 11.67.

- **No visible damage** – joints tight, teeth marks in face, no loose material present, half-barrels visible when pre-splitting and a well-defined toe and crest.
- **Slight damage** – joints opened up, crest loss <1 m, few half-barrels visible when pre-splitting, excavation possible for 1 m beyond designed batter location.
- **Moderate damage** – blocks dislodged, crest loss 1–3 m, excavation possible for 1–3 m beyond designed limit.

Figure 11.64: No visible blast damage

Figure 11.65: Slight blast damage

- **Severe damage** – face shattered, blocks dislodged and rotated, excavation possible for more than 2 m from designed limit.

A detailed record should be made of the post excavation performance of the batter face. A form for this type of analysis is shown in Figure 11.68. Alternative forms for the geotechnical evaluation of blasting performance are illustrated in Figures 11.69 and 11.70.

11.3.9 Design refinement

Wall control blast design refinement should be considered a never-ending process. Designs should be continuously refined based on the changing conditions of the rock mass and slope design requirements. Each design modification should be considered as a separate refinement and evaluated one at a time.

It is important that the magnitude and cause of damage is identified before the design is refined. One way to quantify design performance is to conduct a 3D survey

Figure 11.66: Moderate blast damage

Figure 11.67: Severe blast damage

of the post excavated face (Figures 11.71 and 11.72). If the face is dug too hard, this information can be very valuable in refining the blast design.

The following five cases provide the logic for design refinement.

11.3.9.1 Case 1: Overbreak along the entire face

In the example shown in Figure 11.73, the blast produced breakage well beyond the desired bench face. This overbreak is related to the strength and structure of the rock mass and is caused primarily by insufficient horizontal relief. Additional relief may be achieved simply by increasing the delay interval between rows. However, caution should be used with delay intervals greater than 13 ms/m of burden due to the increased potential for explosive column cutoffs. If the face was cleaned up as much as possible and the proper delay was used between rows, the explosive energy adjacent to the wall must be reduced. Initially, this should be done by moving the entire design farther away from the wall. The distance moved should be basically equal to the amount of back break observed.

11.3.9.2 Case 2: Overbreak at the top of the bench

In Case 2 (Figure 11.74), only the upper portion of the bench is damaged by the blast. Assuming that the delay interval between rows was sufficient to allow horizontal relief, this damage is typically caused by excessive subdrilling on the bench above, adverse geological structures and/or excessive energy in the inner and outer buffer rows. Since the toe of the slope is in the proper location, the trim row load and location should not be modified. In this case, the overbreak can be reduced by:

- adding an air deck to the inner and outer buffer rows;
- reducing the charge in the inner buffer row;
- moving the inner and outer buffer rows farther away from the slope;
- splitting the charge in the inner and outer rows into two independently delayed decks and detonating the top charge 25 ms before the bottom charge.

11.3.9.3 Case 3: Underbreak at the toe

In Case 3 (Figure 11.75), the distance between the inner and outer buffer rows is correct. However, the trim row should be modified by:

Wall Control Blast Performance Analysis

Pit Sector:

Blast Design

Bench Height:		Double Benched:	
Bench Face Angle:		Overall Bench Height:	
Catch Bench Width:		Inter Ramp Slope Angle:	
Number of Rows:		Rock Density:	
Staggered Pattern (y/n):		Water Conditions:	

	Toe Row	Row	Row	Row	Row	Face Row
Hole Diameter:						
Hole Angle:						
Toe Offset:						
Top Back Break:					Crest Offset of Hole Collar:	
					Toe Burden:	
Burden:						
Spacing:						
Stemming:						
Airdeck:						
Subdrill:						
Top Charge Type:						
Top Charge Density:						
Top Charge Diameter:						
Top Charge Energy (AWS):						
Top Charge Weight:						
Bottom Charge Type:						
Bottom Charge Density:						
Bottom Charge Diameter:						
Bottom Charge Energy (AWS):						
Bottom Charge Weight:						
In Hole Delay:						

Timing Configuration

show surface delay location and dominant structure orientation

Recommended Design Modifications

Page 2

Figure 11.68: Wall control blast evaluation form

Wall Control Blast Performance Analysis

Date: ______
Blast ID: ______

Pre-Blast Analysis

Select dip and strike and direction of initiation

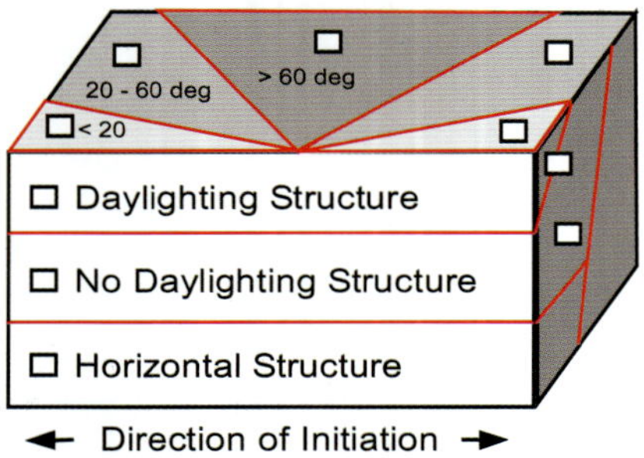

Blastability Matrix: select design

Increasing Block Size →

Increasing Hardness ↓

Design 1	Design 2	Design 3
Design 4	Design 5	Design 6
Design 7	Design 8	Design 9

Confinement:

Two Free Faces	One Free Face	No Free Face	Shot w/ Production

Comments: ______

Blast Performance Monitoring

Video Analysis

	Poor	Average	Good
Horizontal Displacement:			
Vertical Displacement:			
Displacement Beyond Dig Limit:			
Slope Raveling:			

Vibration Analysis

	Distance	PPV	Dom. Freq.
S1:			
S2:			
S3:			
S4:			

Post Blast Damage Analysis

	None	Minor	Significant
Cracks Parallel to Crest:			
Contact Shift Angled to Crest:			
Block Heave Beyond Crest:			

	Depth
Power Trough:	

Excavation Analysis

	None	Minor	Significant
Oversize in Top Half of Bench:			
Oversize in Lower Half of Bench:			
Scaling Required:			
Difficulty Achieving Toe:			
Difficulty Maintaining Crest:			

Excavator:	
Operator:	
Average Productivity (tph):	

Post Excavation Analysis

	Underbreak	On Design	Overbreak
Toe Location (m):			
Crest Location (m):			

Damage Classification: ______

None:	joints tight, teeth marks in face, no loose material present, well defined toe and crest
Slight:	joints opened up, crest loss < 1m, few half barrels visible excavation possible for 1 m beyond designed limit
Moderate:	blocks dislodged, crest loss between 1 to 3 m, excavation possible for 1 - 2 m beyond designed limit
Severe:	face shattered, blocks dislodges and rotated, excavation possible > 2 m beyond designed limit

Damage Diagram

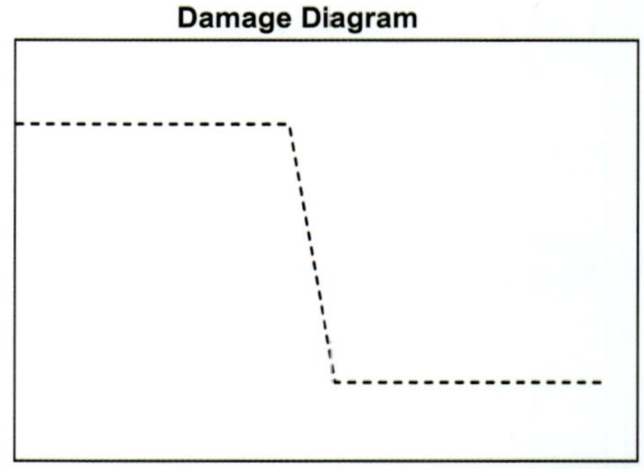

Comments: ______

Reviewer: ______

Page 1

Figure 11.69: Wall control blast evaluation form

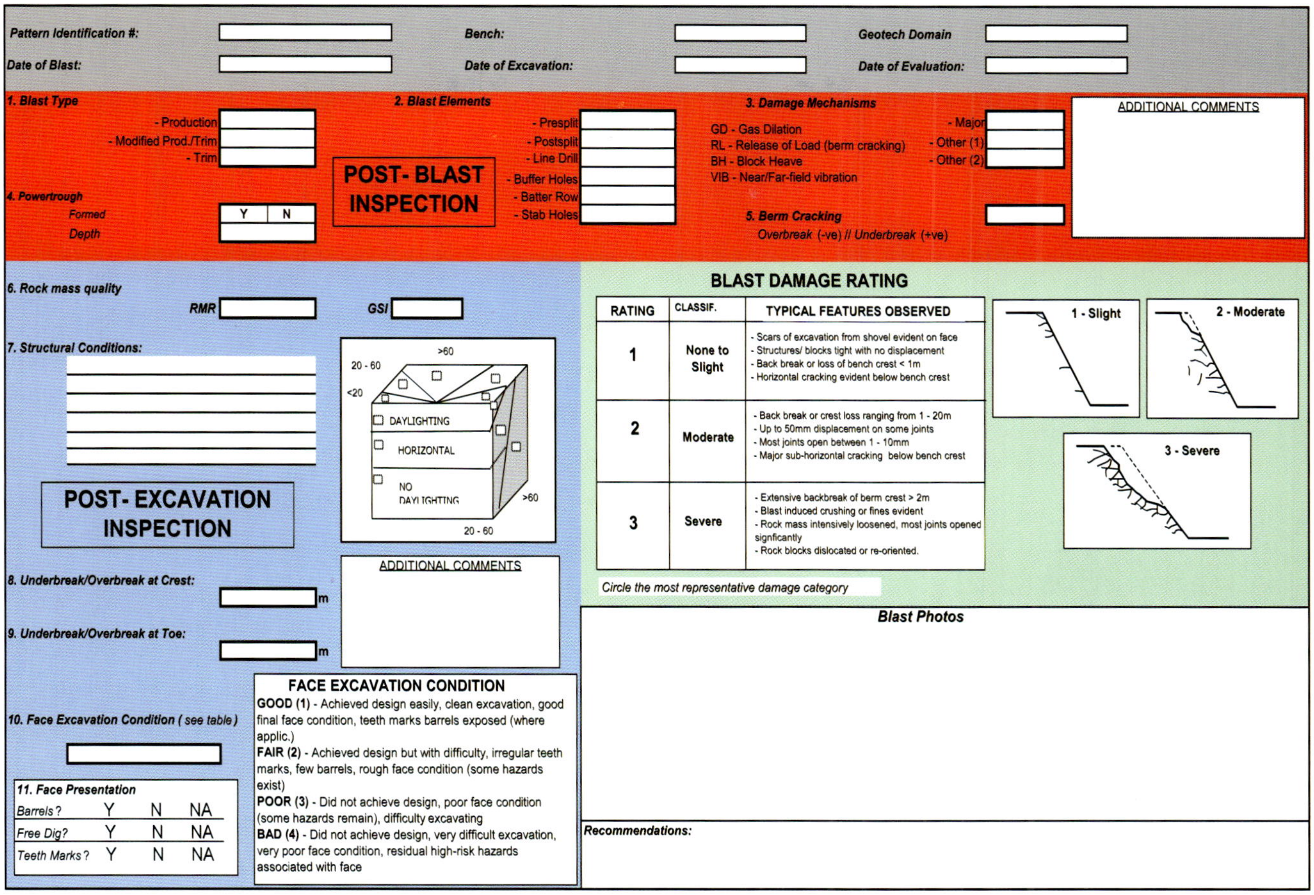

Figure 11.70: Geotechnical evaluation of blasting performance

Figure 11.71: 3D batter face survey

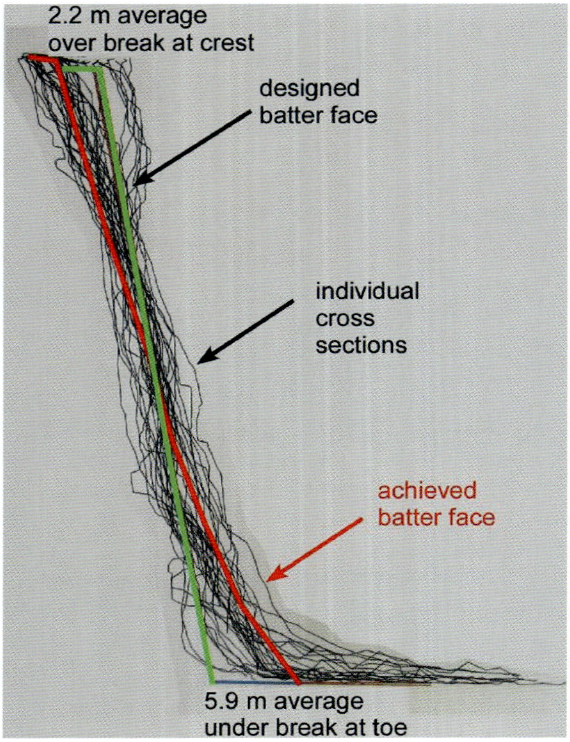

Figure 11.72: Batter profile analysis

- increasing the load in the trim row;
- decreasing the spacing of the trim row;
- moving the trim row closer to the bench face.

11.3.9.4 Case 4: Overbreak at the toe

In this type of wall damage (Figure 11.76), the charge or location of the trim row and perhaps the inner buffer row should be modified. The possible refinements include:

- air deck the trim row;
- air deck the inner buffer row;

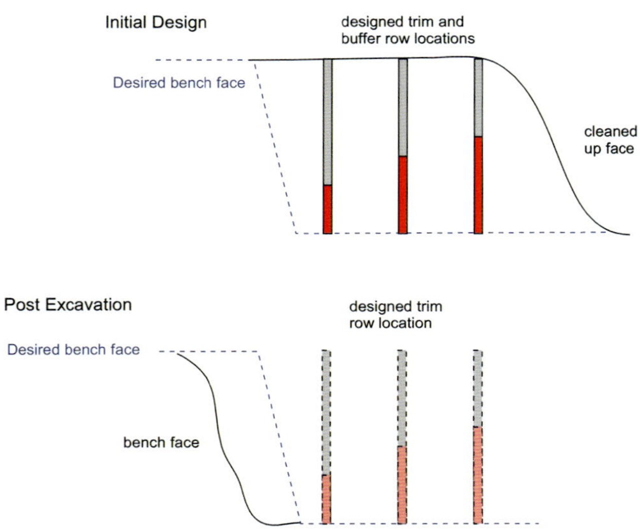

Figure 11.73: Overbreak along entire batter

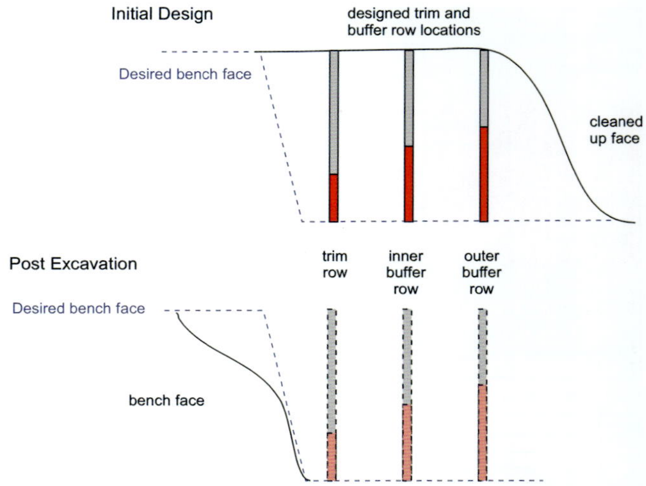

Figure 11.74: Overbreak along crest

- reduce the charge in the trim row;
- reduce the charge in the inner buffer row;
- reduce the trim row's charge and spacing.

11.3.9.5 Case 5: Overbreak at the crest, underbreak at the toe

Case 5 (Figure 11.77) illustrates a design that needs to be modified to improve toe breakage while limiting crest damage. Assuming that the proper delay intervals are used, the following steps should improve design performance:

- reduce the spacing on the trim row;
- increase the charge in the trim row;
- air deck the inner and outer buffer rows;
- move the inner and outer buffer rows farther from the face;
- split the charge in the inner and outer rows into two independently delayed decks and detonate the top charge 25 ms before the bottom charge.

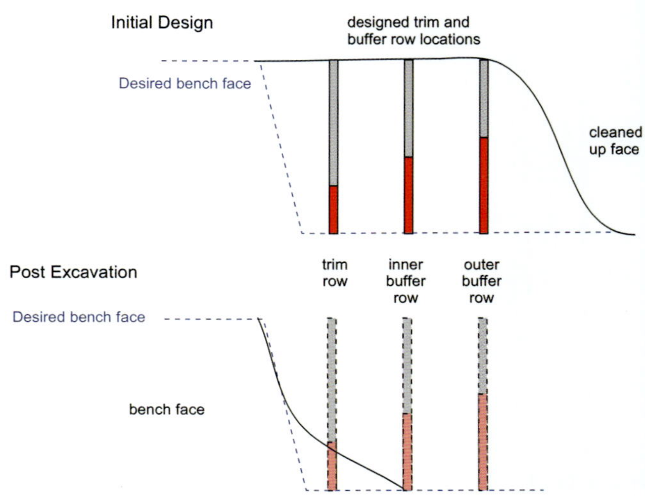

Figure 11.75: Underbreak at toe

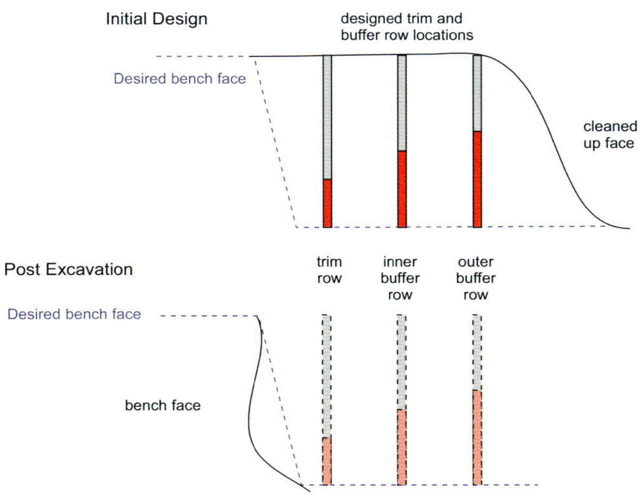

Figure 11.76: Overbreak at toe

When pre-split techniques are used, the damage beyond the pre-split is controlled by site-specific geology and the location and charge of the inner and outer buffer rows.

The modifications shown in Cases 1 to 5 also apply to reducing damage to a pre-split face. It is recommended that the location and charge of the inner and outer buffer rows be dictated by the current performance of blasts in similar conditions. For example, if the inner buffer row is breaking back 1.5 m from the bottom of the hole, it should be initially placed 1.5 m away from the bottom of the pre-split (Figure 11.78). In some cases, hard rock masses require a stab hole to improve the breakage at the crest (Figure 11.79).

11.3.10 Design platform

The design, implementation evaluation and refinement of the wall control blasts outlined above require a group approach with input from many specialists at the site. It is important that the designs be based on actual field conditions. Important inputs are the delineation of

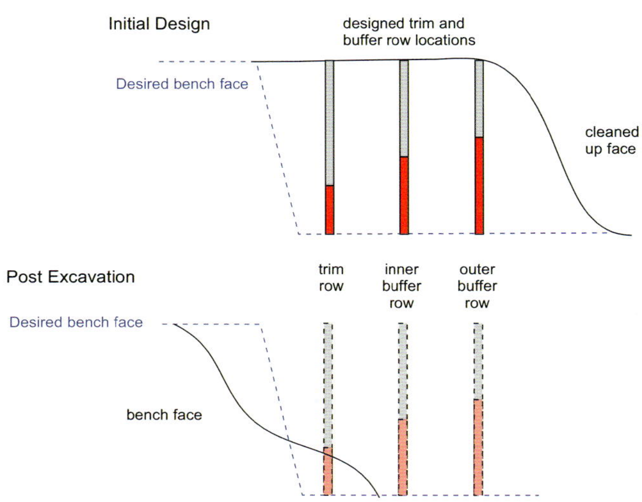

Figure 11.77: Overbreak along crest, underbreak along toe

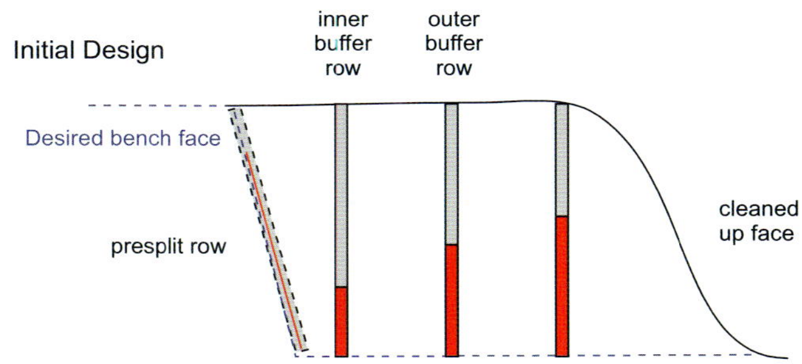

Figure 11.78: Pre-split cross-section

blasting domains, an appreciation of the geometric and mining constraints, and the mining 'tool kit'. This tool kit comprises the drills, the explosives and the initiation systems that are available for use in any blast. These provide the platform from which the designs can be derived to achieve the given objectives.

Although the properties of the rock mass have a controlling influence on blast performance, these properties are seldom used explicitly in blast design. However, some basic rock mass characteristics such as rock substance strength, fracture frequency and density can be used to guide design decisions (section 11.3.4). Blasting domains are zones within a rock mass that have a similar response to blasting. The definition of domains can be based on lithology, structure, alteration or any other property that is indicative of changes in blasting performance. Figure 11.80 shows an example from the Kalgoorlie Consolidated Gold Mines in Western Australia.

Blasting domains may or may not coincide with domains delineated for geotechnical purposes so they should be jointly delineated by the geology, geotechnical and blasting departments. It is important that the geotechnical data are stored in a central and auditable database with systems to evaluate data reliability. While there is still some debate as to the most appropriate geotechnical or rock mass properties to describe the influence of a rock mass on blast performance (Scott et al. 2006), it is generally agreed that the key parameters should include:

- stiffness parameters, which control the distortion of the blast hole wall and hence the pressure developed in the blast hole;

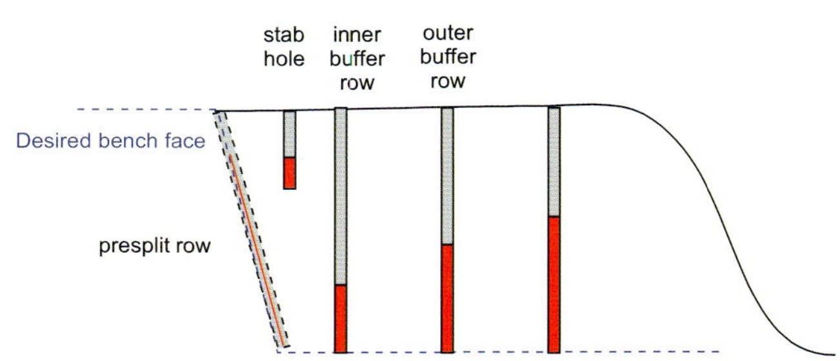

Figure 11.79: Pre-split with stab hole

Figure 11.80: Delineation of geotechnical and blasting domains and the KCGM mine, Australia
Source: Brunton (2001)

- strength parameters, which control the crushing at the blast hole wall and failure of the rock from the reflection of stress waves from surfaces;
- attenuation properties, which determine how far the stress wave travels before its energy falls below the levels that cause primary breakage;
- in situ fracture frequency, orientation and character, which together define the in situ block size distribution and influence the attenuation of the shock wave and the migration of explosion gases;
- the density of the rock mass, which affects its inertial response to the forces imposed on it during blasting.

Mining geometric conditions and constraints are parameters defined by long- and short-term planning requirements. They include bench geometry and the pit wall and bench angles. Mining constraints may include access, proximity to sensitive structures, presence of water or equipment issues. These can have a major impact on blast design and results.

11.3.11 Planning and optimisation cycle

The optimisation cycle has three major components:

1 design and analysis;
2 implementation;
3 performance measurement.

11.3.11.1 Design and analysis

Blast design and analysis may utilise design rules or blastability indices, supplemented by models and/or simulations.

In the case of deep pit mining, the use of effective models needs to become an integral part of the blast engineering process. The approaches currently used in mining are empirical or mechanistic and numerical; each is briefly discussed below.

Empirical methods

Empirical relationships can be used to predict fragmentation, muckpile characteristics, ore movement, blast vibrations and damage for given blast design parameters. A range of fragmentation and damage models have been reported in the literature (e.g. Da Gama 1983; Cunningham 1987; Lu & Latham 1998; Djordjevic 1999). The extent to which they are used in practice is difficult to quantify.

Empirical models can provide indicative results or be used for comparative purposes such as studying the gross effects of changes to blast designs. This is a useful way to determine the effects of proposed changes to a blast design. However, empirical models cannot reliably quantify blast performance to the accuracy required by some sectors of the industry.

Numerical and mechanistic methods

Mechanistic models generally combine mathematical relationships describing average rock behaviour, with empirical rules. A good example is the mine to mill fragmentation model described by Kanchibotla et al. (1998). Numerical models attempt to explicitly mimic the rock damage, breakage, fragmentation and movement processes through first principles based on rock mechanics and/or detonation physics. Examples include those described by Minchinton and Lynch (1996) and Yang and Wang (1996).

Cundall et al. (2001) reviewed available numerical methods and identified their strengths and limitations in terms of their ability to model the complete blasting process. The review highlighted the need for blasting codes that can completely model the blasting process from detonation to rock fracture, and finally to rock heave.

The most recent models of the complete blasting process include the HSBM (hybrid stress blasting model, Cundall et al. 2001) and the ELFEN/MBM/SoH model (Minchinton 2006). These offer the most promising platform for studying 'what if' scenarios for new and anticipated mining conditions in large and deep open pits. To date, their application has been constrained by their computing requirements, which have resulted in modelling of relatively small and often unrepresentative blast volumes. This is expected to change with improvements in computing power and more efficient computational techniques and algorithms.

HSBM

The HSBM constitutes a state-of-the-art modelling framework of the complete blasting process (Guest 2005). It is a single framework which dynamically links a non-ideal explosives detonation code to a geomechanical rock or material model. The detonation code is based on Vixen (variable ideality explosives energetics, Cunningham 2004). The material or geomechanical code is based on the Itasca code PFC3D, hence the term 'hybrid stress blasting model'.

The HSBM offers the potential to model and predict the dynamic forces at the rock/explosive interface for a given explosive and rock mass conditions, followed by rock breakage (micro- and macro-damage causing fracturing), gas flow through the fracture network (Guest 2005), fragmentations and displacement.

In its current form (2006), the HSBM is better suited to studying or evaluating what-if scenarios using limited but representative blast volumes. The principal limitations are the memory requirements, the calculation time and the lack of validation of the code's macroscopic behaviour. These limitations are being addressed. A second phase of the HSBM is in progress, with the objective of developing the code so that it is an effective medium- to long-term planning tool with the ability to provide input to short-term (day-to-day) blast design requirements.

ELFEN/MBM/SoH

The MBM (Minchinton & Lynch 1996) is an explicit finite element–discrete element code that dynamically models the complete blast process. This includes pressure loading from the detonation of non-ideal explosives in the blast hole, stress wave propagation, strain-rate dependent rock fracture, coupled gas flow through the fracture network and bulk motion of rock fragments.

The MBM is the Orica component of the ELFEN package developed by Rockfield Software, Swansea, Wales, which acts as the computational engine and provides pre- and post-processing facilities. Another variant of the code, SoH, uses quadrilateral rigid elements tessellated in a defined region, thereby avoiding the fracture phase. This can be used to study bulk motion or flow processes in blasting (Minchinton & Dare-Bryan 2005).

Both the MBM and SoH are used by Orica to study fundamental aspects of customer problems and to provide guidance for blast designs involving considerations of air blast, ground vibration, timing, fragmentation and muckpile expression.

Like most codes of this type, the computing requirements (memory, calculation time and data storage) can be high, especially for 3D calculations with fractures. A recent parallel version of SoH (2D and 3D) is expected to address these issues; efficient parallelisation of the fracture code remains a computational research issue.

11.3.11.2 Implementation and performance measurement

The implementation component requires formalised procedures for quality assurance and quality control (QA/QC). This is to ensure that what is done in the field (section 11.3.8) is consistent with the intended design.

The quantification of blasting outcomes is a crucial component of the optimisation process. There are a number of technologies available to assist with this process, some of which are listed in Table 11.9.

An important requirement, and therefore a component of the blast engineering methodology, is to assess and quantify the value of the blasting results to the operation. Blasting is not just a process that enables excavation to proceed; it can add or destroy value for the operation. A proper accounting of this value would allow blast designs to achieve their target objectives.

The final stage of the methodology is to compare the blasting outcomes with the original objectives. The cycle of design optimisation is a continuous process with changing inputs as blasting encounters different rock mass or mining conditions. It contains a learning component as the database of experience refines design rules or improves the calibration of models.

Just as the design rules and models can vary from basic relationships to quite sophisticated numerical tools, the measurements used to quantify blasting outcomes may be simple provided they are compatible with the blasting objectives and can be routinely captured. Examples include the ability to measure productivity indicators such as instantaneous loading productivity on a blast-by-blast basis (see Figure 11.81).

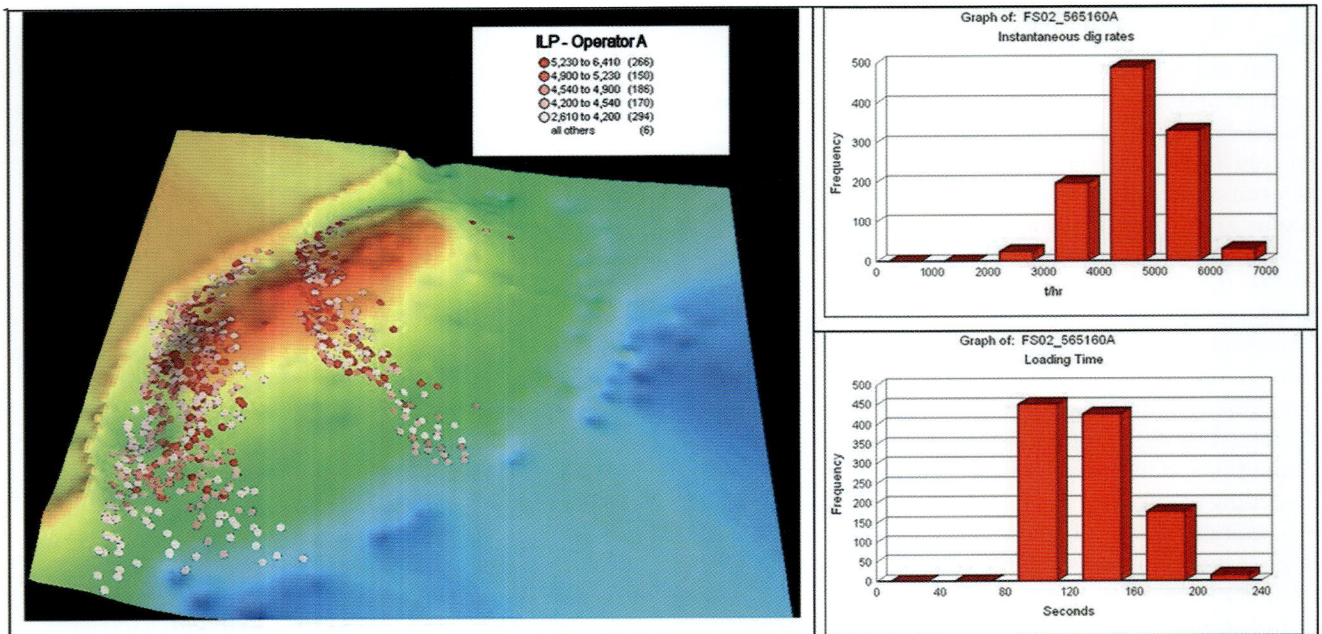

This plot shows instantaneous shovel productivity for operator "A". Statistics indicate an average ILP of 4641 t/h and a standard deviation of 768 t/hr. Note that higher ILP rates (4540-6410 t/hr) concentrate in the upper section of the muckpile with below average rates in corner and flatter regions of the muckpile.

Figure 11.81: Instantaneous loading productivity measurements on a muckpile using HP-GPS data from modular mining
Source: Onederra et al. (2004)

Table 11.9: Technologies for drill and blast optimisation

Tools	Application	Status	Expertise required
Geotechnical data collection and analysis			
Digital photogrammetry	Structural data and analysis	Commercial	Moderate
JointStats	Geotechnical data management. Basic and advanced analysis of jointing	Commercial	Moderate/high
Measure while drilling (MWD) systems	Drill hole monitoring for rock hardness	Proven research tool/commercial	Moderate/high
Blast layout and basic analysis			
JKSimBlast	Data management, design and analysis	Commercial	Low/moderate
ORICA packages	Design, analysis and implementation	Proprietary	Moderate
AEL packages	Design, analysis and implementation	Proprietary	Moderate
Mine planning packages	Pattern layout and data management	Commercial	Moderate/high
Blast monitoring and assessment			
Digital image analysis systems	Fragmentation assessment	Commercial	Moderate
High-speed video	Movement	Commercial	Moderate/high
Vibration monitoring systems	Near field damage Environmental compliance	Commercial	Moderate
VOD monitoring systems	In situ explosive performance	Commercial	Moderate/high
Laser profiling systems	Muckpile shape Face profile Toe and crest profiles	Commercial	Moderate/high
Blasting accessories			
Electronic detonators	Ore, waste production and limit blasting	Commercial	Moderate
Equipment monitoring and fleet management systems			
Modular Mining	Equipment monitoring and management	Commercial	Moderate/high

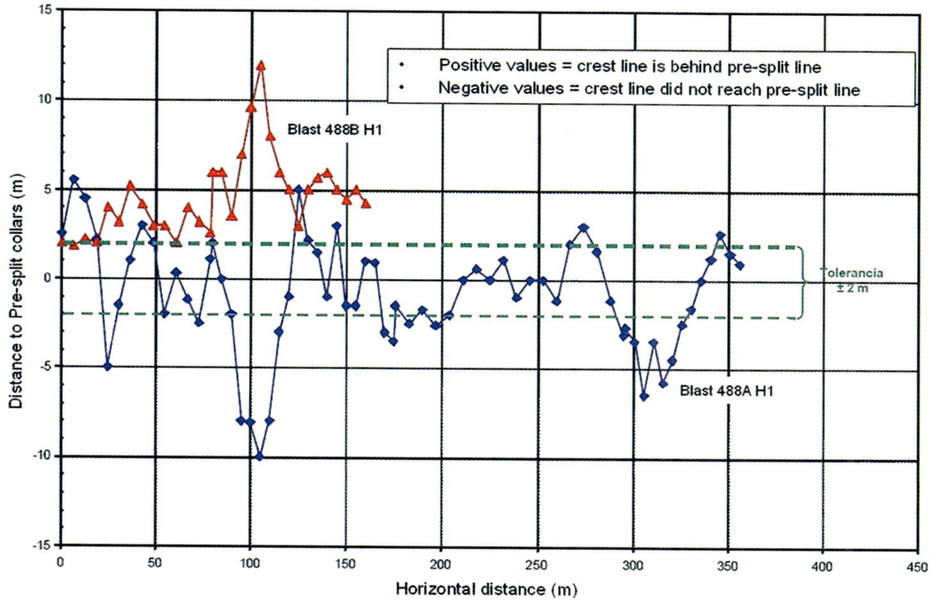

Figure 11.82: Comparison between pre-split line and actual mean overbreak crest line from two blasts
Source: After Flores (2001)

A key blasting objective may be to reliably excavate to the designed pit wall alignment. The cost implications of leaving toe in the wall that prevents access for the drilling of the next bench, or overbreak that denies access to a berm, can be severe. These issues are magnified as pits become deeper.

Figure 11.82 shows a pre-split line and the actual crest line achieved by different blasts and indicates significant overbreak for the blast, plotted in blue.

The implementation and performance measurement methodology described here is simply one approach that has proven successful in major mining operations. A number of variations may be applied with equal success, provided the process suits the operation to which it is applied. Similar systems have been proposed and successfully applied, for example a system developed by Bye (2005) for in-house research at an Anglo Platinum operation in South Africa (Figure 11.83).

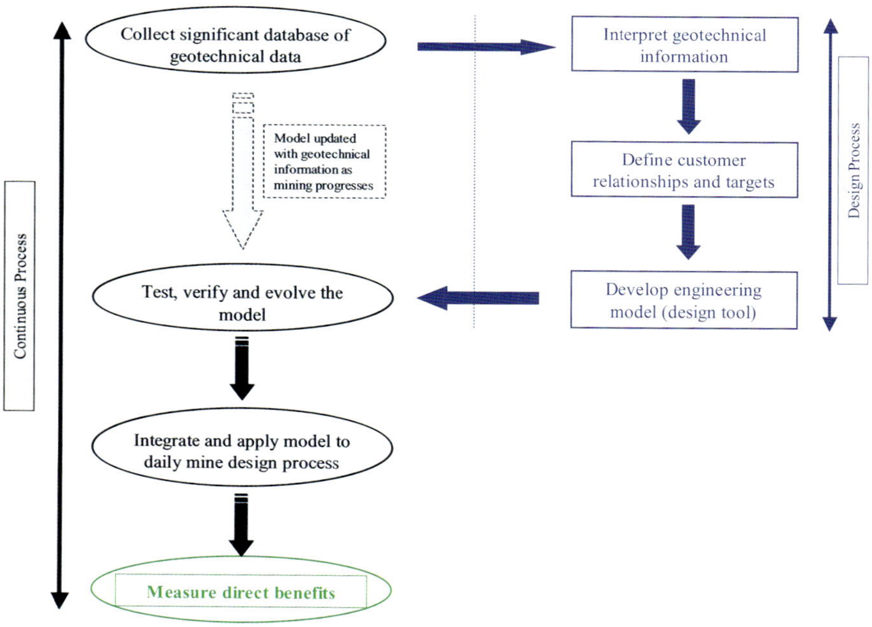

Figure 11.83: Engineering design process used for Anglo Platinum research
Source: After Bye (2005)

The system shown in Figure 11.84 had the following attributes (Bye 2005):

- a significant database of geotechnical information was collected from exploration drill holes, in-pit face mapping and rock strength testing;
- an interpolated 3D geotechnical model was developed from the data;
- the model was tested and adjusted until the model predictions were representative of the field conditions;
- the customer relationships and design targets with the largest economic benefit for the company as a whole were defined;
- the 3D geotechnical model was used as a platform to develop engineering design tools for mine optimisation. These tools were developed as fragmentation and slope design models;
- a simple user-interface (front-end) to the model was designed, which enabled mine planners and engineers to use the model;
- the model was applied over an 18 month period and the efficiency and financial benefits were recorded.

The 3D geotechnical model provided information well ahead of the mining face, and could be used for rock quality prediction, production optimisation, slope evaluation and design as well as planning and costing.

11.3.12 Conclusion

Clearly, the design, implementation evaluation and refinement of wall control blasts require a group approach with input from many specialists at the site. It is important that the designs be based on actual field conditions. However, a good blast design has little value unless it is implemented correctly in the field. Procedures should therefore be established to quantify the degree of accuracy during the implementation of the blast.

During the excavation process, the operators and supervisors must pay close attention to achieving the designed toe and crest locations and make sure all loose material is removed from the berm prior to moving to another location in the pit. This is discussed further in section 11.4.

Blast design refinement begins with identification of the source of blast damage such as shock, crack extension and block heave. Once the source has been defined, the design can be modified to improve blast performance. Design refinement should always be considered an ongoing process to match any changing geological conditions and/or mining requirements.

11.4 Excavation and scaling

Excavation and subsequent scaling of the bench faces are a crucial step in the achievement of safe and optimum slopes in all open pits. It is therefore important that production personnel understand the need to define the toe and crest of the bench. An operator's performance is usually judged in terms of productivity. However, for the mining of final and phase walls the primary performance criteria should involve achieving the design batter face angle and bench width, as well as minimising rockfall hazards from the bench faces.

With the increasing size of loading equipment required for efficient operations in large open pits, a conflict has arisen between production excavation and the requirement for stable clean bench faces. Large electric rope shovels are not well-suited for scaling bench faces, nor are they cost-effective in that role. In some mines, specific teams are used for phase and ultimate slope excavation, cleanup and scaling.

This section briefly discusses the methods of achieving the design bench configuration goals and assessing the results.

11.4.1 Excavation

In most large open pit mines the primary means of excavating the blasted material is a rope shovel (Figure 11.85) or, less frequently, hydraulic excavators operating in a front shovel configuration. In smaller open pits, excavators operating in either a front shovel or backhoe configuration or wheeled front-end loaders may be used.

The operating bench height is generally regulated to be no more than 1.5 m above the reach of the excavating equipment. Therefore, with large shovels a maximum height of 15–16 m is normal, whereas with smaller equipment 10–12 m is the norm.

The approach used for excavating bench faces to design specifications depends on the nature of the material and the mine's operating capability. Several authors have established general relationships between uniaxial compressive strength or point load strength, fracture frequency and appropriate excavation techniques. An example is the suggested guide for ease of excavation developed by the US Department of the Navy Naval Facilities Engineering Command (Figure 11.84).

Overburden and weathered or weaker rocks can often be free-dug with large equipment, as shown in Figure 11.86. Alternatively, the materials can be ripped and dozed. Using this approach, experienced operators can achieve steep benches (Figure 11.87). Similarly, ripping and dozing is often used to establish footwall configurations in coal mines and on slopes where the benches are controlled by strongly developed bedding or foliation (Figure 11.88).

A critical factor in bench excavation in weak rocks, particularly in wet climates, is to cut the slopes as steeply as possible to encourage runoff. This approach must be combined with well-designed and well-maintained surface water control measures, as discussed in Chapter 6 (section 6.5.5.2).

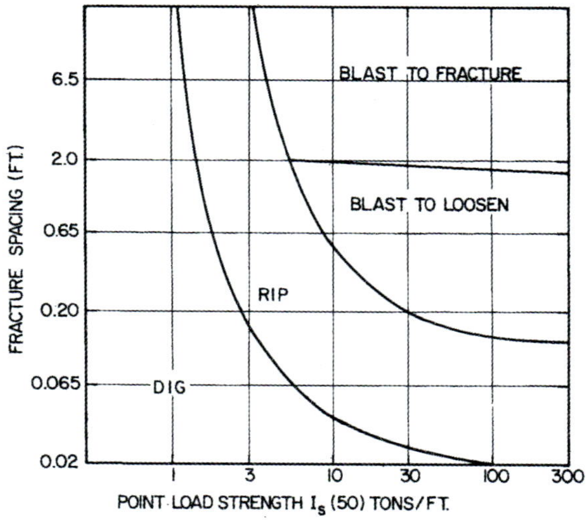

Figure 11.84: Suggested guide for ease of excavation (point load strength corrected to a reference diameter of 50 mm)
Source: NAVFAC DM-7.2 (May 1982)

Figure 11.85: Bench excavation using a rope shovel

Figure 11.86: Shovel-excavated overburden slope

Figure 11.87: Steep dozer-cut face in weak rock

For strong rocks, drilling and blasting to fragment material is normally required prior to excavation by the primary loading equipment. The large equipment is very powerful and there is a strong potential of overdigging benches, particularly where there is blast damage. Further, rope shovels cannot mine efficiently along the crest of benches without a significant overspill.

One method used to control overdigging of the bench face is to cut a crest trench with an excavator (Figure 11.89). This clarifies the limit to which the shovel operator should dig and, where the bench height is close to the operating limit of the excavating equipment, reduces the tendency to create an overhang at the crest of the bench.

In large open pits with steep slopes, overspill between concurrently operated expansion cuts on the same wall is increasingly problematic. The requirement for optimum slopes for the respective phases, combined with high production rates, can result in the catchment on the wall above the lower cut being filled. This results in a rockfall

Figure 11.88: Dozer-cleaned face in mountain coal mine
Source: Courtesy Elk Valley Coal Corp.

Figure 11.89: Defining future bench crest with backhoe

Figure 11.91: Chain for scaling bench faces

hazard close to the toe of the wall above the lower cut. In extreme cases, it may be necessary to step out to create a wider bench; this generally results in a reduction in ore recovery from the lower cut.

Mining procedures have been developed to minimise overspill. These include mining a slot along the bench on the upper cut, leaving a rib of material along the crest (Figure 11.90). The rib is subsequently blasted into the slot and the material along the crest is then mined by an excavator configured as a backhoe, which can pull the blasted muck back from the crest.

11.4.2 Scaling and bench cleanup

Scaling to remove loose rock from the bench faces, followed by cleanup of loose material along the bench toe, should be considered as the final stages of bench excavation on phase and ultimate walls. This 'good housekeeping' significantly reduces the rockfall hazards.

Figure 11.90: Slot mining on an upper cut where two expansions are being mined on the same wall

11.4.2.1 Scaling

Scaling of the bench crest and face following excavation is an important component of the excavation cycle, but is sometimes overlooked or ignored. Scaling is intended to remove loose blocks and slabs that may form rockfalls or small failures, creating potential safety hazards. It is particularly important where benches are being stacked to a double-benched configuration, since rockfall is a major hazard for the crews drilling and blasting the lower cut. Scaling also minimises the amount of debris that collects on the bench following excavation, thus preserving valuable catchment volume.

Primary scaling is usually conducted on a second pass along the face by the shovel or excavator that removed the original blasted muck. Depending on the nature of the rock mass, the effective bench height, the size of the shovel/excavator, operator experience and design catchment berm width, this may be sufficient. However, in many circumstances secondary scaling of the bench faces using specialised equipment and techniques is required for optimum results. Secondary scaling of bench faces can be performed from the bench above or by equipment standing on the bench floor. It must be accomplished before access becomes difficult or is lost, and before final cleanup at the toe of the bench.

Scaling from the bench above is normally done by chaining the face using a large chain (ship's anchor chain) with or without attached dozer track plates (Figure 11.91). The chain can be dragged along the face by a dozer or backhoe. In no circumstances should the backhoe be used to scale the face from the bench above, as large rocks may pull the machine off-balance.

Scaling from the bench below is generally performed by an excavator configured as a backhoe (Figure 11.92). Most manufacturers offer specialised units equipped with long booms holding small buckets and/or rock picks.

Figure 11.92: Scaling bench face with specially equipped excavator

11.4.2.2 Toe cleanup

Cleanup of debris that accumulates at the toe of the bench should be done immediately after scaling, before access to the toe is lost. In some circumstances, supplementary cleaning or redistribution of debris that has accumulated on the bench may be necessary to maintain adequate catchment. Supplementary bench cleaning will depend on access and the service life of the slope. Periodic bench inspections should identify bench sectors that require cleaning. The cleanup ensures maximum catchment on the bench (Figure 11.93) and should be recorded on a bench maintenance plan.

11.4.3 Evaluation of bench design achievement

Several systems are available for evaluating bench design achievement. One such system, which is modified from work at Chuquicamata in Chile, involves a matrix, which includes:

- design achievement (Df) for the bench configuration;
- face condition (Fc).

Figure 11.93: Bench clean-up

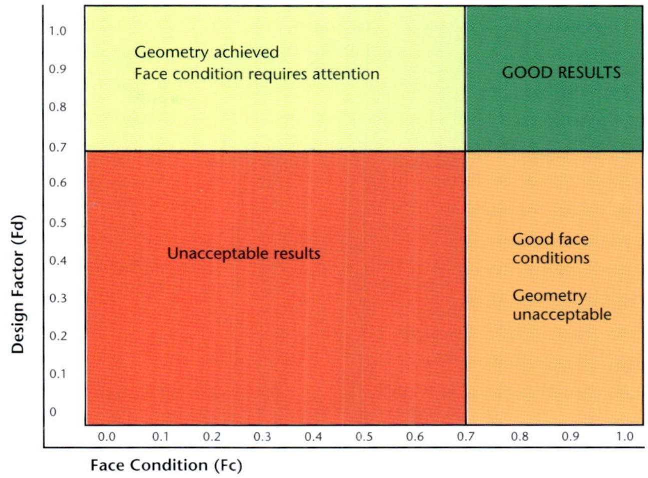

Figure 11.94: Controlled blast evaluation chart

The components of these two sets of parameters are outlined in Table 11.10, together with the respective ranges of values applied in the matrix.

The total of assigned values for each component in the two factors should be reduced to a factor between 0 and 1 and plotted in the matrix shown in Figure 11.94.

In general, double-benching with steep bench faces should only be considered if the results of the controlled blasting at a single bench scale fall within the green area to the upper right in Figure 11.94, where both the design factor and face condition factors are greater than 0.7.

This system is general and should be modified for specific site conditions.

11.5 Artificial support

This section discusses the use of artificial support and construction of stabilising structures in open pit mining applications. It briefly describes the basic approaches in determining suitability of support systems, design considerations and some reinforcement measures.

11.5.1 Basic approaches

The purpose of stabilising structures and placing artificial support is to increase the forces resisting slope failure. This can be done by increasing or enhancing the shear strength of the in situ material, changing the geometry of the slope or providing additional shear resistance along a potential failure surface.

Artificial support to rock slopes is relatively common practice in civil engineering applications where excavations are of moderate dimensions and the costs of structures (roads, bridges, high-rise buildings) are high compared to the excavation and support costs. Techniques range from the use of diaphragm walling or secant piling (that provide vertical walls for foundations) to chemical or physical treatment (vibro-flotation,

Table 11.10: Blast evaluation components

Component (weighting)	Assigned values		Comments
Design achievement (Df)			
Component (weighting)	**Assigned values**		**Comments**
Bench face angle (50%)	≥ Design	50	Achieved overall bench face angle relative to design
	Design –3°	25	
	Design –5°	10	
	Design –10°	0	
Bench width (40%)	≥ Design	40	Achieved average bench width relative to design
	Design –1 m	35	
	Design –2 m	25	
	Design –3 m	15	
	Design –5 m	0	
Toe position (10%)	On design	10	Is design toe is being achieved?
	Design –1 m	8	
	Design – m	5	
	Design –3 m	0	
Face condition (Fc)			
Component (weighting)	**Assigned values**		**Comments**
Half-barrels visible (20%)	≥80%	20	If half-barrels only visible in lower part of bench reduce by 5–10 points
	70–80%	15	
	60–70%	12	
	50–60%	8	
	30–50%	5	
	10–30%	2	
	<10%	0	
Intact rock breakage (15%)	<1/m3	15	Subjective evaluation, interpolate between 0–15
	>5/ m3	0	
Open joints (10%)	All closed – 10	10	Subjective evaluation, interpolate 0–10
	Many open – 0	0	
Loose material on face (20%)	No blocks	20	Assess in terms of rockfall hazard
	Few small blocks	15	
	Large blocks	10	
	Many blocks	0	
Face profile (20%)	Straight	20	Shape of face and basis for variations
	Hard toe	10	
	Overhang crest	5	
	Irregular face	0	
Crest condition(15%)	Achieved	15	For loose rock on crest deduct 0–5 points more
	<1 m loss	12	
	1–2 m loss	10	
	2–3 m loss	5	
	>3 m loss	0	

freezing, grouting) of soils for stabilisation of retaining walls. These are practical in civil engineering applications, given the limited slope heights and the economic value and service life of the infrastructure.

Artificial support in mining presents a different range of issues and challenges from those in civil engineering. The length and height of slopes in mining applications are often much greater, the service life is often short (particularly where a number of different cutbacks are involved) and the economics and practicality of artificial support are affected by the larger volumes of rock to be supported.

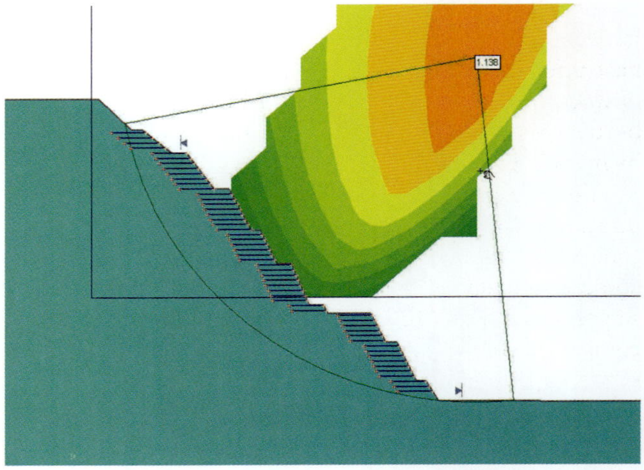

Figure 11.95: Critical slip circles and rock volumes that need support

The use of rock support and reinforcement in Australian open pit mines gained momentum during the 1970s but has now waned. In the 1980s and early 1990s there was a trend to use cable bolting as a blanket form of support over all final walls. This was to allow the extraction of more ore by mining steeper slopes than could normally be achieved without rock reinforcement.

However, it became apparent that large-scale failures were difficult to control with this design approach. Experience showed that slopes approximately 100 m high were the maximum that could be supported with 30 m long cable bolts. Beyond that height, failure occurs behind the supported volume, creating larger deeper-seated masses which are more difficult to control.

This can be understood by looking at an inter-ramp slope of 200 m, a common height in a medium size pit. For this slope, simple geotechnical analysis of a range of materials would indicate critical slip circles extending about 50 m into the slope. Placement of support to that depth is often impracticable unless there is underground access. Further, the magnitude of forces being generated for such a slope can be significantly higher than the support capacities of cable bolts (Figure 11.95). The result is that rock reinforcement is now used mostly for stabilisation of batter-scale wedges/blocks of rock or other localised instability within pit walls.

Even though the volume of wall rock being reinforced at individual mines has reduced, it can still involve large amounts of material. Therefore, it is essential that each rock reinforcement element be designed and installed correctly.

11.5.2 Stabilisation, repair and support methods

A distinction needs to be drawn between stabilisation, repair and artificial support methods.

'Stabilisation' of rock slopes refers to rock slopes that have experienced movement and may be approaching or have undergone some failure. The most common method of stabilising structures include placing a graded rockfill buttress at and beyond the toe and providing drainage behind the buttress.

'Repair' refers to soil or rock slopes that have undergone some failure and can be repaired by removing some of the failed material and replacing it with more competent material. Most methods involve benching slopes through the failure surface and replacement with compacted competent materials or the provision of a shear key to improve shear resistance.

'Artificial support' methods may include retaining walls, placement of rock or cable bolts, or structures such as drilled or cast in-place piles, earth and rock anchors, reinforced earth including the use of geotextile, and protection against erosion. The measures are generally aimed at preventing instability.

11.5.3 Design considerations

In providing artificial support in large open pits, the engineer has to match the design of the rock support system and reinforcement to the ground conditions. In general, there is less benefit from providing a stiff reinforcement in soft or plastic rock than in a strong stiff rock. Achieving large and stable slopes with aggressive wall angles through global wall reinforcement is difficult.

As with any engineering material, the basic approach to artificial support in open pit mining must consider the capacity of the support system, the desired factors of safety, the service life of the support, the timing for installation and quality control/quality assurance programs. These are discussed below.

11.5.3.1 Design of support

Design of any support system must take into account the rock properties, the properties of the support systems, potential failure surfaces and the appropriate factors of safety. Any design method has to consider:

- the function of the support, e.g. to prevent rockfall, slope failure or rock slide;
- geological structure in and around the slope;
- in situ rock mass strength;
- groundwater regime;
- groundwater chemistry;
- behaviour of the rock support or reinforcement system under load;
- rock stress levels and the changes in rock stress during the life of the excavation;
- the potential for seismic events (earthquake or blasting).

Rock mass strengths can be categorised using RMR, Q or GSI ratings. Geological structures which create unstable blocks or 3D wedges often need to be defined through field mapping or commercial digital photogrammetric systems. Given the non-homogenous nature of rock masses, it is often difficult to define these parameters and statistical properties may be assigned.

In contrast, the behaviour and capacity of the artificial support or reinforcement system under load can be defined through standard engineering calculation. Manufacturers of ground support quote breaking or ultimate tensile strengths of support installations, but the engineer must also take into account other possible failure modes such as bending, shear or bearing failure.

Many rock engineering design softwares now allow inclusion of support systems such as end anchored, micro-piling, soil nails, grouted tiebacks, rock bolts, cable bolts and geotextiles. The engineer needs to specify the capacities of such systems. Ideally, testing of support in field scale trials would be used for calculating the capacities.

As with any design process, it is vital to check that increasing the shear strength of a critical failure surface does not make another deeper-seated surface, that extends beyond the reinforced failure surface, approach the critical level.

11.5.3.2 Timing of support installation

In general, the earlier that ground support is installed the more effective it is. The exception is swelling or squeezing ground where, if support is installed too early, the forces on the support may cause it to fail; a delay in installation would allow the ground to squeeze and redistribute stresses so that the support capacity is better suited to the ground conditions.

The timing of installation of ground support and reinforcement should be an integral part of the design implementation, to limit the potential for unravelling of the rock mass. In areas requiring reinforcement, the delay in installation of ground support should be minimised as far as reasonably practicable.

In mining, it is recognised that it may take several days from the firing of a blast until the blasted area is clear of debris and ready for the installation of ground support and reinforcement. However, extended delays (e.g. weeks) may jeopardise the effectiveness of the ground control because of reduced access, and the general loosening (and weakening) of the rock mass.

Ideally, identified wedges or blocks in pit walls that have the potential to daylight or prove unstable should be secured as mining continues, with the support being installed progressively.

11.5.3.3 Corrosion and service life

Corrosion is an important factor in the design and selection of the rock support and reinforcement. The influence of corrosion means that virtually none of the conventional forms of rock support and reinforcement will last indefinitely; they all have a finite design life. Two causes of corrosion are oxidation of the steel elements, and galvanic consumption of iron by more inert metals, e.g. copper.

Caution must be exercised when installing grouted dowels in highly corrosive rocks or potentially acid-forming rocks. Care is required around the collars of grouted cable bolts and dowels, where the sulphides are oxidising with continual weathering of the surface. In highly acidic rock, the collars and plates of cable bolts may have a service life of only months.

11.5.3.4 Quality control

It should be recognised that the various levels of rock support and reinforcement, together with their surface fittings, combine to form an overall ground support and reinforcement system that consists of different layers. Each layer makes a unique contribution to the success of the system. Each element/layer of support and reinforcement must be combined in such a manner that the overall system is well-matched to the ground conditions for the design life of the excavation.

It is therefore important that mine management develop a quality control procedure that ensures that the standard of installation and reinforcement elements meets that required by the design criteria for all ground conditions in the mine.

11.5.3.5 Limitations of design procedures

All engineering design procedures are based on various assumptions that may restrict the application of a particular design procedure. These limitations must be taken into account when applying design procedures in geotechnical engineering. By using appropriate design factors of safety, the limitations can be managed.

11.5.3.6 Alternatives to artificial support

A careful study of the geologic structures must be performed to select the proper reinforcement (i.e. length of bolts or cables, thickness of shotcrete). Rock bolts that are too short will do little to prevent slope stability problems. In some cases, reinforcement may tie several small failures together and create a larger failure Other potential solutions to stop or delay a slope failure are to build a buttress at the toe or unload the top, dewater or reprofile the slope.

11.5.4 Economic considerations

Whenever artificial support is being considered in open pit mining, economic considerations play a crucial part in the decision-making process. The cost and timing delays associated with artificial support must be less than the cost of flattening the batters, dewatering the slope or stepping out the pit walls, to justify its use.

In determining whether artificial ground support is suitable, the following cost–benefit study should be carried out:

- the provision and installation cost of the support, together with delays of mining productivity through allowing access for support installation, are the cost of the system;
- the benefits result from achieving steeper walls and therefore a lesser volume to be mined, and from increased ore recovery.

The cost–benefit analysis may not be well-defined if ground support is to be used to control short sections of a wall to allow an overall steeper angle to be achieved, without ‘noses’ developing in the walls in areas of poorer ground.

Where unstable wedges intersect a pit wall, a cost–benefit analysis should assess the benefits of removing that wedge against those of supporting the wedge. The study should take into account the possibility that removing a wedge may affect another part of the wall, possibly destabilising or undercutting it.

The economic assessments described above apply to decision-making for pit walls at the design stage. During excavation, cost–benefit studies may be required if bench scale (or larger scale) failures occur. If the failure is in a non-critical area of the pit, the easiest response may be to leave the failed material in place. Mining can continue at a controlled rate if the velocity of the failure is low and predictable and the mechanism of the failure is understood. However, if there is any question about subsequent stability, an effort should be made to remove the material. Large-scale failures can be difficult and costly to clean up.

A mining company may choose to leave a step-out in the mine design to contain the failed material and continue mining beneath the step-out. The value of the lost ore needs to be evaluated against the costs of cleanup to determine if this is a feasible solution. The size of blasts may also need to be reduced, to minimise impacts on the unstable zone. To prevent small-scale failures from reaching the bottom of the pit, the number of catch benches and their width can be increased. Catch fences have been installed at some operations to contain falling material.

If allowing the instability to fail is not an option, artificially supporting the failure may be a good solution. Artificial support can be expensive but if the overall angle of the pit slope can be steepened and cleanup costs are reduced, the added expense of reinforcement can be justified.

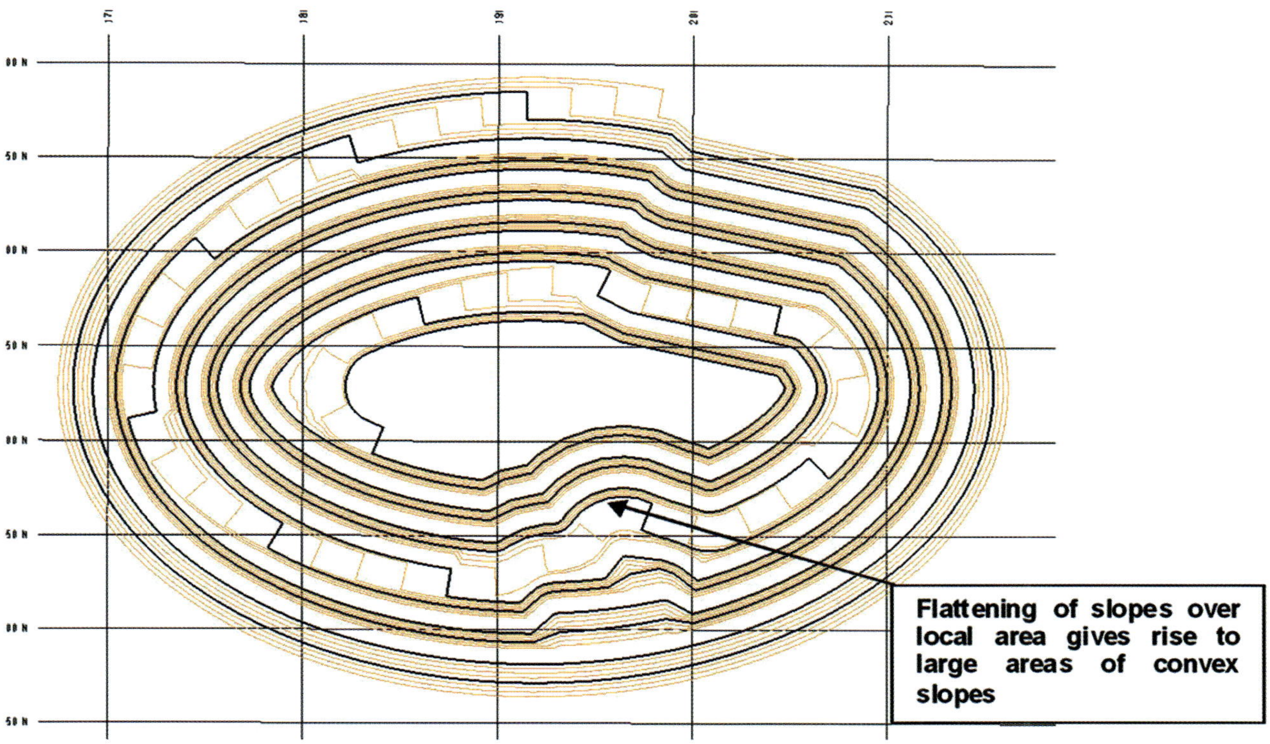

Figure 11.96: Effect of noses on pit geometry

11.5.5 Safety considerations

Installation of artificial support exposes personnel to hazards. Although drilling of rock bolt or cable bolt holes can be mechanised and the operator remains a reasonable distance away from the pit wall, this does not apply when pushing the rock bolt or cable bolt into the hole and plating and tensioning the cable bolt. In that situation, personnel are exposed to rockfall hazards much greater than usual. These risks may be increased if the working bench is lower than the level where the cable or rock bolts are being installed. Such work may need to be carried out from elevated platforms or cages suspended from cranes.

Selective artificial support may target specifical poorer-quality rock such as heavily jointed or sheared zones, faults and clay-filled gouge. Again, personnel operating in these areas are exposed to a higher rockfall risk than in more competent ground.

This does not mean that the risks cannot be managed. As with any hazard, appropriate management can minimise, isolate or eliminate the risk by employing job safety and hazard assessment or similar tools. These may include installing support only in dry weather, use of robotic monitoring, use of personnel as spotters, use of cages to protect personnel, and rescaling faces prior to installing support.

11.5.6 Specific situations

Artificial support may be justified where localised changes in geology occur along or within a face slope or where historic underground workings intersect the pit walls.

11.5.6.1 Changes in geology

From a pit design perspective, it is often poor practice to have sudden changes in pit slope angles over short lengths of pit slopes. This will result in pit walls becoming locally convex and, in extreme cases, noses developing in the pit wall. In these areas the rock often loses confinement and may become unstable. Where a change in geology occurs it is possible to go from a hard rock to a soft rock or soil-like material very quickly. Appropriate face slopes in the harder rock may be 70°, and 45° in softer rock. The effect of a sudden transition is shown in Figure 11.96. In these circumstances it is often appropriate to artificially support the local area to sustain steeper slopes through the weaker material, so as to maintain a uniform wall.

11.5.6.2 Historic workings

Historic underground workings (stopes) which intersect the pit wall may give rise to local wall instability. Stopes that were developed by cut and fill methods will result in fill exposures in the pit wall. The fill will be lower strength than the rock, but confining the rock walls means that the

Figure 11.97: Filled and open stopes intersecting pit wall

hanging wall or footwall to the stope should be in a reasonable condition. In this case the fill should be retained as much as possible but this is difficult if the overall inter-ramp wall angle is greater than the angle of repose of the fill material. Arching of the fill material between the stope walls may enhance stability. Artificial support in the form of retaining walls is often needed to retain the fill material. Where stopes are narrow (e.g. up to 5 m) they are significantly easier to stabilise than are wider filled stopes. It is often more expedient from a geotechnical perspective, and more cost-effective at the pit design stage, to minimise intersections with historic workings and to minimise the width of such intersections where they do occur.

Open stopes that intersect pit walls (Figure 11.97) pose a different set of problems. The surrounding rock is not confined and will probably have loosened. The rock may be stable around the stope but, by introducing another degree of freedom for movement towards the pit wall excavation, large areas of the wall may become distressed and unstable. Consideration may need to be given to filling the voids or supporting the stope walls where they intersect the pit wall.

11.5.7 Reinforcement measures

In this section, reinforcement measures and artificial support to slopes have been categorised into four main groups:

- rock bolting systems;
- retaining type structures;
- surface treatments;
- buttressing.

There may be combinations of these groups. Only a brief description of reinforcement measures is provided.

11.5.7.1 Rock bolting systems

Rock bolting systems typically fall into three categories: rock bolts, dowels (shear pins) and cable bolts. Rock bolting systems can be enhanced by connecting individual components by welded mesh or strapping.

Rock bolts

Rock bolts consist of plain steel rods with a mechanical or chemical anchor at one end and a face plate and nut at the other end, which can be tensioned. They work mainly in tension but can also be used to provide shear resistance. They transfer stress to a more competent rock mass and thus confine the rock mass, thereby increasing its strength. The rock bolt can also apply additional stress on joints or discontinuities to increase friction and resistance to movement. Rock bolts provide active resistance where the bolt is tensioned in situ (bolts) or passive resistance where ground deformation places tension on the bolt (dowels).

All bolts or dowels share the same installation specifications:

- bolts should be installed perpendicular to the wall of the excavation most of the time, but perpendicular to the rock fabric if the bolts are supporting potential structural failures;
- bolts should be long enough to extend into more confined rock away from the excavation;
- when bolting up blocks, it is important to anchor well past the discontinuities and deep enough in the competent rock that bolts do not pull out.

Commercially available rock bolts include cone and shell, grouted and chemically (resin) anchored rebar.

Rock dowels

When installation of support can be carried out very close to the excavated face or in anticipation of stress changes that will occur at a later excavation stage, dowels can be used instead of rock bolts. The dowel depends upon movement in the rock to activate the reinforcing action.

The simplest form of dowel in use is the cement-grouted dowel (Figure 11.98). Other types of dowel are the friction stabiliser (split set) and Swellex dowels. Different load capacities are given by these types of support, depending on diameter and length. Resin-grouted and cement-grouted dowels, if installed correctly, can achieve 16–18 t capacity

Figure 11.98: Rock dowels and mesh to slope

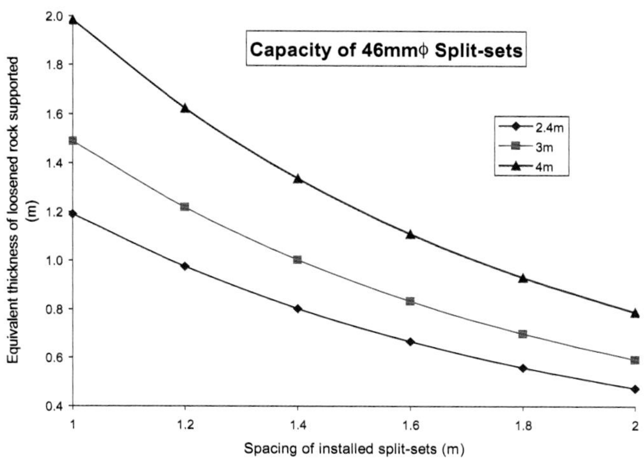

Figure 11.99: Capacity of rock dowels and mesh to slope

whereas the friction stabiliser/swellex can achieve 5–10 t capacity. Capacities can be significantly reduced if the bond between the bolt and the ground, or between the grout and the ground, is inadequate.

Capacities of these support systems are the minimum of:

- the tensile strength of the reinforcing element;
- the friction between the reinforcing element and the rock, and/or grout if used.

The results of pull-out tests show that it is preferable to plot the support system capacity by spacing and length of bolt as shown in Figure 11.99.

Cable bolting

Typically, cable bolts are installed to prevent sliding of blocks on discontinuities dipping out of the face. They are designed to act principally in tension and therefore provide both a normal force to resist sliding, and a shear force component. Cable bolts may be fully or partially grouted. Where higher loads are required, rock anchors in the form of thread-bar rock anchors or multi-strand tendon cable anchors can be used.

A comprehensive review of cable support in underground mining was given by Hutchinson and Diederichs (1996). The methods of installation, selection of cable type and design applications are very appropriate to open pit mining.

The cable bolt consists of one or more steel reinforcing strands placed in a hole drilled in rock, with cement or other grout pumped into the hole over the full length of the cable. A steel face plate, in contact with the bench face, is usually attached to the cable by a barrel and wedge anchor. The cable may be tensioned or untensioned. The steel rope may be plain strand or modified to achieve the appropriate load transfer from the grout and steel strand to the rock mass.

The capacities of cable bolts typically range from 25 t for plain single strand (15.2 mm) 7 wire to 30 t for double bird-caged strand (14 wire). The bolts are prepared for installation by attaching a breather tube to the full length of the bolt, with approximately 2 m to protrude from the hole mouth. Approximately 1 m of grout tube is attached to the bottom of the cable with sufficient tail to connect to the grout pump. The cable bolt is inserted into the hole and the hole mouth sealed to eliminate loss of grout when pumping. The grout tube is connected to the pump and the air bleed (breather) tube is placed into a container of water and pumping commenced. Air bubbles exhaust from the hole during pumping and are visible in the water container. When the hole is full of grout the air bubbles cease. After the grout has cured the cable bolt can be tensioned, provided there is sufficient free cable.

Engineering considerations in achieving full cable bolt capacities are the bond strength, grout quality, installation method and expected service life.

For many wedge, planar type or rotational type slide analyses, the resisting forces of the cable bolts can be easily accommodated in the FoS calculation. In terms of potential planar type failures the FoS is given by the relationship between the resisting forces (S) and the driving forces (T) resolved parallel to the plane of weakness (excluding seismic and groundwater forces):

$$\text{FoS} = \frac{S}{T} = \frac{cA + W\cos\alpha\tan\phi}{W\sin\alpha} \qquad \text{(eqn 11.1)}$$

where

c = cohesion
W = weight of wedge
A = surface area of failure plane
α = angle of slide plane from horizontal.

Modifying this equation for the effects of cable bolts is a relatively simple process, shown below for a passive support system (dowels) and an active support (tensioned) system. A general assumption is that active cable bolt support reduces the driving forces, whereas passive cable bolt support increases the resisting forces.

Passive support:

$$\text{FoS} = \frac{S}{T} = \frac{cA + W\cos\alpha\tan\phi + Ts}{W\sin\alpha} \qquad \text{(eqn 11.2)}$$

Active support:

$$\text{FoS} = \frac{S}{T} = \frac{cA + W\cos\alpha\tan\phi + Tn\tan\phi}{W\sin\alpha - Ts} \qquad \text{(eqn 11.3)}$$

where

$Ts = F\cos(\alpha + i)$ is the shear component of the force applied to base of slide plane
$Ts = F\sin(\alpha + i)$ is the normal component of the force applied to base of slide plane

PLANAR SLIDE WITH CABLE SUPPORT		
Geometric properties	**Symbol**	
Volume of wedge /m		181.7 m3 / m
Surface area of slide plane /m		23.1 m2 / m
Angle of slide plane ά	α	30 degrees
Face Angle	β	45 degrees
Material properties		
cohesion c	c	15 kPa
friction angle Φ	Φ	28 degrees
Density t /m3	γ	25 kN / m3
Factor of safety		
Weight of wedge	W	4543 kN / m
S = cA + W cos α tan Φ	S	2438 kN / m
T = W sin α	T	2271 kN / m
F = S / T		**1.07**
Cable Bolt Properties		
Cable bolt capacity	C'	250 kN
No installed in face slope		4
Centres along face		2 m
Installed angle to horizontal	i	-5
Angle to slide plane	(α+ i)	35
Tn = C sin (α+i)		287 kN/m
Ts = C cos (α+i)		410 kN/m
Active Support		
S = cA+Wcosα.tanΦ+ Tn.tanΦ		2591
T = W sin α - Ts		1862
F = S / T		**1.39**
Passive Support		
S = cA+Wcosά.tanΦ+ Tn.tanΦ+ Ts		3000
T = W sin ά		2271
F = S / T		**1.32**

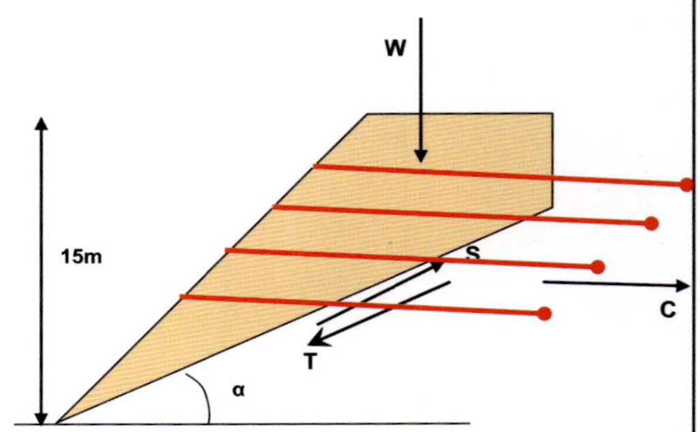

Figure 11.100: Typical patterned cable bolting calculation

i = the angle of the cable bolt(s) from horizontal
F = the ultimate strength of the cable bolt.

A typical example of a 2D planar type wedge calculation is shown in Figure 11.100. The study for a 15 m high bench face at 45° with discontinuity at 30° shows that the wedge has a FoS of 1.07 for the estimated material properties of the discontinuity. The installation of 4 ×25 t capacity cable bolts at 2 m centres beyond the discontinuity improves the FoS to 1.3–1.4, depending on whether an active or passive support is used.

Shear pins

Shear pins are reinforcement bars or larger steel, concrete or post sections that may be grouted in situ. They are designed to be oriented perpendicular to a particular discontinuity and to act mainly in shear. The support provided by the shear pin is equal to the shear strength of the steel bar and possibly the cohesion of the rock/concrete surface. The shear key primarily acts as a resisting force in the limit equilibrium equation. Although shear pins are installed principally perpendicular to the potential slide plane, other applications have involved horizontal installations to provide shear support to blocks defined by flat-lying underground workings intersecting the pit wall or unstable clay seams within an eroded rock wall.

Meshing and W straps

Where bolting alone is insufficient and support is required for small fractured material, welded or arc mesh secured to the rock bolts, dowels or cable bolts is a suitable form of support. Usually a 100 × 100 mm mesh is used, but the size is determined by the desired bag strength. The use of mesh in very blocky ground reduces the potential for unravelling and can be a very useful ground support method. Figure 11.101 shows the use of mesh to protect personnel working under an almost vertical rock face where a pump station has been located inside the pit wall.

W strapping is used to connect the collars of rock bolts. They are nominally 2–3 mm thick and 200–300 mm wide, and can be bolted to follow the contours of the rock face. Support tension can be exerted between bolt sets through the strap (Figure 11.102).

11.5.7.2 Retaining walls

Retaining walls fall into the categories of gravity (mass walls) or cantilevered walls. They are typically formed from precast concrete or in situ poured concrete, steel sheet piling or bored piles. Walls can be reinforced or un-reinforced and can be tied back with tendons into the rock (Wu 1975).

Walls or large-diameter piling can be used to stabilise relatively small slides in the direction of movement or to retain steep toe slopes so that failure will not extend back into a larger mass.

Proper drainage behind the wall is critical to the performance of retaining walls. Drainage material will reduce or eliminate the hydraulic pressure and increase the stability of the fill material behind the wall.

Figure 11.101: Rock bolts and meshing around pumping station

Mass retaining walls

Gravity walls are made from a large mass of stone, concrete or composite material. They depend on the size and weight of the wall mass to resist pressures from behind. A good footing is required for gravity type walls.

Analytical techniques normally require that the resultant force must pass through the middle third of the base of the gravity wall. Examples of mass retaining walls in mining are gabion basket structures, stacked tyres and reinforced earth.

Figure 11.102: Patterned cable bolting and strapping at Kalgoorlie, Western Australia

Gabions

Gabion walls are a traditional effective and practical means of stabilising cuts and slopes. As they apply a surcharge load to the underlying pit wall, they must be installed upwards from a location where there is strong enough rock for a suitable foundation. In this way each bench provides a base for its overlying bench or, from another perspective, each berm provides an effective buttress for overlying benches.

Construction of gabions is labour-intensive and requires appropriately sized rock that is high-quality and not prone to long-term degradation. Hydrothermally altered rock is not often appropriate. Each basket is about 1 m high and can be stair-stepped up each batter at the batter slope angle. Construction will require a cutback of batters to form appropriate stair-steps.

Stacked tyres

Stacked tyres (Figure 11.103) are an alternative to gabions, and may be simpler and cheaper to install. Tyres are stacked in walls along a batter to heights of 1–2 m stepping up slope. Each stack is secured with a coaxial post rail line or drill rod grouted into 4–5 m deep holes. The insides of the stacks are filled with concrete, encasing the post and securing each stack of tyres. Effectively, tyres are formwork for a concrete wall.

Tyres or disused tyres are cheaper than other forms of stabilisation and do not corrode or degrade. If posts are adequately protected, the system is appropriate for long-term stabilisation. It is a popular way of stabilising embankments at mine sites.

The combination of embedding posts and concrete-filled tyres connected to the posts provides shear and normal resisting forces. Stacked tyre walls only need to be applied to the lower half of each batter.

Reinforced earth

Reinforced earth retaining walls (Figure 11.104) are gravity structures consisting of alternating layers of

Figure 11.103: Stacked tyre retaining wall

Figure 11.104: Reinforced earth retaining wall around crusher slot

Figure 11.105: Tied-back wall consisting of 12 m cable bolts, straps, mesh and sprayed concrete

granular backfill and reinforcing strips with a modular precast concrete facing. They are used extensively in transportation and other civil engineering applications but rarely in mining. Because of its high load-carrying capacity, reinforced earth is ideal for very high or heavy-loaded retaining walls.

The inherent flexibility of the composite material makes it possible to build on compressible foundation soils or unstable slopes. These performance advantages, combined with low materials volume and a rapid, predictable and easy construction process, make reinforced earth an extremely cost-effective solution compared with conventional retaining structures.

However, the application of reinforced earth in mining is limited because of the need to excavate material then backfill to fairly tight tolerances. It does, however, have applications in the construction of retaining walls associated with crushing and conveying installations.

Cantilevered retaining walls

Cantilevered walls are made from a relatively thin stem of steel-reinforced, cast-in-place concrete or mortared masonry, often in the shape of an inverted T. These walls cantilever loads to a large structural footing, converting horizontal pressures from behind the wall to vertical pressures on the ground below. This type of wall uses much less material than a traditional gravity wall but requires a very firm footing, particularly at the toe of the wall where high bearing pressures and rotational moments are imposed.

Tied-back walls

Tied-back walls generally comprise a concrete wall, often reinforced with mesh or reinforcement bars tied back into the rock face using cable bolts, or rock bolts in smaller structures. These walls are particularly suited to mining applications where they can be constructed progressively as the benches are mined, using cable bolting meshing and a shotcrete application (Figure 11.105).

Secant piling

Secant piling can be undertaken in soft rock or soils. Secant piles are constructed so that there is an intersection of one pile with another. The usual practice is to construct alternative piles along the line of the wall, leaving a clear space a little less than the diameter of the required intermediate piles. The exact spacing is determined by the achievable construction tolerances. The initial piles do not have to be constructed to the same depth as the following intermediate piles, depending on the way the wall has been designed and reinforced.

Concrete is added and, before it has fully set, the intermediate holes are drilled along a parallel but slightly offset line so that the holes cut into the first piles. The intermediate piles are placed through a heavy casing with toothed cutting edges. This enables the casing to cut into the concrete of the initial piles on either side. Subsequent concreting results in the continuous wall. The concrete often has a slow rate of setting, to ease the problem of cutting one pile into another.

Secant piles can be bored to depths of 30 m. However, the difficulties of construction increase as the pile depth goes below 20 m.

Steel sheet piling

Sheet pile walls are often used in soft soils and tight spaces. They are made out of steel, vinyl, fibreglass or plastic sheet piles driven into the ground.

Structural design methods for this type of wall are readily available. As a rule of thumb, one-third of the pile should be above-ground and two-thirds below-ground. However, structural checks are required for the bending and shear properties of the pile cross-section. Taller sheet pile walls usually require a tie-back anchor 'dead man' tied to the wall face, placed in the soil some distance behind the wall face. The anchor is usually tied by a cable or a rod. Anchors must be placed behind the potential failure plane in the soil.

11.5.7.3 Surface treatments

Shotcrete

A shotcrete lining provides ground support and can lock key blocks into place. It also protects the rock against erosion by water and weathering. To protect water-sensitive ground, the shotcrete should be continuous and crack-free and reinforced with a wire mesh or fibres.

Shotcrete and gunite (pneumatically applied mortar) are two common terms for concrete sprayed through a hose and applied pneumatically at high velocity onto a surface. Shotcrete undergoes placement and compaction at the same time due to the force with which it is projected from the nozzle. Two methods of application are the dry mix method and the wet mix method.

The dry shotcrete method uses a dry cement mix batched into a mixer/pump arrangement, which is then pumped by compressed air along a hose to a nozzle where the mix is ejected and water added. The addition of water is at the discretion of the nozzleman, who monitors addition according to prevailing conditions and perceived mix quality.

The wet shotcrete method involves the batching and wet mixing of the shotcrete under controlled conditions in a mixer. The mix is pumped along a hose and compressed air is added at the nozzle to facilitate spraying the mix onto the required surface.

Until recently, dry shotcrete was the dominant method. However, the wet shotcreting method is rapidly gaining market share because it allows a better working environment, it follows health and safety regulations and yields a higher and more consistent quality plus higher production.

Sprayed concrete is reinforced by conventional steel rods, steel mesh and/or fibres. Fibre reinforcement (steel or synthetic) is also used for stabilisation, and often replaces steel mesh.

There are a few mechanisms through which shotcrete could fail under a rock load. The first is the loss of adhesion between the shotcrete and rock surface. Failure by direct shear is the controlling mechanism, if adhesion is not lost. If adhesion is lost, the shotcrete typically fails in flexure (by bending as it spans between rock bolts) or may fail by punching through the bolt plates. Figure 11.106 shows these failure mechanisms. It is typical that adhesion is lost, or it cannot be relied upon and the shotcrete fails in flexure.

Fibrecrete

Steel fibre reinforced shotcrete was introduced in the 1970s and has gained worldwide acceptance as a replacement for traditional wire mesh reinforced plain shotcrete.

Steel fibre reinforced shotcrete (SFRS) is defined as a concrete containing discontinuous discrete steel fibres, which are pneumatically projected at high velocity onto a surface. Steel fibres are incorporated in the shotcrete to improve its crack resistance, ductility, energy absorption and impact resistance characteristics. Properly designed, SFRS can reduce or eliminate cracking, a common problem in plain shotcrete.

The most important aspects controlling the performance of steel fibres in shotcrete are the:

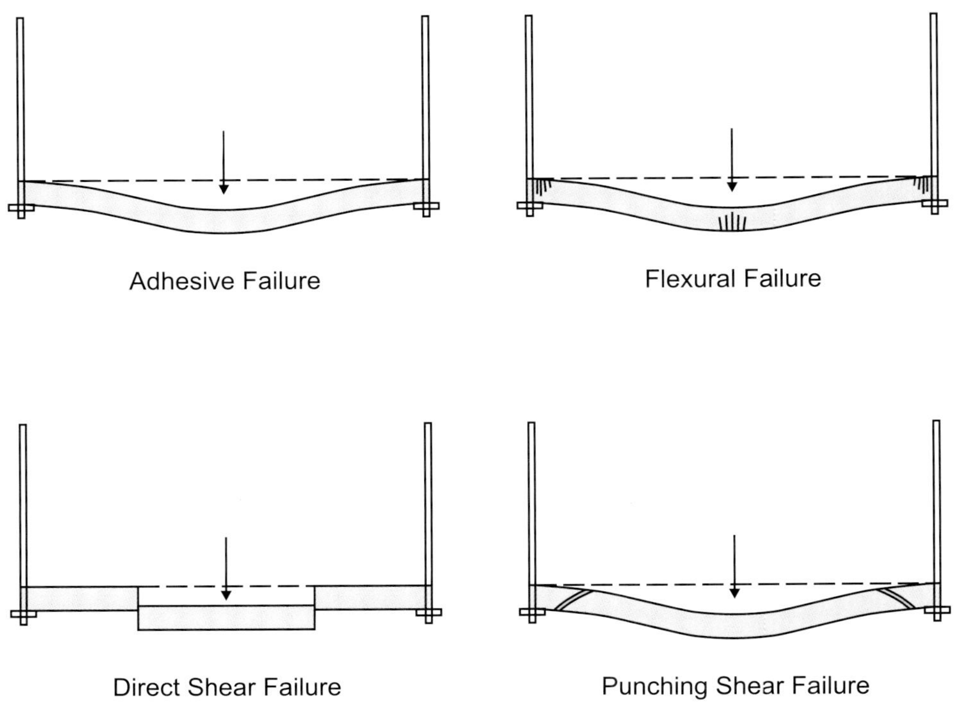

Figure 11.106: Possible failure modes for shotcrete

- aspect ratio;
- volume concentration;
- geometrical shape.

Generally, the higher the aspect ratio and volume concentration of the fibre the better the performance of the SFRS with respect to flexural strength, impact resistance, toughness, ductility and crack resistance. Unfortunately, the higher the aspect ratio and volume concentration of the fibre, the more difficult the shotcrete becomes to mix, convey and spray. Thus there are practical limits to the amount of single fibres which can be added to SFRS, the amount varying with the different geometrical characteristics of the several fibre types. Loose steel fibres with a high l/d aspect ratio, which is essential for good reinforcement, are difficult to add to the concrete and to spread evenly in the mixture.

Slope erosion protective measures

Slopes which are highly susceptible to erosion by rain and wind should be protected from those elements. Protection is also required for slopes that may be subjected to wave action, which occurs where the open pit is expected to fill with water and there is a discharge from the pit lake at a set level.

The usual protection against erosion by wind and rainfall is a layer of rock, cobbles or sod. Protection from wave action may be provided by rock riprap (dry dumped or hand placed), concrete pavement, pre-cast concrete blocks, soil-cement, fabric and wood. Typical specifications for slope erosion measures are given below.

- Stone cover – a rock or cobbles cover of 300 mm thickness is sufficient to protect against wind and rain. A major consideration is whether the slope is flat enough to retain the rock cover.
- Grasses suitable for a given locality should be selected with provision for fertilising and uniform watering. Again, a major consideration is whether the slope is flat enough to retain the soil cover. Hydroseeding is a popular method of quickly establishing grasses on steep batters. Coconut fibre mats containing seeds is a popular method in South America.
- Dumped rock riprap provides the best protection against wave action. It consists of coarse, competent rock fragments dumped on a properly graded filter. Rock should be hard, dense and durable against weathering. It should also be heavy enough to resist displacement by wave action.
- Concrete paving – as a successful protection against wave action, concrete paving should be monolithic and of high durability. Underlying materials should be pervious to prevent development of uplift water pressure. The paving should be a minimum thickness of 150 mm with joints kept to a minimum and sealed. The slab should be reinforced at mid depth in both directions with continuous reinforcement through the construction joints.
- Gabions – slopes can be protected by gabions (section 11.5.7.2).

Figure 11.107: Construction of stabilising buttress halfway up pit wall

11.5.7.4 Buttressing

Construction

A simple method of increasing slope stability is to increase the weight of material at the toe, creating a counterforce that resists failure (Figure 11.107). A berm or buttress of earth or rockfill can simply be dumped onto the toe of the slope. Buttresses across the pit can consist of in situ rock left unmined or constructed from a waste rock or, temporarily, ore.

Broken rock or riprap is preferred to overburden because it has a greater frictional resistance to shear and is free draining, reducing problems with plugging groundwater flow.

Rockfill buttresses are most effective when the natural ground is excavated below the potential failure surface and the excavation backfilled with the rock. This forces the failure circle to occur through the stronger rockfill or along a deeper failure surface that is more resistant to failure.

This method requires considerable volumes of fill. The volume of rockfilled berms or buttresses should be at least one-quarter to one-third of the unstable soil/rock mass. If earth fills are used, the volume should be increased to between one-third and one-half the potential failure volume.

Shear trenches

Shear trenches or shear keys provide increased shearing resistance to failure and also serve as a subsurface drain. A shear trench is frequently a good supplement to flattened slopes and berms. Shear trenches should extend the full length of the slope.

Backfilling of voids

Open pits may be developed around historic underground workings. Existing mined stopes which have limited vertical or horizontal continuity (less than about 50 m) are unlikely to have a significant negative impact on large-scale pit instability but can be a real problem at batter scale. However, stopes with substantial continuity which intersect the final pit walls could pose a significant risk to large-scale instability upslope of the daylight intersection. The level or risk depends on pre-existing natural stress fields and any intersecting major discontinuities.

Remedial measures for old mine workings beneath open pits can be conveniently subdivided into stope bridging or filling solutions. Bridging solutions involve capping with a concrete or roller-compacted type concrete, whereas filling schemes involve gravity filling, pneumatic filling or hydraulic backfilling.

The bridging solutions can provide support to the pit walls and some confinement to the rock at the surface. Bridging solutions do not provide stability to large-scale block movement that may occur as the pit is developed.

Where the stopes are filled, the risk of local instability due to historic slopes can be alleviated by constructing slots or buttresses of cemented fill within the old stopes. The plugs need to extend a reasonable distance below the intersection with the pit face and be of a reasonable width or thickness and not particularly stiff (UCS ~ 5 MPa). Other solutions may be retaining walls (tyres) or tied-back walls.

Backfilling of old mine voids is used to reduce the effective void and limit roof or sidewall caving. Steed et al. (1991) summarise costs for this method, which are normally prohibitive unless a ready source of deslimed tailings or other waste is available.

Wide stopes can be backfilled with run of mine waste. Narrow stopes (less than 10 m width) will tend to choke off, arch or bridge.

11.5.8 Rockfall protection measures

Open pit slope design uses a combination of face slope angles, bench heights and berm widths to achieve the desired inter-ramp angles. The principal objective of the berm is to contain small failures or rockfall hazards from the face slope. In localised areas of particularly weak rock, the berm may become compromised in width and there is potential for rockfalls from the face slope to travel long distances. Where artificial support cannot be adequately installed to control this risk, measures can be taken to control or manage the potential for rocks to bounce, often at high velocities, into working areas. Three main methods of controlling rockfall hazards are ditches, netting or mesh and fences.

11.5.8.1 Ditches and bunds

Catch ditches or bunds on benches, along ramps and at the toe of slopes are often a cost-effective means of stopping rockfalls provided there is adequate space at the toe of the slope. In mining applications, the catch bund can often be placed adjacent to the haul road and a single-lane travelway be maintained. The required dimensions of the bund or ditch can be determined from work field tests by Ritchie (1963), provided the slope height is less than 40 m (section 10.2.1.2). However, that work was primarily intended for highway cuts where high factors of safety are required. This degree of conservatism may not be warranted for a mine excavation. For a 20 m high face at 80° the design chart suggests that a catch width of 7 m and depth of 1.5 m is adequate. Beyond this dimension, rockfall modelling can be done with proprietary software. Bunds can be constructed of waste rock or from gabions.

11.5.8.2 Mesh

Wire mesh hung down the face of a rock slope can be an effective method of containing small rockfalls and ravelling close to the face and preventing them from bouncing. The mesh absorbs some of the energy of the falling rock. A chainlink mesh is suitable for small face heights and small rocks. As face heights increase and rockfall dimensions increase, the mesh can be reinforced with lengths of wire rope and stitched together.

The mesh is anchored securely at the top but is not anchored at the bottom. It should be kept as close as possible to the unstable area, as shown in Figure 11.108. The opening at the base allows rock to work its way down to the bottom rather than accumulating behind the mesh.

11.5.8.3 Catch fences

A catch fence or rock-restraining net is a device engineered to stop large rockfalls. The system consists of a net

Figure 11.108: Avalanche mesh to rock slope

vertically supported by steel posts with energy dissipators able to dissipate high kinetic energy by large displacements.

These fences are usually placed to protect against rockfalls in mountainous areas, but have been successfully used in mining applications. Figure 11.109 shows a commercially available system installed at the Kalgoorlie Consolidated Gold Mines open pit in Western Australia. Dynamic rockfall protection nets were developed decades ago in Switzerland and designs have been continually researched and modified to catch and retain rockfall, debris flow and small snow avalanches. The standard design rockfall barriers cover an impact energy range of 250–3000 kJ. Impacts are calculated from consideration of fall trajectories, slope profile and maximum boulder size.

These types of catch fences have been installed at the Sons of Gwalia open pit at Leonora, Western Australia, and at the Los Bronces copper mine in Chile to provide for slopes with no berms. They have also been installed at the El Soldado mine in Chile to increase bench height and decrease berm width. Other applications are protection of haul roads and ramps and other facilities at the base of a major failure.

Figure 11.109: Catch fence, Kalgoorlie, Western Australia

12 PERFORMANCE ASSESSMENT AND MONITORING

Mark Hawley, Scott Marisett, Geoff Beale and Peter Stacey

12.1 Assessing slope performance

12.1.1 Introduction

In most open pit mines, the physical environment in which slope designs must be developed and implemented is extremely complex. Models that depict the distribution and characteristics of the soil, rock and groundwater throughout the deposit (Figure 12.1) are often based on incomplete information and subjective interpretations. Confidence in these models is affected by limited exploration and investigation budgets, tight project schedules and the practical limitations of investigative tools and techniques. This problem is particularly acute during the pre-development stages of new projects that have little or no previous mining history, where surface exposures are scarce, drill holes are widely spaced and instrumentation is lacking.

While understanding the geological and structural framework of the deposit is fundamental, other factors, including the excavation technique, mine planning and sequencing considerations, and surface water and groundwater management requirements, also influence slope stability and present another layer of complexity for the designer. Because of this complexity, and despite the growing sophistication of the computer-supported analytical techniques outlined in the preceding chapters of this book, it would still be impossible to explicitly model all the factors and interactions that influence the behaviour of a pit slope. Therefore, although modelling is clearly an important tool that helps us understand and hopefully bracket the expected behaviour of the pit slopes, pit slope design relies to varying degrees on empirical science, the design philosophy implemented at the mine (section 11.2.2), the judgment of the designer and feedback based on actual performance.

Given this reality, ongoing empirical calibrations of the key factors that influence the design, and validation of the design methodology and criteria, are necessary components of any rational slope engineering program. In this context, slope design must be considered as an iterative process whereby:

- design criteria are based on the best available information and a defendable methodology;
- slope designs are implemented according to the established criteria;
- actual geologic conditions, as-built slope geometry and slope behaviour are monitored and documented;
- documented versus predicted conditions and behaviour are compared and slope design criteria are modified accordingly, completing the cycle.

This process requires systematic monitoring and documentation of the geological conditions and slope performance, and periodic reviews and updates of the design criteria and mine plan as the mine is developed. Techniques for assessing the performance of the slope design are described in the remainder of section 12.1. Slope monitoring techniques are described in section 12.2 and ground control management plans are outlined in section 12.3.

12.1.2 Geotechnical model validation and refinement

The geotechnical model developed during the project feasibility study is based primarily on the interpretation of limited exposures such as surface outcrops, test pits, trenches, adits, declines and limited drilling (section 8.5.1 and Table 8.1. The positions and attitudes of major geologic structures such as faults, folds and contacts that form critical model boundaries may not be well-established. In addition, interpretation uncertainties and biases may be introduced by the nature of the data, such as line bias due to the orientations of drill holes or underground workings, or differences in scale that may occur when mapping structures in surface outcrops versus oriented core. Therefore, it is important that the

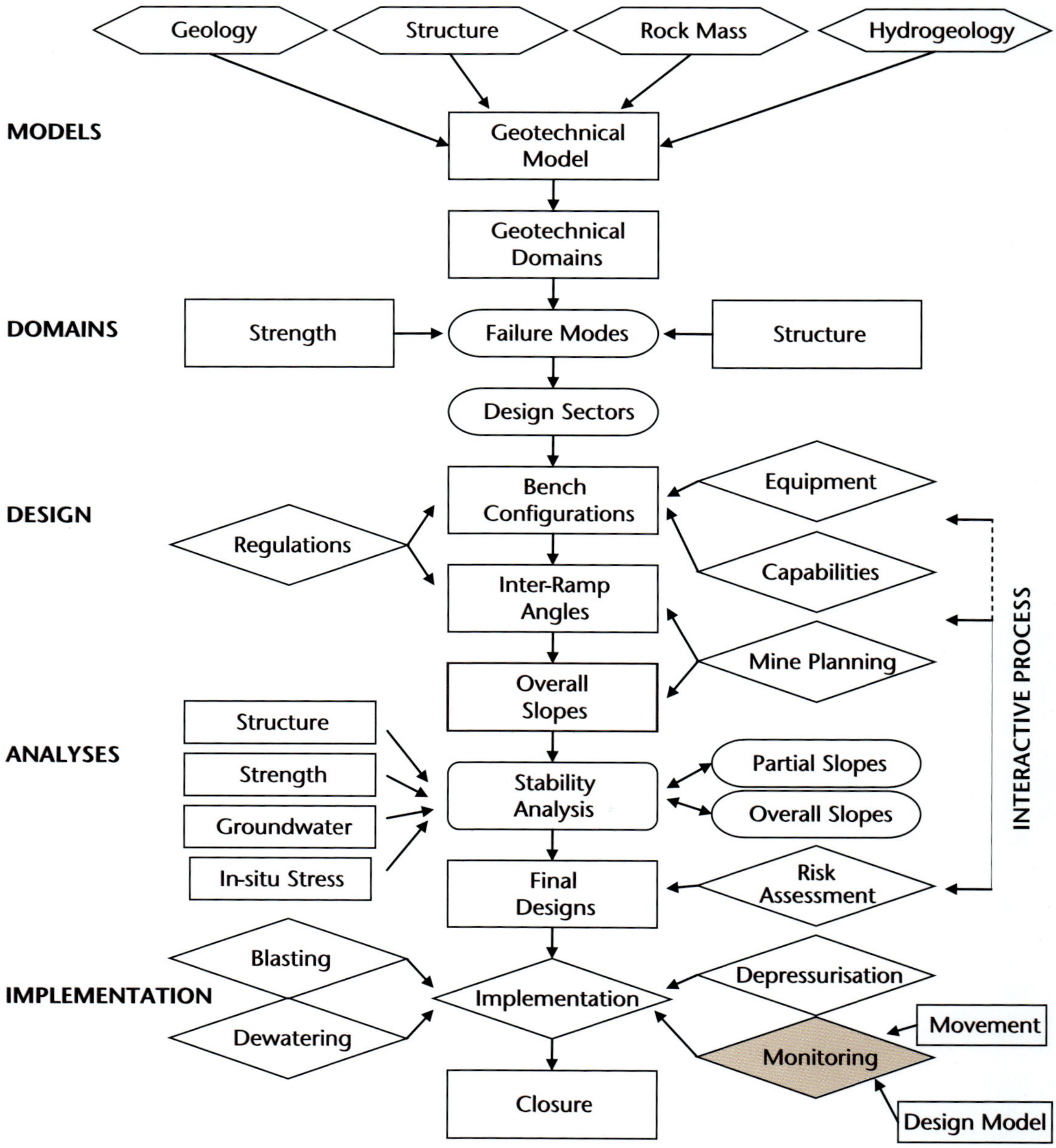

Figure 12.1: Slope design process

geotechnical model (Table 8.1) be validated through mapping and drilling programs as mining progresses.

12.1.2.1 Geological mapping

As the deposit is mined and the geology is progressively exposed, there are opportunities for supplementary geological mapping and interpretation, and ongoing validation and updating of the geotechnical model. Optimisation of the mine design requires that operators maintain up-to-date as-built geology plans, sections and 3D models that detail and correlate all major structures, alteration and weathering zones and mineralisation boundaries. Particular attention needs to be given to documenting the location and nature of boundaries between structural domains, geotechnical and hydrogeological units and groundwater flow compartments. Geotechnical and hydrogeological plans, sections and 3D models also need to be regularly updated according to the most current as-built information to ensure that interpreted boundaries are represented as accurately as possible.

12.1.2.2 Bench mapping

In addition to facilitating mapping and correlating the major structures and geological boundaries, bench exposures allow detailed mapping of the structural fabric and geomechanical characteristics of the rock mass. A comprehensive understanding of the nature of the structural fabric and competency of the rock mass is required for rational design of benches and inter-ramp slopes.

Systematic bench mapping programs can provide detailed information on the characteristics of individual discontinuities and discontinuity sets. Discontinuity orientation data obtained from structural fabric mapping on a bench scale can also be used to validate and refine structural domain boundaries. Slope designs can be sensitive to the orientation and intensity of specific discontinuity sets, and detailed stereographic analysis of the mapping data can help to identify important shifts in discontinuity orientations or degrees of development. Bench mapping programs can also improve confidence in statistical design approaches by increasing the population and reliability of the database.

The spacing, continuity and form of discontinuity sets also influences slope behaviour. Low persistent, widely spaced, undulating or stepped discontinuity sets may have little impact, whereas sets that are closely spaced, continuous and planar may control design. Data on the spacing, continuity and planarity of individual discontinuity sets, and their interrelationships, are often difficult to obtain in the pre-development stage of a mining project when data sources may be limited to weathered outcrops and drill core. Initial bench exposures are often the first opportunity for accurately documenting these important parameters. In addition, as the pit is developed, the characteristics of specific discontinuity sets may change in response to changing stress conditions, changes in overall rock mass competency and other factors. Blasting can also affect continuity and spacing, and this factor should be taken into account in the mapping process.

The shear strength characteristics of discontinuities are related to the roughness and hardness of the discontinuity surfaces, the aperture or width of the opening between the surfaces and the nature and thickness of the infilling materials. Documentation of these parameters and correlation with the results of laboratory and field testing can be used to validate and refine shear strength assumptions used for stability analysis and design. As for other discontinuity parameters, the shear strength characteristics can vary by discontinuity type and set, and may be affected by alteration, weathering and mineralisation of the rock mass.

Bench scale discontinuity data can be collected using a variety of techniques, as discussed in detail in Chapter 2, section 2.2.3. Conventional line or window mapping requires access to the benches and can be labour-intensive and hazardous. Alternative techniques that use laser scanners or photogrammetry with sophisticated computer data-processing may be more safe, efficient and accurate and can be done remotely; however, field validation is still recommended. In addition, some of the key characteristics of discontinuities, such as joint roughness, wall hardness and infilling characteristics, can only be reliably obtained by close physical examination.

12.1.2.3 *Supplementary drilling*

As the mine develops and the understanding of the geological framework of the deposit evolves, questions regarding the accuracy of the underlying interpretation often emerge. Specific geotechnical and hydrogeological issues may require further investigation and analysis. Areas that may have been inaccessible or too deep or costly to investigate earlier in the mine life may become more accessible or less costly to investigate as the deposit is mined. Geotechnical and hydrogeological issues may have been identified but not fully investigated because potential impacts would not present until later in mine life. In these cases, a targeted supplementary geotechnical and hydrogeological drilling and instrumentation program may be a key component of an ongoing slope performance assessment program.

Supplementary core drilling programs should consider opportunities for detailed geological and geomechanical core logging and core orientation, as well as in situ hydraulic conductivity (packer) or stress testing. Opportunities for drill hole instrumentation, such as piezometers, inclinometers or TDRs, drill hole extensometers and microseismic sensors should also be considered. Rotary drilling programs that do not produce cores are generally less desirable from a geotechnical perspective, but may be preferred for certain types of hydrogeological investigations (e.g. air lift or pump tests). Data obtained from supplementary drilling and instrumentation programs should be used to validate and update the geological and geotechnical models.

12.1.3 Bench performance

12.1.3.1 *Bench documentation*

Systematic documentation and evaluation of the performance of benches is an important component of any slope assessment program. Benches are the fundamental building blocks of the pit slope, and their geometry and behaviour often controls the inter-ramp and hence overall slope design. In addition to discontinuity specific data, which can be captured using various bench mapping techniques as described above, information on the geomechanical characteristics of the rock mass and the geometry and general behaviour of the bench is required.

Bench documentation should be customised to the project needs and the information collected referenced to specific design sectors at the mine site. Figure 12.2 shows a documentation format designed for collecting bench performance data directly in the field using a documentation window approach.

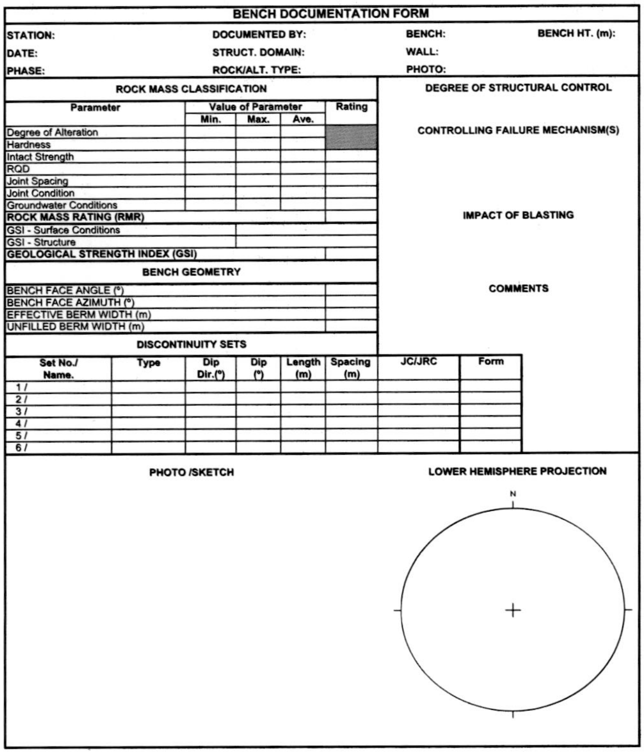

BENCH DOCUMENTATION FORM

STATION: | DOCUMENTED BY: | BENCH: | BENCH HT. (m):
DATE: | STRUCT. DOMAIN: | WALL:
PHASE: | ROCK/ALT. TYPE: | PHOTO:

ROCK MASS CLASSIFICATION

Parameter	Value of Parameter			Rating
	Min.	Max.	Ave.	
Degree of Alteration				
Hardness				
Intact Strength				
RQD				
Joint Spacing				
Joint Condition				
Groundwater Conditions				
ROCK MASS RATING (RMR)				
GSI - Surface Conditions				
GSI - Structure				
GEOLOGICAL STRENGTH INDEX (GSI)				

BENCH GEOMETRY

BENCH FACE ANGLE (°)	
BENCH FACE AZIMUTH (°)	
EFFECTIVE BERM WIDTH (m)	
UNFILLED BERM WIDTH (m)	

DEGREE OF STRUCTURAL CONTROL

CONTROLLING FAILURE MECHANISM(S)

IMPACT OF BLASTING

COMMENTS

DISCONTINUITY SETS

Set No./ Name.	Type	Dip Dir.(°)	Dip (°)	Length (m)	Spacing (m)	JC/JRC	Form
1 /							
2 /							
3 /							
4 /							
5 /							
6 /							

PHOTO /SKETCH

LOWER HEMISPHERE PROJECTION

N

Figure 12.2: A bench documentation data collection form

The type of information that should be collected includes:

- general geologic characteristics such as lithology, alteration, weathering and mineralisation;
- geomechanical characteristics such as the hardness of the rock, intensity of fracturing and condition of discontinuities;
- the as-built geometry of the bench, including the orientation and breakback angle of the bench face, bench height and bench width;
- factors controlling the achieved geometry, whether geological/geotechnical (e.g. fabric and material strength) or mechanical (e.g. mining equipment);
- observations of blast damage and the effectiveness of controlled blasting;
- observed seepage and groundwater conditions;
- the nature and effectiveness of surface runoff controls;
- the quality and effectiveness of toe cleanup, scaling and maintenance of catchment;
- observations regarding specific failure mechanisms and the overall importance of structure and kinematic controls.

12.1.3.2 Rock mass classification

Bench documentation and mapping should facilitate collection of all the data required to classify the rock mass using one of the industry standard rock mass classification systems outlined in Chapter 5 (GSI, RMR, MRMR) or a system customised to the specific site. The system should be compatible with the underlying rock mass model so that the classification data can be used to validate and refine both the model and the design assumptions concerning rock mass competency and strength. Systematic comparison of documented rock mass competency in a given geotechnical unit with the competency predicted by the rock mass model will indicate the reliability of the model. Over time, as more documentation data are incorporated into the model, reliability should improve.

Figure 12.3 shows a statistical comparison of Bieniawski's 1976 version of Rock Mass Rating (RMR_{76}) for a given rock mass unit based exclusively on data obtained from core drilling prior to slope development, and RMR_{76} data documented on benches developed in the same rock mass unit. In this case, the average RMR_{76} from the bench documentation is slightly lower than indicated in the core. This discrepancy might be the result of blasting disturbance, biases due to the orientation of the core holes, spatial variations in the distribution of rock mass competency or a combination of these factors. Depending on the design's sensitivity to changes in rock mass competency or rock mass strength, modification of the design criteria may be necessary.

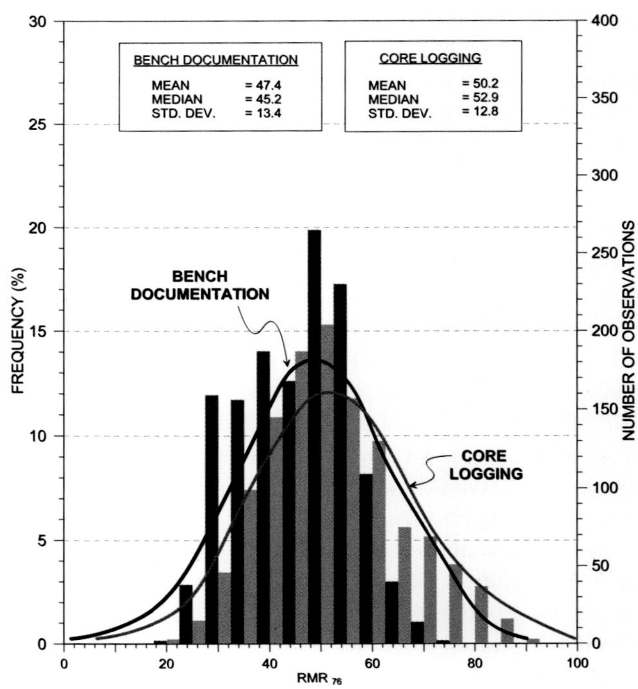

Figure 12.3: RMR obtained from drill core vs bench documentation

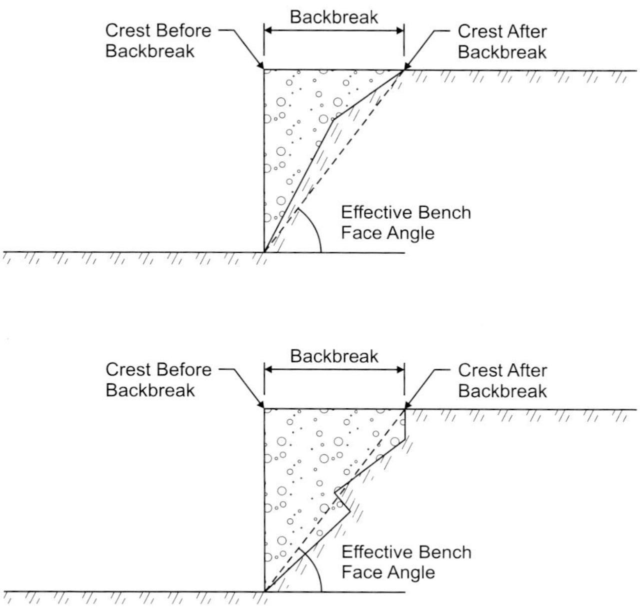

Figure 12.4: Definition of backbreak and effective bench face angle
Source: Ryan & Pryor (2000)

12.1.3.3 Bench face angle

Most bench and inter-ramp slope design methodologies rely heavily on predicting how benches will backbreak in response to mining. One of the key parameters used to define the geometry of a bench is the effective bench face angle, which is the angle to which the bench face is expected to backbreak during mining (Figure 12.4).

The bench face angle is a complex function of many factors, including:

- the structural fabric (orientation, spacing, continuity of discontinuities);
- the condition of the discontinuities (shear strength, planarity);
- the competency of the rock mass;
- the excavation technique (drilling, blasting, excavation sequence and equipment type);
- the type and intensity of scaling;
- environmental factors (freeze–thaw, precipitation);
- service life of the slope.

In massive competent rock masses, the bench face angle is often strongly influenced by blasting. In this case, expected bench face angles may be estimated for design purposes based on the type and layout of the blast pattern and experience with similar blast designs in comparable rock masses. In blocky to moderately fractured competent rock masses, the occurrence of wedge, planar or toppling failures involving discontinuities is often key. Kinematic and stability analyses can identify potential failure modes based on the structural fabric and discontinuity characteristics in a given structural domain and rock mass unit (section 10.2.1.3). Consideration is also given to discontinuity spacing and persistence. Results of these analyses may be evaluated using various statistical, semi-empirical or judgment-based approaches to estimate the expected bench face angle. In intensely fractured rock masses, breakback may be controlled by general degradation and ravelling of the rock mass. Expected bench face angles in these cases are often difficult to assess with confidence using conventional analytical techniques, and provisional designs are often based on experience or judgment. In very weak incompetent rock masses, breakback may be controlled by failure modes more typically associated with excavations in soils. In these cases, conventional slope stability analysis techniques used for the design of soil cuts should be used to evaluate the bench face angle.

Because it is impossible to explicitly account for all the complex factors that influence the performance of the benches, validation and empirical calibration of the design methodology used to estimate breakback is a critical component of a slope performance assessment program. The most direct method of calibration is to compare the as-built bench face angles with the expected bench face angles. Where the structural fabric influences breakback, comparisons are strictly valid only within a given design sector. Broader comparisons may be valid in cases where the structural fabric has limited influence and other factors (e.g. rock mass competency) are at least as important. If the as-built and expected bench face angles are comparable, confidence in the design methodology is reinforced and modification of the design approach may not be required. However, material differences between the as-built and expected bench face angles may indicate a flaw in the design methodology that requires correction.

There are several approaches for documenting the as-built bench face angle. The most basic method is to back-calculate the angle using survey or mine status plans that depict bench toes and crests. However, this method typically returns angles that are significantly flatter (2–10°) than the actual angles because mine survey crews are hesitant (and may not be permitted) to get close to the crest or the toe of the bench. Topographic plans based on aerial photography can be used, but their accuracy is limited by resolution and contour interval. Direct measurement of as-built bench face angles using a Brunton compass, or indirect measurements using a clinometer (Figure 12.5) are typically more accurate (±2–5°). Both these methods require access to the bench. Digital photogrammetry and laser scanning methods are probably the most accurate and detailed method for obtaining as-built breakback data.

Histogram plots and cumulative frequency analysis are convenient ways of presenting and evaluating achieved

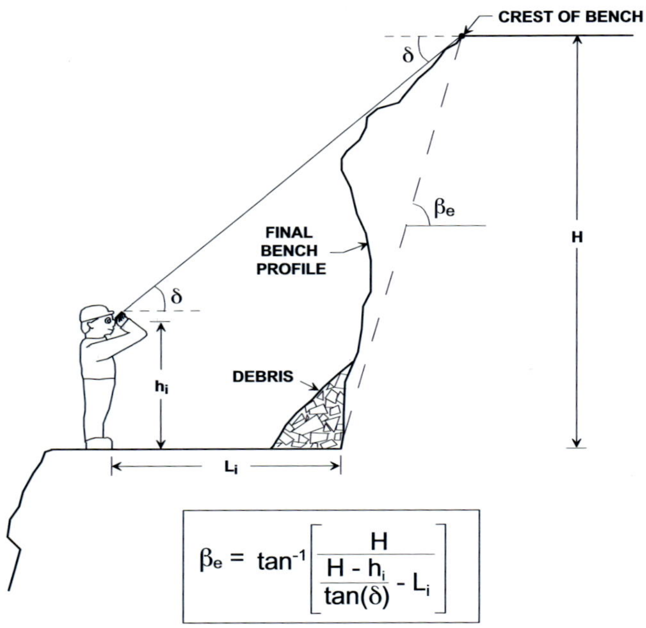

Figure 12.5: Measuring the bench face angle using a clinometer

bench face angle data. Depending on the design methodology, direct comparison of the as-built and expected angles may be possible, as illustrated in Figure 12.6.

Cumulative frequency analysis is also a convenient tool for objectively evaluating bench performance based on established acceptability criteria, or for assessing the sensitivity of the acceptability criteria to breakback. For example, if unacceptable performance is defined as an as-built bench face angle that is flatter than the expected angle, acceptability criteria for a given design case could be based on a specific cumulative frequency threshold, taking into account risk tolerance and consequences of failure. This is analogous to a probability of failure (PoF) analysis, where failure is defined as unacceptable performance of the design. In a critical design case where the consequences of failure are high and risk tolerance is low, a cumulative frequency or PoF of less than 20% might be appropriate. In practical terms this means that it is acceptable for up to 20% of the bench face to breakback to an angle that is flatter than the expected or design breakback angle. Alternatively, where the consequences of failure are low and risk tolerance is high, a cumulative frequency/PoF of 50% might be acceptable. In either case, if sufficient as-built breakback data are available, bench performance can be evaluated objectively using a cumulative frequency analysis (Figure 12.7).

In Figure 12.7, as-built breakback data from two different design sectors are represented by Curves A and B. In both cases, evaluation of acceptable performance is

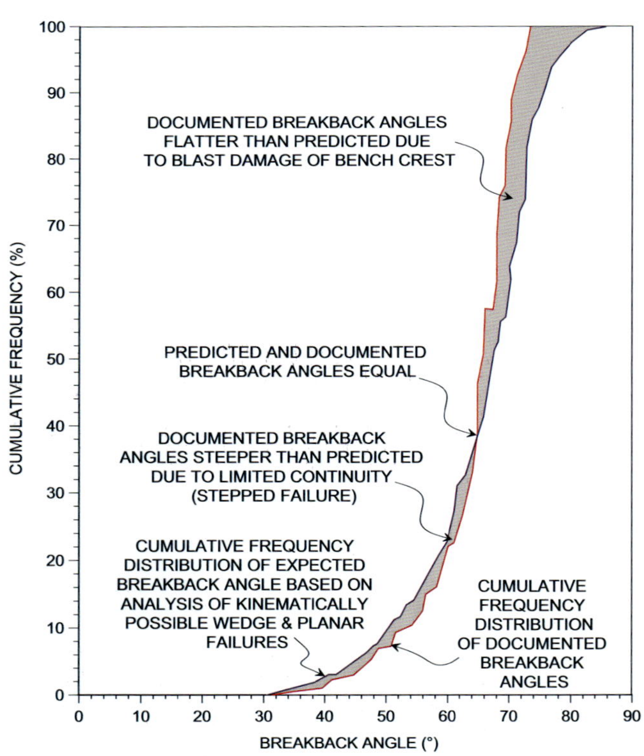

Figure 12.6: Expected (predicted) breakback vs documented breakback using cumulative frequency analysis

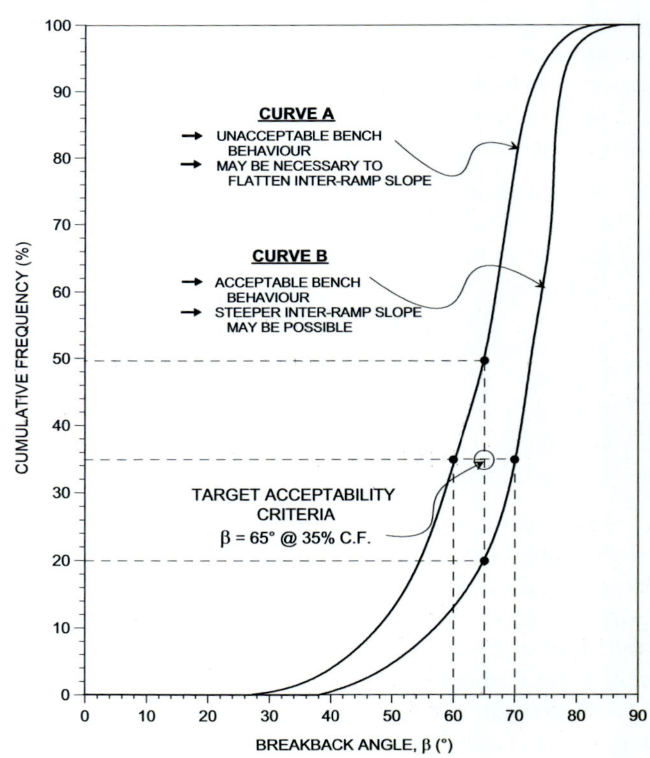

Figure 12.7: Direct assessment of bench performance based on cumulative frequency analysis of documented bench face angle

based on an expected bench face angle of 65° and a design cumulative frequency of 35%. Curve A indicates an as-built bench face angle of 60° at the target cumulative frequency of 35% and a cumulative frequency of 50% for the design bench face angle of 65°. Based on the established acceptability criteria, the design sector represented by Curve A is not performing acceptably, and a revision to the design criteria (e.g. flattening of the inter-ramp slope and/or modification of the excavation method) is required. Curve B indicates an as-built breakback angle of 70° at the target cumulative frequency of 35% and a cumulative frequency of 20% for the design breakback angle of 65°. Based on the established acceptability criteria, the design sector represented by Curve B is performing acceptably and a revision to the design criteria is not required. Alternatively, the data represented by Curve B suggest that it may be feasible to steepen the design or relax excavation controls (and reduce costs).

Backbreak measurements are typically made as part of geotechnical or geological mapping routines. Where window (cell) mapping techniques are used, one bench face angle is recorded for every cell (section 2.2.3.2). If continuous survey data from digital imaging (e.g. air photos, photogrammetry) or laser scanning is used, sampling routines can be written to extract bench face angles for whatever spacing is desired.

Another important facet of assessing bench performance is understanding the specific mechanisms that influence breakback. This is particularly important in cases where as-built and expected breakback angles are materially different. Field recognition and validation of the various discontinuity sets, as well as comments about their relative degree of development, continuity and spacing, and which sets actually form failures that are important to design, can be very useful. Discontinuity sets that were initially thought to be weakly developed or absent in a particular structural domain may have a stronger influence than expected. In the absence of reliable data on the continuity and spacing of discontinuity sets, theoretical assessments of potential failure modes may over- or underestimate the relative importance of a specific failure mode. Failure modes that, due to lack of data, were conservatively assumed in the original design to be continuous for the full bench height, may instead have only limited continuity and form smaller-scale stepped failures instead of larger continuous failures. Alternatively, blasting could extend structures that were originally thought to be tight, healed or discontinuous. Failure modes observed in excavated bench faces should be critically compared to those assumed during design to control stability. In addition, qualitative assessments of the degree to which structure controls breakback should be conducted and compared to the original design expectation. The bench documentation format illustrated in Figure 12.2 is designed to capture these types of comments, as well as illustrative sketches, simple kinematic projections and representative photographs to help understand the various failure mechanisms and their relative importance.

12.1.3.4 Bench height and width

In addition to the expected bench face angle, bench height and width are required to define the design geometry of a bench. Bench height is usually fixed, based on equipment specifications (section 10.2.1.1). In the context of this discussion, the effective bench (or berm) width is defined as the width of the catchment bench or safety berm that remains after excavation, scaling and cleanup of the bench. Bench width (section 10.2.1.2) may be based on a variety of objectives and criteria, including:

- competency of the rock mass;
- height of the bench;
- volume of potential bench scale failures;
- rockfall catchment criteria;
- performance and condition of overlying benches;
- service life of the slope;
- inter-ramp slope angle and height;
- access requirements;
- general risk tolerance.

Comparison of as-built versus design bench widths is another opportunity for objectively assessing the performance of the slope. Several approaches are available for documenting the as-built bench width. The easiest methods involve measuring the bench width using survey or mine status plans that depict bench toes and crests, or topographic plans based on ground surveys or aerial photography. These approaches suffer from the same accuracy issues as indicated for bench face angles. Direct measurement of the as-built bench width using a tape measure is typically more accurate, but is labour-intensive and requires access to the bench. Digital photogrammetry and laser scanning methods provide the most accurate and detailed methods for obtaining as-built bench width data. Remote survey methods can be easily repeated and used to assess changes to the bench width (and bench face angle) that may occur over time.

Histogram plots and cumulative frequency analysis also provide convenient methods for presenting and evaluating bench width data. Using the same approach as described for bench face angles, unacceptable performance (or design failure) could be defined as an as-built bench width that is narrower than the design bench width. Acceptability criteria for a given design case could be based on a specific cumulative frequency/PoF threshold, taking into account risk tolerance and design objectives.

12.1.3.5 *Effectiveness of catchment*

While comparisons of design versus as-built bench face angle and bench width can be used to evaluate whether the design intent is being met, slope performance assessments should also evaluate the effectiveness of the design. The primary purpose of benches is to provide catchment for bench scale failures, ravelling debris and rockfalls, and thus a safe working environment for equipment and personnel. Benches also establish platforms for electrical infrastructure (power poles, electrical substations, cables), dewatering infrastructure (wells, horizontal drain holes, collection pipes and sumps, booster pumps), instrumentation (prisms and GPS monuments, piezometers, extensometers and crack monitors, inclinometers and TDRs, seismic sensors) and ditches and surface water diversions. Benches provide access for visual inspection of slope and bench conditions.

Unrestricted access to benches is vital for monitoring activities; this must be part of the mine operations standard operating plans. Many mines do not routinely design access in to their ramps systems due to space restraints, which is poor mining practice.

Benches should be inspected periodically to assess their effectiveness. Inspections should focus on the following questions.

- Are bench scale failures and rockfalls being caught and controlled on berms, or are they spilling over onto subsequent benches?
- Are bench faces generating frequent rockfalls?
- Is there adequate access for inspection and monitoring?
- What portion of the design catchment has been filled with debris from failures, ravelling and rockfalls?
- Does accumulated debris need to be removed or levelled to maintain access or adequate catchment?
- How is the bench performing over time? Is available catchment being sustained, or is it decreasing over time due to ongoing failures and/or general degradation of the bench?

A database of inspection reports should maintained for each bench in each design sector. This information should be periodically reviewed and used for modifying the bench design.

12.1.3.6 *Blasting*

Blasting can have a significant and often controlling influence on bench performance. Consequently, the impact of blasting on bench stability must be included in a slope design performance assessment and optimisation program. Unlike the underlying structural geology and rock mass characteristics, which are fixed, the impact of blasting on wall stability can be controlled by careful design and implementation (subject to practical limitations). Comprehensive blasting trials, including detailed documentation of blast vibrations and systematic modification of key design parameters, are usually required for optimal design and to minimise disturbance to the rock mass (section 11.3). Methods of inspecting, documenting and analysing the post blast damage to the benches are outlined in section 11.3.8.2 and section 11.4.3.

12.1.3.7 *Documentation and back analysis of bench failures*

In addition to periodic inspections, any significant failures that occur after the bench has been excavated should be documented and back analysed. Personnel working in the pit should be encouraged to report all rockfalls. Significant rockfall events should be documented, plotted on a plan and captured in a database. Documentation should include the following information:

- a unique identifier;
- location, with figure;
- date the bench was completed;
- date of occurrence;
- estimated volume and/or mass of the failure;
- type of failure (e.g. wedge, plane, stepped failure, toppling, slump, ravelling);
- orientation, spacing and continuity of discontinuity sets involved;
- planarity, roughness and infilling characteristics of the discontinuities involved;
- catchment bench width;
- width of bench filled by failure/catchment width remaining;
- additional pertinent comments, such as apparent association with blasting or precipitation, or any precursors such as tension cracks or rockfalls/ravelling.

Detailed rockfall documentation data is essential for the calibration of analytical rockfall models such as CRSP (Jones et al. 2000) and RockFall (Rocscience Inc. 2002), which can provide valuable insight into the mechanics of rockfalls and their sensitivity to slope geometry and mitigative measures. Photographs and local plans, sections, sketches and stereographic representations of the failure help to more clearly identify and understand the failure mechanism.

An up-to-date status plan showing the locations of all documented failures and highlighting where failure debris has spilled over the catchment bench should be maintained. This plan can be used to visually assess the distribution of failures and identify potential problem areas where the design may need adjustment.

Back analysis of bench failures may be useful to help confirm and refine initial shear strength assumptions. Most bench scale wedge, planar and toppling-type failures that result in breakback of the bench face occur in conjunction with the initial mining, scaling and cleanup of the bench. Wedges and planes that have steep plunges or dips and are undercut by the bench face typically have low initial stability and fail spontaneously as the bench is mined. Likewise, toppling failures on strongly developed, closely spaced discontinuities that dip steeply into the slope are often initiated by blasting and bulk excavation of the bench. These types of spontaneous failures are difficult to back analyse using conventional deterministic analysis techniques because they were not in a condition of limiting equilibrium at the time of failure. Hence, even though the geometrical constraints may be well-understood, the effective factor of safety (FoS) is not known, and there is no unique solution to the back analysis. Sensitivity analyses based on the results of multiple back analyses may be helpful in bracketing shear strength characteristics and numerical back analyses may help in understanding failure mechanics and sensitivities, but rigorous analysis of these types of failures is usually not worthwhile.

Kinematically possible failures that occur some time after the initial excavation of the bench are usually of more interest. These types of failure typically involve planes or wedges with dips or plunges that are steeper than their basic friction angles, but not so steep that they fail spontaneously when they are undercut. These failures 'hang up' when exposed on the bench face as a result of interlocking or cohesion. Over time, blasting vibrations, freeze–thaw, seasonal groundwater pressures, chemical weathering and general degradation of the rock mass may reduce the effective shear strength acting on the discontinuities to the point of limiting equilibrium, at which time failure occurs. Failures of this type provide excellent opportunities to back calculate the effective shear strength and assess sensitivities using deterministic analysis techniques.

12.1.3.8 Rockfall hazards and catchment

Establishing and maintaining sufficient bench width to catch and control structural failures and provide safe and reliable access to the slope is important, but bench design may ultimately be dictated by the need to control rockfalls. This is often the case where bench face angles and/or inter-ramp slopes are steep, or where bench heights are high. Steep bench faces tend to generate more rockfalls. Rockfalls generated on high steep benches attain higher velocities and kinetic energies than those generated on low flat benches. Rockfalls on steep inter-ramp slopes are more likely to bounce and sustain their momentum; on flatter slopes, rockfalls tend to bounce less and roll more, maintaining more contact with the slope and dissipating more energy.

Observations made during periodic bench inspections can generally only indicate the effectiveness of the rockfall catchment along a given bench, and identify source areas. Sectors where rockfalls are not being adequately controlled can be identified and plotted on a rockfall activity/hazard plan. This plan should be periodically reviewed to assess the overall effectiveness of rockfall controls and whether mitigative measures or modifications to the bench geometry are required. Mitigation could include passive measures such as buffer zones, hazard warnings or temporary sector closures intended to limit exposure. Active mitigation measures could involve construction of impact berms or rockfall fences, supplementary scaling or installation of draped mesh (section 11.5.8). Design modifications could include increasing the bench width or flattening the bench face angle, either of which results in a flattening of the inter-ramp slope angle. Where effective rockfall catchment has been lost on multiple benches and the residual risk is deemed unacceptable, the only viable option may be to step in and establish a wide catchment bench. Active mitigative measures and design modifications such as these are intended to reduce the hazard.

A well-maintained rockfall activity/hazard plan is a valuable tool in mitigating risk to personnel and equipment, and planning and prioritising remedial measures. For example, a comprehensive rockfall hazard identification and mapping system was developed and systematically implemented at the Antamina Mine in Peru. The system sequentially evaluates a number of parameters to arrive at one of five qualitative hazard levels (Very Low, Low, Moderate, High or Very High). The system is illustrated in Tables 12.1 and 12.2.

Table 12.1 is applied on slope segments where the effective slope height above the working level is less than or equal to 60 m (i.e. two double benches). Table 12.2 is used where the effective slope height is greater than 60 m.

The parameters considered most important for identifying rockfall hazards in the Antamina open pit are:

- slope height;
- mining activities above the slope;
- condition of catchment benches;
- degree of scaling and condition of bench faces;
- presence and effectiveness of impact berms.

The rockfall hazard levels identified throughout the pit using these criteria are displayed on a current mine status plan, such as shown in Figure 12.8. This hazard level status plan is posted in key locations to convey the information to mine personnel and is used to facilitate discussions and directives during safety, operations and planning meetings.

Implementation of the system at Antamina has been very successful. Calibration and future modifications could extend application of such a plan to include sensitivities to various remedial measures and an exposure factor that will enable semi-quantitative assessments of relative risk.

DETERMINING ROCKFALL HAZARD FOR: *LOW SLOPES*				
Is there mining above?		***Condition of benches and bench faces?*** SELECT WORST CASE OF CONDITION A OR B	***Condition of rockfall impact berm?***	***Hazard Level***
NO or greater than 40m from crest	A	Benches intact and clean	Good	Very Low
			Moderate	Low
	B	Good scaling, no adverse structure	Poor	Moderate
	A	Benches deteriorating and/or up to half full	Good	Low
			Moderate	Moderate
	B	Moderate scaling, some adverse or open structure	Poor	High
	A	Benches lost and/or completely full	Good	Moderate
			Moderate	High
	B	Poor scaling, bench-scale failures, blocks perched on crest	Poor	High
YES = 15m high muck pile at crest, and within 40m of crest	A	Benches intact and clean	Good	Low
			Moderate	Low
	B	Good scaling, no adverse structure	Poor	Moderate
	A	Benches deteriorating and/or up to half full	Good	Low
			Moderate	Moderate
	B	Moderate scaling, some adverse or open structure	Poor	High
	A	Benches lost and/or completely full	Good	Moderate
			Moderate	High
	B	Poor scaling, bench-scale failures, blocks perched on crest	Poor	Very High
YES < 15m high muck pile at crest, or cleaning crest	A	Benches intact and clean	Good	Moderate
			Moderate	High
	B	Good scaling, no adverse structure	Poor	Very High
	A	Benches deteriorating and/or up to half full	Good	High
			Moderate	Very High
	B	Moderate scaling, some adverse or open structure	Poor	Very High
	A	Benches lost and/or completely full	Good	Very High
			Moderate	Very High
	B	Poor scaling, bench-scale failures, blocks perched on crest	Poor	Very High

Low slopes are slopes less than or equal to 60 m, or two double benches, high.

Start in the left column, and select the appropriate parameters to arrive at a hazard level on the right.

When the crest height of the muck pile starts coming down due to mining, an increase in rockfall occurs at the crest. It is estimated that this occurs approximately 30 m from the crest. Increased rockfall has also been observed prior this point, due to disturbance of the muck pile from digging activities; 40 m is an arbitrary estimate.

Scaling
good – no loose material, clean crest
moderate – some loose material, oversteepened crest
poor – abundant loose material, ragged & loose crest

Rockfall Impact Berm Condition
good – berm > 2 m high and > 5 m from toe
moderate – berm > 2 m high and 1 to 5 m from toe, or berm < 2 m high and > 5 m from toe
poor – no or discontinuous berm, or berm < 0.5 m high, or berm < 1 m from toe

Table 12.1: Rockfall hazard rating system for low slopes
Source: Courtesy Compañia Minera Antamina S.A.

DETERMINING ROCKFALL HAZARD FOR: *HIGH SLOPES*				
Is there mining above?		***Condition of benches and bench faces?*** SELECT WORST CASE OF CONDITION A OR B	***Condition of rockfall impact berm?***	***Hazard Level***
NO or greater than 40m from crest	A	Benches intact and clean	Good	Low
			Moderate	Low
	B	Good scaling, no adverse structure	Poor	Moderate
	A	Benches deteriorating and/or up to half full	Good	Low
			Moderate	Moderate
	B	Moderate scaling, some adverse or open structure	Poor	High
	A	Benches lost and/or completely full	Good	Moderate
			Moderate	High
	B	Poor scaling, bench-scale failures, blocks perched on crest	Poor	Very High
YES = 15m high muck pile at crest, and within 40m of crest	A	Benches intact and clean	Good	Low
			Moderate	Moderate
	B	Good scaling, no adverse structure	Poor	High
	A	Benches deteriorating and/or up to half full	Good	Moderate
			Moderate	High
	B	Moderate scaling, some adverse or open structure	Poor	Very High
	A	Benches lost and/or completely full	Good	High
			Moderate	Very High
	B	Poor scaling, bench-scale failures, blocks perched on crest	Poor	Very High
YES < 15m high muck pile at crest, or cleaning crest	A	Benches intact and clean	Good	High
			Moderate	Very High
	B	Good scaling, no adverse structure	Poor	Very High
	A	Benches deteriorating and/or up to half full	Good	Very High
			Moderate	Very High
	B	Moderate scaling, some adverse or open structure	Poor	Very High
	A	Benches lost and/or completely full	Good	Very High
			Moderate	Very High
	B	Poor scaling, bench-scale failures, blocks perched on crest	Poor	Very High

High slopes are slopes greater than 60 m, or two double benches, high.

Start in the left column, and select the appropriate parameters to arrive at a hazard level on the right.

When the crest height of the muck pile starts coming down due to mining, an increase in rockfall occurs at the crest. It is estimated that this occurs approximately 30 m from the crest. Increased rockfall has also been observed prior this point, due to disturbance of the muck pile from digging activities; 40 m is an arbitrary estimate.

Scaling
good – no loose material, clean crest
moderate – some loose material, oversteepened crest
poor – abundant loose material, ragged & loose crest

Rockfall Impact Berm Condition
good – berm > 3 m high and > 8 m from toe
moderate – berm > 3 m high and 3 to 8 m from toe, or berm < 3 m high and > 8 m from toe
poor – no or discontinuous berm, or berm < 2 m high, or berm < 3 m from toe

Table 12.2: Rockfall hazard rating system for high slopes
Source: Courtesy Compañia Minera Antamina S.A.

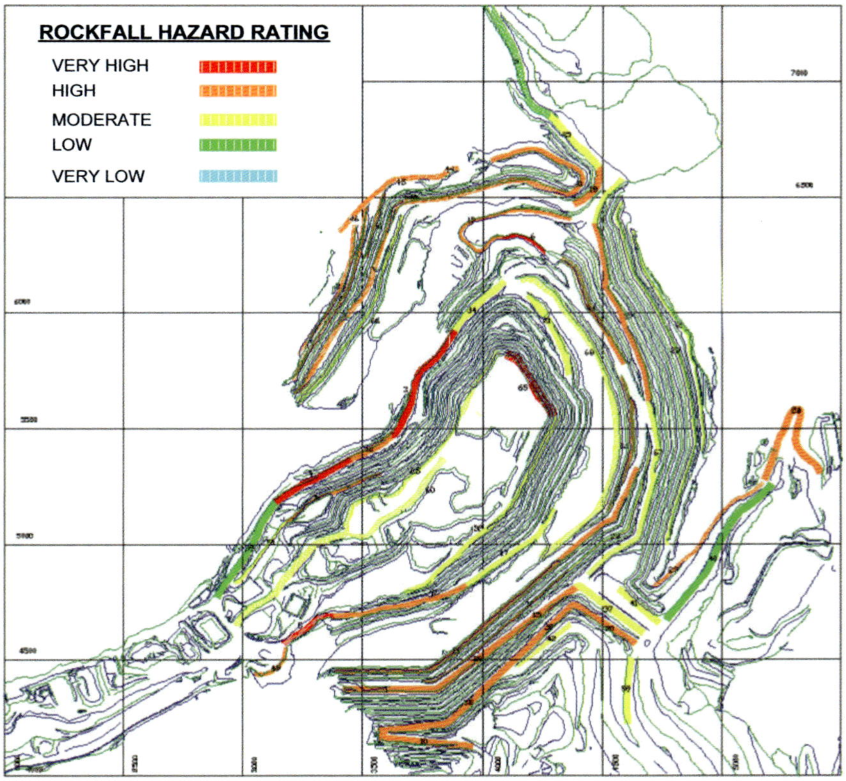

Figure 12.8: Rockfall hazard plan
Source: Courtesy Compañia Minera Antamina S.A.

12.1.4 Inter-ramp slope performance

12.1.4.1 As-built vs design inter-ramp slope geometry

Inter-ramp slope design criteria typically include specification of the inter-ramp slope angle and the maximum slope height between ramps, haul roads or planned step-out benches (also known as the bench stack height) (Figure 12.9).

The inter-ramp slope angle is often based on the results of bench geometry assessments but should also consider the potential for larger-scale failures involving continuous structures such as major joints or faults (section 10.2.2). In weak or incompetent rock masses, the potential for rock mass failure must be considered. Many large inter-ramp slope failures involve combinations of structural and rock mass failure. The inter-ramp slope height may be limited, based on a need to control multi-bench failures, to provide secure access to the slope at intervals to facilitate monitoring or the application of remedial measures, or to control overall slope stability.

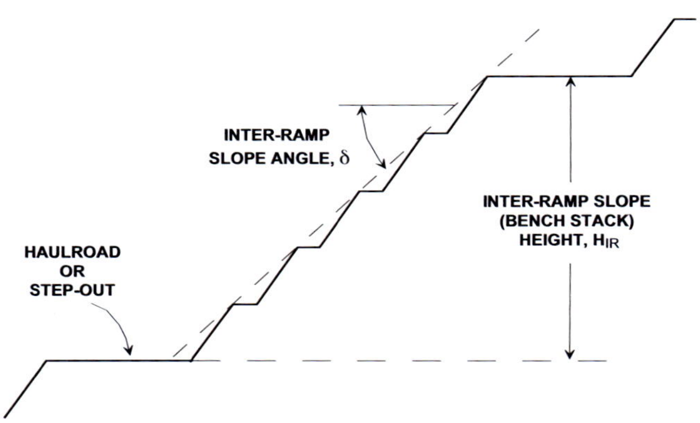

Figure 12.9: Inter-ramp slope geometry parameters

Slope performance assessments should include systematic and periodic documentation of as-built inter-ramp slopes to verify that the design criteria appropriate for the ground conditions are being applied by the mine planners and are correctly implemented by mine operations. Basic as-built inter-ramp slope geometry data can be collected directly in the field using a clinometer. Alternatively, regularly spaced sections, drawn perpendicular to the slope using mine survey plans, that show the locations and elevations of bench crests and toes can be used to measure inter-ramp slope angle and height. If detailed topographic plans or scans of the slope are available, statistical sampling techniques can be used to collected detailed data.

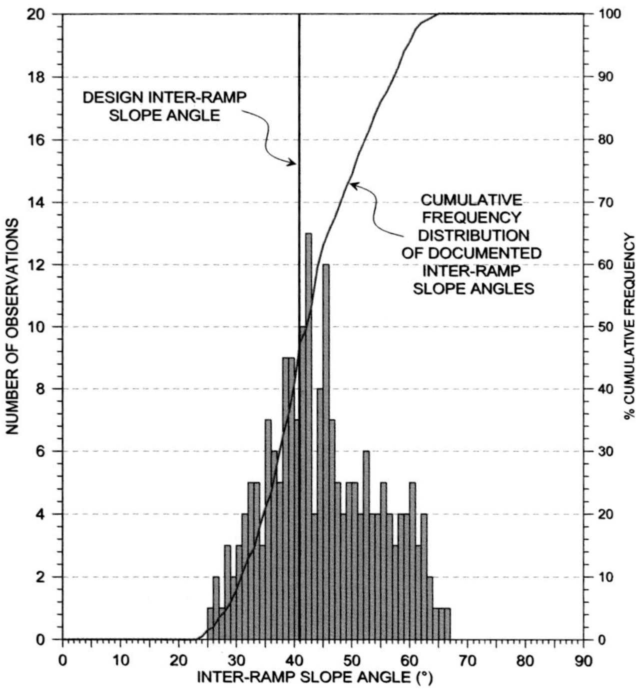

Figure 12.10: Comparison of as-built and design inter-ramp angle

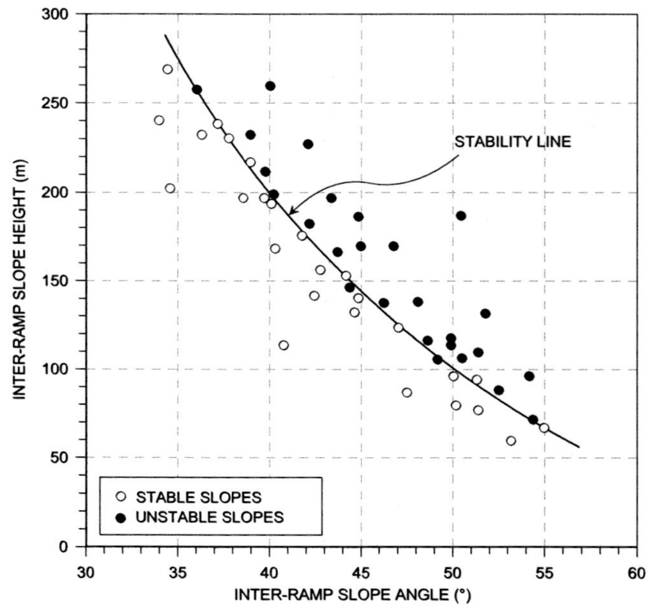

Figure 12.11: Inter-ramp slope height vs inter-ramp slope angle

Simple statistical techniques, such as the cumulative frequency analysis technique outlined in section 12.1.3.3, can be used to evaluate the distribution of inter-ramp slope angles in comparison to the design criteria (Figure 12.10).

The data can also be used to examine the relationship between inter-ramp slope angle and inter-ramp slope height. As illustrated in Figure 12.11, plotting inter-ramp slope angle versus height, and identifying data points that represent sections where multi-bench or inter-ramp slope instabilities have occurred versus stable sections, may provide additional insight into slope behaviour and highlight opportunities for further optimising slope geometry. For example, where as-built inter-ramp slope heights are consistently lower than permitted by the design criteria and excavated slopes are performing well, consideration might be given to steepening the inter-ramp slope. This situation might also result from the shape of the ore body, or for mine planning reasons.

In addition to assessing the degree of compliance with the design criteria, slope performance assessments should evaluate the effectiveness of the inter-ramp slope design. Does the design meet expectations in terms of preventing or controlling inter-ramp scale instability? Is the design too conservative or too aggressive? In some cases, inter-ramp slope performance can be evaluated in terms of objective acceptability criteria such as the frequency or size of multi-bench failures, the cost of cleanup and remedial measures or the frequency and length of disruptions to production. However, these types of assessments typically require detailed and statistically reliable historical records that are often not available. In most cases it is necessary to apply more subjective criteria, such as qualitative assessments about the overall effectiveness of catchment, the accessibility of the slope and whether multi-bench failures are being adequately controlled on the slope.

It is usually easier to identify inter-ramp slope segments that are not performing adequately because they are too steep or too high. The solution to this type of problem could be to flatten (lay back) the slope, either by reducing the height of the bench stack or by increasing the step-out width, both of which affect the stripping ratio or push the toe of the slope off design, reducing the quantity of recoverable ore. Suboptimal inter-ramp slope performance, where the design is too conservative, is more difficult to identify because positive slope performance does not necessarily indicate an inappropriate or overly conservative design. One way of assessing opportunities for steepening is to establish trial slopes where the inter-ramp slope angles and/or heights are incrementally steeper and/or higher than the design. Such trials are best suited for interim or temporary slopes where the consequences of instability are not significant and can be controlled. This may necessitate the inclusion of contingency plans in the trial design.

12.1.4.2 Documentation and back analysis of multi-bench instabilities

For the purposes of this section, multi-bench instabilities include failures that involve more than one bench but are limited to a single inter-ramp section and do not involve the loss or potential loss of a haul road (Figure 12.12).

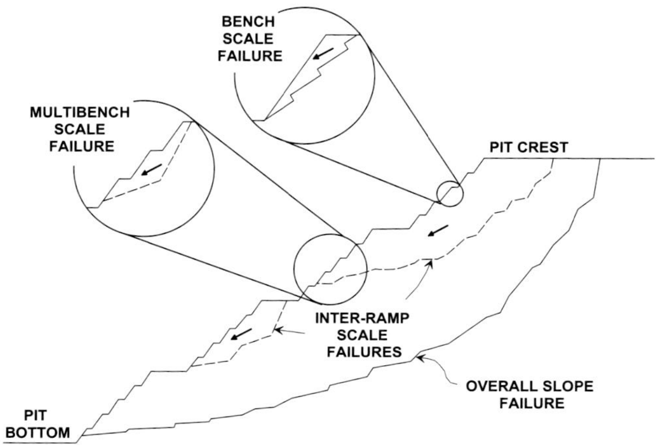

Figure 12.12: Scope and scale of instabilities

Larger-scale failures involving multiple inter-ramp sections or the overall slope are discussed later.

Depending on the height of the inter-ramp slope, the definition typically includes instabilities that range in size from tens of metres to a few hundred metres. Because of this order of magnitude range and the consequent wide range of potential impacts, a one-size-fits-all approach to documentation and back analysis of multi-bench instabilities is inappropriate. Rather, the level of documentation and back analysis should depend on the size and potential impact of a given failure. Small-scale multi-bench failures involving a few benches that do not affect critical infrastructure might be considered using the approach suggested for bench scale failures. Large-scale failures involving the full inter-ramp slope height or affecting critical haul roads or infrastructure require a much higher level of investigation and study.

In addition to basic information identifying the location, date, geometry and type of failure, and the characteristics of any discontinuities involved, documentation of multi-bench failures should include compilation and review of deformation monitoring, bench inspection, blasting and precipitation records. A suggested checklist of things to consider when documenting multi-bench failures is given in Table 12.3.

Depending on their size, the failure mechanism and the nature of the rock mass, multi-bench instabilities may exhibit precursors to failure, such as tension cracks or scarps, accelerating deformation rates or an increase in the frequency or size of rockfalls or ravelling. Ideally, monitoring programs should recognise any multi-bench failures that might present a significant risk to the mining operation far enough ahead of failure that effective mitigative measures can be implemented. Good monitoring records also provide a valuable source of information that may help in understanding the failure mechanism and triggering events.

Stereographic projections, plans, sections and 3D representations that illustrate the geometry of the failure and structural controls may be useful in understanding the mechanism. For more complex, structurally controlled failure mechanisms, simple planar representations of the key structures may be inadequate and it may be necessary to develop and interpret detailed structural contour plans of individual discontinuities. Before-and-after photographs and scans can help visualise and quantify the failure.

Back analysis of multi-bench failures can help to calibrate and refine discontinuity shear strength assumptions. Failures that involve a component of shearing through the rock mass may provide unique opportunities to calibrate rock mass shear strength. The type and extent of back analysis depends on the size and nature of the failure and the amount and reliability of documentation data. Simple limit equilibrium analysis techniques may be sufficient for small-scale or mechanically simple failures, whereas large complex failures may require sophisticated numerical modelling. In some cases, multiple approaches may be needed when assessing the reliability of an analysis technique. Results should be expressed in terms of sensitivity to key input parameters and compared to previous back analyses to validate and refine critical assumptions.

If back analysis reveals significant variations in key shear strength assumptions or material changes to the underlying geological interpretation, it may be necessary to review and revise the slope design criteria.

12.1.5 Overall slope performance

12.1.5.1 Slope documentation and expected behaviour

Overall slopes are usually limited by the inter-ramp design criteria, the shape of the ore body, haul road access, other mine planning considerations or a combination of factors. Overall slope performance assessments may be limited to documenting the as-built slope geometries to ensure that they are in compliance with the design, and routine monitoring to warn of unanticipated deformations, geological complications or developing adverse pore pressures. If there are adverse structural conditions or the rock mass is weak, overall slope stability issues may also control design. In these situations, overall slope performance evaluation is more critical and may require specific instrumentation, more vigilant monitoring and supplementary investigations to confirm the design as the slope is developed.

To evaluate overall slope performance based on instrumentation monitoring, it is important to consider the expected response of the slope to mining, and to set thresholds or targets against which performance can be compared. The expected response will depend on the

Table 12.3: Checklist for documentation of multi-bench failures

General Information
- Identification Number or Code
- Location (Phase, Structural Domain, Wall, Design Sector)
- Limits (Height, Width, Depth, Volume, Mass)
- Date First Recognised
- Initial Manifestations (Cracking, Settlement, Heave, Rockfalls/Ravelling)

Survey and Monitoring Data
- Before and After Photographs
- Before and After Topographic Surveys or Scans
- Crack Maps
- Slope Inspection Records
- Monitoring Plan (Instrument Locations)
- Movement Monitoring Records (Prisms, Extensometers, Inclinometers/TDRs, SSR)
- Piezometer Monitoring Records
- Dewatering Records
- Precipitation Records

Geologic Information
- Drillhole Locations and Logs
- Geologic Maps, Sections and 3D Models (Lithology, Alteration, Structure, Mineralisation)
- Bench Mapping Data
- Stereographic Projections of Structural Fabric
- Contour Plans of Structures Controlling Instability

Geotechnical Information
- Slope Documentation Records
- Geotechnical Maps, Sections and 3D Models
- Discontinuity Shear Strength Criteria
- Rock Mass Classification Information
- Rock Mass Strength Criteria
- In Situ Stress Conditions

Blasting and Seismic Information
- Design Details and Dates for Recent Proximal Blasts
- Blast Monitoring Records
- Blast Damage Reports or Surveys
- Seismic Event Records

Slope Design and Performance
- Bench and Interramp Design Criteria
- Analytical Basis for Design
- Slope Performance Records
- Previous Instability History

Assessment of Failure Mechanism
- Nature of Deformations (Deformation Rate, Movement History, Triggering Mechanisms)
- Structural and Kinematic Controls
- Rock Mass Controls
- Results of Back Analyses

Risk Assessment and Remedial Measures
- Current Status (Stable, Metastable, Unstable)
- Impact on Operations (Current, Future)
- Alternative Remedial Measures
- Cost–Benefit Analyses

geology, the nature of the rock mass and the height and steepness of the slope. Initial predictions of expected response should be prepared during the design stage of the project. For modest slopes, initial predictions might be based on experience with similar slopes or simple modelling. Large slopes or slopes in complicated geological environments may require sophisticated numerical modelling. Such models typically require detailed calibration based on documented response over time. Regardless of the approach used to estimate expected response, adjustments will be required as the slope is developed, to reflect documented slope response.

Maintaining detailed and representative as-built geotechnical sections through each main pit slope is fundamental. The sections can be used to compare the as-built slope geometry to the overall slope design criteria and to qualitatively evaluate the potential impact of variances. Key instruments and monitoring results (e.g. prisms, inclinometers and piezometers) should be shown. Monitoring results should be shown on appropriately scaled plans or 3D visualisations so that the location and extent of zones of abnormal or unexpected response can be identified early. Such plans might illustrate total or component movement vectors, incremental or cumulative deformation magnitudes or rates, piezometric pressure contours or pore pressure dissipation rates and the distribution of microseismic events.

A periodic photographic record of the slope as it develops is strongly advised. Comparison of photographs of overall slopes taken from strategic vantage points can reveal subtle variations in slope behaviour over time that may not be apparent in other types of monitoring. Individual and time-series photographs can help communicate specific performance issues. Targeted video surveillance may be useful in specific cases where instability is anticipated within a reasonably narrow time frame. Some operations run continuous video surveillance of pit walls.

12.1.5.2 Large-scale slope instabilities

Large-scale slope instabilities that involve multiple inter-ramp slope segments or the overall slope can threaten the economic or social viability of a mine, and in rare cases may result in fatalities. It could be argued that the most important objective of any ongoing slope performance assessment program ought to be early recognition of large-scale instability. Early recognition allows mitigative measures to be designed and implemented in time, and human and economic risks to be appropriately managed.

If large-scale instability does develop, documenting the progression is key to understanding the failure mechanism and to developing mitigative or remedial plans. Detailed monitoring and photographic records are critical for reliable calibration and validation of the stability analysis models necessary for rational response plans. Understanding the mode of failure and triggering mechanisms may require detailed analysis of the mining sequence. The impact of blasting, pore pressures, in situ stresses and other factors should be considered, and supplementary investigations may be needed to fill knowledge gaps and validate models.

Understanding, predicting and managing potential large-scale instabilities, and developing effective mitigative and remedial plans, requires a comprehensive holistic approach that considers all factors and is unique to each situation. Specialist advice and external reviews are strongly advised, to ensure that all appropriate steps are taken.

12.1.5.3 Slope depressurisation and pit dewatering

The design of most large slopes requires at least a general understanding of the potential impact of groundwater on the mining operations. Large slopes may be sensitive to piezometric pressures and designs may anticipate natural or enhanced depressurisation of the walls, as outlined in Chapter 6. It is important to monitor changes in piezometric pressure as the slopes develop to ensure that depressurisation targets are met. If targets are not being achieved, planned depressurisation may have to be advanced or increased, or slopes may have to be flattened to maintain stability. If the groundwater flow system is reasonably well-understood and the slopes are appropriately instrumented with piezometers, depressurisation rates can be tracked and results plotted on plans or sections and compared to projections and targets.

Some open pits require only pit bottom sumps with modest pumps to maintain a dry excavation. However, many open pits require some form of in-pit or pit rim well dewatering. Installation of deep dewatering wells can be very expensive, and delaying their installation for as long as possible is usually attractive. This approach requires careful monitoring of groundwater levels and pore pressures to ensure that wells and drain holes are installed and replaced as needed to keep up with mining. Monitoring usually includes sealed piezometers and open standpipes, as well as observations of seeps and local ponding. Hydrographs that track groundwater levels and dewatering rates over time and in relation to pit development, in combination with appropriate plans, sections or 3D representations showing the current water table or piezometric surface contours and profiles, are convenient ways of evaluating performance.

At mines where pit slope dewatering performance is a critical part of the slope management program, it is important that the geotechnical engineer validate the dewatering data to ensure that dewatering targets are being achieved. This usually involves visually checking piezometric levels of the most recent data against critical geotechnical stability sections. If numerous discrepancies exist between actual data and target water levels, a new hydrogeological interpretation should be initiated, or management should be notified that the dewatering system is in non-compliance of target levels (section 6.5).

12.1.6 Summary and conclusions

Open pit slope design is an iterative and often highly empirical process whereby slope designs are based on the available information, implemented according to the established criteria, systematically documented and evaluated, and modified in accordance with observed performance. Slope performance assessments must be customised to suit the unique environment and available resources of each operation, and should include slope documentation and monitoring. Slope documentation programs should provide the information needed to progressively validate and refine the geotechnical models, in particular the underlying geological model. Documentation must address each of the key scale components of the design – benches, inter-ramp slopes and overall slopes. This information should be reviewed periodically and the design criteria and mine plan updated as the mine is developed. The frequency of design reviews will depend on the complexity of the rock mass, the mining rate and the overall performance of the slope.

Our ability to model and understand the mechanics of slope stability has advanced dramatically since the introduction of modern pit slope design methodologies in the 1970s. However, large-scale slope instabilities still occur. In most cases these involve structures or mechanisms that were unanticipated or misunderstood, often because the underlying interpretive geological model was incomplete or incorrect. Comprehensive and rigorous slope monitoring remains the most important tool for assessing overall slope performance.

Where possible, slope performance should be evaluated objectively against expected responses and clear acceptability criteria. However, due to the complexity of the geologic environment and mining process, the slope design and performance evaluation cycle will continue to involve subjective judgment. In this context, there is no substitute for experience and continuity.

12.2 Slope monitoring

12.2.1 Introduction

Practical limitations may force a mine to establish interim and final slope limits and develop final walls with incomplete information. At the same time, as discussed elsewhere in this book, the potential stability of high pit slopes is difficult to predict with available investigation data and analysis techniques, particularly at the design stage. The result is a strong reliance on slope management systems, of which a comprehensive monitoring system must form an integral part.

Monitoring is an invaluable tool for assessing design performance and failure risk, and for aiding risk minimisation. In today's environment, mining companies have a moral and financial obligation to eliminate the potential for accidents, and a legal obligation to protect the workforce. Legislation stipulates that employers must take every precaution practicable to provide a safe working environment. Failure to identify potential hazards and manage the associated risks could result in fines or imprisonment or both. The presence of monitoring instrumentation not only aids hazard and risk identification but reduces any workforce anxiety by confirming that ground conditions are being monitored by experienced and competent personnel.

When the need for a monitoring system is correctly established and the program is properly planned, cost savings may be a direct result. However, the justification for monitoring is not primarily that of cost reduction. In some cases, the program can be valuable in proving that the design is correct and viable. In other cases, instrumentation might show that the design is inadequate, resulting in slope design modification and associated increased mining costs. In all cases, the indirect value of added safety and the avoidance of failure (and remedial costs) will make the instrumentation program cost-effective.

The main objectives of the slope monitoring program can be summarised as:

- maintaining safe operating conditions to protect personnel and equipment;
- providing advance notice of zones of potentially unstable ground so that mine plans can be modified to minimise the impact of slope displacement;
- providing geotechnical information for analysing any slope instability mechanism that develops, designing appropriate remedial action plans and conducting future slope design;
- assessing the performance of the implemented slope design.

A displacement monitoring system should be established as soon as possible during the early stages of mining and maintained throughout the operating life of the open pit. In many cases a monitoring system may be required beyond closure of the pit.

Elements of the program should be aimed at the following basic goals:

- detecting and recording any slope movement as a basis for:
  - → ensuring safety of the operation;

- → establishing the basis for the movement (failure mode);
- → managing instability;
- investigating failures and ongoing instability. Monitoring of instability assists in identifying failure mechanisms, providing crucial data for back analysis and defining appropriate remedial work;
- confirming the design model or providing a basis for assessing and modifying the designs, including specific elements:
  - → geology, including rock type distribution and alteration;
  - → structural model, with consideration of major and minor structures;
  - → rock properties;
  - → groundwater pressures;
  - → in situ stress levels, particularly for high slopes;
- ensuring that the slope design criteria are being achieved in terms of operating procedures.

Monitoring systems and procedures can be designed to meet these objectives. Pre-excavation instrument installation can provide important benchmark data for subsequent monitoring during mining, to validate design assumptions and modify future designs as required. In practice, such long-term instrumentation programs must withstand the ordeals associated with large-scale production mining.

Movement monitoring systems are usually operated by in-house personnel, but occasionally part or all of the system may be contracted out. The strength of a monitoring program depends on the capabilities of the equipment and techniques, and on the people driving the program. The success of the monitoring also depends on support from higher levels of mine management.

12.2.2 Movement monitoring systems

12.2.2.1 Introduction

In near-surface, low-stress environments (pit slopes) where gravitational failures dominate, the large movements associated with rock instability are nearly always preceded by smaller ones that can be detected by sensitive instruments. Thus, movement monitoring gives the most direct and useful measurement of impending instability. However, in highly stressed, massive and brittle ground, displacements up to the point of failure can be small and harder to detect.

Delay intervals between occurrence and detection of movement, and between detection and collapse, depend on the characteristics of the ground and on the sensitivity of the monitoring instruments. In most cases, a warning period of several hours to weeks can be achieved.

A golden rule for the installation of a geotechnical movement monitoring program is that every instrument installed on a project should be selected and placed to assist in answering a specific question. Following this simple rule is the key to successful field instrumentation.

The approach to planning a movement monitoring program should involve the following steps:

- definition of project conditions;
- prediction of all potential mechanisms that could control instability;
- determinination of parameters to be monitored and potential magnitudes;
- establishment of suitable monitoring systems, including instrumentation and location;
- formulation of measurement procedures, including frequency, data collection, processing, interpretation and reporting;
- assignment of tasks for design, construction and operation of systems;
- planning of regular calibration and maintenance;
- establishment of trigger action response plans (TARPs) and associated accountabilities for action to minimise impacts of ground movement.

Guidelines for the design and execution of monitoring programs are presented in section 12.2.3.

Monitoring methods for open pit slopes can be divided into surface and subsurface, with further subdivision into qualitative and quantitative systems. All have their place in specific open pit mine environments and are often related to the potential failure size.

Qualitative systems can include visual inspections – human observations are subjective but can prove invaluable. These can be general inspections to detect the onset of instability (e.g. tension cracks, excessive rockfalls) or be part of the safety aspects of difficult mining situations (spotting for rockfalls). Training operations staff in hazard identification is extremely valuable in slope failure detection and management.

Qualitative assessments by pit production supervisors of the working conditions are often required by regulating bodies (e.g. MSHA) in the form of pre-shift work inspections. Observations of inspections should be documented and passed on to the next shift using a 'red book' or similar documentation method (section 12.3.2.7).

Quantitative systems usually involve instruments measuring surface or subsurface displacements. Typical components are listed below in order of increasing complexity.

Surface displacement:

- visual inspection;
- cross-crack measurements, either manual or by wireline extensometer;
- survey monitoring;
- GPS;

- photogrammetry;
- laser scanning;
- radar, both ground-based and satellite-based (InSAR);
- tiltmeters and electrolevels.

Subsurface components (instruments typically installed in drill holes) can include:

- inclinometers;
- shear strips and time domain reflectometer (TDR) cables;
- extensometers;
- thermistors;
- micro-seismic;
- piezometers.

A slope monitoring system for a large open pit mine generally includes a combination of these, one of which provides the primary monitoring that forms the basis for slope management. For any system, acceleration of displacement is generally the key precursor to slope collapse.

12.2.2.2 Types of instruments

Monitoring instruments usually consists of three basic components:

- a transducer or sensor to measure the property of interest;
- a transmitting system, e.g. rods, electrical cables or telemetry devices to transmit the information to the readout location;
- a readout unit, such as a dial gauge, to give a digital or graphical display of the measured quantity.

There are many variations on this theme. For example, the sensor may be fixed in the ground or built into a portable probe to allow measurements at more than one location. The readout may consist of a display only, or may include facilities for recording data on paper, magnetic tape or disk. The recording may be continuous, in response to a command signal or at preset intervals. The readout may be located close to each instrument station or remote, transmitting the readings to a central control house by a pneumatic, hydraulic or electrical system.

Instruments in common use are not nearly as complex as they might appear, and rely on a few simple measuring sensors. For example, the electrical resistance strain gauge can be used to measure not only strains but also water and ground pressures and the tensions in rock bolts, by including the gauge as part of an appropriately designed measuring transducer.

Mechanical instruments for measuring movement, called displacement or strain gauges, include a simple steel measuring tape, wire or rod fixed to the rock at one end and in contact with a dial gauge micrometer at the other. Mechanical systems are usually the most reliable, as they are simple and free from drift and other sources of inaccuracy that often impair the performance of more complicated devices.

Apart from the human eye, the best known optical instrument is the telescope. Although not a measuring device, a telescope is an essential component in traditional survey theodolites and levels. Magnification of a graduated scale is a direct and simple way of measuring movements to accuracy better than a millimetre over substantial distances. The principle of optical interferometry is employed over much greater distances by an electro-optic distance measuring (EDM) device. This compares transmitted and reflected laser or infrared beam signals and measures distances in terms of the wavelength of the light source.

Temperature measurements are required in geomechanics for temperature compensation to be applied to other types of measurement and thus improve precision, or for more direct reasons, including the definition of permafrost. Measurement is by thermocouple, thermistor or resistance temperature device (RTD).

12.2.2.3 Surface monitoring methods

Slope monitoring for surface expressions of displacement should extend beyond the limit of the possible movement zone to areas known to be stable. In this way, possible surface rebound can be monitored in advance of cracking.

The following is a summary of typical surface monitoring methods.

Visual inspection

An essential part of any slope monitoring system is visual observation, which is qualitative but has the advantage of involving the experience of people working in a particular environment. It has the benefit that all people working in the pit can and should be involved to their degree of capability. Thus, equipment operators should be required to report rockfalls, field supervisors should be involved in the slope inspection process and geotechnical staff should be involved in a regular detailed inspection process.

Simple visual observations may be sufficient. A record of the pattern of cracking in rock around the crest of an open pit mine gives very useful information on the mechanisms and directions of movements. Cracks should be marked with spray paint so that at each subsequent date of observation it will be possible to recognise new cracks and measure the elongation of old ones. Crack patterns and the history of cracking should be recorded on plans and cross-sections just as carefully as more complex types of measurement.

A further step is to fix 'tell-tales' such as glass microscope slides attached to the rock surface with an epoxy adhesive across selected cracks. Cracked tell-tales confirm the continuation of movement and give a useful indication of movement direction and amount. Cement or plaster tell-tales can be more convenient than glass plates when the surface is irregular, but care must be taken to distinguish between cracks caused by ground movements and those caused by plaster shrinkage.

In difficult mining situations, such as mining below unstable ground, it is often prudent to employ one or more spotters in radio contact with the operators working below the slope. The role of the spotter is to warn of the onset of any rockfall or apparent changes in the condition of the slope, and thereby allow personnel to be evacuated from the area. This type of work requires well-trained personnel, since the situation usually involves a high degree of risk. There should also be a clear evacuation plan for personnel working in the area of the slope.

Any visual monitoring program must be supported by instrumentation to provide a quantitative basis for defining any movement.

Cross-crack measurements

Tension cracks at the crest of the slope may be the first sign of instability. If cracks appear at the crest of the slope or elsewhere, their lengths, widths and vertical offsets should be monitored. Crack measurements give clues to the behavior of the entire slope and the direction of movement may often be inferred from the pattern of cracking, particularly by matching the irregular edges of the cracks.

Because crack movements are much easier to observe on a shotcreted surface than on the rock, a thin coating of shotcrete, concrete or plaster can be applied to aid monitoring of ground movements.

Measurements between deeply embedded anchor targets are better in soils and weathered or weak rocks where the surface is friable. Target separation and embedment must be sufficient that the bolts remain securely fixed in the rock or soil. Measurements can be made with a portable tape extensometer. The same bolts can also be used with various types of fixed-in-place electrical displacement or strain gauge, as discussed below. A portable clinometer can be used to measure differential vertical movements to supplement measurements of changes in distance between targets. This consists of a spirit-level bubble mounted on a bar with micrometer adjustment at one end.

Crack measuring pins

A steel pin or timber peg is fixed firmly on each side of a tension crack at selected locations and the distance between the two pins/pegs is measured periodically, using a measuring tape, a vernier calliper or a micrometer to determine the progress of the crack (Figure 12.13).

The technique is easy, straightforward and economical but it is not feasible to take a large number of measurements because of its relatively low efficiency and high labour intensity, especially at sites that are not easy to access. Nevertheless, this technique has proven to be an effective method of gathering data under certain conditions.

Figure 12.13: Cross-crack measurement

An automated system for cross-crack measurement is shown in Figure 12.14. That system involves a vibrating-wire extensometer designed to measure displacement across joints and cracks in concrete, rock, soil and structural members. The switch end and the shaft are attached firmly to steel or timber pegs on each side of the crack. Movement of the shaft changes the tension in an internal wire, causing a corresponding change in its frequency of vibration. The instrument has high accuracy and resolution, and displacement beyond a preset threshold can trigger a visual or audio warning alarm activated through an alarm controller.

Wireline extensometers

A system commonly used for cross-crack measurements in open pits, particularly on waste dumps, is the wireline

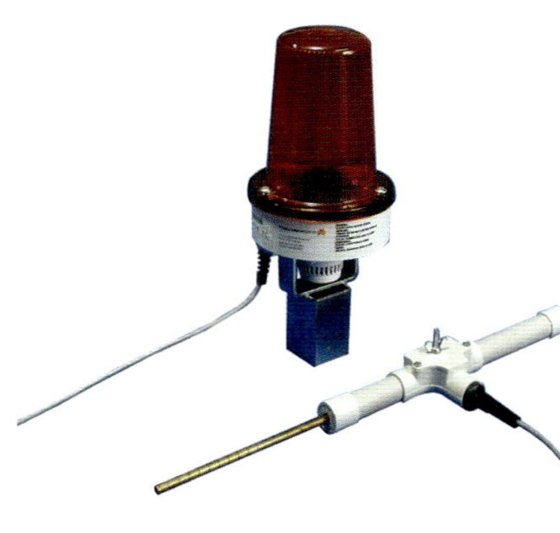

Figure 12.14: Automated crackmeter system
Source: Courtesy Geotechnical Systems Australia Pty Ltd.

Figure 12.15: Wireline extensometers, manual (left) and automated units

extensometer. A wire firmly anchored on one side of a crack or cracked area (normally at or immediately below the slope crest) is passed over a pulley on a tripod and tensioned using a weight (Figure 12.15). The displacement of the weight or movement of the pulley is then measured to assess the displacement across the crack. This technique has the advantage that it can be automated and fitted with an alarm and/or the movement data transmitted to a central control.

The basic instrument for manual reading can be easily fabricated in a mine workshop, and a limit switch can be added to activate a visual or audible alarm if there is more than a specific amount of displacement. If an alarm is activated, mine operators should be given clear instructions on evacuation routes and procedures.

Manual reading of the extensometer can be performed by mine supervision rather than geotechnical staff, as long as they have specific instructions on the assessment of displacement rates and associated trigger levels for alarms. This involves the operations staff in the monitoring program.

Commercially available automated wireline extensometers provide the basis for real-time measurement of displacement as well as data transmission to the central dispatch office or geotechnical staff. Excessive movement rates can be flagged for further investigation or mine evacuation. Such a unit is shown on the right in Figure 12.15.

Survey monitoring

Geodetic survey remains the standard method for monitoring large open pit slopes. The techniques used are traditional methods of general survey measurement and positioning. Mine surveyors are therefore usually more familiar with these techniques than are other professionals such as engineers and geologists.

Depending on the methods and procedures, geodetic techniques can be used to determine the absolute position and the positional variations of selected points on the surface of a pit slope in one to three dimensions.

When using survey techniques, survey instruments such as levels, theodolites, total stations, GPS receivers, photogrammetric cameras or a combination of these instruments are used to collect field data. These data are processed to determine the positions of the surveyed points in a given reference frame such as the reduced level coordinate system. Displacement of the surveyed points can be determined by comparing the coordinates from two or more survey periods.

Conventional survey techniques have proven very useful for monitoring open pit mine slopes and ground subsidence; they also have application for monitoring in underground mines.

Monitoring by geodetic methods requires measurement of horizontal and vertical angles and of distances to a series of targets. These measurements are used to compute the 3D or x, y, z coordinates of points in and around the deforming ground. When measured at intervals, the displacements in 3D can be calculated in relation to reference points in stable ground and provide rates and direction of movement.

Triangulation with precision theodolites was initially used, but the introduction of infrared and laser-based electro-optic distance measuring (EDM) devices in the early 1970s led to the use of trilateration, the principal survey-based slope monitoring system. Instrumentation normally involves a high-precision (one second) theodolite and an EDM, which can be combined into a single unit (total station). For maximum accuracy, the system must be carefully planned and designed by an experienced survey engineer.

Instruments

A wealth of technical literature is available on survey instruments, especially from the manufacturers. The following points are intended to highlight only the major applications of each main type to slope displacement monitoring.

- Levels are occasionally used to determine height variations along the crests, berms or ramps of a pit. In relatively small survey areas, levelling data can provide the best height survey accuracy. The main disadvantage of levelling measurement is that human access is required to all the surveyed points. This can be difficult or impossible in the open pit mining environment.
- Optical mechanical theodolites are now rarely used for slope monitoring, mainly due to their low efficiency and susceptibility to significant atmospheric refraction

errors in observed angles. However, optical electronic theodolites form an integral part of total station units when combined with an EDM.

- EDM equipment is used for measurement of distances, either for direct determination of distance change or for determination of lateral position change by trilateration. Over the last 20 years, increasingly reliable and accurate EDM equipment has radically changed conventional surveying practices. EDM devices require fewer personnel than conventional optical instruments, are faster to use and are more accurate.

  EDM equipment uses the velocity of electromagnetic radiation to measure the distance between the instrument and a reflector prism placed at the measuring point. Some equipment uses microwave radiation (laser), others use infrared or visible light. Since air density has an effect on the velocity of light, air temperature, pressure and humidity must be monitored. In modern EDMs, these factors are monitored and processed internally.

  Precise measurements over distances greater than a few tens of metres (measurements of up to 60 km are possible) require targets in the form of retro-directive optical reflectors/prisms (glass 'corner cubes' usually 60 mm diameter). These are best fixed permanently to the ground. For shorter distances, it is often sufficient to use the cheaper alternatives of truck reflectors or adhesive reflecting tape. Problems can develop if these cheaper targets rotate even slightly from the line of sight or become coated by dust.

  Depending on the model, an EDM can have a range of up to several kilometres. Most instruments have two components of error: a random error plus a small percentage of the sight length. For example, the unit shown in Figure 12.16 has a manufacturer-quoted accuracy of 1 mm + 1 ppm. However, this depends on atmospheric conditions.

  To achieve high precision with EDM instruments, they must be frequently calibrated for zero and scale (frequency of modulation) corrections. The zero correction (an additive constant) given by the manufacturer usually changes with time, and may also be a function of the intensity of the reflected signal. In some older EDM instruments, the zero correction may demonstrate phase-dependent cyclical changes. Each instrument should be calibrated at least twice a year, following special procedures described by the manufacturer. The calibration must account for all combinations of EDM reflector pairings, because each reflector may also have a different additive constant correction.

  Reflectorless EDM systems have been developed, primarily for volume measurements. While at present this type of system does not appear to have the accuracy required for slope monitoring, there is potential for such an application.
- Electronic total stations are the most commonly used survey instruments for pit slope monitoring. The instrument is designed to survey the 3D coordinates of reflective prisms located around the slope being monitored. Readings (distances and/or angles) are usually taken from a fixed instrument station on the crest of a pit to all prisms in view. The prism locations and movements are then computed from the readings. A typical total station instrument incorporating the EDM and an electronic theodolite into a single unit is shown in Figure 12.16.
- Servo-driven electronic total stations are increasingly popular in the mining industry for their high survey efficiency and accuracy. These can be programmed to perform preset surveying schedules with automatic detection of target prisms distributed across a pit wall. Therefore, much less human involvement is required in the field. This is a great advantage as manual observations are usually repetitive, labour-intensive and time-consuming. The system can operate continuously, if required.

The EDM instruments often include an automatic data capture and storage system. The data can be manually downloaded or automatically transmitted to a computer adjacent to the unit or by telemetry to a central server. Software is available from manufacturers or specific

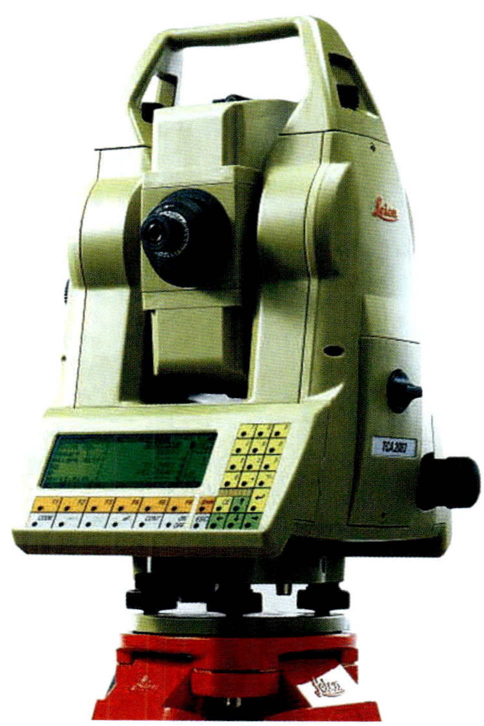

Figure 12.16: Leica TCA2300 universal total station

Figure 12.17: Monitoring shack containing total station unit

Figure 12.18: GPS survey equipment

suppliers that can analyse the data and trigger an alarm or notify key personnel (the mine dispatcher and/or geotechnical engineer) if movement beyond a specified level is detected. Depending upon the environment, it is usual for the total station unit to be housed in a protective building immediately behind a pit crest (Figure 12.17).

Atmospheric influences on survey observations are a major source of errors for survey monitoring of open pit slopes. Atmospheric refraction errors are introduced when a line of sight passes through the atmosphere with an uneven density distribution. The errors vary with temperature, pressure and humidity, and can be adjusted at the instrument. However, where the line of sight crosses a pit, it is difficult to compensate for factors such as temperature inversions and dust. Typically, distance measurements are less affected by atmospheric conditions than angles, and horizontal angles are usually less affected than vertical angles. This must be taken into account when assessing indicated movements, as discussed below.

GPS surveying systems

Global positioning systems (GPS) based on satellites orbiting the earth can be used for real-time positioning at any location 24 hours a day in any weather. The positioning is accomplished through the use of timing signals transmitted by the satellites to ground receivers.

With two or more receivers working simultaneously in a so-called differential mode, relative positions (3D coordinate differences) between the receivers can be measured with an accuracy of a few millimetres to about 20 mm over distances up to several kilometres, and about 1 ppm over distances up to several hundred kilometres.

A GPS receiver requires an unobstructed view of at least four satellites. It requires three satellites to determine its horizontal (2D) position, and a fourth satellite to determine its altitude (3D).

Unlike conventional survey techniques such as those using EDM, total stations and levels, GPS does not require a direct line of sight between survey stations. This is useful in the open pit mining environment as sight to stable reference stations is often unavailable. GPS is not affected by local atmospheric conditions when the GPS baseline length is within 1 km, so GPS is usually more efficient and accurate, and requires less labour than conventional survey techniques. In consequence, GPS has been adopted as the general surveying technique in many mines. The advantages also make it an ideal tool for setting up control surveys for slope monitoring.

There are certain limitations to GPS that affect its application to routine slope monitoring, including the following.

- GPS surveys require survey personnel to physically access all the surveyed points unless GPS receivers are permanently affixed on the surface of the monitored pit slope. In some cases slope surfaces can be difficult to access at various stages of the pit life. Therefore GPS receivers on a pit slope may not be practical due to the very high cost of the equipment.
- Surveys using GPS may be slower than using a total station system if travel to each survey point is required.
- GPS receivers require direct line of sight to satellites, which may be blocked, thus restricting satellite geometry and affecting GPS positioning accuracy.
- A pit surface may introduce multi-path errors.

Multi-path is when a GPS satellite signal bounces off an object before reaching a receiver. This makes the time of signal arrival different from what it would have been with no reflection. Multi-path causes degraded accuracy because GPS receivers base their readings on the timing of the signals from the satellites.

The main applications for GPS systems in an open pit are therefore typically in the areas of:

- monitoring waste dumps,
- providing high accuracy control for survey monitoring base stations.

For standard static GPS surveys, the effect of poor GPS satellite geometry and multi-path problems can be largely compensated by increasing the time over which field observations are made. However, when more efficient techniques such as the fast static method are used, the decrease in the survey accuracy can become very obvious. Tests have shown that multi-path effects can easily contribute to a few millimetres of positioning errors.

Achieving greater accuracies, such as those necessary for mining applications, requires the use of a relative positioning technique such as differential GPS or DGPS. The principle of DGPS involves a GPS receiver placed at a known fixed position and allowed to capture GPS satellite signals. The receiver's computed position is compared to its known position to determine the measurement inaccuracies. Error corrections, known as differential corrections, are then broadcast by radio link to the receivers on the rovers (in mining: trucks, shovels, drills, etc.) to fix their calculations and bring accuracy to less than a metre.

Obtaining even higher precision requires similar but more complex techniques which allow accuracy in the order of a few millimetres, with good satellite geometries. Real-time kinematic (RTK) systems achieve these accuracies on-the-fly. Because the associated errors are not constant, GPS systems must broadcast DGPS or RTK corrections at a certain minimum rate to achieve the desired accuracy. For precision of 1–2 m, corrections every 10–20 sec are adequate. Systems designed to achieve accuracy in the range of 1–2 cm require corrections each second.

The distance from the base station to the rover may also be an issue. For a properly installed RTK system with a completely accurate base station antenna position, the error is usually 1–10 ppm of horizontal distance, with a limit of the baseline between base and rover of 10–50 km. Thus, at 10 km the error will be 1–10 cm in addition to any other errors due to geometry.

At least one remote GPS system is available. These units use solar panels for power and telemetry to send results to a base computer, which can also provide alarms to a control. It works well, but requires frequent sunlight for power and must be accessible so that the solar panels can be cleaned regularly.

Photogrammetry

Traditionally, photogrammetric methods used film cameras and analytic instruments and were slow and expensive. Photogrammetric methods for mapping structure and measuring structural movements evolved rapidly with the development of digital cameras and better software and computers.

The following discussion of traditional methods is included only for completeness.

In the past, precise photography for measuring structural movements employed photo-theodolites to take successive photographs from a fixed station along a fixed baseline. Movements were identified in a stereocomparator by steroscopic advance or recession of pairs of photographic images in relation to stable background elements. The procedure defined components of movement taking place in the plane of the photograph.

Digital photogrammetry has effectively replaced the traditional analog and analytical photogrammetric techniques. Digital photogrammetry can greatly reduce the turnaround time of photogrammetric measurements, enabling on-line real or near-real time 3D coordinate measurements. Modern digital photogrammetry systems produce dense 3D point clouds that are integrated with the image data to deliver 3D images which can be visualised and used for measurements on standard personal computers.

Digital photogrammetry has the advantage that hundreds of thousands of potential movements may be recorded on a single stereo photographic pair, and many times that in multiple overlapping stereo pairs, allowing appraisal of the overall displacement pattern over an extended area in minimum time.

The methods employed in photogrammetry are based on triangulation and involve the intersection of lines of sight. To calculate the relative position of a point, the focal length, position and orientation of the camera must be determined for each image, as well as the elevation of the ends and the length of the baseline. Much detail of the slope surface can be surveyed quickly as the technique does not require significant time.

Measurement accuracy depends on many factors. The baseline should be as long as topography permits, not less than approximately one-eighth the sight distance and as near to perpendicular to the line of sight as practicable. Disparity measurements are usually made by the software to micron accuracy within an image.

When using photogrammetry the user must be aware of the difference between accuracy and precision in the measurements:

- accuracy is a quantitative estimate of the difference between a measurement of a parameter and the true value of the parameter;

- precision is a measure of the random variation of the measurement of a parameter, often expressed as the root mean square (rms) deviation about a mean value, usually assumed to be the true value. Repeatability is sometimes used with precision or in the same context.

Accuracy is a function of many factors ranging from the quality of survey data to the calibration of the camera. Precision, sometimes referred to as standard error, is primarily a function of camera resolution and the length of baseline relative to the sight distance. Both degrade rapidly once the baseline length is reduced to less than about one-eighth of the sight distance.

Both accuracy and precision are highly dependent on the focal length of the lens.

In general, the precision/standard error of measurement vary, but with calibrated high-resolution digital SLR cameras precision of 1:10 000 to 1:20 000 is routinely achievable. This corresponds to 5–10 mm at 100 m (0.016– 0.033 ft at 330 feet) with a 35 mm lens.

Typical accuracy of stereo photogrammetric measurement varies normally from 1:5000 to 1:100 000 of the sight distance. Recent research shows that the accuracy of digital photogrammetry routinely exceeds 1:50 000 and is fast approaching 1:100 000. Recent work with aerial photogrammetry has demonstrated mean positional errors on a set of control points of 1:100 000 at a flying height of 1500 m.

Photogrammetric methods offer very high field efficiency, i.e. a great number of points can be surveyed simultaneously. In addition, photographic images can be stored permanently for future reference and analysis. One of the major advances of digital photogrammetry is that the special equipment and skills necessary for traditional photogrammetric methods are not necessary for photogrammetric measurement. The extended times required for analog and analytical techniques to process the photographs and to carry out the necessary measurements area are a thing of the past, with digitisation. These are among the reasons why the use of photogrammetric techniques in pit slope monitoring was very limited. With recent developments, digital photogrammetry could have a much wider role in slope monitoring.

Photogrammetric methods of surface contouring can be as precise as conventional surveying or electro-optic distance measurements, and have the great advantage of covering a complete field of view rather than a set of prelocated targets. Hence, it is not necessary to forecast the locations or directions of potential movements. A sequence of photographs taken at suitable time intervals can be compared to detect movements wherever they might develop. Inaccessible faces can be surveyed without interfering with mining. Ground topography can be quantified as a basis for geotechanical mapping. Useful permanent records are obtained, to permit back-analysis of slides and rockfalls.

Determination of absolute positions requires a control survey using conventional methods to locate selected points in the photograph. This can be avoided if movement measurements alone are sufficient, provided that the camera stations are stable and permanent.

Laser scanning (LiDAR)

Light detection and ranging (LiDAR) technology using terrestrial laser scanning (TLS) is finding application in the mapping of the topography of pit faces and in slope monitoring. Laser scanning has similarities to radar scanning but, because of the difference in the wavelength of the signal and repeatability of measurement, is less accurate in a reflectorless mode. However, when calibrated against known reference targets that have been located by first order surveys, a 3D image with accuracy in the centimetre range can be achieved at a range of up to 2000 m. Longer range scanners (up to 6 km) are also available, but the accuracy drops somewhat.

Several survey equipment manufacturers produce laser scanning systems. A typical instrument is shown in Figure 12.19.

Because of the lower accuracy compared to radar, laser scanning is generally not used for slope monitoring. However, it is used increasingly for defining the topography of inaccessible slopes. For example, a failure

Figure 12.19: Laser scanning unit (LPM-2K in use for volcano monitoring)

can be scanned remotely to the degree of accuracy required both to establish the size of the failure and to define the topography of the failed mass. LiDAR also has application for monitoring large waste dumps post closure, as well as natural landslides.

As the technology improves and accuracy increases it is possible that laser scanning could find more extensive application in slope monitoring.

Radar

Radar has been adopted as a tool for monitoring slope movements and is now used in many open pit mines around the world. The system has the advantage that an entire specified section of wall can be monitored remotely in near-real time without the use of reflectors and regardless of atmospheric conditions. Movements can be detected to millimetre accuracy, depending on the range.

Basic units with a range of approximately 800 m (and providing submillimetre displacement accuracy) are in use in many open pit mines. Further developments have provided units with enhanced range of up to 1800 m. Figure 12.20 shows a typical unit.

The system works by recording the time taken for a signal of known amplitude, wavelength and velocity to be sent, reflected and received from the subject wall. Interferometry, a signal processing technique, assesses the phase information inherent in the reflected radar signal in order to achieve the desired level of accuracy.

Scanning of a target area defined on a slope takes approximately 15 min or less. The size of the pixels scanned in each pass depends partly on the distance of the radar unit from the face; typically they are a maximum of a few metres.

The data from the radar unit are transmitted to computers in a central office via telemetry, where they are reduced and plotted for immediate visual review. Data may also be viewed at the unit itself. By comparing the signals received from each radar pass, movement in the line of sight is determined and reduced to plots of the face showing the amounts of relative displacement. Displacement plots over specified periods of time can also be generated (Figure 12.21).

The main use of the system to date has been for operational safety (production-critical monitoring). Its rapid response and total area coverage makes it ideal for monitoring areas of slope instability or anticipated instability with mining operations below. The software for operating the radar and reducing the data can trigger alarms at levels pre-specified by the mine geotechnical personnel.

Because the 3D aspects of movement are not monitored the system is less useful for defining the mode of instability, although it can provide a more accurate determination of the extent of the moving mass. Radar is therefore frequently used in combination with a survey monitoring system, which can not only define the sense of displacement but can also cover a larger area from a greater distance.

Figure 12.20: Slope stability radar unit

Satellite imaging subsidence monitoring (InSAR)

Interferometric synthetic aperture radar analysis (InSAR) is a technique that uses the differences in phase between successive SAR images, which can be acquired by aircraft or satellite.

SAR sensors send out signals of microwave energy at a specific wavelength. Some of this energy is absorbed, but most bounces off a surface and is recorded back at the sensor. Data are then downloaded and analysed.

When phase information is compared between subsequent images, an interferogram is produced to show the differences. Phase differences include changes in elevation or motion at the earth's surface, baseline effects (differences in the positions of the satellite at the time of image acquisition) and atmospheric noise components.

This system works continuously and in all weather conditions. It can be used to detect very small (cm to mm) movements (Figure 12.22).

There are two approaches to InSAR analysis:

- differential InSAR (DInSAR);
- permanent Scatterer InSAR (PSInSAR).

Differential InSAR (DInSAR) infers a measurement of surface motion, by accurately measuring the phase differences for each data pixel in two successive images. This approach can be used to produce a high-resolution motion field. DInSAR requires high coherence between image acquisitions, making it well-suited to areas with a stable surface environment. Arid regions are ideal for DInSAR applications.

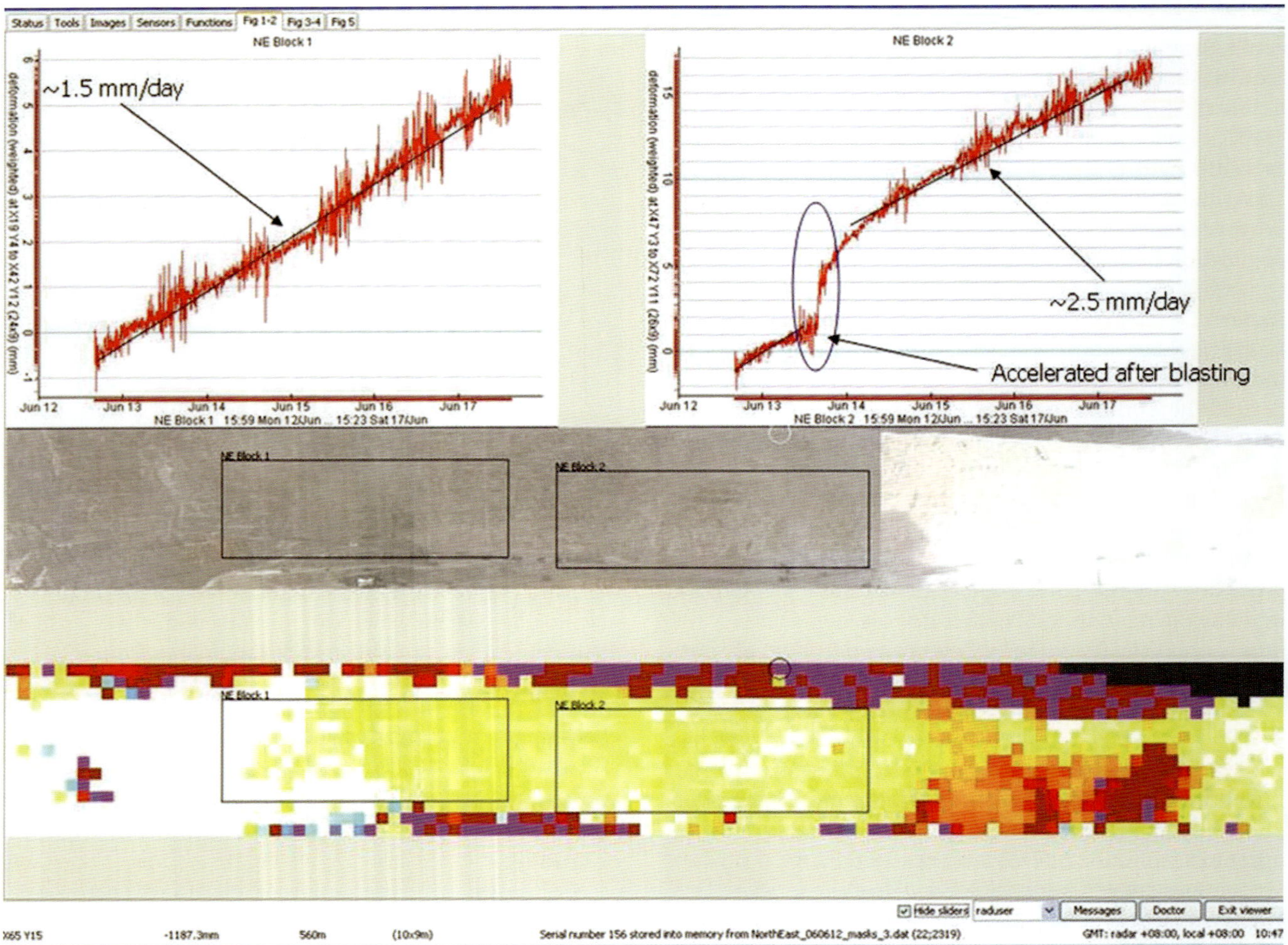

Figure 12.21: Output from slope monitoring radar showing areas of movement and displacement plots

DInSAR requires an accurate DEM to remove the topographic phase component, and is subject to layover and shadow. These factors can limit its utility over steep terrain. It cannot be used to remove the atmospheric noise component. Its effectiveness is limited over areas of poor coherence, such as surfaces constantly changing with vegetation or snow.

Permanent Scatterer InSAR (PSInSAR) provides a measurement of phase differences of permanent scatterers that produce a stable SAR signal return over time. PSInSAR is ideal for areas that have a high density of permanent scatterers, typically man-made structures such as buildings, dams, and towers. Certain natural features, such as exposed rock, may also be used.

PSInSAR is effective in areas with poor coherence and can provide millimetre accuracy in relative measurement. However, to eliminate atmospheric noise and improve accuracy, PSInSAR requires more images than DInSAR, usually a 'stack' of about 15, or the use of known stable reference points. Stable local reference points can be used to eliminate local atmospheric noise and achieve millimetre accuracy with as few as five images.

Other limitations of InSAR include the following:

- InSAR is not particularly effective at determining deformation/subsidence over areas less than approximately 100 metres square.
- if the slope pitch is too steep, InSAR may not provide good results. The SAR satellites are in polar orbits and therefore they are more effective at assessing movement where the direction is not directly north/south facing.
- the satellite orbit periods vary and the method depends on comparison of consecutive images so the data is not real time.
- InSAR is very useful technique to detect a small surface motion, however limiting factors include:
  - → changes in ground cover (cultivation, irrigation, snow cover and vegetation growth) and occasionally atmospheric issues between data acquisitions.
  - → significant ground movement. InSAR detects ground movement through the comparison of two radar images by converting phase difference between the radar signatures. If the total amount of deformation between two images is in excess of

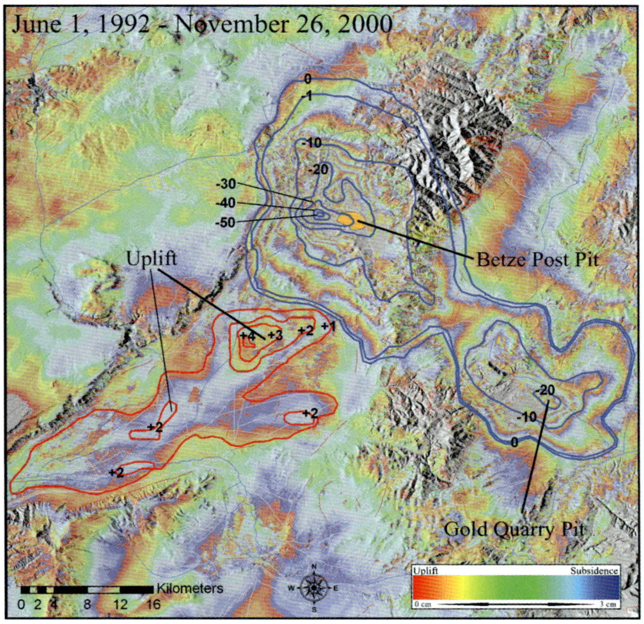

Figure 12.22: A SAR-generated interferogram is overlain on a height-displacement map with aerial photo as background

5.5 cm or 23 cm (depending on the satellite used) then InSAR may not be able to quantify deformation.

These limitations restrict the use of InSAR for slope monitoring. However, the system has found extensive use for monitoring such factors as ground subsidence and post closure movement. Figure 12.22 shows ground subsidence (blue contours) and ground uplift (pink contours) due to the water management activities associated with the open pit mining activities in the Carlin Trend of Nevada. Subsidence is associated with groundwater withdrawal from the carbonate aquifer and uplift is associated with groundwater re-injection and the irrigation program in lower Boulder Valley. The InSAR image not only confirms the field water level monitoring results but also the projection of the regional groundwater flow model. It clearly identifies the maximum areal extent of subsidence by mine dewatering. There is no drawdown beyond the 0 subsidence contour.

Tiltmeters and electrolevels

For slopes in rock, monitoring the tilt of critical blocks can provide an assessment of stability if the deformation has a rotational component. Tiltmeters with electrolytic level transducers provide the most precise data, and the high precision allows trends to be determined in a minimum time period. Multi-point liquid level gauges have been installed on benches of large excavations in rock where there is concern for a wedge failure; the instruments are used to monitor vertical deformation and are intended to forewarn of any instability.

Tiltmeters are very sensitive instruments for determining the change in elevation between two points. Resolutions down to 1 sec of arc are possible. In geomechanics applications, portable tiltmeters usually make use of servo-accelerometer sensors identical to those in the better known drill hole inclinometer instruments described below.

An array of tiltmeters may successfully replace precise geodetic levelling in ground subsidence studies.

There are three types based on physical operation principles:

- electrolytic tiltmeters (also called electrolevels), which use the spirit level principle. The spirit level vial contains three electrodes linked in an AC-energised Wheatstone bridge type circuit;
- vertical pendula;
- accelerometers in which the electric signal is proportional to the sine of the angle of tilt.

In all three types, the tilt is converted into an electrical signal with a typical voltage output of a ± 5 V range.

There are many factors affecting the accuracy of tilt sensing, besides the resolution of the readout. A temperature change produces dimensional changes of the mechanical components and changes in the viscosity of liquid in the electrolevels and the damping oil in the pendulum type tiltmeters. Also, the electrical characteristics change with temperature. Drifts of tilt indications and fluctuations of the readout may occur. Most of the errors may be compensated for a certain cost or their effect may be made linear, thus allowing easy calibration.

The tiltmeter instrumentation is generally intended for semi-permanent installation to monitor angular deformations of the ground. These sensors can be rapidly deployed in drill holes or attached to surfaces or structures to monitor ground movements that manifest as angular rotations over time.

12.2.2.4 Subsurface monitoring methods

Surface monitoring systems are less expensive than subsurface systems, primarily due to the requirement for drilling. However, where it is necessary to locate the sliding surface or examine propagation of subsurface movement, subsurface instrumentation gives additional, more reliable and more detailed information. Subsurface deformation measurements may also be required where it is anticipated that movement may not be detected by surface instrumentation.

For slopes in soil and overburden, inclinometers are the instruments of choice for subsurface measurements, although shear plane indicators can be used for crude measurements. Slope extensometers may be preferred if deformation is predicted to occur along thin shear zones.

Critical movements for slopes in rock are often smaller than those for movements of slopes in soil, and are typically confined to specific surfaces. Therefore instrumentation to detect shearing movement can be more appropriate and the required accuracy of deformation measurements is generally greater. These requirements tend to favour TDR cables and other shear indicators, fixed drill hole extensometers and in-place inclinometers, although removable inclinometers are also used, particularly where rock mass failures are anticipated.

Multiple extensometers, TDR cables and in-place inclinometers can provide real-time monitoring of subsurface deformation, and can be connected to alarms if required.

A discussion of the types of subsurface monitoring systems in use in open pit slopes is presented in the following sections.

Inclinometers

Inclinometers are used to measure the subsurface lateral displacement of discontinuities within soil or rock, as intersected in a drill hole. Two types are available, one of which involves a portable probe that is lowered down a grooved casing installed in a drill hole and measures deformation over the length of the hole relative to the bottom of the hole. These units can operate in holes inclined as much as 50° from the vertical. The other consists of one or more probes that are left in place in a drill hole and record lateral displacement to a remote sensor.

Portable inclinometer probes

The most common type of inclinometer used in open pit slope monitoring is the probe inclinometer, which must be inserted in the hole each time a set of readings is required (Figure 12.23). The probe travels along guide grooves in an aluminium or plastic tube grouted into a drill hole. Probe inclinometers generally operate in holes inclined up to 30° to the vertical; horizontal versions are also available. They can detect differential movements of 0.5–1.0 mm per 10 m length of hole.

The portable inclinometer probe is usually lowered through a guide casing to the base of the drill hole. The probe is then pulled up while the inclination information of the probe in two orthogonal planes is registered at certain intervals. From this information,

Figure 12.23: Portable inclinometer probe

profiles of the drill hole in the two planes can be derived and reviewed graphically. The lateral displacements of the drill hole can be determined by comparing the measured profiles of the bore hole obtained at different times. Drill holes up to 300 m or greater in depth can be measured with inclinometers.

In practice it is usual to extend a drill hole into stable ground to have a common reference point to compare drill hole profiles for determining displacements. In addition, for accurate, absolute readings to indicate direction of movement, the casing must be measured for spiral deflection, which is used to adjust the subsequent inclinometer readings.

A series of drill hole surveys over a period of time gives an indication of the deflection of the drill hole and surrounding ground relative to the base of the hole (Figure 12.24).

Inclinometers are ideal to measure lateral displacements within a pit slope. However, measurement with inclinometers cannot be fully automated and it is difficult to link them to alarm systems. In addition, where the displacement is taking place on a single plane, the inclinometer casing will shear at that point, cutting off access for the probe below that level.

Where movement is expected across a single plane, a simple form of probe inclinometer can be used (Figure 12.25). A thin-wall PVC pipe is installed in a nominally vertical drill hole and the shearing depth is determined by inserting a rigid rod and measuring the depths at which the rod stops. A rod will stop when it cannot pass a bend or break in the pipe resulting from horizontal movement. The device is normally used as a failure plane indicator, and is referred to as a shear probe.

In-place inclinometers

Real-time changes in the inclination or curvature of a drill hole can be measured by an in-place inclinometer system. These are normally lowered into the hole in grooved casing; multiple units can be installed in the same casing to provide relative displacements over sections of the hole. It should be noted that, unless previously modified, in-place inclinometers usually work well only in a near-vertical mode. Readings are interpreted in terms of displacements perpendicular to the drill hole axis.

One type of fixed-in-place inclinometer, called a chain deflectometer, consists of a chain of pivoted rods. Rotations are measured at the pivots between each pair of rods, by means of resistance strain gauges on flexible steel strips, or inductance or capacitance transducers. Another type uses a series of gravity-sensing transducers.

Shear strips and time domain reflectometers (TDRs)

Both of these devices detect the location, but not the magnitude, of deformation in a drill hole. The shear strip

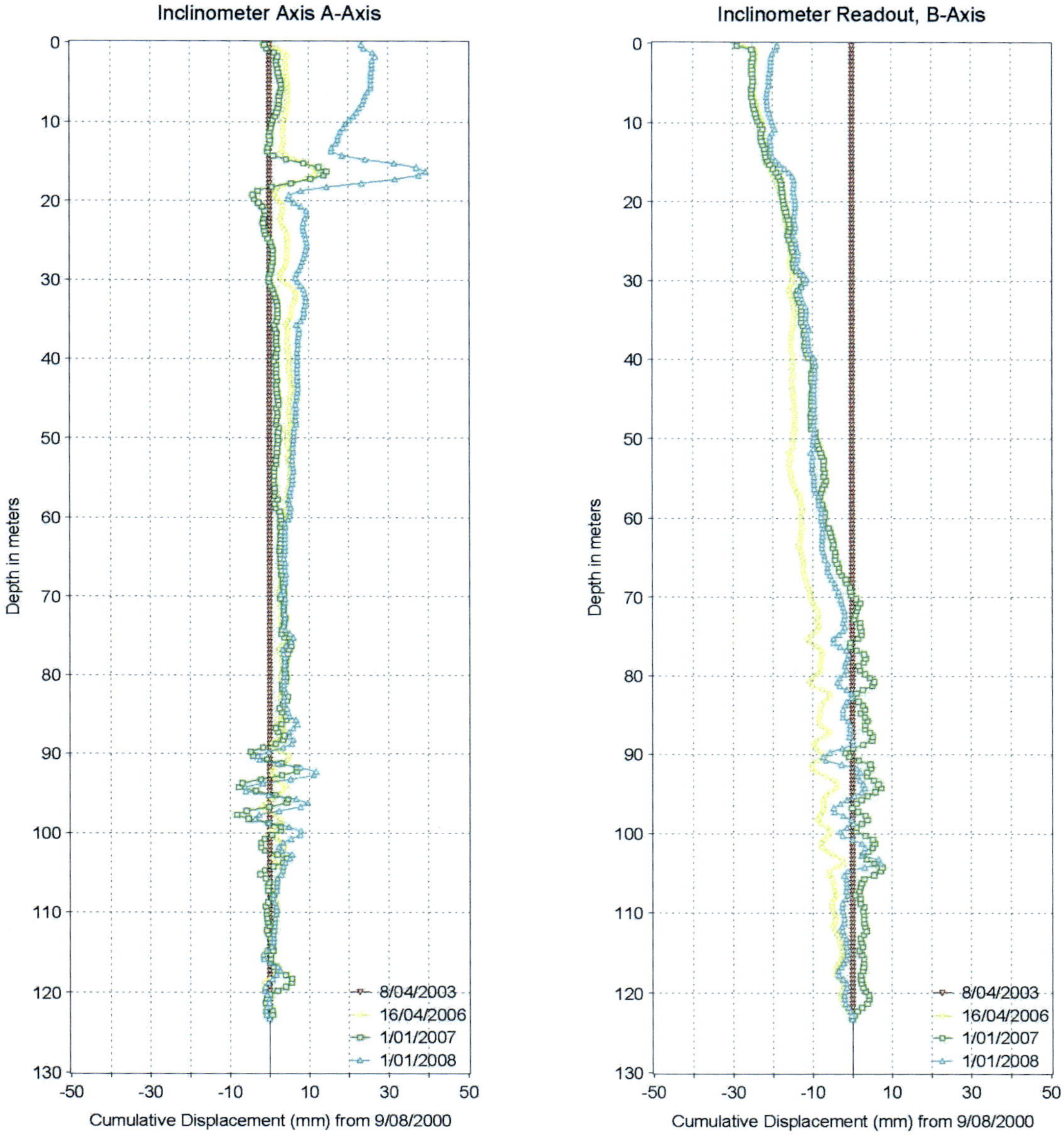

Figure 12.24: Typical inclinometer survey of drill hole

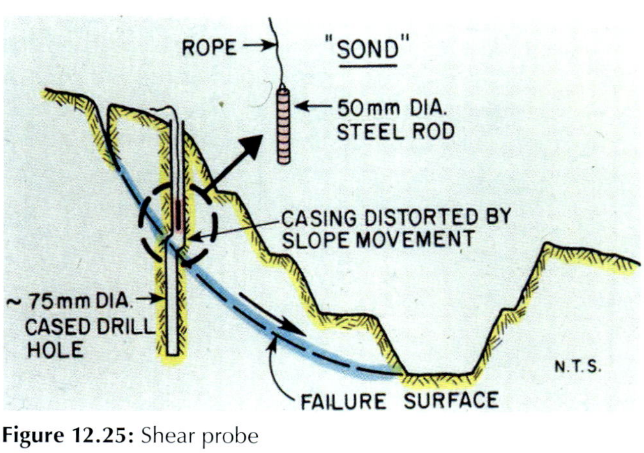

Figure 12.25: Shear probe

is a string of electrical resistors connected in parallel at regular intervals along the length of a printed circuit conductor, and grouted in the hole. Localised movement breaks the strip.

The location of the break is determined from the change in resistance measured at the drill hole collar. Typical applications are to monitor the migration of caving upward from the roof of a mine stope, or the depth of shearing beneath a landslide.

Time domain reflectometry (TDR), which is similar in concept to a shear strip but more versatile and less expensive, is an electrical pulse-testing technique developed to locate breaks in power transmission cables. Coaxial cables are installed in drill holes or along tunnels or mine roadways and anchored or grouted to the rock so that any movement will pinch or break the cable. Indications are that cable damage occurs at about 2.8 mm of strain per 30 m length of cable between anchors. Cable defects such as crimps, short circuits or breaks are detected by transmitting voltage pulses from a cable tester. The distance to the cable fault is proportional to the elapsed time between transmission and arrival of a reflected signal. Accuracy is about 2% of the distance between the tester and the cable break. This can be improved by pre-crimping the cable at regular, typically 10–20 m intervals: the crimps distort the signal and act as markers on the arrival waveform, to which any further distortions or breaks may be related.

For holes longer than 30 m, TDR cables are less expensive and easier to install than multiple point extensometers. Systems can be combined with dataloggers and multiplexers to allow remote readings of multiple cables. TDR cables can also be fitted with an alarm system.

Fixed extensometers

Fixed extensometers are installed in drill holes to measure the movements in the rock mass. In contrast to inclinometers, which measure movements in directions perpendicular to the hole, they measure movement along the axis of the hole (e.g. settlement when the hole is vertical). Simple single-position extensometers usually consist of an end-anchored rod equipped with a measuring facility at the hole collar. Multiple-position extensometers record differential movements of anchor points installed at various depths from which either rods (Figure 12.26) or stainless steel wire ropes (Figure 12.27a) extend to the measuring facility at the hole collar. The anchors are grouted into position, with the connecting rods or wires being sheathed in polythene tubing so that they are free to move within their sleeves. Usually, multiple installations are made sufficiently deep for the deepest anchor to act as a stable reference. As a further reference check, the movement of the drill hole collar is monitored with respect to an external datum. Rod extensometers are untensioned, but wire extensometers are tensioned either by a suspended weight or from a spring-loaded headframe at the hole collar (Figure 12.27b). The measuring precision of wire extensometers can be limited by loss of tension and the potential for creep and kinking, so they are often used mainly for economy where considerable movement is expected and some loss of precision is expected.

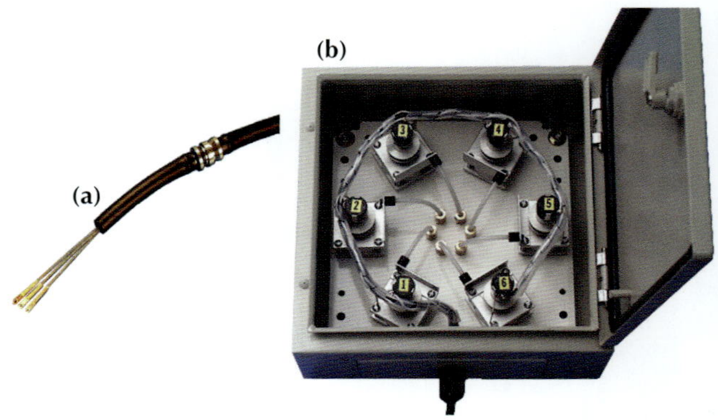

Figure 12.27: (a) Multiple point wire extensometer; (b) Multiple point wire extensometer headframe
Source: Courtesy Geotechnical Systems Australia Pty Ltd.

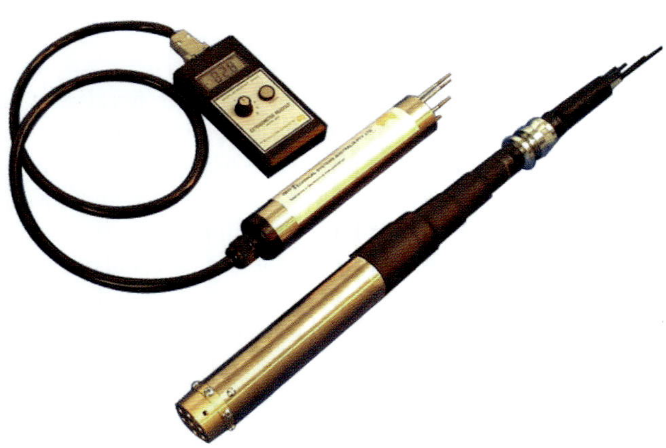

Figure 12.26: Multiple point rod extensometer with digital reading measurement indicator
Source: Courtesy Geotechnical Systems Australia Pty Ltd.

Rod extensometers can be read with a portable dial gauge or digital indicator (Figure 12.26, upper) or a fixed electrical transducer if continuous monitoring and automatic warning is required. Wire extensometers can be read from a remote site, data logged or read at the collar using a digital multi-channel readout.

Multiple electrical extensometers (Figure 12.28) are an alternative to both multiple rod and wire extensometers; they have the advantages of high accuracy and permitting a large number of measuring points in a single drill hole. The instrument consists of four main components: the mechanical anchor, the measurement module, the extension tubing and the centralisers. Unlike multiple rod and wire types, the instrument measures relative displacement between adjacent anchors rather than transmitting all anchor displacements to a collar reference head.

Resistance wire extensometers

CSIRO in Australia developed a resistance wire extensometer (RWE) to measure rock deformation across sheared or weak seams and zones (Figure 12.29, lower). This is a 1 m long strain gauge anchored or grouted into a drill hole. Characteristics of range, sensitivity, resistance and gauge factor are similar to those of a normal strain

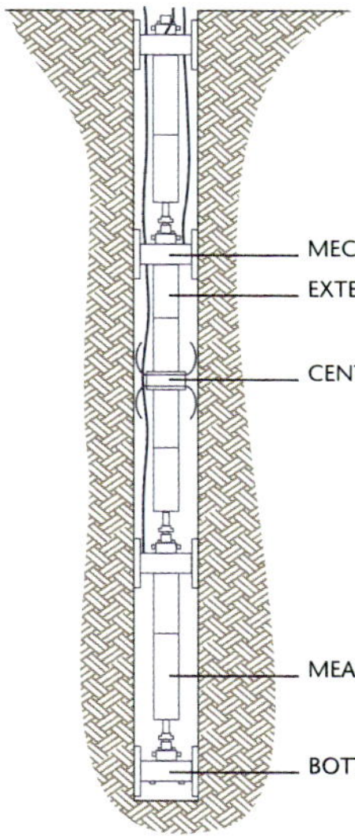

Figure 12.28: Multiple-point single-tube extensometer

gauge, and the RWE has a high-frequency response and good long-term stability.

There are several variants of the RWE, which may be cemented or resin grouted, or simply clamped by expansion anchors into drill holes. When grouted into a drill hole, the string can measure a profile of incremental strains.

The hollow RWE consists of a hollow tube containing a resistive-wire element suspended under a preset tension. Positive anchorage is effected by enlarged flanges at each end. Pure axial displacement is recorded free from the effects of transverse shears.

A flexible variant of the RWE (Figure 12.29, upper) comprises a single resistive wire coated with a plastic sheath, 1–4 m long. It registers the gross deformation of jointed rock including shear displacements and has been

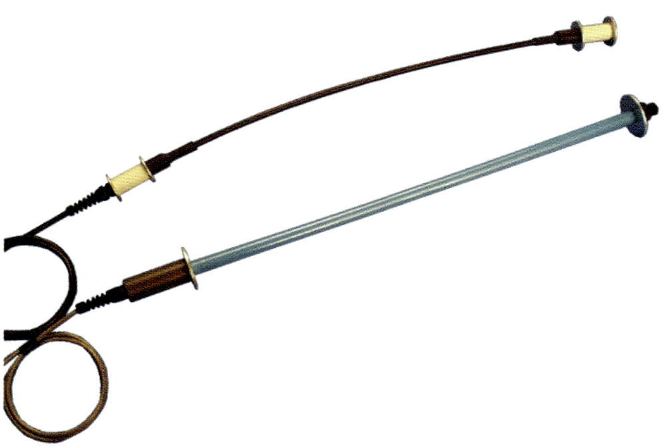

Figure 12.29: Rigid RWE resistance wire extensometer
Source: Courtesy Geotechnical Systems Australia Pty Ltd.

used to monitor the deformation of the grout in the same hole as a reinforcement cable.

Probe extensometers

The multiple-magnet extensometer comprises a series of ring-magnets that slide along the outside of a plastic guide tube and are fixed to the rock by grout or with mechanical anchors. A probe containing a reed switch is inserted in the guide tube. When it enters the field of a magnet the reed switch closes, activating a lamp or buzzer in the readout. The version used to record movement in mine backfill, tailings or overburden soil makes use of magnet targets fixed in the drill hole by spider springs. The probe is usually suspended by its steel measuring tape, taking readings as the probe is pulled to the surface. Accuracy is 1–2 mm.

Most rock applications, particularly those involving inclined holes, require the greater accuracy of grouted magnets and a rod-mounted probe containing two reed switches, which detect a pair of adjacent magnets. Magnet separation is measured by a micrometer at the hole collar. The repeatability of the instrument is ±0.5–±5 mm.

Thermistors

Thermistors are used to define ground temperatures, particularly in cold areas where the presence of permafrost can affect slope performance. The normal application is for strings of thermistors to be installed in drill holes to define the extent of frozen ground and the depth of the active layer affected by seasonal thawing.

Thermistor cables for drill hole installation should be assembled and calibrated in a laboratory prior to shipment to the field. Specific depths for the thermistors are defined prior to assembly, which involves each thermistor being attached to a cable pair in a multi-strand cable. The area of the thermistor is then resealed to prevent water entering the cable. While factory-calibrated thermistors are available, it is normal practice to recalibrate the assembled cable. Prior to installation in the drill hole the cable should be checked in an ice bath to ensure that the sensors are working.

The hole that contains the thermistor cable may also contain other instrumentation such as vibrating-wire piezometers and a TDR cable. It should be backfilled with material appropriate to the overall application.

An alternative to thermistor cables is a vibrating-wire piezometer tip that includes temperature sensors. These have particular application where water temperature rather than ground temperature is important.

Micro-seismic monitoring

Routine real-time micro-seismic monitoring in the open pit environment can provide 3D data where rock breakage or movement is occurring. These data can be used to augment surface monitoring systems in identifying potential instability and the associated failure mode.

Brittle fractures in rock radiate seismic waves. If these waves can be recorded sufficiently clearly as seismograms by a number of sensors, the seismic event's origin time, location and source parameters such as radiated seismic energy and inelastic co-seismic deformation can be estimated (Mendecki et al. 1997). The technique is commonly used in underground mining operations and has recently been applied in the open pit environment. Seismic events can be located and quantified in the 3D volume of the rock slope, opening the door to a deeper understanding of how mining is affecting the slope. Analysis of recorded data indicates that it may be possible to infer significant surface movements from seismic data before it is observed in surface deformation data. It may also be possible to determine if known geological structures are becoming seismically active and to indicate previously unknown planes of weakness within the slope.

Micro-seismic arrays

For seismic events to be reliably located by an array of seismic sensors, the sensors should surround the volume of rock being monitored. In an open pit environment, this means that sensors must be located near to the surface as well as at the bottom of the monitored volume. The sensors are usually installed into a combination of long inclined and short vertical holes. The entire pit is not monitored, rather the slope suspected of potential instability.

Some open pit seismic arrays are shown in Figures 12.30 and 12.31. The typical distances between sensors would be 100–200 m and the dimensions of the monitored volumes would be about 400 × 200 × 150 m. The near-surface sensors can be installed into short (10 m) vertical holes and would be 4.5 Hz geophones. Since these sensors cannot be installed into holes inclined beyond 2° from vertical, 14 Hz omni-directional geophones are used for long (100–300 m) inclined drill holes. Geophones are preferred to accelerometers since the typical frequencies recorded from slope seismic events are in the 10–400 Hz range, and geophones are more sensitive in this band. In addition, accelerometers are less reliable and, since the sensors are permanently grouted into these long expensive drill holes, reliability is a serious issue.

Micro-seismic data acquisition units

Signals from the geophones are monitored by seismic stations. Since the signals can be very low amplitude (peak ground motions of 1 μm/sec are common), a wide dynamic range is required to accurately monitor micro-seismicity in slopes. The signals are typically sampled at 6000 Hz with 24-bit or 32-bit analog-to-digital (A/D) converters, and time is synchronised across the network via GPS timing signals. To minimise the risk of lightning-induced damage to the sensitive A/Ds, one station is usually sited at the top of a long drill hole and exposed sensor cable runs are limited by only monitoring the sensors in that drill hole. Digital radios enable real-time two-way communication with the central

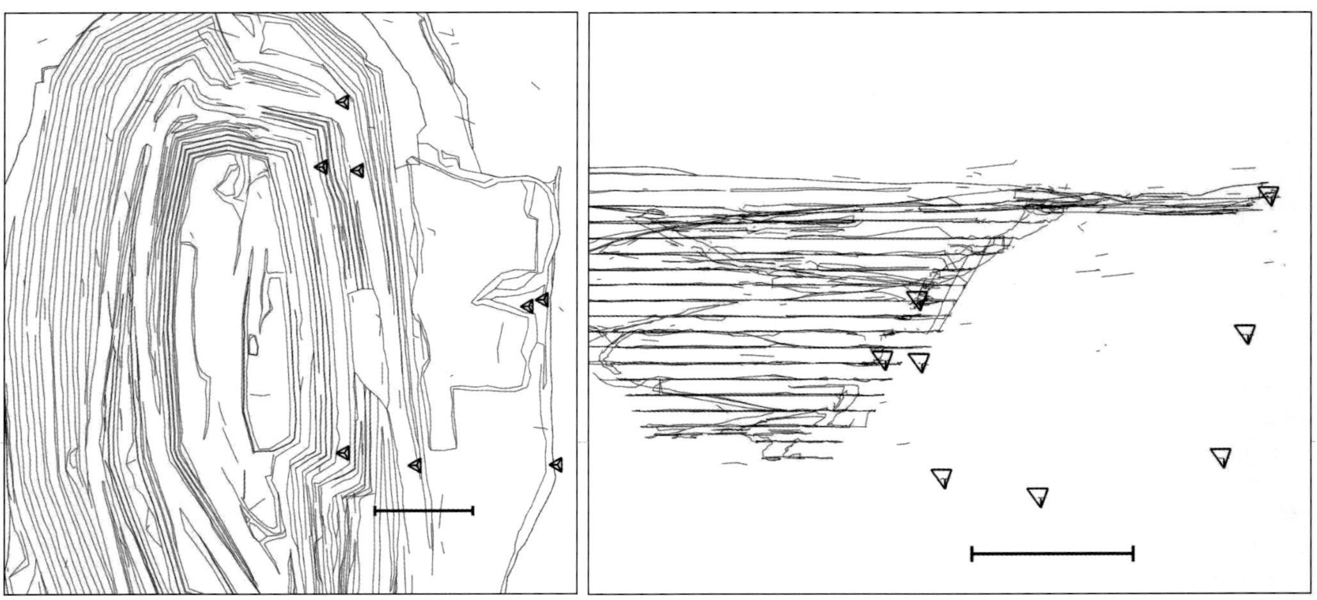

Figure 12.30: A plan view (left) and side view looking north (right) of the seismic array at Navachab gold mine in Namibia. A scale bar 100 m long is visible in each picture. The triaxial geophones are indicated by triangles
Source: Courtesy ISS International

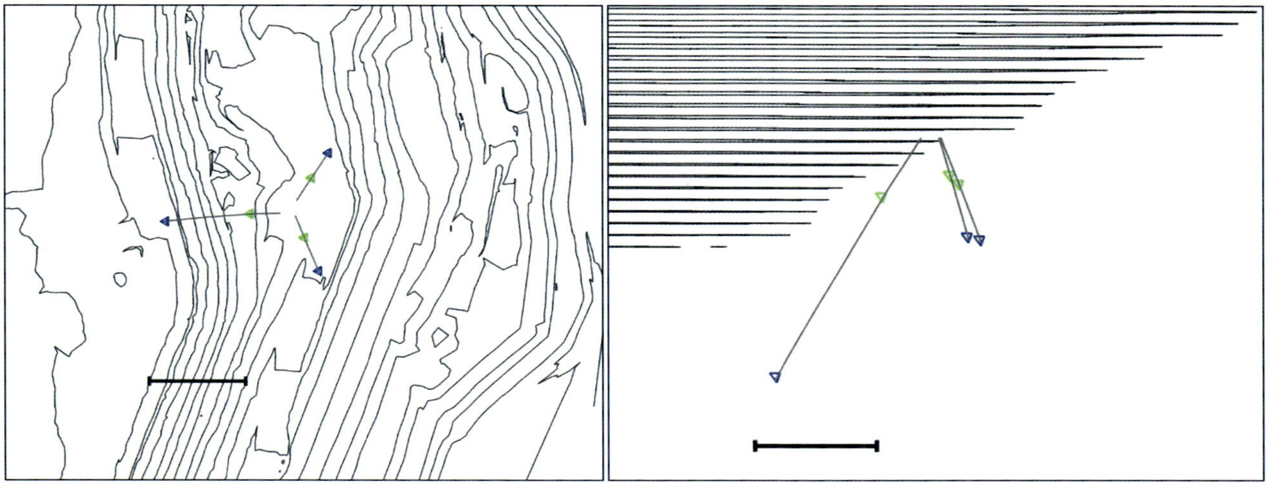

Figure 12.31: A plan view (left) and side view looking north (right) of the seismic array at Sunrise Dam gold mine in Australia. A scale bar 100 m long is visible in each picture. The triaxial geophones are indicated by blue triangles, uniaxial geophones are indicated by green triangles
Source: Courtesy ISS International

computer in the geotechnical offices, and 100 W solar panels should be sufficient to power the seismic station. The stations report each trigger; if several stations have triggered in a consistent manner the seismograms are requested and transferred to the central computer for processing and storage.

Micro-seismic data processing, analysis and reporting
Fast internet connections allow the seismic data to be automatically sent to a remote central facility for processing and analysis. The origin time, spatial location and estimates of source parameters are computed for each seismic event. After routine analysis of the seismic event data, regular reports are sent back to the mine geotechnical engineers for interpretation along with other geotechnical data. This kind of data processing and analysis has been used by mines in Australia, South Africa and Namibia, and circumvents the need for advanced seismology training for the on-site geotechnical engineer.

Micro-seismic event data
Seismic events typically occur as far behind open pit slopes as the pit is deep. The events are typically very small and are not felt by miners. However, the data provides a sensitive measure of how mining is affecting the slope. For example, a case study has shown how fracturing behind the slope is more influenced by the removal of broken rock (unloading) at the toe of a slope than by actual blasting (Figure 12.32).

Two other case studies have shown that seismic data may be used to infer surface movements in places on the slope where movements were then detected later. The cause was a series of deep tensile cracks that did not cause

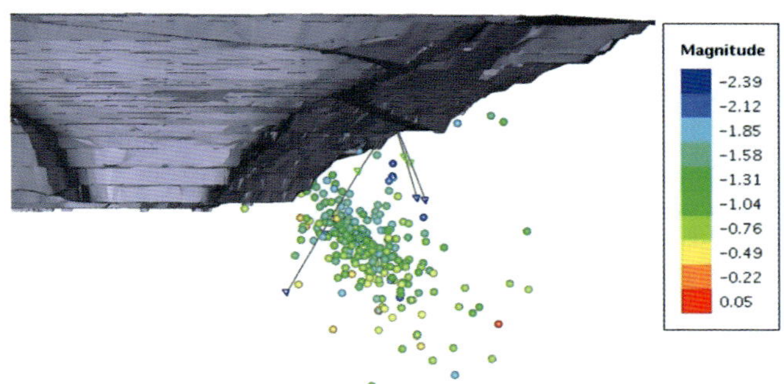

Figure 12.32: Side view (looking north) of the located micro-seismic events recorded during seismic response to removal of broken rock after pit blasts at the toe of this slope. The local magnitude scale is also shown. Most events occur beneath the pit floor level
Source: Courtesy ISS International

slope instability, although significant slope displacements were observed (see Figure 12.33). In one case the located seismic events tended to lie on planes of weakness which matched the known geology at that mine, explaining the mechanism of the slope response to mining.

Table 12.4 summarises the pros and cons of micro-seismic monitoring of open pit slopes.

12.2.2.5 Monitoring of groundwater pressure

If the actual pit slope design is predicated on achieving a given future pore pressure profile, it is important that year-by-year pore pressure targets are developed to ensure that depressurisation is occurring at the desired rate.

The final slope design must include piezometer installations in the most critical areas for slope performance. Target pressures are then developed for each piezometer, for

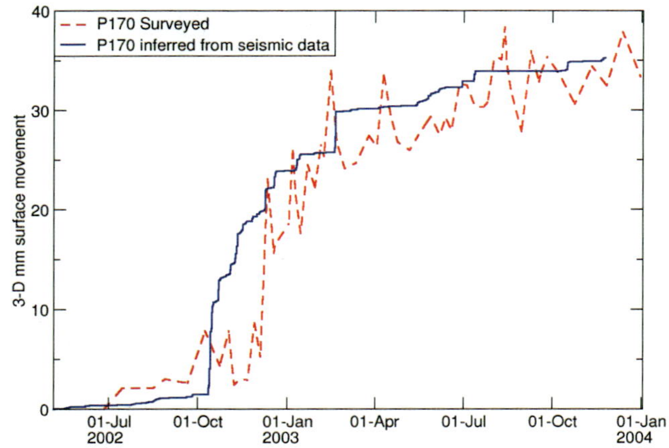

Figure 12.33: The 3D movement of a measuring point on the slope at Navachab gold mine, as surveyed (red dashed line) and inferred from seismic data (blue solid line).
Source: Courtesy ISS International

each year of mine operation. The components of a groundwater monitoring are outlined below.

Monitoring components

Operational slope monitoring for hydrogeology includes the following.

Piezometers

- Weekly water levels or pore pressures in standpipe piezometers.
- More frequent automated pore pressure monitoring of vibrating-wire piezometers, or transducer installations in standpipe piezometers.
- Weekly pressure monitoring for each horizontal piezometer using a portable manual pressure gauge, or more frequent automated monitoring using a pressure transducer sealed into the collar of the drain holes.
- If appropriate, regular (e.g. quarterly) chemistry monitoring of standpipe piezometers can be carried out for a limited suite of indicator parameters to provide further information on changes to the groundwater flow system (the requirement for well purging prior to sampling would need to be evaluated).
- Collation of all monitoring data on a monthly basis.

Table 12.4: Positives and negatives of micro-seismic monitoring of open pit slopes

Potential beneficial aspects	Negative aspects
3D data for imaging inside the slope	Requires long (few hundred metre) drill holes
Real-time detection of even very small fracturing for quantified slope response to mining	Sensors are permanently installed and only surface equipment can be moved later – only recommended for slopes that must be monitored for at least a few years
Stability of known geological structures can be monitored	Off-site processing of seismogram data is required
Detection of previously unknown geological structures, to within the resolution of the array	Real time warnings are not possible
Medium-term indications of where the slope is expected to move	

Horizontal drain flows

- Monitoring of the flow rate from each new horizontal drain hole during drilling advancement, with the flow rate recorded at the end of drilling each pipe.
- Regular monitoring of the flow rate from each completed horizontal drain. A typical monitoring schedule may be:
  - → daily recording of the flow rate for 7 days after completion;
  - → weekly recording the flow rate for 2–3 months after completion;
  - → monthly recording of the flow rate as long as access permits.

If appropriate, regular chemistry monitoring of horizontal drain flows can be carried out, typically for a limited suite of indicator parameters.

Dewatering well discharges

- Normally, the discharge rate is monitored at the wellhead using an in-line flow meter and the instantaneous and cumulative discharge is recorded during a daily inspection. A sounding tube should be installed in the well to allow measurement of the non-pumping and pumping water level to track changes to the specific capacity of the well with time. A typical monitoring schedule would include:
  - → a daily visit to the wellhead to record the instantaneous discharge rate and carry out a general inspection;
  - → weekly measurement of pumping water level, cumulative discharge, amperage and line-to-line voltage;
  - → monthly collation of data.
- Regular (e.g. quarterly) chemistry monitoring of completed well discharges is done, for a limited suite of indicator parameters.

Slope conditions

- Visual or photographic mapping of pit wall seeps.
- Visual inspection of runoff areas, paying particular attention to the potential for surface water to enter tension cracks.
- Periodic measurement of the flow in any seep that can be easily accessed.
- Periodic monitoring of the chemistry of any prominent seeps.

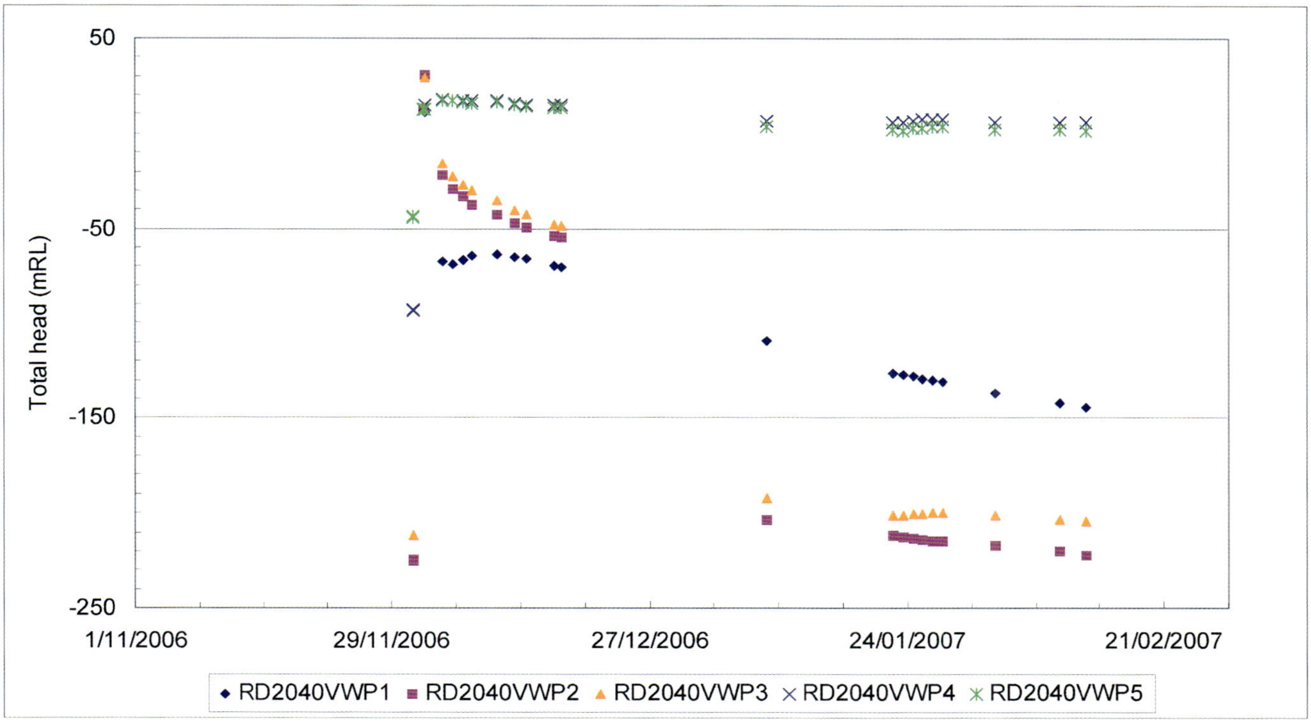

Figure 12.34: Typical hydrograph of multiple piezometer installation

- Monitoring of the flow rate from each in-pit sump. The flow rate is typically monitored at the pressure side of the sump pump using an in-line flow meter, and the instantaneous and cumulative flow is recorded during daily inspection.
- Regular inspection and cleaning of surface water diversion systems, particularly prior to the onset of the rainy season (wet climates) and for winter snow accumulation (cold climates).

Presentation of monitoring results

Pore pressure monitoring results may be presented as follows.

- Piezometer hydrographs showing total head plotted against time. Records from multiple piezometers, particularly vertically discretised piezometers, may be included on a single plot (Figure 12.34). Correction of the data for atmospheric pressure may be required. Pore pressure targets may be included as end points on the plots. Piezometer plots are usually grouped by their physical location within the pit but may be grouped by hydrogeological unit.
- Horizontal drain plots showing flow rate plotted against time. Multiple drains from each sector of the slope may be included on a single plot, with the total flow rate from all the drains in the sector shown on the plot. Drain flow plots are usually grouped by their physical location within each pit sector but may be grouped by hydrogeological unit.
- Vertical well flow plots showing flow rate plotted against time. Wells from each sector of the slope may be included on a single plot, with the total flow rate from all wells in each sector also shown. The flow rate plots are grouped by their physical location within each pit sector and the wells may also be grouped by their hydrogeological unit. Plots showing the variation of specific yield (discharge rate divided by drawdown) over time may be useful for certain hydrogeological units, or for the system as a whole.
- Composite plots, which may show:
  - → changes in pore pressure shown with changes in the flow rate from drains or wells;
  - → changes in horizontal drain flows or pore pressure shown with changes in the rate of movement of prisms or other slope monitoring instruments (Figure 12.35). It is often useful to annotate the plots with dates where new wells came online, where a pushback was mined in a particular slope sector, of high precipitation events and other factors that may have influenced the observed hydrograph response.
- Water level maps which show total head or pore pressure at a given moment or which show change in total head or pore pressure over a given time period. To assist with data interpretation, it is normal that all

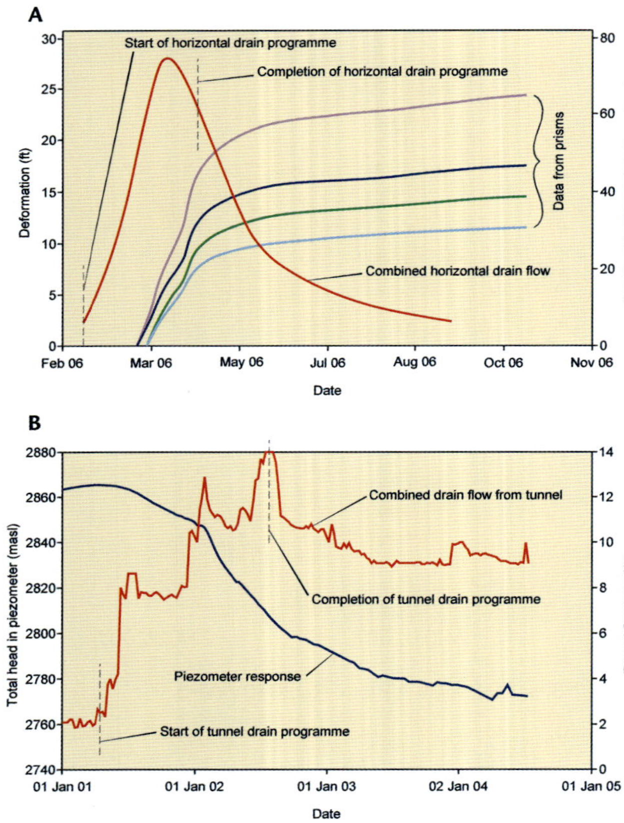

Figure 12.35: (a) Plot of prism data vs horizontal drain flow, for the SE wall, Chino Mine, New Mexico, USA; (b) Plot of piezometer response vs drain flows from tunnel, for the NE wall, Escondida Mine, Chile.

piezometer and pore pressure data are plotted as pore pressure elevations (total head) onto a base map showing the key geological structures, and possibly lithology and/or alteration. Vertical differences in pressures can be presented by colour-coding the data from vertically discretised piezometers.

- Maps showing the distribution of drain hole and production well flows, which can include:
  - → drain hole locations, with drain holes colour-coded by their initial flow rate;
  - → pumping well locations with initial and current flow rates. Again, it is normal that all information is plotted onto a base map showing the key geological structures, and possibly lithology and/or alteration. Depending on the amount of data available, the flow rate map can be integrated with one or more of the water level maps.
- Hydrogeological cross-sections which show:
  - → vertically discretised piezometers plotted along the section;
  - → changes of the water table or pore pressure distribution with time. Cross-sections normally include lithological, alteration and structural information.
- Numerical model cross-sections that show actual monitored pressures compared with model output at a given point in time. An example is given in Figure 12.36.

Monitoring of transient pore pressures

In mines in wet climatic zones or where local recharge occurs seasonally, pore pressures can vary considerably throughout the year. Infiltration of precipitation into the material above the crest of the pit or the slope itself can cause transient pore pressures to develop in material that is dry for much of the year. The largest seasonal fluctuations are normally observed at shallower locations within the slope. The transient pressures are often related to water moving down the blast-damaged (overbreak) zone, or in permeable zones at shallow depths beneath the slope.

Because of the seasonal nature of the recharge and hence the transient nature of the pore pressure, it is often difficult to install effective measures to remove the water. The preferred method is to intercept and manage the recharge at source, although this is clearly not possible where the transient pressures are caused by precipitation and runoff from the slope itself. Good surface water removal and management is invariably beneficial.

Fluctuating pore pressures can also be difficult to monitor. Normally they are characterised using a combination of:

- monitoring shallow piezometers within the slope, which may be dry and have zero pore pressure for much of the year;
- observing locations and flow rates from prominent seasonal seepage faces on the pit slope;
- observing seasonal increases in the flow rate from horizontal drains.

12.2.2.6 Data acquisition systems

With increasingly complex instrumentation in mines, it has become impractical in many instances to take readings manually. The extra capability of a datalogged system is attractive and hard to ignore, even for more conservative instrument users. Modern electronics systems scan many channels of sensors quickly and reliably. Intelligent data acquisition (DA) systems can not only acquire, analyse and store data but can also initiate actions (e.g. sound alarms, start recorders) on the basis of the data acquired in real time, and the trigger levels defined by the geotechnical engineer.

A basic DA system comprises six stages: signal acquisition, signal conditioning, AID (analog to digital) conversion, signal transmission, signal processing and data storage. Such systems are powerful tools because they can exercise intelligence in the collection of routine data, for example by automatically changing, when appropriate, the

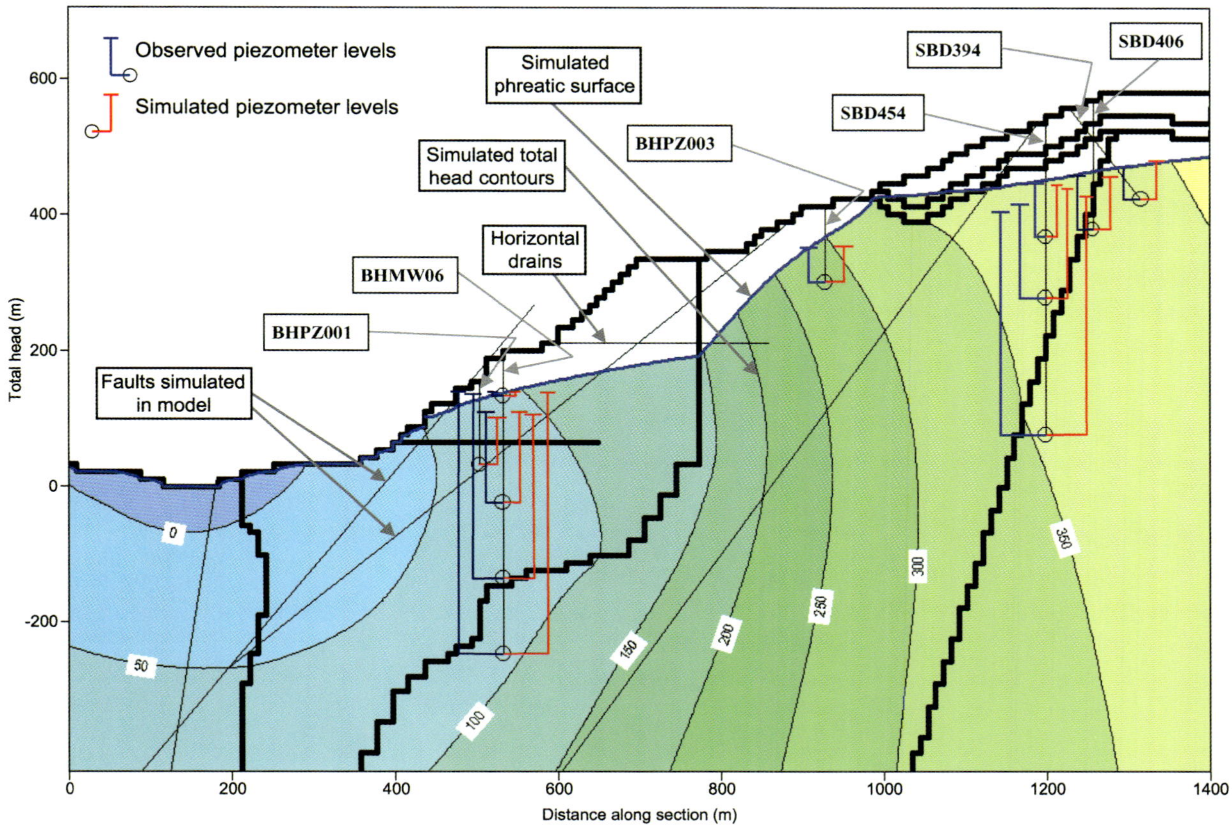

Figure 12.36: Plotting actual monitoring data on model output

frequency and number of sensors read, and they can act immediately on the basis of data received (e.g. by warning of faulty sensors or by triggering alarms when thresholds are exceeded).

The advent of cheap portable microcomputers that can perform reliably even with power interruptions makes it possible to build sophisticated DA systems for in situ geomechanics monitoring.

All DA systems consist of two distinct components that are strongly interdependent: hardware (microprocessors, sensors, amplifiers, counters) and software (protocols, instructions). Although good software can compensate for many hardware deficiencies, excellent hardware can be crippled by inadequate software, which can limit the performance of some commercially available systems.

In designing a DA system for geomechanics monitoring it is necessary to specify exactly what is required of the system, then choose the most cost-effective solution which meets (but does not grossly exceed) the design specifications. Because many of the elements used to build the system (transducers, signal conditioners, dumb loggers) may be available from previous projects, a great deal of expense can be saved by making full use of existing resources.

12.2.3 Guidelines on the execution of monitoring programs

A slope monitoring program should form an integral part of the engineering of any open pit mine. The methods and monitoring requirements for open pit applications depend on the type and scale of mine slopes being considered. Although the common range of pit depths is 100–500 m, much higher slopes are being developed, particularly in copper mines in western Canada, South America and Indonesia. As more efficient and larger equipment becomes available, increasingly deeper open pits may be proposed.

12.2.3.1 Preparatory work

Many open pit mines do not install a monitoring system in the early phases of mine development because interim mining slopes are often of lower heights and/or relatively flat and there is a lower risk of major instability, compared with the risk in more aggressive final slopes. However, without these initial readings, if movements are observed at a later stage of mine development it is often difficult to evaluate their significance in terms of absolute measurements.

Installation of a monitoring system on a pit slope that has already started to move will provide valuable

Table 12.5: Summary of pit wall stability monitoring procedures

Procedure	Areas	Frequency	Activities	Personnel	Reporting and actions
Daily inspections	All current active mining areas, high-risk areas	Daily	Visual checks for cracking, dilation and scaling requirements	Geotechnical geologist	Wall inspection book, discussed at daily meeting, geotechnical superintendent and operations superintendent (if problems exist). Batter and berm inspection forms to be filled and signed off at daily production meeting as required
Periodic visual inspections of pit perimeter and berms	Pit perimeter and all accessible berms	Weekly to fortnightly, and after heavy rainfall	Visual checks for tension cracking, other signs of slope movement and rockfalls	Geotechnical geologist, technician	Berm walk over inspection form, slopes and berms overlay to pit plan, cracks to be painted and surveyed, advise geotechnical geologist, operations superintendent, issue hazard alert as appropriate, production meetings
Tension crack monitoring	Cracked areas on pit perimeter and berms	Initially daily and after heavy rainfall. Frequency to be adjusted depending on rates of opening	Measurements of crack widths and visual checks of other signs of slope movement and rockfalls	Technician	Spreadsheets. Geotechnical geologist to be notified if accelerations noted in opening, production meetings. Hazard alert to be issued as appropriate
EDM monitoring	Cracked areas around perimeter and on berms, and other designated pit areas	ATS and semi-automatic. Frequency to be adjusted depending on rates of prism movement	Survey of changes in prism northings, eastings and elevations	Surveyor	Quickslope database and graphs. Autoslope alarm system. Geotechnical geologist and operations superintendent to be notified of acceleration in movement. Hazard alert to be issued as appropriate. Prepare monthly report
Slope stability radar, ground probe	High-risk rockfall hazards	Ongoing, continuous	Daily checks of data and instrument status	Technician, geotechnical geologist	Dispatcher to notify the E Zone controller, geotechnical geologist if alarm is triggered. E Zone evacuation initiated for all alarms
Slope photography	All portions of walls	Ongoing over pit life	Monthly to quarterly	Geotechnical geologist	Geotechnical superintendent, file
Slope failure records (hazard alert and incident reports)	Any portion of walls where a rockfall has occurred into a working area	As required	Complete hazard alert and incident report for the rockfall or failure event	Geotechnical geologist	Management and senior mine operations personnel

Source: Geita Gold Mine (2006)

information regarding safety, but may be difficult to evaluate in terms of overall movement and planning of remedial works. Therefore, all open pits should have the instruments for and capability of installing and executing a basic monitoring program, which should be instituted in the initial stages of operation. The basic slope monitoring system should be established by the mine engineers in collaboration with geotechnical engineers, taking into account the relative importance and potential scale of possible failures. Thus, the design of the initial system must be based on an assessment of the potential modes of instability.

12.2.3.2 Monitoring program

The monitoring system for an open pit mine normally comprises a number of elements, starting with a surveillance system to evaluate slope performance and detect the onset of unexpected movement. Such a system typically involves visual observation and geodetic surveying of prisms distributed at a relatively wide spacing around the pit crest and down the slopes. If specific unstable conditions are anticipated or if movement is indicated, a more extensive and possibly more sophisticated program may be required, building upon the basic system.

The initial system can involve relatively simple and inexpensive techniques. However, the program must be able to form the basis for detailed monitoring systems in specific areas, if movements are detected. It will also be influenced by such factors as the size of the pit, access and the anticipate size of failures.

Table 12.5 provides an example of the procedures typically associated with the monitoring of pit wall stability in an operating mine. Table 12.6 shows applications of different procedures and instrument types

Table 12.6: Summary of monitoring methods by potential failure size and implication

Block size (m^3)	Speed of failure	Implications	Monitoring for detection	Typical remedial
1–10	Immediate	Rockfall – safety	**Visual**	**Catchment**
10–1000	Very rapid to rapid	Safety	**Visual** Radar	**Catchment**
1000–100 000	Rapid to slow	Operational	Visual Surveying **Radar** Seismic (?)	**Manage** Modify slope (step-out)
100 000–1 000 000	Moderate to slow	Operational/financial	**Surveying** **Radar** TDR/inclinometer Seismic	**Manage** *Modify slope (step-out)* Recut (?)
>1 000 000	Slow to moderate	Force majeur	**Surveying** TDR/inclinometer Seismic Radar	**Modify slope (recut)** Mine closure (>10 Mm^3) Manage

Speed of failure
Very rapid = immediate to minutes
Rapid = minutes to hours
Moderate = days to weeks
Slow = weeks to months

Monitoring and typical remedial
Bold = most likely
Italic = alternative or support system

used to detect instabilities of different sizes and potential implications.

This section discusses how the various methods described above are applied to an overall slope monitoring program.

Visual inspection

A basic element of a slope monitoring program should be visual inspection by the mine geotechnical engineer and members of the engineering staff, combined with observations by all personnel working in the mine. This qualitative, but extremely important aspect of the program should be maintained throughout the life of the mine.

Monitoring of rockfalls

Besides the visual inspections by the mine geotechnical staff, operating personnel's observations of rockfalls or unusual conditions can be extremely important and should be encouraged through a specific reporting system.

In many pits, small failures, rockfalls or general ravelling may develop because of poor natural or induced ground conditions. Prediction is difficult, although rockfalls and small failures may cause major operational problems in terms of safety, maintenance of berms and protection of haul roads. The overall slope may be stable, but the design of the slope may be controlled by the requirement to control numerous possible single-bench failures or rockfalls. Hence, procedures for evaluating and monitoring rockfalls and small failures are mandatory for safety and the preparation of a rational design.

Probably the most effective method of monitoring rockfalls is by visual inspection and documentation of historic and recent slope behavior and rockfall incidents. Slope behaviour is documented by recording the relevant slope geometry parameters including bench face angle, berm width, bench breakback and the volume of debris on berms, as described in section 12.1.3. Larger failures should be documented more rigorously. These parameters, when compared to the mine design parameters, provide a statistical evaluation of the effectiveness of the slope design and the significance of small failures. The parameters can be recorded at regular intervals (10–30 m) on all benches around the pit or in strategic structural domains to enable statistical assessment. Chronological photographs are useful to monitor the rate of rockfalls and build-up of debris on berms over time.

Surface extensometers and crack monitoring

If evidence of movement is detected from visual inspection, the first step in augmenting the monitoring program might be simple crack monitoring systems. Results of visual inspection and crack monitoring are a useful guide when selecting additional secondary monitoring points for detailed survey assessments.

Crack monitoring techniques typically consist of:

- regular detailed mapping of location, depth, width of cracks, rate of extension and opening;
- installation of targets on opposite sides of cracks to monitor rates of opening;
- installation of surface (wireline) extensometers;
- installation of picket lines or lines of targets that can be monitored using theodolites or precise levels to detect changes in alignment, location or elevation along a given crack or the crest of the slope.

Many mines have applied surface extensometers for monitoring displacement local wall failures and dump

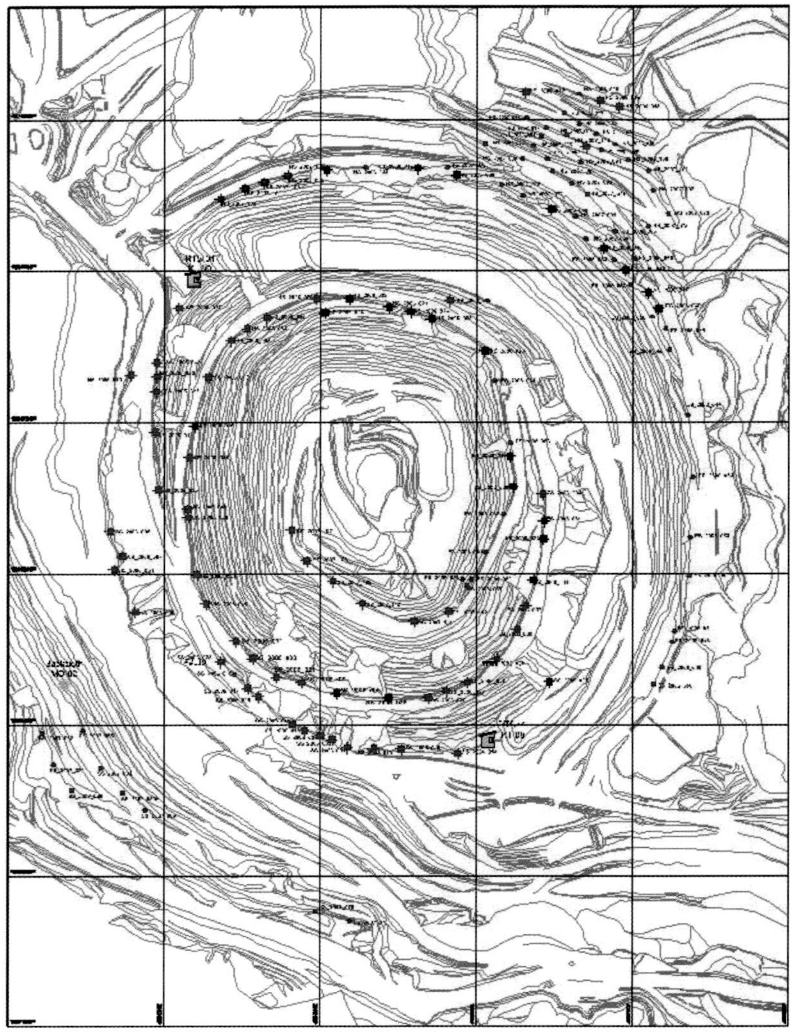

Figure 12.37: Typical survey monitoring points in large open pit which is experiencing some instability

displacement. These devices, which can be easily constructed in the mine shop, provide a rugged practical system of movement monitoring that can be inspected and interpreted on a regular basis by mine operations personnel. They can also be equipped with automatic devices such as lights or sirens to provide warning of excessive movement. More sophisticated units can provide real-time indication of movement to remote locations (dispatch or engineering office) through a telemetric link.

Terrestrial geodetic surveys

The most reliable and complete measurements of the 3D movements associated with initial movement could be obtained from conventional geodetic survey techniques using precise theodolite and EDM combinations or total stations. These systems can be installed by mine survey personnel, generally with survey equipment in regular use at the mine.

Geodetic surveys should start by installing a survey network of stable instrument stations and primary monitoring points around the pit perimeter. This network should be tied to at least three stable reference stations well behind the pit crest. Monitoring points (prisms) are established at regular intervals on each wall of the pit and in areas identified from geotechnical investigations as more susceptible to instability. If instability is detected, additional secondary monitoring points may be established in the area to determine the size, failure geometry and movement rates, and to assist in the planning of remedial measures. A typical arrangement of primary and secondary monitoring points for a large open pit copper mine is shown in Figure 12.37.

Primary monitoring points should be surveyed at regular intervals consistent with the type of rock and expected rates of movement. Surveillance monitoring frequencies vary from weekly to quarterly depending on conditions such as the stage of mining, mining rate, changes in piezometric surface and climatic variations.

The individual aspects of a typical system are as follows.

- Control points for the system should consist of the instrument stations near the crest of a pit slope and reference stations located away (100 m to 3 km) from mining activities. Control points are usually established by conducting a first-order survey, using conventional survey techniques such as triangulation, trilateration or triangulateration, or GPS.
- GPS is much more efficient, accurate and less labour-intensive than the conventional survey techniques when used for control surveys, especially when the network covers a relatively large area. The main requirements are a differential system and good-quality equipment.
- The stability of instrument stations can be checked by resurveying the control network or reference stations each time the instrument station is used. Care must be taken to ensure sufficient observations are made to all reference stations on a regular basis.

Data from the survey monitoring should be plotted and assessed after each set of readings. If movement is detected, monitoring frequency of secondary points will depend on the size of the failure and movement rates and could be hourly to weekly. Methods of assessing the movement monitoring data and establishing trigger levels for remedial action are discussed in section 12.2.3.3.

Radar

Where more extensive areas of movement are detected, radar enables real-time monitoring of the displacements to help ensure the safety of operators working below the slope. Radar units are used by many mines in conjunction with geodetic surveying, because they effectively provide real-time warning of displacement and accelerations.

It is important that the radar system does not become the sole basis for monitoring. Further, it is essential to maintain a degree of conservatism in determining when to withdraw personnel from below a moving slope, even if it is being monitored by radar. Even small rockfalls resulting from the deformation can have serious safety consequences and may not be detected by radar.

Subsurface techniques

Although the costs of subsurface techniques are greater than those for surface instrumentation, they can be modest if available drilling equipment is used and mine personnel perform the installation after instruction from specialists or the instrument supplier. Inclinometers and TDR cables, for example, give very valuable and precise information on the locations of deep-seated slide surfaces and on rates of movement, without which remedial work cannot be adequately planned.

Subsurface instrumentation is normally installed only if drill holes are available from other programs or underground access is already available. A typical layout for a large slope from an underground drainage gallery and exploration gallery is shown in Figure 12.38.

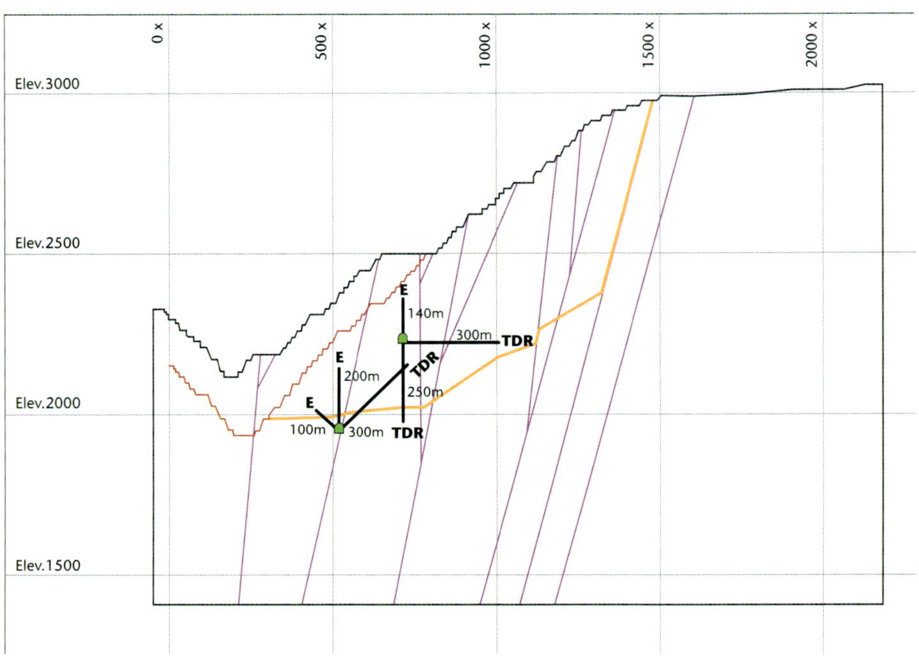

Figure 12.38: Subsurface monitoring from underground access drives (E = extensometer)

12.2.3.3 *Collection, processing, presentation, interpretation and reporting of instrumentation data*

A detailed draft of monitoring and reporting procedures should be prepared during the planning phase and finalised after the instruments have been installed. At that time responsible personnel will be familiar with operation of instruments and specific site considerations. These procedures should include:

- a list of data collection;
- equipment specifications, including servicing requirements;
- processing and presentation procedures;
- interpretation procedures, including alarm levels.

This information must be included in a comprehensive instruction manual, but users should recognise that, although manufacturers provide basic information, they are not familiar with specific site conditions. Users must therefore prepare their own procedures.

Collection of instrumentation data

Responsibility for collection of instrumentation data is determined during the planning phase and should be under the direct control of the mine geotechnical engineer or instrumentation specialists selected by the mine.

Processing and presentation of instrumentation data

The primary aim of data processing and presentation is a rapid assessment of information to detect changes that require immediate action. A secondary function is to summarise and present the data to show trends and compare observed with predicted behaviour so that any necessary action can be initiated.

Monitoring data should be presented in a format that is easy to read and identify problem areas quickly. Items of interest at most open pit mines include slope distance, velocity, wander plots, and inverse velocity. An example graph of a typical format for presenting survey prism data is shown in Figure 12.39. Additionally, maps showing the geographic location of monitoring points with respect to current and recent instabilities are very useful to mine operators. These maps generally show movement vectors with current velocities, along with systematically named instabilities so as not to create confusion with location.

Responsibility for processing and presentation of instrumentation data is determined during the planning

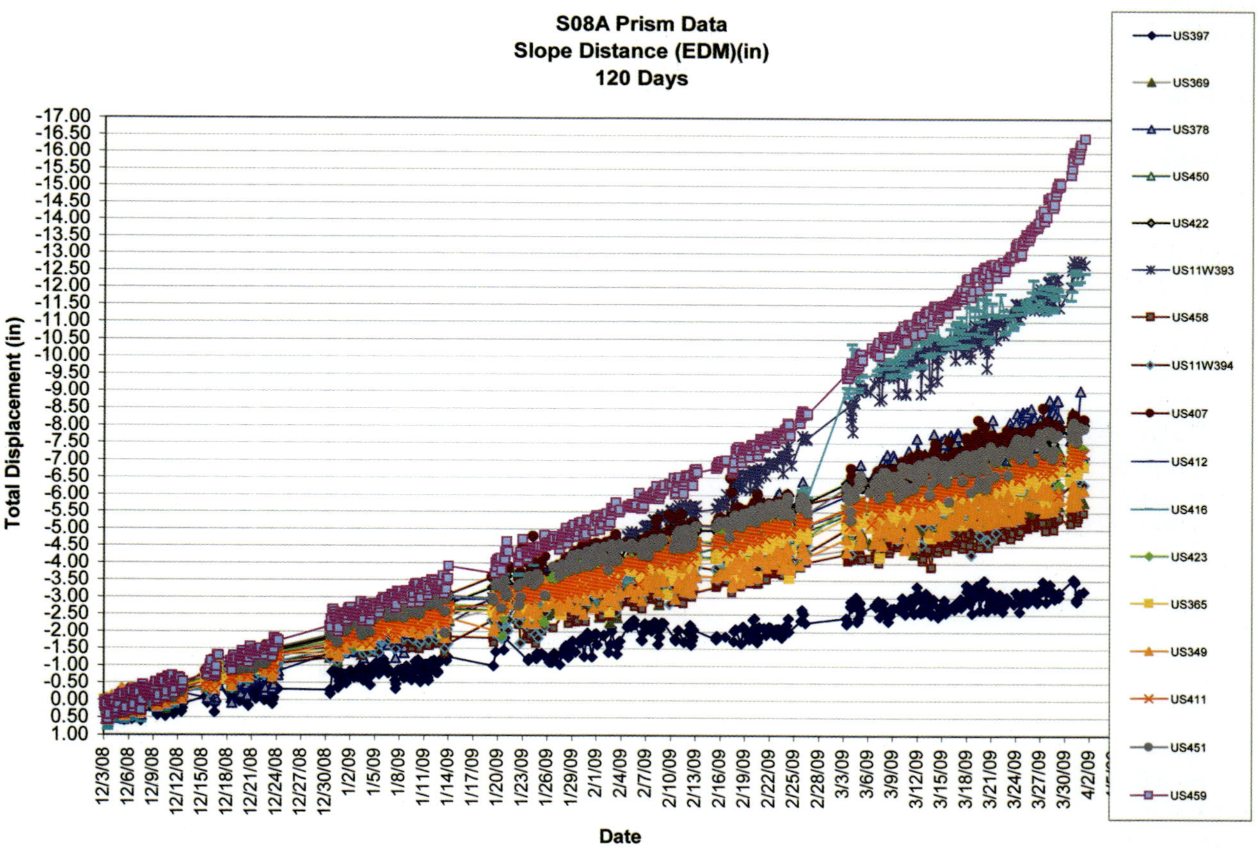

Figure 12.39: Typical survey prism monitoring data
Source: Courtesy Barrick Goldstrike

phase and should be under the direct control of the geotechnical engineer on site or, in special cases, consultants who have immediate 24-hour access to the data. Personnel requirements for these tasks are frequently underestimated, resulting in the accumulation of unprocessed data and failure to take appropriate action. Similarly, experienced geotechnical engineers may use much of their time in support of monitoring systems instead of delegating these responsibilities to technicians and undertaking the required technical analysis to minimise and/or manage the impacts of potential slope failures. The time required for data processing and presentation is usually similar to, and may even exceed, the time required to collect data.

Data processing and presentation depends on the specific monitoring system. For surveillance monitoring and for small pits it can often be performed with standard spreadsheet packages. More comprehensive monitoring programs may require commercial survey reduction and GIS programs.

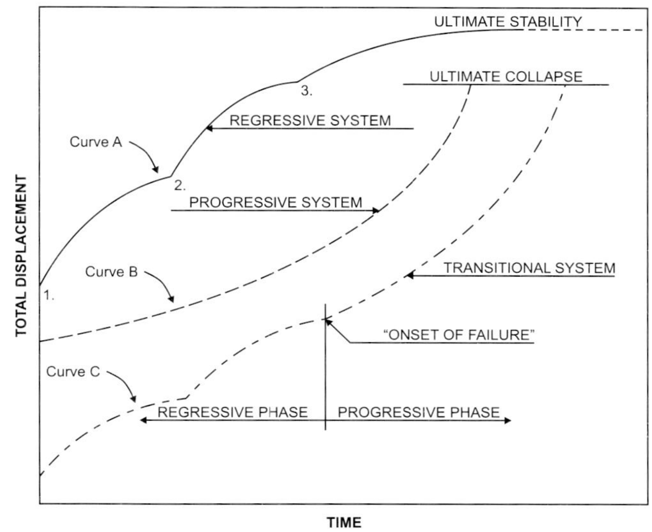

Figure 12.40: Typical regressive/progressive stage displacement curves
Source: Broadbent & Zavodni (1982)

Interpretation of instrumentation data

Monitoring programs have failed because the data generated were never used. If there is a clear sense of purpose for a monitoring program, this will guide the data interpretation. Without a purpose there can be no clear interpretation.

Early data interpretation steps must be aimed at determining the accuracy of the monitoring system. For example, atmospheric changes may result in diurnal variations of several times the manufacturer's quoted accuracy for EDM and total station units, particularly in desert climates where there are significant temperature differences between day and night, and in arctic climates where temperature inversions can develop in a pit overnight. These survey accuracy variations must be filtered out as part of the interpretation process, either by setting wider bands before alarms are triggered or by putting emphasis on readings taken at the same time of day.

The purpose of subsequent data interpretation steps is to correlate the instrument readings with other factors (cause and effect relationships) and to study the deviation of the readings from the predicted behaviour. By its very nature, interpretation of data is a labour-intensive activity and no technique has yet been developed for complete automatic interpretation.

Interpretation of data from movement monitoring systems primarily involves assessing the onset of changes in the movement rate. This is generally reflected by acceleration but, where a slope is already moving, deceleration may also occur. Typical movement patterns associated with instability are summarised in Figure 12.40.

Normally a series of trigger points or trigger action responses (TARPS) are established for each monitoring method. These should take into account changes in movement rates above the survey accuracy background. A typical system of trigger points might be as follows.

- The initial trigger point for concern with the monitoring data should be if the movement rate is double the survey accuracy from the last reading. In this case the reading should be repeated as soon as possible. If the reading is proven correct, additional readings should be taken at an increased frequency.
- The second trigger point would be if the movement rates double over two consecutive readings. In this case, the area of the moving prisms should be inspected. If the cause of movement cannot be determined, mining in the area should be reduced to day shift only or suspended and the reading frequency increased. Continued acceleration of movement should require closure of the pit floor below the moving area until the situation has been fully investigated.
- If an increase in movement greater than four times the survey error is recorded for *any* reading when there has been no previous accelerations noted on a prism, the operations staff should be informed immediately and the area below cleared until the point has been resurveyed. If the reading is confirmed, the area should remain cleared until the situation has been investigated.

The reporting procedure in the event of any TARP should be clearly defined and understood by all. This is discussed further in section 12.3.

The interpretation of data to predict future responses depends on many factors, including rainfall and mining

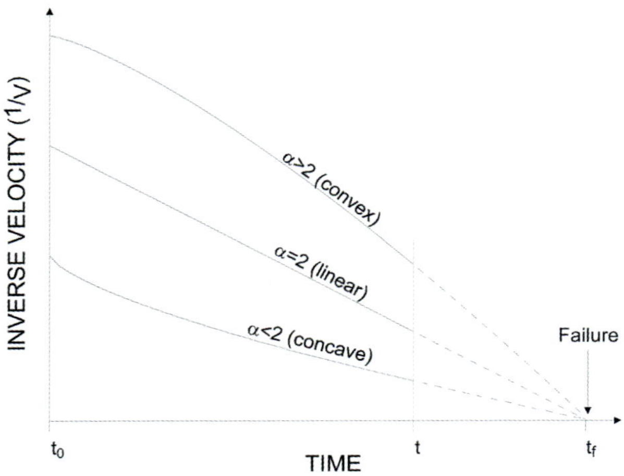

Figure 12.41: Inverse velocity vs time relationships preceding slope failure
Source: After Fukuzono (1985)

activity, both of which can result in further accelerations, and the onset of freezing conditions, which can slow movement. Several papers have discussed methods of predicting failure. The inverse velocity method (Fukuzono 1985) provides a useful tool for interpreting instrument monitoring data with the eventual objective of anticipating or predicting slope failure.

The concept of inverse velocity is based upon observations from large-scale well-instrumented laboratory tests simulating rain-induced landslides in soil. The conditions simulated in the laboratory were considered to be characteristic of accelerating creep (i.e. slow continuous deformation) under gravity loading. When the inverse of observed displacement time rate (inverse velocity) was plotted against time, its values approached zero as velocity increased asymptotically towards failure. A trend-line through values of inverse velocity versus time could be projected to the zero value on the abscissa (x -axis), predicting the approximate time of failure, as shown in Figure 12.41.

Based on retrospective assessment of slope monitoring data from slope failures in the late 1990s (Rose 2002), the assumption was made that linear trend fits of inverse velocity data could be applied to long-term trends as well as the short-term forecasts proposed by Fukuzono (1985). Figure 12.40 is a plot of inverse velocity versus time showing the trends of nine survey prisms located at various elevations on the south-east wall of the Betze-Post open pit, over the six weeks preceding slope failure in August 2001. Filtering (data smoothing) of two-hour robotic total station prism monitoring measurement data was achieved by calculating six-day average (incremental) slope distance velocities to reduce the effects of instrument error in low-level velocity values. As displacement rates increased, a clear inverse velocity trend developed and began to converge on a failure time of 29 August 2001. Linear regression was applied to the inverse velocity values, which defined regression coefficients (R^2) of 99% for all nine prisms. The failure occurred on the predicted date, encompassing an overall slope height of 550 m and an estimated 47 million tonnes. The failure lasted several hours as a series of nested wedge failures and rock avalanches.

Two subsequent slope failure prediction case histories presented by Rose and Hungr (2007) utilised shorter time durations in data filtering of average prism and wireline extensometer data. Time increments were reduced to one or two days to avoid the observed divergence from the linear trends in inverse velocity data from the 2001 Betze-Post open pit case history, in the days leading up to failure (Figure 12.42). The appropriate amount of data filtering with the inverse velocity method varies with each application and depends on the type of monitoring system and the associated error or noise in the data. As velocities increase and inverse velocities correspondingly decrease, reductions in the filtered time increment may be required, to provide more accurate predictions of failure time. A review of the method and its potential limitations is given in Rose and Hungr (2007).

Forecasting rock slope failure is a complex problem involving observations, analysis and experienced judgment. In many cases, the inverse velocity method is a simple but powerful tool to aid the process.

Reporting conclusions

After each set of data has been interpreted, conclusions must be reported in the form of an interim monitoring report and submitted to personnel responsible for implementation of remedial actions indicated by the data. At the very least, management should be supplied with a monthly summary report of the results from the monitoring program, even if no movement is detected. A final report of the monitoring program is often required, and a technical paper may be prepared.

12.3 Ground control management plans

12.3.1 Introduction

Ground control management plans are vital to the safe conduct of mining operations in that they facilitate an effective risk management process. The plans document the geotechnical responsibilities at the mine and the basis for the slope designs, their implementation and the associated monitoring systems. They provide a form of communication and corporate governance reinforcing current practice, and are often required by corporate

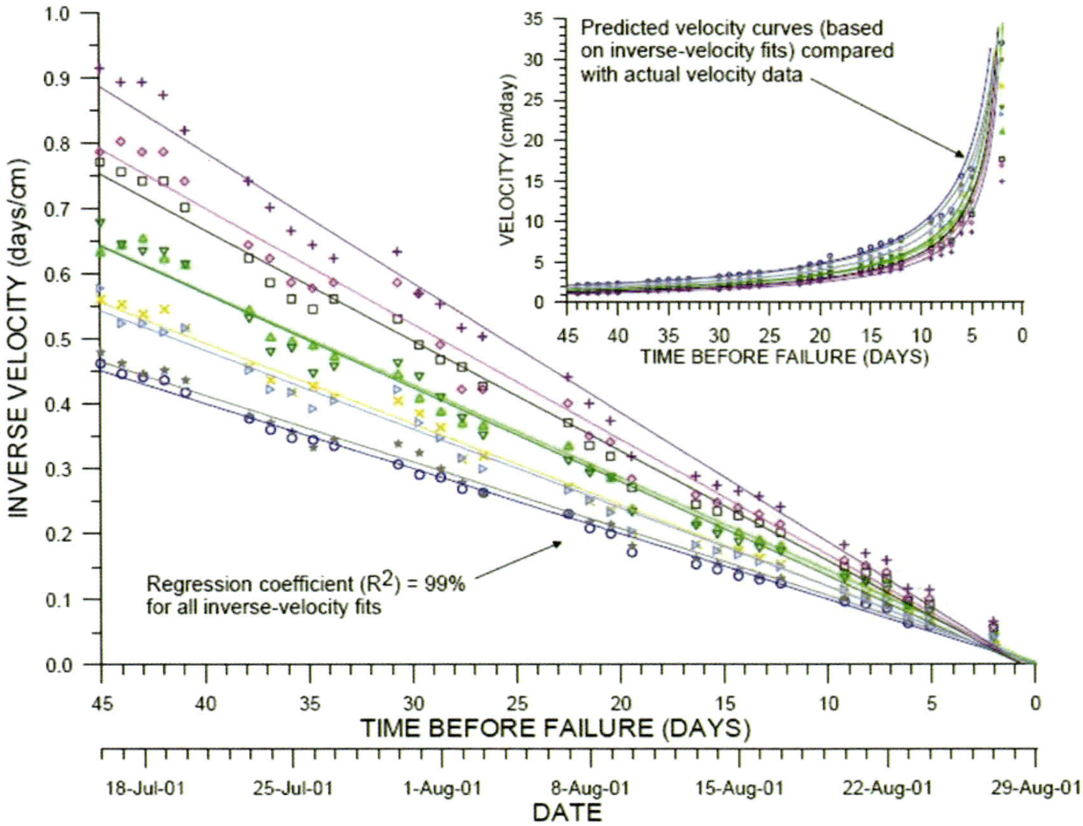

Figure 12.42: Plot of six-day average slope distance inverse velocity and velocity (predicted curves vs actual values on inset graph) vs time for nine prisms (time 0 was the observed time of failure)
Source: After Rose & Hungr (2007)

management and/or by regulators in the mining jurisdiction. Table 12.7 shows the type of information covered under geotechnical management plans used in open pit mines.

While the management plan should contain site-specific information, there is commonality among corporations in the conduct of safety, production and geotechnical management of open pit mines. The majority of geotechnical management plans are produced at the completion of the planning and design phase and form part of an overarching document.

12.3.2 Hazard management plan

Some companies have a slope stability (hazard management) plan which may be a subset of an overarching ground control management plan or may exist as a separate entity. A typical plan may include but not be limited to the following information:

- introduction;
- hazard inventory;
- risk reduction;
- trigger action responses (TARPs);
- emergency response procedures;
- roles and responsibilities;
- records, communication, training, document control;
- performance reviews.

Important features of this list are outlined in the sections below.

12.3.2.1 Introduction

The introduction should contain general statements as to the objective and scope of the plan, plus any relevant historic information used to generate the plan. Because of differences in regulatory requirements, ground conditions and mining methods, this section describes ground control management issues in general terms and outlines the suggested technical content of the plan. Based on information presented in this section, each individual operation should formulate and implement their site specific plan in accordance with local requirements and needs, and in some cases base parts of their plans on historic precedent.

Table 12.7: Type of information in a ground control management plan

Typical Content of Ground Control Management Plans	
Information	**Typical table of contents**

• Legislative requirements
• Hazard identification and risk management process
• Basic geotechnical domain model based on geological and geotechnical history of the operation and mining lease
  ▪ Regional and structural geology site characterisation
  ▪ Rock mass characterisation
  ▪ Groundwater distribution
  ▪ Stress and seismicity
• Ground control management plan and related standard operating procedures
  ▪ Data collection
  ▪ Modelling, analysis and pit slope and ground support design
  ▪ Excavation and performance monitoring
  ▪ Mitigation options and remedial measures
• Inspection and monitoring program
• Trigger action response plans
• Duties and responsibilities
• Training requirements
• Communication
• Audit, review and feedback process
	1 Introduction
	  ▪ Objective
	  ▪ Scope
	  ▪ Definitions
	  ▪ References
	  ▪ Mining environment description and characteristics
	  ▪ Relevant mine history
	2 Identified hazards
	3 Control procedures
	  ▪ Risk management system
	  ▪ Identification of high-risk environments
	  ▪ Communication protocols
	  ▪ Permits
	  ▪ Sampling and monitoring
	  ▪ Accepted movement threshold values
	  ▪ Formalised controls
	4 Roles and responsibilities
	5 Resources required
	  ▪ Training
	  ▪ Communication
	  ▪ Review and audits
	6 Trigger action response plans
	7 Communication
	8 Training
	9 Corrective actions
	  ▪ Audit (internal/external) or reviews
	  ▪ Verbal communications, incident reports
	  ▪ Non-conformance reporting
	10 Review/audit
	  ▪ Not exceeding 12 months, or when a significant change occurs
	11 Document dontrol
	12 Records
	  ▪ Safety management plans
	  ▪ Hazard management plans
	  ▪ TARPs
	  ▪ SOPs

In addition to the open pit, geotechnical engineers are often charged with the responsibility to design and operate all geotechnical aspects a mine property including tailings disposal, waste rock storage facilities and other civil geotechnical infrastructure. The ground control management plan approach can be used to encompass all mine facilities.

In order to maintain a consistent high standard, an effective plan should be based on corporate risk acceptance guidelines, local ground conditions and local regulatory requirements.

12.3.2.2 *Hazard inventory*

The geologic hazard inventory should describe the geologic hazards at the mine site. These are not limited to the open pit but at some mines should include other geologic hazards. Typical in-pit hazards include:

- rock fall;
- bench scale failure;
- underground workings (voids);
- inter-ramp pit wall instability;
- overall pit wall instability.

Other examples of geologic hazards which can affect mining operations include:

- liquefaction;
- floods;
- debris flows;
- natural rock fall;
- snow avalanche;
- slope creep related to plastic deformation within permafrost;
- rock slides or landslides (particularly fragmental rock fall);
- periodic seismic activity with the potential to trigger unwanted events.

Extreme weather at some mine sites (wind, snow, rain, fog) are often categorised using geologic hazard systems and managed in a similar fashion.

A useful means of communicating geotechnical hazards to mine personnel is a hazard identification plan, which illustrates areas of instability, rock fall hazards and other hazards (Figure 12.43). This type of plan should be developed by the geotechnical engineers in consultation with the mine operations group. It can be posted in the crew rooms and copies given to the mine field supervisors. The hazards map should be regularly maintained and should include:

- locations of all instabilities within the mine;
  - → active,
  - → non-active,
  - → under remediation.
- locations of geotechnical monitoring equipment in the mine;
  - → prisms (with most recent movement vectors and velocities),
  - → piezometers,
  - → inclinometers,
  - → extensometers.
- dewatering infrastructure (wells and drains);

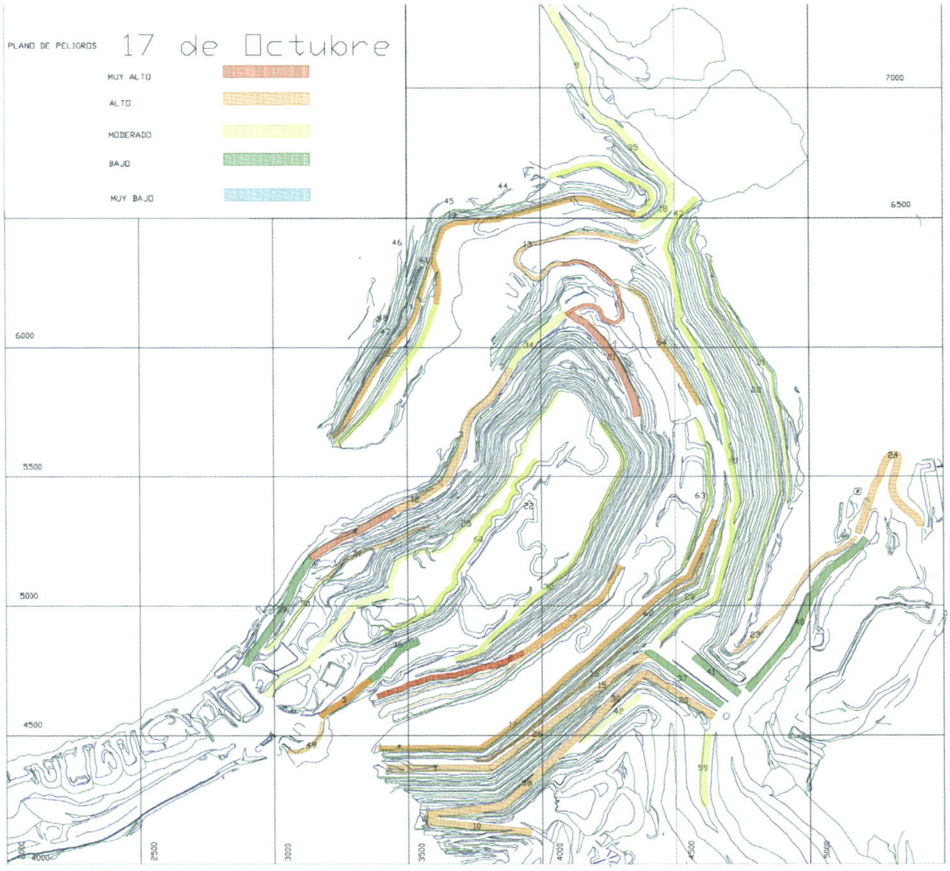

Figure 12.43: Hazard identification plan
Source: Courtesy Compañia Minera Antamina S.A

- surface water controls;
- other instrumentation.

Other items to locate on the map include:

- rock fall hazard areas;
- bench clean-up areas;
- voids;
- critical geotechnical sections.

12.3.2.3 Risk reduction

This section of the ground control management plan should describe methods utilised by the mine to reduce risks identified in the geologic hazard inventory. Typical open pit risk reductions methods employed include:

- detailed slope designs (specified slope angles catch benches, rock fall berms, geotechnical step-outs);
- groundwater depressurisation;
- slope performance monitoring (as-built slope configuration);
- bench clean-up and wall scaling;
- geologic and geotechnical mapping and geologic model reconciliation;
- regular visual monitoring (pre-shift inspections);
- slope deformation monitoring of surface and subsurface monitoring and analysis of conditions in real time;
- slope reinforcement (particularly around critical in-pit infrastructure);
- surface water management plans;
- blast energy monitoring;
- seismic monitoring;
- risk management plans and resulting geotechnical contingency or trigger response plans.

Risk management is discussed further in Chapter 13.

12.3.2.4 Trigger action responses

Most ground control management plans include a set of planned responses to likely events. These are commonly called trigger action responses or TARPs. Table 12.8 shows

Table 12.8: A TARP for pit slope condition

Alert	Green	Yellow	Orange	Red
Condition of pit slope	Long-term cracks	Existing cracks opening with new tensional cracks developing around crest	Rapid opening and slumping of cracks around crest or toe	Slope failing
	Minor floor heave	Floor heave requiring road maintenance	Rapid onset of floor heave	Imminent failure of slope indicated by rapid opening of cracks, floor heave, constant movement of spoil material
	<X mm day convergence	Rate of movement >X mm/hr and <Y mm/hr increasing rate over 12 hr period	Rate of movement Y mm/hr and <Z mm/hr increasing rate over 12 hr period	Rate of movement >Z mm/hr increasing rate over 12 hr period
	Deceleration in monitoring over X weeks (3 data sets)	Consistence acceleration in monitoring over 1 week (3 data sets)	Rapid acceleration in monitoring over 1 week (3 data sets)	Slope failure imminent
Responses				
Control			Contact Emergency Response Committee, mine geotechnical engineer and notify personnel of orange alarm level	Contact Emergency Response Committee, mine geotechnical engineer and notify personnel of red alarm level
Mine Manager		Monitor situation as required	Prepare to evacuate pit. Monitor situation as required	Evacuate pit, agree on recovery plan, notify corporate, mine inspectors, emergency services and monitor situation as required
Mine Superintendent	Monitor production activities	Monitor production activities. Communicate with mine geotechnical engineer	Liaise with shift supervisor, assess situation and inspect as required. Communicate with mine geotechnical engineer. Notify mine manager of the situation as appropriate	Inspect area from outside the failure zone and report to mine manager. Implement recovery plan once formulated (risk assessment required)
Shift Supervisor	Report with daily production plan	Monitor slope conditions throughout shift. Report any noticeable change in conditions to the mine geotechnical engineer. Report any change of conditions or change in TARP level to the next shift	Communicate with workforce that an orange level has been reached. Closely monitor slope conditions throughout shift. Report any noticeable change in conditions to the mine geotechnical engineer. Report any change of conditions or change in TARP level to the next shift	Communicate with workforce that a red level has been reached and withdraw personnel and equipment to a safe location. Secure to prevent entry. Inspect area from outside the failure zone and report to mine superintendent and mine manager immediately. Implement recovery plan once formulated (risk assessment required)
Geotechnical Engineer	Routine mapping and monitoring	Assess area. Determine frequency of inspections, monitoring and remedial work. Notify management of any change. Communicate with mine workers the location, nature and expected conditions associated with the failure	Evaluate the monitoring data and provide recommendation for TARP level advance. Assess area. Determine frequency of inspections, increased monitoring and remedial work. Notify management of any change	Inspect, investigate and formulate recovery plan (formal risk assessment required). Report findings to mine management
Mine Worker	Report with daily production plan	Become familiar with location and potential change in pit slope condition during shift. Report any significant change in conditions to shift supervisor. Report with daily production plan	Elevate level of awareness and monitor pit slope conditions during shift. Minimise 2-way chatter and provide feedback on pit slope conditions	Comply with emergency evacuation procedures and withdraw to a safe location

Source: Modified from Pisters (2005)

a basic TARP triggered by changes in the condition of the pit slope and lists what should be done by whom in response to the changed condition. Values for the X, Y and Z components noted in the table are site-specific and should be tailored to each mine site. They vary according to the rock mass characterisation, potential failure modes, rainfall and seismic activity.

12.3.2.5 Emergency response procedures

Despite the best practice of corporate governance and a robust risk and safety management system, it is still possible for an unplanned or low-probability/high-consequence event to occur. An emergency response plan can prevent an emergency turning into a disaster by minimising panic through training in procedures, effective communication to the workforce, implementation of evacuation protocols and having sufficient qualified resources on hand to minimise any impact on operations. Each site will be different and have unique risks to assess. It is important to assess the risks not only associated with evacuating mine site personnel and property critical to operations but also the downstream impacts of remediation, commodity supply and processing to ensure resilient business continuity in terms of a business continuity plan and effective business continuity management (BCM).

A BCM flowchart is outlined in Figure 12.44, taken from HB 293-2006 Executive Guide to Business Continuity Management, and provides a good overview of the BCM process and integrated responses.

In general, most BCMs involve some form of the following three response strategies.

1 Emergency response – first 48 hours of an emergency (e.g. TARP of the slope management plan was exceeded by a slope failure which enacted the emergency response plan and evacuated the pit).
2 Continuity response – impact on business and what is required to meet output demands and achieve business objectives (e.g. evaluate remediation cost–benefit analysis vs mine redesign, expected downtime and impact to business, production in another part of the pit).
3 Recovery response – predetermined management responses to a particular event based on maximum outages and recovery objectives (e.g. bench failure due to blasting results in remedial excavation back to a predetermined face angle, or mine an alternate access roadway).

Once developed, an emergency response plan and business continuity plan (including recovery) should be scenario tested, communicated to the workforce, resourced, personnel trained, practised and reviewed at regular intervals.

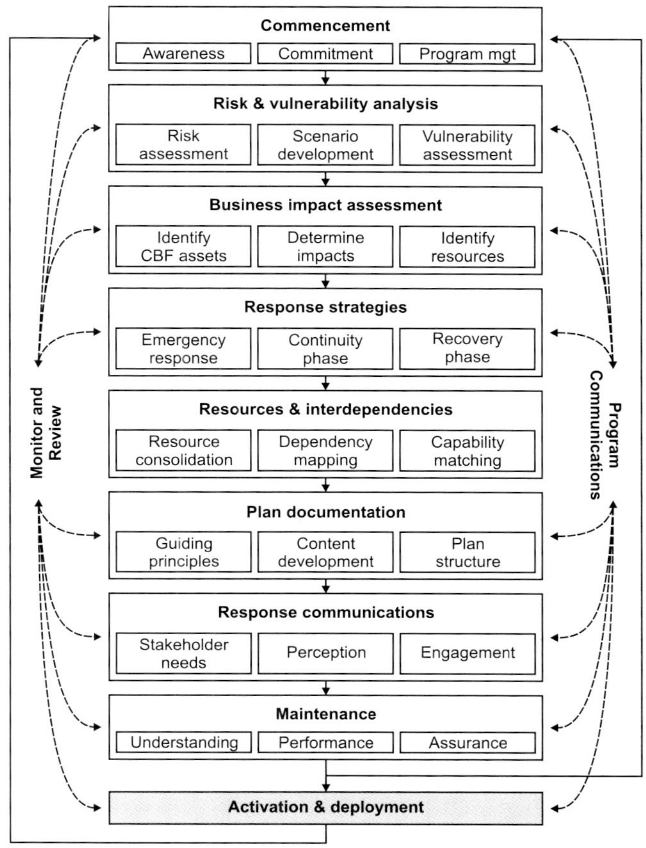

Figure 12.44: The BCM process (HB 293-2006)

The geotechnical engineer may be asked to provide input into the BCM process in the following areas.

In the area of emergency response:

- stability of emergency escape routes and safe locations;
- stability of access roadways for emergency response personnel;
- slope/water monitoring and pit slope management;
- slope failure investigation.

In the area of continuity response:

- cost–benefit analysis of remediation options;
- long-term impact to business objectives, life of open pit and stakeholders' interest (loss of ore recovery and potential loss of shareholder confidence);
- reassess the geotechnical model;
- pit redesign – stability models;
- operational rescheduling – alternate production sites and impact on mine design;
- slope/water monitoring and performance monitoring and management;
- slope failure investigation and management review.

Table 12.9: Contingency responses to slope movement

Approach	Rationale
Leave the unstable area alone	This option can be taken when instability is in an abandoned or inactive area, or the cost of remediation is excessive
Continue mining without changing the mine plan	If the velocity is low and predictable and the area must be mined, living with the displacement while continuing to mine may be an acceptable option. This course of action requires an effective and well-understood monitoring system, as well as a good understanding of the failure mechanics and historical information on local slope behaviour
Unload the slide through additional waste removal	Unloading is a common response and is usually a successful remediation option. However, there are situations involving high water pressure where unloading can decrease stability. The failure geometry and the failure mechanics must be understood to ensure that unloading will stabilise the failure
Leave a step-out; buttress	Step-outs have been used successfully. The choice between step-out and cleanup is determined by the trade-off between the value of the ore lost, ore deferred and the cost of cleanup. Buttressing to add passive resistance may be required
Do a partial cleanup	Partial cleanup may be the best choice where a failure mass blocks a haul road or fails onto the mining platform. Only the material necessary to get back into operation or to optimise the mine plan needs to be cleaned up
Mine out the failure	Where the failure occurs along a specific structure and there is competent rock behind the structure, mining out the failure mass may be the optimum choice
Support the unstable ground – ground support and/or buttressing	Mechanical ground support may be the most cost-effective option when a crusher, conveyor or haul road must be protected. Ground support is usually not a remediation option in a weak/deformable rock mass; buttressing the failed zone is often effective where toe support is required
Dewater the unstable area	Where pore pressures exist, dewatering/depressurisation is an effective method of stabilisation that may be used in conjunction with other options. There are few cases where it is cost-effective to live with the effects of high groundwater pressure

Source: Adapted from KCGM (2004)

In the area of recovery response:

- slope/water monitoring and management;
- slope stability assessment post failure – additional support requirements;
- geotechnical considerations for alternate access roadways;
- remediation of slope failure.

Table 12.9 gives a general list of possible options for a managing a moving slope.

12.3.2.6 Operational geotechnical roles and responsibilities

It is important to understand the roles and responsibilities of mine site personnel and how they interact as a team to provide effective geotechnical management. This example of roles and responsibilities for operational geotechnical management has been based on information provided by KGCM (2004).

Mine manager

Ensure that:

- suitably trained and qualified persons are formally appointed to key ground control positions: planning superintendent, senior geotechnical engineer, mine production superintendent;
- legislation is complied with;
- all ground-related hazards are identified and controlled to tolerable levels by the appointed staff, management systems and a ground control plan;
- resources are made available to achieve a high-quality ground control performance.

Planning superintendent

Ensure that:

- geotechnical department is provided with adequate staffing levels;
- adequate training is given to all ground control personnel;
- suitable equipment and monitoring instruments are supplied and maintained to the specification required;
- regulations are complied with;
- standard operating procedures are implemented and work practice regularly monitored;
- audit, review and quality assurance programs are carried out regularly and documented.

Mine production superintendent

Ensure that:

- operations are conducted in accordance with the relevant regulations;

- standard operating procedures are geotechnically sound and are implemented and work practices are regularly monitored;
- regular liaison occurs with mine planning, geology and geotechnical engineering staff.

Senior geotechnical engineer

Ensure that:

- the required geotechnical data is collected, analysed and interpreted;
- ground and slope performance monitoring systems are used and maintained;
- ground control plan is developed and implemented;
- all credible slope failure modes are considered;
- all significant ground-related incidents are inspected, evaluated and reported appropriately;
- all work activities and plans comply with regulatory requirements;
- liaison with technical services, geology and mine production occurs regularly;
- monthly reports are provided to management;
- audit, review and quality assurance programs are carried out regularly and documented.

Geotechnical engineers/geologists

Ensure that:

- standard operating procedures and work instructions are followed;
- ground conditions at active faces are inspected and monitored regularly;
- the required geotechnical data is collected, analysed and interpreted;
- all ground stabilisation measures are designed and implemented, based on geotechnical analysis;
- monitoring results are analysed and any anomalies reported;
- assist with pit design modifications based on new void or structural information;
- all significant slope failures are investigated and reported;
- liaise with production supervisors on daily basis;
- supervise slope stabilisation work.

All employees/contractors

Ensure that:

- no work is undertaken without an authorised plan;
- safe operating procedures are followed;
- all ground-related hazards are identified and reported to supervisor and/or geotechnical staff;
- ground conditions are inspected prior to and during work activities.

12.3.2.7 Recording events, communication, training and control documents

Recording events

Record keeping is an important component of a ground control management system for two reasons:

- open pit mining operations are required not only to consider the geotechnical aspects of the operation and to exercise due diligence in safety aspects, but are also required to be able to demonstrate that they have fulfilled these obligations;
- records are the only means by which personnel can measure the performance of the operation and therefore, can identify potential problems and deficiencies.

Recording slope instabilities is an essential prerequisite for the development of an optimised ground control system. Records of remedial actions and geotechnical instructions enable the mine to assess the effectiveness of the mine design and provide information for future mine development. During the process of developing or updating the ground control management plan, new standards, programs or procedures may be generated. All these new tasks or documents require additional resources. Therefore, commitments from the mine management and other relevant departments are critical in formulating the final plan, and more importantly, its implementation.

To identify potential ground control problems as they develop, so that they can be mitigated promptly, geotechnical personnel should make regular open pit walkover inspections of all safely accessible benches and pit crests that could be affected by or influence mining activity. Inspection records are also a source of input data for model calibration, establishment of the ground conditions model and back analysis. Additionally, many jurisdictions require mine operations to perform inspections of high wall conditions before the start of each shift (e.g. MSHA 30 CFR § 56.3401).

It is essential that all inspections be properly recorded and that an effective system is in place to transfer inspection recommendations into remedial works, preventive works, or detailed investigation. Where a data base exists, the inspection reports should be added to produce an historical record of the behavior of the pit slopes. The inspecting personnel should always check that the last sets of recommendations have been fully implemented. The 'red book' concept, or shift pass-on document, is a time tested method to communicate conditions on to the next shift.

All significant uncontrolled rock falls (e.g. size $>50m^3$), or any conditions where rock fall due to raveling is an ongoing concern, should be record regardless of whether injury or equipment damage has occurred. These include events such as debris-covered slopes in rainy or freeze-thaw conditions. Digital photography is essential in completing this task, making a fast and easily distributable document of conditions just after the event occurred.

Technical descriptions of all damaging incidents will assist the analysis of these events and develop a proactive response to them. A record of the analysis demonstrates an effort to understand and prevent these occurrences. In some cases the analysis will be incomplete or incorrect or it may not be possible to come to a definitive conclusion. However, over time, the compilation and evaluation of such records will result in a better understanding of the rock mass behaviour and will support optimisation of slope designs.

Some jurisdictions require incidents of specified magnitude to be reported. There are normally standard forms for such purpose so that records of these reportable incidents will be available at each mine site. For incidents not reportable to regulators, mines should have a mechanism in place to record and analyse them. Some mines record incidents in the shift logbook (or 'red book') and the geotechnical personnel sign and date each incident report to indicate that it has been read. A better system would be to maintain a database recording all incidents with defined parameters and additional comments. All reportable incidents have detailed geotechnical or investigation reports associated with them while non-reportable incidents are simply entries in the database. The database is easy to set up and useful for future data analysis such as performing back-analysis, evaluating the engineering geology model, and calibrating numerical models.

Communication

Communicating problems, challenges and solutions to the workforce is important for obtaining their active participation and cooperation. Simply making the effort to present the information emphasises the company's concern over these issues and reinforces the importance of reporting problems as they arise. Upon approval of the ground control management plan at a mine site, all new procedures, protocols, tasks and standards have to be effectively communicated to the workforce. Processes related to ground control should be periodically reviewed and audited against the existing plan and corrective measures taken if areas of improvement are identified. The plan has to be a living document requiring updates at least once a year.

Common means of communication are safety cards, shift logs ('red book'), shift change briefings, regular production meetings, and information posting and meetings. Some mines use Ground Control/Occurrence Report forms specifically for such purposes. In addition to these formal means, information exchange often happens at active workplaces during regular inspections and this type of information should be kept in the inspection records. Mines should select the best methods for communication with due consideration for their local workforce and regulations.

Training

The workforce should be trained to recognise indications of potential slope instability to enable them to inform technical staff of observed indicators and implement corrective adjustments to operations. This process creates a workforce that is actively engaged in problem recognition and resolution. Attendance of ground control training courses by all mine workers is often a jurisdictional requirements (MSHA, for example, in the USA requires this of all mine workers). The ground control management plan at each mine should include a training program involving the geotechnical staff and safety professionals in order to inform all workers how to recognise ground control hazards, who to contact should a hazard be identified, and what to do about the hazard.

Training content for employees in open pit mining is regulated in some jurisdictions and standardised training packages are developed by regulatory authorities for such purposes. Ground control components in training packages may include the general knowledge of ground hazard recognition, geological controls on stability, and support systems. In jurisdictions where training is not explicitly regulated, mines develop their own training packages with content similar to that in the standardised training packages.

In addition to the general knowledge on ground control aspects, most mines now have training programs that give employees a training period including both theory and practice of individual tasks, using manufacturer's training manuals if use of equipment is involved, and Standard Work/Operating Procedures (SWP/SOP) are developed in accordance with a job safety analysis. Annual refresher training on ground control for all employees is mandated in some jurisdictions and is provided annually or semi-annually at many operations where it is not mandated.

Document control

All pertinent documents and regular reports should be filed in a common directory controlled by the senior geotechnical engineers. Most modern mine sites have network serves on which these documents can be stored securely. Only the geotechnical engineer (or designee) should have rights to edit or add documents. Document and data read rights should be distributed to mine management or even external consultants on an as needed basis. All monthly reports and data files from monitoring and mapping should similarly be stored on these mine servers. It is important that managers realise up front the storage size requirements to support geotechnical data and document requirements.

12.3.2.8 Performance reviews

Where geotechnical staff are employed at site or consultants are used on a regular basis, it is suggested that external geotechnical performance reviews should be undertaken on an annual basis. This frequency of review is a nominal industry standard. To facilitate an objective process, it is recommended that an in-house and/or outside expert reviews the geotechnical performance of the mine. In this way a fresh pair of eyes is used to assess whether further geotechnical hazards are present, or likely to occur as part of the developing operation.

Reviews and audits are similar but distinct in the way they assess the geotechnical performance at a mine site. A review addresses the ability of the management practices to adequately address the hazards in the mine. An audit addresses whether or not the mine is following the procedures to reduce the risk in the identified hazard.

Geotechnical performance reviews should address both review and audit issues (section 1.6.2) as they are both related to overall geotechnical performance. This section uses the term review, but both review and audit concepts are included in the performance review concept.

Performance reviews must provide value to the mine operation, including not only the identification of unacceptable risk exposure and other performance gaps, but also to develop a mutually agreeable corrective mechanism. The geotechnical specialists (section 1.6.3) should form a partnership with on-site personnel, who have the ultimate responsibility for the successful operation of the facility in accordance with its production and cost targets.

The review process

In order to objectively assess an operation's compliance with the company-wide strategy of achieving and maintaining adequate ground control management plans, a means of auditing the operation has been developed. The objective of the review process is not only to evaluate the content and presentation of the ground control management plan, but also to assess the implementation of the plan into the routine activity at the mine.

For the review process to be successful, it has to provide something of value to those who are audited. This means that the process must be conducted in a constructive atmosphere that is respectful of the needs of both the operation and the reviewer.

The review should identify any discrepancies and gaps that exist in the system so that a clear and practical plan can be devised to address the outstanding issues in a reasonable timeframe.

Methodology – assessment approach

By adopting a practical and less rigorous process, ground control issues at the mine can be identified in several days. The review will typically be performed by a team of no more than three professionals (internal and/or external).

The ground control management plan should form the basis of the review questionnaire. There should be three aspects of the review; the presentation of the ground control plan as a document, compliance with ground control activities summarised in the plan, and the effectiveness of the plan in terms of reducing the potential for ground instability incidents.

It should be recognised that many mines continue to rely on empirical evidence and personal experience rather than analytical results and detailed engineering. These two extremes are not mutually exclusive and the primary objective of the audit is to identify issues that need to be improved so that the operation can fulfill its own production, cost and safety targets, corporate objectives, and local regulatory requirements.

The performance of a mine should be evaluated through subjective assessment criteria. Although subjective, these criteria have proven remarkably robust and two auditors seldom disagree by more than one category. Discussion of the reason for the discrepancy can usually result in agreement.

At the end of the review the findings should be discussed with the site management team and a work program agreed and documented.

13 RISK MANAGEMENT

Ted Brown and Alison Booth

13.1 Introduction

13.1.1 Background

Risk is inherent in most human activity. It is an ever-present feature of business and engineering undertakings and of modern life. At the general level, significant risks may be associated with accidents of several types (domestic, industrial, traffic), natural hazards (earthquakes, landslides, hurricanes), disease, business or financial undertakings, security and the risk of conflict, and failures of engineered structures and systems. At a personal level, risks can result in loss of property, financial loss and personal injury or, ultimately, the loss of life. Hambly and Hambly (1994) provide a concise overview of the levels of risk associated with a range of everyday and engineering activities.

In the past few decades, formal procedures have been developed for the assessment and management of risk for a wide range of purposes, including safety, in almost all industries, business and financial management and government services and agencies. Formal risk assessment and management has a longer history in the aviation, military, nuclear, petrochemical and space industries, for example, than it does in the minerals and some other industries. Given that mining has always been considered a high-risk business from both safety and economic aspects (section 1.4.8), it is only logical that risk management should be adopted as a standard procedure by the mining industry.

In general, the development and implementation of proactive approaches to risk management in place of the former 'fix it when it breaks' approach was triggered by major disasters that entered the public domain (Joy & Griffiths 2005). There are now national standards for risk management, codes of practice for a range of applications and formal requirements for the implementation of risk management procedures as part of corporate governance processes. Risk assessment and risk management involve specialist university courses, books, research journals, conferences, professional organisations and specialist practitioners.

If used effectively, risk management can be a powerful decision-making and management tool. However, it is not a panacea. To be effective it requires perception and detailed understanding of the range of risks involved in an undertaking, the development and implementation of risk assessment and risk management procedures, the leadership of management in developing a risk culture, documented procedures, high levels of communication and consultation, and personnel training and commitment.

13.1.2 Purpose and content of this chapter

The purpose of this chapter is to show how general risk management concepts and processes can be applied to the geotechnical risks associated with each stage of the open pit slope design process (Figure 13.1). In section 13.2, terminology will be defined, the general risk management process will be outlined and illustrated and the wider use of risk management in the minerals industry will be discussed for background and context. Section 13.3 provides an overview of how to apply risk assessment and management procedures to the evaluation and management of the geotechnical risks associated with large open pit slopes. Details of applicable risk assessment methodologies, including risk identification, risk analysis processes and tools, the role of data uncertainty and risk evaluation are given in section 13.4. Finally, section 13.5 discusses the treatment, control and management of geotechnical risk in large open pit slopes.

This chapter draws together and illustrates how the concepts and procedures discussed elsewhere in this book can be applied to open pit slope design and management. In particular, it complements, and has direct links to, Chapter 1: Fundamentals of slope design, Chapter 8: Data uncertainty, Chapter 9: Acceptance criteria, Chapter 11:

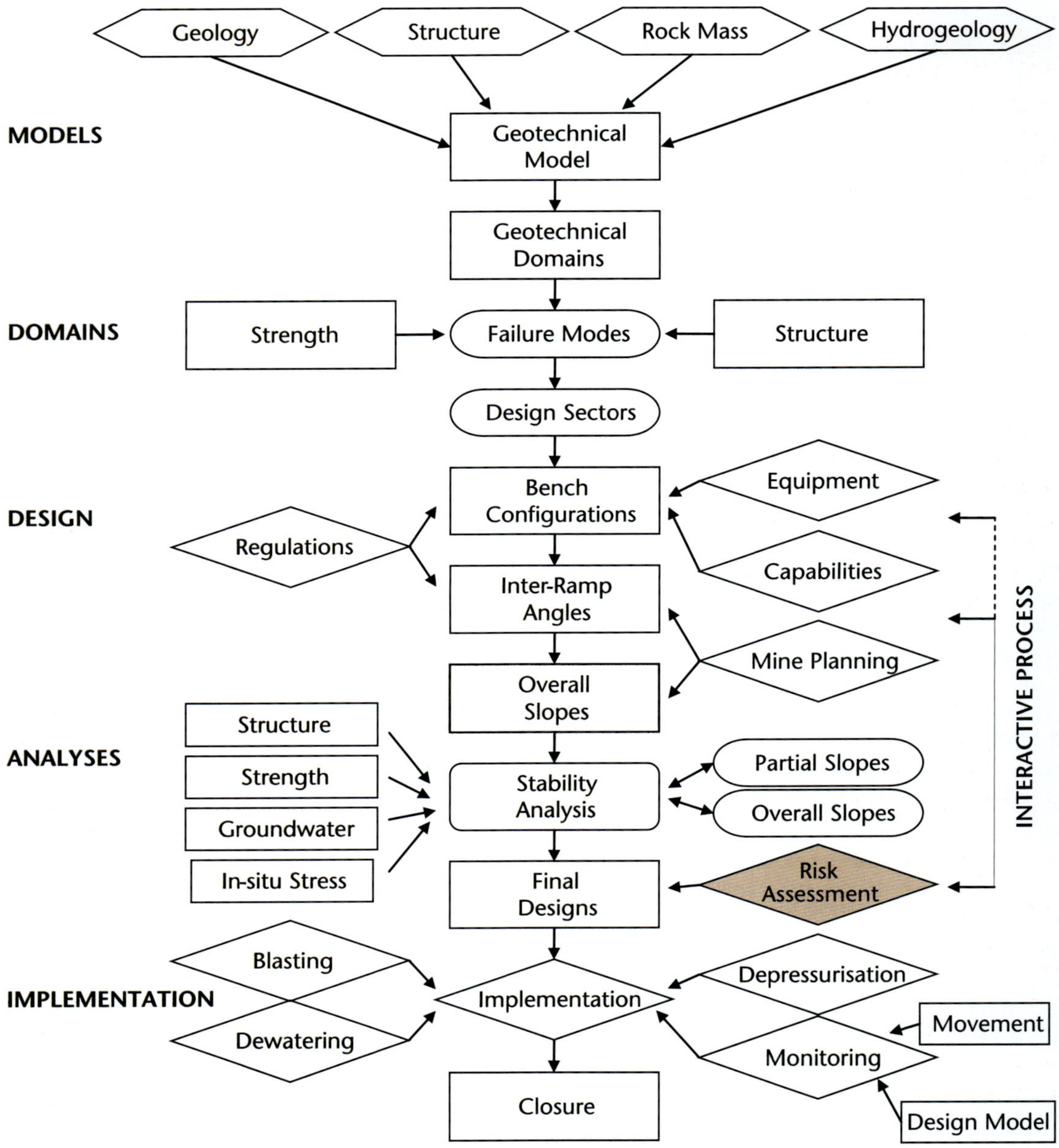

Figure 13.1: Slope design process

Design implementation and Chapter 12: Performance assessment and monitoring. Material in those chapters will be referred to, but not repeated, in this chapter.

13.1.3 Sources of information

This chapter draws on a wide range of published and unpublished information. The published information is referenced in the text (e.g. Aven 2003; Calderón & Tapia 2006; Joy & Griffiths 2005; Karzulovic 2004; Standards Australia 2004). Details of the publications are given in the References at the end of this book. Unpublished information was provided by a number of individuals who contributed to the preparation of this chapter and by a number of sponsor companies of the Large Open Pit Project. Where possible, these contributions are acknowledged in the text and in the References.

As indicated in section 13.1.1, risk management is now applied for a number of purposes in a wide range of enterprises of almost every conceivable type. This has spawned a burgeoning literature on the subject. For example, an internet search on risk management on the websites of a few major international publishers produced hundreds of items. Some of the range may be of interest to readers who require more detail on particular issues and techniques touched on in this chapter. Relevant specialist journals include *Journal of Risk and Uncertainty, Reliability Engineering and System Safety, Risk Analysis, Risk, Decision and Policy* and *Risk Management*. Textbooks with detailed information include those by Aven (2003), Bedford and Cooke (2001), van Steveren (2006) and Vose (2000). There is also a wide range of electronic sources.

13.2 Overview of risk management

13.2.1 Definitions

A large number of terms are associated with risk management and the processes it involves. The detailed definitions of these terms may differ depending on the context and country from which they emanate, but there is broad agreement about the meanings of the major terms used in this chapter. The following definitions of key terms are based on those of the International Organisation for Standardisation's (2002) ISO/IEC Guide 73 on risk management vocabulary, and the Australian and New Zealand Standard on Risk Management, AS/NZS 4360: 2004 (Standards Australia 2004):

- consequence – the outcome or impact of an event;
- hazard – a source of potential harm; a potential occurrence or condition that could lead to injury, damage to the environment, delay or economic loss;
- likelihood – the probability or frequency of occurrence of an event, described in qualitative or quantitative terms;
- risk – the chance of something happening that will have an impact on objectives;
- risk analysis – a systematic process to understand the nature of and deduce the level of risk;
- risk assessment – the overall process of risk identification, risk analysis and risk evaluation;
- risk criteria – the terms of reference by which the significance of risk is assessed;
- risk evaluation – the process of comparing the level of risk against risk criteria;
- risk identification – the process of determining what, where, when, why and how something could happen;
- risk management – the culture, processes and structures directed towards realising potential opportunities while managing adverse effects;
- risk treatment – the process of selecting and implementing measures to modify risk.

13.2.2 General risk management process

The general risk management process to be developed and applied here is that used in AS/NZ 4360: 2004 (Standards Australia 2004), illustrated in Figure 13.2. Other risk management processes, such as that developed by the Institute of Risk Management in the UK (Institute of Risk Management 2002) are similar but not necessarily identical to that used here.

As shown in Figure 13.2, the process follows a number of clearly defined and inter-related steps:

- establish the context – establish the external, internal and risk management contexts in which the rest of the process will take place. Establish the criteria against which risk will be evaluated and define the structure of the analysis;
- identify the risks – identify where, when, why and how events could prevent, degrade, delay or enhance achievement of the objectives;
- analyse the risks – identify and evaluate the existing controls. Determine the consequences and likelihoods of particular occurrences and therefore the associated levels of risk, considering the range of potential consequences and how these could occur. Generally, the risk is quantified as the product of the likelihood and consequence of the particular occurrence;
- evaluate the risks – compare estimated levels of risk against the pre-established criteria and consider the balance between potential benefits and adverse outcomes. This enables decisions to be made about the extent and nature of treatments required and their priorities;
- treat the risks – develop and implement specific cost-effective strategies and action plans for increasing potential benefits and reducing potential costs or adverse effects;
- monitor and review – it is necessary to monitor and review progress and the effectiveness of all steps in the risk management process to ensure continuous improvement and that the risk management plan is implemented effectively and remains relevant.

The first three steps are regarded as comprising risk analysis, while risk assessment involves those steps plus risk evaluation. Figure 13.2 shows that communication and consultation is required at every stage in the process, and that monitoring and review create feedback loops that may require modifications to earlier results.

It will be necessary to adapt this general procedure to take account of the special features and factors involved in a particular risk management exercise. For example, Figure 13.3 illustrates the adaptation of the general process to the risk management of landslides undertaken after a disastrous landslide in the Snowy Mountains region of New South Wales, Australia, in 1997. Full details of the landslide risk management concepts and guidelines developed are given by the Australian Geomechanics Society (AGS) Subcommittee on Landslide Risk Management (2000). Practice note guidelines are given by the AGS Landslide Taskforce Practice Note Working

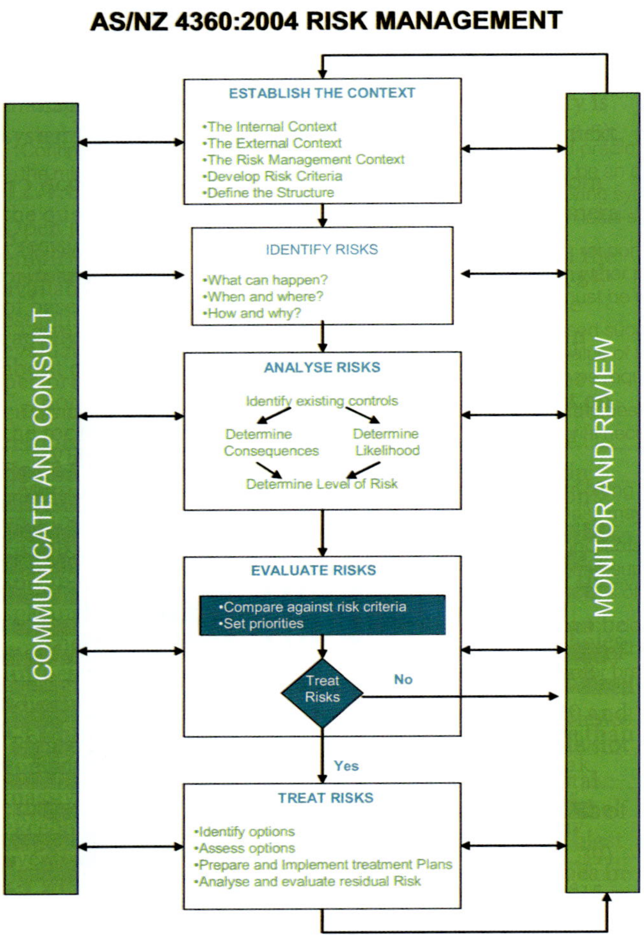

Figure 13.2: Risk management process
Source: Standards Australia (2004)

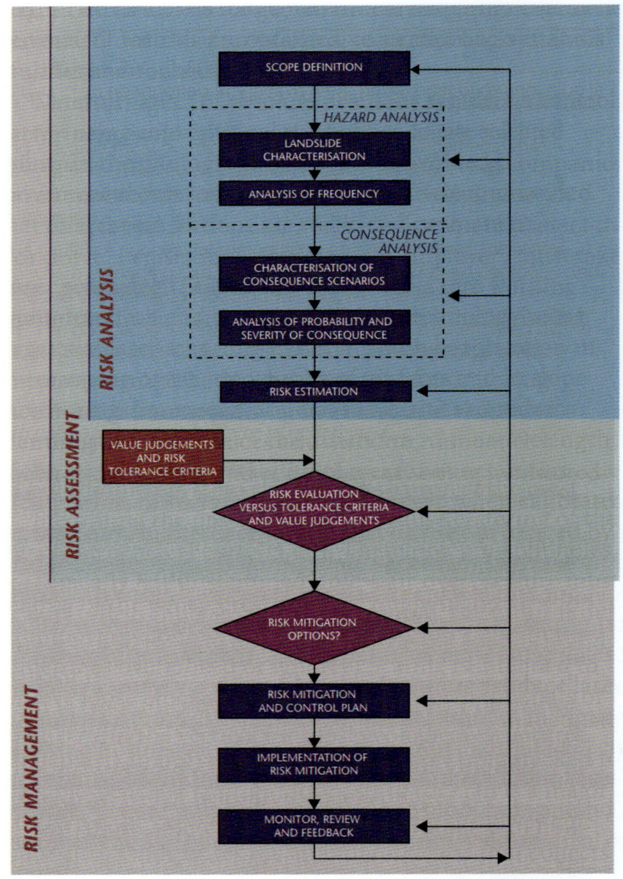

Figure 13.3: Flowchart for landslide risk management
Source: Fell et al. (2005), Leventhal (2007)

Group (2007), while overviews of the framework are given by Fell et al. (2005) and Leventhal (2007).

13.2.3 Risk management in the minerals industry

Throughout its long history, the mining industry has been plagued by the economic failure of mining ventures through various causes and by damage to mining infrastructure, and injury to and loss of life, arising from hazards such as water inrushes, rockfalls, rock bursts and gas outbursts. Safety issues associated with the use mechanical equipment in surface and underground mining have long been of concern. It is not surprising that the international mining industry is now using formal and systematic risk assessment and management procedures in business and operational applications.

Joy and Griffiths (2005) noted that the Australian industry has applied these procedures since 1990. However, they also concluded that 'the growth of methods and competency has been erratic in many ways leading to issues with the quality of risk assessment application'. The Australian *National Minerals Industry Safety and Health Risk Assessment Guidelines* (Joy & Griffiths 2005) was developed partly as a result of this perception. Earlier, in New South Wales a *Risk Management Handbook for the Mining Industry* was developed by the NSW Department of Mineral Resources (1997). More specialised approaches to risk assessment and management address the hazards and categories of risk in particular types of mining, such as block and panel caving (Brown 2007). Systematic safety assessment techniques have a longer history in the international minerals industry than the more broadly based formal risk management methods discussed in this chapter (Joy 2004).

Most companies in the minerals industry have corporate risk management requirements and procedures, as do the operating sites. These cover most aspects of the corporate governance, business continuity, exploration, project development and mining and minerals processing operations. Quinlivan and Lewis (2007) cited a multinational mining company's use of the overall risk

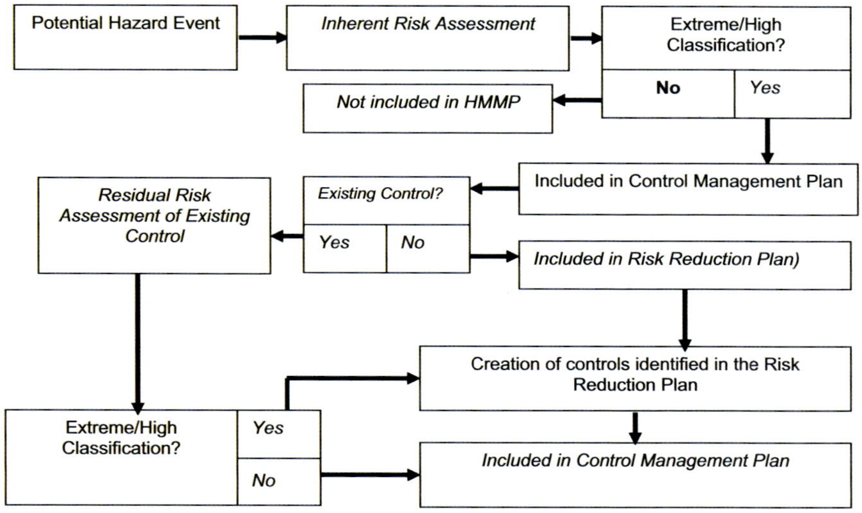

Figure 13.4: An operational risk management decision-making process
Source: Courtesy Newcrest Mining Limited

management approach discussed here. At the project level, the assessment and management of risk are now important components of the studies at all levels of the project, from conceptual studies (Level 1) to operations (Level 5). They are particularly applied to the management of safety and health risks in the minerals industry (Joy & Griffiths, 2005; MIRMgate 2007). Related approaches, generically known as system safety accident investigation (SSAI) techniques, are used in the investigation of accidents and other incidents (Gibb et al. 2004, Joy 2004). Figure 13.4 shows the risk management decision process adopted by Newcrest Mining's Cadia Valley Operations (CVO 2003). This approach provided the basis for development of a range of operational risk and hazard management plans, including the air inrush hazard management plan described by Logan and Tyler (2004).

Not all companies and operations use risk management techniques to the same degree. Management must take responsibility for the development of a risk-averse culture and the use of risk management approaches within their organisations. The minerals industry risk management maturity chart shown in Table 13.1 (Joy 2005) illustrates how a company's risk culture can improve and mature by increasing employees' awareness of risk and introducing risk assessment and management procedures.

13.3 Geotechnical risk management for open pit slopes

This section looks at the application of the principles and general risk management procedures already introduced, to the assessment and management of the geotechnical hazards and risks arising in the various stages of a large open pit project. It represents the first stage, 'Establish the context', in Figure 13.2.

In the risk management approach in AS/NZS 4360: 2004 (Standards Australia 2004), establishing the context involves a number of steps – establishing the external context, establishing the internal context, establishing the risk management context, developing risk criteria and defining the rest of the process. In terms of the geotechnical risks associated with large open pit slopes, the first four steps may be interpreted in the following ways.

1 Establishing the external context involves defining the external environment in which the open pit operates and the relationship between the mining organisation and that external environment. This involves considerations of the key business drivers, external stakeholders and their perceptions of the mine and the organisation, and the business, social, regulatory, cultural, competitive, financial and political environments in which the open pit operates. Many of these issues are not normally regarded as within the purview of geotechnical or slope engineering but, when the risk of overall slope failure in a large open pit is being evaluated, they need to be considered.
2 Establishing the internal context requires an understanding of the business and other (e.g. social, environmental) goals and objectives of the project or operation, and the strategies in place to achieve them. It also involves an understanding of the internal stakeholders, the culture of the organisation (including its tolerance of risk) and the structure and capabilities of the organisation in terms of people, systems, processes and access to capital. These may influence

Table 13.1: Minerals industry risk management maturity chart

Minerals Industry Risk Management (MIRM) Maturity Chart

• No care culture
  • Apathy/resistance
  • Near misses not considered
  • Negligence
  • Dishonesty
  • Hiding of incidents
• No or little training
• Poor or no communication
	• Blame culture
	  • Accept need to care
	  • Some near miss reporting
	  • Some window dressing e.g. pre-inspection cleanups and light duty
	  • Disciplinary action
	• Minimum/inconsistent training
	• Some communication on a need-to-know basis
		• Compliance culture
		  • Some participation
		  • Near miss discussions
		• Acceptable training/awarenes s
		• Established and good communication channels
		• Regular people involvement and focus
			• Ownership culture
			  • Involvement at all levels
			  • Near miss involvement
			• High level of training/awareness
			• Communication at a high level hiding nothing
				• Way of life
				  • Comes natural
				  • Personal involvement by all to prevent incidents
				• Complete understanding
				• All informed at all times about everything

Accept that incidents happen	**Prevent a similar incident**	**Prevent incidents before they occur**	**Improve the systems**	**Way we do business**
Vulnerable	**Reactive**	**Compliant**	**Proactive**	**Resilient**

• Reactive approach
• No systems
• **No risk assessment**
• Legal non compliance
• Accept equipment / process decay
• Superficial incident investigation
• **Poor investigation**
• **No monitoring/audits**
• Permit non-compliance
• Potential illegal practices
	• Administrator driven
	• Loose systems, elements of a HS Management System
	• **Re-active risk assessment**
	• Minimum legal compliance
	• Apply PPE as a way of eliminating exposure
	• **Incident investigation but limited analysis**
	  • Focus on what happened
	  • No systems focus
	  • Human fault focus
	• **Ad hoc monitoring/ audits**
	• No occupational hygiene or health initiatives
	• Reactive medical monitoring
	• Monitoring as per regulations
		• OH&S Coord. driven
		• OH&S stds system and ISO 9002 or equivalent
		• **Risk assessment through existing systems**
		• Total legal compliance
		• Strictly enforce the use of PPE where required (knowing risk)
		• **Causal incident analysis based on event potential**
		• Info sharing from events
		• Planned occupational hygiene /environmental monitor ing
		• Periodical medical examinations
		• **Planned monitoring/audits**
		• Safety meetings & talks
		• Some task observations
			• Line driven systems improvement
			• ISO 14001 and OHSAS 18000 or equivalent
			• **Pro-active formal risk assmt**
			• Beyond legal compliance
			• Seek to actively engineer out process/equipment inadequacies
			• **Incident learnings shared with all levels**
			• Well designed plans/procedures
			• **Focus on adhering to site plans and procedures**
			• **Integrated audits**
			• Peer evaluation and discussion
				• Individually internalised
				• Integrated management systems
				• **Risk assessment integrated into all systems**
				• Self regulating style
				• **Eliminate problems before they occur**
				• All threats considered in decision-making
				• **Systems enhancement through external evaluation/auditing**

the controls or treatments used to limit the likelihood or consequences of slope failures, for example.

3 Establishing the risk management context sets the objectives, boundaries and scope of the risk management process for the activity or part of the organisation concerned. In the case of large open pit slopes, the objectives are usually to maximise both safety and economic returns. This stage also involves the definition of the roles and responsibilities of various parts of the organisation and the individuals participating in the risk management process, the resources required and the records to be kept.

4 Developing risk criteria identifies the criteria (operational, technical, financial, social, legal, environmental) against which geotechnical risk is to be evaluated. In a large open pit, the geotechnical risks are mainly but not only safety and economic risks. Acceptability criteria for these risks are discussed in Chapter 9. In open pit slope design, economic risk criteria may be based on the results of risk–return or cost–benefit analyses carried out at each stage of the project.

The geotechnical risks associated with large open pit slopes mostly arise from the uncertainties inherent in the qualitative and quantitative descriptions of the rock masses concerned (i.e. the uncertainties in the geotechnical model) and from the uncertainties associated with analyses carried out using that model. The first type of uncertainty arises particularly from inherent variability and measurement error, while the second is sometimes known as transformation uncertainty (Phoon & Kulhawy 1999).

In earlier chapters, it was shown how the geotechnical model for an open pit slope is based on geological, structural, rock mass and hydrogeological models. It was also noted, as illustrated in Table 8.1, that the levels of confidence associated with these constituent models and with the resulting geotechnical model increase as the project moves from the conceptual (Level 1) to the pre-feasibility (Level 2), feasibility (Level 3), design and construction (Level 4) to operational (Level 5) stages. It follows that updated risk assessments and risk management plans will be required at each stage.

Figure 13.5 shows a generic flowchart for the open pit design process (Steffen et al. 2006). It illustrates the interacting roles of the geotechnical, mine planning and management teams in the overall process. Figure 13.5 also indicates that slope designs may be based on factor of safety (FoS) or probability of failure (PoF) criteria and their corresponding acceptance levels as discussed in detail in Chapters 8 and 9. As the geotechnical model is refined through the successive project stages, these criteria and acceptance levels, and the acceptable levels of economic and safety risk, also have to be refined. In other words, new designs and risk assessments must be produced at each stage of the project.

Using a risk-based approach to geotechnical slope design throughout the life of an open pit has a number of advantages over more deterministic approaches. It provides probabilistically established likelihoods and consequences of slope failures of various scales and types. It gives management the opportunity and responsibility to define acceptable levels of safety and economic risk at each stage. It quantifies the levels of risk associated with different slope configurations and provides a basis for the development of geotechnical and slope management plans. This involves quantifying the economic value added with increased levels of risk. Examples of risk-based approaches to geotechnical slope design are given by Calderón and Tapia (2006), Johnson et al. (2007), Pothitos and Li (2007) and Tapia et al. (2007).

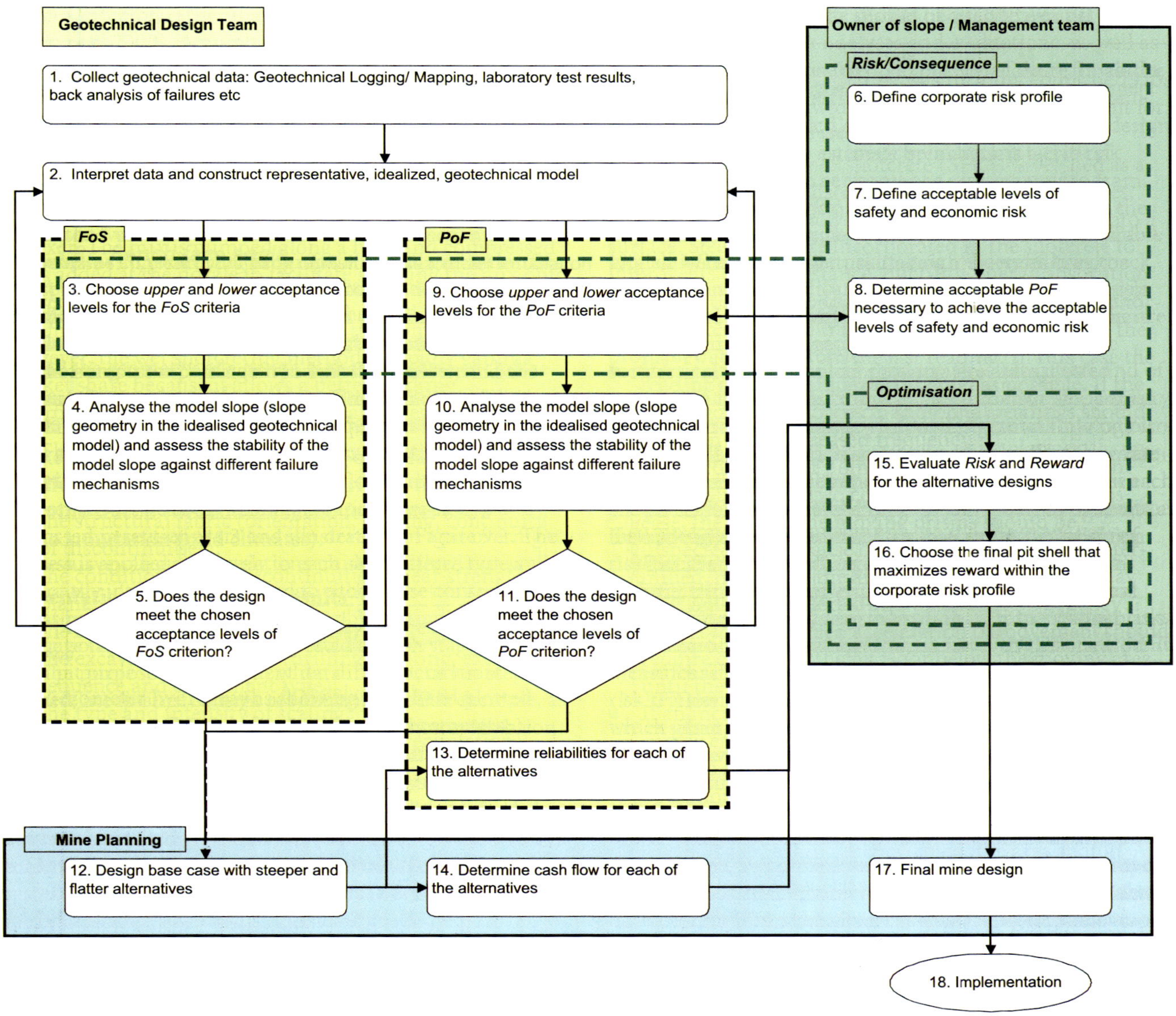

Figure 13.5: Flowchart for the open pit slope design process
Source: Steffen et al. (2006)

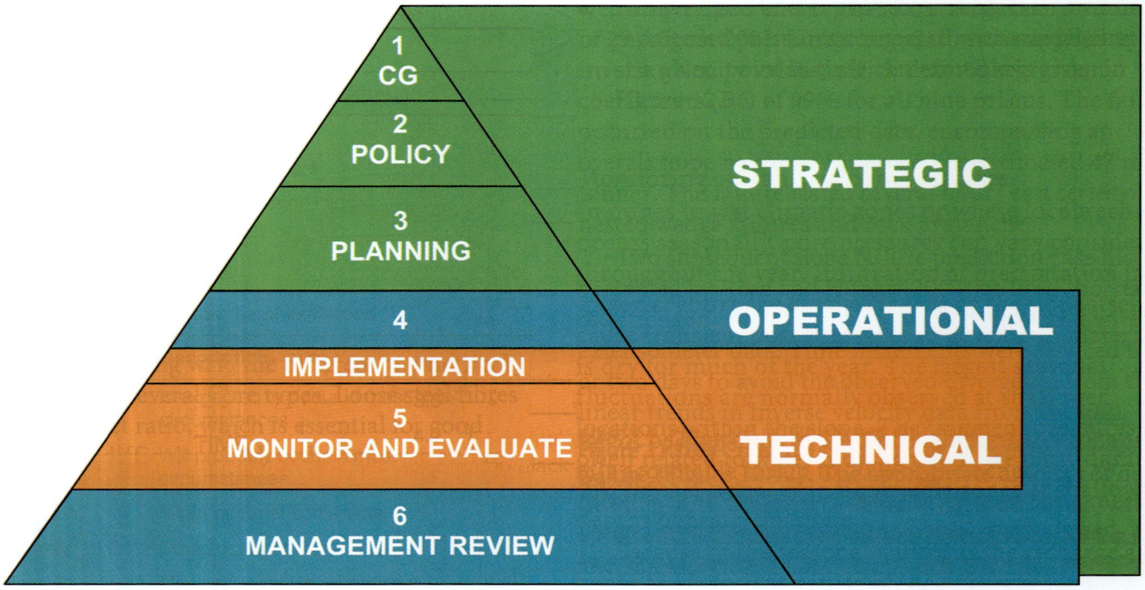

1	**Corporate Governance** – Establish the overarching Risk or Safety Management System.
2	**Policy** – Company directed policy on occupational health and safety (OHS) and acceptance criteria for geotechnical considerations in the life of the open pit
3	**Planning** – covering Stages 1-5 (Conceptual through Operations) 3.1 Risk management and generic business continuity management 3.2 Legal requirements and compliance with standards 3.3 Business objectives, targets, plans, risk benefit analysis, geotechnical model, pit design.
4	**Implementation** of plans, procedures, standard operating procedures (SOPs) and records at the design, operational and closure/transitional levels 4.1 Organisational structure, roles and responsibility 4.2 Operational risk management – geotechnical model, pit design, implementation, slope management plan, geotechnical procedures, eg, mapping and monitoring 4.3 Business continuity management 4.4 Consultation, communication and reporting 4.3 Training and competency 4.5 Documentation and data control
5	**Monitor and Evaluate** – stability management, mine to design, design performance 5.1 Monitoring and measurement 5.2 Incident investigation, corrective action and preventative action 5.3 Records and record management 5.4 Audit - internal and external
6	**Management Review**

Figure 13.6: Overarching methodology for managing the risk of open pit slope failure

Figure 13.6 shows a flowchart for the overall management of the risk of slope failure, developed from information on corporate processes provided by a number of sponsors and a generic occupational health and safety management system (Monash University 2006). It shows how strategic, operational and technical considerations are involved at the corporate governance, policy, planning, implementation, monitoring and evaluation and management review levels of the overall risk management process.

The focus in this book is on the geotechnical risks associated with large open pit slopes. However, just as there is uncertainty about the geotechnical model at the various stages of an open pit project, similar uncertainties about the resource model can have major impacts on the open pit and slope designs. Risk assessment and management techniques of the type discussed in this chapter may be used to address such risks throughout the open pit evaluation and design process (Steffen 1997, 2007).

13.4 Risk assessment methodologies

13.4.1 Approaches to risk assessment

Risk assessment is the overall process of risk identification, risk analysis and risk evaluation. These three stages of the risk assessment process will be discussed in the context of the geotechnical risks associated with large open pit slopes.

Generally, risk assessment methods, in particular the methods used in the risk analysis stage, may be categorised as qualitative, semi-quantitative or quantitative. The following brief accounts of these approaches are based on those given in AS/NZS 4360: 2004 (Standards Australia 2004).

Qualitative methods use verbal descriptions of the likelihoods that particular events will occur and of the magnitudes of the potential consequences of those events. This approach may be used in the initial screening to identify risks requiring more detailed analysis (where it is appropriate for the decisions required) or where the data or resources available are inadequate for quantitative analyses.

Semi-quantitative methods apply weightings or scales to the qualitative descriptions of likelihoods and/or consequences in order to produce more detailed ranking scales than those usually obtained by qualitative analyses. Care must be taken because the scales and numbers chosen may not properly reflect relativities, leading to inconsistent, anomalous or inappropriate outcomes. Semi-quantitative analysis may not differentiate properly between risks, particularly when the likelihoods or consequences are extreme. In some cases, it may be possible to partly overcome these difficulties by using qualitative consequence descriptions in conjunction with quantitative likelihood data based on probabilistic analysis. A semi-quantitative approach may be useful in ranking and prioritising risks.

Quantitative methods use numerical values, not descriptive scales, for the likelihood and consequences of an occurrence. This approach depends on the accuracy and completeness of the numerical data available and the validity of the models. The consequences may be expressed in terms of monetary, technical, operational or human impact (e.g. safety) criteria. Different criteria, values and ways of combining likelihood and consequence may be required at the different stages of a project. The uncertainties in the likelihood and consequences should be considered in the assessment and communicated effectively. This approach provides the clearest and most useful outcomes when valid tools and data are used.

Table 13.2 ranks the suitability of these approaches in assessing risk for a range of purposes. The quantitative approach is most useful for risk assessments of large open pit slopes, particularly in the operational stage (Level 5). Calderón and Tapia (2006), Karzulovic (2004) and Tapia et al. (2007) provide examples of its application in large open pits in Chile. Calderón and Tapia (2006) argue that a quantified risk assessment procedure offers advantages in that it:

Table 13.2: Suitability of risk assessment approaches

Suitability	Qualitative	Semi-quantitative	Quantitative
To improve an actual problem	Satisfactory	Good	Not necessary
To plan for change	Satisfactory	Good	Not necessary
To select the best option	Poor	Satisfactory	Good if data are available
To decide on acceptability of risk	Inadequate	Satisfactory if the acceptability rank is set	Good if data are available

Source: Lilly (2005)

- defines acceptable risk levels in terms of working safety and economics;
- allows meaningful comparisons of competing slope designs;
- provides detailed information for use in managing risk.

Calderón and Tapia (2004) also noted that the use of quantified risk assessment can guide:

- geotechnical engineers where to collect more information, where to improve slope monitoring and where to improve bench excavation procedures;
- hydrogeologists where to collect more information and where to improve drainage;
- mine planners where to provide flexibility in slope design.

13.4.2 Risk identification

This step identifies the risks that have to be analysed and managed. The comprehensive identification of risks using a well-structured systematic process is critical because a risk not identified at this stage may be excluded from further analysis. The objective is to identify what, where, when, why and how something might happen that represents a hazard or a risk.

Generally, in a mining context, the identification of risk begins with the identification of hazards or sources of potential harm. Generically, the identification of hazards is typically addressed in terms of potential energy sources. Joy and Griffiths (2005) identify ten potential energy sources of interest – gravity, electrical, mechanical, chemical, pressure, noise, thermal, radiation, body mechanics and biological.

In open pits, slope designers and geotechnical engineers are primarily concerned with identifying and controlling gravitational energies to prevent and manage rockfalls and slope failures at various scales. Other energy sources which may trigger or contribute to the likelihood of gravitational energies inducing falls or failures include:

- pressure energies in terms of elevated pore or joint water pressures and/or an elevated stress environment which may require slope depressurisation;
- thermal energies in more extreme climates where diurnal temperature variations could cause frost jacking or exfoliation of the rock mass;
- chemical energies in explosives which may cause significant blast damage to the rock face, or in acidic mine waters which may increase weathering rates or corrode secondary support and reinforcement systems.

Once the hazards have been identified, it is necessary to determine the associated risks in terms of what can happen, when and where, how events affect the objectives of the operation and why. The following list of information sources and possible approaches, modified from UNSW (2003), may provide a useful starting point:

- draw on past experience/history of the open pit, neighbouring pits, other locations within the industry, previous records, industry-wide information;
- check compliance with standards and/or regulatory requirements;
- build a database to maintain a record and associated statistics. Access industry databases such as MIRMgate (http://www.mirmgate.com/browse-hazard.asp);
- draw on personal experience, brainstorming with individuals familiar with each process stage and/or consultants with specific areas of expertise;
- consider the steps in the process, the inherent risk at each stage and how failures could occur – a process system map may be useful;
- use checklists as a guide. Each open pit is unique and a given pit may generate more or less risk than the industry norm. Suggested geotechnical checklists of information required through the life of an open pit are given in Appendix 4;
- audit checks for monitor and review;
- use systems/scenario analyses, system engineering techniques and formal techniques such as job safety/hazard analysis (JSA/JHA), energy barrier analysis (EBA), preliminary hazard analysis/workplace risk assessment and control (PHA/WRAC), hazard and operability studies (HAZOPS), failure modes and effects analysis (FMEA), failure modes, effects and criticality Aanalysis (FMECA), fault tree analysis (FTA), event tree analysis (ETA) and level of protection analysis (LOPA). Outlines of some techniques will be given in section 13.4.3.2.

Some common hazards and risks associated with the inputs to the geotechnical model as defined in this book are listed below.

Geology risk: Potential slope failures and uneconomical returns because of a poorly defined geological model and a lower than expected level of confidence for the relevant project level. This may arise from:

1 human error because of:
  - → inaccurate sampling or data collection;
  - → lack of experience;
  - → non-compliance with established procedures;

2 insufficient mapping;
3 insufficient drilling.

Structure hazard/risk: Slope failure resulting from previously unknown structures during operations and lower than expected level of confidence of the structural model for the relevant project level because of:

1 insufficient structural data/drilling;
2 insufficient/incorrect laboratory testing;
3 human error arising from:
  - → structure missed and/or inaccurate mapping in data collection, leading to potential for unfavourable discontinuity orientation and frequency to create unstable slopes;
  - → lack of experience to adequately consider all modes of structural failure;
  - → non-compliance with established procedures;

4 incorrect/limitations of methods of analysis such as kinematic analyses, empirical assessments and computer programs (e.g. 2D vs 3D).

Rock mass hazard/risks 1: Rock mass weaker than expected leading to inadequate design and slope failure arising from:

1 insufficient or inadequate rock mass data collected during exploration or before proceeding to mine, resulting in an incomplete geotechnical model. This has the potential for the modelling and mine design calculations to be based on information that does not adequately represent the geotechnical domain being mined and the overestimation of stable slope angles;
2 lower than expected level of confidence in rock mass model;
3 from human error because of:
  - → inaccurate sampling, data collection, inaccurate/incorrect rock testing or calculation of RMR or GSI;
  - → lack of experience;
  - → non-compliance with established procedures;

4 poor blast design, with the potential to use too much explosive to achieve the desired fragmentation pattern, leading to blast damage creating less stable slopes.

Rock mass hazard/risks 2: Reduced revenue if the rock mass is stronger than expected and the mine is overdesigned because of:

1 insufficient data collected during exploration or before proceeding to mine, resulting in an incomplete geotechnical model. This has the potential for modelling and mine design calculations to be based on information that is not representative of the geotechnical domain being mined and the underdesign of stable slope angles;
2 the potential to use more explosives in the blast design to achieve the desired fragmentation pattern. This may lead to increased costs in remediation and scaling to achieve the designed slope angles;
3 human error, as above.

Hydrogeology hazard/risks: The amount of water present exceeds the anticipated pumping requirements, inducing pit slope failure because of:

1 a high-frequency rainfall event and/or an uncontrolled water event;
2 groundwater level being higher than expected and the slope not being depressurised or dewatered;
3 increased expenditure for extra pumping capacity;
4 groundwater monitoring or dewatering failed;
5 human error arising from:
   → inaccurate sampling, data collection, calculation or modelling of mine water processes and systems;
   → lack of experience of the analyst;
   → non-compliance with procedures.

Geotechnical model hazard/risks: Slope failure at bench, inter-ramp or overall slope scales because of:

1 incomplete or inaccurate geotechnical model for the project level, leading to incorrect definitions of geotechnical domains, computer modelling and subsequent mine design;
2 the limitations of the computer models and other methods used in mine design not being fully understood or accounted for in slope design;
3 human error based on:
   → inaccurate sampling, data collection, calculation, modelling or monitoring, inadequate review and continual improvement of geotechnical and mine processes/systems;
   → lack of experience of the analyst;
   → poor communication of slope design or implementation requirements
   → non-compliance with procedures or with the mine design;
4 failure of existing control measures such as monitoring, dewatering, mine systems and procedures.

To record and manage the identified hazards and risks, a comprehensive hazard/risk register should be compiled and maintained throughout the life of the open pit. A hazard or risk register is likely to group geotechnical hazards and risks under headings such as those used for Chapters 3 to 7. The entry for each hazard/risk should list its causes, the risk rating and what controls exist or are proposed to eliminate the risk or, more usually, limit its likelihood of occurrence and/or consequences. The hazard/risk register may also identify the safety or business objectives affected by particular risks. The register should be updated at each stage of the project as the results of further investigations and studies become available. The implementation of controls, or otherwise, should be recorded in Levels 4 and 5.

13.4.3 Risk analysis

13.4.3.1 Risk analysis process

Risk analysis is the process of developing an understanding of each risk. It provides an input to decisions on whether risks need to be treated and the most appropriate and cost-effective risk treatment strategies. Risk analysis involves consideration of the sources of risk, their consequences and the likelihood of those consequences occurring. Risks are usually analysed by combining their likelihoods and consequences. The analyses may be qualitative, semi-quantitative or quantitative. In most circumstances, existing controls are taken into account.

A preliminary analysis may be carried out so that similar risks are combined or low-impact risks are excluded from detailed analysis. Where possible, the risks excluded from further analysis should be listed (e.g. in the risk register) to demonstrate the completeness of the risk analysis (Standards Australia 2004).

In the case of large open pit slopes, the major sources of geotechnical risk are slope failures on bench, inter-ramp and overall slope scales. As discussed in more detail in section 9.4.1, the consequences of these failures can be regarded as having safety or economic impacts and can be categorised as falling into one or more of the following groups (Tapia et al. 2007):

- injury or fatalities to personnel;
- damage to equipment;
- the economic impact on production (e.g. loss of production, cleanup costs, temporary or longer-term sterilisation of part of the pit);
- force majeure (a major economic impact, e.g. an overall slope failure);
- industrial action (leading to loss of production);
- adverse public and stakeholder relations, including impacts on permissions to mine.

The safety consequences of an event may be expressed through measures such as the number of lost time injuries (LTI) or fatalities over a given time or for a particular tonnage mined. Economic consequences may be expressed as the tonnages involved in a particular failure or the cleanup costs, the costs of equipment repair and/or replacement and the costs of lost production. The likelihoods of events occurring, or the probabilities of failure (PoF), are established using the methods discussed in Chapters 8, 9 and 10. In the quantitative risk analysis of open pit slopes, the level of risk is usually quantified as:

$$\text{Risk} = \text{PoF} \times \text{consequence of the event}$$

13.4.3.2 Risk analysis tools

A wide range of risk analysis tools is available for various stages of the risk analysis process. Each has a specific purpose and outcomes as summarised in a mineral industry context by Rasche (2001). Details of the tools are given by Joy and Griffiths (2005) and in specialist texts such as those by Aven (2003) and Bedford and Cooke (2001). The following summaries of some major tools are mainly based on the accounts given by Joy and Griffiths (2005) and Rasche (2001).

'Bowtie analysis' shows how a range of controls may eliminate or minimise the likelihood of occurrence of specific initiating events that may generate a risk, or reduce the consequences of an event once it has occurred. The results are represented in a bowtie diagram (Figure 13.7). Bowtie diagrams originated as a technique for analysing safety incidents, but are also useful for analysing other types of complex risks and communicating key risks and critical controls (Quinlivan & Lewis 2007). For risks such as the geotechnical risks associated with open pit slopes, where the causes of risk and the required controls are relatively obvious, bowtie analysis may be neither required nor effective.

Consequence or cause–consequence analysis is a combination of fault tree analysis and event tree analysis (see below). The outcome is a diagram which displays the relationships between the causes and the consequences or outcomes of an incident. This technique is used most commonly when the failure logic is simple, since a diagram combining both fault and event trees can become quite complex.

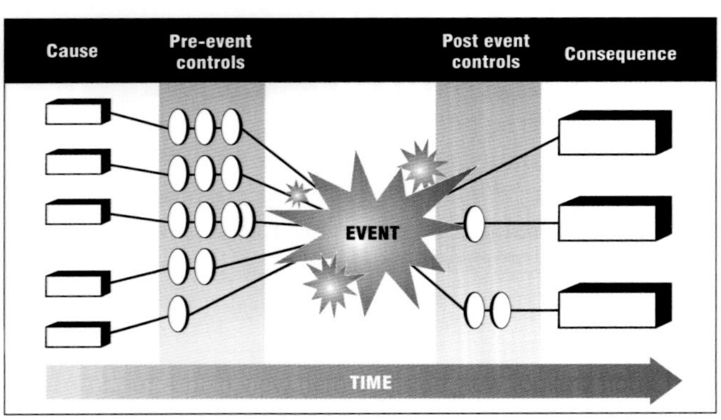

Figure 13.7: A bowtie diagram

Construction hazard assessment and implication review (CHAIR) is a structured, facilitated discussion process involving designers, constructors and other key stakeholders that is used to make final design changes to construction projects by accounting for probable construction methods. It uses a series of guidewords or prompts and may be used at the conceptual, final design, operation and closure or demolition stages of projects. Although it was developed for the construction industry by WorkCover NSW (2001) it has obvious application to construction issues in the minerals industry, including open pit projects.

Energy barrier analysis (EBA) is a qualitative process used to identify hazards by tracing energy flow into, through and out of a system. The technique identifies not only the energy source but also the barriers or controls in place to prevent release of the energy or control its consequences. As discussed in section 13.4.2, the energies involved in open pit slope failures are mainly gravitational energies which may be augmented by water pressure, thermal and chemical energies. Where the relationships are reasonably well-understood, the use of energy barrier analysis may add little value.

Event tree analysis (ETA) is a tool that provides a systematic mapping of realistic event scenarios with potential to result in a major incident, and of the relationships, dependencies and potential escalation of events with time. It also provides numerical estimates of the likelihoods of occurrence of the component events and of an escalated event. Event tree analysis is a classical quantitative risk analysis tool that has been used in the nuclear and process industries, in particular, to model catastrophic risk. The application of ETA to the evaluation of economic consequences of slope failure is discussed in section 9.5.2.2. Figure 13.8 shows an event tree diagram evaluating the economic impact of slope failures in a large copper open pit (Tapia et al. 2007).

Failure modes and effects analysis (FMEA) or failure modes, effects and criticality analysis (FMECA) addresses the basic question of what is the consequence of the failure of a component of a system or of a piece of equipment. It identifies the causes, consequences and criticality of possible component failures, often as input to fault tree analyses. It is usually used after the finalisation of a design or as part of an accident investigation.

Fault tree analysis (FTA) identifies, quantifies and represents the faults or failures, and the combinations of faults or failures, which can lead to a major hazard or event. It is an excellent tool for use in complex systems

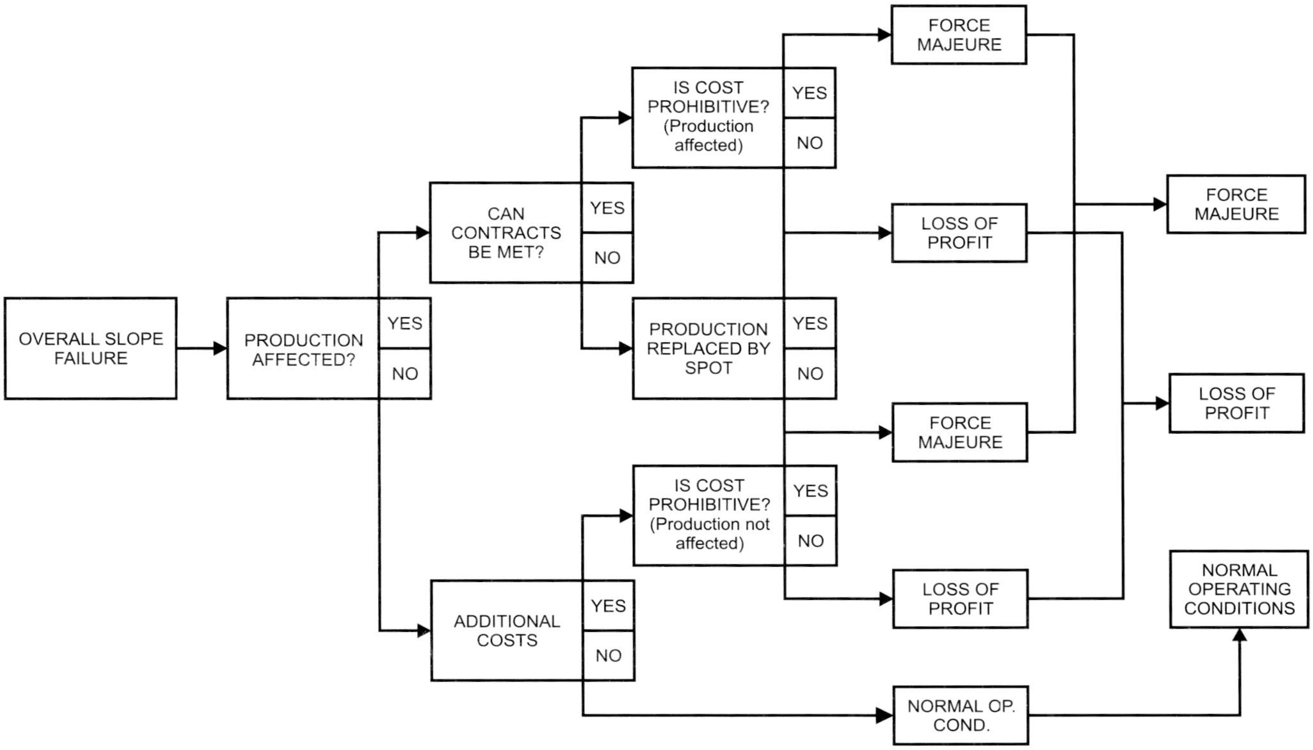

Figure 13.8: Event tree diagram evaluating the economic impact of open pit slope failures
Source: Tapia et al. 2007

where the interaction and combination of events and faults or failures need to be considered. FTA is another classical quantitative risk analysis tool used widely in a range of industries, often in combination with other techniques such as ETA and FMEA/FMECA. Figure 9.3 shows a simple example of its application to evaluate the probability of an open pit slope failure in a particular geotechnical domain.

Hazard and operability studies (HAZOP) is a structured brainstorming approach to hazard analysis developed in the chemical and processing industries in the 1970s. Its main purposes are to identify process hazards and operability problems and their consequences, and to evaluate the existing or proposed controls at the design stage.

Human error analysis (HEA) is a qualitative or quantitative approach to the identification and management of human errors that could lead or contribute to significant hazards. The ultimate objective is to develop changes to the system that will eliminate, or at least minimise, the probabilities of occurrence and impacts of these errors.

Job safety/hazard analysis (JSA/JHA) is a task-oriented qualitative risk assessment carried out by a work team on site to guide the development of safe working procedures for potentially hazardous tasks. It represents an important part of the total risk management process but is used at a different level from the risk analysis techniques discussed here.

Preliminary hazard analysis (PHA) or hazard analysis (HAZAN) provides an initial listing of hazards that must be addressed in the design process to ensure that they are managed adequately. This approach is likely to be useful in identifying potential hazards andrisks in Levels 1 and 2 of a large open pit project. Such an assessment might produce a table listing the hazards or risks identified and the causes, major effects and corrective or preventative measures or strategies required, as in a risk/hazard register.

Obviously, it is important to select the right tools for each particular analysis or stage of the analysis. Table 13.3 suggests some of the risk analysis tools that are suitable for analysing geotechnical risk in the various stages of a large open pit project. The risk categories used in Table 13.3 are similar to those illustrated in Figure 13.6 and are more broadly based than the categorisation introduced in section 13.4.3.1. Table 13.3 also introduces some financial and economic tools (e.g. NPV, risk vs return) and, at the strategic level, SWOT (strengths, weaknesses, opportunities and threats) analyses that are not included in the summaries of risk analysis tools.

If a risk assessment is to be performed at the design (Level 4) and operational (Level 5) stages, a quantitative approach should be used wherever possible. Points to

Table 13.3: Suitability of risk analysis tools for the stages of an open pit project

Life of open pit project stage	Risk category	Recommended type of risk analysis
Desk or conceptual study Pre-feasibility Feasibility (Levels 1–3)	Economic/financial/project	Qualitative/semi-quantitative NPV, risk vs return, bowtie, determine level of uncertainty
	Strategic	Qualitative semi-quantitative SWOT, bowtie, determine level of uncertainty
	Technical/design	Semi-quantitative/quantitative Risk vs consequence, bowtie, FTA, FMEA/FMECA, determine level of uncertainty
	Operational	Semi-quantitative/quantitative Risk vs consequence, bowtie, FTA, FMEA/FMECA, determine level of uncertainty
Detailed (or final) design Operating stage Closure/transition to underground (Levels 4–5)	Economic/financial/project	Qualitative/semi-quantitative NPV, risk vs return, bowtie, FTA, determine level of uncertainty
	Strategic	Qualitative/semi-quantitative SWOT, risk vs consequence, bowtie, FoS, PoF, determine level of uncertainty
	Technical/design	Majority quantitative (some semi-quantitative in the absence of data) Risk vs consequence, FTA, FMEA/FMECA, FoS, PoF, determine level of uncertainty
	Operational	Majority quantitative (some semi-qualitative in the absence of data) Risk vs consequence, FTA, FMEA/FMECA, determine level of uncertainty

consider when selecting the best tool or method for an assessment include:

- the objective or purpose of the risk assessment;
- the estimated nature of the risk;
- available supporting data and the adequacy of those data;
- the expertise and resources required or available;
- site and industry history of incidence;
- any constraints or limitations on the process;
- the socio-political context for the conduct of the risk assessment;
- the assumptions used to support a particular tool;
- the levels of uncertainty involved (see section 13.4.3.3).

13.4.3.3 Data uncertainty

There are always varying degrees of uncertainty with the input data and range of estimates in the overall risk assessment process. It is essential that this inherent uncertainty be recognised and, wherever possible, recorded at each stage of the analysis using confidence levels. Generally, it might be expected that the higher the levels of uncertainty with the data and assumptions used in analysing the consequences and likelihoods of events, the more conservative will be the decision-making process and the risk acceptance criteria.

The issue of uncertainty and input data used in geotechnical analyses for open pit slopes has been discussed in general terms in Chapter 1 and in detail in Chapter 8. The varying target levels of confidence associated with the geological, structural, rock mass, hydrogeological and geotechnical models used at Levels 1 to 5 of a large open pit project are given in Table 8.1. In general, a project should not proceed to the next stage until these target confidence levels have been achieved. Concepts of uncertainty in the context of geotechnical risk management for open pit slopes were also introduced in section 13.3. As an example of data uncertainty, consider the volume and mass of a wedge of rock that could slide from an open pit bench. The spacings, orientations and persistences of the planes forming the faces of the wedge are very rarely deterministic values. They are more likely to be the means, best estimates or central tendencies of data sets that contain limited populations and may be subject to biases of some type. In this case, there is uncertainty about the mass of a wedge of rock that may slide out of the bench face.

In the standard or classical approach to risk and risk analysis, it is assumed that risk exists objectively and can be expressed by probabilities and expected values. A probability of occurrence is interpreted in the classical statistical sense as the relative fraction of times the event would occur if the situation were repeated an infinite number of times. Event and fault trees may aid in estimating risk. The probabilities of the component events or faults occurring are estimated using hard data and expert opinions. Uncertainty may be measured by the statistical variation in the hard data, for example as reflected by confidence intervals (Aven & Kristensen 2005).

In advanced and comprehensive studies, a more thorough analysis of uncertainty may be performed using a combination of this classical approach and a Bayesian approach in which probability is used as a means of expressing uncertainty. In the Bayesian approach, there are no true objective probabilities on which to base risk calculations. However, an objective probability may be introduced though the concept of chance which reflects variation in a population (e.g. of uniaxial compressive strength values). A second component of probability is the analyst's subjective assessment of the uncertainty associated with this chance. This is often referred to as the probability of frequency approach (Aven & Kristensen 2005; Kaplan 1992). An alternative or predictive Bayesian approach focuses on the use of observable or measurable quantities to assign probabilities of occurrence. Models such as those developed from event and fault trees may be used as tools to express the uncertainties associated with the probabilities of occurrence of the observable quantities (Aven 2003).

Monte Carlo simulation is one technique that can be used to develop models of uncertainty. It can provide a range and distribution of possible values for each unknown factor, rather than a single discrete average (Baczynski 2000). A number of software packages can perform the simulations and provide the required outputs.

13.4.4 Risk evaluation

13.4.4.1 Risk evaluation process

The risk evaluation process involves comparing the level of risk derived from the risk analysis with the risk criteria established when the context for the risk management process was considered. The purpose of risk evaluation is to use the outcomes of risk analysis to decide which risks require treatment, and the treatment priorities.

As discussed in Chapter 9 and earlier in this chapter, the risks associated with an open pit slope failure may be expressed quantitatively as the product of the probability of that failure occurring and the consequences of the failure. The resulting risk is evaluated against the acceptance criteria for each established type of risk, as described in section 9.4.4. The overall risk evaluation process is discussed in section 9.4.3 and illustrated in Figure 9.2. The process is applied separately to each slope failure type and previously identified risks, and to each of the consequences as outlined in sections 9.4.1 and 13.4.3.1. Usually, the risk evaluation process has to be repeated at each stage of an open pit project (Levels 1–5) when different, or more refined, acceptance criteria may be applied. For example, as more geological and geotechnical data become available and the levels of confidence in the geological and geotechnical models increase, the risk analysis and risk evaluation processes may be applied separately to a number of geotechnical domains in the overall open pit slope (Swan & Sepúlveda 2000; Tapia et al. 2007).

13.4.4.2 Risk matrices

The results of the analyses of the likelihood of occurrence of particular risks and their consequences or impacts, and of their evaluation using acceptance or severity criteria, may be represented using risk matrices. Risk matrices are particularly useful as a means of communicating the main results of a risk assessment exercise. Risk matrices may be

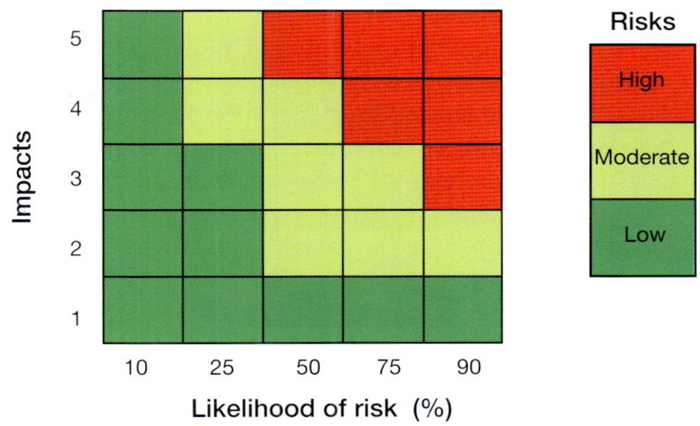

Figure 13.9: A semi-quantitative risk matrix

used to communicate the results of qualitative and semi-quantitative risk analyses and evaluations, as well as quantitative. In the case of open pit slope failure, it would be usual to develop risk matrices for safety and economic consequences of particular risks.

Figure 13.9 shows a generic semi-quantitative risk matrix in which the likelihood of a risk occurring is plotted as a percentage on the horizontal axis and the impact is plotted on a 1–5 scale on the vertical axis. Table 13.4 shows a qualitative risk matrix representating the personnel safety implications of a particular risk. Levels of consequence and likelihood with verbal descriptors are established and the possible likelihood–consequence combinations are assigned qualitative levels of risk and of acceptability. The resulting levels of risk and their severity or acceptability are colour-coded with the warmer colours used for the higher or more extreme levels of risk.

The measures that may be used to treat or control geotechnical risks will be discussed in section 13.5. In some instances, these measures may entirely eliminate a given risk but more usually they reduce the likelihood of occurrence of the event generating the risk and/or the consequences of that occurrence. In this case, residual risks will remain following implementation of the risk treatment measures. These residual risks should be listed in a residual risk register and may be depicted in residual risk matrices, which generally take the same forms as the examples above.

Strategic risk matrices focus on a company's long-term objectives. They may be used to depict the results of the risk–benefit analyses used in the tollgating process for project selection, for example, and must be tailored to the company's strategic requirements. Table 13.5 shows semi-quantitative strategic residual risk matrices for the influence of slope design and the impact of incorrect geotechnical design on capital expenditure (Capex), operational expenditure (Opex) and safety at the scoping or conceptual, pre-feasibility study, feasibility study, operational and closure stages of a particular open pit project.

Another form of risk matrix is a risk– return matrix depicting the results of a risk evaluation carried out using

Table 13.4: Qualitative risk matrix for personnel safety

Consequences for personal injury		
Level	**Descriptor**	**Example of description**
1	Insignificant	No injuries
2	Minor	First aid treatment
3	Moderate	Medically treated injury
4	Major	Extensive injuries/permanent disability
5	Catastrophic	Fatality

Qualitative measures of likelihood		
Level	**Descriptor**	**Description**
A	Almost certain	Is expected to occur in most circumstances
B	Likely	Will probably occur in most circumstances
C	Possible	Might occur at some time
D	Unlikely	Could occur at some time
E	Rare	May occur only in exceptional circumstances

Qualitative risk matrix

	Consequences				
Likelihood	**Insignificant**	**Minor**	**Moderate**	**Major**	**Catastrophic**
A	H	H	E	E	E
B	M	H	H	E	E
C	L	M	H	E	E
D	L	L	M	H	E
E	L	L	M	H	H

E: Extreme risk – immediate action required; unacceptable risk
H: High risk – Senior management attention required; unacceptable risk without action
M: Moderate risk – management responsibility; acceptable with control measures
L: Low risk – manage by routine procedures; acceptable risk

a risk–return, risk–benefit or cost–benefit analysis. Table 13.6 shows a risk–return matrix for the financial consequences of the geotechnical risk associated with the operational stage of an open pit project.

One mining company follows the steps below when carrying out analyses that lead to risk matrices.

1 Determine the possible slope failure modes based on empirical, FoS or PoF calculations made using the current geotechnical model.
2 Establish the triggering events (e.g. groundwater, rainfall, blasting, excavation).
3 Determine the scale of potential slope failures at the bench, inter-ramp and overall slope:
   - → calculate the percentage of ramp or berm loss;
   - → calculate the potential tonnage of material displaced and compare with acceptability criteria for FoS and/or PoF.
4 Determine the cost of remediation for each event:
   - → calculate the cost of removal and remediation work;
   - → k (tonnes/hour);
   - → calculate the cost of operational downtime;
   - → calculate the impact cost on the life of mine plan and potential redesign. Alternatively, if it is not practicable to remove the failed mass, calculate the cost of leaving it in place using a step out (loss of ore).
5 Management should then be able to provide acceptable limits based on the probability and costs of a certain sized slope failure.

13.5 Risk mitigation

13.5.1 Overview

Risk mitigation is the final stage in the general approach to risk management illustrated in Figure 13.2 and discussed

Table 13.5: Semi-quantitative strategic residual risk profile

	Scoping	Pre-feasibility	Feasibility	Operational	Closure
Strategic influence of slope design					
Capex	Low	Low	High	Moderate	Low
Opex	Low	Low	Low	Moderate	Low
Safety	Low	Low	Moderate	High	High
Impact of incorrect geotechnical design					
Capex	Low	Moderate	High	High	Low
Opex	Low	Low	Moderate	Moderate	Low
Safety	Low	Low	Low	High	High

$ value Opex/Capex	
Low	<10 million
Moderate	10–100 million
High	>100 million

in this chapter. It involves identifying the range of options for treating risks, assessing those options and preparing and implementing treatment plans. Options for treating risks that could lead to negative outcomes (as in the case of the geotechnical risks associated with excavating large open pit slopes) include:

- avoiding the risk by not starting or continuing the activity giving rise to the risk;
- changing the likelihood of occurrence of the event that generates the risk (e.g. change the inter-ramp slope angle) so as to reduce the likelihood of negative outcomes;
- changing the consequences of the event to reduce the extent of injuries or losses (e.g. changing the potential size of the failure by reducing the slope height);
- sharing the risk with other parties through contractual arrangements, insurance and structural arrangements such as partnerships and joint ventures;
- retaining the risk and seeking to manage it.

Control measures are generally implemented to reduce the inherent risk. The initial risk is then reassessed with the control measures in place, to determine the level of remaining or residual risk. This process of adding control measures and reassessing risk can be carried out a number of times to reduce the residual risk until it falls within the accepted criteria or management accepts the level of residual risk.

It is important to recognise that not all risks are equally amenable to mitigation. Therefore, it is necessary to consider the level of mitigation of each risk compared with others. This can be achieved by normalising the levels of risk established for each issue with a variable risk mitigation multiplier:

$$\text{Risk mitigation} = \text{risk} \times \text{mitigation multiplier}$$

Brown (2007) gives examples of the mitigation multipliers applied to a range of risk items associated with underground mining by block and panel caving.

Table 13.6: A risk–return matrix for the financial consequences of geotechnical risk in the operating stage of an open pit project

Cost of remediation ($US)	Production delays	Lost production (t)	Possible consequences	Acceptability of design criteria
>$10 million	Pit closed	> 2 000 000	Total wall failure	Unacceptable Total redesign required
$5–10 million	< 6 months closure	500 000–1 000 000	Wall failure; lost revenue	Seek management and stakeholder approvals Minor modifications to life of mine plan
$1–5 million	<1 week	100 000–500 000	Remedial work required	Acceptable with management approval Ground support and/or pit drainage required
<$1 million	1–2 shifts	< 500 000	Minor damage to mine equipment	Acceptable Remedial work required

Table 13.7: Hierarchy of controls for a risk endangering personnel safety

1	Elimination	If possible, eliminate the risk so it doesn't reach the worker
2	Substitution	Can the job be reorganised or a different area mined to reduce the risk?
3	Isolation	Perform the task using remote mining methods
4	Engineering controls	Change the design or use guards/ monitoring to make the process safer
5	Administration	General procedures including JSA/ JHA, competency training, standard operating procedures (SOPs), monitoring with trigger action response plans (TARPs)
6	Personal protective equipment	Safety boots, hard hats, gloves, ear muffs etc. (last resort; relies on worker behaviour to reduce risk)

13.5.2 Hierarchy of controls

It is common practice to require that unacceptable risks be eliminated or controlled. The hierarchy of controls is the preferred order of implementation of control measures aimed at eliminating or minimising the occurrence and/or impact or consequences of a particular risk. Table 13.7 shows a hierarchy of controls that might be used to address a risk that endangers personnel safety.

It is important to reassess the risks with control measures in place to determine if the residual risks are acceptable. The values or weightings applied to particular control measures may be subjective and become the subject of debate.

Joy and Griffiths (2005) suggest a general six-point strategy to be followed in seeking to control the hazards associated with the release of unwanted energies (Table 13.8). In this example, the approach has been adapted for an open pit slope failure triggered by the release of gravitational energy.

In all cases it should always be established how effectively a control has been implemented. There are often controls in place that are not used in the manner that they were designed for and thus provide less protection against the risk than was originally intended.

13.5.3 Geotechnical control measures

Table 13.8 introduces some of the measures that may be used to control the geotechnical hazards and risks associated with open pit slopes. Table 13.9 groups the common geotechnical control measures into categories of detecting slope movement, reducing the impact of rockfalls and controlling slope movements. Most of the geotechnical control measures listed in Tables 13.8 and 13.9 have been discussed in detail in Chapters 10 and 11 (controlled blasting, slope protection, slope support and reinforcement) and Chapter 12. Karzulovic and Sepúlveda (2007) give examples of the effective treatment of slope instability problems in Chilean open pit mines through controlled blasting, slope depressurisation, pit dewatering, retaining dikes, buttresses, removal of material from unstable slopes and filling of pre-existing underground excavations and surface craters.

Table 13.8: Strategies to prevent and manage the effects of gravitational energy triggering a slope failure

Hazard: Gravitational energies triggering slope failure potentially causing loss of life and / or property damage		
No.	**Strategy**	**Examples**
1	Prevent the marshalling of the energy	Don't mine
2	Reduce the amount of energy marshalled	Use flatter slope angles Use dewatering measures to depressurise the slope
3	Prevent the release of energy	Robust mine design and procedures Leave a buttress at the toe of the slope
4	Modify the rate of release or spatial distribution of the energy	Slow the rate of extraction within the mine schedule Increase the rate of dewatering
5	Separate the energy release and the susceptible structure in time or space	Allow a settling period after blasting Monitoring, geotechnical management plans and evacuation procedures
6	Separate the energy release from the susceptible structure by a barrier	Install catch fences and safety berms Increase the berm widths on benches

Source: After Joy & Griffiths (2005)

Table 13.10 provides a general checklist of questions that could be used as a starting point in establishing what geotechnical control measures are likely to be workable in a given case. Different weightings will apply to each project. For the example in Table 13.10, nine (64%) of the

Table 13.9: Common open pit slope geotechnical control measures

	Hazard/risk	Control measures
1	Detect slope movement	Slope monitoring, radar, mapping and performance monitoring, extensometers, wall prisms, piezometers
2	Reduce impact of rockfalls	Increased berm width, catch fences, reduced slope angle, bunding at base of slope, secondary support – mesh
3	Control slope movement	Depressurisation, good blasting, rock buttresses, rock reinforcement

Table 13.10: Checklist of possible open pit slope geotechnical control measures

Questions	Yes	No	Weight
Can increased monitoring mitigate safety risk?	1		1
Can we modify work practices or schedules?	1		1
Do we know the mechanism and severity of the failure?	1		1
Are the technical and management skills available?	1		1
Can local slope be buttressed?		1	1
Can we use catch berms?	1		1
Can we use catch fences?		1	1
Can we use catch benches?	1		1
Can we install ground support?		1	1
Can we dewater or depressurise the slope?	1		1
Can we modify or improve blasting?	1		1
Can we accelerate or induce the failure?		1	1
Can we mine out the failure zone?		1	1
Can the design be changed?	1		1
Total	**9**	**5**	**14**
Suggested probability	64	36	
Selected probability	64	36	

14 possible measures are evaluated as being able to be implemented successfully.

Sometimes, selecting suitable geotechnical control measures from the range of alternatives will be relatively straightforward. In other cases the optimal solution will not be readily apparent and will require consideration of the advantages and disadvantages of the alternatives. For example, a wide range of surface protection measures including drains of several types, rock buttresses, geofabrics or geogrids, shotcrete or fibrecrete and bitumen mix coatings are available for treating erodable ground in open pits. Each measure has advantages, disadvantages, limitations and probable residual risks.

At each stage of an open pit project, the pit slope design must be reviewed in terms of its impact on the overall mine design, geotechnical risk and the adequacy of existing or planned control measures. It is necessary to identify levels of geotechnical risk and provide options for managing the risks in each project stage. The range of acceptability criteria and tolerable or defendable risk varies at each project development stage in accordance with the levels of uncertainty associated with the existing geotechnical and slope design models. The level of geotechnical uncertainty decreases as more data, eventually including monitoring data and excavation experience, become available as the project progresses. Each open pit and each slope in the pit should be subject to this evaluation.

Upon completion of the initial risk analysis and the evaluation and selection of control measures, the geotechnical risks need to be reassessed with the control measures in place and the residual risk assessed. It is important to highlight the level of uncertainty associated with the new residual risk ranking. This information is generally communicated to management through a mitigation plan.

13.5.4 Mitigation plans

The results of risk assessment and the selection of control measures and assessment of their likely effects are communicated through a mitigation plan. Such a plan is usually based on a hazard identification plan and a hazard/risk register (section 12.3). For each geotechnical hazard (or the inherent risk ranking) it tabulates the existing level of data uncertainty, the existing control measures, the new data available, the new control measures proposed, the residual risk ranking and the new level of data uncertainty, and assign responsibility for further action on that hazard or risk. This process must be repeated at the various stages of the open pit project.

Generally, the mitigation plan will feed into slope management plans which themselves may be part of broader geotechnical management plans. The slope monitoring procedures discussed in Chapter 12 form central parts of these plans. Although slope monitoring results will not become available until the operational stages, it is necessary that the various plans be developed and updated at each project stage. Table 13.11 provides a synthesis of the geotechnical risk management activities undertaken by several sponsor companies during the design and construction, operation, and closure or transition to underground mining stages of an open pit project.

The development and use of the ground control management plans discussed here and in Chapter 12 are critical to the safe and profitable conduct of mining operations. They are part of the corporate governance processes and provide a form of communication that records elements of required processes and mining practices. They also provide a basis for an effective monitoring and review stage of the overall slope risk management process. While each management plan reflects site-specific conditions, there is a significant degree of commonality in the conduct of safety, production and geotechnical management of large open pit mines. The majority of detailed ground control management plans are produced at the completion of the planning and design phase, and form part of an overarching design document prepared for implementation. The geotechnical and slope management

Table 13.11: Geotechnical risk management activities during the design, operation and closure stages of an open pit project

Summary of geotechnical risk management activities	
Geotechnical model and mine design	Update geotechnical risk assessment and mine design
Starter pit	Initial risk assessment (use data from feasibility stage and update geotechnical model and mine design) Slope management plan – controlling the risks/hazards Monitor and review
Mine to design	Update geotechnical model and mine design with data from starter pit and exploration Update risk assessment Update slope management plan – controlling the risks/hazards Monitor and reassess designs based on performance experience Slope or geomechanics management plan
Operations	Update geotechnical model with data collected from monitoring operations and exploration Update risk assessment Update mine design Update slope management plan – controlling the risks/hazards Update trigger action responses Update emergency evacuation procedures
Closure/transition to underground	Update geotechnical model Risk assessment based on end use, transition to underground or rehabilitation Conduct remedial work Slope management plan

plans required for large open pit slopes can be detailed and have links to a number of other management processes.

13.5.5 Monitoring, review and feedback

Monitoring, review and feedback are essential parts of the overall risk management process and its adaptation to the management of geotechnical risks in large open pit slopes. Ongoing review is essential to ensure that the geotechnical and slope management plans remain up-to-date, relevant and useful. As an open pit project progresses, factors that may affect the likelihoods of occurrences and their consequences may change, as may the factors that affect the suitability or cost of particular treatment options. Therefore, it is necessary to repeat the risk management cycle regularly, most notably as part of each stage or level of the open pit project.

The monitoring and recording of actual progress against risk mitigation plans provides an important performance measure and should be incorporated into the organisation's performance management, measurement and reporting systems. Monitoring and review should also capture lessons from the risk mitigation process through reviews of events, the treatment plans and their outcomes. Section 12.1 provides more information on geotechnical performance reviews.

Details of the technologies used and guidelines for the design and execution of open pit slope performance monitoring are documented in Chapter 12 and will not be repeated here. An important element of these plans is the definition of a set of actions required in response to likely events, slope conditions or monitoring results, known as trigger action responses (TARPs), as outlined in section 12.3.2.1. Displacement rates and the rate of change in those rates are important indicators of developing slope instabilities. Pisters (2005) provides a good example of the TARPs used and a tabulation of who should do what in response to certain changes in slope conditions in a group of open pit coal mines. Calderón et al. (2002) illustrate how the intensive monitoring of a slope over eight years allowed the definition of suitable threshold values for displacement rates.

Even with the adoption of best practice in corporate governance and the implementation of a robust risk mitigation system, low probability/high consequence slope instabilities may still occur in large open pits. The implementation of an effective Emergency Response Plan (ERP) can prevent an emergency from becoming a disaster by ensuring that training in procedures has been carried out, that those procedures are implemented, and that sufficient resources are on hand to minimise the failure's impact on operations. Section 12.3.2.2 provides information regarding typical emergency response plans for slope instability. In addition to an emergency response plan, it is also necessary to have recovery and business continuity response plans to minimise the downstream effects of corrective remediation and the impacts on processing, commodity supply and business continuity.

14 OPEN PIT CLOSURE

Dirk van Zyl

14.1 Introduction

This chapter deals with the final stage of the open pit slope design process by providing the reader with an understanding of pit closure (Figure 14.1).

In this chapter, mine closure is defined as the activities that take place before and after operations cease. Active closure planning and implementation during operations can reduce the amount of work required at the end of mining operations. This can result in cost savings at the end of operations and a better outcome. It can be accomplished by developing and implementing a closure plan for the open pit (and other facilities) during operations. Such closure plans should be updated throughout the mine life and the detail of the closure plan increased as the mine approaches the later stages of its economic life.

Open pit mines close when the ore reserves are exhausted. There may still be resources at the site but closure occurs when they cannot be recovered economically, although they may become reserves in the future with the development of new technology or changing metal prices.

There are many large open pit base and precious metal mines that operate for long times, often many decades. However, many smaller open pit mining projects are planned for time horizons of less than 10 years, in some cases as little as five years. The resulting open pits are not necessarily 'large' but many of the issues remain the same. The closure of multiple smaller open pit mines since the late 1990s reveal many of the issues that large open pit mines will face when they close. However, the size of large open pits and the volumes of waste generated limit some of the options that are feasible at smaller properties.

The focus on mine closure became central to mine planning and operations in the 1990s. Prior to that time, disturbed land, waste rock dumps and tailings impoundments were often closed on a reclamation basis. 'Reclamation' refers to regrading, topsoil placement and revegetation. However, the closure of open pits was seldom included as a specific consideration in the original mine planning process. The result was that historic open pits, such as the Berkeley Pit in Butte, Montana, were abandoned when open pit mining ended.

The focus of this book is large open pit mines. Few large open pit mines have closed since the 1990s and there is very little experience related to such closures. There are large differences between closure of historic mines and new mines. New mine designs should include closure as an integral activity to mining. This is not always true for existing mines, especially those without well-developed closure plans. The next section discusses this issue.

Closure of an open pit includes the whole mine site and results in a variety of activities, e.g. removal (or transfer) of buildings and other infrastructure, management of remaining fluids, such as tailings supernatant and heap leach effluent, and reclamation of disturbed land. Specific closure approaches and techniques are developing for mine facilities such as tailings impoundments, heap leach facilities and open pits. New technologies are continually being developed for mine closure; these include water treatment technologies, cover design approaches and monitoring

Sustainability of nearby communities and the sustainable long-term post-mining land use are important considerations in developing and implementing mine closure plans. Such considerations also decide potential future economic uses of mine sites and the various facilities, for example golf courses on mine waste facilities, wind farms on mine waste rock dumps, and the use of pit lakes for recreational facilities.

Development of a complete mine closure plan is an extensive task involving a multidisciplinary team of experienced professionals. The team can include mining

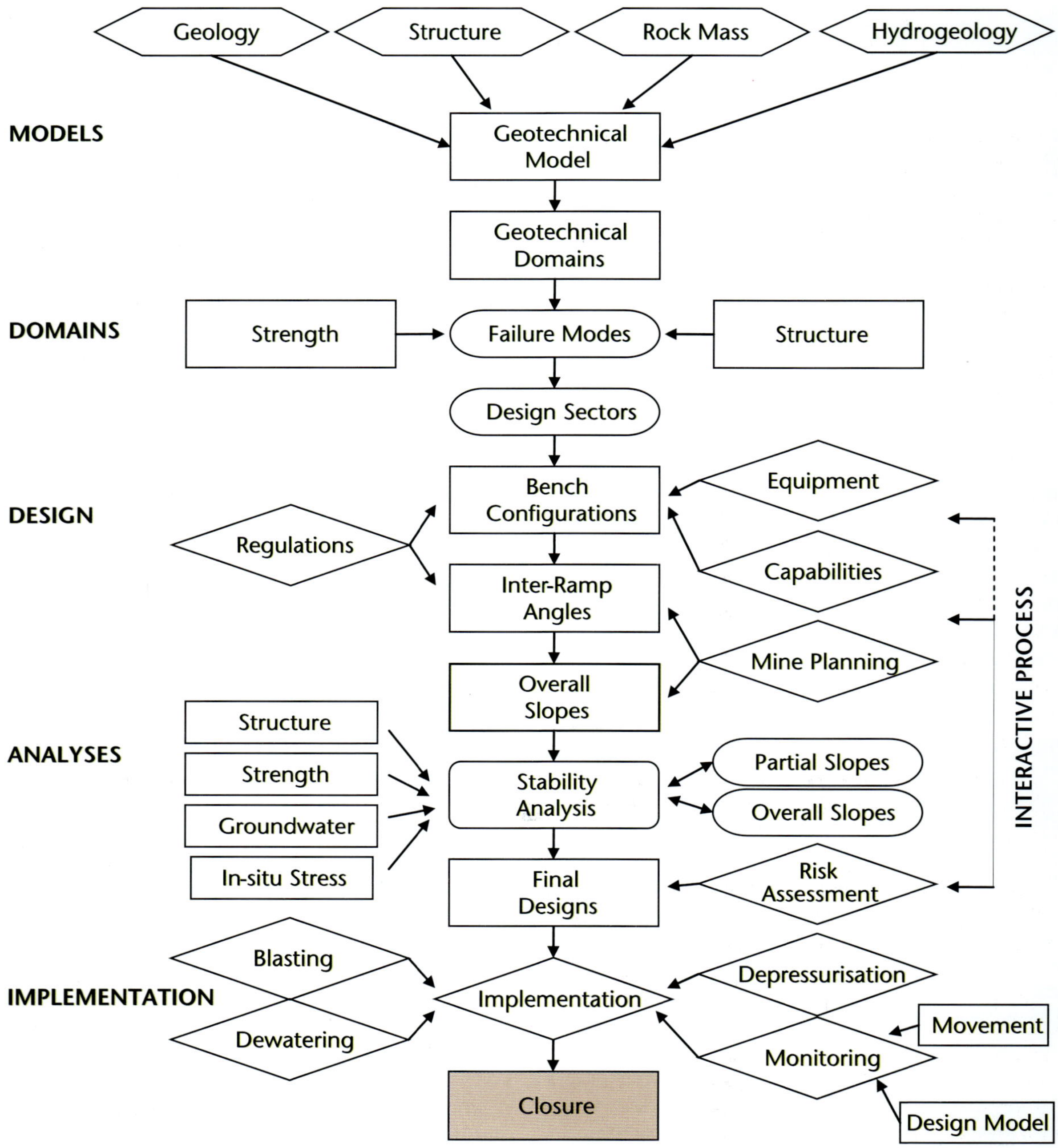

Figure 14.1: Slope design process

and geotechnical engineers, geologists, hydrogeologists, geochemists, reclamation specialists, community engagement and development specialists and human resources specialists.

Complete mine closure is a complex and involved process. This chapter focuses on closure planning and implementation with respect to open pit slopes. The activities examined, which are discussed in more detail in section 14.3, include:

- site characterisation;
- geochemical evaluations;
- surface and groundwater hydrological considerations;
- access to the pit;
- pit wall stability.

Socio-economic issues and issues such as the closure of tailings impoundments and waste rock are beyond the scope of this book and are not discussed.

14.2 Mine closure planning for open pits

14.2.1 Introduction

The mine closure planning process involves many steps for which the 2000 *Strategic Planning for Mine Closure* by ANZMEC/MCA is a very useful guide. The publication offers an objective and principles for each of the following topics (Table 14.1):

1 stakeholder involvement;
2 planning;
3 financial provision;
4 implementation;
5 standards;
6 relinquishment.

These objectives and principles were compiled in specific response to Australian and New Zealand regulatory frameworks and conditions. They may be a useful guide for other jurisdictions, even if the laws and regulations are different.

The closure objectives and principles of ANZMEC/MCA (2000) provide a checklist of closure planning activities for all mines. They can be used for the closure of open pits and should be included as part of the planning process.

- Stakeholder involvement – stakeholders in open pit closure include the mine closure team as well as the communities in the mine area. The engagement and community specialist is responsible for much of this task. However, the mine planning and geotechnical departments at the mine must participate in the discussions to make sure that solutions are feasible.
- Planning – closure planning is an ongoing process of refinement, often required by changes in mine plans and technology as the mine approaches the closure target date. Mine closure planning includes a large number of issues (described in section 14.3); the principles in Table 14.1 form an important guide for the planning process.
- Financial provision – not all jurisdictions have financial provisions in place at this time. However, it is a large part of sustainable mining practices and is expected to become universal in the future. An important component of this objective is the development and updating of an accurate closure cost estimate as it relates to open pits.
- Implementation – this objective requires an implementation plan that identifies accountability, also during operations.
- Standards – closure planning must have concrete targets, both for planning and implementation. This issue is discussed in the next section as part of closure goals and criteria.
- Relinquishment – this is the ideal end point when everybody agrees that the closure is successful and has met the agreed goals. For open pits, the timing for relinquishment depends on whether it is a dry pit or whether a lake will form. If the latter, it may take a long time to accomplish relinquishment because pit filling and water quality stabilisation may take decades.

14.2.2 Closure planning for new mines

For new mines, a closure plan should be developed as part of the initial mine design and permitting process and be completed by at least Level 3 (Feasibility Level, Table 1.2) of the project. This plan should be updated on a regular basis, ideally every two to three years. Ongoing data collection will result in improvement of the closure plan over time. Some regulatory agencies require that a final closure plan be submitted two years before its implementation. The typical steps in developing an open pit closure plan are:

- develop closure goals and criteria, making sure that local communities and other stakeholders are engaged at this point;
- incorporation of closure planning into the mine geologic and hydrogeologic models to assess the potential final condition of the pit walls, surface water, groundwater and surrounding areas;
- perform site characterisation to address the closure goals and criteria and to address data gaps;
- develop the closure plan with specific considerations such as potential pit lake formation and water quality, geotechnical stability of the pit walls and public access;
- identify areas where further data must be collected or research done to provide information for the final closure plan;
- implement identified closure activities during mining;
- review and update the closure plan and closure cost on a regular basis;
- prepare final closure plan for implementation at least 2–8 years before closure (this range of times reflect various corporate philosophies and mine life scenarios).

14.2.3 Closure planning for existing mines

Where closure plans are not in place for large open pit mines, it is a high priority to develop a closure plan for these sites to:

- understand the closure costs and the implications of closure for ongoing operations;
- focus operations towards eventual closure;
- work towards developing a final closure plan and cost, following the steps above.

Table 14.1: Summary of objectives and principles for mine closure planning

Stakeholder involvement

Objective

To enable all stakeholders to have their interests considered during the mine closure process.

Principles

1 Identification of stakeholders and interested parties is an important part of the closure process.

2 Effective consultation is an inclusive process which encompasses all parties and should occur throughout the life of the mine.

3 A targeted communication strategy should reflect the needs of stakeholder groups and interested parties.

4 Adequate resources should be allocated to ensure the effectiveness of the consultation process.

5 Wherever practical, work with communities to manage the potential impacts of mine closure.

Planning

Objective

To ensure the process of closure occurs in an orderly, cost-effective and timely manner.

Principles

1 Mine closure should be integral to the whole of mine life plan.

2 A risk-based approach to planning should reduce both cost and uncertainty.

3 Closure plans should be developed to reflect the status of the project or operation.

4 Closure planning is required to ensure that closure is technically, economically and socially feasible.

5 The dynamic nature of closure planning requires regular and critical review to reflect changing circumstances.

Financial Provision

Objective

To ensure the cost of closure is adequately represented in company accounts and that the community is not left with a liability.

Principles

1 A cost estimate for closure should be developed from the closure plan.

2 Closure cost estimates should be reviewed regularly to reflect changing circumstances.

3 The financial provision for closure should reflect the real cost.

4 Accepted accounting standards should be the basis for the financial provision.

5 Adequate securities should protect the community from closure liabilities.

Implementation

Objective

To ensure there is clear accountability and adequate resources for implementation of the closure plan.

Principles

1 The accountability for resourcing and implementing the closure plan should be clearly identified.

2 Adequate resources must be provided to ensure conformance with the closure plan.

3 The ongoing management and monitoring requirements after closure should be assessed and adequately provided for.

4 A closure business plan provides the basis for implementing the closure plan.

5 The implementation of the closure plan should reflect the status of the operation.

Standards

Objective

To establish a set of indicators, which will demonstrate the successful completion of the closure process.

Principles

1 Legislation should provide a broad regulatory framework for the closure process.

2 It is in the interest of all stakeholders to develop standards that are both acceptable and achievable.

3 Completion criteria are specific to the mine being closed and should reflect its unique set of environmental, social and economic circumstances.

4 An agreed set of indicators should be developed to demonstrate successful rehabilitation of a site.

5 Targeted research will assist both government and industry in making better and more informed decisions.

Relinquishment

Objective

To reach a point where the company has met agreed completion criteria to the satisfaction of the responsible authority.

Principles

1 A responsible authority should be identified and held accountable to make the final decision on accepting closure.

2 Once the completion criteria have been met, the company may relinquish its interest.

3 Records of the history of a closed site should be preserved to facilitate future land use planning.

Source: ANZMEC/MCA (2000)

These plans are important for mining as a whole so that it can develop a positive reputation for sustainability and land use.

14.2.4 Risk assessment and management

Risk assessment and management is an underlying approach in all closure planning. The risks associated with the facility and those remaining after closure planning and mitigation must be understood. Ongoing data collection and closure design is undertaken to manage the risks so that the final risks remaining at closure are acceptable to the corporation. The risks are related to technical, safety, environmental, political and financial impacts and many other aspects of open pits.

A risk assessment identifies the hazards (probability of occurrence, cause and pathway) and the consequences of an event. There are many tools available for such analyses and a formal failure modes and effects analysis (FMEA) is a powerful technique to evaluate the risks associated with closure of a mine and risks remaining after mitigation. It can be used as a screening tool before a more detailed probabilistic risk assessment is implemented.

14.3 Open pit closure planning

Specific issues, consequences and options for reducing the impacts of closing open pits have been proposed (Table 14.2) in a recent Australian publication (Australian Government 2006). The issues are listed under the following headings:

1 acid rock drainage (ARD) and leachate production from exposed walls;
2 void stability (in this chapter 'pit slope stability' is used);
3 public and fauna safety;
4 aesthetics;
5 post-mining land use;
6 long-term viability of rehabilitation.

Options and techniques to address each issue are presented in Table 14.2. Many will be discussed (under different headings) in later sections.

In the long term, open pit closure must focus on:

- safety concerns for people that may gain access to the closed mine;
- environmental concerns to humans, avian life and other fauna. This includes water quality issues related to the pit lake (if one forms) and the surrounding groundwater as it may influence human health and environmental risks to fauna.

This section describes the details of open pit closure planning, such as closure goals and criteria (including

Table 14.2: Open pits – issues, consequences and options for reducing impacts

Issues and consequences	Options and techniques
Acid rock drainage and leachate production from exposed walls Poor groundwater quality	Backfill above predicted recovered groundwater level Maintain water quality during mining Treat water (lime) Seal potential ARD-generating surfaces Refill pit with water (e.g. stream diversion and/or groundwater recovery)
Void (wall) stability Slumping Wall failures	Bench high wall and reshape low wall to a stable slope angle Batter or blast high wall to safe and stable angle Backfill to support internal walls
Public and fauna safety Injury or death	Hostile materials may need immediate covering (e.g. possible spontaneous combustion) Barrier to discourage human access Abandonment bunds of competent rock (where possible) and located outside area of wall instability Fencing and signage
Aesthetics High visual impact Industry reputation Negative public reaction	Stakeholder engagement to identify community view Revegetate void surroundings Screening Create wetlands Backfill or collapse and revegetate berms
Post-mining land use	Stakeholder engagement to determine possible uses Aquaculture Recreational facilities Educational areas Water storage Domestic and/or hazardous waste disposal
Long-term viability of rehabilitation	If infilled, weed control and revegetation

Source: Australian Government (2006)

regulatory criteria) and technical aspects related to closure planning. The range of topics clearly illustrates the multidisciplinary requirements of the closure planning team. It is important to emphasise that open pit closure planning must focus on site-specific conditions. Every site and mine is different and must be approached in this light.

14.3.1 Closure goals and criteria

The first step in developing a closure plan is to establish closure goals and criteria. Ideally, this process should be implemented early in the project planning. 'Goal' is defined in Webster's Dictionary as 'the end or final purpose; the end to which a design tends or which a person aims to reach or accomplish' and 'criterion' is defined as 'a test, means of judging, a standard of judgment, any established law, rule, principle, or fact by which a correct judgment may be formed'.

Mine closure goals, or the targeted outcomes of the mine closure design, must be based on site-specific conditions and be set with input from the stakeholders. Typical open pit closure goals include:

- divert all upstream water around the open pit;
- eliminate acid drainage from open pit walls;
- prevent access to pit;
- meet groundwater discharge criteria at a compliance point;
- develop wildlife habitat.

Closure criteria or standards must also be set on a site-specific basis. Mine closure criteria can originate from many sources, including:

- prescribed criteria (mostly regulatory criteria and decisions);
- corporate criteria;
- non-regulatory stakeholder desires;
- closure performance criteria.

14.3.1.1 Prescribed criteria

Prescribed criteria include the commitments made in the EIS for the mine, or other permitting requirements. All regulations that are applicable to a site for all media (air, water and soil) are prescribed criteria. It is essential that a complete list of prescribed criteria be prepared before advancing the mine closure plan. This section reviews closure requirements from Nevada and California and from British Columbia (Canada).

Regulatory programs to address mine closure have been developed in many jurisdictions. They may contain specific requirements for the closure of waste disposal facilities to protect the quality of groundwater and surface water, but are typically less clear for open pits. Appendix 5 cites the appropriate sections from these regulations to show typical details of requirements. Regulations change, so updates should be obtained at the time of closure planning. For example, Californian regulations were recently changed to require backfilling of open pits after metal mining. This is a very unrealistic set of regulations intended to discourage metal mining in that state.

Many regulations were developed in the late 1980s or early 1990s and consider many aspects of mine closure. The sections dealing with open pits form a small part of the overall regulations. The regulatory requirements may be summarised in terms of the following goals.

- Maintain physical stability – the open pit walls should remain stable in the long term. For example, there are a few instances where progressive failure of the pit slopes resulted in the necessity to expand the land ownership around a mine.
- Maintain chemical stability – the long-term water quality from the exposed pit walls and from potential seeps into pit must not cause contamination of surface or groundwater systems.
- Select an appropriate future use for the site – this depends on the pit's proximity to communities and the presence of avian species and other wildlife. Potential future uses for the open pit should be discussed with stakeholders and may include items such as an inaccessible pit lake because of expected poor water quality, an accessible pit lake for water sports or fishing, preservation of historic operation and the disposal of solid waste.

Based on a review of the requirements in Appendix 5, the specific regulatory issues with respect to open pit closure are as follows:

1. pit walls do not have to be vegetated at the time of closure;
2. stability of the pit walls is important for safety and environmental controls;
3. water quality in the pits and surrounding groundwater must not affect human health, avian life or other fauna.

14.3.1.2 Corporate criteria

It is important to develop a list of corporate closure criteria. Some corporations have very well-developed and documented closure procedures and criteria, while other companies are only beginning to develop them. Commitments in corporate health, safety, environmental and community statements must be collected and honoured in the mine closure planning. There may be other corporate policies and criteria that must be incorporated in the closure planning. Many mining companies accept World Bank or IFC guidelines[1] for general environmental and social controls; these can be considered part of the corporate closure criteria. The IFC criteria are applicable for private financing by about 40 financial institutions who are signatories to the Equator Principles.[2] These requirements are enforced with loans above $10 million from those lending institutions.

14.3.1.3 Non-regulatory stakeholder and closure performance criteria

Non-regulatory stakeholders such as nearby communities and NGOs may demand that certain closure criteria be honoured, e.g. community agreements or land swap criteria. Closure desires from non-regulatory stakeholders must be given appropriate priority. It is important for the mine to establish good communication with NGOs so that goals are understood and compromises worked out well before actual closure.

Many open pit mines are located near local communities. This does not mean that the local communities live next to the mine; more distant communities where employees live may feel a close attachment to a mine. As part of the mine closure planning and design, these communities should be engaged to provide input on closure goals and objectives. It is the communities that must live with the closed mine. These stakeholders must be engaged to reduce the impacts of the boom-and-bust cycle of historic mining activities.

The local community may have preferred closure objectives. For example, if a lake forms they may see an opportunity to have access to a water body where they can swim, fish and boat. It will be necessary to ensure that the expected water quality in the pit will be consistent with such uses (chemical stability). It will also be necessary to review the stability of the walls (physical stability) if the lake is used for aquatic activities.

Successful operations usually have effective engagement and communication processes throughout the mine life. If such processes are not in place, they have to be established for the closure planning.

An example where engagement processes played an important role in the selection of a closure plan is at the Martha Mine located in the city of Waihi, New Zealand.[3] Public engagement started before open pit mine operations commenced in 1987 and was ongoing during operations and closure planning. Through the engagement process the community supported the idea that the eventual pit lake could be used for recreation. On closure of the pit a lake will form in the long term and this may be used for recreational purposes (Castendyk & Webster-Brown 2007).

14.3.2 Site characterisation

Site characterisation activities will be based on the closure goals and criteria. Ongoing characterisation and incorporation of environmental/chemical data in the mine model will result in a more complete closure database at the time of closure. The following aspects must be characterised.

14.3.2.1 Site climate

The site climate, including the distribution of precipitation and evaporation and the magnitude of storm events, must be available on a site-specific basis. Mining operations exist in all types of climatic conditions from dry desert environments (e.g. the Atacama Desert in Chile) to tropical rainforests (e.g. Indonesia). A mine in the south-western USA may only be subjected to large rain events but one in the Philippines may also be subjected to cyclones (typhoons).

Precipitation and evaporation data may not be readily available during the design of a large open pit in a remote area. However, they should be available for existing mines as it is common to install meteorological stations at prospective mines, at least at the pre-feasibility stage (Level 2, Table 1.2). The location of the meteorological station is important if the data are to be used for closure design, as there are climatic changes with elevation and micro-climatic effects must be understood.

Table 14.3: Aridity index scale

Classification	Aridity index (AI)	Global land area
Hyperarid	AI < 0.05	7.5%
Arid	0.05 < AI < 0.20	12.1%
Semi-arid	0.20 < AI < 0.50	17.7%
Dry sub-humid	0.50 < AI < 0.65	9.9%
Humid	AI > 0.65	
Boreal/polar		

Source: UNEP (1992)

The ratio of annual precipitation to potential evapotranspiration (aridity index) can be used to classify climate according to the scale in Table 14.3 (UNEP 1992). Open pits in hyperarid and arid regions may be dry because the rainfall is extremely low and the evaporation and evapotranspiration are high. Infiltration is typically low in these climatic regions and therefore the groundwater level may have been significantly modified by the mining activities.

In semi-arid areas, infiltration is typically high enough to recharge the groundwater and, depending on the geologic conditions, it is possible to form a pit lake after closure. Table 14.3 indicates that about 37% of the global land area is in semi-arid or drier regions.

14.3.2.2 Site hydrology

Site hydrology includes the surface water and groundwater conditions. Surface water hydrology characterisation includes the watershed characteristics upstream from the pit and the expected surface water runoff from precipitation events. The surface water hydrology information is used to design surface water collection and/or diversion systems during and after operations (the latter is described below). Water quality information on the surface water upstream, in the pit itself and downstream must be collected prior to mining and regularly during mining.

Site hydrogeology information includes the groundwater level, the groundwater flow direction, water quality and aquifer characteristics such as hydraulic conductivity. It is essential to have good baseline data on groundwater conditions at all mine sites, specifically in the pit area. The hydrogeological information is used to design a dewatering or depressurisation system to maintain stability of the pit (Hall 2003). Large-scale dewatering operations may require that groundwater be treated and released to the environment. Such treatment can include chemical (e.g. metal removal) or physical (e.g. raising the temperature) activities.

After cessation of dewatering or depressurisation operations, free water may collect in the bottom of the pit or it may all evaporate. If there is sufficient inflow the water level in the pit will rise and a permanent pit lake will form. Hydrogeological calculations, combined with a complete water balance (discussed below), can be used to estimate the time it would take to fill the pit to a final static water level. In many instances that will be a very long time, as the inflow gradient reduces as the pit fills. The effects of gradual concentration of chemical constituents in the pit due to evaporation and other processes must be clearly understood (see section 14.3.5).

14.3.2.3 Site geology and geochemistry

Site geological and geochemical conditions must be characterised, including:

- the geologic units;
- weak zones in the pit walls that may degrade or erode;
- the presence of sulphides;
- major metals;
- leach characteristics;
- water quality of runoff and seepage into the pit;
- water quality in potentially mined-out pits in similar settings.

The geology and geochemistry of the walls may have a significant impact on the closure of open pits. The physical and chemical stability of the pit walls and the water quality of a potential pit lake will be determined by the site geology and geochemistry. It is critical to characterise the geological conditions very carefully before developing the final closure design. Ongoing rock characterisation is part of this task.

The concept of geo-environmental models for various ore bodies developed in the 1990s. A geo-environmental model is an integration of economic geology, process engineering and metallurgy, geochemistry, environmental conditions and historic environmental behaviour. It is applied to a deposit type with consistent characteristics with the objective of predicting the future and long-term environmental behaviour of a mine site (du Bray 1995). Geo-environmental models have been developed for many types of deposits where large open pits are possible, including porphyry copper, gold-rich porphyry copper and porphyry molybdenum. Understanding the geo-environmental characteristics of a specific deposit can provide insight into pit wall conditions (lithology, structural geology, geochemistry) at the time of closure. Ongoing sampling, characterisation and test work during the mine life can help to solidify the knowledge about the model and therefore the environmental behaviour at closure.

Geologic and geochemical information is used to derive the water quality of flows from or off the exposed pit walls. This information is used to predict the water quality in a pit lake if one forms.

The geologic, geotechnical and geochemical information is also used by mine planners to develop block models that allow segregation of waste into that which may produce acid and that which is non-acid producing. With this information, pit wall maps of the ultimate pit configuration can show the location of various formations and rock characteristics, both physical and chemical.

14.3.2.4 Geotechnical conditions

Geotechnical conditions, including the shear strength and degradation conditions of the intact rock, joints and fractures for the various geologic units (geotechnical domains), mine scale geological structures and their potential influence on pit wall stability, regional structures and operational experience are inputs to evaluating the long-term stability.

14.3.2.5 Other characterisations

Other characterisations are determined by the goals and criteria. For example, if other process fluids will be discharged to the pit, e.g. heap leach effluent, the quality and quantity of such flows must be known.

14.3.3 Ore body characteristics and mining approach

The shape of an open pit at the time of closure very much depends on the characteristics of the ore body. This shape will, in turn, determine whether any waste rock can be backfilled into the pit, thereby reducing its effective size.

While it is difficult to predict how the concepts of waste minimisation can be effectively applied to a particular open pit, it is important to carefully consider mine planning options that could result in some waste rock placement in the pits. The following three potential pit shapes provide the range.

14.3.3.1 Single conical pit

A single conical pit is a typical shape for smaller open pits. The final ore is removed from the bottom of the pit and there is no room for placing waste rock in the pit during operations as it will affect access to the ore at the bottom of the pit.

14.3.3.2 Elongated pit with some waste backfill

If the ore body is shaped so that an elongated pit can be formed, it is possible to place some waste in the pit as the mine develops. The backfill can be selected such that reactive waste will ultimately be covered by a pit lake.

14.3.3.3 Multiple pits sequentially mined and backfilled

In some mines, ore bodies occur in separate locations and separate pits are excavated. If these pits are developed in sequence it is possible to place waste materials in the initial pits, depending on groundwater and geochemical characteristics of the pit walls and the waste. The waste may include waste rock or tailings, or co-disposal of the two waste streams. Only the final pit may remain open at the end of the mine life.

There are compelling reasons for putting waste rock in a pit during operations, including haul distance and reducing surface impacts. It is important not to sterilise potential future resources that may remain in the bottom of the pit by covering them with waste rock.

14.3.4 Surface water diversion

Surface water diversion systems are required if an open pit is located in the valley of a major drainage or if significant upstream watersheds are intercepted by the pit. It is necessary to develop containment and/or diversion systems for the surface runoff. It may occasionally be possible to route the surface water through the pit, both during operations and for closure. In other cases, diversion structures must be developed around the mine and an assessment made of whether these systems become a permanent fixture or whether the surface stream is routed back through the pit.

The return storm used for designing the surface water diversion systems depends on the life of the mine and the consequences of structure failure. For example, if it is expected that the mine will have a 30 year life and that failure of the diversion system may result in a large volume of water flowing into the pit and interrupting operations, it may be best to use a 1 in 500 year or less frequent return period, e.g. 1 in 1000 years, for the design of such a diversion. This will reduce the risks of a potential diversion failure. The decision must be made on a site-specific basis. For closure, both the return period and consequences of a hydrologic event may change. A risk assessment may be an appropriate tool to assess both probability and consequences of various closure scenarios at various return periods.

Runoff into the pit, even from small watersheds, may result in sediments washing into the bottom of the pit. This can affect operations as well as long-term water quality in a pit lake or smaller amounts of water collected in a pit after closure.

14.3.5 Pit water balance

A full water balance of the open pit must be developed to account for groundwater inflow, surface water inflow and evaporation. The change in storage in the pit (ΔS), can be calculated from:

$$\Delta S = \text{inflows} - \text{outflows}$$

Where:

– inflows can include:

- precipitation;
- surface runoff;
- seepage into the pit;
- inflows from other sources.

– outflows can include:

- evaporation;
- groundwater flows away from the pit;
- outflows to other sources, e.g. use of water for other purposes.

A pit water balance is used to evaluate the time dependent rise in pit lake level after closure. In semi-arid and drier climates where the net evaporation is high, an open pit will often behave as a sink, i.e. groundwater will flow to the pit from all directions and the evaporation is high enough that no water will leave the pit into the groundwater. Original 'downstream' flows will be reversed. The final surface elevation of the lake may be lower than the original groundwater level as a result of evaporation. Under these conditions evapo-concentration of metals and other constituents can occur in the pit.

In wet climates water may be added to the pits as a result of precipitation. Consequently, the pit may be a flow-through system (a gradient may exist across the pit) and in some cases there could be sufficient additional water that the pit will overflow.

14.3.6 Pit lake water quality

The water quality of a pit lake or seeps into a pit is determined by the quality of the inflowing groundwater, the geochemical interactions of the water with the rocks through which it flows and the geochemical characteristics of the materials on the pit walls. The last may contain significant amounts of weathering products and may have a large impact on the initial water quality of a pit lake. Blast damage of the final wall face can increase near-face permeability and affect these interactions. Walder et al. (2006) stated that 'Closure of large-scale open pit mines is problematic when sulfide minerals are associated with low-grade, un-mineable wall rocks'.

In the presence of oxygen and water, sulphide minerals can generate acid. Depending on the types of sulphide minerals present, the initial pH may remain high while sulphate is produced. Over time the pH reduces, sulphate is produced and metals are dissolved, resulting in poor water quality. These reactions are exothermic and many waste dumps generate heat for extended periods. This contributes to low water quality in leachate and runoff from exposed pit walls containing sulphide minerals. Geochemical specialists should be consulted to evaluate site-specific conditions.

Limnological aspects of open pit lakes must be well-understood when predicting the long-term water quality in a pit lake. Vertical mixing of the water column (turnover) can result in changes in pit lake chemistry.

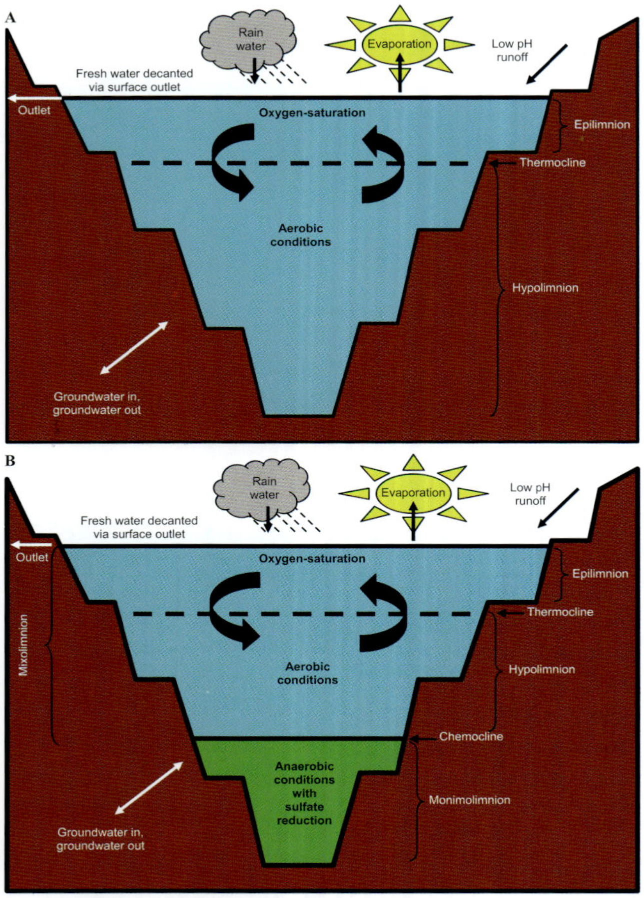

Figure 14.2: Conceptual models of (a) holomictic and (b) meromictic pit lakes. Large arrows indicate layers that circulate during turnover
Source: Castendyk & Webster-Brown (2007)

Figure 14.2 shows conceptual models of holomictic and meromictic lakes (Castendyk & Webster-Brown 2007).

Holomictic lakes undergo complete annual turnover (Figure 14.2a) and meromictic lakes undergo partial annual turnover (Figure 14.2b). The epilimnion and hypolimnion layers become chemically homogenised during turnover events. Epilimnion layers are in chemically oxidising conditions as there is continuous dissolution of oxygen across the air–water interface. The hypolimnion may have lower dissolved oxygen concentrations and redox potentials before turnover because of the decay of organic matter and oxidation reactions. The third layer in meromictic lakes, the monimolimnion, does not circulate annually. This layer has reducing characteristics and high total dissolved solids (TDS). The turnover of a meromictic lake could result in acute environmental impacts (Castendyk & Webster-Brown 2007).

Boehrer & Schultze (2006) summarised the characteristics of 20 stratified open pit lakes. The stratification can be due to chemical enrichment of deep layers resulting from geochemical reactions on the pit walls, inflow of high-salinity water and/or initial pit filling with saline water. The last was described by Fisher and Lawrence (2006) at the Island Copper mine on Vancouver Island.

14.3.7 Ecological risk assessment

In Nevada, when a mine is located on public lands managed by the Bureau of Land Management there are specific requirements for ecological risk assessment (ERA) for open pit mine lakes (BLM 2004). The ERA is used for additional analysis when the predicted pit water chemistry identifies a potential problem with the future pit lake.

The factors to be considered in an ERA include predicted pit water chemical constituents, toxicity benchmark values for avian and terrestrial receptor species for the site, possible exposure pathways for humans, terrestrial or avian wildlife, potential for development of wildlife habitat or aquatic habitat at the pit lake and potential pit lake uses such as recreation. Risk characterisation is based on exposure and effects assessments.

14.3.8 Pit wall stability

Pit slopes are designed to be stable under operating conditions. They are typically optimised to provide sufficient stability for safe working conditions during operations. The slopes may deteriorate as operations proceed, but safety remains the major concern. A large part of this book discusses slope stability.

After closure, there are many changes in boundary conditions that influence the geotechnical stability of the pit walls. The pore pressures change as the dewatering or depressurisation activities cease. New joint surfaces may become inundated and thereby change the shear strength of surfaces. Alteration (weathering) of stresses in the pit walls may result in shear strength changes. The overall effect is that the pit walls may ravel or fail over time, resulting in a much larger surface area and depositing mineralised materials into the pit, often into a lake. The pit water quality may be affected if newly exposed mineralised surfaces are highly reactive.

In extreme cases a major failure of the pit wall may create a wave in the pit lake. Where the pit lake fills the void and has an outlet, the wave could overtop the outlet and travel downstream from the mine area. Some mines (e.g. Brenda in British Columbia, Canada) have constructed a permeable buttress across the outlet to allow continuing seepage of the lake overflow and contain the project maximum wave originating from a post-closure failure.

The geotechnical stability of the pit walls after closure is influenced by:

Table 14.4: Open pit stability conditions related to hydrogeology

Purpose of dewatering	Conditions during mining operations	Possible actions and conditions at closure	Consequences
Depressurisation	Low flow, reduction in pore pressure to stabilise slopes	Remove pumps and allow pore pressure to increase	Increased pore pressure that can influence slope stability, very little flow into the pit
Reduce groundwater table	Medium to large dewatering rates, water can be treated and discharged	Stop pumping, large volumes of inflow	Increased pore pressure that can influence slope stability, pit filling up
		Fill pit rapidly from surface water inflows, e.g. river diversion	Pit is filled in short period, phreatic surface outside pit not recovered and pore pressure conditions can be complicated
Collect water from concentrated flow, e.g. a fault zone	Local controls of groundwater flows and pore pressure	Stop water collection	Water could collect in bottom of pit and rapid infilling of pit

- hydrogeological changes;
- weathering and slaking of certain soft rocks;
- debris flows;
- filling in of benches;
- loss of access to the pit due to instability;
- loss of surface drainage (ditches) and surface water controls;
- undercutting of the pit wall by the pit lake erosion processes;
- increased rockfall hazard;
- stress relief or relaxation;
- seismicity;
- shear strength changes of pit wall materials (including intact materials, fractures and structures).

Activities during operations also contribute to site conditions that can influence the long-term behaviour of the pit at the time of closure. For example, aggressive blasting practices can lead to significant rock damage in the pit walls (section 11.3). This can result in bench level failures and an overall ravelling of the pit slope (Holley et al. 2003).

Hydrogeological changes at the time of closure can be dramatic. For example, in the Sleeper mine Nevada, and Kori Kollo mine, Bolivia, there was no rebound of the phreatic surface as the pits were rapidly filled with groundwater and surface water. Pore pressure conditions therefore resulted in seepage into the pit walls until the phreatic surface rebound was complete. Table 14.4 explores a number of pit stability conditions that depend on hydrogeological conditions.

Closure also extends the seismic exposure period and slopes that are designed to industry standard factors of safety and probabilities of failure are now exposed to longer seismic return periods and resulting higher seismic accelerations. The impacts of seismic activity on long-term slope stability should be assessed; a risk analysis methodology may be an appropriate tool.

Although there is not extensive research about the potential long term effects of stress relief on slope stability, stress relief or relaxation can affect slope stability and result in ravelling of the surface and/or rock fracture in the vicinity of the toe of the slope, either in the slope or the floor of the pit (Stacey & Xianbin 2004). Whilst this is a genuine concern during mining, it is reasonable to assume that post closure the effects of stress will have dissipated. However, as pit walls are increasing in height, these uncertainties must be addressed. Consequently, additional research into the potential effects of stress relief and strain on slope stability has been incorporated into the ongoing LOP Project research plan.

The strength of the pit wall materials may change because of reduced normal stress on the materials near the pit walls or as a result of water flow along fractures or geologic structures. Last, the pit lake may cause erosion of softer deposits, causing undercutting and overhanging zones to form. Small pit slope failures are therefore not unusual during the pit filling process and may also occur once the final lake level has been established. Where this process is possible, some form of protection must be designed into the slopes, for example by rip-rapping the benches and bench faces around the anticipated lake level, taking into account any seasonal fluctuations.

The major concerns with geotechnical stability are:

- safety concerns for people accessing the pit, whether dry or with a pit lake;
- potential impacts on pit lake water quality as pit wall failures occur;
- enlarging the top area of the pit over time so that it is very difficult to establish the ultimate pit limits for access control by bunds or fences.

Regulatory agencies appear to focus on pit lake water quality, not the stability of the pit walls. For example, the author could not find any publicly available reports from mining companies in Nevada that evaluate the post-closure stability of open pits. It is an area of large uncertainty and will require future research.

14.3.9 Pit access

Careful consideration must be given to the issue of public access to a pit after closure. As mentioned above, open pits are designed for operational safety but are not designed to the same requirements as slopes in civil engineering, where public safety is the prime driver.

Artisanal and small-scale miners may forceably intrude the operation access so that they can mine the remaining minerals. This may result in considerable danger, and in most cases such access is discouraged or prohibited.

In the case of a pit lake with good water quality, the public may want access for recreational purposes such as fishing and boating. To allow such access, a beach area must be established – this will require pre-planning and targeted earthworks. Two sites where this is being considered are the pit lakes at the Sleeper and Lone Tree mines in Nevada. Concern remains about the safety of the public, especially if there is potential for pit slope failures after formation of the lake. Such failures can result in rockfalls, large waves (if a sudden release of pit wall rock falls into the lake) and changes in water quality (if sulphide surfaces with oxidation products are exposed following the failure).

The long-term liability will probably remain with the mine owner and/or the land owner. This can affect financial assurance release (relinquishment) and must be addressed by mining companies and regulatory agencies.

14.3.10 Reality of open pit closure

Experience in the closure of smaller pits and a few larger pits has highlighted two very important realities related to mine closure.

1 No walking away – open pits must be monitored after mine closure for stability and hydrological and geochemical conditions such as pit lake formation and water quality. The length of time that monitoring will be required depends on site-specific conditions. Even very successful closure projects have required continued monitoring and maintenance.
2 Historic significance – large open pits are significant man-made structures. It is expected that closed open pits will attract many visitors and there is a potential for the creation of visitor centres at many mines. An excellent example is the visitor centre at the Bingham Canyon mine near Salt Lake City in Utah.

14.4 Open pit closure activities and post-closure monitoring

14.4.1 Closure activities

When all operations cease at an open pit, closure activities can commence. These may include:

- removing all equipment and materials from the pit, including all chemicals and oils and similar materials;
- maintaining a monitoring program for pit slope stability;
- halting dewatering operations and allowing water to flow into the pit;
- filling the pit from other sources, such as river diversions and groundwater pumping;
- potential treatment of the pit water following filling;
- establishing access controls such as blocking the access road and placing a fence or bund around the pit;
- grading and recontouring part of the pit wall to allow safe access to the future pit lake.

14.4.2 Post-closure monitoring

Extensive monitoring of pit walls during operations is an industry standard of care. Similarly, the industry must develop a standard of care for the post-closure monitoring of open pits.

A post-closure monitoring plan must be implemented to obtain information on a regular basis about the pit lake water level and quality. It may be necessary to sample from boats and samples are typically taken at different depths to obtain the pit lake water quality profile. Water temperatures are also measured. Parshley and Bowell (2004) gave details of monitoring procedures during the development of a pit lake.

The stability of the slopes must also be monitored by regular inspection and potentially by using surveying, laser scanning (LiDAR) or photogrammetry, piezometers and other approaches. Pit slope failures may prevent effective monitoring if entry into the pit lake may pose a hazard and when approaching an unstable slope. Recently, satellite-based techniques such as InSAR and other photogrammetric techniques have been employed to evaluate long-term deformations of pit walls. Monitoring is becoming more automated and can be operated remotely. Such techniques should be considered when developing closure plans, especially for large pits and waste dumps.

14.5 Conclusions

Considerable experience is developing in the closure of smaller and medium sized open pits. There have been a number of successful and well-publicised case histories where closure goals have been achieved, especially in the mines that were designed and developed in the late 1980s and 1990s. In those projects, pit lakes have successfully formed and mine waste has been placed back into the pit void, resulting in physically and chemically stable land forms. The challenge that awaits the mining community is to apply the principles from these smaller closures to large open pits. Economics will dictate what can be achieved.

Some of the techniques used on small pits, such as complete pit backfilling, are not economical for larger pits and other solutions are required. Each site has different physical, geologic, hydrogeologic and geochemical conditions. There must be adequate characterisation of these conditions to develop an initial closure plan. Plans must be communicated to stakeholders and appropriate modifications incorporated into the final closure.

The essential considerations for open pit mine closure include:

1 development and communication of the closure goals and criteria with stakeholders and interested parties;
2 maintaining safety for personnel who need access to the closed pit (e.g. monitoring) or are given access for other opportunities (e.g. recreation);
3 addressing environmental concerns of humans, avian species and other fauna. This includes water quality issues related to the pit lake and the surrounding groundwater as it may influence human health and pose environmental risks;
4 evaluating and understanding the physical stability of the pit over the long term and developing appropriate monitoring plans.

Much is to be learned about the closure of large open pits as few have been closed since the 1990s. Setting the closure goals and criteria is the first step in the mine pit closure design. The regulatory framework specifically related to open pits is not extensive and the mining industry has the opportunity to develop economically sound and environmentally sensitive solutions.

Endnotes

1 http://www.ifc.org/ifcest/enviro.nsf/Content/EnvironmentalGuidelines, accessed June 2007.
2 http://www/equator-principles.com, accessed October 2007.
3 http://www.marthamine.co.nz/.
4 http://www.newmont.com/en/pdf/nowandbeyond/NB2003-KoriKollo.pdf, accessed January 2007.

APPENDIX 1
Groundwater data collection

Attachment A: Data collection procedures in RC drill holes

The following is a list of data that may be collected from drill holes as part of the hydrogeological data collection process. The supervising geologist/hydrogeologist (or the driller) should record the required information on a standard hydrogeological log sheet. An example of typical log sheet is given in Table A1.1.

Geological data

The following geological data should be recorded, as appropriate:

1 depth;
2 lithology;
3 mineralisation;
4 degree of fracturing;
5 alteration (e.g. potassic, argillic, propylytic, carbonatitic, silicification, veining);
6 clay content;
7 grain size and degree of sorting.

Depth

Record the depth to which the description relates. This may be at the end of each drill rod.

Penetration rate

Penetration rate is controlled by the driller to optimise the speed of drilling and the life of the drill bit. It is generally a function of bit type (hammer or tricone) and size, the weight on the bit (length and number of drill collars, number of drill pipes and pull-down pressure), the amount of cuttings and who is drilling. An increase or decrease in the penetration rate may be related to the degree to which the formation is fractured.

Record the start time, finish time and time for the interval to be drilled. The penetration rate is related, to a certain extent, to the degree of fracturing or competence of the strata being drilled. It is also highly dependent on the weight on the bit and hence on the size of the bit, number of drill rods and drill collars being used.

Unloading pressure

This is the highest pressure reached on the pressure gauge that monitors the air compressor on the drill after a connection is made (another length of pipe is added). The pressure is partly related to the water level in the hole. This pressure is to be recorded only after the hole begins to produce water. After a joint of pipe has been added and the driller has engaged his compressor clutch, the gauge will rise to a high point then begin to fall. Shortly after it begins to fall the hole will 'unload' and water will be produced at the surface. Record the highest pressure reached by the gauge.

Bit type

Note the type of bit used to drill the hole: H for hammer or T for a tricone bit.

Degree of fracturing

An estimate of fracture intensity should be made using the following scale.

0 = None
1 = Weak
2 = Moderate
3 = Moderately strong
4 = Strong
5 = Intense

Identification of fractures can be site-specific. The following guidelines may be followed:

- rely on the driller and let the driller determine when the ground is fractured;
- look for larger than average chips that are broken on oxide-coated faces;

Table A1.1: Typical hydrogeological data log sheet

Log sheet for hydrogeological data	
Project/ location	
Northing	
Easting	
Elevation	
Hole number/ name	
Contractor	
Rig type	
Date started	
Date completed	

Hole depth (m)	Penetration rate		Airlift discharge (L/sec)	Water temp/pH (°C/pH units)	Electrical conductivity (μS/cm)	Unloading pressure (kPa)	Drilling air pressure (kPa)	Degree of fracturing (0–5)	Clay content (0–5)	Bit type H/T	Comments/ water levels
	Start time	End time									

- more than average cuttings circulated to the surface;
- increase in water airlifted to the surface;
- erratic and difficult drilling.

The following criteria may be generally indicative of or associated with open fracturing bedrock, individually or in combination.

1. Oxidation – rocks with any amount of pyrite adjacent to oxidising water-bearing fractures may contain iron oxide disseminations and coating on chip faces. Bleaching may also be observed, particularly in carbonaceous rocks.
2. Drill cuttings – an increase in the size of the cuttings (or chips) may be due to rock fracturing or to lithological changes. Variations in chip size may be related to changes in texture, mineralogy or hardness. Bit type and condition also play a role; there may be a gradual reduction of chip size with depth as a bit wears out.
3. Voids – voids may be recognised by a sudden drop of the drill rods. Evidence of voids will not appear in a sample tray either because the entire interval is void or all cuttings are lost into the void (no sample).
4. Motion of drill rods – jerky rotation or hanging, and jumping of the drill rods often indicate interception of a fracture zone. It is important to pay close attention to rod advance to accurately record the location and width of the fracture zone, and to relate the occurrence to the returned cuttings and other data. In some cases, a single fracture zone may account for all the groundwater ingress to the drill hole.
5. Volume of cuttings return – fracture zones may cave as they are drilled, resulting in a hole that is much greater in diameter than the bit. The amount of cuttings from the drilled interval may therefore be noticeably greater than normal.

6 Chip shape (flat vs irregular faces) – drill cuttings are normally of similar size and irregularly shaped in homogeneous unfractured rocks (e.g. young intrusive rocks). Flat chip faces may be due to jointing or faulting. Iron oxides may be present on chip faces.
7 Clay content – clay gouge is frequently associated with fault zones. Hydrothermal alteration frequently results in argillisation of rocks.
8 Changes in groundwater chemistry/temperature – a sudden change in groundwater chemistry or temperature often indicates influx of water from a different source, e.g. a water-bearing fracture.

Clay content

A record of clay content in the column can be made using the following scale. It is useful to record the percentage of the sample that is in the form of clay balls.

0 = None
1 = Trace
2 = Minor
3 = Moderate
4 = Mostly
5 = All

Airlift flow rate

This is one of the most important data types collected during drilling. Airlifted flow indicates the amount of water a particular zone of the formation is capable of transmitting when water is forced out of the hole through the annulus of the drill stem. This should be measured by timing how long it takes to fill a container of known volume. In situations where the flow is small, make sure the driller has turned off the water injection pump and wait several minutes for injected water to stop. Flow rates should be measured at the end of each drill rod after water has been encountered.

Airlift flow rate is influenced by several factors, the most important of which are:

- the amount of submergence of the drill rods below the static water level;
- whether the annulus between the drill rod and the hole is open, and the length of hole above the bit in which the annulus is open;
- the size of the annular space (flow around the bit is more restricted with a small annulus than a large annulus);
- new water-bearing fractures encountered as the hole is advanced.

Running air pressure

The air pressure should be recorded at regular intervals during drilling (usually just before the compressor is shut off to allow additional drill rods to be added). The pressure recorded is the pressure required to lift the water to the surface (i.e. the height of water above the drill bit). The readings can be used to estimate the depth to water (pumping water level) during drilling.

Water level

The water level should be measured at least once (preferably more frequently) during the drilling of each hole. The most accurate water level readings will be obtained after any period during which the drill stands idle over the drill hole.

The water level can be easily measured through the centre tube in RC pipe using an electric water level sounder. It is common practice to measure the depth to water in a drill hole at the start of the day before drilling commences and whenever drilling is interrupted, e.g. when adding drill rods and during breakdowns. It may be necessary to periodically interrupt drilling for short periods (10–30 minutes) to measure the water level.

Water quality (field parameters)

Temperature, electrical conductivity and pH can be measured using field meters at the end of each drill rod. The driller should pull the drill bit off the bottom of the hole and continue to run the airlift for several minutes to obtain a clearer water sample.

Water samples

Full water samples can be collected from each hole, during and after completion of drilling, by collecting water from the airlift discharge pipe. Prior to water sampling, the hole would be purged for a standard length of time or until consistent values of pH, conductivity and temperature of the water are measured consecutively. However, it should be noted that samples collected by airlift pumping are often unsuitable for analysis of metals because of their relatively rapid oxidation

Attachment B: Drill stem injection tests

Injection testing provides an easy, low-cost method of investigating the hydraulic response of fracture zones during RC drilling.

The purpose of the injection test is to provide quantitative specific capacity information for the fracture zones that are being drilled. In holes that encounter large fracture zones, the water level may recover very rapidly and hence an injection test rather than slug test (see Attachment E) may be more beneficial for characterisation of fractures.

Ideally, multiple tests should be carried out to assess the main fracture zones. The number of tests performed will depend on the depth of the hole and the characteristics of the formations penetrated during drilling.

The test takes advantage of the design of the RC drill pipe. Water is injected down the annulus of the drill pipe and the resultant rise in head in the drill hole is measured in the inner tube.

The test requires that a special subadapter be fabricated to connect a water supply to the dual reverse circulation (DTRC) drill rods. This can be easily made by the drilling contractor and should introduce the water supply into the annular space between the inner and outer tubes of the drill string. In addition, a 50 mm in-line flow meter, a 3–5 KW trash pump, a water source, a water level sounder and a stopwatch will be required. The typical setup for injection testing is shown in Figure A1.1.

The general procedure for injection testing is as follows.

- Stop drilling when the drill bit passes through a significant fracture zone (this may be indicated by an increase in water airlifted to the surface or jerky motion of the drill rods). Lift the drill bit off the bottom of the hole, to just above the zone of interest. Circulate air to clean the hole and clear cuttings from the drill pipe.

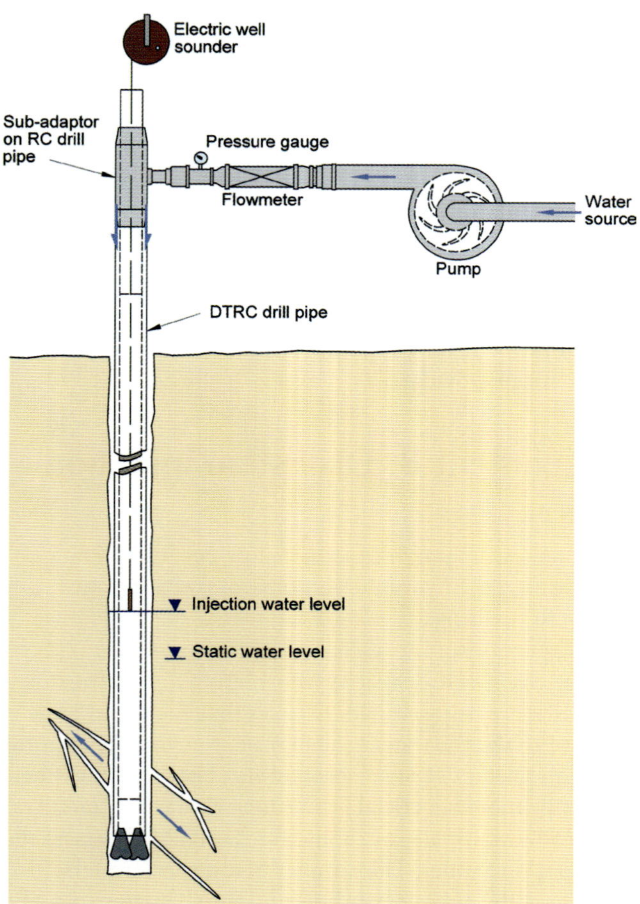

Figure A1.1: Injection testing through DTRC pipe

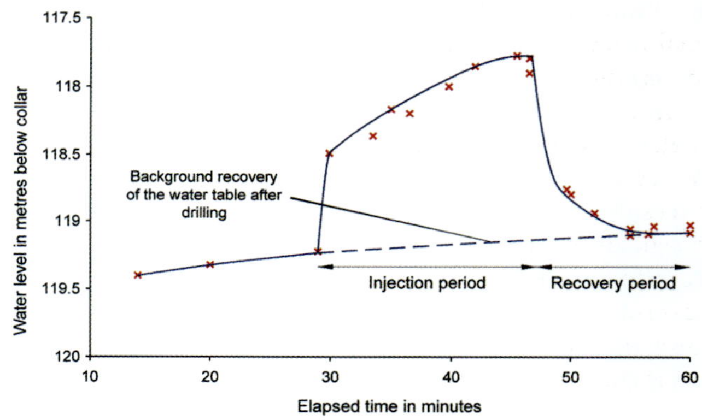

Figure A1.2: Typical response to injection test in drill hole

- Allow the water level to stabilise for a short period (15–30 min), then measure and record the static water level inside the RC drill pipe using a water level sounder.
- While waiting for water level to stabilise, set up the equipment for the test.
- Pump water into the hole at a constant rate via the subadapter.
- Record the change in water level using a water level sounder, and flow rate using an in-line flow meter or by recording the time taken to empty a tank of known volume.
- Pumping should be continued until the water level stabilises. Note that in lower-permeability formations, water may rise to the top of the hole without stabilising. If this occurs, discontinue the test when water reaches the surface.
- At the end of the injection period, continue to monitor and record the fall (recovery) in water level after the pump is turned off.

Figure A1.2 shows an example of the data obtained from an injection test. In this case the response to injection is superimposed on a background rise in water levels due to recovery after completion of drilling.

Injection tests can be run for as long as required, but valuable data on the hole and hydraulic characteristics of fracture zone can be obtained from a test lasting 20–30 minutes.

It is important that the hole is airlifted prior to testing to make sure that the bit and the inner tube are clean. Air in the formation subsequent to drilling can produce spurious results. Better results are usually obtained if the bit is pulled back from the bottom of the hole for the test; if the bit can be pulled back to above the level of the inflow zone, the friction losses due to up-hole flow can be eliminated.

Testing at frequent depth intervals, as the hole is being drilled, allows the depth of groundwater inflow zones to be determined. A rise in specific capacity is observed after each inflow zone is penetrated.

When interpreting the data from injection testing, the main assumption is that as the water level in the hole rises in response to injection, significant loss of fluid out of the hole does not occur in the unsaturated zone above the water table. For this reason, the test procedure is better suited to fractured rock aquifers where groundwater flow takes place in discrete zones. Careful reference to the log of the hole will help these flow zones to be identified, minimising the possibility of misinterpretation. The test is less suited to porous intergranular formations, unless the formation is confined and the rise in water level is through an impermeable confining layer.

Injection testing generally involves a small amount of rig standing time. It can be carried out with minimal cost on as many exploration or test holes as required. It allows a wide areal spread of groundwater data to be built up, data which cannot be obtained for fractured rock aquifers from isolated conventional pumping tests.

The main limitation of the injection test is that it cannot be conducted through a down-the-hole hammer as the check valve located within the hammer prevents water being pumped down the annulus at low pressure. Clay zones and boots around the drill pipe may also affect interpretation of the data.

Attachment C: Guidelines and procedures for piezometer installation

Standpipe piezometers

A typical standpipe piezometer installation may be as follows.

1 Drill the hole to the required depth of the piezometer. Flush the hole with water or biodegradable drilling fluid.
2 If the hole extends below the zone of interest it may need to be partly backfilled. Tremmie cement, bentonite grout, or bentonite into the base of the hole until the desired piezometer level is reached. Place a small amount of sand or gravel above the bottom seal.
3 Lower the standpipe into the hole until the perforated section is adjacent to the zone of interest. The pipe should be suspended from a clamp at the surface throughout the operation. At no time should it be allowed to sit unsupported on the bottom of the hole as this could result in damage to the pipe, particularly in deep installations.
4 Install a filter pack, below and around the piezometer, using a tremmie pipe. Choose a filter pack (sand or gravel) of a grade that will not enter the filter section of the piezometer. The pack is typically continued for a few (2–5) metres above the perforated section of the piezometer.
5 Install the upper seal. Place a bentonite seal above the filter pack, using bentonite chips (for shallow installations) or a bentonite grout (for deeper installations). The seal should be at least 1–2 m thick but can be longer depending on the nature of the formation. The bentonite seal typically requires 2–3 hours to set; refer to the bentonite instructions for exact times. Keep the hole filled with water to fully hydrate the bentonite.
6 Cement the upper hole. Above the upper bentonite seal, the annulus between the hole wall and the piezometer pipe should be filled with a cement/bentonite grout. The use of bentonite causes the grout to swell and prevents the cement from cracking, thereby reducing the potential for movement of water behind the standpipe.

If the diameter of the hole is sufficient, a second (and possibly a third) piezometer can be installed at shallower depths. However, there is a risk that hydraulic connectivity will occur between such nested piezometers if intervening bentonite seals are not well-formed. The risk is eliminated if multi-level piezometers are installed in separate drill holes.

Vibrating-wire piezometers

The vibrating-wire instrument utilises a sensitive stainless steel diaphragm to which a vibrating-wire element is connected. Changing pressures cause the diaphragm to deflect, and this deflection is measured as a change in tension and frequency of vibration of the vibrating-wire element.

A piezometer installation must satisfy the following two basic criteria:

- the measured pressure must be sufficiently close to the pore water pressure of the target formation;
- the piezometer must quickly equilibrate with the pore water pressure of the target formation.

Vibrating-wire piezometers installed using a conventional sand pack and cement/bentonite seals can satisfy these requirements. Vibrating-wire piezometers can also be installed as fully grouted installations, which also satisfy the requirements, as long as the grout is less permeable than the surrounding formation. The grouted-in method is a rapid way to install multiple piezometers in one hole, or to install piezometers in holes with inclinometers or other geotechnical instruments.

Installation methods for vibrating-wire piezometers vary, but a typical procedure is as follows.

1 Prepare the piezometer. The filter stone should be saturated and the space between it and the diaphragm filled with water. After drilling and cleaning the hole, submerge the piezometer in a bucket of clean water, pull off the filter to allow air to escape from the piezometer

then replace the filter. Hold the piezometer with the filter end up to prevent water from draining out.

2 Prepare the steel grout pipe by cutting slots in a number of pipe lengths. Typically, one length of slotted pipe should be installed every 20 m below the water level.
3 The depth of the hole should be checked prior to installation to ensure that there has been no collapse and the sensor can be installed in the open hole. The level of water or drilling fluid in the hole should be recorded.
4 The serial number of each piezometer sensor to be installed (deep, intermediate, shallow) should be recorded prior to installation and the corresponding cable marked accordingly.
5 The sensor should be attached approximately 2 m above the bottom of the first length of pipe to be installed in the hole. This pipe should be slotted.
6 The sensor should be securely attached to the pipe with insulating or duct tape.
7 If multiple piezometer sensors are to be installed in a single hole, the length of pipe to be installed between each sensor should be calculated prior to installation. Intermediate and shallow sensors are attached to the pipe at the correct intervals as it is lowered.
8 The piezometer should be tied in its own signal cable so it can be lowered, filter end up, into the hole.
9 As the sensor is lowered into the hole with the pipe, the cable should be attached to the pipe with tape at regular intervals, normally at the top and bottom of each length of pipe.
10 Once the installation depth has been reached, the pipe can be suspended in the hole with a clamp. The pipe should remain suspended during the entire grouting process.
11 The hole should be backfilled with a cement/bentonite grout using the steel guide pipe as a tremmie pipe. The amount of bentonite should be adjusted to produce a grout with a consistency of heavy cream. If the grout is too thin, the solids and water will separate. If the grout is too thick, it will be difficult to pump. The aim is to achieve a grout that has permeability up to two orders of magnitude lower than the formation permeability. For deep piezometer installations, grouting may need to occur in several stages to avoid excessive pressure on the instrument. The installer should be aware that there may be a lag time between when the transducer measures the fluid pressure of the liquid grout (a function of its specific gravity and the height of the column installed) and when it measures the formation pressure, which presumably is in equilibrium with the pore pressure in the cement after it solidifies and becomes a porous media. Observed lag times have ranged from 7–14 days to 3 months.
12 The vibrating-wire cables should be secured at the surface and terminated above ground level in a waterproof enclosure or with a waterproof connector. The installation should be marked with a high visibility stake.

Once installed, the sensor readings should be checked. Each set of cables should be connected in turn to the readout box and a stabilised reading taken, including digits and temperature. The measured head above each sensor can be calculated by comparing the reading to the initial zero reading recorded prior to installation of each sensor.

If the hole is unstable, the vibrating-wire piezometers can be installed through the drilling rods using the following procedure.

1 Remove the rods from the hole and change the diamond bit for a casing shoe.
2 Lower the drilling rods to the bottom of the hole and flush the hole with clean water.
3 Raise the rods to approximately 5–6 m above the installation zone for the deepest vibrating-wire sensor.
4 Lower the sensors on the steel guide pipe through the rods. The sensors should be connected to the guide pipe in a fashion which allows the casing shoe to slip over the sensors without damaging them.
5 Prepare the grout as appropriate and, using the drill rods as the tremmie, backfill the hole with grout. Remove the rods as the hole is grouted.
6 The rods should not be used to tag the grout level during this process due to the risk of damaging the cable and vibrating-wire sensors.

Check the manufacturer' instructions on how to set up and calibrate vibrating-wire instruments. Further information can be found on relevant websites (e.g. www.Geokon.com and www.slopeindicator.com).

Attachment D: Falling or rising head tests (slug tests)

Rising or falling head tests (slug tests) are a rapid and practical method of obtaining permeability data for the formation using open holes or standpipe piezometer installations. In a slug test, a small volume (slug) of water is quickly removed from (rising head) or introduced into (falling head) an open hole or piezometer, then the rate of water level change in the piezometer is measured. From these measurements, the permeability of the formation in the immediate vicinity of the hole can be determined.

Notes on test procedures

Unless the water level is near the top of the piezometer casing, adding a slug of water may be less satisfactory because the water does not arrive instantaneously at the water surface; some trickles slowly down the hole and is

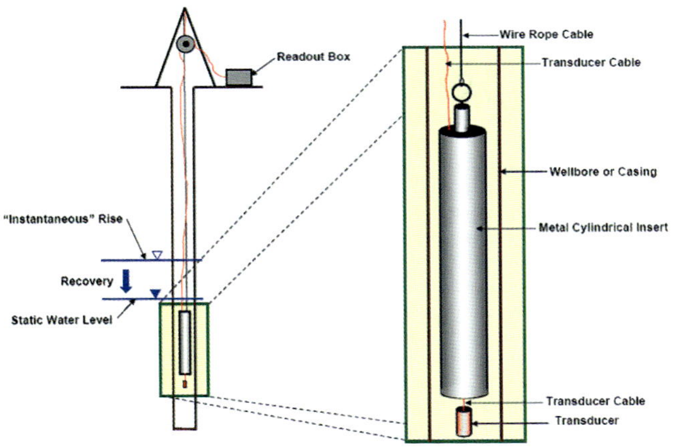

Figure A1.3: Using a transducer to record water level changes in slug tests

still adding to the water level after the water level has begun to fall. A better method is to lower a closed heavy container to just above the water level then lower it quickly into the water, causing a very rapid rise of level. This allows a falling head test to be carried out. When the water level has returned to its initial position the slug can be removed quickly, causing the water level to drop for a rising head test.

Slug tests are very simple to carry out and require strict attention to only two factors:

- efficient removal/addition of the slug;
- measurement and recording of water level changes as quickly as possible after the slug is removed.

As in all hydraulic tests, the water levels in the well change at a much greater rate in the early part of the test. It is therefore important that the slug is introduced into, or removed from, the piezometer as quickly as possible. The water level change in removing or adding the slug should generate an initial water level change of at least 10–50 cm, if possible. The bailer or closed cylinder should therefore be designed just smaller than the internal diameter of the piezometer, to induce the maximum possible change in water level. An effective way of measuring and recording the change in water level is is to install a transducer immediately below the container, as illustrated in Figure A1.3.

The rate of water level change upon commencement of the test will depend on the hydraulic conductivity of the formation around the piezometer. A method of measurement should be designed to produce water level measurements at a time interval in keeping with this rate of change. If the hydraulic conductivity is higher than around 10^{-5} m/sec, the water level will recover too quickly for accurate manual measurements and an automatic recording device (e.g. pressure transducer and data logger) will be needed.

Automatic loggers and pressure transducers can provide water level data every second and are preferable to manual measurement. Automatic loggers reduce measurement error and are particularly useful at the start of the test. The practicality of simultaneously removing a slug and commencing intensive monitoring is limited, unless automated devices are employed.

Table A1.2: Measurement intervals for slug tests

Time since start	Measurement interval
0–2 min	10 sec
2–5 min	30 sec
5– 15 min	60 sec

Manual measurements are suitable where the rate of water level change is slow enough to allow the accurate measurement and recording of water level data at a suitable frequency. Measurements can be made manually with a water level sounder. Water levels must be measured to a datum, which is usually the top of the piezometer casing at an indicated point.

If an automated logger is used, it is also advisable to confirm the water level with a manual measurement at various times during the test. This is more important in longer-duration tests but is good practice for any test.

If water levels are being measured manually, the rate at which they must be taken should mirror the rate of water level change. Suitable measurement intervals may be as shown in Table A1.2.

As slug tests are quick and simple to carry out, a number of different tests on each hole or piezometer can be undertaken. In doing so, any inaccuracies or introduced errors can be averaged and minimised. The number of tests should depend on the consistency of the results and the confidence in the method.

Although slug tests form a very useful component of the hydrogeological investigation program, it should be noted that there are a number of limitations that must be understood in reviewing the results. Any hydraulic data collected through a slug test will pertain to a highly localised portion of the formation. As the head change in the piezometer or well is usually less than 1 m, flow towards the piezometer will be initiated within a few metres radius of the hole, hence the results are relevant for only this volume of the formation. As for any hydraulic testing, the wall of the drill hole may still bear the effects of drilling (mud caking or drilling-induced fracturing) so errors in results can be significant and the data need sensible interpretation.

Method of analysis

There are a number of methods for analysing slug test data. Slug test analysis was originally devised by Hvorslev (1951). The most common methods in use are perhaps those developed by Cooper et al. (1967), Papadopulos et al. (1973) for confined aquifers and Bower and Rice (1976) for unconfined aquifers.

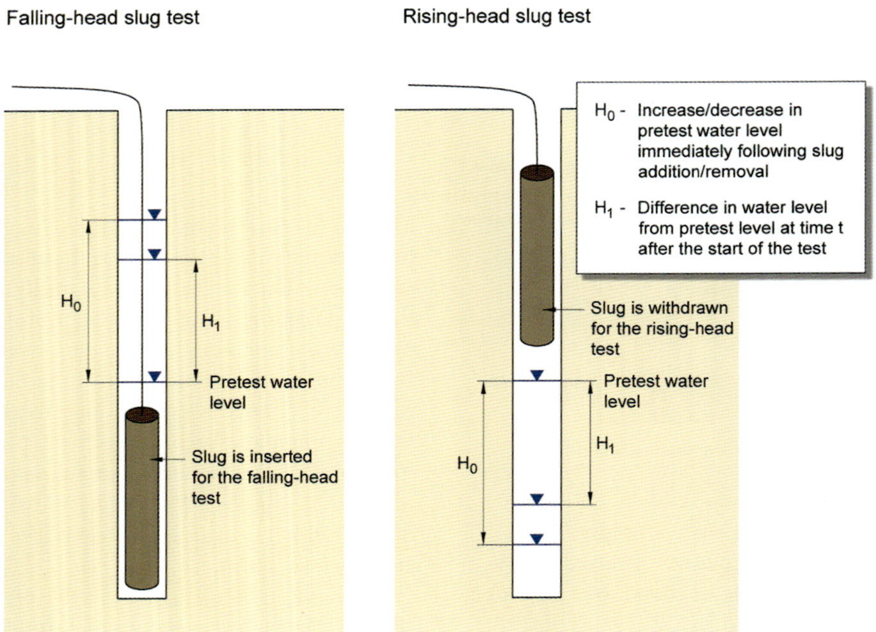

Figure A1.4: Falling head and rising head slug tests

Whatever method is employed, the procedures for analysis are generally straightforward. The original water level in the hole is first noted and recorded. For slug injection, water is instantaneously added to the hole by insertion of a closed cylinder below the water level, raising the water level above its static level. This is termed H_o and is the height of the slug (water level) at time zero (Figure A1.4).

The water level will then start to decay as water flows out of the hole and into the formation. This change in head over time is noted, where H_1 is the height of the slug for any given time. Therefore, at time equal to zero, $H_1/H_o = 1$. As time increases, H_1/H_o approaches zero, i.e. the slug has been completely dissipated and the water level has returned to its original position.

A detailed description of slug test analysis is beyond the scope of this appendix – further information is readily available. A good summary of the various analytical methods is given in Kruseman and Ridder (1991).

Attachment E: Packer test guidelines and procedures

Hydraulic packer testing can be performed to characterise the hydraulic properties of the formation. Conventionally, the procedure consists of a series of injection (constant head) tests using a single packer or double (straddle packer) system. The packer tests can be carried out as the hole is advanced or at the end of the drilling. In fractured rock, test intervals are based on lithology alteration or structure identified from logging of the drill core. They are typically performed below the water level. In many settings, best results are obtained using a single packer setup and a series of constant head injection tests on a number of intervals in each core hole.

Packer configuration, equipment and assembly

The following packer testing equipment is normally used:

- centrifugal pump;
- right-angle fitting with gate valve;
- in-line flow meter and pressure gauge;
- wireline inflatable packers and fittings;
- ¼ inch diameter nylon hose inside composite cabling;
- compressed nitrogen bottle and pressure regulator assembly;
- stuffing box.

Figure A1.5: Electric winch and composite cable

Figure A1.6: Injection manifold with three pressure gauges, each accurate within a certain range of pressure

Figure A1.8: Nitrogen bottle showing shut-in inflation pressure

Figures A1.5 to A1.10 are a series of photographs showing some of these components.

The packer assembly is typically lowered through the drill rods. Single packer setups often include an additional packer, which is inflated inside the end of the drill rods to create a watertight seal between the packer section and the drill rods. A watertight seal is created at the surface where the inflation tube and wireline exit, using a stuffing box. The packers are inflated via a ¼ inch diameter nylon hose using the bottle of compressed nitrogen and pressure regulator assembly.

During testing, fluid is pumped into the test interval through the drill rods with the centrifugal pump. The in-line flow meter and pressure gauge measure flow rate and pressure at the surface. The flow rate is controlled using a gate valve located upstream of the flow meter and pressure gauge.

The centrifugal pump should preferably be a turbine pump to distribute an even flow throughout the injection process. The pump must be able to deliver flows up to the highest anticipated pressure, which could be as high as 200–250 psi. The pump should be able to pump as little as 0.1 L/sec to as much as 5 L/sec.

Good flow meters and pressure gauges that are sized for expected pressure on a manifold with proper sized

Figure A1.7: Manifold showing bypass line

Figure A1.9: Running packer down HQ drill string

Figure A1.10: Tightening down packing

piping are required. The flow meter should be able to accurately measure flow across the range of flow rates described above. The pressure gauges should be sized appropriately for the wide range of pressures in the testing. In some cases, installation of multiple pressure gauges, each with a specific range and accuracy, may be needed for the best possible data.

The cable and airline that support the packer should be run into the hole together as one. The most practical form of cable and air line is called 'composite', where the airline and steel wireline are held together inside a rubber tube. These arrive on a reel that can be raised and lowered using an electric motor or the hydraulics from the drill rig.

Spare packers and nitrogen cylinders should be maintained on site to keep the test work and drilling progressing in the case of a packer failure.

Soap is required for checking leaks in connections (brass or stainless steel) before placing the packer down-hole. A 2% solution of chlorine (Clorox) or hypochlorite may be needed to clean the additives in the drill hole before the test.

Important pretest calculations

Maximum surface test pressure

The maximum surface test pressure (P_{max}) is calculated prior to testing to avoid inducing hydraulic fracturing of the formation during the packer tests. This could lead to an overestimation of the hydraulic properties of the rock formation. Specifically, P_{max} corresponds to the highest allowable injection pressure during testing as measured by the pressure gauge mounted on the surface pipe.

Typically, P_{max} is calculated by taking the distance from the ground surface (adjusted for angle holes) to the top of the test interval and multiplying it by a factor of 0.98–1.64 psi/m. For example, if the top of the test intervals below the ground surface for a 45° inclined hole are 121.9 m and 271. m, the corresponding P_{max} values are 85 psi and 18 psi respectively, based on the following calculation:

Cosine 45° $= Z/121.9$ m → $Z = 86$ m; and
$Z \times 0.985$ psi/m → 85 psi
Cosine 45° $= Z/271.3$ m → $Z = 192$ m; and
$Z \times 0.985$ psi/m → 189 psi

where Z = vertical depth in metres.

It may be appropriate to subtract a 15% safety factor from the pressure calculations at depths below about 100 m. If this advice is followed, the values for P_{max} would be reduced to about 72 psi and 161 psi. In any event, P_{max} should not be exceeded. To be technically correct the pressure values should be expressed in SI units, but as psi gauges are used almost universally by drillers, psi values are used in this example.

Calibration of equipment for head loss calculations

In most cases, it is assumed that injection losses due to friction in the drill rods will not be significant. Additionally, although friction losses through the packer assembly flow pipe may need to be considered, the short length involved reduces this impact and so it will be ignored in subsequent calculations.

The pressure loss within the delivery pipe due to friction can be empirically determined by laying out lengths of drill rod on the ground, inflating the packer setup inside the rods and measuring the pressure differential for different flow rates on either side of the inflated packer. The information is used to plot a head loss vs flow curve.

Packer configuration integrity

Once the packer assembly is installed at the desired depth interval and inflated, its integrity should be checked to ensure that the test zone is sealed. The packers must be inflated to the working pressure specified by the manufacturer to ensure a proper seal. This is normally up to 250 psi but, because the packers are inflated down-hole, hydrostatic pressure (pressure from overlying water column) must be considered. If the water column is assumed to be approximately 1.42 psi/m of water (based on the density of fresh water), the inflation pressure (gauge pressure at surface) required can be calculated as follows:

$$Pi = Pw + Hwc \times 1.42 \text{ psi/m}$$

where

Pi = packer inflation pressure
Pw = packer working pressure
Hwc = height (vertical) of water column above packer (adjusted for angled holes).

Systematic testing procedures

The systematic steps for preparing for a packer test are outlined below. The procedures apply to testing both non-flowing and flowing (artesian) drill holes.

1 Prepare the packer assembly: two packers with open bottom for nominal single packer testing or three packers with perforated middle pipe section and closed cap on the bottom for straddle packer test.
2 Check inflation line connecting the packers and fittings: do not overtighten, which will strip the threads. Make sure there is adequate slack in the inflation line between the packers to accommodate displacement. Normally, the packers will pull away from each other during inflation (but double-check the specific configuration of the system).
3 Check the packer assembly for leakage. Inflate to the maximum gland working pressure in an appropriate length and diameter of drill casing or drilling rods. The packer must not be inflated beyond the manufacturer's recommended pressure (the recommended pressure is usually stamped on the metal band at one end of the packer).
4 Check the wireline connectors on the packer assembly and stuffing box components, especially the seals.
5 Prepare and check the water supply system including the tank, supply, pump, connection hoses, pressure gauges, valves and flow meter.
6 Design the test parameters including the depth and length of test intervals, the bit depth (double-check the driller's record by counting the rods in the hole), the position of packers, the inflation pressure and the water pressure for three stages.
7 Prepare the the test interval by removing any drilling mud and cuttings by flushing with clear water and/or Clorox (2% solution) or hypochlorite.
8 Pull the rods up to locate the drill bit at selected depth.
9 Prepare the wireline winch.
10 Install the stuffing box on the drill rods.
11 Measure the groundwater level several times prior to installing the packer system to assess the static level (or measure pressure, if the hole is flowing under artesian conditions), once the packer assembly and stuffing box have been installed.
12 Lift the packer assembly using the wireline and lower to the landing ring above the drill bit and check that it seats on the landing ring by listening to rods using a wrench. A click is sometimes audible when the packer has been seated properly.
13 If possible, check the depth marking on the wireline if this has been marked for the expected depth.
14 If the hole is flowing under artesian conditions, install stuffing box seals around wireline and inflation lines. If not, go to Step 16.
15 If the flow is artesian, measure the shut-in pressure. Wait for the pressure to stabilise and record the pressure (a shut-in test can be carried out during the pressure stabilisation as described below).
16 Inflate the packer slowly (e.g. in increments of about 50 psi) until the required working pressure is reached. This will require inflating to the working pressure plus the calculated hydrostatic pressure.
17 After inflation is complete, monitor the packer inflation line pressure for at least 2 minutes to see if the system is leaking.
18 After it is determined that the packer is not leaking, measure the water level within the drilling rods and record the height of the pressure gauge above the ground surface.
19 Tighten the stuffing box gradually as water is slowly injected through the rods. All trapped air must be eliminated from the drilling rods before the stuffing box is fully sealed. Any remaining air is bled from the system by a release valve once the stuffing box is fully sealed.
20 Seal the stuffing box cap and attach the water supply system.
21 Check the inflation lines and inflation pressure to ensure that no leaks occur, and check the water feeding system. The packer system is now ready for testing.
22 Prepare the field test form.

Injection test data collection

The injection tests are generally conducted at five consecutive pressures, three ramping up and two ramping down, at values that are 50%, 75%, 100%, 75% and 50% of P_{max} (referred to as P_1, P_2, P_3, P_4 and P_5). The expected pressure range will be based on the estimated permeability of the rock and the expected intake of injected water. These will have to be assessed on a hole-by-hole basis and matched with the pumping equipment available. If insufficient or excessive pressures are used as the initial starting point (50% of P_{max}), the test can be stopped and restarted after adjustments are made in the field.

Flow rates are adjusted as necessary during the test using the upstream gate valve. Time, flow rate and pressure data are collected every 2 minutes during each test step. Injection is continued at each step until the flow rate and injection pressure has stabilised over the course of three consecutive readings.

When performing the tests, always confirm that the packer is sitting on the crown of the diamond drill bit.

Basic test procedures

Non-flowing conditions

1 Open the water feeding system valve and maintain constant initial pressure until it appears to have stabilised.
2 During this time, record the elapsed time and total volume of consumed water every 30 seconds for the first 2 minutes of the test stage, then every minute thereafter.
3 After the pressure has stabilised for three consecutive readings, increase the pressure to P_2.
4 Record the time vs flow rate until the step has stabilised for three consecutive readings.
5 Increase the pressure to P_3 and continue recording data until the step has stabilised for three consecutive readings.
6 Decrease the pressure to approximately P_4 and continue recording data until the step has stabilised for three consecutive readings.
7 Decrease the pressure to approximately P_5 and continue recording data until the step has stabilised for three consecutive readings.
8 After conducting the last injection step, conduct a recovery test by shutting the feed valve and recording pressure (decreasing) vs time for about 10–15 minutes, or until 90% recovery has occurred.
9 Deflate the packer assembly and remove the stuffing box cap and seal.
10 Wait until all the nitrogen escapes from the packer cells, wait an additional 5 minutes, then pull the assembly carefully to top of drill rods, watching for a previously placed marker flag to avoid pulling the assembly into the overhead sheave.
11 Measure the groundwater levels after the test several times to assess water level recovery.

These steps are only a guide. Tests can be modified and made shorter or longer and at several intervals in each hole before proceeding to Steps 10 and 11.

Flowing-artesian conditions

To carry out a discharge test under flowing artesian conditions, the packer is located at the drill bit in the same manner as for the injection test. The procedures for testing are fairly simple, but care must be taken to ensure a good packer seal since pressure response curves will not indicate leaking seals as they do in an injection test.

The following procedures outline the basic test setup.

1 Locate the drill bit and packer assembly as in an injection test.
2 Close the flow valve at the top of drill rods and monitor the pressure.
3 After the pressure has stabilised (i.e. a steady reading is achieved for at least 3 minutes), record the shut-in pressure.
4 Start the timer and record pressure and flow rate (Q) vs time at intervals of 0.5, 1.0, 1.5, 2.0, 2.5, 3.0, 3.5, 4.0, 4.5, 5.0, 6, 7, 8, 9, 10, 12, 14, 16, 18, 20, 25, 30, 35, 40, 50, 60, 80, 100, 120, 150, 180, 210, and 240 minutes.

A recovery test should be carried out in conjunction with a discharge test. Following the discharge period, the time is recorded and the valve is again shut. This will allow the formation pressure to recover. The relationship of time vs pressure is recorded for this test on the same intervals as those listed for the discharge test.

To minimise standing time for the drill, the test duration may be shortened to 120 minutes of discharge and 120 minutes of recovery. Ideally, the test time should never be less than 60 minutes of discharge and 60 minutes of recovery.

Data analysis

A typical procedure is to calculate permeability from the measurements of total excess injection head (hydraulic head at centre of the test interval) and flow rate for each pressure step, using the following formula for steady-state flow conditions:

$$K = Q\log_e(L/R)2\pi LH$$

where:

K = permeability (m/day)
Q = flow rate (m^3/day)
L = length of test interval (m)
H = hydraulic head at centre of test interval (m of water)
R = drill hole radius (m).

Data from a discharge test can be plotted as psi vs flow rate (Q) on a logarithmic scale. There are other formulae in the literature for calculating permeability from packer test data based on slightly different assumptions regarding the nature of the flow from or into the test interval. They all give approximately the same results.

Routine quality control checks

To ensure that the data collected are accurate and, more importantly, representative of the zone of interest, the tester must verify that the test assembly is not leaking. Leaks through the supply line or rods, or past the packers, will have the effect of apparently increasing the permeability of the test zone. This is because water pumped during the test will be assumed to be flowing into the test zone, but will instead be a combination of zone uptake and leakage. This will become more significant as the permeability of the zone decreases and/or the injection pressure increases towards $P_{max.}$

A common area for leaks is past the packer, between the expanded gland and the wall of the hole. Incomplete inflation, irregularities on the hole wall and tears in the outer gland material are likely reasons. Leakage can often be difficult to assess as flow past the packer cannot be distinguished at the surface. To determine if the packers are sealing properly, various procedures can be followed:

- prior to the packer being run into the hole, it should be inflated inside a length of plain pipe. A check should then be made to ensure that there are no leaks in the nitrogen pressurising system;
- during the test, the packer pressure as indicated by the gauge on the nitrogen bottle should remain constant. Check for any drops in the packer inflation pressure during the test. A drop in pressure will indicate that the packer is deflating or the supply line is leaking;
- check for bubbles if the water level is visible at the top of the drill rods;
- note unexpected flow vs pressure relationships within a single test or as compared to other test zones of similar rock (determined from the drill core). This could suggest leakage;
- increase the inflation pressure slightly. If a good seal has been achieved, this should make no difference to the flow/pressure relationship.

If any of these measures suggest leakage, it will be necessary to remove the packer assembly, test it and rerun the test to verify that the data collected are representative and accurate.

Another area where leaks can occur is at the joints in the drill string. These leaks can be stopped by wrapping a string or wicking material around the rod threads before connecting the rods. The leakage may be greatly reduced, but may still have a significant effect when the cumulative leakage is taken into account.

To test for apparent leakage, a blind packer assembly (that is plugged at the bottom) can be lowered to just above the bit and inflated. Water is then pumped into the rods and the flow vs pressure response is recorded. If it is assumed that the packers are sealing the rods and that water is not flowing through the bit, then any flow will be the cumulative joint leakage. The pump pressure should correspond to the expected test pressures. In performing this test, the pressure should not exceed 80% of the packer inflation pressure as this could force water past the gland regardless of proper inflation.

Attachment F: General pumping test procedures

Principles

Pumping tests are a method of characterising the groundwater system over a wider area than slug tests. The action of pumping a well creates a cone of depression in the potentiometric surface, whose shape and rate of expansion depend on the hydraulic properties of the formation. By monitoring the way the area of drawdown expands, it is possible to derive information about the hydraulic properties of the area surrounding the well. In a typical pumping test, a larger-diameter well is pumped and observation wells or piezometers installed within the area of pumping influence are monitored for drawdown response.

Pumping tests are usually undertaken in permeable settings, where it is possible to pump at significant rates from wells. Thus, pumping tests are more applicable for dewatering evaluations. In many situations involving slope depressurisation, the permeability of the materials may be too low to allow conventional pumping tests. However, it is often possible to carry out tests at a low pumping rate or by airlifting, with an array of closely spaced piezometers monitoring the drawdown in the vicinity of the pumping well.

Types of pumping tests and methods of analysis are numerous and form a principal method of characterising any groundwater system. There are various documents on aquifer test procedures. One of the most complete texts on pumping test setup and interpretation is Kruseman and de Ridder (1991).

If low-volume pumping tests are conducted specifically for pit slope evaluations, the empirical data obtained from the tests can be as valuable as the analysis for hydraulic properties.

Constant-rate pumping tests

Pumping tests have a much larger radius of influence than slug tests. The results of a long-duration constant-rate pumping test should therefore provide hydraulic data that are more fully representative of the groundwater system as a whole, broadly averaging out any localised heterogeneities.

A constant-rate test (constant-discharge test) is the most common type of pumping test and can provide values of permeability and storage coefficient. A constant-rate test involves pumping a hole at a constant rate over a number of days, weeks or months, and measuring the groundwater response in the pumping hole and in nearby observation wells. The time–drawdown data can be analysed by a suitable analytical equation or numerical model to provide values of the formation's hydraulic parameters.

Pre-test observations and considerations

A number of outside influences can affect water levels in the formation and should be monitored prior to and during the pumping test so that corrections can be made, if required, during analysis. These include background changes or trends in the groundwater system due to nearby pumping or natural influences, including seasonal changes, barometric effects, earth tides in strongly confined settings and other more site-specific effects such as pumping from other wells or from the pit area. To correct for these effects, there must be a period of monitoring groundwater levels in the observation wells or piezometers that will be monitored during the test itself.

Rainfall should be measured before and during the pumping test so that the groundwater response to any such events can be determined and corrected for. However, if the groundwater is at significant depth it is unlikely that rainfall will create any groundwater recharge effects if the test duration is relatively short.

Any groundwater discharges (e.g. from springs and nearby pumping wells) should be monitored before and during the pumping test so that their effects on groundwater levels may be taken into account during analysis of the test results. Ideally, other pumped wells in the vicinity of the test site should be held at a pumping rate that is as constant as possible during the pretest observations and during the test. The test data may need to be corrected if pumping from neighbouring wells is stopped during the course of the test.

Selection of pumping rate

Selection of the pumping rate during the test is important. The rate should be within the bounds of the sustainable yield of the well, so that it can be maintained as steadily as possible for the duration of the test.

Knowledge of well yield is therefore important for planning the test. If no such information is available, a short pumping test or step-drawdown test to establish the yield of the well, or its specific capacity, may be advisable.

Measurement of discharge

The discharge rate is liable to vary during pumping tests, particularly during the early stages where a drop in the groundwater level in the pumping well increases the head on the pump. To ensure a constant rate is maintained, the discharge rate must be monitored to allow adjustments to the pump as the head increases.

The most reliable method of measuring pumped discharge is to use a weir tank with the pumped water discharging over a rectangular or V-notch weir, or through an orifice plate. Weir tanks need to be carefully set up and levelled and care must be taken that filling the tank does not cause it to settle and tilt.

Modern flow meters can be just as accurate and can be linked directly to data loggers. During the test, discharge readings should be taken at regular intervals (approximately every 30 min), and more frequently near the start of the pumping test when the rate is most likely to change because of the increase in head.

Measurement of groundwater levels and test duration

As for slug tests, pressure transducers with loggers or manual dip readings are both suitable as long as resolution of measurement in the test well is 10 mm or better. If the former, it is essential to carefully consider the choice of transducer as it must be capable of measuring the whole of the anticipated range of drawdown at the required resolution. If water levels are to be measuremented in observation wells, the resolution of these measurements should be 5 mm or better. If automated loggers and pressure transducers are to be used throughout the test, it is advisable that manual measurements be taken at various stages of drawdown during the test to confirm transducer accuracy.

A time–drawdown graph will be used during analysis, with time on a logarithmic axis. To allow an even spread of data on the log scale, the data should be collected at increasing intervals throughout the test. A suitable measurement schedule may be as follows:

- immediately before discharge is started, stopped or changed;
- every 30 seconds for the first 10 minutes;
- every minute for next 20 minutes;
- every 10 minutes thereafter until the completion of 2 hours of pumping;
- every 30 minutes thereafter until the completion of 8 hours of pumping;
- every 1 hour thereafter until the completion of 18 hours of pumping;
- every 2 hours thereafter until the completion of 48 hours of pumping;
- every 4 hours thereafter until the completion of 96 hours of pumping;
- every 8 hours thereafter until the completion of 168 hours of pumping;
- and so on…

The total length of the test should be such that an approximate steady state is reached (i.e. no significant further drawdown can be identified). In an unconfined aquifer, the test should be long enough for delayed yield effects to be recognised (i.e. a decrease in the rate of drawdown followed by an increase in the rate). The length of the test is usually dependent on the development of the cone of depression and cannot always be planned in advance. Approximate predictions can be made using the Cooper-Jacob equation (Kruseman & de Ridder 1991).

Discharge of pumped water

The pumped water should be discharged at the surface into a drain or other feature and directed away from the pumping well and out of the area of influence of the pumping test. This is to ensure that percolation of discharge waters does not recharge the groundwater system and recirculate to the well during the test period.

Analysis of constant-rate pumping tests

The analysis of pumping test data is usually performed with analytical equations based on a number of simplifications and assumptions. The accuracy of the estimates of hydraulic properties depends on the fit of the tested aquifer with the idealised aquifer assumed in the chosen analytical solution.

For most analytical solutions, the measured water level is referenced to some radial distance from the pumping well, not the water level in the well itself. The pumping water level in the test well itself is generally not representative of conditions in the formation adjacent to the well because of the head losses as the inflowing water velocity increases immediately around the well.

Most methods of analysis require the test data to be fitted to a type-curve which represents the response in the idealised aquifer. Several software packages allow the data to be plotted and compared with the ideal case; some even take the input direct from a data logger. As with all analytical solutions, the estimates are only as good as the match between the tested system and the idealised aquifer (and all the stated assumptions) upon which the solution is based.

The purpose of most pumping test analyses is to provide values of the transmissivity and storage coefficient. For detailed procedures on analysis methods of pumping tests, see standard texts such as Kruseman and de Ridder (1991). However, in a mine setting, where high hydraulic stresses are involved and where complex fracture-flow conditions often occur, the practical use of transmissivity and storage coefficient values can be somewhat limited and a more empirical approach to pumping test analysis is often more appropriate.

APPENDIX 2
Essential statistical and probability theory

With the permission of the author, much of the text and examples that follow have been abstracted directly from *Reliability-based design in civil engineering* (Harr 1996), to provide a brief introduction to the essential concepts of probability and statistics.

1 Axioms of probability

The starting points in representing uncertainty and reliability are the two primary definitions of the concept of probability: relative frequency and subjective interpretation.

The concept of relative frequency was introduced by Laplace (1812): if an outcome A occurs T times in N equally likely trials, the probability of its outcome P[A] is:

$$\mathrm{P}[\mathrm{A}] = \frac{T}{N}$$
$$= \text{favourable outcomes/total possible outcomes} \qquad \text{(eqn A2.1)}$$

Example: Find the probability of drawing a red card from an ordinary, well-shuffled deck of cards.

Solution: Of the 52 likely outcomes there are 26 favourable outcomes (red cards). Hence:

$$\mathrm{P}\left[\text{drawing red card}\right] = \frac{26}{52} = \frac{1}{2} = 50\%$$

It is understood in the example that if the process was repeated a very large number of times, a red card would appear in half the trials.

As an example of the concept of subjective interpretation, take the statement 'The probability of failure of a proposed structure is 1% (= 0.01)'. Here, the concept of repeated trials is meaningless. Within the sense of the example, the structure will be built only once and it will either fail or be successful during its lifetime. It cannot do both. This is an example of the subjective interpretation of probability. It is a measure of information as to the likelihood of the occurrence of an outcome.

Although subjective probability is generally more useful than the relative frequency concept, both are governed by the same rules. These rules are encapsulated with three axioms.

Axiom 1: The probability of an outcome A ranges between zero and unity:

$$0 \leq \mathrm{P}[\mathrm{A}] \leq 1 \qquad \text{(eqn A2.2a)}$$

Axiom 2: The certainty of an outcome C is a probability of unity:

$$\mathrm{P}[\mathrm{C}] = 1 \qquad \text{(eqn A2.2b)}$$

Axiom 3: The third axiom requires the concept of mutually exclusive outcomes. Two outcomes are mutually exclusive if they cannot occur simultaneously. The third axiom states that the probability of the occurrence of the sum of a number of mutually exclusive outcomes A_1, A_2 … A_N is the sum of their individual probabilities. That is:

$$\mathrm{P}\left[\mathrm{A}_1 + \mathrm{A}_2 + \ldots + \mathrm{A}_N\right] = \mathrm{P}\left[\mathrm{A}_1\right] + \mathrm{P}\left[\mathrm{A}_2\right] + \ldots + \mathrm{P}\left[\mathrm{A}_N\right] \qquad \text{(eqn A2.2c)}$$

A particularly important application of these axioms is that after the construction of a structure, only one of two outcomes can occur in the absolute structural sense: either it is successful or it fails. These are mutually exclusive outcomes. They are also exhaustive in that, within the sense of the example, no other outcomes are possible. Hence, the second axiom (Equation A2.2b) requires that:

$$\mathrm{P}[\text{success} + \text{failure}] = 1$$

Since they are mutually exclusive, the third axiom specifies that:

$$\mathrm{P}[\text{success}] + \mathrm{P}[\text{failure}] = 1$$

The reliability, R, of a structure is its probability of success. Designating the probability of failure as $p(f)$, it follows that:

$$R + p(f) = 1 \text{ or} \qquad \text{(eqn A2.3a)}$$

$$R = 1 - p(f) \qquad \text{(eqn A2.3b)}$$

2 Probability distributions

Measures of probability are of two types: discrete or continuous. An example of a discrete variate is an earthquake, which either will occur or will not. In contrast, a variate such as the frictional strength (Ø) of a joint exhibits a continuous range of possible values between specified lower and upper limits.

The two widely used discrete probability distributions are the binomial and the Poisson distributions.

2.1 Binomial distribution

The binomial distribution:

$$b(x; N, R) = \frac{N!}{x!(N-x)!} R^x p^{N-x} \qquad \text{(eqn A2.4)}$$

provides the probabilistic measure of the likelihood of success (or failure) of a number of experiments N repeated x times for which the following properties hold.

1. The experiment is repeated a number of times N, with each outcome independent of all others.
2. Only two mutually exclusive outcomes are possible, success or failure. x is the number of successes.
3. The probability of occurrence (success) in each trial, denoted R (reliability), remains the same for all trials. p is the probability of non-occurrence (failure): $R + p = 1$.
4. The experiment is performed under the same conditions for all N trials.
5. Interest is in the number of successes x in the N trials and not in the order in which they occur.

Example: A basketball player made two-thirds of his foul shots through the regular playing season. If he maintained the same probability of success in a post-season game, what is the probability that he would make only one foul shot in five attempts?

Solution: The probability of success is $R = \frac{2}{3}$, $p = \frac{1}{3}$, $N = 5$, $x = 1$ and, from Equation A2.4:

$$P[x \mid N, R] = P\left[1 \mid 5, \tfrac{2}{3}\right] = \frac{5!}{1!4!}\left(\frac{2}{3}\right)\left(\frac{1}{3}\right)^4 = 0.041$$

The probability that he would make only one foul shot from five attempts is approximately 4%.

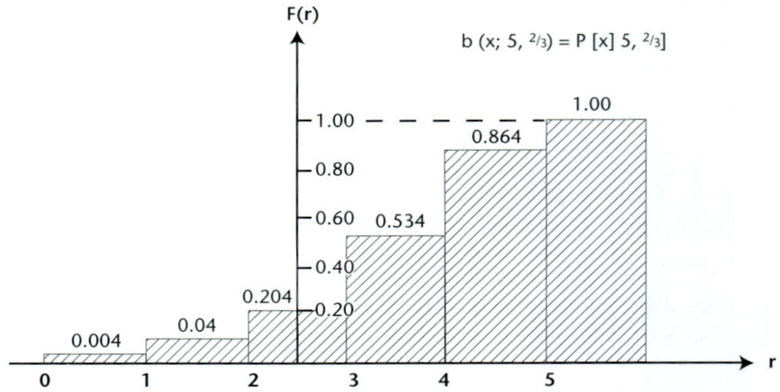

Figure A2.1: CDF for binomial distribution function $P\left[1 \mid 5, \frac{2}{3}\right]$

2.1.1 *Cumulative distribution function*

The probability that a random variable will take on a value less than or equal to a particular numerical value or will take a range of values can be provided by the cumulative distribution function (CDF). Thus, if x is a random variable and r is a real number, then the CDF, designated as $F(r)$, is the probability that x will take on values equal or less than r.

For the binomial distribution this can be written as:

$$F(r) = P[x \leq 1] = \sum_{\text{all } x_i \leq r} b(x_i; N, R) \qquad \text{(eqn A2.5)}$$

For the example of the basketball player, the CDF for the probability function $P\left[1 \mid 5, \frac{2}{3}\right]$ is given in Figure A2.1.

2.1.2 *Expected value*

The expected value is the measure of the central tendency of a probability distribution for a random variable (x):

$$E[x] = \bar{x} = \sum_{i=1}^{N} x_i f_i \qquad \text{(eqn A2.6)}$$

The expected value is closely related to the arithmetic mean, or mean, of a collection of numbers, but a difference between the two should be noted. The arithmetic mean is a measure of the central tendency of a collection of sample observations, with each sample having an equal probability of occurring. The expected value is obtained from the probability distribution function of a random variable wherein the individual outcomes may have very different probabilities of occurring.

Example: What is the expected value of the number of dots that will appear if a fair die is tossed?

Solution: Each of the random variables 1, 2, 3, 4, 5 and 6 has the equal probability $f_i = 1/6$ of appearing. Hence, from Equation A2.6:

$$\begin{aligned} E[\text{toss of fair die}] &= 1(1/6) + 2(1/6) + 3(1/6) \\ &\quad + 4(1/6) + 5(1/6) + 6(1/6) \\ &= 3.5 \end{aligned}$$

Note that even though 3.5 is the expected value, it is an impossible outcome. There is no face on the die that will show 3.5 dots. It simply represents the value about which scatter can be expected if repeated experiments are conducted. Thus, if a fair die is tossed over and over again, on average 3.5 dots per toss will be observed. Also, as all $f_i = 1/6$, for this case the expected value is the same as the arithmetic mean.

For a binomial distribution, the expected value of the success of each trial is R and it remains constant. Hence:

$$E[x \mid \text{binomial distribution}] = E[x_1] + E[x_2] + \ldots E[x_N] = NR \quad \text{(eqn A2.7)}$$

2.1.3 *Variance*

The variance is the measure of scatter (or variability) of a random variable:

$$V[x_i] = E[x_i^2] - (E[x_i])^2 \quad \text{(eqn A2.8)}$$

2.1.4 *Standard deviation*

The standard deviation is a more meaningful measure of dispersion and is the positive square root of the variance:

$$\sigma[x_i] = \sqrt{V[x_i]} \quad \text{(eqn A2.9)}$$

2.1.5 *Coefficient of variation*

The coefficient of variation is a relative measure of the scatter of a random variable. It is usually designated as a percentage and uses parentheses to differentiate it from the similarly appearing variance V[x], which is given as a decimal:

$$V(x) = \frac{\sigma[x]}{E[x]} \times 100\% \quad \text{(eqn A2.10)}$$

The coefficient of variation expresses a measure of the reliability of the central tendency (expected value). For example, for an expected value of 10 a coefficient of variation 10% would indicate a standard deviation of 1, whereas the same expected value with a coefficient of variation of 20% would specify a standard deviation of 2. As a rule of thumb, coefficients of variation below 10% are thought to be low, between 15% and 30% moderate, and greater than 30% high.

2.1.6 *Sigma units*

Another useful measure is obtained by expressing the difference between a value of the variate and its expectation as a number of standard deviations, called the number of sigma units (Figure A2.2), which has valuable relationships with Chebyshev's inequality and Gauss' inequality.

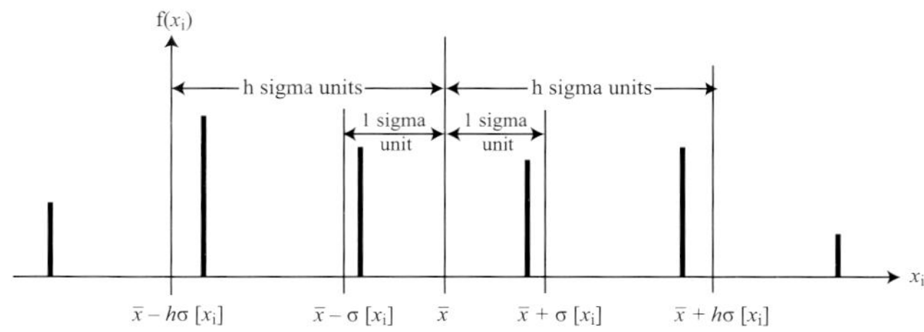

Figure A2.2: Sigma units, discrete probability distribution

Chebyshev's inequality states that if x_i is a random variable with an expected value $\bar{x}$ and a standard deviation σ, then for h sigma units $(h\sigma)$ to each side of $\bar{x}$:

$$P\left[\left| x_i - \bar{x} \right| \geq h\sigma\right] \leq \frac{1}{h^2} \quad \text{(eqn A2.11)}$$

In other words, for any probability distribution the probability that random values of the variate will lie within the range of h standard deviations on each side of the expected value is at least $1 - 1/h^2$. This is also true for continuous probability distributions (Figure A2.3).

Gauss' inequality states that if any distribution function is unimodal (i.e. symmetrical with respect to its expected value and the expected value is also its maximum value), then:

$$P\left[\left| x_i - \bar{x} \right| \geq h\sigma\right] \leq \frac{4}{9h^2} \quad \text{(eqn A2.12)}$$

Probabilities for a range of expected values $\pm h$ sigma units for the Chebyshev and Gauss' inequalities, together with exact values for the exponential, normal and uniform distributions, are given in Table A2.1.

The information in Table A2.1 is extremely valuable. Specifically, for the normal distribution, which is the most frequently used distribution, it tells us that 68% of the information contained in the distribution is within one standard deviation of the expected value ($\bar{x}$) and 96% within two standard deviations of $\bar{x}$.

2.2 Poisson distribution

The Poisson distribution is an adaptation of the binomial distribution to cater for the probability of occurrence of rare events.

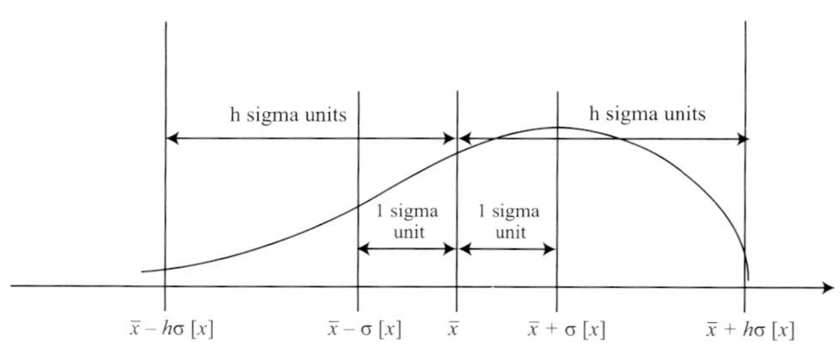

Figure A2.3: Sigma units, continuous probability distribution

Table A2.1: Probabilities for range of expected values ±*h* sigma units

H	Chebyshev's inequality	Gauss' inequality	Exponential distribution	Normal distribution	Uniform distribution
0.5	0	0	0.780	0.380	0.29
1	0	0.56	0.860	0.680	0.58
2	0.75	0.89	0.950	0.960	1.00
3	0.89	0.95	0.982	0.997	1.00
4	0.94	0.97	0.993	0.999	1.00

The binomial distribution Equation A2.4 can be rewritten as

$$b(x; N, p(f)) = \frac{N!}{x!(N-x)!} p(f)^x R^{N-x} \quad \text{(eqn A2.13)}$$

where $p(f)$ is the probability of a failure occurring in the present context and is small compared with the probability of a failure not happening (= success = R).

The expected value for the binomial distribution is $Np(f)$. Suppose that $Np(f)$ is held constant, e.g. $Np(f) = \mu$ and that N approaches infinityi.e. there are a very large number of trials. If the occurrence of a failure is a rare event, very many trials are required to find one. In these circumstances, i.e. $N \to \infty$ and $Np(f) = \mu$, the distribution can be written in the form of the Poisson distribution (Harr 1996):

$$b(x_i; N, p(f)) = f(x_i) = \frac{\mu^{x_i} e^{-\mu}}{x_i!} \quad x_i = 0, 1, 2, \quad \text{(eqn A2.14)}$$

The expected value and variance of the distribution are both equal to μ.

2.3 Normal distribution

The most frequently used continuous probability distribution is the normal distribution.

$$f(x) = N(\bar{x}, \sigma) = \frac{1}{\sigma\sqrt{2\pi}} \exp\left[\frac{-(x-\bar{x})^2}{2\sigma^2}\right] \quad \text{(eqn A2.15)}$$

where

$\bar{x} = E[x]$, $\sigma = \sigma[x]$ and the range is $x(a) = -\infty$ to $x(b) = +\infty$

and

$$E[x] = \int_{x(a)}^{x(b)} x f(x)\,dx \quad \text{(eqn A2.16a)}$$

As before:

$$V[x] = E[x^2] - (E[x])^2 \quad \text{(eqn A2.16b)}$$

and

$$\sigma[x] = \sqrt{V[x]} \quad \text{(eqn A2.16c)}$$

The distribution can be viewed as as a probability distribution function (PDF), which is the well-known symmetrical bell-shaped curve (Figure A2.4), or as a cumulative distribution function (CDF).

Probabilities associated with the distribution are widely tabulated in the literature and are readily available in many computer software applications, so examples are not given here.

Other useful continuous probability distributions include the uniform, exponential, lognormal and beta distributions.

2.4 Uniform distribution

The uniform distribution is the continuous probability distribution $f(x) = C$ where C is a constant between $x = a$ and $x = b$.

For the uniform distribution:

$$f(x) = \frac{1}{b-a} \quad a \le x \le b \quad \text{(eqn A2.17)}$$

$$E[x] = \frac{a+b}{2} \quad \text{(eqn A2.18a)}$$

$$V[x] = \frac{(b-1)^2}{12} \quad \text{(eqn A2.18b)}$$

and

$$\sigma[x] = \frac{b-a}{2\sqrt{3}} \quad \text{(eqn A2.18c)}$$

2.5 Exponential distribution

The exponential distribution is the continuous probability distribution $f(x) = \lambda \exp(-\lambda x)$ where $z \ge 0$ and λ is a constant.

$$E[x] = \frac{1}{\lambda} \quad \text{(eqn A2.19a)}$$

and

$$V[x] = \frac{1}{\lambda^2} \quad \text{(eqn A2.19b)}$$

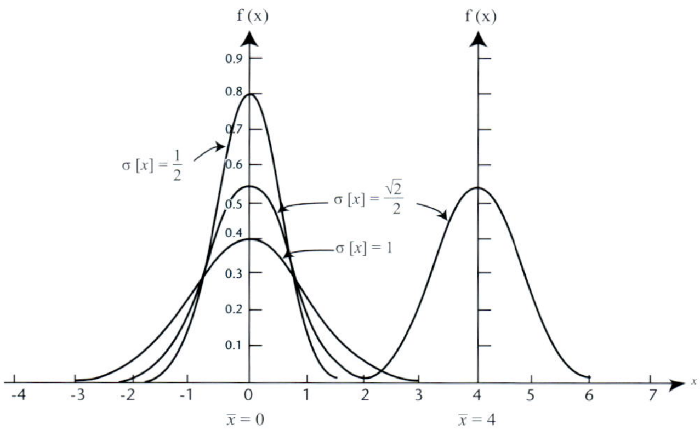

Figure A2.4: Examples of normal probability distribution functions

2.6 Lognormal distribution

The lognormal distribution is the continuous probability distribution where, if $x = \ln y$ or $y = \exp(x)$, then y is said to have a lognormal distribution. If the lognormal is bounded between $y(a)$ and $y(b)$, the distribution has a minimum value of zero, $y(a) = \exp(-\infty) = 0$, and is unbounded above, $y(b) = +\infty$.

The probabilities associated with lognormal variates can be quickly obtained from those of mathematically corresponding normal variates. Thus, if $E(y)$ and $V(y)$ are the expected value and coefficient of variation of a lognormal variate y, the corresponding normal variate x will have the expected value and standard deviation given by Equation A2.20.

$$(\sigma[x])^2 = \ln\left\{1 = [v(y)]^2\right\} \qquad \text{(eqn A2.20a)}$$

$$E(x) = \ln E[y] - \frac{(\sigma[x])^2}{2} \qquad \text{(eqn A2.20b)}$$

2.7 Beta distribution

The beta probability distribution is defined over the range $[a, b]$ where a is the minimum value of the variate x, b is the maximum value of the variate x and:

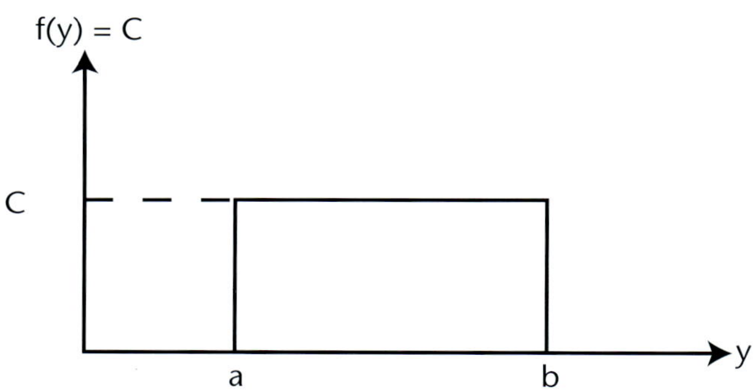

Figure A2.5: Uniform probability distribution

$$f(x) = C(x-a)^{\alpha}(b-x)^{\beta} \qquad \text{(eqn A2.21)}$$

where $\alpha > -1$ and $\beta > -1$, and the normalising constant C is:

$$C = \frac{(\alpha+\beta+1)!}{\alpha!\beta!(b-a)^{\alpha+\beta+1}} \qquad \text{(eqn A2.22)}$$

The parameters α and β can be obtained from the expected value $E[x] = \bar{x}$ and the standard deviation $\sigma = \sigma[x]$ of the variate from:

$$\alpha = \frac{X^2}{Y^2}(1-X) - (1+X) \qquad \text{(eqn A2.23a)}$$

$$\beta = \frac{\alpha+1}{X} - (\alpha+2) \qquad \text{(eqn A2.23b)}$$

where $X = (\bar{x} - a)/(b-a)$ and $Y = \sigma/(b-a)$.

For further details of the beta distribution, including curves that yield α and β as a function of X and Y and some schematic probability distributions as functions of α and β, see Harr (1996, pp.79–85).

3 Information and distributions

If nothing is known about a probability of distribution except that it is one, then:

$$\sum p_i = 1 \qquad \text{(eqn A2.24)}$$

This is all the information that is available. What can be assumed about the distribution of p? What does experience say? Suppose we come to a fork in a previously untravelled road, only one branch of which leads to our desired destination. What is the probability of success (or failure) in reaching our destination, having no additional information? Intuition would consider both to be equally likely, hence each is equal to ½. If we extend the number of forks to N, the same reasoning demonstrates each of the probabilities to be equal to $1/N$. Hence, the p_i are uniformly distributed. Our experience added enough information to Equation A2.24 to provide the distribution of probabilities.

This is an example of the principle of insufficient reason (Hume 1740), framed more succinctly by Bayes (1763), who stated that if we know of no reason for the probabilities to be different, we should consider them to be equal. Jaynes (1957, 1978) generalised the principle of insufficient reason into the principle of maximum entropy, declaring the assignment of a probability to be that which maximises entropy subject to the additional constraints imposed by the given information.

Applying Jaynes' principles, for the range of constraints most frequently encountered in our daily endeavours, the distributions that maximise entropy are those listed in Table A2.2.

Table A2.2: Maximum entropy probability distributions

Given constraints	Assigned probability distribution
$\int_a^b f(x)dx = 1$	Uniform
$\int_a^b f(x)dx = 1$ Expected value	Exponential
$\int_a^b f(x)dx = 1$ Expected value, standard deviation	Normal
$\int_a^b f(x)dx = 1$ Expected value, standard deviation, range (minimum and maximum values)	Beta
$\int_a^b f(x)dx = 1$ Mean occurrence rate between arrivals of individual events	Poisson

4 Confidence limits

The general expression for the symmetrical upper (U) and lower (L) limits that will satisfy the condition $P[x \geq U] = P[x \leq L] = \acute{a}/2$ for a normal variate x is:

$$P[E[x] - h\sigma[x] \leq x \leq E[x] + h\sigma[x]] = 1 - \acute{a} \quad \text{(eqn A2.25a)}$$

or

$$P[L \leq x \leq U] = 1 - \acute{a} \quad \text{(eqn A2.25b)}$$

Table A2.3: Confidence coefficients for normal distribution

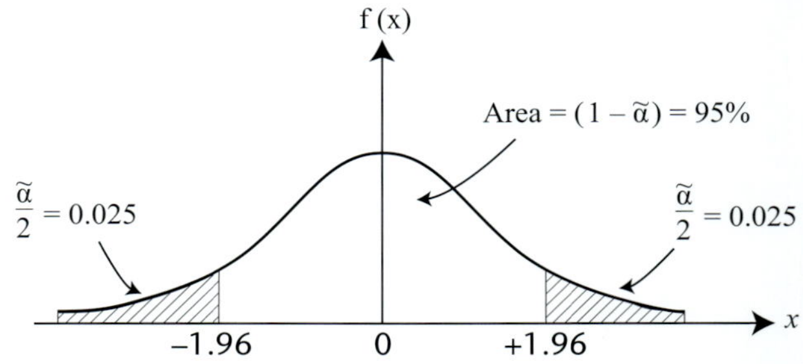

Confidence level $(1 - \tilde{\alpha})$, %	$h_{\tilde{\alpha}}/2$
80	1.28
90	1.64
95	1.96
98	2.33
99	2.58
99.5	2.81
99.99	3.89

The range $[L, U]$ is the confidence interval of the variate and $(1 - \acute{a})$ is the confidence level. As such, the probability $(1 - \acute{a})$ provides a measure of belief that the confidence limits $[L, U]$ will contain x.

A typical statement of this type is that 'the 95% confidence limit is the interval 25 to 35'. The choice of the confidence level $(1 - \acute{a})$ is a matter of the risk that one is willing to take in accepting an erroneous value of the variate under consideration.

Values of $h\left(= \left|\frac{\bar{x} - x}{\sigma[x]}\right|\right)$ can be readily obtained from any set of normal probability tables. Some example values are given in Table A2.3.

APPENDIX 3
Influence of in situ stresses on open pit design

Evert Hoek, Jean Hutchinson, Kathy Kalenchuk and Mark Diederichs

1 Introduction

All rock masses are subjected to in situ stresses resulting from the weight of the overlying strata and from locked in stresses of tectonic origin. The excavation of an open pit results in disruption of the existing stress field with some of the stresses being relieved, while others are intensified. These changes in the in situ stresses can have very important consequences in terms of the behaviour of the rock mass surrounding underground excavations. However, the question to be considered in this appendix is whether these stress changes are important in the design of the slopes of an open pit mine.

As the open pit is excavated, the rock mass deforms and the stresses in the rock mass immediately behind the slopes are relieved. It is generally assumed that the rock mass in which the slopes are excavated is loaded by gravitational stresses only, and that these stresses are not high enough to exceed the rock mass strength and cause failure. Is this true or are there conditions in which the in situ stresses have to be considered?

Open pit design practice, as presented throughout the rest of this book, holds that the effect of in situ stress is an issue only when the stresses induced in pit wall slopes are substantial enough to exceed the rock mass strength. This leads to rock mass damage, producing an enlarged zone of weakened rock which may subsequently fail under gravity loading.

After a brief discussion on the origins and characteristics of in situ stresses, the typical stress fields surrounding open pit mines are examined in this appendix. This is not a comprehensive discussion on in situ stresses, as this is a very large topic in its own right. Rather, the purpose is to make the reader aware of some of the basic issues related to in situ stresses and of the special circumstances under which these stresses may be important in open pit design.

2 In situ stresses

Consider an element of rock at a depth of 1000 m below the surface. The weight of the vertical column of rock resting on this element is the product of the depth and the unit weight of the overlying rock mass. (typically about 2.7 tonnes/m3 or 0.027 MN/m3). Hence the vertical stress on the element is 2700 tonnes/m2 or 27 MPa. This stress is estimated from the simple relationship:

$$\sigma_v = \gamma z \qquad \text{(eqn A3.1)}$$

Where

σ_v = vertical stress
γ = unit weight of the rock and
z = depth below surface.

Measurements of vertical stress at various mining and civil engineering sites around the world confirm that this relationship is generally valid, although there is a significant amount of scatter in the measurements, as shown in Figure A3.1.

The horizontal stresses acting on an element of rock at a depth z below the surface are much more difficult to estimate than the vertical stresses. Normally, the ratio of the average horizontal stress to the vertical stress is denoted by the letter k such that:

$$\sigma_h = k\sigma_v = k\gamma z \qquad \text{(eqn A3.2)}$$

Terzaghi and Richart (1952) suggested that, for a gravitationally loaded rock mass in which no lateral strain was permitted during formation of the overlying strata, the value of k is independent of depth and is given by $k = \nu/(1-\nu)$, where ν is the Poisson's ratio of the rock mass. This relationship was widely used in the early days of rock mechanics but, in reality, applies only to normally consolidated soils and recent sedimentary rocks that have not been subjected to tectonic stresses or significant

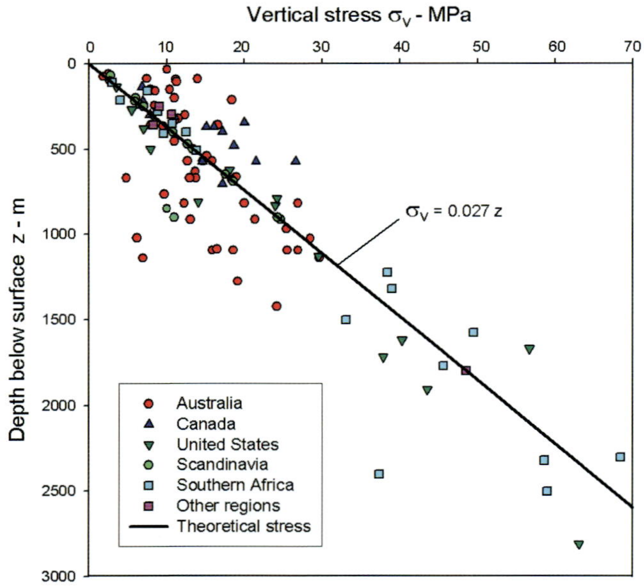

Figure A3.1: Measurements of vertical stress in different regions of the world
Source: Hoek & Brown (1980b)

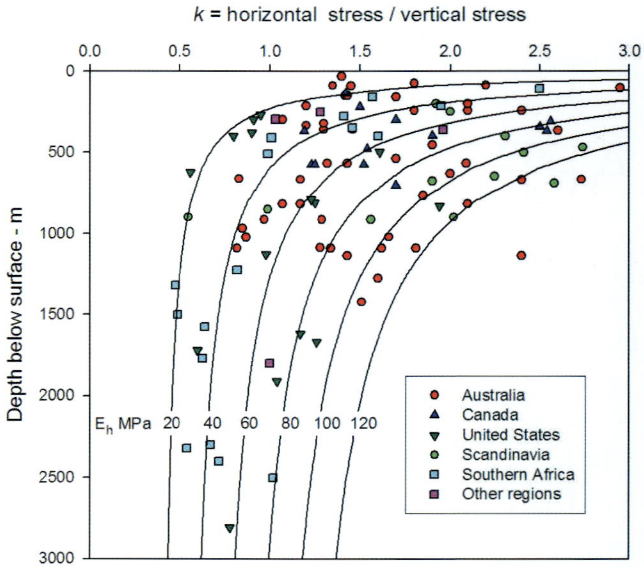

Figure A3.2: Measurements of horizontal stress in different regions of the world
Source: After Hoek & Brown (1980b)

unloading due to surface erosion or deglaciation. For most crustal environments, this equation is inaccurate and is seldom used today.

Measurements of horizontal stresses at civil and mining sites around the world show that the ratio k tends to be high at shallow depth and that it decreases with increasing depth (Hoek and Brown 1980b). In order to understand the reason for these horizontal stress variations it is necessary to consider the problem on a much larger scale than that of a single site.

Sheorey (1994) developed an elasto-static thermal stress model of the earth. This model considers curvature of the crust and variation of elastic constants, density and thermal expansion coefficients through the crust and mantle. A detailed discussion on Sheorey's model is beyond the scope of this appendix, but he did provide a simplified equation which can be used for making an initial estimate of the horizontal to vertical stress ratio k. This equation is:

$$k = 0.25 + 7E_h\left(0.001 + \frac{1}{z}\right) \qquad \text{(eqn A3.3)}$$

where z (m) is the depth below surface and Eh (GPa) is the average deformation modulus of the upper part of the earth's crust measured in a horizontal direction. This direction of measurement is important, particularly in layered sedimentary rocks in which the deformation modulus may be significantly different in different directions.

Figure A3.2 gives a plot of measured horizontal to vertical stress ratios for different parts of the world and, superimposed on this plot, are curves generated from equation A3.3 for different rock mass moduli.

While the measured and theoretical horizontal to vertical stress ratios show similar trends with depth, it is interesting to note that measurements from different continents occupy different parts of the plot. Hence, measurements from Southern Africa exhibit significantly lower k values than those from Australia and Scandinavia. It is difficult to believe that the deformation modulus of African rocks, particularly in the deep level gold mines, is one fifth of that of the Australian and Scandinavian rocks. This suggests that Sheorey's model is over simplified and that the tectonic history of the area also plays an important role in determining the in situ stress conditions.

Hence, while equation A3.3 provides a first estimate of the value of k, it is prudent to check this value against other evidence, where possible.

3 Geological factors

The tectonic history of the rock mass has a major influence upon the in situ stresses in that mass. This is demonstrated by the World Stress Map reproduced in Figure A3.3 (from Zoback, 1992) which shows that the orientation of maximum horizontal stresses is dependent upon their location in tectonic plates. The stresses on this map are colour coded according to the stress ratio measured: NF, or normal fault (red), SS, or strike slip fault (green) and TF, or thrust fault (blue), and can be related to the state of stress acting upon a fault as defined by Anderson in 1951 (Figure A3.4).

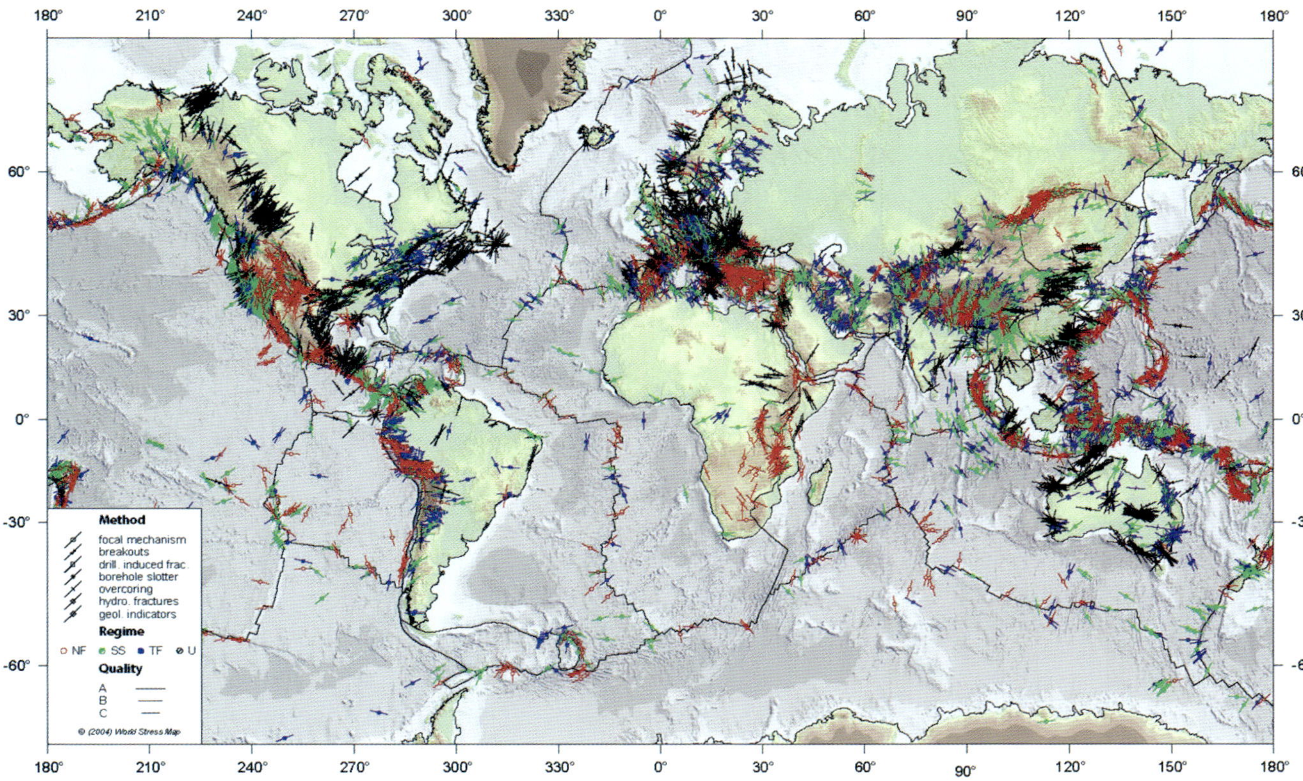

Figure A3.3: World Stress Map (2005 edition) displays the orientations of the maximum horizontal compressive stress. The length of the stress symbols represents the data quality with the longest lines corresponding to the highest quality. The World Stress Map project, completed in July 1992, was carried out under the auspices of the International Lithosphere Project (Zoback, 1992). The WSM is now maintained and it has been extended by the Geophysical Institute of Karlsruhe University as a research project of the Heidelberg Academy of Sciences and Humanities (www.world-stress-map.org).

Generally, k ratios of 0.5 or less are encountered only in extensional zones such as the south-west USA, while k ratios of 0.5 to 1 are encountered in sedimentary basins such as the South African gold fields, where the stresses are disassociated with the stresses in the underlying basement rock. Stress ratios of much greater than 1 are found in compression zones or in continental shield rocks.

Local variations are possible, however, particularly in alpine regions where *k* can vary spatially and with depth due to complex tectonic strain distribution. Due to the substantial variation in the possible range of locked in horizontal stresses near surface, these rules of thumb should only be applied to the evaluation of deep stresses.

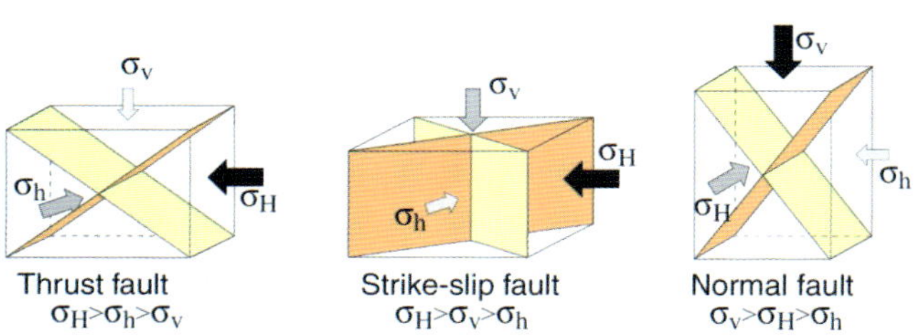

Figure A3.4: State of stress determined from the type of fault present
Source: After Anderson (1951)

The state of regional or local stress can sometimes be determined by the orientation and sense of slip along currently active faults as shown in Figure A3.4. These fault stress relationships hold for the creation of new faults by current or contemporaneous stress fields. Where ancient faults are reactivated by more recent stress fields, the relationships in Figure A3.4 may not apply, as the fault represents a weak zone within the system.

These stresses are also influenced by the thickness of the earth's crust and by past and present topography of the site. For example, melting of ice sheets or erosion of the surface will relieve the vertical stress, but leave the horizontal stresses locked in place. In alpine regions or areas with high topographic relief, the stresses at a site located above the regional topographic baseline may be decoupled from the tectonic stresses and therefore may be under-stressed, with low horizontal stresses in relation to the lithostatic stress (Doe et al. 2006). A review of geological and geophysical literature and discussions with a competent geologist to create a suitable geological and conceptual geotechnical model will usually reveal many of the reasons behind a local in situ stress anomaly.

4 In situ stress measurement

Determination of the in situ stresses from the surface is a difficult, time consuming and expensive operation at the depths of interest in deep open pit mines. The two stress measurement techniques commonly used in mining and civil engineering projects are overcoring and hydraulic fracturing. Another method involves acoustic emission testing to monitor when core is loaded to the point where new damage occurs (Villeascusa et al. 2002). This point is related to the previous damage due to stress, and the data from six pieces of core provides a full stress tensor, as is described elsewhere in this book (section 3.6). This method has not yet gained wide acceptance by the international geotechnical community.

Overcoring involves drilling a large diameter borehole (usually about 100 mm in diameter) to the point at which the stresses are to be measured. A smaller diameter hole is then drilled from the end of the large hole and a measuring instrument containing strain gauges is bonded into this smaller hole. Once the adhesive has cured, an initial set of strain measurements is taken. Strain measurements are recorded as the large diameter hole is continued, using a thin wall diamond bit, thereby releasing the stresses on the overcored piece of rock. The piece of core containing the overcored instrument is removed, and is loaded biaxially to measure the elastic constants for the rock with the gauges. Analysis of the measurements taken before and during overcoring allows the three dimensional stresses in the rock to be calculated.

While the maximum depth at which overcoring techniques can be used is less than 500 m, overcoring in highly stressed rock is very problematic, as it depends on the assumption of linear elastic rock behaviour and when stress magnitudes are elevated, this condition is violated by the development of micro-fractures within the rock mass (Doe et al. 2006). In addition, at this time, there is only one contractor (Vattenfall, Sweden) in the world that has the technology to do stress measurements at these depths.

If underground access is available from which holes can be drilled into the zone of interest, the problem becomes much simpler, since the overcoring method can be used with a higher probability of success. However, the technique requires great care in drilling and preparing the holes, in bonding the instruments in place and in monitoring the overcoring process. In addition, interpretation of the results is a complex process which requires considerable experience. Consequently, it is usually advisable to hire a specialist contractor to perform these measurements rather than attempting to carry out the task in house.

The hydraulic fracturing method utilises high pressure water in a packed off section of a borehole to induce and then test fractures in the rock mass. Careful measurement of the pressure and flow of the water, and knowledge of the tensile strength of the rock, provides the information required to estimate the stress level at which the cracks open and this, in turn, permits estimates to be made of the minimum principal in situ stress. Care must be taken to measure the orientation of the fracture generated during the test, which will be normal to the orientation of the minimum stress, and to select appropriate values of factors accounting for the poroelasticity of the material in the interpretation of the data (Doe et al. 2006).

If the minimum principal stress is horizontal, then the technique provides a reasonable estimate of its magnitude. Since only one stress component is measured using this method, it is therefore assumed that the vertical stress is equal to the product of the depth below surface and the unit weight of the rock mass.

If the horizontal stress is higher than the vertical stress, which is frequently the case as discussed above, then the hydraulic fracturing technique will produce a horizontal fracture and therefore measures the weight of the overburden. Hence, no measurement of the horizontal stress is possible in this case.

As noted by Doe et al. (2006), measuring stress in a high horizontal stress environment is extremely challenging, and is seldom successful when done from surface drilled holes. Where determination of the stress magnitude is important, it is advisable to use a combination of different stress measurement techniques.

5 Stresses in open pit slopes

It is well known that rock 'noses' or slopes that are convex in plan are less stable than concave slopes. This is generally because of the lack of confinement in convex slopes and the beneficial effects of confinement in concave slopes. These observations provide practical evidence that lateral stresses, in the rock in which slopes are excavated, can have an important influence on slope stability.

In the current state of practice in rock slope design, these lateral stresses are usually ignored or are dealt with in a very simplistic manner. In fact, all limit equilibrium models are based upon gravity loading only, and lateral stresses are excluded from any slope stability analysis that uses these models. Numerical models can incorporate lateral stresses but most analyses, using these models, are based upon a simple approximation in which the horizontal stress applied to the model is some proportion of the vertical stress.

Measurement of the in situ stress field in the vicinity of large slopes is seldom carried out, because the horizontal stresses are generally considered to be of minor

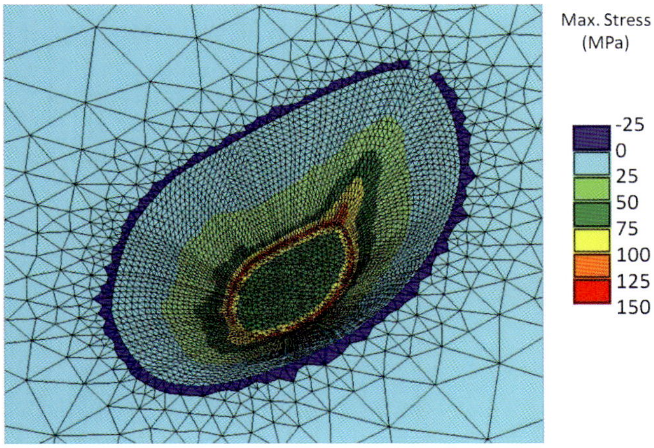

Figure A3.5: The effect of geometry on stresses in the rock mass surrounding a 1000 m deep open pit. The in situ stresses are 2H:1V in this case

significance. It is assumed that the in situ or regional stresses do not have an influence on the pit wall stability because failures are gravity controlled. The failure is controlled by the orientation of structural features such as joints, planes of bedding or foliation, sheared zones, faults, or zones of weaker rock, and by the influence of groundwater pressures. This assumption may be adequate for many slopes, where failure is gravitationally driven, but it needs to be questioned for the design of very large slopes, as is noted elsewhere in this book (section 10.3.4.3).

The three dimensional geometry of an open pit influences the stresses and displacements induced in the surrounding rock mass as illustrated in Figures A3.5 and A3.6. Figures A3.7 and A3.8 show that pre-mining in situ stresses also influence the stresses and displacements induced in the rock mass in which an open pit is excavated.

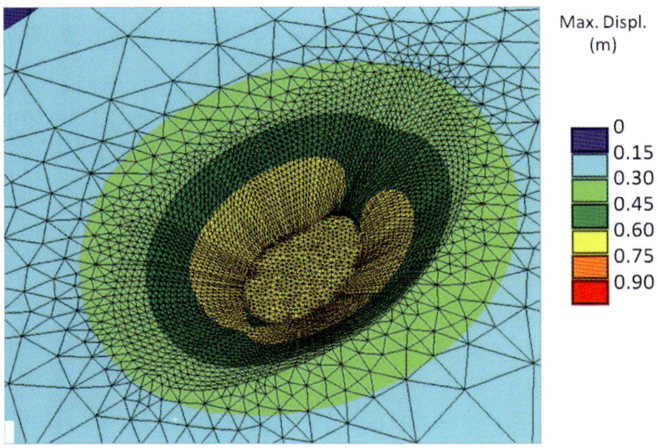

Figure A3.6: The effect of geometry on displacements in the rock mass surrounding a 1000m deep open pit in an in situ stress field of 2H:1V, and a 45° slope, in a rock mass with a modulus of 10 GPa

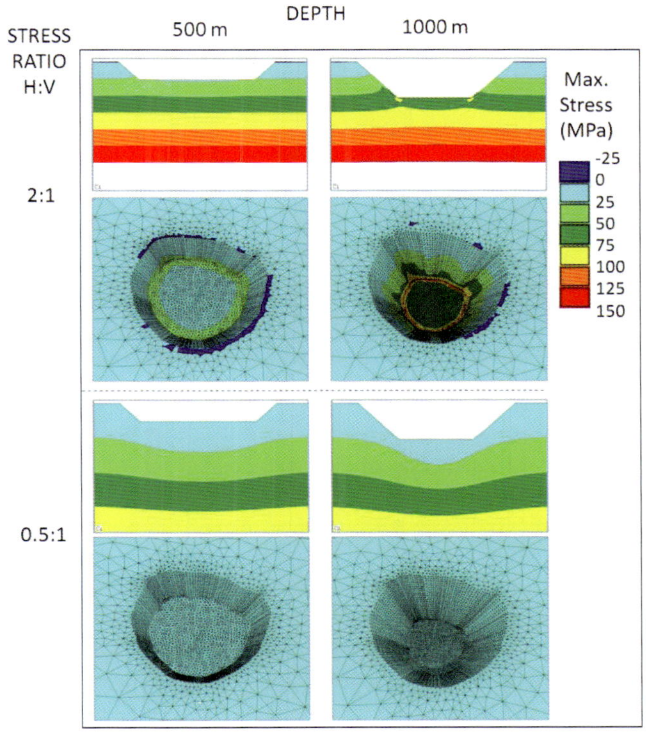

Figure A3.7: 3D modelling showing the effect of pit depth and stress ratio on the stresses induced in the rock mass surrounding an open pit with 45° slopes

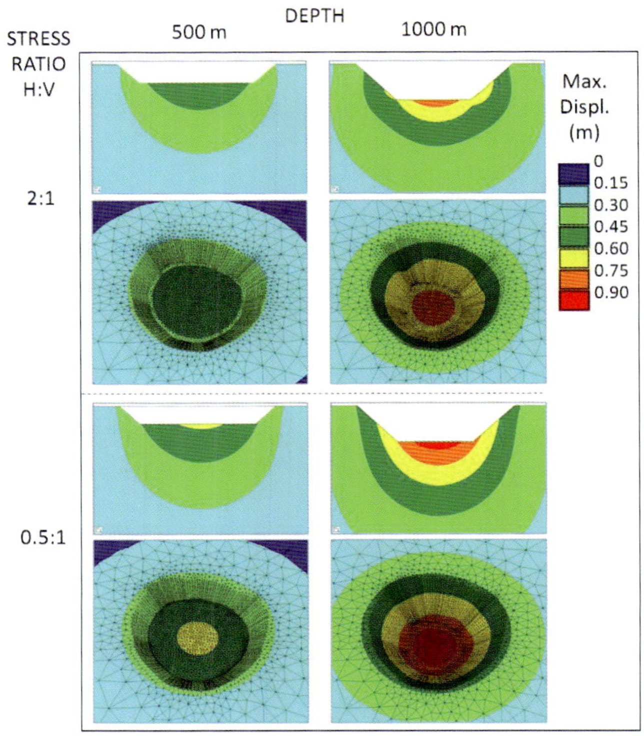

Figure A3.8: 3D modelling showing the effect of pit depth and stress ratio on displacements in the rock mass surrounding an open pit with 45° slopes for rock mass modulus of 10 GPa

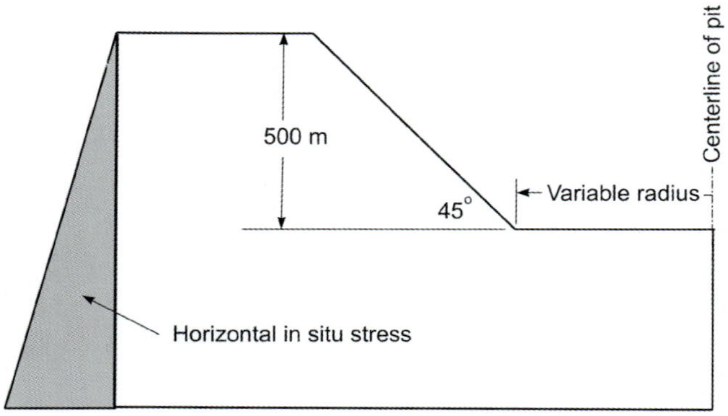

Figure A3.9: Slope geometry and in situ stresses

In order to understand the role of each of these factors, a series of simple numerical analyses have been carried out in which the pit geometry and in situ stresses are investigated separately. Two dimensional finite models were used to investigate the influence of pre-mining in situ stresses and axisymmetric analyses were used to study the influence of pit geometry. The model geometry is shown in Figure A3.9 and results of these analyses are presented in Figures A3.10 to A3.15.

The 45° slope is 500 m high and it has been excavated in 40 m steps in a dry rock mass with the following properties:

- Peak friction angle $\phi = 37°$
- Peak cohesive strength c = 0.7 MPa
- Dilation angle $\alpha = 15°$
- Residual friction angle $\phi r = 30°$
- Residual cohesive strength cr = 0.3 MPa
- Deformation modulus E = 1000 MPa
- Rock unit weight $\gamma = 0.027$ MN/m^3

Note that the rock mass defined by these parameters is weaker and more deformable than that encountered in

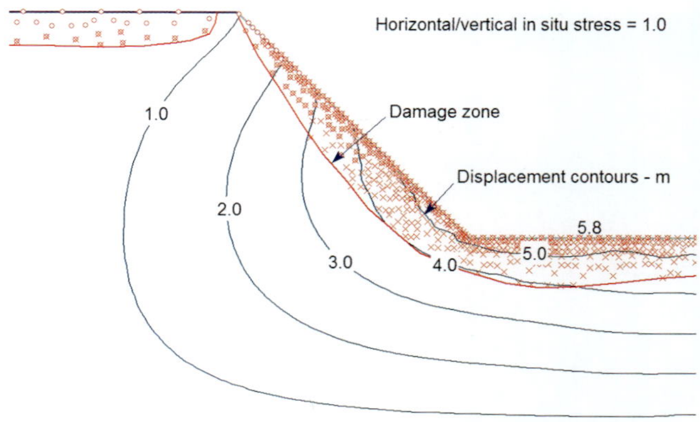

Figure A3.11: Damage zone and displacements in a 500 m high 45° slope in a homogeneous rock mass with a horizontal to vertical in situ stress ratio of 1.0

many open pit mines around the world. These properties have been chosen deliberately in order to demonstrate the influence of in situ stress changes.

Figures A3.10 to A3.12 illustrate the extent of the zones of damaged (drop from peak to residual strength) rock and the induced displacements in the rock mass for pre-mining in situ stress ratios of 0.5, 1.0 and 2.0. These analyses were all carried out using a plane strain model which represents a cross-section through a long slope with an infinite radius of curvature in plan. Figures A3.13 to A3.15 present the results of axi-symmetric analyses in which the model represents a vertical slice through a circular open pit. These analyses assume that the horizontal and vertical in situ stresses are equal and analyses were carried out for ratios of pit bottom radius to slope height of 0.2, 0.5 and 1.0.

It can be seen that the damage zones and displacements are very similar for all of these cases with the exception of the results given in Figure A3.12 for a horizontal to vertical

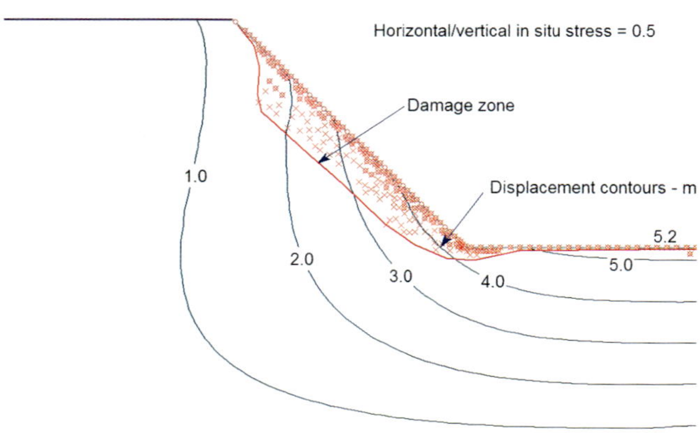

Figure A3.10: Damage zone and displacements in a 500 m high 45° slope in a homogeneous rock mass with a horizontal to vertical in situ stress ratio of 0.5.

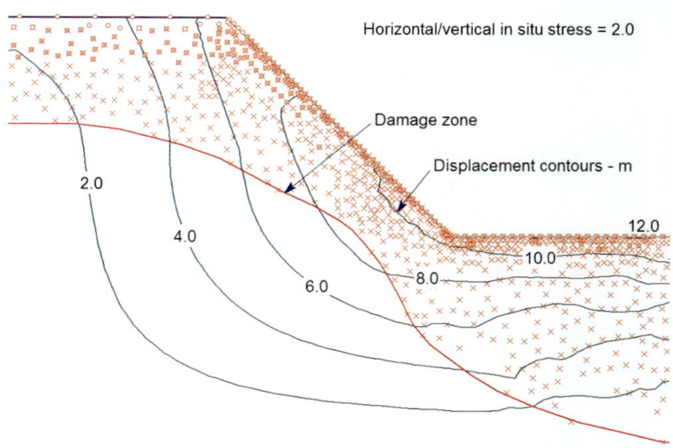

Figure A3.12: Damage zone and displacements in a 500 m high 45° slope in a homogeneous rock mass with a horizontal to vertical in situ stress ratio of 2.0

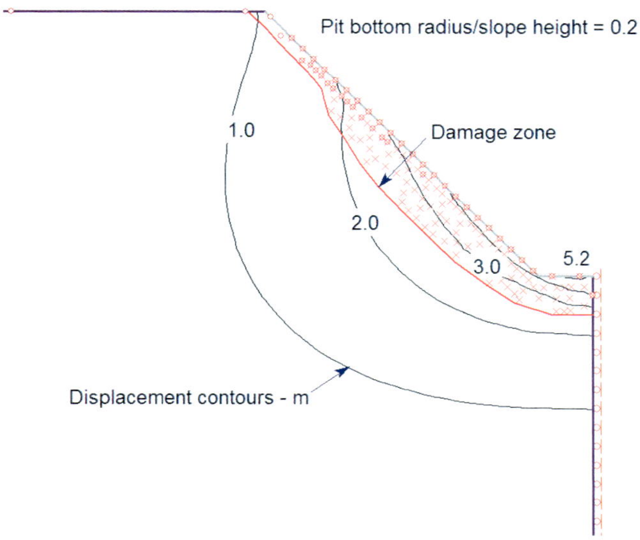

Figure A3.13: Damage zone and displacements in a 500 m high 45° slope in a homogeneous rock mass with a pit bottom/slope height radius of 0.2

in situ stress ratio of 2. This suggests that, except for very high horizontal in situ stresses, the influence of in situ stresses and pit geometry may not be significant when considered in relation to the uncertainties in the geological model, the strength and deformation characteristics of the rock units making up the rock mass, and the influence of groundwater pressures in the rock mass.

While the extent of the damage zone and the displacement contours provide a useful qualitative assessment of the influence of in situ stresses and pit geometry, many readers may prefer to have a more quantitative assessment for a factor of safety. (FoS).

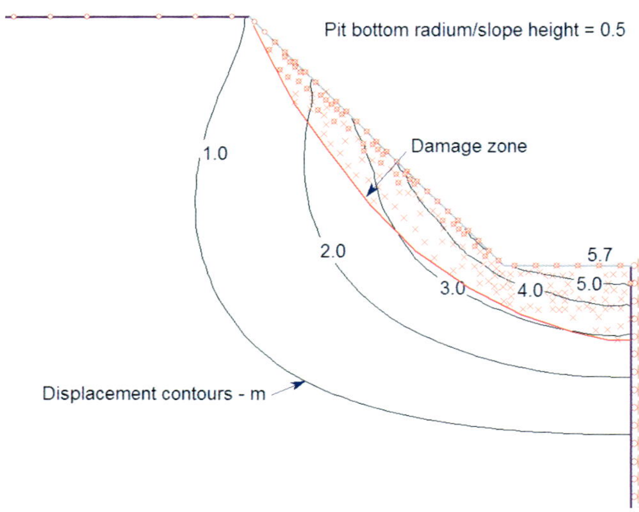

Figure A3.14: Damage zone and displacements in a 500 m high 45° slope in a homogeneous rock mass with a pit bottom/slope height radius of 0.5

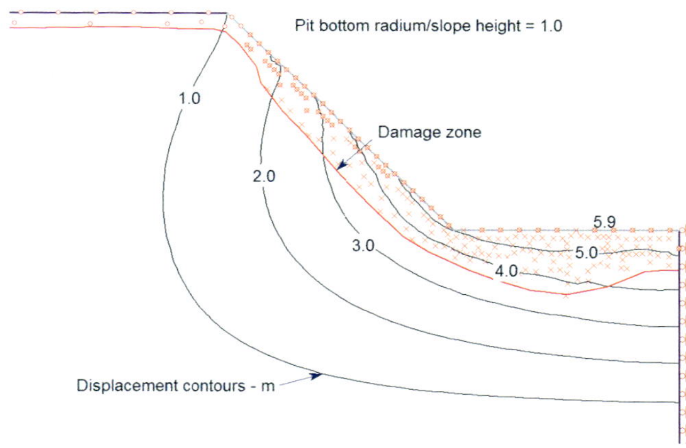

Figure A3.15: Damage zone and displacements in a 500 m high 45° slope in a homogeneous rock mass with a pit bottom/slope height radius of 1.0

Conventional limit equilibrium analyses do not consider non-gravitational in situ stresses or the three dimensional plan geometry of the slope. However, by utilising the extent of the damage zone from the finite element models, a two material slope model can be created and analysed in a limit equilibrium model, as illustrated in Figures A3.16 and A3.17. The damaged zone in this model is assigned the residual shear strength parameters (ϕ_r = 30° and c_r = 0.3 MPa) discussed earlier.

For the analysis illustrated in Figure A3.17, the optimised non-circular failure surface search option in the RocScience program Slide has been used to calculate the factor of safety. The figure also shows the failure surfaces with factors of safety of less than 1.3.

Evaluation of the limit equilibrium analysis results should include consideration of the zone of instability that develops. The width of the zone containing surfaces with FoS<1.3, for example, represents the zone of possible slumping or cracking and should be a consideration in determining setbacks for infrastructure, and in the

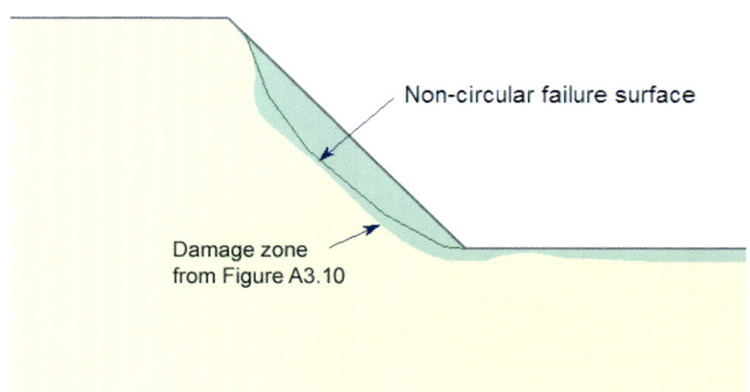

Figure A3.16: Two-material slope model created by incorporation of the damage zone from a finite element model. In this case the example given in Figure A3.10 has been used to define the second material

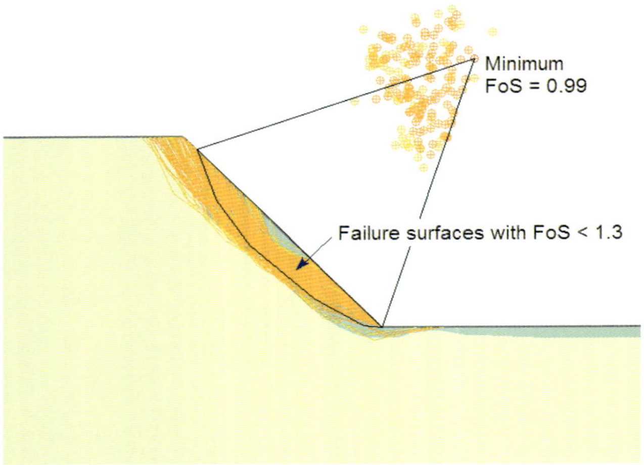

Figure A3.17: Optimised non-circular search for a minimum factor of safety in the slope

evaluation of potential instability of the pit walls. The development of the zone of instability, depending on the in situ stress ratio, is illustrated in Figure A3.18.

An alternative method for estimating the factor of safety of the slope is the Shear Strength Reduction (SSR) method (Dawson et al. 1999; Diederichs et al. 2007), which is discussed in section 10.3.4.3 of this book. This method was used by Lorig (1999) to investigate the influence of open pit geometry and in situ stresses.

The technique utilises an iterative modification of the shear strength parameters of the rock mass to determine the factor of safety. This process also changes the size of the damage zone with each iteration, and the final results may not accurately reflect the factor of safety for a gravity driven failure. However, the method has the advantage of incorporating the horizontal in situ stresses directly and it can also be applied to the axisymmetric model used to study the effect of pit geometry.

The results obtained from both limit equilibrium and shear strength reduction calculations, for the range of conditions included in Figures A3.10 to A3.15, are presented in Table A3.1. The limit equilibrium results show no significant difference between any of the models

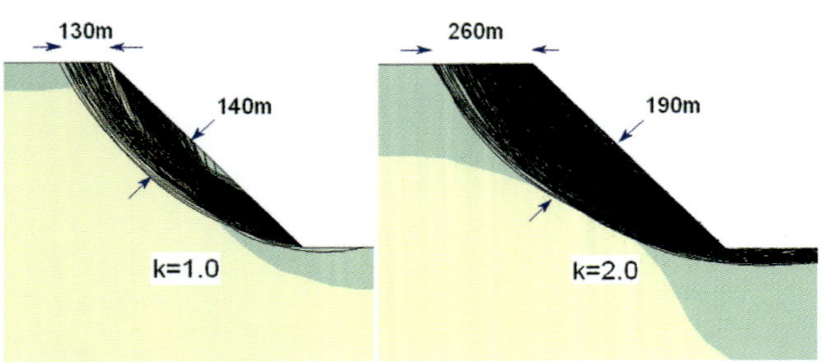

Figure A3.18: Slip surfaces with FoS less than 1.3, for the case in Figures A3.10 and A3.12, for differing in situ stress ratios.

Table A3.1: Results of approximate factor of safety calculations for different in situ stresses and open pit geometries.

In situ stress ratio	Pit bottom radius – m	FoS – Limit Equilibrium	FoS – SSR
0.5	∞	0.99	1.04
1.0	∞	0.97	1.04
2.0	∞	0.93	1.03
1.0	100	1.00	1.25
1.0	250	0.98	1.21
1.0	500	0.98	1.15

with the exception of marginally lower values for the case where the horizontal to vertical stress ratio is 2.0. The SSR results show no change in the factor of safety for changes in the in situ stress ratio but a small increase in stability with decreasing pit bottom radius. The SSR results exhibit similar trends to those obtained by Lorig (1999), who also used the SSR technique (with different rock mass properties).

While the minimum factor of safety may or may not be sensitive to the inclusion of stress induced damage, the overall sensitivity of the slope should be considered by comparing the extent of possible slip surfaces or shear zones with a factor of safety less than a design threshold (e.g. FoS<1.3) as illustrated in Figure A3.18. This comparison may be important for assessing the overall risk to surface and in-pit infrastructure.

6 Conclusion

It has long been accepted that slopes that are concave in plan are marginally more stable than straight or convex slopes. It has also commonly assumed that pre-mining in situ stresses do not have a significant influence on slope stability and that stability calculations based on the assumption of gravitational loading only are adequate.

The studies reported in this appendix support these traditional views. The evidence presented suggests that the levels of uncertainty normally associated with geological models, rock mass properties and groundwater conditions would swamp any of the minor changes associated with in situ stress variations or pit geometry.

Exceptions that would require more careful consideration are cases where high horizontal stresses may occur, where mining of two adjacent open pits may create a highly stressed pillar, or where underground excavations are located close to the pit walls.

It is recommended that the following procedures should be followed when it is possible that the in situ stresses may have a significant impact on the stability of mine excavations.

- From the depth below the surface and the estimated horizontal deformation modulus, estimate the vertical and horizontal stresses as discussed in this appendix, ensuring that the values calculated are considered in light of geological evidence regarding the expected stress conditions due to the tectonic regime, as recorded in a general sense on the World Stress Map. The general trend of the stress ratio, whether extensional, compressional or pseudo-hydrostatic, should be determined in consultation with geological experts. As noted by Doe et al. (2006), heterogeneity introduced by the influence of topography and litho-mechanical variability should also be considered.
- Using the simplest numerical tools, evaluate the influence of pit geometry and regional stresses on the induced stresses and displacements around the pit as described in the preceding section. The amount of elastic rebound of the pit walls, as the pit develops, may be quite substantial, and should be considered in the interpretation of long term displacement monitoring data recorded during pit deepening or widening. Estimate the extent of the damage zone behind the pit walls as described, but bear in mind that there may also be a significant amount of blasting damage in these walls. Estimate the factors of safety of the slopes using the simple tools described and take these into account, together with all the other information available, in deciding whether more detailed studies are justified.

These more detailed studies may involve in situ stress measurement from surface or from available underground excavations, as well as more sophisticated three dimensional analyses. Such studies are particularly relevant when considering very large deep open pits, simultaneous mining in adjacent pits or a pit and underground mine or mining in massive brittle rock in which spalling and rock bursts may be triggered by very high stresses.

APPENDIX 4

Risk management: geotechnical hazard checklists

(Courtesy Anglo Technical Division, Anglo American plc)

Geotechnical hazard checklists provide an example of the types of questions that should be asked throughout the life of an open pit project. These are not inclusive lists and questions should be tailored to each site's specific requirements. The following example checklists have been provided by sponsor companies.

The colour-coded legend shows the importance of the information to the decision-making process at each project stage concerned. Note that closure does not end liability for property.

Required	Required
Advisable	Advisable
Unnecessary	Unnecessary

	Procedural/code of practice/data availability	**Conceptual**	**Pre-feasibility**	**Feasibility**	**Operational**	**Closure**
1	Is there a company geotechnical policy that needs to be adhered to?	Required	Required	Required	Required	Required
2	Are guidelines available to compile a suitable Code of Practice?	Unnecessary	Unnecessary	Advisable	Required	Unnecessary
3	Are the guidelines clear in the requirements for a Code of Practice?	Unnecessary	Unnecessary	Advisable	Required	Unnecessary
4	Has an introduction stating the origin, review, drafting committee and implementation plan been completed?	Unnecessary	Unnecessary	Advisable	Required	Unnecessary
5	Has a section been completed to describe the mine environment in terms of locality, geology, hydrology, rock mass and geotechnical domains?	Advisable	Advisable	Required	Required	Unnecessary
6	Are records available on rockfall and failure incidents?	Unnecessary	Unnecessary	Unnecessary	Required	Required
7	Have these records been included in the Code of Practice?	Unnecessary	Unnecessary	Unnecessary	Required	Required
8	Has rock-related hazard identification been done?	Required	Required	Required	Required	Required
9	Are there strategies to manage rock-related risks?	Required	Required	Required	Required	Required
10	Does this include a strategy on geotechnical properties and geological structure?	Unnecessary	Advisable	Required	Required	Required
11	Does this include a strategy on hydrology?	Unnecessary	Advisable	Required	Required	Required
12	Does this include a strategy on monitoring?	Unnecessary	Advisable	Required	Required	Required
13	Does this include a strategy on control measures?	Unnecessary	Advisable	Required	Required	Required
14	Does this include a strategy on proactive wall design through projection of data?	Unnecessary	Advisable	Required	Required	Required
15	Does this include a strategy on overall mine stability?	Unnecessary	Advisable	Required	Required	Required

		Required	Required (colour key)			
		Advisable	Advisable (colour key)			
		Unnecessary	Unnecessary (colour key)			
	Procedural/code of practice/data availability	**Conceptual**	**Pre-feasibility**	**Feasibility**	**Operational**	**Closure**
16	Does this include a strategy on planning for the total life of mine?	Unnecessary	Advisable	Required	Required	Required
17	Does this include a strategy on special areas?	Unnecessary	Advisable	Required	Required	Required
18	Have these strategies been included in a Code of Practice?	Unnecessary	Unnecessary	Required	Required	Required
19	Are there procedures to ensure adherence to the strategies to manage rock-related risks?	Unnecessary	Unnecessary	Required	Required	Required
20	Are there procedures on how to report hazards?	Unnecessary	Unnecessary	Required	Required	Required
21	Is there a procedure for geotechnical sign-off and input into the mine plan?	Required	Required	Required	Required	Required
22	Is there a procedure for geotechnical sign-off into blasting input?	Unnecessary	Unnecessary	Required	Required	Required
23	Is there a procedure for geotechnical sign-off into structural geology modelling?	Unnecessary	Advisable	Required	Required	Required
24	Does the daily workplace inspection include geotechnical aspects and is this signed off?	Unnecessary	Unnecessary	Unnecessary	Required	Unnecessary
25	Are there procedures for the inspection interval and assessment requirements of the haul road?	Unnecessary	Unnecessary	Unnecessary	Required	Unnecessary
26	Are there procedures for failure data collection?	Unnecessary	Unnecessary	Unnecessary	Required	Required
27	Are there procedures on what/when/why/whom to report geotechnical incidents?	Unnecessary	Unnecessary	Unnecessary	Required	Required
28	Are there procedures to address a criticality arising from the monitoring trends?	Unnecessary	Unnecessary	Unnecessary	Required	Required
29	Is there a follow-up procedure for reports (e.g. return/delivery receipt)?	Unnecessary	Advisable	Advisable	Required	Required
30	Are the monitoring criteria defined in procedures?	Unnecessary	Advisable	Advisable	Required	Required
31	Are there procedures for obtaining monitoring data?	Unnecessary	Advisable	Advisable	Required	Required
32	Are there management procedures in place once the critical points are reached (warning systems, evacuation procedures etc.)?	Unnecessary	Unnecessary	Unnecessary	Required	Required
33	Once evacuation has taken place, are re-entry procedures available?	Unnecessary	Unnecessary	Unnecessary	Required	Required
34	Are there procedures on data collection and database maintenance?	Unnecessary	Advisable	Advisable	Required	Required
35	Are there appropriate procedures in place to establish and monitor the extent of existing mine workings/cavities?	Unnecessary	Advisable	Required	Required	Required
36	Are there procedures in place to monitor surface water?	Unnecessary	Unnecessary	Unnecessary	Required	Required
37	Do procedures exist for installing equipment for, and monitoring the groundwater?	Unnecessary	Advisable	Required	Required	Required
38	Are there procedures for data collection, analysis and validation?	Unnecessary	Required	Required	Required	Advisable
39	Did all relevant parties contribute to the Code of Practice?	Unnecessary	Unnecessary	Unnecessary	Required	Required
40	Has this Code of Practice been implemented?	Unnecessary	Unnecessary	Unnecessary	Required	Required
41	Has a review been done for this Code of Practice?	Unnecessary	Unnecessary	Unnecessary	Required	Required
42	Are there statements of qualification for geotechnical consultants?	Required	Required	Required	Required	Required

		Required	Required			
		Advisable	Advisable			
		Unnecessary	Unnecessary			
	Procedural/code of practice/data availability	**Conceptual**	**Pre-feasibility**	**Feasibility**	**Operational**	**Closure**
Database						
43	Are data available in electronic format database?	Unnecessary	Required	Required	Required	Required
44	Are data stored electronically for groundwater (piezos, pump testing etc.)?	Unnecessary	Required	Required	Required	Required
45	Are data stored electronically for slope monitoring (survey, extensometers etc.)?	Unnecessary	Unnecessary	Unnecessary	Required	Required
46	Are data stored electronically for face mapping (MRMR, Dipsdata etc.)?	Unnecessary	Advisable	Required	Required	Required
47	Are data stored electronically for failures (dimensions, photos, source mechanism etc.)?	Unnecessary	Advisable	Required	Required	Required
48	Are data stored electronically for blast evaluation (drill direction, effectiveness etc.)?	Unnecessary	Unnecessary	Unnecessary	Required	Required
49	Are records stored within a recognised database format (Excel, Access, QPro, Word etc.)?	Unnecessary	Required	Required	Required	Required
50	Are there query/search capabilities in this database?	Unnecessary	Required	Required	Required	Required
51	Are data records cross-referenced (photo, docs, DIPS etc.)?	Unnecessary	Required	Required	Required	Required
52	Can data be exported/imported to/from other database formats?	Unnecessary	Required	Required	Required	Required
53	How efficient is the layout of the electronic database (low, moderate, high)?	Unnecessary	Advisable	Required	Required	Required
54	How effectively are reports and data retrieved from the electronic database (little, moderately, highly)?	Unnecessary	Advisable	Required	Required	Required
55	Are there filing categories for all the geotechnical categories and associated data/information?	Unnecessary	Advisable	Required	Required	Required
56	Are the records verified in some way (checked physically, statistically, visually)?	Unnecessary	Advisable	Required	Required	Required
57	Is the database up to date?	Unnecessary	Required	Required	Required	Required
58	Are there available backups of the electronic database?	Unnecessary	Required	Required	Required	Required
59	How regularly are backups made of the database (low = yearly, moderate = monthly, high = ad hoc)?	Unnecessary	Required	Required	Required	Required
60	Is the current database status sufficient to draw up a geotechnical model?	Unnecessary	Unnecessary	Required	Required	Required
61	Has this geotechnical model been compiled?	Unnecessary	Unnecessary	Required	Required	Required
62	Are data on groundwater (piezos, pumptesting etc.) available in the geotechnical model?	Unnecessary	Unnecessary	Unnecessary	Required	Required
63	Are data on slope monitoring (survey, extensometers etc.) available in the geotechnical model?	Unnecessary	Unnecessary	Unnecessary	Required	Required
64	Are data on face mapping (MRMR, Dipsdata etc.) available in the geotechnical model?	Unnecessary	Unnecessary	Unnecessary	Required	Required
65	Are data on failures (dimensions, photos, source mechanism etc.) available in the geotechnical model?	Unnecessary	Unnecessary	Unnecessary	Required	Required
66	Are data on blast evaluation (drill direction, effectiveness etc.) available in the geotechnical model?	Unnecessary	Unnecessary	Unnecessary	Required	Required
67	Is this model up to date?	Unnecessary	Unnecessary	Required	Required	Required
68	Has this model been used to aid in designs?	Unnecessary	Unnecessary	Required	Required	Required
69	Can different combinations of data be selected and viewed in the model?	Unnecessary	Unnecessary	Required	Required	Required

	Procedural/code of practice/data availability	Conceptual	Pre-feasibility	Feasibility	Operational	Closure
	Key: Required / Advisable / Unnecessary					
70	Are there procedures in place on data collection and database maintenance?	Unnecessary	Required	Required	Required	Required
Data collection						
71	Is there a statement of coordinate systems, system units, local mine grid, etc.?	Required	Required	Required	Required	Required
72	Is the data stored in a secure database?	Advisable	Advisable	Required	Required	Required
73	Is data referenced to coordinate systems?	Required	Required	Required	Required	Required
74	Is data referenced in time?	Required	Required	Required	Required	Required
Geology						
75	Does a lithological model of the operation/potential operation exist?	Advisable	Advisable	Required	Required	Unnecessary
76	Is there knowledge on whether the pit will intersect multiple rock types?	Advisable	Advisable	Required	Required	Unnecessary
77	Are dykes, faults or sills present in the rock mass?	Advisable	Advisable	Required	Required	Unnecessary
78	Could their frequency of distribution be termed as low, moderate or high?	Advisable	Advisable	Required	Required	Unnecessary
79	Could the geology of the area be described as complex?	Advisable	Advisable	Required	Required	Unnecessary
80	Are all major geological features considered to be mapped?	Advisable	Advisable	Required	Required	Unnecessary
81	Is the material weathered?	Advisable	Advisable	Required	Required	Unnecessary
82	What is the degree of weathering (low, moderate, high)?	Advisable	Advisable	Required	Required	Unnecessary
83	Is there a clearly defined weathered–fresh rock interface?	Advisable	Advisable	Required	Required	Unnecessary
84	Are they displaced by other dykes or faults?	Advisable	Advisable	Required	Required	Unnecessary
85	Are geological discontinuities associated with localided rock fracture, metamorphism or other factors which could reduce the rock mass strength?	Advisable	Advisable	Required	Required	Unnecessary
86	Is a structural model available?	Advisable	Advisable	Required	Required	Unnecessary
87	Does structural verification take place?	Advisable	Advisable	Required	Required	Unnecessary
88	How functional is the structural model (low, moderate, high)?	Advisable	Advisable	Required	Required	Unnecessary
86	Does the average dip of the major structures lie between 35° and 70°?	Advisable	Advisable	Required	Required	Unnecessary
89	Do the major structures intersect planned walls at less than 25° (strike)?	Advisable	Advisable	Required	Required	Unnecessary
90	Has the initial exploration drilling for the pit been completed?	Advisable	Required	Required	Required	Unnecessary
91	Was the core logged geologically by a geologist?	Advisable	Required	Required	Required	Unnecessary
92	Can the standard of geological core logging be described as low, moderate or high?	Advisable	Required	Required	Required	Unnecessary
93	Was any of the core relogged geotechnically by an engineering geologist?	Advisable	Required	Required	Required	Unnecessary
94	Can the standard of geotechnical core logging be described as low, moderate or high?	Advisable	Required	Required	Required	Unnecessary
95	Are all cores stored in a core shed?	Advisable	Required	Required	Required	Unnecessary
96	Are records kept of all core logging?	Advisable	Required	Required	Required	Unnecessary

		Required				
		Advisable				
		Unnecessary				
	Procedural/code of practice/data availability	**Conceptual**	**Pre-feasibility**	**Feasibility**	**Operational**	**Closure**
97	Has an exploration geological model been constructed?					
98	Has an operational geological model been constructed?					
99	Have the differences been reconciled?					
100	Is the model available in an approved electronic format?					
101	Does the model reflect the current pit status?					
102	How functional is the geological model (low, moderate, high)?	Low	Moderate	High	High	
Geotechnical						
103	Has any geotechnical drilling been completed for the pit?					
104	How adequate is this geotechnical drilling database (low, moderate, high)?		Moderate			
105	Was this core logged geotechnically by an engineering geologist?					
106	What is the standard of geotechnical logging (low, moderate, high)?		Moderate			
107	Has oriented core been obtained?					
108	Has geotechnical classifcation been carried out?					
109	Are all cores stored in a core shed?					
110	Are records kept of all core logging?					
111	Are procedures in place for data collection, analysis and validation?					
112	Are these procedures adhered to?					
113	Has any analysis been done to establish joint populations and failure mechanisms?					
114	How well are these failure mechanisms understood (little, moderate, high)?					
115	Has any geotechnical testing been done on intact soil/rock samples?					
116	Have intact strength parameters been derived (Hoek-Brown etc.)?					
117	Has any geotechnical testing been done on features within soil/rock samples?					
118	Have typical joint parameters been established?					
119	How adequate is this geotechnical testing database (low, moderate, high)?					
120	Has any geotechnical face mapping been done for the pit?		Natural outcrop	Trial mine	50%	
121	What is the standard of geotechnical face mapping (low, moderate, high)?				High	
122	Is it necessary to analyse face mapping data regularly?					
123	Is the face mapping data analysed regularly?					
124	Is the design and geotechnical parameters validated by the in-pit face mapping?					
125	Has an operational geotechnical model been constructed?					

		Required				
		Advisable				
		Unnecessary				
	Procedural/code of practice/data availability	**Conceptual**	**Pre-feasibility**	**Feasibility**	**Operational**	**Closure**
126	Is the model available in an approved electronic format?					
127	Does the model reflect the current pit status?					
128	How functional is the geotechnical model (low, moderate, high)?			Moderate	High	
Groundwater						
129	Do old workings (shafts, tunnels, stopes) which may contain accumulations of water occur near the existing or planned open pit?					
130	Do any other accumulations of water lie above or adjacent to the existing or planned open pit?					
131	Could the groundwater level rise significantly with rainfall recharge?	Assume				
132	Is the groundwater flow structurally controlled?	Assume				
133	Could circular failure occur in the rock mass (high percentage soil)?	Assume				
134	Could groundwater flow cause liquefaction within the rock mass (high percentage soils, major fault gouge)?	Assume				
135	Has a hydrological study been completed for the area?		Low	High	High	
136	Has the permeability of the rock mass (low, medium, high) been established?					
137	Has the storativity of the rock mass (low, medium, high) been determined?					
138	Has the recharge rate of the groundwater (low, medium, high) been determined?					
139	Has it been determined if perched aquifers exist or form?					
140	Has it been determined if closed compartments of groundwater exist along the pit perimeter?					
141	Has the necessity for dewatering pumping be required?					
142	Have these results been incorporated into the slope design?					
143	Does a system exist which is capable of monitoring the groundwater level and inflow?					
144	Are records kept of the groundwater levels and inflow?					
145	Does a system exist to handle the groundwater and associated inflow?					
146	Do procedures exist for installing, maintaining and monitoring the system?					
147	Are records kept of the pumping, storage and inflow?					
148	Are the groundwater procedures adhered to?					
149	Are water-bearing features identified and mapped on an ongoing basis?					
150	Are there any measurements/modelling of the current inflow/recharge?					
151	Is the potential of the current dewatering system known?					
152	Could any inflow from a permeable horizon/aquifer be accommodated easily by existing water handling systems?					

		Required	(dark orange)			
		Advisable	(light orange)			
		Unnecessary	(green)			
	Procedural/code of practice/data availability	**Conceptual**	**Pre-feasibility**	**Feasibility**	**Operational**	**Closure**
153	Would additional water handling equipment/systems be needed to handle any increased inflow?	Unnecessary	Unnecessary	Unnecessary	Required	Required
154	Does the groundwater serve as a source of water to the mine?	Unnecessary	Unnecessary	Unnecessary	Required	Required
155	Is there a groundwater management plan?	Unnecessary	Unnecessary	Advisable	Required	Required
156	Is the groundwater management plan adhered to?	Unnecessary	Unnecessary	Unnecessary	Required	Required
Surface water handling						
157	Is there any water source in the form of a spring, stream, river or pipeline?	Advisable	Advisable	Required	Required	Required
158	Is there more than one of these sources present within the mining area?	Advisable	Advisable	Required	Required	Required
159	Is the source small (cause minor wetting), medium (flood sections in pit) or large (flood whole pit) relative to the size of mine workings?	Advisable	Advisable	Required	Required	Required
160	Is the source temporary/seasonal (e.g. a dried-up river during summer)?	Advisable	Advisable	Required	Required	Required
161	Is the source on/near the crest of the pit?	Advisable	Advisable	Required	Required	Required
162	Could the volume of the source increase rapidly (e.g. during periods of heavy rainfall)?	Advisable	Advisable	Required	Required	Required
163	Does any connection exist or could one develop between the source and planned or existing mine workings?	Advisable	Advisable	Required	Required	Required
164	Is there any accumulation in the form of a lake, dam, reservoir, vlei or pan?	Advisable	Advisable	Required	Required	Required
165	Is there more than one of these accumulations present within the mining area?	Advisable	Advisable	Required	Required	Required
166	Is the accumulation small (cause minor wetting), medium (flood sections in pit) or large (flood whole pit) relative to the size of mine workings?	Advisable	Advisable	Required	Required	Required
167	Is the accumulation temporary/seasonal (e.g. a seasonal pan)?	Advisable	Advisable	Required	Required	Required
168	Could the accumulation generate pore water pressures in strata adjacent to the workings?	Advisable	Advisable	Required	Required	Required
169	Has the potential inflow of these sources been determined (small = <25 L/sec, moderate = 25–125 L/sec, large = >125 L/sec)	Advisable	Advisable	Required	Required	Required
170	Does a system exist which is capable of continuously monitoring the state of these accumulations?	Advisable	Advisable	Required	Required	Required
171	How effective is this system (low, medium or high)?	Advisable	Advisable	Required	Required	Required
172	Could any inflow from the source be accommodated easily with existing water handling systems?	Advisable	Advisable	Required	Required	Required
173	Is regular maintenance undertaken or planned for stormwater drainage facilities?	Unnecessary	Unnecessary	Unnecessary	Required	Required
Surface features/structures						
173	Are there any surface structures within/close to the mining area (dams, dumps, slopes, embankments, railways, buildings, powerlines etc.)?	Advisable	Advisable	Required	Required	Required
174	If required, are these structures monitored during the mining operation?	Unnecessary	Unnecessary	Unnecessary	Required	Required
175	Do any of the structures require a risk assessment?	Required	Required	Required	Required	Required

		Required				
		Advisable				
		Unnecessary				
	Procedural/code of practice/data availability	**Conceptual**	**Pre-feasibility**	**Feasibility**	**Operational**	**Closure**
176	Have the required risk assessments been completed for the structures?					
177	Are emergency operational procedures in place for any structure?					
178	Are personnel familiar with these procedures?					
179	How efficient is the data available on surface structures (little, moderate or high)?	Little	Moderate	High	High	High
180	Are there any dam/slimes dams close to the mining area?					
181	Are there buildings within 300 m of the pit perimeter?					
182	Is there a railroad/public road within 300 m of the pit perimeter?					
183	Are there powerlines/pipelines within 300 m of the pit perimeter?					
184	Does any unnatural land (restored land, landfill areas or waste dumps) overlie all or part of the planned/ existing workings?					
185	Could any of these structures influence slope stability?					
186	If necessary, could any of these structures be moved?					
187	Have the requisite risk assessments been carried out?					
188	Could the threat(s) imposed by any of these structures be eliminated?					
189	Have the long-term breakback angles for the slopes been determined?					
190	Are waste dumps planned more than 300 m from the pit rim?					
191	Is backfilling planned within this operation (Y/N)?					
192	Have the angles of repose been determined for the waste material?					
193	Is segregated dumping practised (fines/soft material separate from hard coarse)?					
194	Has the time-dependent failure nature of the waste material been determined?					
	Design elements					
195	Is there a slope design available (empirical, limit equilibrium, numerical)?	Empirical	Limit	Numerical	Numerical	
196	What is the validity of the design (low, moderate, high)?	Empirical	Limit	Numerical	Numerical	
197	Has the design been reviewed by a consulting company?				Numerical	
198	Has the design been audited by an external company?				Numerical	
199	Does the slope design conform to corporate guidelines for PoF/FoS?				Numerical	
200	Was any geotechnical drilling done?					
201	Were the holes planned with aid from a geological model?					
202	Were all the planned holes drilled?					
203	Has all the core been logged?					
204	Have all the core data been analysed?					
205	Have joint sets been identified?					

	Procedural/code of practice/data availability	Conceptual	Pre-feasibility	Feasibility	Operational	Closure
		Required				
		Advisable				
		Unnecessary				
206	How competent is this analysis data (little, moderate, high)?					
207	Could major features intersecting the core be identified?					
208	Has this data been incorporated in the design?					
209	Have rock/soil samples been selected for testing?					
210	Were shear and triaxial tests performed to obtain an estimated c and phi of the rock mass?	Assume				
211	Were shear and triaxial tests performed to obtain an estimated c and phi of the joints and other geological weaknesses?	Assume				
212	Were shear and triaxial tests performed to obtain an estimated c and phi of the intact rock?	Assume				
213	What is the bias variation of the test data (low, moderate, high)?	High				
214	How competent is the analysis data (little, moderate, high)?	Low				
215	Have these data been incorporated in the design?					
216	Has a groundwater study been completed?	Assume	Assume			
217	Have available groundwater monitoring data been analysed?					
218	How competent are the analysis data (little, moderate, high)?	Low	Low			
219	Have the groundwater data been incorporated in the design?					
220	Has any in-pit face mapping been done?		Outcrop	Trial		
221	Have these data been analysed and projected?					
222	Has any validation of design been done on exposed faces and failures?					
223	How competent are the analysis data (little, moderate, high)?	Low				
224	Have these data been incorporated in the design?					
225	Have domains been selected according to the data obtained above?					
226	Is the chosen mining method considered to reflect acceptable industry practice?					
227	Have appropriate numerical modelling methods been used to establish slope angles and other slope design criteria?					
228	What is the standard of the numerical modelling that has been done (low, moderate, high)?	Low	Low	Moderate	High	
229	If there are underground workings/cavities in the area, has appropriate numerical modelling been done to assess the interaction?					
230	What is the standard of the numerical modelling that has been done (low, moderate, high)?	Low	Low	Moderate	High	
231	Was the numerical design calibrated using planned geometry and geotechnical properties from test results?					
232	Has this numerical design been updated with current validated geotechnical data?					

Key	
Required	(dark red)
Advisable	(orange)
Unnecessary	(green)

	Procedural/code of practice/data availability	Conceptual	Pre-feasibility	Feasibility	Operational	Closure
233	Are controls in place to ameliorate the effects of failures in unstable domains?	Unnecessary	Advisable	Required	Required	Unnecessary
234	Are identified geological features/weak areas supported (or to be supported) with appropriate reinforcement (when necessary)?	Unnecessary	Advisable	Required	Required	Unnecessary
235	Have support systems been designed in accordance with unstable blocks/zones determined from numerical modelling?	Unnecessary	Advisable	Required	Required	Unnecessary
236	Is there any history of slope failures or similar massive rock movements when mining with this geometry?	Unnecessary	Advisable	Required	Required	Unnecessary
237	Has any other mining taken place (adjacent or underground mining) which could influence slope stability?	Unnecessary	Advisable	Required	Required	Unnecessary
238	Does the design allow for changes later in the life of the mine?	Unnecessary	Advisable	Required	Required	Unnecessary
239	How regularly is the design updated with new data (low = yearly, moderate = quarterly, high = ad hoc)?	Unnecessary	Unnecessary	Unnecessary	Required	Unnecessary
240	How regularly is the design validated with new face and failure data (low = quarterly, moderate = yearly, high = ad hoc)?	Unnecessary	Unnecessary	Unnecessary	Required	Unnecessary
241	Does the mine planning process involve geotechnical personnel (Whittle/Practical Pit)	Unnecessary	Advisable	Required	Required	Unnecessary
242	Are slope angles signed-off by geotechnical personnel prior to implementation?	Unnecessary	Unnecessary	Required	Required	Unnecessary
243	Are changes in slopes (short-term mining requirements) authorised by geotechnical department?	Unnecessary	Unnecessary	Unnecessary	Unnecessary	Unnecessary
Personnel and training						
244	Have mine personnel who work in the pit been inducted?	Unnecessary	Unnecessary	Unnecessary	Required	Unnecessary
245	Does induction include a geotechnical section?	Unnecessary	Unnecessary	Unnecessary	Required	Unnecessary
246	Are mine personnel informed on issues pertaining to safety on the mine?	Unnecessary	Unnecessary	Unnecessary	Required	Unnecessary
247	What is the requirement for geotechnical aides on the mine (little, moderate, high)?	Unnecessary	Unnecessary	Advisable	Required	Unnecessary
248	What is the requirement for geotechnical technicians on the mine (little, moderate, high)?	Unnecessary	Unnecessary	Advisable	Required	Unnecessary
249	What is the requirement for geotechnical engineers on the mine (little, moderate, high)?	Unnecessary	Unnecessary	Advisable	Required	Unnecessary
250	Is the geotechnical department staffed with enough fully trained personnel?	Unnecessary	Unnecessary	Advisable	Required	Unnecessary
251	Are geotechnical aides sufficiently computer-literate (spreadsheet, word processor and analysis programs)?	Unnecessary	Unnecessary	Unnecessary	Required	Unnecessary
252	Do geotechnical aides require any further training?	Unnecessary	Unnecessary	Unnecessary	Required	Unnecessary
253	How competent are the geotechnical aides in their work (little, moderate, high)?	Unnecessary	Unnecessary	Unnecessary	Required	Unnecessary
254	Are there personnel dedicated to gathering geotechnical data?	Unnecessary	Unnecessary	Advisable	Required	Unnecessary
255	Have they been fully trained by a recognised institution (consultant, in-house, university)?	Unnecessary	Unnecessary	Advisable	Required	Unnecessary

Required	(red)	
Advisable	(orange)	
Unnecessary	(green)	

	Procedural/code of practice/data availability	**Conceptual**	**Pre-feasibility**	**Feasibility**	**Operational**	**Closure**
256	Are the aides tested to evaluate their training competency (appraisals, exams, audits)?	Unnecessary	Unnecessary	Unnecessary	Required	Unnecessary
257	How competent are the geotechnical technicians in their work (little, moderate, high)?	Unnecessary	Unnecessary	Advisable	Required	Unnecessary
258	Are there personnel dedicated to analysing geotechnical data?	Unnecessary	Unnecessary	Advisable	Required	Unnecessary
259	Have they been fully trained by a recognised institution (consultant, in-house, university)?	Unnecessary	Unnecessary	Advisable	Required	Unnecessary
260	Are the technicians tested to evaluate their training (appraisals, exams, audits)?	Unnecessary	Unnecessary	Advisable	Required	Unnecessary
261	How competent are the geotechnical engineers in their work (little, moderate, high)?	Unnecessary	Unnecessary	Advisable	Required	Unnecessary
262	Are development plans in place for further training?	Unnecessary	Unnecessary	Advisable	Required	Unnecessary
263	Are geotechnical personnel aware of hazards in the mine environment?	Unnecessary	Unnecessary	Unnecessary	Required	Unnecessary
264	Do all geotechnical personnel know how to address hazards?	Unnecessary	Unnecessary	Advisable	Required	Unnecessary
265	Is there a procedure to report hazards?	Unnecessary	Unnecessary	Advisable	Required	Unnecessary
266	Are job descriptions and responsibilities available for each position?	Unnecessary	Unnecessary	Advisable	Required	Unnecessary
267	Are systems in place to measure work performance?	Unnecessary	Unnecessary	Unnecessary	Required	Unnecessary
268	What is the standard of the work done by the geotechnical aide (low, moderate, high?	Unnecessary	Unnecessary	Advisable	Required	Unnecessary
269	What is the standard of the work done by the geotechnical technician (low, moderate, high)?	Unnecessary	Unnecessary	Advisable	Required	Unnecessary
270	What is the standard of the work done by the geotechnical engineer (low, moderate, high)?	Unnecessary	Unnecessary	Advisable	Required	Unnecessary
271	Are regular appraisals completed for geotechnical personnel (low = yearly, moderate = quarterly, high = ad hoc)?	Unnecessary	Unnecessary	Advisable	Required	Unnecessary
272	Is work reviewed on a regular basis by senior personnel?	Unnecessary	Unnecessary	Advisable	Required	Unnecessary
273	When personnel leave (resign/holiday) can their work be continued by another?	Unnecessary	Unnecessary	Advisable	Required	Unnecessary
274	What is the level of competence of the persons continuing the work (low, moderate, high)?	Unnecessary	Unnecessary	Advisable	Required	Unnecessary
275	Is some form of handover completed after geotechnical personnel return from leave?	Unnecessary	Unnecessary	Advisable	Required	Unnecessary
276	Is outside technical support available (consultants, in-house)?	Unnecessary	Unnecessary	Advisable	Required	Unnecessary

APPENDIX 5

Example regulations for open pit closure

Examples of regulations applicable to open pit mines are provided in this appendix. Applicable regulations for Nevada and California in the USA and British Columbia in Canada are listed below.

Nevada

The Nevada Water Pollution Control regulations are contained in Nevada Administrative Code (NAC) 445A.350 to 445A.447. The following is directly applicable to open pit mines.

A NAC 445A.429. Procedures required to prevent release of contaminants; requirements concerning impoundments.

1 The holder of the permit must institute appropriate procedures to ensure that all mined areas do not release contaminants that have the potential to degrade the waters of the state.
2 Open pit mines must, to the extent practicable, be free-draining or left in a manner which minimises the impoundment of surface drainage and the potential for contaminants to be transported and degrade the waters of the state.
3 Bodies of water which are a result of mine pits penetrating the water table must not create an impoundment which:
   (a) has the potential to degrade the ground waters of the state;
   (b) has the potential to affect adversely the health of human, terrestrial or avian life.
4 The holder of a permit may apply to the commission to establish a beneficial use with a level of protection less than that required by paragraph (b) of subsection 3 for water impounded in a specific mine pit.

The reclamation of mines in Nevada is regulated through NAC 519A.010 to 635. The following aspects are directly applicable to open pit mines.

B NAC 519A.250. Exemption of open pits and rock faces from requirements.

1 An operator may request in writing that the Division grant an exception to the requirements for reclamation for open pits and rock faces which may not be feasible to reclaim.
2 If the operator proves to the satisfaction of the Division that reclamation is not feasible, the Division shall exempt an open pit or rock face from the requirements for reclamation of NAC 519A.010 to 519A.415, inclusive.
3 The Division shall base its determination of the feasibility of reclaiming open pits and rock faces on the technological and economic practicability of achieving a safe and stable condition suitable for a productive post-mining land use. The Division shall consider, without limitation, the:
   (a) topography of the site;
   (b) geology and stability of the site;
   (c) time required to complete reclamation;
   (d) consumption of resources required to complete reclamation;
   (e) potential adverse environmental impacts to the quality of the air and water associated with the activities for reclamation;
   (f) future access to mineral resources.
4 Upon request by the applicant, the return of material to the open pit from which it was extracted shall be considered to be not feasible for the purposes of reclamation.
5 If an open pit or rock face is exempted from reclamation, public safety must be provided for by means other than reclamation, including, but not limited to, restrictions on access to the site or restrictions on the deed to the property.

C NAC 519A.315. Manner for abandonment of site; selection of appropriate activities for reclamation of site.

1 The abandonment of a site must be conducted in a manner which ensures public safety, encourages techniques to minimise adverse visual effects and establishes a safe and stable condition suitable for the productive post-mining use of the land.
2 In selecting appropriate activities for reclamation for a particular site, techniques which minimise adverse visual impact must be considered.
3 As used in this section, 'ensures public safety' includes minimising hazards in areas to which the public may have legal access by, if applicable:
   (a) removing or burying structures, equipment, reagents or scrap;
   (b) sealing or securing shafts, tunnels and adits pursuant toNAC 513.390;
   (c) plugging drill holes;
   (d) leaving slopes in a structurally stable condition;
   (e) restricting access to areas which cannot practicably be made safe.
4 As used in this section, 'stable condition' means a condition that is resistant to excessive erosion and is structurally competent to withstand normal geologic and climatic conditions without significant failure that would be a threat to public safety and the environment.

D NAC 519A.345. Authority of Division to require operator of mining operation to perform certain types of reclamation. The Division may, if appropriate, require an operator of a mining operation to reclaim.

9 Open pit mines by:
   (a) Performing activities that will provide for public safety;
   (b) Stabilising pit walls or rock faces where required for public safety;
   (c) Constructing and maintaining berms, fences or other means of restricting access;
   (d) Creating a lake for recreational use, wildlife or other uses;
   (e) Revegetation.

Reclamation of open pits or rock faces does not require backfilling although backfilling in whole or in part with waste rock from an adjacent mining operation may be encouraged if backfilling is feasible and does not create additional negative environmental impacts.

California

Changes in 2003 to the California Surface Mining and Reclamation Act and Associated Regulations require that open pit mines for specific metal mines be backfilled after operations. This regulatory framework is effective in discouraging exploration and development of new open pit metal mining operations. The specific stipulations are as follows.

§ 3704.1 Performance standards for backfilling excavations and lands disturbed by open pit surface mining operations for metallic minerals.

Notwithstanding the provisions of Section 3700(b) of the Article, no reclamation plan, including any reclamation plan in which the end use is for wildlife habitat, wildland conservation, or open space, or financial assurance for a surface mining operation subject to the provisions of this section, shall be approved by a lead agency unless the reclamation plan meets the provisions of this section. Financial assurances must be maintained in an amount sufficient to provide for the backfilling and contour grading of the mined lands as required in this section.

(a) An open pit excavation created by surface mining activities for the production of metallic minerals shall be backfilled to achieve not less than the original surface elevation, unless the circumstances under subsection (h) are determine by the lead agency to exist.
(b) Backfilling shall be engineered, and backfilled materials shall be treated, if necessary, to meet all of the provisions of Title 27, California Code of Regulations, Division 2, Chapter 7, Subchapter 1, Mining Waste Management, commencing with Section 22470, and the applicable Regional Water Quality Control Board's Water Quality Control Plan.
(c) Excavated materials remaining in overburden piles, waste rock piles, and processed or leached ore piles not used in the backfilling process and remaining on the mine site shall be graded and contoured to create a final surface that is consistent with the original topography of the area. Care shall be taken to avoid the creation of unnatural topographic features, impediments to natural drainage or conditions hazardous to human life and wildlife.
(d) Backfilling, recontouring and revegetation activities shall be performed in clearly defined phases to the engineering and geologic standards required for the end use of the site as stipulated in the approved reclamation plan. All fills and fill slopes shall be designed to protect groundwater quality, to prevent surface water ponding, to facilitate revegetation, to convey runoff in a non-erosive manner and to account for long-term settlement.
(e) The requirements of subsections (a), (b), (c), and (d) notwithstanding, no final reclaimed fill slopes shall exceed 2:1 (horizontal:vertical), nor shall the resultant topography exceed in height the pre-mining surface contour elevations by more than 25 feet. Final fill slopes shall have static and dynamic factors of safety as

determined by an engineer licensed in California that are suitable for the proposed end use of the site and meet or exceed the requirements of applicable building or grading codes, ordinances, statutes and regulations. Final slopes must be capable of being revegetated, and shall blend in visually with the local topography. Surface soil shall be salvaged, stored and reapplied to facilitate revegetation of recontoured material in accordance with the requirements of Section 3711 of this Article.

(f) For the purposes of this section, a metallic mine is defined as one where more than 10% of the mining operation's gross annual revenues as averaged over the last five years are derived from the production of, or any combination of, the following metallic minerals by the open pit extraction method: precious metals (gold, silver, platinum); iron; nickel; copper; lead; tin; ferro-alloy metals (tungsten, chromium, manganese); mercury; uranium and thorium; minor metals including rubidium, strontium and cesium; niobium and tantalum.

(g) For the purposes of this regulation, an open pit mine is the same as an open pit quarry, opencast mine or open cut mine, and is defined as a mine working or excavation that is open to the surface and in which the opening is approximately the full size of the excavation.

(h) The requirement to backfill an open pit excavation to the surface pursuant to this section using materials mined on site shall not apply if there remains on the mined lands at the conclusion of mining activities, in the form of overburden piles, waste rock piles and processed or leached ore piles, an insufficient volume of materials to completely backfill the open pit excavation to the surface, and where, in addition, none of the mined materials has been removed from the mined lands in violation of the approved reclamation plan. In such case, the open pit excavation shall be backfilled in accordance with subsections (b) and (d) to an elevation that utilises all of the available material remaining as overburden, waste rock and processed or leached ore.

(i) This regulation does not apply to any surface mining operation as defined in Public Resources Code Section 2735(a) and (b) for which the lead agency has issued final approval of a reclamation plan and a financial assurance prior to December 18, 2002.

Note

Authority cited: Sections 2755 and 2756, Public Resources Code. Reference: Sections 2733, 2772 and 2773, Public Resources Code.

History

1 New section filed 12-18-2002 as an emergency; operative 12-18-2002 (Register 2002, No. 51). A Certificate of Compliance must be transmitted to OAL by 4-17-2003 or emergency language will be repealed by operation of law on the following day.

2 New section refiled 4-15-2003 as an emergency; operative 4-15-2003 (Register 2003, No. 16). A Certificate of Compliance must be transmitted to OAL by 8-13-2003 or emergency language will be repealed by operation of law on the following day.

3 Certificate of Compliance as to 4-15-2003 order, including repealer and new section, transmitted to OAL 4-18-2003 and filed 5-30-2003 (Register 2003, No. 22).

British Columbia

The Province of British Columbia in Canada published its mine closure regulations in 1996. The specific regulations related to open pit closure are the BC Mines Act (RSBC 1996), Chapter 293, Part 10 Reclamation and Closure.

Securing of openings

10.6.5 When a mine is closed for an indefinite period, or otherwise left unattended for any length of time, the owner, agent or manager shall take all practible measures to prevent inadvertent access to mine entrances, pits and openings that are dangerous by reason of their depth or otherwise, by unauthorised persons and ensure that the mine workings and fixtures remain secure.

Open pits

10.7.13 Pit walls constructed in overburden shall be reclaimed in the same manner as dumps unless an inspector is satisfied that to do so would be unsafe or conflict with other proposed land uses.

10.7.14 Pit walls including benches constructed in rock, and/or steeply sloping footwalls, are not required to be revegetated.

10.7.15 Where the pit floor is free from water, and safely accessible, vegetation shall be established.

10.7.16 Where the pit floor will impound water and it is not part of a permanent water treatment system, provision must be made to create a body of water where use and productivity objectives are achieved.

Terminology and definitions

SI system of units

As far as practicable, the SI or International System of Units for physical measurement and quantities is used throughout this book. Derived SI units most commonly used by geotechnical engineers are listed in Table 1.

- The prefixes most commonly used to indicate multiples and submultiples are G (giga or 10^9), M (mega or 10^6) and k (kilo or 10^3).
- Good SI practice suggests that multiple and submultiple metric units should be used in increments of 1000 (e.g. mm, m, km). The use of the centimetre (cm) is incorrect.
- For derived combinational units such as pressure or stress (pascals or newtons per square metre), multiples and sub-multiples of the basic metric units should be avoided. Hence, N/cm^2 or N/mm^2 is wrong. Correctly, the expressions should be kN/m^2 or MN/m^2, i.e. the appropriate prefix should used with the numerator to indicate larger or smaller quantities.
- Although the SI system is the world's most widely used system of units, some countries (e.g. the USA and UK) and some cultures (e.g. diamond drilling contractors) still recognise non-SI units. In this context, non-SI units will be encountered in some parts of this book (e.g. Appendix 1, Attachment E).

Symbols and abbreviations

W = sample weight
M = sample mass
V_T = total volume of sample
V_V = volume of voids
γ = unit weight of intact rock
γ_w = unit weight of water
n = porosity
n_p = porosity of non-fissured rock
e = void ratio
u = pore pressure
S_y = specific yield
S_r = specific retention
S_s = specific storage
ρ = density of intact rock
G_s = specific gravity of intact rock
σ_n = normal stress
σ_n' = effective normal stress
σ_1 = major principal stress at failure
σ_1' = major principal effective stress at failure
σ_3 = minor principal stress at failure
σ_3' = minor effective principal stress at failure
m = Hoek-Brown dimensionless material constant for rock
m_b = Hoek-Brown dimensionless material constant for broken rock

Table 1: Derived SI units commonly used in geotechnical engineering

Quantity	Unit	SI symbol	Formula
acceleration	metre per second squared	m/s^2	–
area	square metre	m^2	–
area	hectare	ha	$hm^2 = 10^4 m^2$
density	kilogram per cubic metre	kg/m^3	–
energy	joule	J	N m
force	newton	N	$kg\ m/s^2$
power	watt	W	J/s
pressure	pascal	Pa	N/m2
stress	pascal	Pa	N/m2
unit weight	newton per cubic metre	N/m^3	$kg/s^2\ m^2$
velocity	metre per second	m/s	–
volume	cubic metre	m^3	–
volume	litre	L	$dm^3 = 10^{-3} m^3$

Source: After Holtz & Kovacs (1981)

m_i = Hoek-Brown dimensionless material constant for intact rock
s = Hoek-Brown dimensionless material constant for rock mass, ranging from 1 for intact rock with tensile strength to 0 for broken rock with zero tensile strength
a = Hoek-Brown dimensional material constant for rock mass
D = Hoek-Brown disturbance factor for rock mass
τ = shear strength
τ' = effective shear strength
c = cohesion
c' = effective cohesion
UCS = uniaxial compressive strength of intact rock
σ_c = uniaxial compressive strength of intact rock
ITS = uniaxial compressive strength of intact rock
RBS = strength of the rock blocks contained within the rock mass
TCS = triaxial compressive strength
TS = tensile strength
σ_t = tensile strength
σ_{tB} = Brazilian tensile strength
I_s = point load strength index
$I_{s(De)}$ = point load strength index for an equivalent core diameter D_e different from 50 mm
D_e = diameter of point load or Brazilian strength test sample
t = thickness of the Brazilian strength test sample (disc)
A = minimum cross-sectional area of point load or Brazilian strength test sample
P = compression load required to break point load or Brazilian strength test sample
E = Young's modulus
E_s = secant Young's modulus
E_{rm} = Young's modulus of rock mass
v = Poisson's ratio
ε_d = dynamic Poisson's ratio
ε = strain
ε_a = axial strain
εr = radial strain
ε_v, = volumetric strain
G = shear modulus
G_d = dynamic shear modulus
K = bulk modulus
V_P = P-wave velocity
V_P^T = theoretical P-wave velocity
$V_{P,i}$ = P-wave velocity of mineral constituent 'i', which has a volume proportion C_i in the rock
ϕ = friction angle
ϕ' = effective friction angle
ϕ_{trial} = reduction friction angle for trial factor of safety
ϕ_j = peak friction angle of joint
ϕ_{jres} = residual friction angle of joint
ϕ_b = basic friction angle of joint
ϕ_{eq} = friction angle of equivalent joint (discontinuity)
ϕ_{tz} = friction angle of transition zone between discontinuity and rock mass
c_{trial} = reduced cohesion for trial factor of safety
c_j = peak cohesion of joint
c_{jres} = residual cohesion of joint
c_{tz} = cohesion of transition zone between discontinuity and rock mass
g = gravitational acceleration
l = unit weight of material
l_j = lengths of discontinuity
l_r = length of rock bridge
k_t = coefficient of transition for transition zone between discontinuity and rock mass
A_c = contact area of planar rock (joint) surface prior to shearing
τ_{max} = peak shear strength of joint
τ_f = shear strength of joint at failure
$u_{s,peak}$ = shear displacement required to reach the peak shear strength of joint
δ_s = relative shear displacement on planar rock (joint) surface
i = roughness angle of asperities on joint surface
σ_{ny} = normal stress that causes the yielding of the asperities
c_{jeq} = 'equivalent' cohesion derived from the asperities
θ = joint dip angle
JS = joint spacing
JC = joint condition
J_r = joint roughness
J_a = joint alteration number
JRC = joint roughness coefficient
JCS = uniaxial compressive strength of the rock wall bounding joint
$R_{n(L)}$ = rebound number of the L-type Schmidt hammer
JRC_F = field value of JRC
JCS_F = field value of JCS
JRC_O = reference value for JRC
JCS_O = reference value for JCS
L_F = length of the structure (joint) in the field
L_O = reference value for joint length
k_n, = normal stiffness
k_{ni} = initial normal stiffness
$k_{ni,mm}$ = initial tangent stiffness for mismatching defects
e_i = initial defect aperture
v_{cmax} = maximum defect closure
k_s = shear stiffness
k_{si} = initial shear stiffness
k_j = stiffness number

n_j = stiffness exponent
α = inclination of defect in the axis of a drill hole
d = spacing between defects in a drill hole
L = length of a drill hole
RMR = rock mass rating (Bieniawski)
IRMR = in situ rock mass rating (Laubscher)
MRMR = mining rock mass rating (Laubscher)
GSI = geological strength index (Hoek-Brown)
RQD = rock quality designation (Deere)
Q = tunnelling index (Barton)
$E[x]$ = expected value
$\sigma[x]$ = standard deviation
$V(x)$ = coefficient of variation
H = slope height
β = slope angle
FoS = factor of safety
CFoS = central factor of safety
FS = static safety factor
f = trial factor of safety
PoF = probability of failure
R = risk (*PoF* × (consequences of failure))
NPV = net present value
BCM = business continuity management
ANZMEC = Australia New Zealand Minerals and Energy Council
MCA = Minerals Council of Australia

Definitions

Bowtie analysis = Analysis showing how a range of controls may eliminate or minimise the likelihood of occurrence of specific initiating events that may generate a risk, or reduce the consequences of an event once it has occurred

Consequence = Outcome or impact of an event

Consequence or cause–consequence analysis = Combination of fault tree analysis and event tree analysis

Construction hazard assessment and implication review (CHAIR) = Structured, facilitated discussion process involving designers, constructors and other key stakeholders used to make final design changes to construction projects by accounting for probable construction methods

Drainable porosity = *see* Porosity, drainable

Effective stress = *see* Stress, effective

Energy barrier analysis (EBA) = Qualitative process used to identify hazards by tracing energy flow into, through and out of a system

Epilimnion = Topmost layer of water in a thermally stratified lake

Event tree analysis (ETA) = Tool that provides a systematic mapping of realistic event scenarios potentially resulting in a major incident, and of the relationships, dependencies and potential escalation of events with time. It also provides numerical estimates of the likelihood of occurrence of the component events and of an escalated event

Failure mode and effects analysis (FEMA)/failure modes, effects and criticality analysis (FMECA) = Addresses the basic question of the consequence of the failure of a system component or piece of equipment. It identifies the causes, consequences and criticality of possible component failures, often as input to fault tree analysis

Fault tree analysis (FTA) = Identifies, quantifies and represents in diagrammatic form the faults or failures, and the combinations of faults or failures, which can lead to a major hazard or event

Groundwater = The water contained in the pore spaces of a rock or soil mass in a saturated state, i.e. below the water table

Hazard = Source of potential harm; a potential occurrence or condition that could lead to injury, damage to the environment, delay or economic loss

Hazard analysis = *see* Preliminary hazard analysis

Hazard and operability studies (HAZOP) = Structured brainstorming approach to hazard analysis used to identify process hazards and operability problems and their consequences, and to evaluate the existing or proposed controls at the design stage

Head, pressure = The height of a column of water that can be supported by the static pressure at a specific point

Head, static = The height above a reference or datum surface at which the water will stand in a piezometer. It represents the sum of the pressure head in the piezometer and the elevation above datum of the measuring point in the piezometer (frequently termed total head)

Human error analysis (HEA) = Qualitative or quantitative approach to the identification and management of human errors that could lead or contribute to significant hazards

Holomictic lake = Lake with uniform temperature and density from top to bottom

Hydraulic conductivity = The volume of groundwater that will move in unit time under a unit hydraulic gradient through a unit area at right angles to the direction of groundwater flow (*see also* Permeability)

Hydraulic gradient = The rate of change of static head with distance

Hypolimnion = Dense, bottom layer of water in a thermally stratified lake

Interconnected porosity = *see* Porosity, interconnected

Job safety/hazard analysis (JSA/JHA) = Task-oriented qualitative risk assessment carried out by a work team on site to guide the development of safe working procedures for potentially hazardous tasks

Kip = 1000 pound-force

Likelihood = Probability or frequency of occurrence of an event, described in qualitative or quantitative terms

Meromictic lake = Lake with layers that do not intermix

Perched groundwater zones = Saturated parts of the rock or soil mass separated by unsaturated material. Perched groundwater zones are normally caused by low-permeability stratigraphic or structural horizons that impede downward percolation of water. The vertical hydraulic gradient between the phreatic levels will be greater than unity

Permeability = A term loosely used to mean intrinsic permeability or hydraulic conductivity. In engineering practice, where there is usually little variation in the viscosity of the pore fluids, permeability is commonly used to mean hydraulic conductivity. It is used in this sense throughout this book

Phreatic water = Groundwater that exists in a saturated state below the vadose zone, i.e. below the water table. The term 'phreatic surface' is used synonymously with water table throughout this book

Pore pressure elevation = Static head above a particular failure surface or zone of weakness

Porosity, drainable = Only part of the interconnected porosity will drain under gravity. Even after prolonged drainage, some water will remain adhering as films (pellicular water) to mineral grains or fracture surfaces, or completely filling the finer pores of the rock or soil. The fraction or percentage of the bulk volume that drains under gravity is termed the specific yield (S_y) and the fraction that remains water-filled is termed the specific retention (S_r). The sum of specific yield and specific retention equals total porosity (i.e. $S_y + S_r = n$)

Porosity, interconnected = The fraction or percentage of the bulk volume of a rock or soil mass occupied by interconnected void space

Porosity, total = The fraction of the total volume of a rock or soil mass occupied by void space, usually expressed as a percentage. These voids may be intergranular spaces or fractures. Total porosity (n) may include pores such as fluid inclusions or isolated fractures that are not connected with other pores

Potentiometric surface = Surface created by contouring static head values. It can apply to confined and unconfined aquifers. The term 'piezometric surface' is used synonymously with 'potentiometric surface' in this book

Preliminary hazard analysis (PHA)/hazard analysis (HAZAN) = Analysis to provide an initial listing of the hazards that must be addressed in the design process to ensure that they are managed adequately

Pressure head = *see* Head, pressure

Risk = Chance of something happening that will have an impact on objectives

Risk analysis = Systematic process to understand the nature of and to deduce the level of risk

Risk assessment = Overall process of risk identification, risk analysis and risk evaluation

Risk criteria = Terms of reference by which the significance of risk is assessed

Risk evaluation = Process of comparing the level of risk against risk criteria

Risk identification = Process of determining what, where, when, why and how something could happen

Risk management = Culture, processes and structures directed towards realising potential opportunities while managing adverse effects

Risk treatment = Process of selection and implementation of measures to modify risk

Specific storage = The volume of water released from storage by a unit volume of saturated aquifer for a unit decline in head. Specific storage, S_s, is related to porosity and compressibility by the equation $S_s = \rho g(\alpha + n\beta)$ where ρ is the density of water, g is acceleration due to gravity, n is the porosity of the rock, α is the compressibility of the rock framework and β is the compressibility of water

Specific yield = Fraction of the total volume of a rock or soil mass that can drain by gravity in response to lowering of the water table

Static head = *see* Head, static

Storage coefficient = Volume of water that a formation releases from storage per unit surface area of the aquifer for unit decline in the component of head normal to that surface. In a confined aquifer, storage coefficient is the product of specific storage and aquifer thickness. In an unconfined aquifer it is the product of specific storage and saturated aquifer thickness plus the specific yield

Stress = Force per unit area. Stress therefore has the same value as pressure, but is normally used for a body's internal response to the externally applied pressure

Stress, effective = Difference between the total stress (the total pressure experienced as a result of the weight of the overlying material) and the pore water pressure. The effective stress is the pressure with which the grains or blocks of the formation are held in contact. If the pore pressure is reduced by lowering the potentiometric surface, the effective stress will increase. If the total stress is reduced (e.g. by removal of overlying material) the effective stress will reduce

Total head = *see* Head, total

Total porosity = *see* Porosity, total

Transmissivity = Measure of the ability of a rock to transmit water; the rate at which water will move under a unit hydraulic gradient through a unit width of aquifer, measured at right angles to the direction of groundwater flow. Transmissivity is the product of hydraulic conductivity and saturated thickness

Vadose zone = Interval between ground surface and the water table where the pore pressure is less than atmospheric pressure. Also referred to as the unsaturated zone

Void ratio = The ratio of pore volume to grain volume. *See also* Porosity, which is the ratio of pore volume to bulk volume

Water table = Surface in a rock or soil mass at which the pore pressure is exactly equal to atmospheric pressure. Frequently used interchangeably with the term 'phreatic surface'

References

Abramson LW, Lee TS, Sharma S & Boyce GM (1996). *Slope Stability and Stabilization Methods*. John Wiley & Sons, New York.

Adhikary DP & Dyskin AV (1998). A continuum model of layered rock masses with non-associative joint plasticity. *International Journal of Numerical Analysis Methods and Geomechanics* **22**(4), 245–261.

Adhikary DP & Guo H (2002). An orthotropic Cosserat elasto-plastic model for layered rocks. *Rock Mechanics and Rock Engineering* **35**(3), 161–170.

Adhikary DP, Dyskin DV, Jewell RJ & Stewart DP (1997). A study of the mechanism of flexural toppling failure of rock slopes. *Rock Mechanics and Rock Engineering* **30**(2), 75–94.

AGS Landslide Taskforce Landslide Practice Note Working Group (2007). Practice note guidelines for landslide risk management. *Australian Geomechanics* **42**(1), 63–114.

AGS Sub-Committee on Landslide Risk Management (2000). Landslide risk management concepts and guidelines. *Australian Geomechanics* **35**(1), 49–92 and **37**(2), 1–44.

Amadei B & Stephansson O (1997). *Rock Stress and its Measurement*. Chapman & Hall, London.

Anderson EM (1951). *The Dynamics of Faulting*. Oliver & Boyd, Edinburgh.

Anglo Gold Ashanti (2005). Geotechnical risk management, risk benefit analysis for slope design. Company presentation.

ANZMEC/MCA (2000). *Strategic Framework for Mine Closure*. Australia & New Zealand Minerals and Energy Council/Minerals Council of Australia, Canberra.

Ashford SA & Sitar N (1997). Analysis of topographic amplification of inclined shear waves in a steep coastal bluff. *Bulletin of the Seismological Society of America* **87**, 692–700.

Ashford SA, Sitar N, Lysmer J & Deng N (1997). Topographic effects on the seismic response of steep slopes. *Bulletin of the Seismological Society of America* **87**, 701–709.

AusIMM (2001). *Field Geologists' Manual*, 4th edn. AusIMM Monograph No. 9.

Australian Govt (2006). *Mine Closure and Completion*. Dept of Industry, Tourism and Resources, Canberra.

Australian Standards (1993). Geotechnical site investigations AS1726-1993.

Aven T (2003). *Foundations of Risk Analysis: A Knowledge and Decision-oriented Perspective*. John Wiley & Sons, New York.

Aven T & Kristensen V (2005). Perspectives on risk: review and discussion of the basis for establishing a unified and holistic approach. *Reliability Engineering and System Safety* **90**(1), 1–14.

Baczynski NRP (2000). STEPSIM4 'step–path' method for slope risks. In *Proceedings of the GeoEng2000 Conference* (ed. MC Erwin), Melbourne, vol. 2, pp. 86–92. Technomic, Lancaster, PA.

Badgley PC (1959). *Structural Methods for the Exploration Geologist*. Harper & Sons, New York.

Baecher GB & Christian JT (2003). *Reliability and Statistics in Geotechnical Engineering*. John Wiley & Sons, West Sussex.

Balmer G (1952). A general analytical solution for Mohr's envelope. *American Society Test Mat* **52**, 1260–1271.

Bandis S (1980). Experimental studies of scale effects on shear strength and deformation of rock joints. PhD thesis. University of Leeds, UK.

Bandis S (1990). Mechanical properties of rock joints. In *Rock Joints. Proceedings of International Symposium* (eds N Barton & O Stephansson), Löen, Norway, pp. 125–140. Balkema, Rotterdam.

Bandis SC (1993). Engineering properties and characterization of rock discontinuities. *Comprehensive Rock Engineering* (eds JA Hudson, ET Brown, C Fairhurst & E Hoek), vol. 1, pp. 155–183. Pergamon Press, Oxford.

Bandis S, Lumsden A & Barton N (1981). Experimental studies on scale effects on the shear behaviour of rock joints. *International Journal of Rock Mechanics and Mining Science and Geomechanics Abstracts* **18**(1), 1–21.

Bandis S, Lumsden A & Barton N (1983). Fundamentals of rock joint deformation. *International Journal of Rock Mechanics and Mining Science and Geomechanics Abstracts* **20**(6), 249–268.

Barton N (1971). A relationship between joint roughness and joint shear strength. Rock fracture. In *Proceedings of International Symposium on Rock Fracture*, Nancy, Paper 1-8.

Barton N (1972). A model study of rock-joint deformation. *International Journal of Rock Mechanics and Mining Science and Geomechanics Abstracts* **9**(5), 579–602.

Barton N (1973). Review of a new shear strength criterion for rock joints. *Engineering Geology* **7**, 287–332.

Barton N (1974). *A Review of the Shear Strength of Filled Discontinuities in Rock*. Norwegian Geotechnical Institute Publication no. 105.

Barton N (1980). Estimation of in situ joint properties, Nasliden Mine. In *Application of Rock Mechanics to Cut and Fill Mining, Proceedings of International Conference* (eds O Stephansson & MJ Jones), Lulea, Sweden. pp. 186–192. IMM, London.

Barton N (1982). Shear strength investigations for surface mining. In *Stability in Surface Mining, Proceedings of 3rd International Conference* (ed. CO Brawner), Vancouver, British Columbia, pp. 171–196. Society of Mining Engineers. AIME, New York.

Barton N (1986). Deformation phenomena in jointed rock. *Geotechnique* **36**(2), 147–167.

Barton N (1987a). *Predicting the Behaviour of Underground Openings in Rock.* Norwegian Geotechnical Institute Publication no. 172.

Barton N (1987b). Discontinuities. In *Ground Engineers' Reference Book* (ed. FG Bell), Chapter 5, 5.1–5.15. Butterworths, London.

Barton N (1993). Application of Q-system and index tests to estimate shear strength and deformability of rock masses. *Workshop on Norwegian Method of Tunnelling*, New Delhi, India, pp. 66–84.

Barton N (2002). Some new Q-value correlations to assist in site characterization and tunnel design. *International Journal of Rock Mechanics and Mining Science and Geomechanics Abstracts,* **39**(2), 185–216.

Barton N & Bandis S (1980). Some effects of scale on the shear strength of joints. *International Journal of Rock Mechanics and Mining Science and Geomechanics Abstracts* **17**(1), 69–73.

Barton N & Bandis S (1982). Effects of block size on the shear behaviour of jointed rock. Issues in rock mechanics. In *Proceedings of 23rd US Symposium of Rock Mechanics* (eds RE Goodman & FE Heuze), Berkeley, California, pp. 739–760. AIME, New York.

Barton N & Bandis S (1990). Review of predictive capabilities of JRC–JCS model in engineering practice. Rock Joints. In *Proceedings of International Symposium* (eds N Barton & O Stephansson), Loen, Norway. pp. 603–610. Balkema, Rotterdam.

Barton N & Choubey V (1977). The shear strength of rock joints in theory and practice. *Rock Mechanics* **12**(1), 1–54.

Barton N, Lien R & Lunde J (1974). Engineering classification of rock masses for design of tunnel support. *Rock Mechanics* **6**(4), 189–236.

Barton N, Loset F, Lien R & Lunde J (1980). Application of the Q-system in design decisions. *Subsurface Space* **2**, 553–561.

Bayes T (1763). An essay towards solving a problem in the doctrine of chances. *Philosophical Transactions of Royal Society of London* **53**, 370–418. Reprinted in *Studies in the History of Statistics and Probability* (eds ES Pearson & MG Kendall) (1970), pp. 131–153. Charles Griffin, London.

Bedford TM & Cooke RM (2001). *Probabilistic Risk Analysis: Foundations and Method.* Cambridge University Press, Cambridge.

Belikov BP, Alexandrov KS & Rysuva TW (1970). Uprugie svoistva porodo–obrasujscich mineralov i gornich porod. *Izdat,* Nauka, Moskva. Science, Oxford.

Bell FG (2000). *Engineering Properties of Soils and Rocks*, 4th edn. Blackwell Scientific, Oxford.

Bickers CF, Dunbar CT, LeJuge GE & Walker PA (2001). Wall control blasting practices at BHP Billiton Iron Ore Mt Whaleback. In *Proceedings of EXPLO 2001*, Hunter Valley, NSW, pp. 93–102. AIMM Publication Series no. 4/2001, Melbourne.

Bieniawski ZT (1973). Engineering classification of jointed rock masses. *Trans South African Institute of Civil Engineering* **15**(12), 335–344.

Bieniawski ZT (1974a). Estimating the strength of rock materials. *Journal of South African Institute of Minerals and Metallurgy* **74**, 312–320.

Bieniawski ZT (1974b). Geomechanics classification of rock masses and its application in tunnelling. In *Proceedings of 3rd Congress of International Society of Rock Mechanics*, Denver, vol. 2A, pp. 27–32. National Academy of Sciences, Washington DC.

Bieniawski ZT (1976). Rock mass classification in rock engineering. In *Exploration for Rock Engineering* (ed. ZT Bieniawski), vol. 1, pp. 97–106. Balkema, Cape Town.

Bieniawski ZT (1978). Determining rock mass deformability: experience from case histories. *International Journal of Rock Mechanics and Mining Science and Geomechanics Abstracts* **15**(5), 237–247.

Bieniawski ZT (1979). The geomechanics classification in rock engineering applications. In *Proceedings of 4th Congress of International Society of Rock Mechanics*, Montreux, vol. 2, pp. 41–48. Balkema, Rotterdam.

Bieniawski ZT (1984). *Rock Mechanics Design in Mining and Tunnelling.* Balkema, Rotterdam/Interscience, New York.

Bieniawski ZT (1989). *Engineering Rock Mass Classifications.* John Wiley & Sons, New York.

Billaux DB, Luvison C & Darcel BP (2006). 3FLO: calculs d'écoulements tridimensionnels. Bases théoretiques. *User's Manual, 3FLO Version 2.31.* Itasca, Ecully, October.

Bird P (2003). An updated digital model of plate boundaries. *Geochemistry Geophysics Geosystems* **4**(3), 1027.

Bishop AW (1955). The use of the slip circle in the stability analysis of earth slopes. *Geotechnique* **5**, 7–17.

Bishop AW & Morgenstern N (1960). Stability coefficients for earth slopes. *Geotechnique* **5**(1), 7–17.

Bjerrum L (1973). Problems of soil mechanics and construction of soft clays and structurally unstable soils. In *Proceedings of 8th International Conference on Soil Mechanics and Foundation Engineering*, Moscow, vol. 3, pp. 111–159.

Blair D & Minchinton A (1997). On the damage zone surrounding a single blasthole. *International Journal of Blasting and Fragmentation (fragblast)* **2**, 59–72.

Blyth FGH & deFreitas MH (1984). *A Geology for Engineers*, 7th edn. Edward Arnold, London.

Bock H (1986). In-situ validation of the borehole slotting stressmeter. In *Proceedings of International Symposium on Rock Stress and Rock Stress Measurement*, Stockholm, pp. 261–270. Centek, Lulea.

Boehrer B & Schultze M (2006). On the relevance of meromixis in mine pit lakes. In *Proceedings of 7th International Conference on Acid Rock Drainage (ICARD)* (ed. RI Barnhisel), pp. 200–213. ASMR, Lexington, Kentucky.

Bower H & Rice RC (1976). A slug test for determining hydraulic conductivity of unconfined aquifers with completely or partially penetrating wells. *Water Resources Research* **12**, 423–428.

Bowles J (1979). *Physical and Geotechnical Properties of Soils.* McGraw-Hill Books, New York.

Brace WF & Riley DK (1972). Static uniaxial deformation of 15 rocks to 30 kb. *International Journal of Rock Mechanics and Mining Science and Geomechanics Abstracts* **9**(2), 271–288.

Brady BHG & Brown ET (2004). *Rock Mechanics for Underground Mining*, 3rd edn. Kluwer, Dordrecht.

Brennan S & Inouye K (1988). Stereo analysis package (unpublished). Golder Associates, Vancouver.

Brent GF (1995). The design of pre-split blasts. In *Proceedings of EXPLO 1995*, pp. 299–305. AUSIMM, Brisbane.

Brinkmann JR (1990). An experimental study of the effects of shock and gas penetration in blasting. In *Proceedings of 3rd International Symposium on Rock Fragmentation by Blasting*, Brisbane, Australia, pp. 55–66.

Broadbent CD & Zavodni ZM (1982). Influence of rock structure on stability. In *Stability in Surface Mining*, vol. 3. Society of Mining Engineers.

Brown ET (2003). *Block Caving Geomechanics*. JKMRC Monograph Series in Mining and Mineral Processing 3. Julius Kruttschnitt Mineral Research Centre, University of Queensland, Brisbane.

Brown ET (2007). *Block Caving Geomechanics*, 2nd edn. JKMRC, Brisbane.

Brunton I (1998). The effect of blast induced transient loads on slope displacements. MEngSci thesis. JKMRC, University of Queensland, Brisbane.

Brunton I (2001). Presentation on blast domain definition at Fimiston Open Pits. JKMRC internal research seminar, Brisbane.

Bureau of Land Management (2004). *Ecological Risk Assessment Guidelines for Open Pit Mine Lakes in Nevada.* Instruction Memorandum no. NV-2004-031, 19 February.

Bye A (2004). Unlocking value through the application of EDD's at Anglo Platinum's, PPRust open pit operations. *1st International Seminar on Strategic vs Tactical Approaches in Mining.* South African Institute of Mining and Metallurgy.

Bye A (2005). The development and application of a 3D geotechnical model for mining optimisation, Sandsloot open pit platinum mine, South Africa. *SME Annual Meeting*, Got Mining–Preprints, 28 Feb.–2 March, Salt Lake City.

Cai M, Kaiser PK, Uno H, Yasaka T & Minami M (2004). Estimation of rock mass deformation modulus and strength of jointed hard rock masses using the GSI system. *International Journal of Rock Mechanics and Mining Science* **41**(4), 2–19.

Calderón AR & Tapia AD (2006). Slope-steepening decision using quantified risk assessment: the Chuquicamata case. In *Fifty Years of Rock Mechanics: Landmarks and Future Challenges. 41st US Symposium on Rock Mechanics* (eds D Yale, S Holtz, C Breeds & U Ozbay), Golden, Colorado. Available on CD. American Rock Mechanics Association, Alexandria, VA.

Calderón A, Catalán A & Karzulovic A (2002). Management of a 15 × 10^6 tons slope failure at Chuquicamata Mine, Chile. In *Proceedings of 12th Panamerican Conference on Soil Mechanics and Geotechnical Engineering and 39th US Symposium on Rock Mechanics* (eds PJ Culligan, HH Einstein & AJ Whittle), Cambridge, vol. 2, pp. 2419–2426. Verlag Gluckauf, Essen.

Call RD, Cicchini PF, Ryan TM & Barkley RC (2000). Managing and analyzing overall pit slopes. In *Slope Stability in Surface Mining* (eds W Hustrulid, MK McCarter & DJA Van Zyl), pp. 39–46. SME, Colorado.

Campbell G (1994). Geophysical contributions to mine-development planning: a risk reduction approach. In *Proceedings of XVth CMMI Congress* (ed. CR Anhaeusser), SAIMM Symposium Series S14, vol. 3, pp. 283–325.

Campbell G & Crotty JH (1990). 3-D seismic mapping for mine planning purposes at the South Deep Prospect. In *Proceedings of International Deep Mining Conference* (eds DAJ Ross-Watt & PDK Robinson), SAIMM Symposium.

CANMET (1977). *Pit Slope Manual,* ch. 5, 'Design'. CANMET Report 77-5. Energy, Mines & Resources Canada, Ottawa.

Carmichael RS (ed.) (1989). *Practical Handbook of Physical Properties of Rocks and Minerals.* CRC Press, Boca Raton.

Carranza-Torres C & Fairhurst C (1999). General formulation of the elasto-plastic response of openings in rock using the Hoek-Brown failure criterion. *International Journal of Rock Mechanics and Mining Science* **36**(6), 777–809.

Carrera J & Guimera J (2000). A comparison of hydraulic and transport parameters measured in low-permeability fractured media. *Journal of Contaminant Hydrology* **41**(3), 261–281.

Carrera J & Neuman SP (1986). Estimation of aquifer parameters under transient and steady state conditions: maximum likelihood method incorporating prior information. *Water Resources Research* **22**, 199–210.

Carter TG, Carvalho JL & Swan G (1993). Towards practical application of ground reaction curves. In *Innovative Mine Design for the 21st Century* (eds WF Bawden & JF Archibald), pp. 151–171. Balkema, Rotterdam.

Carvalho JL, Kennard DT & Lorig L (2002). Numerical analysis of the east wall of Toquepala mine, southern Andes of Peru. In *EUROCK 2002, ISRM International Symposium on Rock Engineering for Mountainous Regions*, Madeira, Portugal, November 2002, pp. 615–625. Sociedade Portuguesa de Geotecnia, Lisbon.

Castendyk D & Webster-Brown JG (2007). Sensitivity analysis in pit lake prediction, Martha Mine, New Zealand. 1: Relationship between turnover and input water density. *Chemical Geology* (accepted for publication).

Chen WF (2007). *Limit Analysis and Soil Plasticity.* J Ross Publishing.

Chen Z, Wang X, Haberfield C, Yin JH & Wang Y (2001). A 3-dimensional slope stability analysis method using the upper bound theorem. Part I: theory and methods. *International Journal of Rock Mechanics and Mining Science* **38**, 369–378.

Chiapetta RF (1982). Pre-splitting and controlled blasting techniques. In *4th High Tech Seminar on State of the Art Blasting Technology. Instrumentation and Explosives Applications.* Blasting Analysis International, Nashville, Tennessee.

Christian (2004). 39th Terzaghi lecture: Geotechnical engineering reliability – how well do we know what we are doing? *Journal of Geotechnical and Geoenvironmental Engineering* **130**(10), 985–1003.

Clark SP (ed.) (1966). Handbook of physical constants, Memoir 97. In *Rock Mechanics Symposium*, Toronto, pp. 35–46. Geological Society of America, Boulder, Colorado.

Cocker A (1997). The influence of jointing on pre-split performance. Research studies for De Beers operations. JKMRC internal reports.

Cocker J, Urosevic M & Evans B (1997). A high resolution seismic survey to assist in mine planning. In *Proceedings of Exploration 97: 4th Decennial International Conference on Mineral Exploration* (ed. AG Gubins), pp. 473–476.

Coon RF & Merritt AH (1970). Predicting in situ modulus of deformation using rock quality indices. In *Determinaton of the In Situ Modulus of Deformation of Rock*, ASTM STP 477, pp. 154–173. ASTM, Ann Arbor, Michigan.

Cooper HH, Bredehoeft JD & Papadopulos SS (1967). Response of a finite diameter well to an instantaneous charge of water. *Water Resources Research* **3**(1), 263–269.

Cosserat E & Cosserat F (1909). *Theorie des corps deformables.* Hermann, Paris.

Cravero M, Iabichino G & Mancini R (1988). The effects of rock blasting with explosives on the stability of a rock face. In *Numerical Methods in Geomechanics*, pp. 1697–1706. Balkema, Rotterdam.

Crempien J (2005). Seismic risk in the Rio Blanco–Sur Sur Region. Unpublished report to A Karzulovic & Asoc., Santiago.

Crook T, Willson S, Yu JG & Owen R (2003). Computational modelling of the localized deformations associated with borehole breakout in quasi-brittle materials. *Journal of Petroleum Science and Engineering* **31**(3–4), 177–186.

Cundall PA (2007). PFC modelling of large slopes in brittle jointed rock. Unpublished confidential report to LOP sponsors, 27 February.

Cundall P, Ruest M, Chitombo G, Esen S & Cunningham C (2001). The hybrid stress blasting model: a feasibility study. Confidential report. JKMRC/ITASCA/AEL.

Cundall P, Carranza-Torres C & Hart R (2003). A new constitutive model based on the Hoek-Brown criterion. In *FLAC and Numerical Modeling in Geomechanics – 2003. Proceedings of 3rd International FLAC Symposium* (eds R Brummer et al.), Sudbury, Ontario, pp. 17–25. Lisse, Balkema.

Cunningham CVB (1987). Fragmentation estimations and the Kuz-Ram model: four years on. In *Proceedings of 2nd International Symposium on Rock Fragmentation by Blasting,* Keystone, Colorado, pp. 475–487.

Cunningham CVB (2003). Evaluation of the need for electronic detonator systems for blasting operations. *Proceedings of EFEE 2nd World Conference on Explosives and Blasting Technique,* Prague, pp. 157–164.

Cunningham C (2004). Detonation velocity: its significance for blast modelling: Report compiled for the HSBM project. Confidential to HSBM project sponsors. African Explosives Ltd.

Da Gama CD (1983). Use of comminution theory to predict fragmentation of jointed rock masses subjected to blasting. *Proceedings of 1st International Symposium on Rock Fragmentation by Blasting,* Lulea, Sweden, pp. 565–579.

Dai SH & Wang MO (1992). *Reliability Analysis in Engineering Applications.* Van Nostrand Reinhold, New York.

Dally JW, Fourney WL & Holloway DC (1975). Influence of containment of the borehole pressures on explosive induced fracture. *International Journal of Rock Mechanics and Mining Science and Geomechanics Abstracts* **12**, 5–12.

Dawson EM & Cundall PA (1996). Slope stability using micropolar plasticity. In *Rock Mechanics Tools and Techniques* (eds M Aubertin et al.), pp. 551–558. Balkema, Rotterdam.

Dawson EM, Roth WH & Drescher A (1999). Slope stability analysis by strength reduction. *Géotechnique* **49**(6), 835–840.

De Ridder GP & De Ridder NA (1991). *Analysis and Evaluation of Pumping Test Data*, 2nd edn. Publication 47, International Institute for Land Reclamation and Improvement, Wageningen.

Deere DU & Deere DW (1988). The rock quality designation (*RQD*) index in practice. In *Rock Classification Systems for Engineering Purposes* (ed. L Kirkaldie), pp. 91–101. ASTM STP 984, American Society for Testing and Materials, Ann Arbor, Michigan.

Deere DU & Miller R (1966). Engineering classification and index properties for intact rock. Technical Report No. AFWL-TR–65-116, Air Force Weapons Lab., Kirtland Base, New Mexico.

Deere DU, Hendron AJ, Patton F & Cording EJ (1967). Design of surface and near surface excavations in rock. In *Proceedings of 8th US Symposium on Rock Mechanics: Failure and Breakage of Rock* (ed. C Fairhurst), pp. 237–302. AIME, New York.

Deschamps R & Yankey G (2006). Limitations in the back-analysis of strength from failures. *Journal of Geotechnical Engineering Div ASCE* **132**(4), 532–536.

Diederichs MS, Lato M, Hammah R & Quinn P (2007). Shear strength reduction approach for slope stability analyses. *Proceedings First CANUS Rock Mechanics Symposium*, Vancouver.

Djordjevic N (1999). Two-component of blast fragmentation. In *Proceedings of 6th International Symposium of Rock Fragmentation by Blasting – FRAGBLAST 6,* Johannesburg, South Africa, pp. 213–219. South African Institute of Mining and Metallurgy.

Djordjevic N, Brunton I, Cepuritis P, Chitombo G & Heslop G (1999). Effect of blast vibration on slope stability. In *Proceedings of EXPLO 1999*, pp. 105–115. AUSIMM, Kalgoorlie, Australia.

Doe TW, Zieger M, Enachescu C & Böhner J (2006). In-situ stress measurements in exploratory boreholes. *Felsbau* **24** (4), 39–47.

Donath FA (1964). Strength variation and deformational behavior in anisotropic rock. In *State of Stress in the Earth's Crust* (ed. W Judd), pp. 281–300. Elsevier, New York.

Donze FV, Bouchez J & Magnier SA (1997). Modelling fractures in rock blasting. *International Journal of Rock Mechanics and Mining Science* **34**, 1153–1163.

Downing CH (1985). *Blast Vibration Monitoring and Control.* Prentice Hall, Englewood Cliffs, NJ.

Downing CH & Gilbert C (1988). Dynamic stability of rock slopes and high frequency travelling waves. *Journal of Geotechnical Engineering* **114**(10), 1069–1088.

Driscoll FG (1989). *Groundwater and Wells.* Johnson Filtration System Inc., Minnesota.

Driscoll FG & Fletcher D (1986). *Groundwater and Wells*, 2nd edn. RG Designs.

du Bray E (ed.) (1995). Preliminary compilation of descriptive geoenvironmental mineral deposit models. USGS Open File report 95-831. Available online at http://pubs.er.usgs.gov/usgspubs/ofr/ofr95831.

Duncan JM (2000). Factors of safety and reliability in geotechnical engineering. *Journal of Geotechnical Engineering Div ASCE* **126**(4), 307–316.

Duncan JM & Chang CY (1970). Non-linear analysis of stress and strain soils. *Journal of Soil Mechanics and Foundation Engineering Div ASCE* **96**, 1629–1655.

Duncan JM & Goodman RE (1968). *Finite Element Analysis of Slopes in Jointed Rock.* US Army Engineering Waterways Exp. Station, Vicksburg, Mississippi.

Duncan JM & Stark TD (1992). Soil strength from back analysis of slope failures. In *Stability and Performance of Slopes and Embankments II. A 25-year Perspective* (eds RB Seed & RW Boulanger), vol. 1, pp. 890–904. ASCE, New York.

Duncan JM & Wright SG (1980). The accuracy of equilibrium methods of slope stability analysis. *Engineering Geology* **16**, 5–17.

Duncan JM & Wright SG (2005). *Soil Strength and Slope Stability.* John Wiley & Sons, New Jersey.

Duncan N (1969). *Engineering Geology and Rock Mechanics,* vol. 1. Leonard Hill, London.

Ebisu S, Aydan O, Komura S & Kawamoto T (1992). Comparative study on various rock mass characterization methods for surface structures. In *Rock Characterization. Proceedings of ISRM Symposium, EUROCK'92* (ed. J Hudson), Chester, UK, pp. 203–208. British Geotechnical Society, London.

Einstein HH, Veneziano D, Baecher GB & O'Reilly KJ (1983). The effect of discontinuity persistence on rock slope stability. *International Journal of Rock Mechanics and Mining Science and Geomechanics Abstracts* **20**(5), 227–236.

Eissa EA & Kazi A (1988). Relation between static and dynamic Young's moduli of rocks. *International Journal of Rock Mechanics and Mining Science and Geomechanics Abstracts* **25**(6), 479–482.

Erban PJ & Gill K (1988). Consideration of the interaction between dam and bedrock in a coupled mechanic-hydraulic FE program. *Rock Mechanics* **21**(2), 99–118.

Evans AM (1993). *Ore Geology and Industrial Minerals: An Introduction.* Blackwell Science, Oxford.

Faccioli E, Vanini M & Frassine L (2002). 'Complex' site effects in earthquake ground motion, including topography. *12th European Conference on Earthquake Engineering.* Paper Reference 844.

Fairhurst C (1964). On the validity of the Brazilian test for brittle materials. *International Journal of Rock Mechanics and Mining Science and Geomechanics Abstracts* **1**(4), 535–546.

Farmer I (1983). *Engineering Behaviour of Rocks,* 2nd edn. Chapman & Hall, London.

Fell R, Ho KKS, Lacasse S & Leroi E (2005). A framework for landslide risk assessment and management. In *Landslide Risk Management. Proceedings of International Conference on Risk Management* (eds O Hungr, R Fell, R Couture & E Eberhardt), Vancouver, pp. 3–26. Taylor & Francis, Leiden.

Fellenius W (1927). *Erdstatische Berenchnungen mit Reibung and Kohaesion.* Ernst, Berlin.

Fellenius W (1936). Calculation of the stability of earth dams. *Transactions of 2nd Congress on Large Dams, Washington DC* **4**, 445–462.

Fisher TSR & Lawrence GA (2006). Treatment of acid rock drainage in a meromictic mine pit lake. *Journal of Environmental Engineering* **132**, 515–526.

Fleming RW, Spencer GS & Banks DC (1970). Empirical study of the behaviour of clay shale slopes. US Army Nuclear Cratering Group Technical Report no. 15.

Flores G (2001). Tronadura de control pared mina chuquicamanta. Report from the Superintendencia de Ingenieria Geotecnia. CODELCO Chile.

Flores G & Karzulovic A (2003). *Geotechnical Guidelines: Geotechnical Characterization*, ICS–II Caving Study, Task 4, JKMRC, Brisbane.

Fourie A & Haines A (2007). Obtaining appropriate design parameters for slopes in weathered saprolites. *Slope Stability 2007. Proceedings of 2007 International Symposium on Rock Slope Stability in Open Pit Mining and Civil Engineering* (ed. Y Potvin), pp. 105–116. Australian Centre for Geomechanics, Perth.

Fourmaintraux D (1976). Characterization of rocks: laboratory tests. *La Mecanique des Roches Appliquee aux Ouvrages du Genie Civil* (ed. M Panet). Ecole Nationale des Ponts et Chausses, Paris.

Fourney WL (1983). Fragmentation studies with small flaws. *Rock Fracture Mechanics, CISM Courses and Lectures no. 275*, pp. 321–340. Springer-Verlag, New York.

Franklin JA (coordinator) (1985). Suggested method for determining point load strength. *International Journal of Rock Mechanics and Mining Science and Geomechanics Abstracts* **22**(2), 51–60.

Franklin JA & Dusseault MB (1989). *Rock Engineering.* McGraw-Hill, New York.

Fredlund DG & Krahn J (1977). Comparison of slope stability methods of analysis. *Canadian Geotechnical Journal* **14**, 429–439.

Freeze AR & Cherry JA (1979). *Groundwater.* Prentice Hall, New York.

Fukuzono T (1985). A new method for predicting the failure time of a slope. In *Proceedings of 4th International Conference and Field Workshop on Landslides*, Tokyo, pp. 145–150.

Fullagar PK & Fallon GN (1997). Geophysics in metalliferous mines for ore body delineation and rock mass characterisation. In *Proceedings of Exploration 97: 4th Decennial International Conference on Mineral Exploration* (ed. AG Gubins), pp. 573–584.

Gardner WS (1987). Design of drilled piers in the Atlantic Piedmont. *Foundations and Excavations in Decomposed Rock of the Piedmont Province* (ed. RE Smith), pp. 62–86. Geotechnical Special Publication 9, ASCE, New York.

Gedney D & Weber W (1978). *Design and Construction of Soil Slopes.* TRB Special Report 176, pp. 172–191.

Giani GP (1992). *Rock Slope Stability Analysis.* Balkema, Rotterdam.

Gibb G, Reason J, De Landre J & Placanica J (2004). The incident cause analysis method (ICAM). *Safety in Australia* **26**(2), 13–19.

Gibson M & Morgenstern N (1962). A note on the stability of cuttings in normally consolidated clays. *Geotechnique* **12**(3), 212–216.

Gilbert RB, Wright SG & Liedtke E (1998). Uncertainty in back-analysis of slopes: Kettleman Hills case history. *Journal of Geotechnical Engineering Div ASCE* **124**(12), 1167–1176.

Glass CE (2000) The influence of seismic events on slope stability. In *Slope Stability in Surface Mining* (eds WA Hustrulid, MK McCarter & DJA Van Zyl), pp. 97–105. SME, Colorado.

Gokceoglu C, Sonmez H & Kayabasi A (2003). Predicting the deformation moduli of rock masses. *International Journal of Rock Mechanics and Mining Science and Geomechanical Abstracts* **40**(5), 701–710.

Golder Associates (2006). FracMan – Interactive discrete fracture network (DFN) analysis for representing key discrete fractures in 3D space, geometric modelling and exploration simulation. Golder Associates Inc., Seattle.

González de Vallejo LI (ed.) (2002). *Ingenieria Geologica.* Prentice-Hall, Madrid.

Goodman RE (1970). Deformability of joints. *Determination of the In Situ Modulus of Deformation of Rock*, pp. 174–196, ASTM STP 477, ASTM, Ann Arbor, Michigan.

Goodman RE (1976). *Methods of Geological Engineering in Discontinuous Rocks.* West Publishing, New York.

Goodman RE (1989). *Introduction to Rock Mechanics*, 2nd edn, Wiley, New York.

Goodman R & Dubois J (1971). *Duplication of Dilatant Behavior in the Analysis of Jointed Rocks.* Dept of Civil Engineering, University of California, Berkeley.

Goodman RE, Taylor R & Brekke T (1968). A model for the mechanics of jointed rock. *Journal of Soil Mechanics and Foundation Engineering Div ASCE* **96**, 637–659.

Grimstad E & Bhasin R (1996). Stress strength relationships and stability in hard rock. In *Recent Advances in Tunnelling Technology. Proceedings of International Conference*, New Delhi, India, vol. 1, pp. 3–8.

Guest A (2005). Dynamic breakage of kimberlite in the near field. PhD thesis. JKMRC, University of Queensland, Brisbane.

Haber E, Ascher U & Oldenburg DW (2004). Inversion of 3D electromagnetic data in frequency and time using an inexact all-at-once approach. *Geophysics* **69**, 1216–1228.

Hack R (2002). An evaluation of slope stability classification. Keynote lecture, *Eurock 2002*, pp. 3–21.

Hagan T & Mercer J (1983). Workshop proceedings: safe and efficient blasting in open pits. ICI Australia Operations, Karratha, 23–25 November.

Haile A (2004). A reporting framework for geotechnical classification of mining projects. *AusIMM Bulletin* **September/October**, 30–37.

Haines A & Terbrugge PJ (1991). Preliminary estimate of rock slope stability using rock mass classification systems. In *7th Congress of International Society of Rock Mechanics*, Aachen, Germany, pp. 887–892.

Hall J (2003). The practical implementation of dewatering and depressurisation in large open pits. In *Proceedings of 5th Large Open Pit Conference* (eds C Workman-Davies & E Chanda), November, pp. 165–176. AIMM.

Hambly EC & Hambly EA (1994). Risk evaluation and realism. In *Proceedings of Institution of Civil Engineers – Civil Engineering*, **102**(May), 64–71.

Hamdi E & du Mouza J (2004). A methodology for rock mass characterisation and classification to improve blast results. *International Journal of Rock Mechanics and Mining Science* **42**, 177–194.

Hamel JV (1970). The Pima Mine slide, Pima County, Arizona. *Geology Society of America*, Abstracts with Programs, **2**(5), 335.

Hamel JV (1971). Kimbley pit slope failure. In *Proceedings of 4th Panamerican Conference on Soil Mechanics and Foundation Engineering*, vol. 2, pp. 117–127. ASCE, New York.

Hamel JV (1972). The slide at Brilliant cut. In *Stability of Rock Slopes, Proceedings of the 13th US Symposium on Rock Mechanics* (ed. EJ Cording), Urbana, Illinois, pp. 487–510. ASCE, New York.

Hansen JB (1967). *The Philosophy of Foundation Design: Design Criteria, Safety Factors and Settlement Limits. Proceedings of the Symposium on Bearing Capacity and Settlement of Foundations*, Duke University, Durham, North Carolina, pp. 9–13.

Harr ME (1987). *Reliability-based Design in Civil Engineering*. McGraw-Hill, New York.

Harr ME (1996). *Reliability-based Design in Civil Engineering*. Dover Publications, New York.

Harries NJ (2001). Rock mass characterisation for cave mining engineering. PhD thesis (unpublished). University of Queensland, Brisbane.

Harrison JP (1999). Selection of the threshold value in RQD assessments. *International Journal of Rock Mechanics and Mining Science* **36**(5), 673–685.

Hatherly PJ (2002). Rock strength assessment from geophysical logging. In *8th International Symposium on Borehole Geophysics for Minerals, Geotechnical and Groundwater Applications*, Toronto, Ontario, 21–23 August.

Hatherly PJ, Medhurst TP & MacGregor SA (2007). A rock mass rating scheme for clastic sediments based on geophysical logs. In *Proceedings of International Workshop on Rock Mass Classification in Underground Mining* (eds C Mark, R Pakalnis & R Tuchman), pp. 57–63. Information Circular 9498, NIOSH, Pittsburgh.

Haverland ML & Slebir EJ (1972). Methods of performing and interpreting in situ shear tests. In *Stability of Rock Slopes. Proceedings of 13th US Symposium on Rock Mechanics* (ed. EJ Cording), Urbana, Illinois, pp. 107–137. ASCE, New York.

Hawkes I & Mellor M (1970). Uniaxial testing in rock mechanics laboratories. *Engineering Geology* **4**(3), 177–285.

Hencher SR & Richards LR (1982). The basic frictional resistance of sheeting joints in Hong Kong granite. *Hong Kong Engineer* **11**(2), 21–25.

Hencher SR & Richards LR (1989). Laboratory direct shear testing of rock discontinuities. *Ground Engineering* **March**, 24–31.

Hendrick N (2006). Converted-wave seismology for coal exploration. In *Recorder: Canadian Society of Exploration Geophysicists* **May**, 27–31.

Henson H Jr & Sexton JL (1991). Premine study of shallow coal seams using high resolution seismic reflection methods. *Geophysics* **56**(9), 1494–1501.

Heuze F, Walton O, Maddix D, Shaffer R & Butkovich T (1990). Analysis of explosions in hard rock: the power of discrete element modelling. In *Mechanics of Jointed and Faulted Rock* (ed. H Rossmanith). Balkema, Rotterdam.

Heuze FE (1980). Scale effects in the determination of rock mass strength and deformability. *Rock Mechanics* **12**(3–4), 167–192.

Hoek E (1970). Estimating the stability of excavated slopes in opencast mines. *Transactions of Institute of Mining and Metallurgy, Section A: Mining Industry* **79**, A109–A132.

Hoek E (1974). Progressive caving induced by mining an inclined orebody. *Transactions of Institute of Mining and Metallurgy, Section A: Mining Industry* **83**, A133–A139.

Hoek E (1983). Strength of jointed rock masses, 23rd Rankine lecture. *Geotechnique* **33**(3), 187–223.

Hoek E (1991). Muller lecture: When is rock engineering design acceptable? In *Proceedings of 7th Congress of International Society of Rock Mechanics*. Aachen, Germany.

Hoek E (1994). Strength of rock and rock masses. *ISRM News Journal* **2**(2), 4–16.

Hoek E (1998). Reliability of Hoek-Brown estimates of rock mass properties and their impact on design, Technical Note. *International Journal of Rock Mechanics and Mining Science* **35**(1), 63–68.

Hoek E (1999). Putting numbers to geology –an engineer's viewpoint: 2nd Glossop lecture. *Quarterly Journal of Engineering Geology and Hydrogeology* **32**(1), 1–19.

Hoek E (2002). *Practical Rock Engineering*. Notes available online at http://www.rocscience.com.

Hoek E (2004). *A Brief History of the Development of the Hoek-Brown Failure Criterion*. Discussion Paper no. 7. Available online at http://www.rocscience.com.

Hoek E (2005). *Uniaxial Compressive Strength versus Global Strength in the Hoek-Brown criterion.* Notes available online at http://www.rocscience.com.

Hoek E & Bray J (1981). *Rock Slope Engineering*, 3rd edn. IMM, London.

Hoek E & Brown ET (1980a). Empirical strength criterion for rock masses. *Journal of Geotechnical Engineering Div ASCE* **106**(GT9), 1013–1035.

Hoek E & Brown ET (1980b). *Underground Excavations in Rock*. IMM, London.

Hoek E & Brown ET (1988). The Hoek-Brown failure criterion: a 1988 update. In *Proceedings of 15th Canadian Rock Mechanics Symposium* (ed. J H Curran), Toronto, pp. 31–38. University of Toronto.

Hoek E & Brown ET (1997). Practical estimates of rock mass strength. *International Journal of Rock Mechanics and Mining Science* **34**(8), 1165–1186.

Hoek E & Diederichs MS (2006). Empirical estimation of rock mass modulus. *International Journal of Rock Mechanics and Mining Science* **43**(2), 203–215.

Hoek E & Imrie AS (1995). Guidelines to establish project consulting boards. *International Water Power & Dam Engineering* **46**(8), 33–44.

Hoek E & Karzulovic A (2000). Rock-mass properties for surface mines. In *Slope Stability in Surface Mining* (eds WA Hustrulid, MK McCarter & DJA Van Zyl), pp. 59–67. SME, Colorado.

Hoek E & Richards LR (1974). *Rock Slope Design Review.* Report to Principal Govt Highway Engineer. Golder Associates, Hong Kong.

Hoek E & Soto CA (1981). Collation of open pit slope behaviour under earthquake loading in Chile. Unpublished report (no. 812–1573) to Bougainville Copper Ltd. Golder Associates, Vancouver.

Hoek E, Wood D & Sha S (1992). A modified Hoek-Brown criterion for jointed rock masses. *Rock Characterization. Proceedings of ISRM Symposium EUROCK'92* (ed. J Hudson), Chester, UK, pp. 209–213. British Geotechnical Society, London.

Hoek E, Kaiser PK & Bawden WF (1995). *Support of Underground Excavations in Hard Rock*. Balkema, Rotterdam.

Hoek E, Marinos P & Benissi M (1998). Applicability of the geological strength index (GSI) classification for very weak and sheared rock masses. The case of the Athens Schist formation. *Bulletin of Engineering Geology Environment* **57**, 151–160.

Hoek E, Read J, Karzulovic A & Chen ZY (2000). Rock slopes in civil and mining engineering, Invited Lecture. *Proceedings of GeoEng2000 Conference* (ed. M C Erwin), Melbourne, vol. 1, pp. 643–658. Technomic, Lancaster, PA.

Hoek E, Carranza-Torres C & Corkum B (2002). Hoek-Brown failure criterion, 2002 edn. In *Mining and Tunnelling Innovation and Opportunity. Proceedings of 5th North American Rock Mechanics Symposium and 17th Tunnelling Association of Canada Conference* (eds R Hammah, W Bawden, J Curran & M Telesnicki), Toronto, vol. 1, pp. 267–273. University of Toronto Press, Toronto.

Hoek E, Marinos P & Marinos VP (2005). Characterization and engineering properties of tectonically undisturbed but lithologically varied sedimentary rocks. *International Journal of Rock Mechanics and Mining Science* **42**, 277–285.

Holley K, McKenzie C & Creighton A (2003). Striking a balance between blasting and geotechnical issues. In *Proceedings of 5th Large Open Pit Conference* (eds C Workman-Davies & E Chanda), November, pp. 225–233. AIMM.

Holmberg R & Persson PA (1980). Design of tunnel perimeter blast hole patterns to prevent rock damage. *Transactions of Institute of Mining and Metallurgy* **89**, A37–A40.

Holtz RD & Kovacs WD (1981). *Introduction to Geotechnical Engineering.* Prentice Hall, Englewood Cliffs, New Jersey.

Hudson JA & Priest SD (1979). Discontinuities and rock mass geometry. *International Journal of Rock Mechanics and Mining Science and Geomechanics Abstracts* **20**(6), 339–362.

Huffman AR (2002). The future of pore pressure prediction using geophysical methods. *Leading Edge* **21**(2), 199–205.

Hume D (1740/1978). *A Treatise of Human Nature,* 2nd edn. Clarendon Press, Oxford.

Hungr O (1987). An extension of Bishop's simplified method of slope stability to three dimensions. *Géotechnique* **37**, 113–117.

Hungr O (2002). *CLARA—W: Slope Stability Analysis in Two and Three Dimensions.* O. Hungr Geotechnical Research, Vancouver.

Hungr O, Salgado FM & Bryne PM (1989). Evaluation of a three-dimensional method of slope stability analysis. *Canadian Geotechnical Journal* **26**, 679–686.

Hunt R (1986). *Geotechnical Engineering Techniques and Practices.* McGraw-Hill, New York.

Hustrulid W (1999). *Blasting Principles for Open Pit Mining.* Vol. 1: General design concepts. Balkema: Rotterdam.

Hustrulid W & Wenbo L (2002). Some general design concepts regarding the control of blast-induced damage during rock slope excavation. *Proceedings of 7th International Symposium on Rock Fragmentation by Blasting–FRAGBLAST 7,* Beijing, pp. 595–603. Metallurgical Industry Press.

Hutchinson DJ & Diederichs MS (1996). *Cable Bolting in Underground Mines.* BiTech Publishing, Richmond, Canada.

Hvorslev MJ (1953). Time lag in soil permeability in ground-water observations. *Waterways Experimental Station Bulletin* **36**, Vicksburg, MS.

Institute of Risk Management (2002). *A Risk Management Standard*. AIRMC, ALARM, IRM, London. Available online at http://www.theirm.org/publications/documents/Risk_Management_Standard_030820.pdf.

International Organization for Standardization (2002). ISO/IEC Guide 73 Risk Management – Vocabulary – Guidelines for use in Standards. ISO, Geneva.

IPA Inc. (2006). *IPA Business Reports*. Available online at http://www.ipa–iba.com.

ISRM (1983). Suggested methods for determining the strength of rock materials in triaxial compression: revised version. *International Journal of Rock Mechanics and Mining Science and Geomechanics Abstracts* **20**(6), 285–290.

ISRM (2007). *The Complete ISRM Suggested Methods for Rock Characterisation, Testing and Monitoring: 1974–2006* (eds R Ulusay & JA Hudson). ISRM Turkish National Group, Ankara, Turkey.

Itasca Consulting Group, Inc. (2003) *3DEC (Three-Dimensional Distinct Element Code)*, Version 3.0. Itasca: Minneapolis.

Itasca Consulting Group, Inc. (2004). *UDEC (Universal Distinct Element Code)*, Version 4.0. Itasca: Minneapolis.

Itasca Consulting Group, Inc. (2005). *FLAC (Fast Lagrangian Analysis of Continua)*, Version 5.0. Itasca: Minneapolis.

Itasca Consulting Group, Inc. (2006). *FLAC3D (Fast Lagrangian Analysis of Continua in 3 Dimensions)*, Version 3.1. Itasca: Minneapolis.

Itasca Consulting Group, Inc (2008a). *PFC2D (Particle Flow Code in 2 Dimensions)*, Version 4.0. Itasca, Minneapolis.

Itasca Consulting Group, Inc (2008b). *PFC3D (Particle Flow Code in 3 Dimensions)*, Version 4.0. Itasca, Minneapolis.

Jaeger JCJ (1960). Shear failure of anisotropic rock. *Geology Magazine*, **97**, 65–72.

Jaeger JCJ & Cook NGW (1979). *Fundamentals of Rock Mechanics*, 3rd edn. Chapman & Hall, London.

Jakubec J & Laubscher DH (2000). The MRMR rock mass rating classification system in mining practice. *Proceedings of MASSMIN 2000* (ed. G Chitombo), Brisbane, pp. 413–422. AIMM, Melbourne.

Janbu N (1954). Application of composite slip surface for stability analysis. In *European Conference on Stability of Earth Slopes*, Stockholm.

Janbu N (1957). Stability analysis of slopes with dimensionless parameters. *Harvard University Soil Mechanics Series*, no. 46.

Janbu N (1968). Slope stability computations. *Soil Mechanics and Foundation Engineering Report*, Technical University of Norway, Trondheim.

Janbu N (1973). *Slope Stability Computations in Embankment Dam Engineering* (eds RC Hirschfield & SJ Poulos), pp. 47–86. Wiley, New York.

Jaynes ET (1957). Information theory and statistical mechanics II. *Phys. Rev.* **108**, 15 October.

Jaynes ET (1978). Where do we stand on maximum entropy? In *The Maximum Entropy Formalism* (eds RD Levine & M Tribus). MIT Press, Cambridge, Mass.

Jefferies M (2006). Calculating the overall slope failure probability (the 'top' probability). Draft manuscript for LOP *Design Guidelines* editors (unpublished).

Jefferies M, Lorig L &Alvarez C (2008). Influence of rock-strength spatial variability on slope stability. In *Proceedings First International FLAC/DEM Symposium on Numerical Modelling*, Minneapolis, USA.

Jennings JE (1970). A mathematical theory for the calculation of the stability of slopes in open cast mines. *Planning Open Pit Mines. Proceedings of International Symposium* (ed. PWJ Van Rensburg), Johannesburg, pp. 87–102. Balkema, Cape Town.

Jennings JE (1972). An approach to the stability of rock slopes based on the theory of limiting equilibrium with a material exhibiting anisotropic shear strength. *Stability of Rock Slopes. Proceedings of 13th US Symposium on Rock Mechanics* (ed. EJ Cording), Urbana, Illinois, pp. 269–302. ASCE, New York.

Jibson RW & Jibson MW (2005). Slope performance during an earthquake. US Geological Survey Open File Report 03–005. Available online at http://earthquake.usgs.gov/resources/softwarer/slop.perf.php.

John KW (1968). Graphical stability analysis of slopes in jointed rocks. *Proceedings of American Society of Civil Engineers*, 497–526.

Johnson JD, Fergusson D & Guy G (2007). Risk-based slope design for opencast coal mines at Rotowaro, Huntly, New Zealand. *Slope Stability 2007. Proceedings of 2007 International Symposium on Rock Slope Stability in Open Pit Mining and Civil Engineering* (ed. Y Potvin), Perth, pp. 157–170. Australian Centre for Geomechanics, Perth.

Johnston IW (1985). Strength of intact geomechanical materials. *Journal of Geotechnical Engineering Div ASCE* **111**, 730–749.

Jones CL, Higgins JD & Andrew RD (2000). Colorado RockFall Simulation Program, Version 4.0. Colorado Dept of Transportation, Denver.

JORC (2004). *Australasian Code for Reporting of Exploration Results, Mineral Resources and Ore Reserves*. AusIMM/MCA/AIG. Available online at http://www.jorc.org.

Joy J (2004). Occupational safety risk management in Australian mining. *Occupational Medicine* **54**(5), 311–315.

Joy J (2005). Personal communication. Minerals Industry Safety and Health Centre, University of Queensland.

Joy J & Griffiths D (2005). *National Minerals Industry Safety and Health Risk Assessment Guidelines*, Version 4, January 2005. Available online at http://nmishrag.mishc.uq.edu.au/NMISHRAG_PDF_Files/NMISHRAG_Contents.pdf.

Jumikis AR (1983). *Rock Mechanics,* 2nd edn. Trans Tech Publications, Clausthal, Germany.

Kaiser J (1953). Erkenntnisse unde Folgerungen aus der Messung von Gerauschen bei Zungbeanspruchung von metallischen Werkstoffen. Archiv. Fur das Eisenhuttenwasen, pp. 43–45.

Kanchibotla SS, Morrell S, Valery W & O'Loughlin P (1998). Exploring the effect of blast design on SAG mill throughput. In *Proceedings of Mine to Mill 1998 Conference*, pp. 153–158. AIMM, Brisbane.

Kanji MA (1970). Shear strength of soil rock interfaces. PhD thesis. University of Illinois, Urbana.

Kaplan S (1992). Formalism for handling phenomenological uncertainties: the concepts of probability, frequency, variability, and probability of frequency. *Nuclear Technology* **102**, 137–142.

Karzulovic A (1988). The use of keyblock theory in the design of linings and supports for tunnels. PhD thesis. University of California, Berkeley.

Karzulovic A (2004). The importance of rock slope engineering in open pit mining business optimisation. *Landslides, Evaluation and Stabilization. Proceedings of 9th International Symposium on Landslides* (eds W Lacerda, M Erlich, SAB Fontoura & ASF Sayao), Rio de Janeiro, vol. 1, pp. 443–456. Taylor & Francis, Leiden.

Karzulovic A (2006). *Fundamentals of Geomechanics* (in Spanish), lecture notes. Universidad de los Andes.

Karzulovic A & Sepúlveda (2007). Open pit problems in engineering practice. *Slope Stability 2007. Proceedings of 2007 International Symposium on Rock Slope Stability in Open Pit Mining and Civil Engineering* (ed. Y Potvin), Perth, pp. 201–211. Australian Centre for Geomechanics, Perth.

Karzulovic A, Brzovic A, Pereira J, Marambio F, Russo A & y Cavieres P (2001). Geomechanical properties of structures in the primary ore of El Teniente mine (in Spanish). *Proceedings of SIMIN 2001.* Dept of Mining Engineering, University of Santiago, Chile.

KCGM (2004). Roles and responsibilities for operational geotechnical management. Internal KCGM document.

King MS (1983). Static and dynamic elastic properties of rock from the Canadian shield. *International Journal of Rock Mechanics and Mining Science and Geomechanics Abstracts* **20**(5), 237–241.

King MS, Pandit BI & Stauffer MR (1978). Quality of rock masses by acoustic borehole logging. In *Proceedings of 3rd International Congress*, International Association of Engineering Geologists, pp. 156–164.

Kirsten HAD (1983). Significance of the probability of failure in slope engineering. *The Civil Engineer in South Africa* **25**(1).

Klerck PA (2000). The finite element modelling of discrete fracture of brittle materials. PhD thesis. University of Wales, Swansea, UK.

Kojovic T, Kanchibotla SS, Poetschka NL & Chapman J (1998). The effect of blast design on the lump:fines ratio at Marandoo iron ore operations. *Proceedings of Mine to Mill Conference,* pp. 149–152. AIMM, Brisbane.

Kozicki J & Donzé FV (2008a). A new open-source software developed for numerical simulations using discrete modelling methods. *Computer Methods in Applied Mechanics and Engineering* **197**, 4429–4443

Kozicki J & Donzé FV (2008b). YADE_OPEN DEM: An open-source software using a discrete element method to simulate granular material. *Engineering Computations* (accepted for publication, 2009).

Krahn J (2004). *Stability Modelling with SLOPE/W,* 1st edn (rev.). GEO-SLOPE/W International, Calgary.

Kramers JD (1979). Lead, uranium, strontium, potassium and rubidium in inclusion-bearing diamonds and mantle xenoliths from southern Africa. *Earth and Planetary Science Letters* **42**(1), 58–70.

Kruseman GP & de Ridder NA (1991). *Analysis and Evaluation of Pumping Test Data,* 2nd edn. Publication 47, International Institute for Land Reclamation and Improvement, Wageningen.

Krynine DP & Judd WR (1957). *Principles of Engineering Geology and Geotechnics.* McGraw-Hill, New York.

Kulhawy F (1975). Stress deformation properties of rock and rock discontinuities. *Engineering Geology* **9**, 327–350.

Kutter HK & Fairhurst C (1971). On the fracture process in blasting. *International Journal of Rock Mechanics and Mining Sciences and Geomechanics Abstracts,* **8**, 181–202.

Lam L & Fredlund DG (1993). A general limit equilibrium model for three-dimensional embankments stability anlaysis. *Canadian Geotechnical Journal* **30**, 905–919.

Lama RD & Vutukuri VS (1978a). *Handbook on Mechanical Properties of Rocks,* vol. II. Trans Tech Publications, Clausthal, Germany.

Lama RD & Vutukuri VS (1978b). *Handbook on Mechanical Properties of Rocks,* vol. III. Trans Tech Publications, Clausthal, Germany.

Lama RD & Vutukuri VS (1978c). *Handbook on Mechanical Properties of Rocks,* vol. IV. Trans Tech Publications, Clausthal, Germany.

Lama RD, Vutukuri VS & Saluja SS (1974). *Handbook on Mechanical Properties of Rocks,* vol. I. Trans Tech Publications, Clausthal, Germany.

Laplace PS (1812). *Theorie analytique des probabilités.* Courcier, Paris.

Laubscher DH (1974). Discussion on engineering classification of jointed rock masses by ZT Bieniawski. *Civil Engineer in South Africa*, July, 239–241.

Laubscher DH (1975). Class distinction in rock masses. *Coal, Gold, Base Minerals of South Africa* **23**(6), 37–50.

Laubscher DH (1977). Geomechanics classification of jointed rock masses – mining applications. *Transactions of Institute of Mining and Metallurgy, Section A: Mining Industry* **86**, A1–A8.

Laubscher DH (1984). Design aspects and effectiveness of support systems in different mining conditions. *Transactions of Institute of Mining and Metallurgy, Section A: Mining Industry* **93**, A70–A81.

Laubscher DH (1990). A geomechanics classification system for the rating of rock mass in mine design. *Journal of South African Institute of Mining and Metallurgy* **90**(10), 279–293.

Laubscher DH (1993). Planning mass mining operations. *Comprehensive Rock Engineering* (eds JA Hudson, ET Brown, C Fairhurst & E Hoek), vol. 2, pp. 547–583. Pergamon Press, Oxford.

Laubscher DH (1994). Cave mining: the state of the art. *Journal of South African Institute of Mining and Metallurgy* **94**(10), 279–293.

Laubscher DH (2000). *Block Caving Manual*. Prepared for International Caving Study. JKMRC/Itasca Consulting Group, Brisbane.

Laubscher DH & Jakubec J (2001). The MRMR rock mass classification for jointed rock masses. *Underground Mining Methods: Engineering Fundamentals and International Case Studies* (eds WA Hustrulid & RL Bullock), pp. 474–481. Society of Mining Engineers, AIME, New York.

Laubscher DH & Taylor HW (1976). The importance of geomechanics classification of jointed rock masses in mining operations. *Exploration for Rock Engineering. Proceedings of International Symposium* (ed. ZT Bieniawski), Johannesburg, vol. 1, pp. 119–128. Balkema, Cape Town.

Lauffer H (1958). Gebirgsklassifizierung für den stollenbau. *Geol Bauwesen* **74**, 46–51.

Leroueil S & Tavenas F (1981). Pitfalls of back-analyses. *Proceedings of 10th International Conference of Soil Mechanics and Foundation Engineering*, Stockholm, Sweden, vol. 1, pp. 185–190. Balkema, Rotterdam.

Leventhal AR (2007). A national landslide risk management framework for Australia. *Australian Geomechanics* **42**(1), 1–11.

Ley GMM (1972). The properties of hydrothermally altered granite and their application to slope stability in open cast mines. MSc thesis. London University.

Li Y & Oldenburg DW (1996). 3D inversion of magnetic data, *Geophysics* **61**, 394–408.

Li Y & Oldenburg DW (1998). 3D inversion of gravity data. *Geophysics* **63**, 109–119.

Li Y & Oldenburg DW (2000). 3D inversion of induced polarization data. *Geophysics* **65**, 1931–1945.

Lilly PA (1986). An empirical method of assessing rock mass blastability. In *Proceedings of AUSIMM–IE Aust. Newman Combined Group, Large Open Pit Mining Conference*, pp. 89–92.

Lilly PA (1992). The use of the blastability index in the design of blasts for open pit mines. In *Proceedings of Western Australian Conference on Mining Geomechanics*, Curtin University of Technology, WA, pp. 421–426.

Lilly PA (2005). Risk analysis and decision making: Lecture presentation. Curtin University of Technology, WA.

Lisle RJ & Leyshon PR (2004). *Stereographic Projection Techniques for Geologists and Civil Engineers*, 2nd edn. Cambridge University Press, Cambridge.

Liu J & Elsworth D (1999). Evaluation of pore water pressure fluctuation around an advancing longwall face. *Advances in Water Resources* **22**(6), 633–644.

Logan A & Tyler D (2004). Air inrush risk assessment for caving mines. In *Proceedings of MassMin 2004* (eds A Karzulovic & M Alfaro), Santiago, pp. 717–721. Chilean Engineering Institute, Santiago.

Londe P (1988). Discussion on the determination of shear failure envelope in rock masses. *Journal of Geotechnical Engineering Div ASCE*, **114**(GT3), 374–376.

Lopez Jimeno C, Lopez Jimeno E & Carcedo F (1995). *Drilling and Blasting of Rocks*. A.A. Balkema Publishers.

Lorig L (1999). Lessons learned from slope stability studies. In *FLAC and Numerical Modeling in Geomechanics (Proceedings of the Conference, Minneapolis, September 1999)* (eds C Detournay & R Hart) pp. 17–21. A. A. Balkema, Rotterdam.

Lu P & Latham JP (1998). A model for the transition of block sizes during fragmentation blasting of rock masses. *International Journal of Blasting and Fragmentation* **2**, 341–368.

Ludvig B (1980). The Nasliden Project: direct shear tests on filled and unfilled joints. In *Application of Rock Mechanics to Cut and Fill Mining. Proceedings of International Conference* (eds O Stephansson & MJ Jones), Lulea, Sweden, pp. 179–185. IMM, London.

Lymbery SC (2001). Hydrodynamic extension of radial fractures by explosive gas loading. PhD thesis. University of Queensland, Brisbane.

Lynch RA & Malovichko D (2006). Seismology and slope stability in open pit mines. In *International Symposium on Stability of Rock Slopes*, Cape Town.

Lynch RA, Smith WB & Cichowicz A (2005). Microseismic monitoring of open pit slopes. In *Rockbursts and Seismicity in Mines 6* (eds Y Potvin & M Hudyma), Australian Centre for Geomechanics, Perth.

Maki K (1985). Shear strength and stiffness of weakness planes created by controlled fracturing of intact specimens. *Fundamentals of Rock Joints. Proceedings of International Symposium* (ed. O Stephansson), Bjorkliden, Sweden, pp. 133–142. Centek Publishers, Luleå.

Makurat A, Barton N, Tunbridge L & Vik G (1990). The measurement of the mechanical and hydraulic properties of rock joints at different scales in the Stripa Project. In *ROCK JOINTS. Proceedings of International Symposium* (eds N Barton & O Stephansson), Loen, Norway. pp. 541–548. Balkema, Rotterdam.

Marinos P & Hoek E (2000). GSI: a geologically friendly tool for rock mass strength estimation. In *Proceedings of GeoEng2000 Conference* (ed. MC Erwin), Melbourne, vol. 1, pp. 1422–1440. Technomic, Lancaster, PA.

Marinos P & Hoek E (2001). Estimating the geotechnical properties of heterogeneous rock masses such as Flysch. *Bulletin of Engineering Geology Environment* **60**, 85–92.

Marinos V, Marinos P & Hoek E (2005). The geological strength index: applications and limitations. *Bulletin of Engineering Geology Environment* **64**, 55–65.

Martel SJ (1999). Analysis of fracture orientation data from boreholes. *Environmental and Engineering Geoscience* **V2**, 213–233.

Martin CD (1997). 17th Canadian Geotechnical Colloquium: the effect of cohesion loss and stress path on brittle rock strength. *Canadian Geotechnical Journal* **34**(5), 698–725.

Martin CD & Chandler NA (1994). The progressive fracture of Lac du Bonnett granite. *International Journal of Rock Mechanics and Mining Science* **31**(6), 643–659.

Martin C, Davison C & Kozak E (1990). Characterizing normal stiffness and hydraulic conductivity of a major shear zone in granite. In *Proceedings of International Symposium* (eds N Barton & O Stephansson), Loen, Norway, pp. 549–556. Balkema, Rotterdam.

Mavko G, Mukerji T & Dvorin J (1998). *The Rock Physics Handbook. Tools for Seismic Analysis in Porous Media.* Cambridge University Press, Cambridge.

McCann DM & Entwisle DC (1992). Determination of Young's modulus of the rock mass from geophysical well logs. *Geological Applications of Wireline Logs II* (eds A Hurst, CM Griffiths & PF Worthington), Special Publication no. 65, pp. 317–325. Geological Society.

McHugh S (1983). Crack extension caused by internal gas pressure compared with extension caused by tensile stresses. *International Journal of Fracture* **21**, 163–176.

McKenna GT (1995). Grouted-in installation of piezometers in boreholes. *Canadian Geotechnical Journal* **32**(1), 355–363.

McKenzie C, Scherpenisse C, Arriagada J & Jones J (1995). Application of computer assisted modelling to final wall blast design. In *Proceedings of EXPLO 95,* pp. 285–292. AIMM, Brisbane.

McLamore RT (1966). Strength-deformation characteristics of anisotropic sedimentary rocks. PhD thesis. University of Texas, Austin.

McMahon BK (1985). Some practical considerations for the estimation of shear strength of joints and other discontinuities. *Fundamentals of Rock Joints. Proceedings of International Symposium* (ed. O Stephansson), Björkliden, Sweden, pp. 475–485. Centek Publishers, Luleå.

McNally GH (1990). The prediction of geotechnical rock properties from sonic and neutron logs, exploration. *Geophysics* **21**, 65–71.

Mellor M & Hawkes I (1971). Measurement of tensile strength by diametral compression of disks and annuli. *Engineering Geology* **5**, 173–225.

Mendecki AJ (ed.) (1997). *Seismic Monitoring in Mines.* Chapman & Hall, Cambridge.

Meunier P, Hovius N & Haines JA (2008). Topographic site effects and the location of earthquake induced landslides. *Earth & Planetary Science Letters* (in press). doi:10.1016/j.epsl.2008.07.020

Meyerhof GG (1970). Safety factors in soil mechanics. *Canadian Geotechnical Journal* **7**(4), 349–355.

Meyerhof GG (1984). Safety factors and limit states in geotechnical engineering. *Canadian Geotechnical Journal* **21(1),** 1–7.

Middlebrook TA (1942). Fort Peck slide. *Proceedings of ASCE*, Paper 2144, **107**, 723.

Miles RE (1972). The random division of space. *Proceedings of Symposium on Statistical and Probabilistic Problems in Metallurgy*, December. In *Advances in Applied Probability* **4,** Supplement, 243–266.

Minchinton A (2006). Personal communication to author (G Chitombo).

Minchinton A & Dare-Bryan P (2005). The application of computer modelling for blasting and flow in sublevel caving operations. In *9th AusIMM Underground Operators' Conference 2005,* Perth, WA, 7–9 March, pp. 65–73.

Minchinton A & Lynch PM (1996). Fragmentation and heave modelling using a coupled discrete element gas flow code. In *Proceedings of 5th International Symposium on Fragmentation by Blasting – Fragblast 5* (ed. B Mohanty), Montreal, 25–29 August, pp. 71–80. AA Balkema, Rotterdam.

MIRMgate (2007). Minerals Industry Risk Management Gateway. Available online at http://www.mirmgate.com/browse–hazard.asp.

Misra KC (2000). *Understanding Mineral Deposits.* Kluwer Academic, Dordrecht.

Mitchell RH (1986). *Kimberlites: Mineralogy, Geochemistry and Petrology.* Plenum Press, New York.

Monash University (2006). Occupational Health and Safety Management System, June 2006. Available online at http://www.adm.monash.edu.au/ohse/assets/docs/ohs–management–systems/ohsms–chart.pdf.

Morgan MG & Henrion M (1990). *Uncertainty: A Guide to Dealing with Uncertainty in Quantitative Risk and Policy Analysis.* Cambridge University Press.

Morgenstern N & Price VE (1965). The analysis of the stability of general slip surfaces. *Geotechnique* **15**, 79–138.

Mosinets V (1966). Mechanism of rock breaking by blasting in relation to its fracturing and elastic constants. *Soviet Mining Science* **5**, 492–499.

Mostyn G & Douglas K (2000). Strength of intact rock and rock masses: Invited Lecture. In *Proceedings of*

GeoEng2000 Conference (ed. MC Erwin), Melbourne, vol. 1, pp. 1389–1421. Technomic, Lancaster, PA.

Muhlhaus HM (1993). Continuum models for layered and blocky rock. In *Comprehensive Rock Engineering*. Invited chapter for vol. II, *Analysis and Design Methods*, pp. 209–230. Pergamon Press.

Murphy W (2008). Earthquake induced landslides. Unpublished report to LOP Sponsors Management Committee.

Mutton AJ (1997). The application of geophysics during evaluation of the Century Zinc deposit. In *Proceedings of Exploration 97: 4th Decennial International Conference on Mineral Exploration* (ed. AG Gubins), pp. 599–614.

Nagaraj TS (1993). *Principles of Testing Soils, Rocks and Concrete*. Elsevier, Amsterdam.

Napier S, Oldenburg D, Haber E & Shekhtman R (2006). 3D inversion of time domain data with application to San Nicolas. SEG 2006 Annual Meeting, New Orleans, pp. 1303–1307

Neuman SP (2005). Nothing older than three years. *Ground Water* **42**(6), 797.

Newmark NM (1965). Effects of earthquakes on dams and embankments: 5th Rankine Lecture of the British Geotechnical Society. *Geotechnique* **15**(2), 137–160.

Nicholson GA & Bieniawski ZT (1990). A nonlinear deformation modulus based on rock mass classification. *International Journal of Mining and Geological Engineering* **8**, 181–202.

NSW Department of Mineral Resources (1997). *Risk Management Handbook for the Mining Industry: How to Conduct a Risk Assessment of Mine Operations and Equipment and How to Manage the Risks*. Publication MDG 1010, May 1997. Department of Mineral Resources, Sydney.

O'Neill A, Dentith M & List R (2003). Full-waveform P-SV reflectivity inversion of surface waves for shallow engineering applications. *Exploration Geophysics* **34**(3), 158–173.

Ohkubo T & Terasaki A (1977). Physical property and seismic wave velocity of rock. OYO Technical Note TN-22, Urawa Research Institute, Japan.

Onederra I, Brunton I, Battista J & Grace J (2004). 'Shot to shovel':understanding the impact of muckpile conditions and operator proficiency on instantaneous shovel productivity. *Proceedings of EXPLO 2004*, Perth, Australia, pp. 205–213.

Oriard L (1981). Influence of blasting on slope stability: state of the art. *3rd Conference on Stability in Surface Mining*, Vancouver.

Ouchterlony F, Olsson M & Bergqvist I (2001). Towards new Swedish recommendations for cautious perimeter blasting. *Proceedings of EXPLO 2001,* Hunter Valley, Australia, pp. 281–303. AIMM, Sydney.

Ouchterlony F, Sjoberg C & Jonsson B (1993). Blast damage predictions from vibration measurements at the SKB underground laboratories at ASPO in Sweden. *Proceedings of 9th Annual Symposium on Explosives and Blasting Research*, ISEE, San Diego, pp. 189–197.

Pajot D (2008). Characterising the rock mass; logging for As, Bs and Ds. Presentation to LOP Project Sponsors (unpublished), Schlumberger Water Services, April 2008.

Pan XD & Hudson JA (1988). A simplified three dimensional Hoek-Brown yield criterion. In *Rock Mechanics and Power Plants* (ed. M Romana), pp. 95–103. Rotterdam, Balkema.

Papadopulos SS, Bredehoeft JD & Cooper HH (1973). On the analysis of slug test data. *Water Resources Research* **9**(4), 1087–1089.

Parshley JV & Bowell RJ (2004). The limnology of Summer Camp Pit Lake: a case study. *Mine Water and the Environment* **23.**

Pariseau WG (2000). Coupled geomechanics–hydrologic approach to slope stability based on finite elements. In *Slope Stability in Surface Mining* (eds WH Hustrulid, MK McCarter & DJA Van Zyl), pp. 107–114. SME, Colorado.

Patton FD (1966). Multiple modes of shear failure in rock. *Proceedings of 1st Congress of International Society of Rock Mechanics*, Lisbon, **1**, 509–513.

Patton FD & Hendron AJ (1974). General report on mass movements, *Proceedings of 2nd International Congress IAEG*, Sao Paulo, pp. V–GR1.

Persson PA, Holmberg R & Lee J (1994). *Rock Blasting and Explosives Engineering*. CRC Press, Boca Raton.

Philley JO (1992). Acceptable risk: an overview. *Plant/ Operations Progress* **11**(4), 218–223

Phillips FC (1960). *The Use of Stereographic Projection in Structural Geology*, 2nd edn. Edward Arnold, London.

Phillips FC (1968). *The Use of Stereographic Projection in Structural Geology,* 2nd edn. Edward Arnold, London.

Phillips N, Oldenburg D, Chen J, Li Y & Routh P (2001). In mineral exploration: applications at San Nicolas. *Leading Edge*, December, 1351–1360.

Phoon K-K & Kulhawy FH (1999). Characterization of geotechnical uncertainty. *Canadian Geotechnical Journal* **36**(4), 612–624.

Pierce M, Cundall P, Potyondy D & Ivars DM (2007). A synthetic rock mass model for jointed rock. In *Proceedings of 1st Canada–US Rock Mechanics Symposium* (eds E Eberhardt, D Stead & T Morrison), 27–31 May, Canada, pp. 341–349. Taylor & Francis, London.

Pinto da Cunha A (ed.) (1990). *Scale Effects in Rock Masses.* Balkema, Rotterdam.

Pisters D (2005). Development of generic guidelines for low wall instability management utilising the slope stability radar: case studies from the Hunter Valley and Bowen Basin. In *Bowen Basin 2005: The Future for Coal – Fuel for Thought* (ed. JW Beeston), pp. 245–252. Available on CD. Geological Society of Australia, Sydney.

Plummer CC, Carlson DH & McGeary D (2007). *Physical Geology,* 11th edn. McGraw Hill, Sydney.

Poropat GV & Elmouttie MK (2006). Structural modelling of open pit mines. In *International Symposium on Stability of Rock Slopes in Open Pit Mining and Civil Engineering Situations*, Cape Town, pp. 125–132. SAIMM, Johannesburg.

Pothitos F & Li T (2007). Slope design criteria for large open pits: case study. In *Slope Stability 2007. Proceedings of 2007 International Symposium on Rock Slope Stability in Open Pit Mining and Civil Engineering* (ed. Y Potvin), Perth, pp. 341–352. Australian Centre for Geomechanics, Perth.

Potyondy JG (1961). Skin friction between various soils and construction materials. *Geotechnique* **11**, 339–353.

Potyondy DO & Cundall PA (2004). A bonded-particle model for rock. *International Journal of Rock Mechanics and Mining Science and Geomechanics Abstracts* **41**, 1329–1364.

Power WL, Edgoose JJ & Enever JR (1991). In situ stresses in the Ok Tedi mine region, Western Province, Papua New Guinea. CSIRO Internal Report (New Series No. 57) to Ok Tedi Mining. Unpublished.

Pretorius CC, Trewick WF & Irons C (1997). Application of 3D-seismics to mine planning at the Vaal Reefs Gold Mine, Number 10 Shaft, Republic of South Africa. *Exploration 97*, 399–408.

Price M (2003). *Agua subterranean* (Spanish translation of *Introducing Groundwater*, translated and with additional material by JJ Carillo-Rivera & A Cardona). Editorial Limusa, Grupo Noriega, Mexico.

Price M & Williams AT (1993). A pumped double-packer system for use in aquifer evaluation and groundwater sampling. *Proceedings of Institution of Civil Engineers, Water, Maritime and Energy Journal* **101**, 85–92.

Priest SD (1993). *Discontinuity Analysis for Rock Engineering*. Chapman & Hall, London.

Priest SD & Brown ET (1983). Probabilistic stability analysis of variable rock slopes. *Transactions of Institution of Mining and Metallurgy, Section A: Mining Industry* **92**, A1–12.

Priest SD & Hudson JA (1976). Discontinuity spacings in rock. *International Journal of Rock Mechanics and Mining Science and Geomechanics Abstracts* **13**(5), 135–148.

Priest SD & Hudson JA (1981). Estimation of discontinuity spacing and trace length using scanline surveys. *International Journal of Rock Mechanics and Mining Science and Geomechanics Abstracts* **18**(2), 73–89.

Proctor RV & White TL (1946). *Rock Tunneling with Steel Supports*. Commercial Shearing, Youngstown, Ohio.

Pyke R (1997). Selection of seismic coefficients for use in pseudo-static slope stability analyses. TAGAsoft. Available online at www.tagasoft.com/opinion/article2.html.

Quinlivan D & Lewis T (2007). A case study of managing and reporting risks in a multi-national company. In *Slope Stability 2007. Proceedings of 2007 International Symposium on Rock Slope Stability in Open Pit Mining and Civil Engineering* (ed. Y Potvin), Perth. pp. 533–541. Australian Centre for Geomechanics, Perth.

Ragan DM (1985). *Structural Geology: An Introduction to Geometrical Techniques*, 3rd edn. John Wiley & Sons, New York.

Ramamurthy T, Rao GV & Rao KSA (1985). A strength criterion for rocks. In *Proceedings of Indian Geotechechnical Conference*, Roorkee, **1**, 59–64.

Rasche T (2001). *Risk Analysis Methods: A Brief Review*. Minerals Industry Safety and Health Centre, University of Queensland, June 2001. Available online at http://www.mishc.uq.edu.au/publications/Risk_Analysis_Methods_a_Brief_Review.pdf.

Read JRL (1994). Risk analysis and uncertainty in open pit mining. In *4th Large Open Pit Mining Conference*, 5–9 September, Perth, pp. 139–143.

Read JRL (2007). Predicting the behaviour and failure of large rock slopes. In *Proceedings of 1st Canada–US Rock Mechanics Symposium* (eds E Eberhardt, D Stead & T Morrison), 27–31 May, Canada, pp. 1237–1243. Taylor & Francis, London.

Read JRL & Lye GN (1983). Pit slope design methods: Bougainville Copper Ltd open cut. In *Proceedings of 5th Congress of International Society of Rock Mechanics*, Melbourne, **1**, C93–C98.

Read JRL & Maconochie AP (1992). The Vancouver Ridge landslide, Ok Tedi mine, Papua New Guinea. In *Proceedings of 6th International Symposium on Landslides* (ed. DH Bell), February, Christchurch, pp. 1317–1321. Balkema, Rotterdam.

Read SAL, Perrin ND & Richards L (2005). Evaluation of the intact properties of weak rocks for use in the Hoek-Brown failure criterion. In *Rock Mechanics for Energy, Mineral and Infrastructure Development in the Northern Regions. Proceedings of 40th US Rock Mechanics Symposium* (eds G Chen, S Huang, W Shou & J Tinucci), Anchorage, paper 05–694. ARMA.

Richards LR, Read SAL & Perrin ND (2001). Comparison of the Hoek-Brown failure criterion with laboratory and field tests results from closely-jointed New Zealand greywacke rocks. In *Rock Mechanics:A Challenge for Society. Proceedings of ISRM Symposium, EUROC 2001* (eds P Särkkä & P Eloranta), Espoo, Finland, pp. 283–288. Balkema, Lisse.

Ritchie AM (1963). Evaluation of rockfall and its control. *Highway Research Record* **17**, 13–28. US Highway Research Board, Washington DC.

Robb L (2005). *Introduction to Ore-forming Processes*. Blackwell Publishing, Malden, MA.

Roberts D & Hoek E (1972). A study of the stability of a disused limestone quarry face in the Mendip Hills, England. In *Stability in Open Pit Mining. Proceedings of 1st International Conference* (eds CO Brawner & V Milligan), Vancouver, pp. 239–256. AIME, New York.

Rockfield (2001). ELFEN. Rockfield Software, UK. http://www.rockfield.co.uk/elfen.htm.

Rocscience (2002). RocFall – Program for risk analysis of falling rocks on steep slopes, Version 4.0. Rocscience Inc., Toronto.

Rocscience (2004a). RocData –program for analysis of rock and soil strength data, and the determination of strength envelopes and other physical parameters, Version 4.0. Rocscience Inc., Toronto.

Rocscience (2004b). RocPlane – Program for performing planar rock slope stability analysis and design. Rocscience Inc., Toronto.

Rocscience (2005a). RocLab – Program for determining rock mass strength parameters, based on the latest version of the generalized Hoek-Brown failure criterion, Version 1.0. Rocscience Inc., Toronto.

Rocscience (2005b). Phase^2 – 2D elasto-plastic finite element stress analysis program for underground or surface excavations in rock or soil. Rocscience Inc., Toronto.

Rocscience (2006). Swedge – Program for evaluating the geometry and stability of surface wedges in rock slopes. Rocscience Inc., Toronto

Rode N, Homand E, Hadadou R & Soukatchoff R (1990). Mechanical behaviour of joints of cliff and open pit, Rock Joints. In *Proceedings of International Symposium* (eds N Barton & O Stephansson), Loen, Norway pp. 693–699. Balkema, Rotterdam.

Romana M (1985). New adjustment ratings for application of Bieniawski classification to slopes. In *Proceedings of International Symposium on Rock Mechanics Excavation Mining Civil Works*, ISRM, Mexico City, pp. 59–68.

Romero SU (1968). In situ direct shear tests on irregular surface joints filled with clayey material. In *Proceedings of ISRM International Symposium of Rock Mechanics*, Madrid, **1**, 189–194.

Roscoe Moss (1990). *Handbook of Ground Water Development*. John Wiley & Sons, New York.

Rose ND (2002). The importance of structural geological mapping and slope monitoring in open pit slope stability. In *Proceedings of Canadian Institute of Mining and Metallurgy Annual Meeting*, Vancouver.

Rose ND & Hungr O (2007). Forecasting potential rock slope failure in open pit mines using the inverse-velocity method. *International Journal of Rock Mechanics and Mining Sciences* **44**, 308–320 (© 2006 with permission from Elsevier).

Ross-Brown DM (1973). Slope design in open cast mines. PhD thesis. London University.

Ross-Brown DM & Walton G (1975). A portable shear box for testing rock joints. *Rock Mechanics* **7**(3), 129–153.

Rossmanith HP, Uenishi K & Kouzniak N (1997). Blast wave propagation in rock mass – Part I: monolithic medium. *International Journal of Blasting and Fragmentation* **1**, 317–359.

Rosso R (1976). A comparison of joint stiffness measurements in direct shear, triaxial compression and in situ. *International Journal of Rock Mechanics and Mining Science and Geomechanics Abstracts* **13**(5), 167–172.

Ruest M (2006). Presentation of gas flow logic at the HSBM meeting. Johannesburg. South Africa. Confidential to sponsors.

Rutqvist J & Stephansson O (2003). The role of hydromechanical coupling in fractured rock engineering. *Hydrogeology Journal* **11**(1), 7–40.

Rutqvist J, Ljunggren C, Stephansson O, Noorishad J & Tsang CF (1990). Theoretical and field investigation of fracture hydromechanical response under fluid injection. Rock Joints. In *Proceedings of International Symposium* (eds N Barton & O Stephansson), Loen, Norway, pp. 557–564. Balkema, Rotterdam.

Ryan TM & Pryor PR (2000). Designing catch benches and interramp slopes. In *Slope Stability in Surface Mining* (eds WA Hustrulid, MK McCarter & DJA Van Zyl), pp. 27–38. SME, Colorado.

SAICE (1989). Code of Practice: Lateral Support in Surface Excavations. South African Institution of Civil Engineers, Geotechnical Division.

Sainsbury D, Pothitos F, Finn D & Silva R (2007). Three-dimensional discontinuum analysis of structurally controlled failure mechanisms at the Cadia Hill open pit. In *Slope Stability 2007 (Proceedings, 2007 International Symposium, Rock Slope Stability in Open Pit Mining and Civil Engineering)* (ed. Y Potvin), Perth, pp. 307–320. Perth, Australian Centre for Geomechanics.

SAMREC (2000). South African Code for Reporting of Mineral Resources and Mineral Reserves. SAInstMM.

Sarma SK (1973). Stability analysis of embankments and slopes. *Geotechnique* **23**(3), 423–433.

Sarma SK (1979). Stability analysis of embankments and slopes. *Journal of Geotechnical Engineering Division*, ASCE 105, GT12, 1511–1523.

Saroglou H, Marinos P & Tslambaos G (2004). Applicability of the Hoek-Brown failure criterion and the effect of the anisotropy of intact rock samples from Athens schist. *Journal of South African Institute of Mining Metallurgy* **104**(4), 209–215.

Savely JP (1972). Orientation and engineering properties of jointing in the Sierra pit, Arizona. MS thesis (unpublished). University of Arizona, Tucson.

Savely JP & Call RD (1981). Clay imprint core orientor manual (unpublished). Call & Nicholas Inc., Tucson.

Schön JH (1996). Physical properties of rocks: fundamentals and principles of petrophysics. In *Handbook of Geophysical Exploration* (eds K Helbig & S Treitel), vol. 18, *Seismic Exploration*. Pergamon, Trowbridge, UK.

Schön JH (2004). *Physical Properties of Rocks: Fundamentals and Principles of Petrophysics*. Elsevier, Amsterdam.

Scott A, David D, Alvarez O & Veloso L (1998). Managing fines generation in the blasting and crushing operations at Cerro Colorado Mine. In *Mine to Mill 1998 Conference,* Brisbane, Australia, pp. 141–148. AIMM.

Scott A, Onederra I & Chitombo G (2006). The suitability of conventional geological and geotechnical data for blast design. In *Proceedings of Fragblast–8,* Santiago, Chile. In publication.

Seed HB (1979). Considerations in the earthquake-resistant design of earth dams. *Geotechnique* **29**(3), 215–263.

Selamat M (1994). Rock slope stability incorporating blast vibrations. PhD thesis. JKMRC, University of Queensland, Brisbane.

Serafim JL & Pereira JP (1983). Considerations of the geomechanics classification of Bieniawski. In *Proceedings of International Symposium of Engineering Geology Underground Construction,* Lisbon, 1: II33–II42. Series S10, 2, pp. 569–597.

Shah S (1992). A study of the behaviour of jointed rock masses. PhD thesis. University of Toronto, Canada.

Sharma S (1992). *Slope analysis with XSTABL.* Technical report, Intermountain Research Station, contract #INT-89416-RJV, June. US Dept of Agriculture – Forest Service, Moscow, Idaho.

Sheorey PR (1994). A theory for in situ stresses in isotropic and transversely isotropic rock. *International Journal of Rock Mechanics and Mining Sciences & Geomechanics Abstracts* **31**(1), 23–34.

Sheorey PR (1997). *Empirical Rock Failure Criteria.* Balkema, Rotterdam.

Sheorey PR, Biswas AK & Choubey VD (1989). An empirical failure criterion for rocks and jointed rock masses. *Engineering Geology* **26**(2), 141–159.

Simons N, Menzies B & Matthews M (2001). *A Short Course in Soil and Rock Slope Engineering.* Tomas Telford, London.

Sjoberg J (1999). Analysis of large scale rock slopes. PhD thesis (unpublished). Lulea University of Technology, Lulea.

Sjoberg J (2000). Failure mechanism for high slopes in hard rock. *Slope Stability in Surface Mining* (eds WA Hustrulid, KM McCarter & DJA Van Zyl), pp. 71–80. SME, Colorado.

Skempton AW & Hutchinson JN (1969). Stability of natural slopes and embankment foundations. State of the art report. In *Proceedings of 7th International Conference of Soil Mechanics and Foundation Engineering,* Mexico City, **1**, 291–340. SMMS, Mexico City.

Sliwa R, Dean P & Poropat G (2007). *ACARP Project C15037 Final Report: Tool for the Rapid and Consistent Capture of Drill Core Observations.* CSIRO Exploration and Mining Report 2007/446.

Snow DT (1968). Rock fracture spacings, openings, and porosities. *Journal of Soil Mechanics and Foundation Engineering Division. Proceedings of the American Society of Civil Engineers* **94**(SM1), 73–91.

Song J & Kim K (1995). Blasting induced fracturing and stress field evolution at fracture tips. In *Proceedings of 35th Symposium in Rock Mechanics,* University of Nevada, Reno, pp. 547–552.

Sonmez H & Ulusay R (1999). Modifications to the geological strength index (GSI) and their applicability to stability of slopes. *International Journal of Rock Mechanics and Mining Science* **36**(6), 743–760.

Sowers GF (1979). *Introductory Soil Mechanics and Foundations, Geotechnical Engineering,* 4th edn. Macmillan, New York.

Spencer E (1967). A method of the analysis of the stability of embankments assuming parallel inter-slice forces. *Geotechnique* **17**, 11–26.

Spencer E (1973). Thrust line criterion in embankment stability analysis. *Geotechnique* **23**, 85–100.

SRK Consulting (2006). Acceptable probabilities of failure, mining rock slopes. Internal SRK Consulting report (unpublished).

Stacey TR & Xianbin Y (2004). Extension in large open pit slopes and possible consequences. In *Proceedings of MassMin 2004, Santiago* (eds A Karzulovic & M Alfaro). Chilean Engineering Institute, Santiago.

Standards Australia (2004). *AS/NZS 4360: 2004 Risk Management.* Standards Australia, Sydney.

Starfield AM & Pugliese JM (1968). Compression waves generated in rock by cylindrical explosive charges: a comparison between computer and field measurements. *International Journal of Rock Mechanics and Mining Science* **5**, 65–77.

Stark TD & Eid HT (1992). Comparison of field and laboratory residual strengths. In *Stability and Performance of Slopes and Embankments II. A 25-year Perspective* (eds RB Seed & RW Boulanger), vol. 1. pp. 876–889. ASCE, New York.

Steed CM, Carter TG & Ingraham PC (1991). Remedial work on abandoned mine hazards. In *Proceedings CIM 10th Underground Operators Conference,* Quebec.

Steffen OKH (1997). Planning of open pit mines on risk basis. *Journal of South African Institute of Mining and Metallurgy* **97**(2), 47–56.

Steffen OKH (2002). Mine planning: its relationship to risk management. In *Slope Stability 2007. Proceedings of 2007 International Symposium on Rock Slope Stability in Open Pit Mining and Civil Engineering* (ed. Y Potvin), Perth. Australian Centre for Geomechanics, Perth.

Steffen OKH (2007). Mine planning: its relationship to risk management. Unpublished keynote address in *Slope Stability 2007. Proceedings of 2007 International Symposium on Rock Slope Stability in Open Pit Mining and Civil Engineering* (ed. Y Potvin), Perth. Australian Centre for Geomechanics, Perth.

Steffen OKH, Terbrugge PJ, Wesseloo J & Venter J (2006). A risk consequence approach to open pit slope design. In *Proceedings of International Symposium on Stability of*

Rock Slopes in Open Pit Mining and Civil Engineering, Cape Town, pp. 81–96. South African Institute of Mining and Metallurgy, Johannesburg.

Stephens RA & Banks DC (1989). Moduli for deformation studies of the foundation and abutments of the Portugues Dam, Puerto Rico: rock mechanics as a guide for efficient utilization of natural resources. In *Proceedings of 30th US Symposium on Rock Mechanics* (ed. AW Khair), Morgantown, Virginia, pp. 31–38. Balkema, Rotterdam.

Stevens K, Watts A & Redko G (2000). In-mine applications of the radio wave method in the Sudbury Igneous Complex: In *70th Annual International Metallurgy, Society of Exploration Geophysics*, Expanded Abstracts, Paper 0398, MIN P1.3.

Stimpson B (1981). A suggested technique for determining the basic friction angle of rock surfaces using core. Technical note. *International Journal of Rock Mechanics and Mining Science and Geomechanics Abstracts* **18**(1), 63–65.

Sugawara K, Faramarzi L & Nakamura N (2006a). Determination of rock mass deformation modulus by means of travelling load tests – Part I: Theory of the travelling load test in an open pit. *International Journal of Rock Mechanics and Mining Science* **43**(2), 179–191.

Sugawara K, Faramarzi L & Nakamura N (2006b). Determination of rock mass deformation modulus by means of travelling load tests – Part II: Travelling load test practice in an open pit. *International Journal of Rock Mechanics and Mining Science* **43**(2), 192–202.

Sullivan TD (2006). Pit slope design and risk – a view of the current state of the art. In *Proceedings of International Symposium on Stability of Rock Slopes in Open Pit Mining and Civil Engineering*, Cape Town. South African Institute of Mining and Metallurgy, Johannesburg.

Swan G & Sepulveda R (2000). Slope stability at Collahausi. In *Slope Stability in Surface Mining* (eds WA Hustrulid, KM McCarter & DJA Van Zyl), pp. 163–170. SME, Colorado.

Tapia A, Contreras LF, Jefferies M & Steffen O (2007). Risk evaluation of slope failure at the Chuquicamata mine. In *Slope Stability 2007. Proceedings of 2007 International Symposium on Rock Slope Stability in Open Pit Mining and Civil Engineering* (ed. Y Potvin), Perth, pp. 477–495. Australian Centre for Geomechanics, Perth.

Terzaghi K (1943). *Theoretical Soil Mechanics.* John Wiley & Sons, New York.

Terzaghi K (1946). Rock defects and loads on tunnel support. In *Rock Tunnelling with Steel Supports* (eds RV Proctor & T White), pp. 15–99. Commercial Shearing, Youngstown, Ohio.

Terzaghi K & Richart FE (1952). Stresses in rock about cavities. *Geotechnique* **3**, 57–90.

Terzaghi RD (1965). Sources of error in joint surveys. *Geotechnique* **15**, 287–304.

Thomson S, Thomson D, Hutton A & Poole G (2003). The field testing of RIM IV technology in Australian underground coal mining conditions. In *Sydney Basin Symposium Conference*, September–October.

Tomlinson MJ (1978). *Foundation Design and Construction*, 3rd edn. Pitman Publishing, London.

Trendall AF & Blockley JG (1970). The iron formations of the Precambrian Hamersley Group, Western Australia, with special reference to the associated crocidolite. *Geological Survey of Western Australia, Bulletin* **119**, 366.

Tse R & Cruden DM (1979). Estimating joint roughness coefficients. *International Journal of Rock Mechanics and Mining Science and Geomechanics Abstracts* **16**(5), 303–307.

Tsiambaos G & Sabatakakis N (2004). Considerations on strength of intact sedimentary rocks. *Engineering Geology* **72**, 261–273.

Tunstall AM, Djordjevic N & Villalobos HA (1997). Assessment of rock mass damage from smooth wall blasting at El Soldado mine, Chile. *Transactions of Institute of Mining and Metallurgy* **106**, A42–A46.

Ucar R (1986). Determination of shear failure envelope in rock masses. *Journal of Geotechnical Engineering Div ASCE*, **112**(GT3), 303–315.

UNEP (1992). *World Atlas of Desertification.*

UNSW (2003). Risk management course notes. Dept of Safety Science, Faculty of Science, University of New South Wales, Sydney.

Urosevic M, Evans B & Vella L (2002). Shallow high-resolution imaging of the Three Springs talc mine, Western Australia, *Leading Edge* **21**, 923–926.

US Bureau of Reclamation (1974). Earth manual, a water resources publication. Designation E–18.

US Dept of the Navy (1971). *Soil Mechanics, Foundations, and Earth Structures*, NAVFAC Design Manual DM-1, Washington DC.

US Dept of the Navy (1982). Naval Facilities Engineering Command, DM-7, May, p. 7-2.

USGS (2007). M7.7 Antofagasta, Chile earthquake of 14–15 November 2007. http//quake.wr.usgs.gov/.

Van Heerden WL (1976). Practical application of the CSIR triaxial strain cell for rock mass measurements. In *Proceedings of ISRM Symposium on Investigation of Stress in Rock, Advances in Stress Measurement*, Sydney, pp. 1–6. Institute of Engineers, Australia.

Van Steveren MT (2006). *Uncertainty and Ground Conditions: A Risk Management Approach.* Elsevier, Amsterdam.

Varona PE & Velasco PP (2000). Criterios para la linealización de la envolvente de rotura de Hoek-Brown. *Ingeopress* **80**, 54–58.

Vasarhelyi B (2003). Some observations regarding the strength and deformability of sandstones in case of dry and saturated conditions. *Bulletin of Engineering Geology Environment* **62**, 245–249.

Vasarhelyi B (2005). Statistical analysis of the influence of water content on the strength of the Miocene limestone. *Rock Mechanics and Rock Engineering* **38**, 69–76.

Vázquez PV (2001). Efectos de escala en el módulo de deformación del Macizo Rocoso. Licentiate thesis, Dept of Mining Engineering, Universidad de Santiago, Chile.

Vick SG (2002). *Degrees of Belief.* ASCE Press, Reston, Virginia.

Villaescusa E (1991). A three-dimensional model of rock jointing. PhD thesis (unpublished). University of Queensland, Brisbane.

Villaescusa E, Li J & Baird G (2002). Stress measurement for oriented core in Australia. In *1st International Seminar on Deep and High Stress Mining*, Australian Centre for Geomechanics, Perth, November.

Vose D (2000). *Risk Analysis: A Quantitative Guide,* 2nd edn. John Wiley & Sons, New York.

Walder IF, Blandford N & Shelly JT (2006). Pit lake water quality modeling calibration of the main pit, Tyrone Mine, New Mexico, USA. In *Proceedings of 7th International Conference on Acid Rock Drainage (ICARD)* (ed. RI Barnhisel), pp. 2240–2261. ASMR, Lexington, Kentucky.

Waltham T (1994). *Foundations of Engineering Geology*, 2nd edn. Spon Press, London.

Wang JSY & Elsworth D (1999). Permeability changes induced by excavation in fractured tuff. In *Proceedings of 37th US Rock Mechanics Symposium, Rock Mechanics for Industry.* vol. 2, pp. 751–757. ARMA, Alexandria, Virginia.

Watry SM & Lade PV (2000). Residual shear strengths of bentonites on Palos Verdes peninsula, California. In *Slope Stability 2000* (eds DV Griffiths, GA Fenton & TR Martin), pp. 323–342. ASCE, Reston, Virginia.

Whitman RV (1983). Evaluating and calculated risk in geotechnical engineering: 7th Terzaghi Lecture. *Journal of Geotechnical Engineering* **110**(2), 145–188.

Whitman RV & Bailey WA (1967). The use of computers in slop stability analysis. *Journal of Soil Mechanics and Foundation Engineering Div, ASCE* **93**(SM4), 475–498.

Williams JH & Johnson CD (2004). Acoustic and optical borehole-wall imaging for fractured rock aquifer studies. *Journal of Applied Geophysics* **55**, 151–159.

Wilson R & Crouch EAC (1987). Risk assessment and comparisons: an introduction. *Science* **236**.

Windsor C & Thompson A (1997). *A Course on Structural Mapping and Structural Analysis.* Rock Technology Pty Ltd, Perth.

Winter JD (2001). *An Introduction to Igneous and Metamorphic Petrology.* Prentice Hall, New Jersey.

Winzer SR & Ritter AP (1985). Role of stress waves and discontinuities in rock fragmentation. In *Proceedings of Fragmentation by Blasting*, 1st edn, pp. 11–23. Society for Experimental Mechanics.

Wittke W (1990). *Rock Mechanics. Theory and Applications with Case Histories.* Springer-Verlag, Berlin.

Wood DF (1991). Estimating Hoek-Brown rock mass strength parameters form rock mass classifications. *Transportation Research Record* **1330**, 22–29.

WorkCover NSW (2001). *CHAIR: Safety in Design Tool.* WorkCover NSW, Sydney. Available online at http://www.workcover.nsw.gov.au/NR/rdonlyres/55F8B8CD–D949–4216–840E–015BD12E2671/0/chair_safe_in_design_tool.

World Stress Map (2005). Release 2005 of the world stress map. Available online at www.world-stress-map.org.

Worotnicki G (1993). CSIRO triaxial stress measurement cell. In *Comprehensive Rock Engineering* (ed. JA Hudson), vol. 13(3), pp. 329–394. Pergamon Press, Oxford.

Wu TH (1975). Retaining walls. In *Foundation Engineering Handbook* (eds HF Winterkorn and H-Y Fang), pp 402–417. VanNostrand Rheinhold Company, New York.

Wyllie DC (1992). *Foundations on Rock.* E & FN Spon, London.

Wyllie DC & Mah CW (2004). *Rock Slope Engineering, Civil and Mining*, 4th edn. Spon Press, London.

Wyllie DC & Munn FJ (1979). Use of movement monitoring to minimize production losses due to pit slope failure. In *Stability in Coal Mining. Proceedings of 1st International Conference* (eds CO Brawner & IPF Dorling), Vancouver, pp. 75–94. Miller Freeman Publications, San Francisco.

Wyllie DC & Norrish NI (1996). Rock strength properties and their measurement. In *Landslides Investigation and Mitigation*, Special Report 247 (eds AK Turner & R Schuster), pp. 372–390. Transportation Research Board, NRC, National Academy Press, Washington DC.

Yang J & Wang S (1996). A new constitutive model for rock fragmentation by blasting: fractal damage model. In *Proceedings of 5th International Symposium on Rock Fragmentation by Blasting, FRAGBLAST–5* (ed. B Mohanty), Montreal, pp. 95–100.

Yu X & Vayssade B (1991). Joint profiles and their roughness parameters. *International Journal of Rock Mechanics and Mining Science and Geomechanics Abstracts* **28**(4), 333–336.

Yu J (1999). A contact interaction framework for numerical simulation of multi-body problems and aspects of damage and fracture for brittle materials. PhD thesis. University of Wales, Swansea.

Yudhbir, Lemanza W & Prinzl F (1983). An empirical criterion for rock masses. In *Proceedings of 5th Congress of International Society of Rock Mechanics*, Melbourne, vol. 1, pp. B1–B8. Balkema, Rotterdam.

Zhang L (2005). *Engineering Properties of Rocks.* Elsevier Geo-Engineering Book Series, vol. 4, Elsevier, Amsterdam.

Zhang L & Einstein HH (2004). Using *RQD* to estimate the deformation modulus of rock masses. *International Journal of Rock Mechanics and Mining Science* **41**(2), 337–341.

Zhou B, Fraser SJ, Borsaru M, Aizawa T, Sliwa R & Hashimoto T (2005). New approaches for rock strength estimation from geophysical logs. In *Bowen Basin Symposium 2005. The future for coal — fuel for thought* (ed. JW Beeston), pp. 151–164. Geological Society of Australia, Coal Geology Group and Bowen Basin Geologists Group, Yeppoon, October 2005, 1–6.

Zienkiewicz OC, Humpheson & Lewis RW (1975). Associated and non-associated visco-plasticity and plasticity in soil mechanics. *Géotechnique* **25**(4), 671–689.

Zoback ML (1992). First- and second-order patterns of stress in the lithosphere: the World Stress Map Project. *Journal of Geophysical Research* **97**(B8), 11761–11782.

Index

acceptance criteria 8–9, 221–36
  factor of safety 221–3
  formulating 232
  probability of failure 223–5
  risk model 225–32
acoustic (ATV) televiewers 29, 36
ACT electronic core orientation tool 29
Agua Rica copper project (Argentina) 43
airlift and recovery tests (RC drilling) 43
anisotropy 152–4
Antamina mine (Peru) 197, 335, 336, 337, 373
artificial support 313–26
  basic approaches 313–14
  design considerations 315–16
  economic considerations 316
  reinforcement measures 318–25
  rockfall protection measures 325–6
  safety considerations 317
  specific situations 317–18
  stabilisation, repair and support methods 314–15

backfilling of voids 325
Bajo de Alumbrera mine (Argentina) 148, 175
Ballmark® system 28
banded iron formation (BIF) deposits 58
Barton-Bandis failure criterion 104–6, 108, 110, 113, 206
batter (Australia) 4
Batu Hijau mine (Indonesia) 197
bench (North America) 4
bench cleanup 312–13
bench design achievement 313
bench documentation 329–30
bench face (North America) 4
bench face angle 241–2, 331–3
bench failure 334–5, 337, 338, 339, 340, 365, 375
bench height 239–40, 333
bench instability 338–9
bench mapping 15, 17, 18, 22, 328–9
bench performance 329–37
bench stack 4
bench toppling 242–4
bench width 240, 333
benches 239–44, 310–13
berm (Australia) 4
berm (North America) 4
Betze-Post mine 59, 60, 370
Bieniawski's Rock Mass Rating (RMR) 117–19, 123, 125, 127, 132, 210, 330
Bingham Canyon mine (USA) 55, 197, 198, 228, 412
blast evaluation components 314
blast-damaged zone 128, 150, 162–3, 164
blast-induced damage, influence of geology 279–82
blasting
  and bench performance 334
  buffer 283
  to open up drainage pathways 192
  post-split 292
  pre- or mid-split 285–92
  trim 283–5
blasting, controlled 276–310
  blast damage mechanisms 278–9
  delay configuration 292–4
  design implementation 294–6
  design platform 305–6
  design refinement 299–305
  design terminology 277–8
  HSBM 307
  performance monitoring and analysis 296–9
  planning and optimisation cycle 306–10
  post blast inspection 298
  post excavation and batter quantification 299
  techniques 282–92
blind zones (core logging) 38–9
block modelling approach 205–6
Bougainville mine (Papua New Guinea) 188, 251
British Colombia and open pit mines 461
bunds (reinforcement measures) 325
buttressing 324–5

cable bolting 319–20
California and open pit mines 460–1
Call & Nicholas Inc. (CNI) criterion 127, 130–2, 240
Candelaria mine (Chile) 56
catch fences 325–6
catchment, effectiveness of 334
Century Zinc deposit 23
Chuquicamata mine (Chile) 55, 80–2, 100, 164, 166, 188, 197–8, 251, 260, 313
classification systems (geotechnical model) 210
clay alteration and weathering 156
closure, pit *see* open pit closure
coal 58–9
coarse-grained soils, field estimates of the relative density of 26
collar surveying 27
communication and hazard management 378
controlled blasting *see* blasting, controlled

Controlled Source Audio-Frequency Magnetotelluric (CSAMT) 23
core barrels 27
core drilling and logging 26–40
collar surveying 27
core barrels 27
core handling and documentation 29–31
core logging 32–9
core orientation 28–9
core sampling, storage and preservation 31–2
downhole geophysical techniques 39–40
downhole surveying 27–8
drill hole location 27
planning and scoping 26–7
process requirements 26
core handling and documentation 29–31
core logging 32–9
blind zones 38–9
determining the orientation of structures 36–8
geotechnical logging of fractures 33–4
geotechnical logging of solid core 32–3
large-scale structures logging 34–6
core orientation 28–9
direct marking techniques 28–9
indirect marking techniques 29
core photography 30–1
core recovery and labelling 29–30
core sampling, storage and preservation 31–2
Cortez Pipeline mine (Nevada) 146, 175
cost–benefit analyses 226–8, 340, 386
Cove Pit (Nevada) 190
crack measuring pins 345
cross-crack measurements 345

Darcy's law 151–2
data, instrumentation 368–70
data acquisition systems (monitoring systems) 362–3
data collection *see* field data collection; groundwater data collection
data management 52
data uncertainty 8, 213–20
assessment criteria checklist 219
causes of 213
impact of 213–14
quantifying 215–16
reporting 216–19
definitions, geotechnical engineering 462–6
degradable rocks 94
depressurisation trials *see* slope depressurisation
design implementation 265–326
artificial support 313–26
controlled blasting 276–310
excavation and scaling 310–13
mine planning aspects 265–76
design rock mass strength (DRMS) 127, 128
dewatering
mine 146–8, 172–3, 180–1, 341–2
pit 341–2
Diavik mine (Canada) 197
digital imaging 21
digital photogrammetry 21, 22, 277, 308, 315, 331, 333, 349–50
digital photography 36, 378
dilation, slope 4–5
Discrete Fracture Network (DFN) modelling 22, 79–80, 139, 172, 254, 260
ditches (reinforcement measures) 325
document control and hazard management 378–9
downhole geophysical techniques 39–40
downhole imaging 36–8
downhole surveying 27–8
magnetic techniques 27
non-magnetic techniques 27–8
drainage pathways, blasting to open up 192
drill hole
cameras 297
location 27
observation well 15, 41, 43, 44, 152, 427, 428
open observation holes 44
piezometer 15, 41, 43, 44, 152, 427, 428
planning and scoping 26–7
drilling mud 47, 48, 50
drill-stem injection tests (RC drilling) 42–3
'Dry' slope approach (pore pressure) 211

earthquakes 62–5, 66, 113, 251–3, 259, 381, 432
ecological risk assessment (pit closure) 410
economic factors 2–3
El Indio mine (Chile) 188
elastic constants 91–2
electrolevels 353
emergency response procedures 375–6
empirical design charts 247–8
energy barrier analysis (EBA) 390, 392
environmental factors 3–4
epithermal deposits 55, 56, 76
Escondida mine (Chile) 54, 55, 188–9, 190, 197, 362
evaluation, slope 10–11
event tree analyses (ETA) 228, 229–30, 231, 390, 392, 393
excavation 310–12
EZY-MarkTM system 28–9

fabric, structural 75–6
joints 76
minor fold structures 75–6

fabric with primary domains 82
Factor of Safety (FoS) 3, 8, 11, 136, 137, 196, 221–3, 235, 238, 241, 255, 259, 335, 387, 443–4
failure, slope *see* slope failure
failure modes and effects analysis (FMEA) 11, 390, 392, 405
failure modes, effects and criticality analysis (FMECA) 390, 392
falling or rising head (slug) tests 49
fatalities and injuries 232–4
fault tree analyses 228, 229, 390, 392
faults 73–4
fibrecrete 323–4
field data collection 15–52
  core drilling and logging 26–40
  data management 52
  digital mapping 21
  general geotechnical logging 17–19
  groundwater data collection 40–52
  line mapping 19
  mapping for structural analyses 19–22
  outcrop mapping and logging 15–23
  overburden soils logging 23–6
  surface geophysical techniques 22–3
  window mapping 19–21
fine-grained soils, field estimates of the strength of 26
fixed extensometers 356
folds 69–73
formulation of slope designs 6–11
  acceptance criteria 8–9
  closure 11
  data uncertainty 8
  design implementation 10
  geotechnical model 6–8
  risk management 11
  slope design methods 9–10
  slope evaluation and monitoring 10–11
FracMan 79, 139, 254, 260

gabion walls 321
geological environments 76–7
  intrusive 76
  metamorphic 77
  sedimentary 76–7
geological mapping 328
geological model 6–7, 53–67
  geotechnical requirements 59–62
  ore body environments 55–9
  physical setting 53–5
  regional seismicity 62–6
  regional stress 66–7
geophysical techniques, downhole 39–40
geotechnical control measures 398–9
geotechnical effort by project stage 13
geotechnical engineering, terminology and definitions 462–6
geotechnical hazard checklists 447–5
geotechnical logging
  of fractures 33–4
  general 17–19
  large-scale structures logging 34–6
  of solid core 32–3
  summary 39
geotechnical model 6–8, 201–12
  applying the model 206–12
  block modelling approach 205–6
  building the model 202–5
  classification systems 210
  constructing the model 201–6
  Hoek-Brown rock mass strength criterion 210–11
  model development 202
  pore pressure considerations 211–12
  required output 201
  scale effects 206–10
  validation and refinement 327–37
geotechnical reporting system 216–19
geotechnical requirements 59–62
geotechnical roles and responsibilities, operational 376–7
Goldstrike mine (Nevada) 146, 166, 368
goniometry 36
GPS surveying systems 348–9
Grasberg mine (Indonesia) 188, 197
ground control management plans 10–11, 266, 327, 370–9, 399
groundwater conditions 6, 42, 148, 166, 172, 234, 250–1, 254, 330, 407, 444
groundwater data collection 40–52, 415–28
  drill stem injection tests 417–19
  drilling mud 48
  falling or rising head tests (slug tests) 420–2
  general pumping test procedures 427–9
  guidance notes: installation of test wells for pit slope depressurisation 47–9
  guidelines and procedures for piezometer installation 419–20
  hydraulic tests 49–51
  Packer test guidelines and procedures 422–7
  phased approach 41–2
  piezometer installation 44–7
  'piggy-backed' data collection 41, 42
  pilot holes 47–8
  procedures in RC drill holes 415–17
  setting up pilot depressurisation trials 51–2
  tests conducted during RC drilling 42–3
  well screen and filter pack, selection of 48–9

groundwater flow 151, 153, 154
numerical modelling 169–72
pore pressure reduction from 163
groundwater hydraulics 151–66
Darcy's law 151–2
fracture-flow groundwater settings 156–61
future research 195–9
groundwater flow 151–4
heterogeneity and anisotropy 152–4
influences on fracturing and groundwater 161–3
mechanisms controlling pore pressure reduction 163–6
porous-medium (intergranular) groundwater settings 154–6
groundwater pressure, monitoring 359–62
dewatering well discharges 360
horizontal drain flows 360
piezometers 360
presentation of monitoring results 361–2
slope conditions 360–1
transient pore pressure 362
groundwater settings
fracture-flow 156–61
porous-medium (intergranular) 154–6

hazard inventory 372–3
hazard management plan 371–9
emergency response procedures 375–6
hazard inventory 372–3
operational geotechnical roles and responsibilities 376–7
performance reviews 379
recording events, communication, training and control documents 377–9
risk reduction 373
trigger action responses (TARPs) 373–5
hazard checklists, geotechnical 447–5
heterogeneity 152–4
Hoek-Brown Geological Strength Index (GSI) 19, 117, 123–7, 210
Hoek-Brown strength criterion 117, 124, 128–30, 132, 138, 139, 201, 206, 207–8, 210–11, 253, 260
horizontal piezometer 46, 360
'Hybrid' approach (pore pressure) 211–12
hybrid stress blasting model (HSBM) 307
hydraulic tests (groundwater data collection) 49–51
hydrogeology and slope engineering 141–51
developing a slope depressurisation program 151
general mine dewatering and localised pore pressure control 146–8
making the decision to depressurise 148–51
open pit stability 411
porosity and pore pressure 141–6
hydrogeology model 8, 141–99
conceptual model 166–8
mine dewatering and slope design 172–3
numerical model 168–80
hydrology, site 407–8
hydrostatic unloading 163, 166

igneous rocks 58, 76, 87, 100, 131, 242
implementation, slope design 10
mine planning 10
operational aspects 10
inclinometers 354
index properties, rock strength 85–8
infilling 19, 21, 22, 97–9, 100, 105, 329, 334
in-pit slope engineering, pore pressure 143–4
in-pit water management 192–3
in-place inclinometers 354
in situ stresses on open pit design 437–46
geological factors 438–40
measurement 440
stresses in open pit slopes 440–4
In-situ Rock Mass Rating system (IRMR) 119–23
instability 4–6
instrumentation data 368–70
instruments for slope monitoring 346–8
inter-ramp angles 10, 237, 244–5, 274, 325
inter-ramp slope performance 337–9
inter-ramp slopes 9, 69, 74, 244–6
inter-ramp toppling 245–6
intrusive activity, igneous and subvolcanic 76

joints 76
JointStats data management 52, 79, 139, 216, 254, 260, 308

Kalgoorlie Consolidated Gold Mines open pit (Western Australia) 305, 321, 326
kimberlites 55, 56–7, 76, 94, 156, 162
kinematic analyses 239–46
benches 239–44
inter-ramp slopes 244–6
Kori Kollo mine (Boliva) 148, 411

laboratory test samples 31–2
Large Open Pit (LOP) Project xiii, 52, 79, 138, 139, 208, 212, 216, 241, 254, 260, 261, 411
laser scanning (LiDAR) 350–1
Laubscher
IRMR and MRMR 117, 119–23, 127, 210
strength criteria 127–8
Letlhakane mine (Botswana) 197

Lihir mine (Papua New Guinea) 53, 54
limited equilibrium methods (slope design) 248–53
  3D methods 251
  design charts 250–1
  incorporating water pressures 249–50
  method of slices 248–9
  seismic analysis 251–3
line drilling 292
line mapping 19
lithostatic unloading 163–6, 212
logging, general geotechnical 17–19
Lone Tree mine (Nevada) 146, 412

Manto Verde mine (Chile) 56
mapping
  bench 15, 17, 18, 22, 328–9
  digital imaging 21
  general geotechnical logging 17–19
  geological 328
  line 19
  practical considerations 21–2
  rock mass cell mapping sheet 18
  for structural analyses 19–23
  window 19–21
  see outcrop mapping and logging
mechanical properties and rock strength 88–92
meshing 320, 325
metamorphic structures 69, 74–5, 77, 80, 87, 100, 131, 142, 201
methods, slope design 9–10
micro-seismic arrays 358
micro-seismic data acquisition units 358–9
micro-seismic data processing, analysis and reporting 359
micro-seismic data event data 359
micro-seismic monitoring 357–8
Mina Sur (Chile) 59, 188
mine closure *see* open pit closure
mine dewatering 146–8, 172–3, 180–1, 341–2
mine design 274
mine planning 10
mine planning aspects of slope design 265–76
  application 268–75
  open pit design philosophy 265–7
  open pit design process 267
mine production planning 274
Mine Safety and Health Administration (MSHA) 3
mine scale dewatering applications (numerical hydrogeological model) 169–73
mineralisation 6, 42, 55–6, 58, 59, 76, 162, 268–70, 328
minerals industry risk management 384–5, 386
Mining Rock Mass Rating system (MRMR) 119–23
minor fold structures 75–6
Mississippi Valley Type (MTV) deposits 58
Mohr-Coulomb failure criterion 90, 94, 95, 96, 106, 117, 127, 132, 138, 139, 143, 243, 246, 248, 250, 253, 259–60
monitoring, slope 10–11
  movement monitoring systems 343–63
  programs, guidelines on the execution of 363–70
monitoring groundwater pressure 359–63
monitoring procedures, pit wall stability 364
monitoring program 5, 364–7
  crack monitoring 365–6
  and radar 367
  rockfalls, monitoring 365
  subsurface techniques 367
  surface extensometers 365–6
  terrestrial geodetic surveys 366
  visual inspection 365
monitoring systems 343–63
  data acquisition systems 362–3
  monitoring of groundwater pressure 359–62
  subsurface monitoring methods 353–9
  types of instruments 344
monitoring systems, surface methods 344–53
  crack measuring pins 345
  cross-crack measurements 345
  GPS surveying systems 348–9
  instruments 346–8
  laser scanning (LiDAR) 350–1
  photogrammetry 349–50
  radar 351
  satellite imaging subsidence monitoring (InSAR) 351–3
  survey monitoring 346
  tiltmeters and electrolevels 353
  visual inspection 344–5
  wireless extensometers 345–6
monitoring systems, subsurface methods 353–9
  fixed extensometers 356
  inclinometers 354
  in-place inclinometers 354
  micro-seismic arrays 358
  micro-seismic data acquisition units 358–9
  micro-seismic data processing, analysis and reporting 359
  micro-seismic data event data 359
  micro-seismic monitoring 357–8
  portable inclinometer probes 354
  probe extensometers 357
  resistance wire extensometers 356–7
  shear strips and time domain reflectometers (TDRs) 354–6

thermistors 357
Morenci mine (Arizona) 150
movement, slope 4–5
mud cake 48

Nchanga mine (Zambia) 188
Nevada and open pit mines 459–60
Nickel West project (Western Australia) 62, 63, 78, 205
numerical hydrogeological models 168–80
coupling pore pressure and geotechnical models 179–80
mine scale dewatering applications 169–73
modelling in 2D 176
modelling in 3D 179
pit slope pore pressure 175–9
pit slope scale 173–5
slice models 176–9
numerical methods (slope design) 253–63
advanced numerical models 259–60
continuum models 253
discontinuum models 253–4
modelling considerations 254–9
research targets 260–3
numerical modelling
groundwater flow 169–72
pit slope pore pressure 175–9
pit slope scale 173–5

observation well 15, 41, 43, 44, 152, 427, 428
Ok Tedi mine (Papua New Guinea) 53, 57, 67, 251
Olympic Dam (South Australia) 45, 59, 178, 197
open pit closure 3, 11, 191, 274–5, 401–13, 461
access 412
activities 412
ecological risk assessment 410
example regulations 459–61
existing mines 403–5
goal and criteria 405–7
mine closure planning 403–5
new mines 403
ore body characteristics 408–9
pit lake quality 409–10
pit water balance 409
planning 405–12
post-closure monitoring 412
reality of 412
risk assessment 405
site characterisation 407–8
surface water diversion 409
open pit design, in situ stresses on 437–45
open pit design philosophy 265–7
open pit design process 267
operational aspects 10
Orapa mine (Botswana) 197
optical (OTV) televiewers 29, 36
ore body environments (geological model) 55–9
outcrop mapping and logging 15–23
general geotechnical logging 17–19
mapping for structural analyses 19–22
surface geophysical techniques 22–3
overburden soils logging 23–6

packer testing 49–51
Penoso mine (Chile) 251
performance reviews and hazard management 379
permafrost 83, 94, 344, 357
photogrammetry 21, 22, 277, 308, 315, 331, 333, 349–50
physical setting (geological model) 53–5
'piggy-backed' data collection 41, 42
piezometer 15, 41, 43, 44, 152, 360, 427, 428
piezometer installation 44–7, 419–20
horizontal 46
installed in drainage tunnels 46
joint piezometer/inclinometer completions 45
open observation holes 44
standpipe 44, 419
standpipe vs grouted vibrating-wire installations 45
terminology 44
vibrating-wire 45, 419–20
Westbay 46–7
pilot holes (groundwater data collection) 47–8
Pinson mine (Nevada) 181, 182
pit access after closure 412
pit closure *see* open pit closure
pit slope depressurisation 165
installation of test wells for 47–9
methods 181, 192
setting up pilot trials 51–2
see also slope depressurisation
pit slope pore pressure, numerical modelling 175–9
pit slope pressures, specific programs 181–92
pit slope scale numerical modelling 173–5
pit slopes
characterisation of local 41–2
geotechnical risk management for open 385–9
hydrogeological model 166–8
pit wall stability
mine closure 410–11
monitoring procedures 364
pit wall terminology 5
point load strength index 85, 89
Poisson's ratio 91–2, 113, 280
pore pressure 141–6
behaviour 195–6
considerations 211–12

coupled 197–9
data 201
localised control 146–8
making the decision to depressurise 148–51
reduction, mechanisms controlling 163–6
standardising the interaction between geotechnical models 196–7
transient 197
porosity (hydrogeology and slope engineering) 141–6
porosity of rock 85–7
porphyry deposits 34, 55–6, 76, 89, 100, 106, 116, 143, 156, 162, 408
portable inclinometer probes 354
primary domain boundaries 82
Probability of Failure (PoF) 8, 196, 221, 223–5, 235, 238, 245, 332, 387, 431, 432
probability theory 431–6
probe extensometers 357
project development 11–12
flowchart 2
project stage, geotechnical effort by 13
pumping tests 51

radar and slope monitoring 351, 367
recording events 377–8
regional seismicity (geological model) 62–6
regional stress (geological model) 66–7
regulatory factors 3–4
reinforcement measures 318–25
buttressing 324–5
retaining walls 320–2
rock bolting systems 318–20
surface treatments 323–4
residual soils 92–3
resistance wire extensometers 356–7
retaining walls 320–2
reverse circulation (RC) drilling 42–3, 186, 418
Rio Narcea project (Spain) 188
risk analysis 391–5
risk assessment
ecological 410
and mine closure 405, 410
risk evaluation 395–6
risk identification 389–91
risk management 11, 381–400
approaches to 389
definitions 383
geotechnical 385–9
geotechnical hazard checklists 447–57
methodologies 389–96
and mine closure 405
minerals industry 384–5, 386
overview 383–5
process 383–4
risk mitigation 396–400
risk model (acceptance criteria) 225–35
risk reduction (ground control management) 373
risk/consequence analyses 11, 228, 229, 230–1, 387
Robinson mine (Nevada) 192
rock, fresh
effect of alteration 17
effect of weathering 17
rock, tensile strength 17, 88–9, 94, 128, 253, 278, 280, 440
rock bolting systems 318–20
rock bolts 318
rock dowels 318–19
rock mass
cell mapping sheet 18
common defects 20
geotechnical model 207–10
rock mass analyses 246–64
empirical methods 246–8
limit equilibrium methods 248–53
numerical methods 253–3
rock mass classification 117–27, 330–1
Hoek-Brown GSI 123–7
Luabscher IRMR and MRMR 119–23
RMR, Bieniawski 117–19
rock mass model 8, 83–139
intact rock strength 83–94
synthetic 127, 138–9, 207, 212
Rock Mass Rating, Bieniawski's (RMR) 117–19, 123, 125, 127, 132, 210, 330
rock mass strength (RMS) 127–39
Call & Nicholas Inc. (CNI) criterion 127, 130–2, 240
design (DRMS) 127, 128
directional 132–8
Hoek-Brown 128–30, 210–11
Laubscher strength criteria 127–8
synthetic rock mass model 127, 138–9, 207, 212
rock mass strength criterion 127
Hoek-Brown 128–30, 210–11
values 201
rock mass unloading 162, 168, 197, 198
rock strength characteristics 280
rock strength, intact 83–94
index properties 85–8
mechanical properties 88–92
point load strength index 85
porosity 85–7
special conditions 92–4
tensile strength 88–9
triaxial compressive strength 90–1
uniaxial compressive strength 89–90
unit weight 87

wave velocity 87–8
rockfall 6
hazards and catchment 335–7
protection measures 325–6
Round Mountain mine (Nevada) 51, 190

safety factors 2, 371–9
San Nicolas deposit 24
satellite imaging subsidence monitoring (InSAR) 351–3
scale effects (geotechnical model) 206–10
scaling 312–13
scoping studies 270
scribe orientation systems 28
secant piling 322
sedimentary environments 76
sediment-hosted ore bodies 58
sedimentary rocks 22, 33, 48, 58, 59, 76, 80, 85, 131, 155–6, 242, 437, 438
seismic analyses 62, 251–3, 259
seismic risk data 65–6
seismic methods 22–3
seismicity, regional 62–6
shear pins 320
shear stiffness 113–17
shear strength, measuring 94–7
shear strips and time domain reflectometers (TDRs) 354–6
shear trenches 324–5
shotcrete 323
SI system of units 462
SIMBLOC 79, 80, 254
site climate 407
site characterisation (mine closure) 407–8
site geology and geochemistry 408
site hydrology 407–8
skarn deposits 57
Sleeper mine (Nevada) 148, 161, 411, 412
slope angle versus slope height charts 246–7
slope angles and levels of confidence 234–5
slope configurations 4
slope depressurisation 341–2
installation of test wells for 47–9
modelling 175–6
methods 181, 192
program 150–1, 180–95
setting up pilot trials 51–2
see also pit slope depressurisation
slope design 1–4
economic factors 2–3
environmental and regulatory factors 3–4
formulation of 6–11
fundamentals 1–14
implementation 10
mine planning aspects 265–76
process (flowchart) 7, 16, 54, 70, 84, 142, 202, 214, 238, 266, 328, 382, 387, 402
requirements by project level 11–12
review 12–14
safety/social factors 2
terminology of 4–6
see also design implementation
slope design methods 9–10
design analyses 238–9
design steps 237–8
kinematic analyses 239–46
rock mass analyses 246–64
slope documentation and expected behaviour 339–41
slope erosion protective measures 324
slope failure
acceptance criteria 8, 221
advanced numerical models 259–60
and artificial support 313
blast-induced damage 282
CNI criterion 132
cost-benefit analysis 226
design considerations 315
design implementation 274, 314–15
discrete fracture network modelling 79
earthquakes 62
emergency response procedures 375
event tree analyses 230, 231
fatalities and injuries 232, 233
forecasting 370
geotechnical risk management 385–8
gravity-driven 254, 343, 398
hydrogeology and slope engineering 141
and instability 5–6, 9
inter-ramp 244–5, 337–9
investigation 375
and mine closure 11
monitoring rockfalls 365, 390, 412
movement monitoring systems 343, 350–1, 368, 370
operational geotechnical roles 377
and pit wall stability 411
and pore pressure 198–9, 212
post-closure monitoring 412
prediction 370
regional stress 66
risk analysis 391–3
risk evaluation 395–6
risk identification 390–1
risk model 225–6, 228, 236
rock mass analyses 246
and rock mass model 83, 93, 99
and slope design 228, 238, 244–5, 246, 253–4, 259–60

and slope monitoring 342, 343
subjective assessment 215
and transient pore pressures 197
see also bench failure; probability of failure (PoF)
slope instabilities, large-scale 341
slope instability 4–6
slope monitoring 10–11, 342–70
movement monitoring systems 343–63
programs, guidelines on the execution of 363–70
slope performance 327–79
assessing 327–42
geotechnical model validation and refinement 327–37
inter-ramp 337–9
overall 339–42
slope steepening, potential impacts of 3
slopes, stresses in open pit 440–4
solid modelling 77
stability *see* pit wall stability
standpipe piezometer 42, 44, 45, 49, 360, 419, 420
statistical theory 431–6
steel sheet piling 322
stereographic projection 77–9
blind zones 78
general guidelines 77–8
Terzaghi correction for joint spacing 79
Terzaghi weighting 79
storativity 144
stratabound deposits 57–9
structural defects, strength of 17, 94–117
Barton-Bandis failure criterion 104–6, 108, 110, 113, 206
defect strength 94–117
effect of defect placement 99–100
effect of surface roughness 100–4
influence of infilling 97–9
measuring shear strength 94–7
scale effects 108–9
shear stiffness 113–17
stress, strain and normal stiffness 109–13
terminology and classification 94
structural domain definition 80–2
fabric with primary domains 82
general guidelines 80
primary domain boundaries 80–2
structural model 7–8, 69–82
fabric 75–6
geological environments 76–7
major structures 69–75
model components 69–76
structural modelling tools 77–80
discrete fracture network modelling 79–80
solid modelling 77
stereographic projection 77–9
structures
determining the orientation of 36–9
large-scale structures logging 34–6
metamorphic 74–5
minor fold 75–6
structures, major 69–76
faults 73–4
folds 69–73
metamorphic structures 74–5
study requirements 12
supplementary drilling 329
surface geophysical techniques 22–3
surface water diversion 409
survey monitoring 346
surveying *see* downhole surveying
synthetic rock mass model 127, 138–9, 207, 212

3FLO 79, 139, 198, 199, 254, 260
televiewers 29, 36
tensile strength of rock 17, 88–9, 94, 128, 253, 278, 280, 440
terminology, geotechnical engineering 462–6
terminology of slope design 4–6
instability 4–6
pit wall 5
rockfall 6
slope configurations 4
terrestrial geodetic surveys 366
Terzaghi classification scheme 117
Terzaghi correction for joint spacing 79
Terzaghi effective stress concept 143
Terzaghi weighting 79
thermistors 357
tied-back walls 322
tiltmeters 353
toe cleanup 313
triaxial compressive strength 90–1
trigger action responses (TARP) (ground control management) 373–5

Ujina open pit (Chile) 224, 225
uniaxial compressive strength (UCS) 33, 39, 85, 89–90, 91, 92, 93, 105–6, 107, 124, 128, 203
field estimates 19
Unified Soils Classification System 19, 23, 25, 32, 34, 35, 94, 97
unit weight of rock 87
unloading response 4

Venetia mine (South Africa) 197
vibrating-wire piezometer 42, 44, 45, 46, 49, 51, 196, 197, 357, 360, 419–20

volcanic massive sulphide (VMS) deposits 57, 76

W straps 320
water, pit lake quality 409–10
water balance, pit 409
water diversion, surface 409
water management and control 192–5
wave velocity in rock 87–8
weak rocks 92–4
weighted core barrel 28
'Wet' slope approach (pore pressure) 211
Westbay piezometer 46–7
window mapping 19–21
windrow (Australia) 4
wireless extensometers 345–6

Young's modulus 91–2

Zambian Copper Belt (ZCB) 58